Lectin Methods and Protocols

METHODS IN MOLECULAR MEDICINE™

John M. Walker, SERIES EDITOR

10. **Herpes Simplex Virus Protocols,** edited by *S. Moira Brown and Alasdair MacLean,* 1997
9. **Lectin Methods and Protocols,** edited by *Jonathan M. Rhodes and Jeremy D. Milton,* 1997
8. ***Helicobacter pylori* Protocols,** edited by *Christopher L. Clayton and Harry L. T. Mobley,* 1997
7. **Gene Therapy Protocols,** edited by *Paul D. Robbins,* 1997
6. **Molecular Diagnosis of Cancer,** edited by *Finbarr Cotter,* 1996
5. **Molecular Diagnosis of Genetic Diseases,** edited by *Rob Elles,* 1996
4. **Vaccine Protocols,** edited by *Andrew Robinson, Graham H. Farrar, and Christopher N. Wiblin,* 1996
3. **Prion Diseases,** edited by *Harry F. Baker and Rosalind M. Ridley,* 1996
2. **Human Cell Culture Protocols,** edited by *Gareth E. Jones,* 1996
1. **Antisense Therapeutics,** edited by *Sudhir Agrawal,* 1996

METHODS IN MOLECULAR MEDICINE™

Lectin Methods and Protocols

Edited by

Jonathan M. Rhodes

and

Jeremy D. Milton

University of Liverpool, UK

Humana Press ✳ Totowa, New Jersey

© 1998 Humana Press Inc.
999 Riverview Drive, Suite 208
Totowa, New Jersey 07512

All rights reserved. No part of this book may be reproduced, stored in a retrieval system, or transmitted in any form or by any means, electronic, mechanical, photocopying, microfilming, recording, or otherwise without written permission from the Publisher. Methods in Molecular Medicine™ is a trademark of The Humana Press Inc.

All authored papers, comments, opinions, conclusions, or recommendations are those of the author(s), and do not necessarily reflect the views of the publisher.

This publication is printed on acid-free paper. ∞
ANSI Z39.48-1984 (American Standards Institute) Permanence of Paper for Printed Library Materials.

Cover illustration: Fig. 3 from Chapter 8, "Electron Microscopy: *Use of Lectin–Gold After Embedding,*" by Rainer Herken and Berti Manshausen.

Cover design by Patricia F. Cleary.

For additional copies, pricing for bulk purchases, and/or information about other Humana titles, contact Humana at the above address or at any of the following numbers: Tel: 973-256-1699; Fax: 973-256-8341; E-mail: humana@mindspring.com or visit our website at http://www.humanapress.com

Photocopy Authorization Policy:
Authorization to photocopy items for internal or personal use, or the internal or personal use of specific clients, is granted by Humana Press Inc., provided that the base fee of US $8.00 per copy, plus US $00.25 per page, is paid directly to the Copyright Clearance Center at 222 Rosewood Drive, Danvers, MA 01923. For those organizations that have been granted a photocopy license from the CCC, a separate system of payment has been arranged and is acceptable to Humana Press Inc. The fee code for users of the Transactional Reporting Service is: [0-89603-396-1/98 $8.00 + $00.25].

Printed in the United States of America. 10 9 8 7 6 5 4 3 2 1

Library of Congress Cataloging in Publication Data

Main entry under title:

Methods in molecular medicine™.

Lectin methods and protocols/edited by Jonathan M. Rhodes and Jeremy D. Milton.
    p. cm.—(Methods in molecular medicine™)
  Includes index.
  ISBN 0-89603-396-1 (alk. paper)
  1. Lectins. 2. Lectins—Analysis.   I. Rhodes, Jonathan M.   II. Milton, Jeremy D.   III. Series.
QP552.L42L413   1998
572'.68—dc21
                                                  98-28305
                                                    CIP

Preface

Lectins have in the past been regarded by many scientists as curious proteins of uncertain structure and specificity that bind to carbohydrates of dubious significance themselves. All this is rapidly changing. The functional importance of glycosylation in cell–cell and cell–pathogen interactions, as well as intracellular events, has been recognized by the explosion of the science of glycobiology. This has been paralleled by the realization that lectins, once they have been well characterized, can be extremely useful tools for examining structural changes in glycosylation and their functional consequences for human pathophysiology.

Different lectins vary considerably in their degree of specificity. Some, such as wheatgerm agglutinin, have fairly broad specificity (for glucosamine or sialic acid), whereas others, such as *Maackia amurensis*, are specific not only for a single carbohydrate, but also for its linkage (2–3 linked sialic acid). Lectins with relatively broad specificity may be very useful as an adjunct to isolation or quantification of soluble glycoproteins, whereas lectins of known, and precise, specificity will be more useful for characterization of carbohydrate structure. We have included an appendix in *Lectin Methods and Protocols* that provides the known specificities of all lectins cited in the text.

Lectin Methods and Protocols describes the use of lectins, mainly of plant origin, in three broad areas of investigation: (1) analysis of carbohydrates, (2) isolation and quantification of glycoproteins, and (3) reaction with living cells. Techniques are described for analysis of carbohydrate structures within tissues by light or electron microscopy, quantitative and qualitative analysis of soluble glycoproteins, separation of human bone marrow cells for clinical use, and assessment of the biological effects of plant lectins.

We hope that these practical accounts of a wide range of applications of lectins will encourage further application and development of this useful and fascinating area of glycobiology.

Jonathan M. Rhodes
Jeremy D. Milton

Contents

Preface v
Contributors xiii

PART I. LECTIN HISTOCHEMISTRY AND CYTOCHEMISTRY

A. LIGHT MICROSCOPY

1 Light Microscopy: *Overview and Basic Methods*
  Anthony J. Leathem and Susan A. Brooks 3

2 Lectin Histochemistry and Cytochemistry—Light Microscopy:
  Avidin–Biotin Amplification on Resin-Embedded Sections
  Robert W. Stoddart and Carolyn J. P. Jones 21

3 Lectin–Gold Histochemistry on Paraffin and Lowicryl K4M
  Sections Using Biotin and Digoxigenin-Conjugated Lectins
  Jürgen Roth, Christian Zuber, Tetsutaro Sata, and Wei-Ping Li ... 41

4 Use of Fluorochrome-Labeled Lectins in Light Microscopy
  Udo Schumacher and Barry S. Mitchell 55

5 The Use of Lectins in Combination with Enzymatic Digestion
  for the Study of Glycoconjugates in Cartilage
  Sibylle Hoedt-Schmidt 65

6 Applications of Lectin Histochemistry and Cytochemistry
  in Diagnosis and Prognosis
  Timothy Richard Helliwell 73

B. ELECTRON MICROSCOPY

7 Electron Microscopy: *Use of Lectin-Peroxidase Prior to Embedding*
  Adi Ellinger 97

8 Electron Microscopy: *Use of Lectin–Gold After Embedding*
  Rainer Herken and Berti Manshausen 111

9 Amplification of Lectin–Gold Histochemistry
  Juan F. Madrid, Francisco Hernández, and José Ballesta 121

10 Electron Microscopic Methods for the Demonstration
   of Lectin-Binding Sites in Cancer Cell Lines
   Barry S. Mitchell and Udo Schumacher 133

PART II. USE OF LECTINS FOR STRUCTURAL ANALYSIS OF OLIGOSACCHARIDE CHAINS

11 Application of Sequential Smith Degradation to Lectin Blots
    Chi Kong Ching 147

12 Blot Analysis with Lectins for the Evaluation of Glycoproteins
    in Cultured Cells and Tissues
    Christian Zuber, Wei-Ping Li, and Jürgen Roth 159

13 Characterization of HIV gp120 Envelope Glycoprotein
    by Lectin Analysis
    Gregers J. Gram and John-Erik Stig Hansen 167

14 Use of Lectins for Characterization of O-Linked Glycans
    of Herpes Simplex Virus Glycoproteins
    Sigvard Olofsson and Anders Bolmstedt 175

PART III. LECTINS FOR DETECTION OF ALTERED GLYCOSYLATION OF CIRCULATING GLYCOPROTEINS

15 Use of Lectin for Detection of Agalactosyl IgG
    Naoyuki Tsuchiya, Tamao Endo, Naohisa Kochibe, Koji Ito, and Akira Kobata 195

16 Lectins for Detection of Altered Glycosylation of Circulating
    Glycoproteins: α-1-Antitrypsin
    Yutaka Aoyagi and Hitoshi Asakura 207

17 Detection of Altered Glycosylation of α-Fetoprotein Using
    Lectin-Affinity Electrophoresis
    Kazuhisa Taketa, Miao Liu, and Hiroko Taga 215

18 Use of Lectin-Affinity Electrophoresis for Quantification
    and Characterization of Glycoforms of α-1 Acid Glycoprotein
    Thorkild C. Bøg-Hansen 227

19 ABO(H) Blood Group Expression on Circulating Glycoproteins
    Taei Matsui and Koiti Titani 235

PART IV. USE OF LECTINS IN QUANTIFICATION OF SOLUBLE GLYCOPROTEINS

20 Lectin/Antibody "Sandwich" ELISA for Quantification of Circulating
    Mucin as a Diagnostic Test for Pancreatic Cancer
    Neil Parker 249

21 Quantification of Intestinal Mucins
    Jeremy D. Milton and Jonathan M. Rhodes 255

Contents

PART V. LECTINS IN AFFINITY PURIFICATION OF SOLUBLE GLYCOPROTEINS

22 Purification and Characterization of Human Serum and Secretory IgA1 and IgA2 Using Jacalin
Michael A. Kerr, Lesley M. Loomes, Brian C. Bonner, Amy B. Hutchings, and Bernard W. Senior 265

23 Use of Lectins in Affinity Purification of HIV and SIV Envelope Glycoproteins
Gustav Gilljam 279

24 T-Cell Receptor Purification
Kelly P. Kearse 291

PART VI. LECTINS IN FLOW CYTOMETRY

25 Use of Monomeric, Monovalent Lectin Derivatives for Flow Cytometric Analysis of Cell Surface Glycoconjugates
Hanae Kaku and Naoto Shibuya 301

26 Analysis of Subcellular Components by Fluorescent-Lectin Binding and Flow Cytometry
Rosa M. Guasch and José-Enrique O'Connor 307

PART VII. LECTINS AS TOOLS FOR CELL PURIFICATION/PURGING

27 Lectins as Tools for the Purification of Liver Endothelial Cells
Daniel E. Gomez and Unnur P. Thorgeirsson 319

28 The Use of Soybean Agglutinin (SBA) for Bone Marrow (BM) Purging and Hematopoietic Progenitor Cell Enrichment in Clinical Bone Marrow Transplantation
Arnon Nagler, Shoshana Morecki, and Shimon Slavin 329

29 Combined Lectin/Monoclonal Antibody Purging of Bone Marrow for Use in Conjunction with Autologous Bone Marrow Transplantation in the Treatment of Multiple Myeloma
Elizabeth G. H. Rhodes 351

PART VIII. EFFECTS OF LECTINS ON MAMMALIAN CELLS

30 Mechanisms and Assessment of Mitogenesis: *An Overview*
David C. Kilpatrick 365

31 Mitogenic Effects of Lectins on Epithelial Cells
Lu-Gang Yu and Jonathan M. Rhodes 379

32	Use of Lectins as Mitogens for Lymphocytes *David C. Kilpatrick* *385*
33	Effect of Lectins on Uptake of Polyamines *Susan Bardocz and Ann White* *393*
34	Effects of Lectins on Cytoskeletal Organization in Mammalian Cells *Paolo Carinci, Ennio Becchetti, and Maria Bodo* *407*
35	Effect of Lectins on Protein Kinase Activity *Kiyonao Sada and Hirohei Yamamura* *423*
36	Lectin-Induced Calcium Mobilization in Human Platelets: *Use of Fluorescent Probes* *Giuseppe Ramaschi and Mauro Torti* *433*
37	Lectin-Triggered Superoxide/H_2O_2 and Granule Enzyme Release from Cells *Alexander V. Timoshenko, Klaus Kayser, and Hans-Joachim Gabius* *441*
38	Cytotoxic Effects of Lectins *Elieser Gorelik* *453*
39	Application of the Lectin-Dependent Cell-Mediated Cytotoxicity Assay to Bronchoalveolar Lavage Fluid and Venous Blood Samples Collected from Canine Lung Allografts *Allan G. L. Lee and Hani A. Shennib* *461*

PART IX. EFFECTS OF LECTINS IN ORGAN CULTURE

40	The Effect of Lectins on Crypt Cell Proliferation in Organ Culture *Stephen D. Ryder* *475*

PART X. EFFECTS OF LECTIN INGESTION

41	Effects of Lectin Ingestion on Animal Growth and Internal Organs *Arpad Pusztai* *485*
42	Lectin Ingestion: *Changes in Mucin Secretion and Bacterial Adhesion to Intestinal Tissue* *Howard Ceri, John G. Banwell, and Rixun Fang* *495*
43	Assessment of Lectin Inactivation by Heat and Digestion *Arpad Pusztai and George Grant* *505*

PART XI. LECTINS IN THE INVESTIGATION OF NEURONAL TRAFFICKING

44	Use of Lectins as Transganglionic Neuronal Tracers in the Study of Unmyelinated Primary Sensory Neurons *Mark B. Plenderleith and Peter J. Snow* *517*

Contents

PART XII. USE OF LECTINS IN THE INVESTIGATION OF PATHOGEN–HOST INTERACTIONS

45 Lectin Inhibition of Bacterial Adhesion to Animal Cells
   Murray W. Stinson and Jen Ren Wang 529

46 Inhibition of HIV Infection by Lectin Binding to CD4
   Jean Favero and Virginie Lafont 539

47 Inhibition of HIV Infection by Lectin Binding to gp120
   Theresa Animashaun and Naheed Mahmood 555

PART XIII. USE OF LECTINS FOR DRUG DELIVERY

48 Absorption Enhancement by Lectin-Mediated Endo- and Transcytosis
   Ellen Haltner, Gerrit Borchard, and Claus-Michael Lehr 567

49 The Use of Lectins for Liposome Targeting in Drug Delivery
   Michael Kaszuba and Malcolm N. Jones 583

Appendix—List of Lectins and Binding Structures 595

Index 599

Contributors

THERESA ANIMASHAUN • *Division of Virology, National Institute for Medical Research, The Ridgeway, Mill Hill, London, UK*

YUTAKA AOYAGI • *Department of Internal Medicine, The Third Division, Niigata University School of Medicine, Niigata, Japan*

HITOSHI ASAKURA • *The Third Division, Department of Internal Medicine, Niigata University School of Medicine, Niigata, Japan*

JOSÉ BALLESTA • *Department of Cell Biology, School of Medicine, University of Murcia, Spain*

JOHN G. BANWELL • *Gastroenterology, Department of Medicine, Case Western Reserve School of Medicine, Cleveland, OH*

SUSAN BARDOCZ • *Food-Gut-Microbial Interaction Group, The Rowett Research Institute, Bucksburn, Aberdeen, Scotland*

ENNIO BECCHETTI • *Histology Section, Department of Experimental Medicine, School of Medicine, University of Perugia, Italy*

MARIA BODO • *Institute of Histology and General Embryology, School of Medicine, University of Ferrara, Italy*

THORKILD C. BØG-HANSEN • *The Protein Laboratory, The Health Science Faculty, University of Copenhagen, Denmark*

ANDERS BOLMSTEDT • *Department of Clinical Virology, University of Göteborg, Guldhedsgatan, Göteborg, Sweden*

BRIAN C. BONNER • *Department of Pathology, University of Dundee, Ninewells Hospital Medical School, Dundee, UK*

GERRIT BORCHARD • *Department of Biopharmaceutics and Pharmaceutical Technology, University of the Saarland, Saarbrücken, Germany*

SUSAN A. BROOKS • *Breast Cancer Research Group, University College London Medical School, London, UK*

PAOLO CARINCI • *Institute of Histology and General Embryology, School of Medicine, University of Ferrara, Italy*

HOWARD CERI • *Biological Sciences and Biofilm Research Group, University of Calgary, Alberta, Canada*

CHI KONG CHING • *Department of Medicine, The University of Hong Kong, Queen Mary Hospital, Pokfulam, Hong Kong*

ADI ELLINGER • *Institute of Histology and Embryology, University of Innsbruck, Austria*
TAMAO ENDO • *Department of Glycobiology, Tokyo Metropolitan Institute of Gerontology, Tokyo, Japan*
RIXUN FANG • *Department of Biological Sciences, University of Calgary, Alberta, Canada*
JEAN FAVERO • *Microbiologie et Pathologie Cellulaire Infectieuse, Institut National de la Sante et de la Recherche Medicale, Universite de Montpellier II, Montpellier, France*
HANS-JOACHIM GABIUS • *Institut für Physiologische Chemie, Tierärztliche Fakultät, Ludwig-Maximilians-Universität, München, Germany*
GUSTAV GILLJAM • *Swedish Institute for Infectious Disease Control, Stockholm, Sweden*
DANIEL E. GOMEZ • *Department of Science and Technology, Universidad Nacional de Quilmes, Buenos Aires, Argentina*
ELIESER GORELIK • *Department of Pathology, School of Medicine, University of Pittsburgh Cancer Institute, University of Pittsburgh, PA*
GREGERS J. GRAM • *Department of Infectious Diseases, University of Copenhagen, Hvidovere Hospital, Copenhagen, Denmark*
GEORGE GRANT • *Division of Nutritional Sciences, The Rowett Research Institute, Bucksburn, Aberdeen, Scotland*
ROSA M. GUASCH • *Departamento de Bioquimica y Biología Molecular, Facultad de Medicina, Universidad de Valencia, Spain*
ELLEN HALTNER • *Department of Biopharmaceutics and Pharmaceutical Technology, University of the Saarland, Saarbrücken, Germany*
JOHN-ERIK STIG HANSEN • *Department of Infectious Diseases, Hvidovre Hospital, University of Copenhagen, Denmark*
TIMOTHY RICHARD HELLIWELL • *Department of Pathology, University of Liverpool, UK*
RAINER HERKEN • *Anatomy Centre, Department of Histology, Kreuzbergring, Göttingen, Germany*
FRANCISCO HERNÁNDEZ • *Department of Cell Biology, School of Medicine, University of Murcia, Spain*
SIBYLLE HOEDT-SCHMIDT • *Institute of Pharmacology and Toxicology, University of Bonn, Germany*
AMY B. HUTCHINGS • *Department of Pathology, Ninewells Hospital Medical School, University of Dundee, UK*
KOJI ITO • *Department of Internal Medicine and Physical Therapy, Faculty of Medicine, University of Tokyo, Bunkyo-ku, Tokyo, Japan*

CAROLYN J. P. JONES • *Department of Pathological Sciences, Stopford Building, University of Manchester, UK*
MALCOLM N. JONES • *School of Biological Sciences, University of Manchester, UK*
HANAE KAKU • *Department of Cell Biology, National Institute of Agrobiological Resources, Ministry of Agriculture, Foresty and Fisheries, Tsukuba, Japan*
MICHAEL KASZUBA • *School of Biological Sciences, University of Manchester, UK*
KLAUS KAYSER • *Abteilung Pathologie, Thoraxklinik de LVA Baden, Heidelberg, Germany*
KELLY P. KEARSE • *Experimental Immunology Branch, National Cancer Institute, National Institutes of Health, Bethesda, MD*
MICHAEL A. KERR • *Department of Pathology, Ninewells Hospital Medical School, University of Dundee, UK*
DAVID C. KILPATRICK • *Department of Transfusion Medicine, Edinburgh and S. E. Scotland Blood Transfusion Service, Edinburgh, Scotland*
AKIRA KOBATA • *Tokyo Metropolitan Institute of Gerontology, Tokyo, Japan*
NAOHISA KOCHIBE • *Department of Biology, Faculty of Education, Gunma University, Maebashi, Gunma, Japan*
VIRGINIE LAFONT • *Microbiologie et Pathologie Cellulaire Infectieuse, Institut National de la Sante et de la Recherche Medicale, Universite de Montpellier II, Montpellier, France*
ANTHONY J. LEATHEM • *Breast Cancer Research Group, University College London Medical School, London, UK*
ALLAN G. L. LEE • *The Joint Marseilles Montreal Lung Transplant Program, The Montreal General Hospital and McGill University, Montreal, Quebec, Canada*
CLAUS-MICHAEL LEHR • *Department of Biopharmaceutics and Pharmaceutical Technology, University of the Saarland, Saarbrücken, Germany*
WEI-PING LI • *Division of Cell and Molecular Pathology, Department of Pathology, University of Zurich, Switzerland*
MIAO LIU • *Department of Public Health, Okayama University School of Medicine, Okayama, Japan*
LESLEY LOOMES • *Department of Pathology, Ninewells Hospital Medical School, University of Dundee, UK*
JUAN F. MADRID • *Department of Cell Biology and Morphological Sciences, School of Medicine and Dentistry, University of the Basque Country, Leioa, Vizcaya, Spain*

NAHEED MAHMOOD • *MRC Collaborative Centre, Mill Hill, London UK*
BERTI MANSHAUSEN • *Anatomy Centre, Department of Histology, Kreuzbergring 36, Gottingen, Germany*
TAEI MATSUI • *Division of Biomedical Polymer Science, Institute for Comprehensive Medical Science, Fujita Health University, Toyoake, Japan*
JEREMY D. MILTON • *Gastroenterology Research Group, Department of Medicine, University of Liverpool, UK*
BARRY S. MITCHELL • *Anglo-European College of Chiropractic, Bournemouth, UK*
SHOSHANA MORECKI • *Department of Bone Marrow Transplantation and Cancer Immunobiology, Hadassah University Hospital, Jerusalem, Israel*
ARNON NAGLER • *Department of Bone Marrow Transplantation and Cancer Immunobiology, Hadassah University Hospital, Jerusalem, Israel*
JOSÉ-ENRIQUE O'CONNOR • *Departamento de Bioquimica, Facultad de Medicina, Universidad de Valencia, Spain*
SIGVARD OLOFSSON • *Department of Clinical Virology, University of Göteborg, Guldhedsgatan, Göteborg, Sweden*
NEIL PARKER • *Cortecs Diagnostics, Deeside Industrial Park, Newtech Square, Deeside, UK*
MARK B. PLENDERLEITH • *School of Life Science, Queensland University of Technology, Queensland, Australia*
ARPAD PUSZTAI • *Division of Nutritional Sciences, The Rowett Research Institute, Bucksburn, Aberdeen, Scotland*
GIUSEPPE RAMASCHI • *Department of Biochemistry, University of Pavia, Italy*
ELIZABETH G. H. RHODES • *Department of Haematology, Countess of Chester Hospital, Chester, UK*
JONATHAN M. RHODES • *Department of Medicine, University of Liverpool, UK*
JÜRGEN ROTH • *Division of Cell and Molecular Pathology, Department of Pathology, University of Zurich, Switzerland*
STEPHEN D. RYDER • *Department of Medicine, Queen's Medical Centre, Nottingham, UK*
KIYONAO SADA • *Department of Biochemistry, Kobe University School of Medicine, Chuo-ku, Kobe, Japan*
TETSUTARO SATA • *Division of Cell and Molecular Pathology, Department of Pathology, University of Zurich, Switzerland*

Udo Schumacher • *Department of Human Morphology, University of Southampton, UK*
Bernard W. Senior • *Department of Medical Microbiology, University of Dundee, Ninewells Hospital Medical School, Dundee, UK*
Hani A. Shennib • *The Joint Marseilles Montreal Lung Transplant Program, The Montreal General Hospital and McGill University, Montreal, Quebec, Canada*
Naoto Shibuya • *Department of Cell Biology, National Institute of Agrobiological Resources, Ministry of Agriculture, Foresty and Fisheries, Tsukobo, Japan*
Shimon Slavin • *Department of Bone Marrow Transplantation and Cancer Immunobiology, Hadassah University Hospital, Jerusalem, Israel*
Peter J. Snow • *Cerebral and Sensory Functions Unit, Department of Anatomy, The University of Queensland, Brisbane, Queensland, Australia*
Murray W. Stinson • *Department of Microbiology, School of Medicine and Biomedical Sciences, State University of New York at Buffalo, NY*
Robert W. Stoddart • *Department of Pathological Sciences, Stopford Building, University of Manchester, UK*
Hiroko Taga • *Department of Public Health, Okayama University School of Medicine, Okayama, Japan*
Kazuhisa Taketa • *Department of Public Health, Okayama University Medical School, Okayama, Japan*
Unnur P. Thorgeirsson • *Division of Cancer Etiology, National Cancer Institute, NIH, Bethesda, MD*
Alexander V. Timoshenko • *Institut für Physiologische Chemie, Tierärztliche Fakultät, Ludwig-Maximilians-Universität, München, Germany*
Koiti Titani • *Division of Biomedical Polymer Science, Institute for Comprehensive Medical Science, Fujita Health University, Toyoake, Japan*
Mauro Torti • *Department of Biochemistry, University of Pavia, Italy*
Naoyuki Tsuchiya • *Department of Internal Medicine and Physical Therapy, Faculty of Medicine, University of Tokyo, Hong, Bunkyo-ku, Tokyo, Japan*
Jen Ren Wang • *Department of Medical Technology, National Cheng Kung University, Medical College, Tainan, Taiwan*
Ann White • *Physiology Division, The Rowett Research Institute, Greenburn Road, Bucksburn, Aberdeen, Scotland*

HIROHEI YAMAMURA • *Department of Biochemistry, Fukui Medical School, Fukui, Japan*
LU-GANG YU • *Gastroenterology Research Group, Department of Medicine, University of Liverpool, UK*
CHRISTIAN ZUBER • *Division of Cell and Molecular Pathology, Department of Pathology, University of Zurich, Switzerland*

I

LECTIN HISTOCHEMISTRY AND CYTOCHEMISTRY

A: Light Microscopy

1

Light Microscopy

Overview and Basic Methods

Anthony J. Leathem and Susan A. Brooks

1. Introduction
1.1. Why Use Lectins?

Lectins are proteins or glycoproteins of nonimmune origin derived from plants, animals, or microorganisms that have specificity for terminal or subterminal carbohydrate residues. They are sensitive, stable, and easy-to-use tools. Lectin histochemistry and cytochemistry can provide an extraordinarily sensitive detection system for changes in glycosylation and carbohydrate expression that may occur during embryogenesis, growth, and disease. Although carbohydrates occur in a vast range of permutations that may be present in and between cells, there is frequently a dominance and conservation of structures to give specific markers of cell types or differentiation. Lectin histochemistry or cytochemistry can reveal subtle alterations in glycosylation between otherwise indistinguishable cells.

Lectins will specifically recognize and bind to carbohydrate structures on the surface of cells, on cytoplasmic and nuclear structures, and in extracellular matrix in cells and tissues from throughout the animal and plant kingdoms, down to bacteria and viruses.

1.2. What Do Lectins Bind to?

Sugar combining specificity of lectins is usually quoted in terms of the monosaccharide that best inhibits its binding to cells; for example, *Helix pomatia* lectin is said to be specific for *N*-acetyl-galactosamine, concanavalin

A for mannose, and so on. However, this is a massive oversimplification, and the exact, complex structures recognized by most lectins remain unknown. Lectins bind approx 10 times more avidly to solid-phase or immobilized sugars as to the same sugars in solution, which probably reflects the association–dissociation equilibrium between the lectin binding sites and the sugar, and the multivalency of lectins. A typical lectin binding site may consist of three monosaccharides, and includes their linkage and adjacent protein or lipid backbone. Furthermore, a lectin will bind increasingly in 10-fold steps between mono-, di-, and trisaccharides.

1.3. Cell and Tissue Samples

Lectin histochemistry to detect alterations in cellular glycosylation can be performed on living cells in suspension, on cell smears, tissue imprints, or fresh cryostat sections. Here, sugars in the extracellular matrix, glycolipids, and glycoproteins can be detected readily. If archival tissues are of interest (e.g., in which cellular glycosylation is to be correlated with disease progression or prognosis in a retrospective study), then sections from formalin-fixed, paraffin-embedded tissues may be preferred. Here, glycolipids will have been lost during tissue processing and glycans may be sequestered during fixation and processing. For this reason, glycans—although still present—may be hidden and unavailable for lectin binding. Such sequestration can usually be at least partially overcome by pretreatment with a digestive enzyme such as trypsin, which reveals hidden structures.

1.4. The Range of Lectins Available

About 100 different lectins are readily available commercially, and hundreds of others have been detected and described, mostly by their agglutination of red cells. Many lectins have not been assessed for their histochemical staining of tissues and cells.

In the United Kingdom, two suppliers with a wide range of lectins are Sigma, Poole, Dorset, UK, and EY Laboratories (UK agents are Bradsure Biologicals, Market Harborough, Leicestershire).

There undoubtedly exist hundreds, thousands, or tens of thousands of lectins in nature that remain to be discovered and purified. It is very simple to produce a crude lectin preparation in the laboratory that will give good results in immunohistochemistry. Excellent sources include virtually any beans or seeds, and most green plant tissue like stems and leaves. The easiest method of detecting lectin binding with such preparations, or with unlabeled commercially purchased lectins in which no antibody is available, is through biotinylation of the lectin.

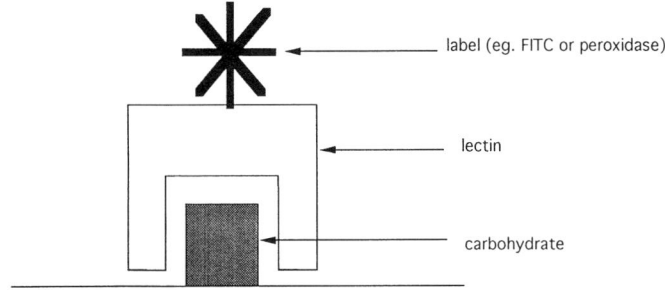

Fig. 1. The direct method.

1.5. Histochemical Methods

1.5.1. The Direct Method

The simplest methods for lectin histochemistry are the so-called direct methods, which rely on detection of lectin binding through the presence of a label conjugated directly onto the lectin molecule. The most commonly used labels are either fluorescent labels such as fluorescein isothiocyanate (FITC), which are visible when viewed under UV light, or enzyme labels like horseradish peroxidase, which yield a visible, colored reaction product (*see* Note 1). The principle of the direct method is illustrated in Fig. 1. The advantages of a direct method is that it is quick, easy, and cheap. The disadvantages are that it is generally less sensitive than the more complex methods (*see* Section 3.6.4.) and that the incorporation of a relatively large labeling molecule, like horseradish peroxidase, may interfere sterically with the combining site of the lectin and therefore slightly alter binding characteristics.

1.5.2. The Indirect Antibody Method

Here, the binding of native, unconjugated lectin is revealed by a second step in which a labeled (again, fluorescent or enzyme label) antibody (usually polyclonal, raised in rabbit) is added. The principle of this technique is illustrated in Fig. 2. The addition of a second layer has the advantage of increasing sensitivity and removes the potential problem of the label molecule interfering with binding specificity. Disadvantages are that the method takes longer, it is slightly more expensive because extra reagents need to be purchased, and there is slightly more potential for nonspecific background staining.

1.5.3. Indirect Sandwich Methods

Further layers can be added. For example, native, unconjugated lectin can be layered with first an unlabeled primary antibody (for example, a polyclonal

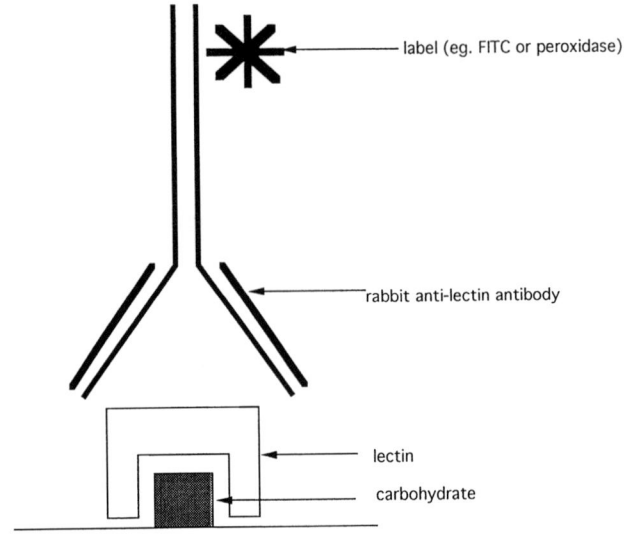

Fig. 2. The indirect antibody method.

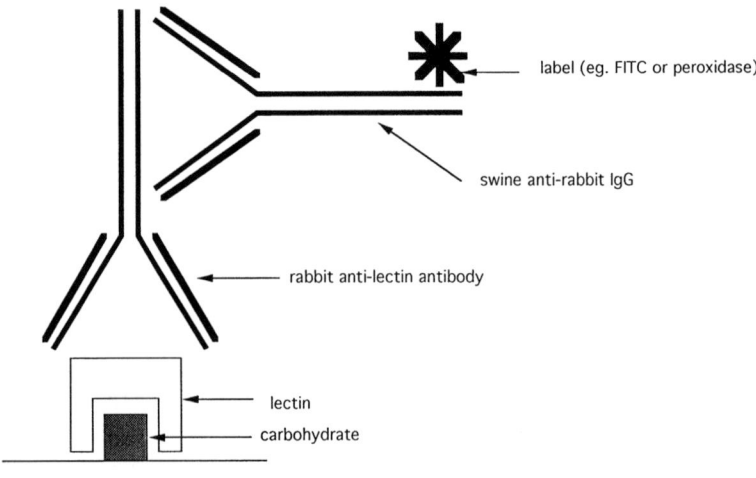

Fig. 3. An indirect sandwich method.

antibody raised in rabbit against the lectin), then a labeled second antibody (for example, a peroxidase labeled polyclonal antibody raised in swine against rabbit immunoglobulins). The principles of this method are illustrated in Fig. 3. Again, the advantage of adding extra layers is increased sensitivity. The disadvantages are, as before, that the method takes longer, is slightly more expensive, and there is slightly more potential for nonspecific background staining.

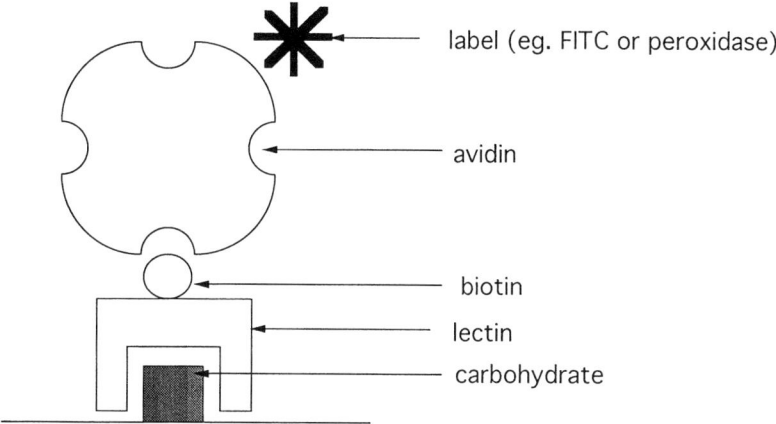

Fig. 4. The simple avidin–biotin method.

1.5.4. Simple Avidin–Biotin Method

Avidin–biotin techniques rely on the great avidity with which avidin (a protein originally derived form egg white) binds to biotin (one of the B group of vitamins). Lectins can be purchased ready labeled with biotin, or can be biotinylated in the laboratory using a simple method (*see* Sections 2.9. and 3.5.). Streptavidin, derived from *Streptomyces avidinii,* has largely replaced avidin, as it gives a cleaner staining result.

The simplest avidin–biotin technique relies on detection of the biotinylated lectin by streptavidin linked to a label such as peroxidase. The principle of this technique is illustrated in Fig. 4.

1.5.5. Multistep Avidin–Biotin Methods

The avidin–biotin interactions can be exploited in more complex multistep techniques. For example, unlabeled lectin is layered with a biotinylated antibody directed against it (e.g., polyclonal antisera raised in rabbit), followed by streptavidin peroxidase. Alternatively, lectin is layered first with unlabeled primary antibody (e.g., polyclonal antisera raised in rabbit) and then a second antibody (e.g., a polyclonal antibody raised in swine to rabbit immunoglobulins) labeled with biotin. This, again, would be followed by streptavidin peroxidase. The principle of these techniques is illustrated in Fig. 5.

1.5.6. Other Methods

Variations on these techniques can be developed to suit individual needs. It is probably better to start with a simple, quick, direct method before going on to more complex systems. Companies such as Dako (High Wycombe, United

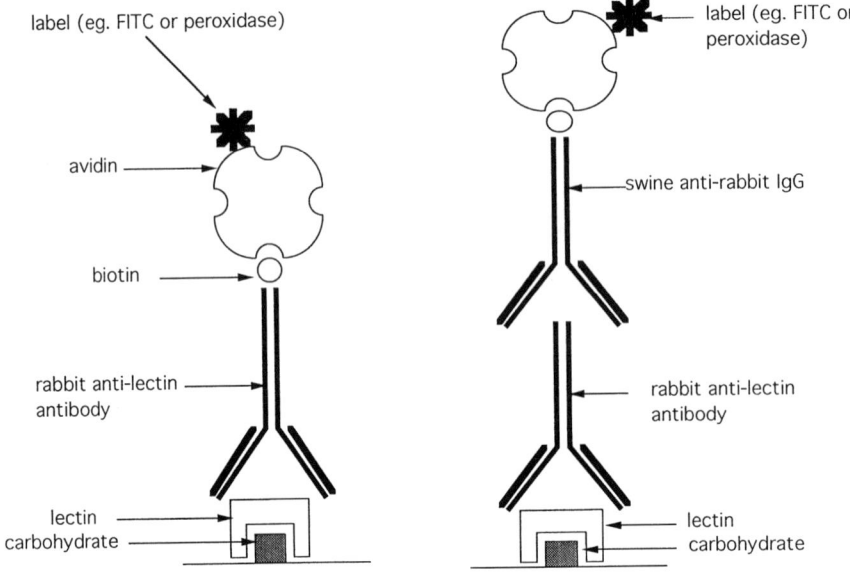

Fig. 5. Examples of multistep avidin–biotin methods.

Kingdom) supply easy-to-use kits of the main steps in some of the more complicated techniques, as well as a comprehensive range of antibodies.

1.6. A Philosophical Point Before Starting

Although numerous lectins have been commercially available for years, remarkably little progress has resulted. Histologists seem reluctant to try something new and the majority of lectin publications report on the same 10 or 12 lectins, frequently purchased as a lectin kit and chosen by the chemical company according to availability rather than a scientific basis. *(So be daring!)*

We recommend that interested researchers:

1. Start by using a well-characterized lectin that is almost certain to work.
2. Progress to try lesser known lectins.
3. Choose a biological phenomenon, a diagnostic problem, or some *a priori* hypothesis to test.
4. Or, select one lectin and explore the staining of a range of tissues; or select one tissue and explore a selection of lectins with a theme, such as galactose-binding or fucose-binding lectins (*see* Note 2).
5. Look for new or little used and uncharacterized lectins.

1.7. Which Lectins to Use and Which to Avoid

To get started, choose a cheap, well-documented lectin that binds to a wide variety of tissues. Concanavalin A is a splendid starter—it is cheap and binds to most cells and tissues. *Ulex europeus* lectin UEAI is interesting in that it binds to blood vessel endothelium. Peanut lectin is also a good starter, as it binds very specifically to a variety of cell types, giving very beautiful staining patterns.

Some lectins are highly toxic, and should be avoided whenever possible. Lectins from *Ricinus communis,* the castor oil bean (also called Ricin); *Abrus precatorius,* the Jaquerty bean (also called Abrin); and *Viscum album* mistletoe, are all extremely toxic.

Toxicity and hazard data are unavailable for most other lectins. Generally, although dust and aerosols should obviously be avoided, no special precautions are needed.

1.8. Which Tissues to Use

To get started, you need a readily available, reproducible source of tissue that binds a wide variety of lectins. Kidney is very good for this—even fresh pig kidney from a supermarket. Epithelial rich tissue is generally successful. Pituitary (for hormone precursors) and salivary glands and gut (different mucins) usually give interesting staining patterns.

1.9. The Way Forward

Lectin histochemistry is really a first step. Binding of lectins to tissue sections can demonstrate differences or changes in glycosylation, but their identification is more difficult. By competitive inhibition of lectin binding by mono- and oligosaccharides and by enzyme digestion of tissue sections before lectin staining, it is possible to get some idea of the structures recognized; and this may be extremely sensitive and clinically useful. But to identify the structure really needs extraction, separation, isolation of the carbohydrate, probably linked to a protein or lipid, then expensive and time-consuming procedures such as mass spectrometry, sequential enzyme digestion, and computer-assisted comparison with other known structures.

1.10. Examples of the Useful Application of Lectin Histochemistry

1.10.1. Ulex europaeus *Lectin UEAI and Endothelial Cells*

Ulex europeus (gorse) has two main lectins—UEAI, which binds to fucose; and UEAII, which binds to *N*-acetyl glucosamine. UEAI is an excellent marker of human endothelial cells in tissue sections and in culture. It is probably a better marker than the more commonly used Von Willebrand factor. UEAI binding gives intense staining, particularly on fresh tissues or after trypsin digest in paraffin sections. Not all endothelial cells stain—for example, those in gut may be negative—but some, like those in capillaries and muscle, do so intensely. The binding is not absolutely specific for endothelial cells—for example, normal intestinal mucosal cells may sometimes stain. Some cancers, including prostate and pancreas, also stain strongly, presumably reflecting a change in fucosylation.

UEAI shows some species specificity, and different lectins may be more suitable for animal endothelium. *Bandeirea simplicifolia* isolectin 1 (BS-I), peanut lectin, and wheat germ lectin all stain endothelium in mouse, rat, pig, and sheep, and work well—though not as well as UEAI—in human tissues too.

1.10.2. Helix pomatia *Lectin (HPA) and Cancer*

The lectin from *Helix pomatia* (the Roman or edible snail) recognizes terminal *N*-acetyl galactosamine, and to a lesser extent *N*-acetyl glucosamine and galactose sugars. It appears to be a very good prognostic marker in breast and other cancers. Approximately 80% of breast cancers will stain strongly for the binding of the lectin, whereas approx 20% are completely negative. The positive cancers tend to be aggressive, to spread to other sites (metastasize), and have a very poor prognosis. The negative cases tend not to spread and have a much more favorable prognosis *(1,2)*. Studies have shown that HPA binding is also a marker of poor prognosis in other cancers, including colorectal *(3,4)*, gastric *(5)*, esophageal *(6)*, and prostatic cancers *(7)*. The HPA appears to recognize complex sugar(s) that are involved in aggressive biological behavior and spread of cancers.

2. Materials
2.1. Lectin Buffer (see Note 3)

1. 60.57 g Tris (Sigma), 87 g NaCl, 2.03 g $MgCl_2$, 1.11 g $CaCl_2$ dissolved in 1 L distilled H_2O.
2. Adjust pH to 7.6 using concentrated HCl.
3. Make up to a total volume of 10 L with distilled water.

2.2. Preparation of Tissue Imprints

1. Piece of fresh tissue approx 1 cm^3 in size.
2. Glass microscope slides.
3. Acetone or methanol.

2.3. Preparation of Cell Smears

1. Cells in suspension (e.g., in tissue culture fluid, any suitable buffer, or in body fluids like blood).
2. Glass microscope slides.
3. Acetone or methanol.

2.4. Preparation of Frozen (Cryostat) Sections

1. Piece of tissue approx 0.5 cm^3 in size.
2. Glass microscope slides.
3. OCT embedding medium (Tissue-Tek, supplied by Raymond A. Lamb).
4. Isopentane or hexane.
5. Liquid nitrogen.
6. Acetone or methanol.

2.5. Preparation of Formalin-Fixed, Paraffin-Embedded Sections

1. Formalin-fixed, paraffin-embedded blocks of tissue.
2. Glass microscope slides.
3. 20% ethanol in distilled water.
4. CNP30 (BDH) or xylene.
5. 99% v/v and 70% v/v ethanol in distilled water.
6. Distilled water.
7. Lectin buffer.

2.6. Trypsinization of Paraffin Sections

Trypsin solution should be made fresh immediately before use. Glassware, buffers, and so on should be prewarmed to 37°C, and trypsinization should be carried out at 37°C either in a water bath or an incubator.

Dissolve 400 mg trypsin (crude type II from porcine pancreas, Sigma) and 400 mg CaCl$_2$ in 400 mL lectin buffer warmed to 37°C. Filter. Immerse slides in the solution at once and incubate at 37°C in an incubator or water bath (*see* Note 4).

2.7. Blocking Endogenous Peroxidase (see Note 5)

3% v/v solution of hydrogen peroxide in methanol. Make fresh as required.

2.8. Preparation of Crude Mushroom Lectin Extract

1. Mushrooms (the white, closed cup variety stocked by most supermarkets work well).
2. Laboratory tissue homogenizer, or domestic food processor or juicer.
3. Lectin buffer or distilled water.

2.9. Labeling with Biotin

1. 0.1M NaHCO$_3$ in distilled water.
2. *N*-hydroxysuccinimidobiotin (NHS-biotin; Sigma) (*see* Note 6).
3. Dimethylformamide or dimethyl sulfoxide.
4. Lectin buffer.

2.10. Biotinylated Mushroom Lectin Binding to Gut Sections

1. Paraffin or frozen sections of human or animal gut (materials listed in Section 2.4. or 2.5. and in 2.7. required).

2. Biotinylated crude mushroom extract (materials listed in Sections 2.8. and 2.9. required).
3. Lectin buffer.
4. Streptavidin conjugated with horseradish peroxidase (Calbiochem, La Jolla, CA) diluted 1/400 in lectin buffer.
5. 3,3-diaminobenzidine tetrahydrochloride (DAB; Sigma). 0.5 mg/mL in lectin buffer. Hydrogen peroxide added to give a concentration of 5% v/v immediately before use (*see* Note 7).
6. Mayers hematoxylin solution (Sigma).
7. 70% v/v, 95% v/v, and 99% v/v ethanol in distilled water.
8. CNP30 (BDH) or xylene.
9. Depex resinous mounting medium (BDH).

2.11. Direct Method to Detect Peroxidase-Labeled Peanut Lectin Binding to Kidney Sections

1. Paraffin or frozen sections of human or animal kidney (materials listed in Section 2.4. or 2.5. required; materials listed in Sections 2.6. and 2.7. required).
2. Peroxidase-labeled peanut lectin (PNA; Sigma) at a concentration of 10 µg/mL in lectin buffer.
3. Lectin buffer.
4. 3,3-diaminobenzidine tetrahydrochloride (DAB; Sigma) 0.5 mg/mL in lectin buffer. Hydrogen peroxide added to give a concentration of 5% v/v immediately before use (*see* Note 7).
5. Mayers hematoxylin solution (Sigma).
6. 70% v/v, 95% v/v, and 99% v/v ethanol in distilled water.
7. CNP 30 (BDH) or xylene.
8. Depex resinous mounting medium (BDH).

2.12. Direct Method to Detect FITC-Conjugated UEAI Binding to Endothelial Cells

1. Paraffin sections, frozen sections, or imprints of human striated muscle, or smears of cultured human endothelial cells (materials listed in Sections 2.2.–2.4., or 2.5. required).
2. Fluorescein isothiocyanate (FITC)-labeled *Ulex europeus* isolectin I (UEAI) (Sigma) at a concentration of 10 µg/mL in lectin buffer.
3. Lectin buffer.

2.13. Indirect Antibody Method to Detect UEAI Binding to Endothelial Cells

1. Paraffin sections, frozen sections, or imprints of human striated muscle, or smears of cultured human endothelial cells (materials listed in Sections 2.2.–2.4., or 2.5. required; materials listed in Sections 2.6. and 2.7. required).
2. *Ulex europeus* isolectin I (UEAI; Sigma) 10 µg/mL in lectin buffer.

Light Microscopy 13

3. Lectin buffer.
4. Horseradish peroxidase-labeled polyclonal rabbit anti-UEAI (Dako, Santa Barbara, CA) diluted 1/200 in lectin buffer (*see* Section 2.1.).
5. 3,3-diaminobenzidine tetrahydrochloride (DAB; Sigma) 0.5 mg/mL in lectin buffer. Hydrogen peroxide added to give a concentration of 5% v/v immediately before use (*see* Note 7).
6. Mayers hematoxylin solution (Sigma).
7. 70% v/v, 95% v/v, and 99% v/v ethanol in distilled water.
8. CNP30 (BDH) or xylene.
9. Depex resinous mounting medium (BDH).

2.14. Helix pomatia *Lectin Binding to Breast Cancers*

1. Paraffin sections of breast cancers (materials listed in Sections 2.5.–2.7. required).
2. *Helix pomatia* lectin (HPA; Sigma) at a concentration of 10 µg/mL in lectin buffer.
3. Lectin buffer (*see* Section 3.1.).
4. Polyclonal rabbit anti-HPA (Serotec, Oxford, UK) diluted 1/200 in lectin buffer.
5. Biotinylated polyclonal swine antibody against rabbit immunoglobulins (Dako) diluted 1/400 in lectin buffer.
6. Streptavidin-labeled horseradish peroxidase (Calbiochem) diluted 1/400 in lectin buffer.
7. 3,3-diaminobenzidine tetrahydrochloride (DAB; Sigma) 0.5 mg/mL in lectin buffer. Hydrogen peroxide added to give a concentration of 5% v/v immediately before use (*see* Note 7).
8. Mayers hematoxylin solution (Sigma).
9. 70% v/v, 95% v/v, and 99% v/v ethanol in distilled water.
10. CNP30 (BDH) or xylene.
11. Depex resinous mounting medium (BDH).

3. Methods
3.1. Preparation of Samples
3.1.1. Tissue Imprints

1. Slice through a piece of fresh tissue using a clean, sharp scalpel blade.
2. Touch the cut surface briefly, gently, and cleanly onto a clean microscope slide. Do not allow the tissue to drag on the surface of the slide and do not press hard—a light touch is enough.
3. Air-dry for 5 min (slides may then be stored wrapped in foil in the freezer until required).
4. Fix by dipping in acetone or methanol for 1 min and then air-dry.

3.1.2. Cell Smears

1. Smear cells onto two clean glass microscope slides by placing a drop of cell suspension near one end of the first slide, placing the second slide on top, and dragging one slide against the other in a swift, smooth movement.

2. Air-dry for 5 min (slides may then be stored wrapped in foil in the freezer until required).
3. Fix by dipping in acetone or methanol for 1 min.

3.1.3. Frozen (Cryostat) Sections

1. Snap freeze a 0.5-cm^3 piece of fresh tissue on an OCT-coated cryostat chuck by immersion in isopentane or hexane, precooled in liquid nitrogen. Allow to equilibrate to the temperature of the cryostat before cutting.
2. Cut 5-μm thick sections using a cryostat.
3. Air-dry for 5 min (slides may then be stored wrapped in foil in the freezer until required).
4. Fix by dipping in acetone or methanol for 1 min.

3.1.4. Formalin-Fixed, Paraffin-Embedded Sections

1. Cool tissue blocks on ice for approx 15 min.
2. Cut 5-μm thick sections by microtome.
3. Float sections on a pool of 20% ethanol in distilled water on a clean glass plate.
4. Carefully float sections from the ethanol solution to the surface of a water bath heated to 40°C.
5. Pick the sections up on clean glass slides. Allow to drain for 5–10 min (*see* Note 8).
6. Dry in an incubator at 37°C overnight, then at 60°C for 20 min. Slides may then be cooled, stacked, or boxed, and stored at room temperature until required.
7. When required, soak in CNP30 or xylene for 20 min to dissolve the wax (*see* Note 9).
8. Transfer through two changes of 99% ethanol; through one change of 70% ethanol and finally to distilled water, agitating the slides for 1–2 min at each stage to equilibriate.
9. Allow to stand for 5 min in lectin buffer.

3.2. Trypsinization of Paraffin Sections (see Note 4)

1. Place sections in a bath of trypsin solution and incubate at 37°C for 5–30 min, depending on the tissues and/or lectins used.
2. Wash in running tap water for 5 min.
3. Allow to stand for 5 min in lectin buffer.

3.3. Blocking Endogenous Peroxidase (see Note 5)

1. Immerse sections in methanol/hydrogen peroxide solution for 20 min.
2. Wash in running tap water for 5 min.
3. Allow to stand for 5 min in lectin buffer.

3.4. Preparation of a Crude Mushroom Lectin Extract (see Note 10)

1. Chop mushrooms roughly with a scalpel blade.
2. Homogenize in a laboratory tissue homogenizer or domestic food processor or juicing machine. A small volume of distilled water or lectin buffer may be added if the extract is too thick or viscous.

Light Microscopy 15

3. Filter.
4. The extract may be stored frozen until required.

3.5. Labeling Lectins with Biotin (see Note 6) (8–10)

1. Dissolve the lectin at a concentration of 1 mg/mL in $0.1M$ $NaHCO_3$, or dialyze crude extract or lectin solution against $0.1M$ $NaHCO_3$ overnight.
2. Remove NHS-biotin from the freezer and allow to thaw for approx 30 min (*see* Notes 6 and 11).
3. Dissolve NHS-biotin in dimethylformamide or dimethylsulfoxide at a concentration of 1 mg/mL.
4. As rapidly as possible after dissolving the NHS biotin, add the biotin solution to the lectin solution in a ratio of approx 120 µL biotin solution:1 mL lectin solution. Mix immediately and well. React at room temperature, preferably with constant gentle mixing, e.g., on an end-over-end mixer, for 2 h or overnight.
5. Dialyze overnight against lectin buffer, and store at 4°C.

3.6. Examples of Staining Methods (see Note 12)

3.6.1. Direct Method to Detect Peroxidase-Labeled Peanut Lectin Binding to Kidney Sections

1. Paraffin sections of human or animal kidney should be used, treated as described in Sections 3.1.4. and 3.3., or fresh, frozen sections treated as described in Sections 3.1.3. and 3.3. Paraffin sections should be trypsinized for 20 min as described in Section 3.2.
2. Incubate sections with peroxidase-labeled PNA for 1 h.
3. Wash in lectin buffer (*see* Note 13).
4. Incubate with DAB for 10 min (*see* Note 7).
5. Wash in running tap water.
6. Counterstain by immersing in Mayers hematoxylin for 2 min.
7. Wash in running tap water for 5–10 min.
8. Pass through 70% ethanol, 95% ethanol, and two changes of 99% ethanol. Agitate the slides at each stage until equilibriated.
9. Agitate for 1–2 min in CNP30 or xylene.
10. Mount in a resinous mountant such as Depex.

3.6.2. Biotinylated Mushroom Lectin Binding to Gut Sections

1. Paraffin sections of human or animal gut should be used, treated as described in Sections 3.1.4. and 3.3., or fresh, frozen sections treated as described in Sections 3.1.3. and 3.3.
2. Incubate with biotinylated mushroom lectin extract, prepared as described in Sections 3.4. and 3.5.
3. Wash in lectin buffer (*see* Note 13).
4. Incubate with strepatavidin peroxidase for 30 min.
5. Wash in lectin buffer.
6. Treat as described in steps 4–10 in Section 3.6.1.

3.6.3. Direct Method to Detect FITC-Conjugated UEAI Binding to Endothelial Cells

1. Paraffin sections of human striated muscle should be used, treated as described in Sections 3.1.4., and trypsinized for 10 min as described in Section 3.2.; or fresh, frozen sections treated as described in Sections 3.1.3.; or imprints treated as described in Section 3.1.1.; or smears of cultured human endothelial cells treated as described in Section 3.1.2.
2. Incubate with FITC-UEAI for 1 h.
3. Wash in lectin buffer (*see* Note 13).
4. View wet under UV light using a fluorescent microscope.

3.6.4. Indirect Antibody Method to Detect UEAI Binding to Endothelial Cells

1. Paraffin sections of human striated muscle should be used, treated as described in Sections 3.1.4. and 3.3., and trypsinized for 10 min as described in Section 3.2.; or fresh, frozen sections treated as described in Sections 3.1.3. and 3.3; or smears of cultured human endothelial cells treated as described in Sections 3.1.2. and 3.3.
2. Incubate with UEAI for 1 h.
3. Wash in lectin buffer (*see* Note 13).
4. Incubate with peroxidase conjugated rabbit anti-UEAI for 1 h.
5. Wash in lectin buffer.
6. Treat as described in steps 4–10 in Section 3.6.1.

3.6.5. Helix pomatia *Lectin Binding to Breast Cancers*

1. Paraffin sections of human breast cancer should be used, treated as described in Sections 3.1.4. and 3.3., and trypsinized for 20 min as described in Section 3.2.
2. Incubate with HPA for 1 h.
3. Wash in lectin buffer (*see* Note 13).
4. Incubate with polyclonal rabbit anti-HPA for 1 h.
5. Wash in lectin buffer.
6. Incubate with biotinylated swine antirabbit for 1 h.
7. Wash in lectin buffer.
8. Incubate with streptavidin peroxidase for 30 min.
9. Wash in lectin buffer.
10. Treat as described in steps 4–10 in Section 3.6.1.

4. Notes

1. Other enzyme labels are available—for example, alkaline phosphatase, alcohol dehydrogenase, and glucose oxidase. Other fluorescent labels are available—for example, Texas red and tetramethylrhodamine isothiocyanate (TRITC). Electron-dense labels such as ferritin and colloidal gold can also be used.
2. Lectins with the same nominal sugar binding specificity—for example, "fucose binding lectins" or "galactose binding lectins"—will actually have quite differ-

ent specificities for complex sugar structures (as described in Section 1.2.) and will thus give very different histochemical staining patterns on the same tissues.
3. The binding of some lectins is dependent on the presence of calcium and other metals. Buffers containing phosphate ions, like phosphate-buffered saline (PBS), are therefore generally unsuitable for lectin histochemistry, as the phosphate ions bind with and sequester the metals. We therefore recommend the use of a general lectin buffer for all dilutions and washes. This is basically a Tris-HCl buffer, pH 7.6, but with added calcium chloride and magnesium chloride.
4. Digestion with an enzyme such as trypsin can often be helpful in revealing carbohydrate structures hidden or damaged during fixation and processing of paraffin-embedded tissues. Different tissues, fixed and processed in different ways, and different lectins, vary enormously in their requirement for trypsinization. We would recommend that when a new method is being tested, the effect of a range of trypsinization times is assessed (e.g., no trypsin, 5 min trypsinization, 10 min trypsinization, 20 min trypsinization, and 40 min trypsinization). Trypsinization can sometimes dramatically enhance staining. It is better to use a fairly crude preparation of trypsin—we recommend the type II crude trypsin from porcine pancreas stocked by Sigma—as sometimes the impurities (like chymotrypsin) actually enhance its effect! Trypsin solutions should be made up fresh when required, as they lose their activity over time. For this reason, it is rarely worth continuing trypsin digestion for more than 30–40 min. Over-trypsinization will result in overdigestion of tissue and loss of morphology. Trypsinization should be carried out at 37°C—all glasswear, buffers, and so on should be prewarmed before use. Trypsinization is only appropriate for formalin-fixed, paraffin-embedded tissues; it is *not* appropriate for cell imprints, smears, or frozen sections.
5. If horseradish peroxidase is to be used as a label, it is necessary to quench any endogenous peroxidase in the tissue sections. Failure to carry out this step will result in high levels of background staining. It is, obviously, not necessary to carry out this step if a fluorescent tag, such as fluorescein isothiocyanate, is used, or if other enzyme labels such as alkaline phosphatase are used. If other enzyme labels are used, it may be helpful to carry out a different blocking step appropriate to that enzyme—for example, if alkaline phosphatase is used, endogenous alkaline phosphatase can be blocked using levamisole.
6. We purchase NHS-biotin in 5-mg aliquots—the smallest supplied by Sigma. For each biotinylation experiment, one 5-mg aliquot is allowed to thaw, then dissolved in 5 mL dimethylformamide. Any unused NHS-biotin is discarded at the end of the procedure. We find this gives consistently successful biotinylation, and is a more satisfactory approach than buying a larger quantity of NHS-biotin, and thawing and weighing out for each experiment.

Biotinylation is a very quick, cheap, and easy way of labeling lectins. For lectins purchased in an unlabeled form in which no antibody is commercially available, or for homemade or obscure lectins, it is often the easiest—sometimes the only—practical way of detecting their histochemical binding.

7. DAB is potentially carcinogenic and should therefore be handled with care. Wear gloves. Avoid spillages and aerosols. Work in a fume cupboard. After use, all equipment, glasswear, and so on should be soaked in diluted bleach overnight before washing; worktops should be swabbed down with dilute bleach and rinsed with plenty of water.

   To limit the risks associated with weighing out, we purchase DAB in 1-g batches. An entire 1-g batch is dissolved in 200 mL distilled water and 1-mL aliquots transferred to 10-mL capped plastic tubes and frozen until required. When required, a tube is thawed, and 9 mL lectin buffer and the appropriate volume of hydrogen peroxide are added.

8. If sections are damaged in any way—for example, if they have tiny scores or tears—or if a very long staining method is to be undertaken, it may be a good idea to precoat the clean glass slides with an adhesive such as poly-L-lysine to help the tissue section stick firmly to the glass. A tiny drop of a 0.1% w/v aqueous solution of poly-L-lysine (Sigma) is applied near one end of a clean glass microscope slide. A second slide is placed against it, and one is dragged against the other in a fast, smooth movement to leave a very thin film on both slides. The slides should be allowed to air dry for about 5–10 min, then sections applied to them as usual.

9. Once paraffin sections have been dewaxed, it it imperative that they not be allowed to dry out at any point, as this will result in high levels of dirty, nonspecific background staining.

10. An extremely crude method for extraction of lectin from mushrooms is given. In spite of its obvious limitations, it works very well and should give excellent histochemical results. This type of approach is equally applicable to many other lectins—for example, a very adequate preparation of peanut lectin can be obtained by soaking peanuts in distilled water or lectin buffer overnight. Experiment!

11. It is critical to exclude water from the NHS-biotin before it mixes with the lectin solution. For this reason, the vial should be thoroughly thawed before the cap is removed in order to discourage condensation of water vapor from the air.

12. The methods are given to act as examples. The scientist is encouraged to try other lectins, other tissues, other staining protocols—the applications are limitless!

13. Thorough washing is critical for good histochemistry results. We recommend washing the sections in at least three changes of lectin buffer. Each wash should last at least 1–2 min. Agitation helps a great deal. We would recommend placing the slides in a metal slide rack and jiggling them up and down in the buffer quite vigorously! If problems with background staining are encountered, one of the major culprits is likely to be insufficiently vigorous washing.

14. If a high level of nonspecific background staining is observed, and this cannot be corrected by vigorous washing (*see* Note 13), a blocking solution may be required. In traditional immunohistochemistry, the approach would be to incorporate 1–5% normal rabbit or swine serum into antibody dilutions and washes. This is *not* appropriate for lectin histochemistry as many lectins bind to glycoproteins in serum. We recommend incorporating 1–5% bovine serum albumin

into the lectin and antibody solutions. This can sometimes dramatically improve staining results.
15. The optimum temperature for lectin binding may be lower than the usual room temperature and, although this should not normally cause any problems, some experimentation is worthwhile. Remarkably, some mammalian lectins appear to bind best at 7 or 8°C, and minimally at 37°C.
16. The optimum binding of lectins may be pH dependent. Again, this should not cause problems, and most lectin histochemistry will work well at pH 7.6. If problems are encountered with unusual or little-used lectins, some experimentation may be helpful.
17. Ten micrograms per milliliter seems to be a good, optimal concentration for most lectins in histochemistry and, if a new lectin is to be tried, it is a sensible concentration to start with. We recommend trying a range of concentrations (e.g., 2.5, 5, 10, 20, and 40 µg/mL) to assess which gives the best results. Some lectins, such as pokeweed, work best at very low concentrations (~1 µg/mL); others, such as *Limulus polyphemus*, require much higher concentrations (~100 µg/mL).
18. Lectins are usually purchased as a dry powder. They can be conveniently stored as a stock solution of 1 mg/mL in lectin buffer at 4°C, and will be stable for months or years. Most native lectins are also reasonably stable, stored at 4°C at their working dilution (e.g., 10 µg/mL in lectin buffer). However, some lectins (such as *Limulus polyphemus*) are remarkably labile.
19. Specificity of lectin binding may be confirmed by competitive inhibition using the appropriate monosaccharide. Lectin is incubated on the tissue section in the presence of $0.1M$ monosaccharide—for example, UEAI in the presence of $0.1M$ fucose. Alternatively, bound lectin may be displaced by subsequent incubation with the sugar solution. With some lectins, competitive inhibition is very strong, giving complete abolition of staining; with other lectins, a significant reduction in staining intensity, but not complete abolition, may be expected.

Suggestion for Further Reading

A publication that is out of print but still full of interesting detail (beautiful electron micrographs and 50 pages of references on lectin cytochemistry and histochemistry) is: Schrevel, J., Gros, D., and Monsigny, M. (1981) Cytochemistry of Cell Glycoconjugates. 250 page monograph in *Progress in Histochemistry and Cytochemistry* **14(2),** Gustav Fischer Verlag, Stuttgart, New York.

References

1. Brooks, S. A. and Leathem, A. J. C. (1991) Prediction of lymph node involvement in breast cancer detection of altered glycosylation in the primary tumour. *Lancet* **338,** 71–74.
2. Walker, R. A. (1993) *Helix pomatia* and breast cancer. *Br. J. Cancer* **68,** 453,454.
3. Ikeda, Y., Mori, M., Adachi, Y., Matsushima, T., and Sugimachi, M. (1994) Prognostic value of the histochemical expression of *Helix pomatia* agglutinin in advanced colorectal cancer. *Dis. Colon Rectum.* **37,** 181–184.

4. Shumacher, U., Higgs, D., Loizidou, M., Pickering, R., Leathem, A., and Taylor, I. (1994) *Helix pomatia* lectin binding as a marker of metastasis and reduced survival in carcinoma of colon and rectum. *Cancer* **74,** 3104–3107.
5. Kakeji, Y., Tsujitani, S., Mori, M., Maehara, Y., and Sugimachi, K. (1991) *Helix pomatia* agglutinin binding activity is a predictor of survival time for patients with gastric carcinoma. *Cancer* **68,** 2438–2442.
6. Yoshida, Y., Okamura, T., and Shirakusa, T. (1993) An immunohistochemical study of *Helix pomatia* agglutinin binding on carcinomas of the oesophagus. *Surg. Gynecol. Obstet.* **177,** 299–302.
7. Shiraishi, T., Atsumi, S., and Yatani, R. (1992) Comparative study of prostatic carcinoma bone metastases among Japanese in Japan and Japanese Americans and whites in Hawaii. *Adv. Exp. Med. Biol.* **324,** 7–16.
8. Bayer, E. A. and Wilchek, M. (1980) The use of avidin-biotin complex as a tool in molecular biology. *Meth. Biochem. Anal.* **26,** 1–45.
9. Bayer, E. A., Wilchek, M., and Skutelsky, E. (1976) Affinity cytochemistry: the localisation of lectin and antibody receptors on erythrocytes via the avidin biotin complex. *FEBS Lett.* **68,** 240–244.
10. Guesdon, J.-L., Ternynck, T., and Avrameas, S. (1979) The use of avidin-biotin interaction of immunoenzymatic techniques. *J. Histochem. Cytochem.* **27,** 1131–1139.

2

Lectin Histochemistry and Cytochemistry—Light Microscopy

Avidin–Biotin Amplification on Resin-Embedded Sections

Robert W. Stoddart and Carolyn J. P. Jones

1. Introduction

Within the last 25 yr, the study of glycans, their structures, and their distribution in tissues has emerged from relative obscurity to become a major theme of molecular and cellular biology: "glycobiology" *(1)*. Glycans are major components of cellular surfaces *(2,3)*, extracellular matrices *(4,5)*, and secretions *(6)*, and play important roles in cell–cell and cell–matrix recognition and adhesion *(7,8)*. They regulate the surface environment of cells by influencing the structure of water *(6,9)*, by modulating diffusion, by sequestering metabolites such as metal ions, by presenting various growth factors to their receptors *(10,11)*, by acting as ligands in recognition-adhesion systems *(11–13)*, and by making major contributions to cell surface charge *(2,14)*. In secretions, they are variously determinants of molecular folding *(15)*, hydration, interaction, and targeting, and they serve in the mechanisms that monitor the aging of glycoconjugates in circulation *(6,9,16,17)*. They are now implicated in mechanisms of molecular targeting and segregation within the endomembrane systems of cells *(18–20)*, in calcium transport in mitochondria *(21)*, the handling of mRNA and its export from the nucleus, the regulation of transcription *(19)*, and in their classical role as energy stores. In all of this, the association of anatomical or ultrastructural localization with biological function is of paramount importance: specific glycans occur in specific places *(22,23)*. The challenge for the histochemist is to reconcile the achievement of sufficiently precise anatomical localization of glycans with the maximum of chemical information about their nature. The lectins have been the main means of accomplishing this, since their first application as fluorochrome-labeled probes to paraffin sections in the early 1970s *(2,24)*. More recently, fluorescence has been largely supplanted by nonfluorescent disclosing systems, such as biotin–avidin–peroxidase and, though paraffin sections are still widely used, resin-embedding of specimens has proved advantageous in terms of resolution and economy of material at the light microscopic level.

1.1. The General Application of Lectins in Glycobiology and Histochemistry

Lectins are carbohydrate-binding proteins that, by definition, have at least two binding sites, are not immunoglobulins, and do not show glycosidase activity *(25)*. Most of those in histochemical use are obtained from plants and the largest group derive from the *Leguminosae*. These show considerable sequence homology, despite a remarkable diversity in the glycans to which they bind. In other orders of plants, related lectins seem less diverse in their ligands.

The majority of lectins in common use have molecular weights in the range 60–130 kDa, are themselves glycoproteins (though they do not bind to their own glycans and seldom do so to glycans of each other), usually contain subunits, are often globular proteins, and are all freely soluble in aqueous buffers in the pH range 6–8. Most are stable to freeze drying, either before or after labeling with biotin, and many require small amounts of calcium for binding. Several also contain transition metals, which may be lost in the presence of the chelating agent EDTA. They are usually slightly smaller proteins than immunoglobulins of class G and mostly show comparable stability and ease of labeling. Both native and labeled forms of many lectins are now commercially available, often at much lower prices than those of monoclonal antibodies (MAbs). In general, lectins can be used in all ways analogous to antibodies, so that histochemical analysis can be supplemented by adsorption studies, the use of blotting, rocket electrophoresis, lectin adsorption assays (analogous to ELISA), and similar techniques.

Hence, though lectins and saccharide-directed MAbs are essentially complementary reagents to each other, there are occasions when a lectin may be the reagent of choice on grounds of cost alone.

With a few exceptions (such as *Helix pomatia* lectin), lectins have extended binding sites and in many plant lectins these are cleft-like. If the cleft is open-ended, the lectin may be able to interact with both terminal and internal saccharide sequences of glycans, but if it is closed at one end, the lectin will only interact with nonreducing terminal sequences *(25)*.

The size and shape of lectin molecules impose restrictions on the largest size of glycan that could be accommodated in a binding site: for a linear glycan this is about 6–7 sugar residues. In practice, most lectins have smaller binding sites than this, with few interacting with linear glycan sequences of more than 3–4 monosaccharide units. MAbs show similar restrictions and tend to have slightly smaller binding sites than lectins.

Most glycans are too big for their entire structure to be accommodated by a lectin or antibody and so, in general, only limited parts of their structures can be defined by the application of a single lectin or antibody probe. However, the use of a set of probes with overlapping binding requirements, particularly in conjunction with a set of exoglycosidases, can enable the analysis of glycans larger than the sequences complementary to single binding sites. This can be done by classical biochemistry, by blotting techniques, and, *in situ*, by histochemical methods.

Histologically, lectins display both specificity and selectivity; it is important to distinguish between these. Specificity defines the minimal chemical structure that can interact with the binding site of the lectin to give a binding of more than a certain, chosen affinity constant. In practice, specificity is often defined operationally in terms of inhibition of binding or red-cell agglutination by a range of saccharides, and the lower cutoff of affinity is in the range 10^3–10^4 moles^{-1}. Occasionally there are X-ray crystallographic or NMR data available to delineate more precisely the chemistry of sugar binding to lectins. Selectivity is a histological term referring to the ability of a lectin to bind to one or more structures in a section of a chosen tissue. It is not directly related to specificity, in that a highly specific lectin may bind to a common sequence present either in many different glycans or on numerous histological features and so may display poor selectivity. However, most highly selective lectins also show fairly high specificity in that they absolutely require specific chemical features in the ligand. These typically include oxygen atoms in various of the hydroxyl groups, in the pyranose rings of the sugars, or in glycosidic linkages, and nitrogen atoms in acetamido- groups of the 2-deoxy, 2-acetamido-sugars. Hence there is great conformational and configurational specificity.

Lectins are used histologically in two general ways. Selectivity may be exploited by using them as markers of particular structures, such as the nuclear cap and acrosomes in developing rat spermatozoa *(26)*. In this type of study, the emphasis is on obtaining maximal resolution, so that the position of a single type of glycan sequence can be defined as closely as possible. This places severe restrictions on section thickness, so that paraffin sections may prove inadequate, but it is not heavily demanding of material.

Specificity is exploited in making general surveys of glycan distribution in tissues. A carefully selected panel of lectins of well-defined specificities is used to stain successive sections and, from their known binding requirements, a series of inferences can be drawn as to where particular sorts of glycan occur *(27–31)*. Such investigations should precede the use of selected lectins of high selectivity as markers in studies of the type indicated above. These general surveys of glycan distribution make heavy demands on material for sectioning, but may not require very high resolution, at least in the first instance. They are often made on paraffin-embedded material, which can have the advantage of a larger sample size than is easily possible with plastic-embedded specimens. However, problems of resolution can occur and re-embedding in plastic may be desirable (*see* Section 1.3.).

The extension of the use of lectin specificity in conjunction with that of enzymes to the partial sequencing of larger glycans is a relatively new development *(32,33)*, and is not yet very widely applied. Like the general surveys above, it is very demanding of material in that it requires relatively large numbers of sections, but is often also demanding in terms of resolution, making semithin sections a useful option.

1.2. The Advantages and Limitations of Resin Embedding

In order to cut semithin (0.5–1.0 µm) sections of good quality, it is necessary to embed suitable well-fixed tissues in plastics that are more rigid and supportive than paraffin wax. This creates five sets of problems, relating to fixation, the choice of embedding agent, the sectioning of the tissue and adherence of the sections to the slide, the achievement of penetration by the lectin and the disclosing system, and the production of a suitable image.

Fixation should be of the best possible quality and, ideally, of a standard that would be acceptable for electron microscopy, since the optical microscopy of semithin sections should yield a resolution comparable with that of lowpower electron microscopy. No general method exists, however, because optimal fixation of human soft tissues, for example, requires conditions different from those needed for the fixation of marine invertebrates. The general rule is that fixatives that are satisfactory at the ultrastructural level can be used with confidence, with the proviso that metals, such as osmium, lead, or mercury should be avoided if possible, as they may alter the appearance of the final image by generating artifacts with the disclosing system. Care should also be taken to avoid powerful oxidizing agents, such as chromates, which may degrade glycans, and strong acids, which can hydrolyze sialyl glycosides on prolonged exposure *(34)*. Aldehyde fixatives based on glutaraldehyde, formaldehyde, or a mixture of the two generally give the best results, but excess aldehyde must be washed away and any free aldehyde residues should be blocked (for example, with saturated glycine in 70% [v/v] alcohol) before exposure to lectin in order to minimize the risk of nonspecific attachment of lectin by crosslinking.

There are two main types of plastic embedding material, and each has its advantages and disadvantages for use in microscopy. Epoxy resin (e.g.,

Lectin Histochemistry and Cytochemistry

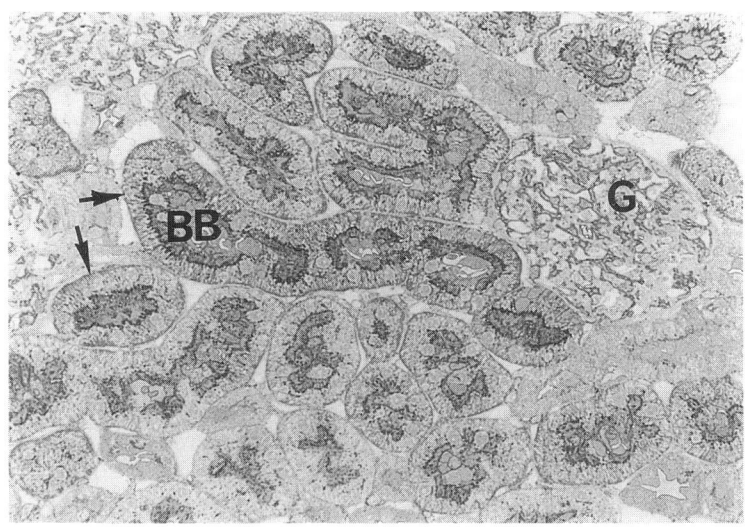

Fig. 1. Semithin epoxy (Taab) resin section of rat kidney stained *(37)* with e-PHA (*Phaseolus vulgaris*, erythroagglutinin, kidney bean lectin) showing binding to the brush border (BB), basal plasma membrane (arrows) and intracellular organelles of proximal tubules, and to components of glomeruli (G) (×350).

Araldite, Epon, Taab embedding resin, Agar 100) has been used for many years for routine electron microscopy and gives excellent results as an embedding medium for use in lectin histochemistry at the light microscopic level (Fig. 1). This is because the resin can be easily removed by a simple etching procedure using alcoholic sodium ethoxide, which, surprisingly, does not degrade glycans nor remove *O*-linked residues by ß-elimination *(35)*, but leaves a thin layer of tissue to be stained as required with the lectins. Alternatively, acrylic resin (e.g., L.R. White, Lowicryl) can be used (Fig. 2), but this resin is not easily removed. In order to visualize the stain at the light microscopic level, when it binds only to the surface molecules of the section and is barely visible even when viewed in the light microscope, stain intensification methods utilizing silver and gold enhancement have to be used *(36)*, some of which we have found to be rather capricious. Acrylic resin has the property, however, of being much more tolerant to water than epoxy resins: L.R. White resin can be infiltrated in a 3:1 mix with 70% (v/v) alcohol, thus avoiding the use of absolute alcohol and the subsequent destruction of sensitive epitopes, which is important when using immunocytochemical techniques to demonstrate protein antigens; glycans are, however, very hardy structures and will tolerate full dehydration during the processing program. Penetration by the disclosing sys-

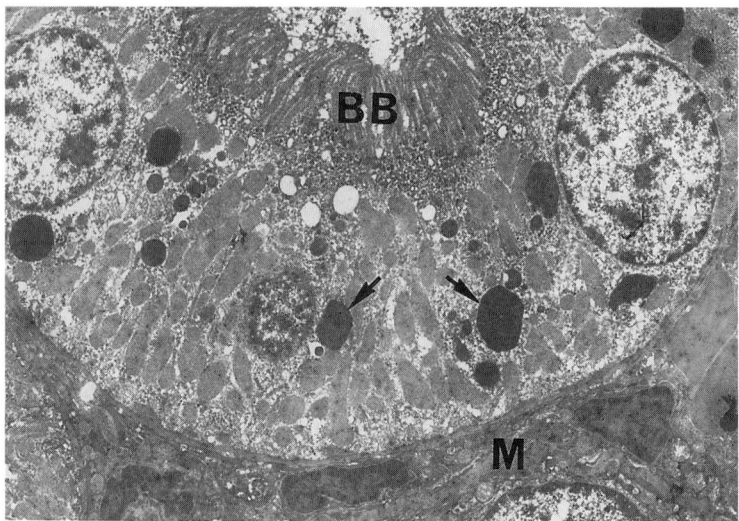

Fig. 2. Ultrathin acrylic (L.R. White) resin section of a rat kidney proximal tubule stained *(37)* with WGA (*Triticum vulgaris*, wheatgerm) lectin showing the strong binding to intracellular granules (arrows), as well to the brush border (BB) and matrix components (M) (×5600).

tem may be limited because of the size of the avidin molecule and its peroxidase conjugates: alternative disclosing systems tend to have similar problems. Acrylic resin will, however, allow even these large molecules to penetrate to a limited extent, unlike epoxy resin, and is the ideal medium for staining at the ultrastructural level. In our laboratory, we routinely embed tissue in epoxy resin (Taab embedding resin) for light microscopic staining with lectins, and in acrylic resin (L.R. White) for ultrastructural examination of the lectin binding properties of tissues. If it is essential to utilize the same block for both light and electron microscopy, the epoxy resin can be gently etched prior to staining grids bearing ultrathin sections, but it is advisable to cut the sections slightly thicker than usual.

It is desirable to control section thickness and the etching conditions carefully, in order to achieve reproducible results and to facilitate comparisons between sections. Because sections may be exposed to a succession of treatments and washings, it is essential to use cleaned and coated glass slides to avoid their detachment.

Because of the thinness of the sections used, the final image must be of relatively greater density than would be required for a paraffin section. Peroxidase-based disclosing systems will produce dense brown images, which can be intensified by treatment with salts of metals such as silver or cobalt *(36,37)*.

Examination or photography of the specimen through a blue or cyan filter is also a useful way of improving contrast. Disclosing systems based on peroxidase yield a highly insoluble colored product that does not diffuse perceptibly at the light microscopic level, whereas those based on alkaline phosphatase are subject to some diffusion. Although this is usually acceptable on paraffin sections, it can be a limitation at the semithin level.

It is important to recognize that the intensity of stain given by the disclosing system will be proportional to the amount of lectin bound, or glycan sequence present, only over a limited range. Hence, very sensitive disclosing systems may show all of the sites at which a particular ligand occurs, but fail to reveal variations in its density that may be better shown by less sensitive systems. Changes in image may occur in switching from paraffin sections to semithin sections, because of this type of effect. Titration of the staining pattern against various concentrations of lectin will often reveal its basis.

1.3. The Adaptability and Applicability of Lectin-Stained Semithin Sections

Lectin histochemistry of paraffin sections often shows features that, although interesting, cannot be resolved sufficiently to show exactly where the stain is located. If all the available tissue has already been processed for paraffin sectioning, as is often the case with pathological material, a solution to the difficulty may be to dewax a portion of the paraffin block and re-embed the tissue in an epoxy resin. This can also be a useful approach in cases in which sampling of an infrequent feature is required and the size of block needed for direct semithin sectioning would be too small to be efficient. A feature found by paraffin sectioning can be excised, re-embedded, and examined at higher resolution and at the electron microscope (EM) level if required. Very often, semithin optical sections will reveal quite sufficient detail to make electron microscopy unnecessary *(38)*.

In going from paraffin sections to successively higher levels of resolution with lectin staining, it is important to ensure, wherever possible, that only one variable is changed at each step. In going to semithin sections from paraffin-embedded sections, the embedding medium is altered, but the staining and disclosing steps can remain essentially unaltered if they are peroxidase-based, so that the images obtained remain comparable. Similarly, the step to low power EM may involve only a change in the microscope system, although sometimes it may be more convenient to use an acrylic resin embedding medium, and the staining procedure again remains unchanged since the peroxidase product is electron-dense. In our laboratory, we routinely use an avidin–biotin–peroxidase system *(37,38)* rather than the avidin–biotin–peroxidase complex (ABC) method of Hsu and Raine *(39)* because the former gives excellent results at both the light and electron optical levels, whereas the large size of the peroxidase complex generated in the latter method makes it difficult to penetrate resin sections. If higher resolution is required, the disclosing system can then be changed from a peroxidase to a gold-based one, allowing the two systems to be compared at lower power, to check their equivalence, before moving to higher magnification. It is this flexibility that makes lectin histochemistry on semithin sections so attractive and powerful a tool.

1.4. Control Procedures

As with other techniques of lectin histochemistry, controls must be applied to its use on semithin sections. These are of two kinds, namely controls on the action of the disclosing systems and controls on the specificity of the lectins used.

Disclosing systems are normally controlled by the omission of reagents at each step, which should abolish all reactions thereafter. Residual staining in the direct avidin–peroxidase system, for example, can arise from the action of endogenous peroxidase, nonspecific adsorption of avidin–peroxidase, the presence of endogenous biotin, or the nonspecific adsorption and oxidation of diaminobenzidine, and these can be separated by appropriate simple controls. More complex disclosing systems require control at each step and are, therefore, more costly of material. Since a major advantage of semithin sections is their economy of valuable material, there is some logic in using the most direct disclosing system possible and using a single disclosing system for a panel of lectins, so that only one set of controls on disclosure is required.

The control of lectin specificity creates more subtle problems. Most studies have used competitive inhibition with free sugars, glycosides, or oligosaccharides to show that the binding of lectin can be selectively abolished. In practice, some lectins such as erythro- and leuko-phytohemagglutinins (e-PHA and l-PHA) have no simple sugar competitors and the oligosaccharides that are their ligands are not sufficiently available to be usable at appropriate competitive concentrations. Similarly, competition with lectins that bind to sialyl glycans, using sialic acids or small sialyl saccharides, cannot be achieved without a concomitant variation in acidity or ionic strength, which will render the controls questionable. In both of these situations, enzymatic controls are to be preferred (*see* Section 3.10. and Note 5).

Complete abolition of lectin binding cannot always be achieved and, since the competition will approximate to pseudo-first-order kinetics, this is to be expected where the affinity constant of the lectin for its target glycan is high relative to that for the free sugar or glycoside. Hence, somewhat paradoxically, it is difficult to prove the specificity of highly specific binding. It should be noted that some lectins show slower, largely irreversible reactions with their ligands, leading to an increase in apparent affinity constant and increasing difficulty in reversing their reaction: to minimize this, the lectin and its competitor should be premixed or placed on the specimen simultaneously.

In cases in which a lectin has two or more subsites within its overall binding site, paradoxical patterns of competitive inhibition may arise. For example, the lectin of *Maclura pomifera* has subsites for Galß1- and GalNAcα1-, and binds with high affinity to Galß1,3GalNAcα1-. However, this binding is inhibited by ß-anomeric glycosides of galactose and α-anomeric glycosides of 2-deoxy,2-acetamido galactose *(40)*, which led, for some time, to confusion as to the real specificity of the lectin. In such cases, controls need to be carefully designed. Even where clear-cut competitive inhibition is shown, it proves only that the lectin is binding via its binding site: for example, concanavalin A binds to α-mannosyl or α-glucosyl termini and α-methyl mannopyranoside will abolish *both* bindings—it does *not* prove that terminal α-mannose is present in the glycan.

For these reasons, the use of competitive inhibition should be treated with care. It is necessary and can be helpful, but can also mislead, confuse, or cause overinterpretation. Hence, it is often desirable to use exoglycosidases, either singly or in panels, to supplement or replace competitive inhibition; they often provide additional structural information as well *(26,32,33)*. Some care must be exercised to show that the enzymes are pure glycosidases, free of proteolytic activity, and remain active during their usage. Also, some glycosidases (especially ß-galactosidase) are very sensitive to the positions of linkages or other structural features of glycans, as well as to anomeric configuration, and may prove unexpectedly unreactive, though this also can be structurally informative *(26)*.

Used in panels, lectins can act as controls on each other, though this demands that their binding requirements have been well-defined by chemical methods.

Discrimination between *O*- and *N*-glycans is possible only with the limited set of glycans that have a requirement for α-mannose in the structures to which they bind; in mammals this sugar is restricted to *N*-glycans and to glycosylated phosphoinositides *(1,15,23,41)*. Whereas some saccharides tend to be more common in *O*- or *N*-glycans, it is dangerous to use the lectins that bind to them as markers of *O*- or *N*-linkage, unless there is additional chemical information to validate this. If, for example, the lectin of *Dolichos biflorus* shows staining that is prevented by pretreatment of the specimen with alkali *(35)* (ß-elimination; *see* Section 3.11.), there are good grounds for attributing its ligand to an *O*-glycan. If staining is unaffected, an *N*-glycan is suggested, but not proven. This requires further chemical analysis.

1.5. Hazardous Lectins

A small number of lectins are liable to be contaminated with plant toxins or to be inherently toxic. Those of *Ricinus communis*, *Abrus precatorius*, and *Viscum album* are the most likely to be encountered because of their commercial availability. If possible, their use should be avoided, but if they *are* used, care should be taken to handle them according to the supplier's recommended instructions and to dispose of wastes safely. All lectins are potentially allergenic and their handling as dry solids should be minimized. Toxic lectins should never be handled in a dry form, but only in solution. They are usually supplied in vials with septum caps to enable them to be made up into solution *in situ*.

2. Materials

2.1. Tissue Fixation

1. 2.5% (v/v) glutaraldehyde in $0.1M$ sodium cacodylate buffer, pH 7.2, prepared by mixing 1 part 25% glutaraldehyde (v/v) (E.M. grade, Taab Laboratories, Aldermaston, UK) with 9 parts buffer immediately before use.
2. 4% (w/v) paraformaldehyde in $0.1M$ phosphate buffer, pH 7.2, or $0.1M$ sodium cacodylate buffer, pH 7.2, dissolved by heat and cleared with a few drops of $1N$ NaOH, to which glutaraldehyde is added to a final concentration of 1% (v/v), freshly prepared.
3. Buffer wash: $0.1M$ sodium cacodylate buffer, pH 7.2, with 3 mM $CaCl_2$.

2.2. Plastic Embedding

1. Analar ethanol, propylene oxide, Taab embedding resin, or epoxy resin equivalent (Taab Laboratories).
2. For coating slides: 3-aminopropyltriethoxysilane (APES) (Sigma, Poole, UK), 3% (v/v) in acetone.
3. To remove resin: 50% (v/v) saturated sodium ethoxide (excess sodium hydroxide pellets in absolute ethanol, left for 3 d minimum, then diluted with an equal volume of ethanol).
4. L.R. White acrylic resin (Taab or London Resin, Reading, UK).

2.3. Lectin Histochemistry

1. Hydrogen peroxide, 30% or 100 vol.
2. $0.05M$ Tris-buffered saline (TBS) (*see* Note 1), pH 7.6, with and without 1 mM $CaCl_2$.
3. Trypsin (Sigma, type II porcine, crude).
4. Biotinylated lectins (Sigma, and EY Laboratories, San Mateo, CA; supplied by Bradsure Biologicals, Loughborough, UK), 10 µg/mL in $0.05M$ TBS with 1 mM $CaCl_2$ (*see* Note 4).
5. Avidin-peroxidase (Sigma), 5 µg/mL in $0.125M$ TBS containing $0.374M$ NaCl, pH 7.6 (made from diluting 1 part $0.5M$ stock TBS with 3 parts distilled water).

6. Diaminobenzidine tetrahydrochloride (Aldrich, Gillingham, UK), 0.05% (w/v) in TBS with 0.015% (v/v) hydrogen peroxide (diluted from 30%, 100 vol).
7. Neuraminidase (Type VI from *Clostridium perfringens*, Sigma), 0.1 U/mL in 0.2M acetate buffer, pH 5.5, with 1% (w/v) $CaCl_2$.
8. Various glycosidases (Boehringer-Mannheim, UK) (*see* Note 5).
9. For ß-elimination: Potassium hydroxide, dimethyl sulfoxide, analytical grade ethanol, 10 mM hydrochloric acid.
10. For the gold–sulfide–silver intensification method of semithin acrylic sections: 0.02% (w/v) gold chloride, neutralized 0.3% (w/v) sodium sulfide, 1% (v/v) acetic acid, Gallyas' solution (1982) made as follows:
    a. 5 g sodium carbonate (anhydrous) in 100 mL distilled water
    b. 0.2 g ammonium nitrate
       0.2 g silver nitrate
       1.0 g tungstosilicic acid
       0.5 mL 37% (v/v) formaldehyde
       100 mL distilled water

    Add reagents in the above order. Glassware must be acid cleaned. For use, add equal parts of a to b, quickly. Apply immediately. If the solution becomes chalky, the glassware has not been adequately cleaned. A faint opalescence is normal. All reagents from BDH (Merck, Poole, Dorset, UK).
11. For gold intensification of peroxidase-stained ultrathin sections: 0.1% (w/v) gold chloride (BDH).
12. Antibiotin gold (10 nm) (Biocell Research, Cardiff, UK) diluted 1:30 in TBS + 1% (w/v) BSA for colloidal gold staining of ultrathin sections.
13. For contrasting ultrathin sections after gold labeling: 2% (w/v) uranyl acetate, and Reynold's lead citrate (1.33 g lead nitrate and 1.76 g trisodium citrate in 30 mL distilled water, shaken for 1 min then left 30 min with intermittent shaking. Add 8 mL 1N NaOH and make up to 50 mL with distilled water). Reagents from BDH (Merck).

3. Methods

3.1. Fixation of Tissues

1. Excise tissue, dice into small pieces with a sharp razor blade, and fix immediately in either 2.5% (v/v) glutaraldehyde in 0.1M sodium cacodylate buffer, pH 7.2, or in 4% (w/v) paraformaldehyde in 0.1M sodium cacodylate buffer or 0.1M phosphate buffer, pH 7.2, with 1% (v/v) glutaraldehyde, at room temperature for 4 h.
2. Wash in 0.1M sodium cacodylate buffer, pH 7.2, with 3 mM $CaCl_2$, three times over 24 h, using a large volume of liquid relative to the size of the tissue. Store in buffer at 4°C until required for further processing.

3.2. Method for Re-Embedding Paraffin Block in Plastic

1. Locate the area of interest and carefully cut it out of the paraffin wax block. Trim away as much excess wax as possible.

2. Immerse in a container of xylene or similar solvent and soak at 60°C for several hours until the wax has melted. Change the xylene two or three times.
3. When there are no more signs of wax around the tissue and the tissue appears translucent, soak in xylene for a further 30 min to ensure there is no residual wax in solution.
4. Transfer the tissue to absolute ethanol for 15–30 min (depending on the size of the tissue). Change and repeat once more.
5. Soak in propylene oxide, two changes of 15 min each. Infiltrate in propylene oxide:epoxy resin (1:1) for 1 h at room temperature.
6. Infiltrate in propylene oxide:resin (1:3) at 4°C overnight.
7. Next day, infiltrate with undiluted resin at 45°C for 1 h, and repeat twice before embedding the tissue either in capsules or planchettes/foil molds and polymerizing the resin as recommended by the manufacturer (24–72 h).

3.3. 3-Aminopropyltriethoxysilane (APES) Coating of Microscope Slides

This is an excellent adhesive for both paraffin and plastic sections, and does not react with the reagents used in lectin histochemistry.

1. Clean slides with acid alcohol or 95% (v/v) ethanol for at least 2 min.
2. Rinse in tap water for 1 min.
3. Rinse in 99% (v/v) ethanol.
4. Air-dry in slide racks.
5. Place in fresh 3% (v/v) 3-aminopropyltriethoxysilane in acetone for 30 s.
6. Rinse in acetone then distilled water.
7. Dry slides and protect from dust.

3.4. Lectin Staining of Paraffin Wax Sections Using the Avidin–Peroxidase Method (42)

1. Cut sections 4 µm thick and mount on APES-coated slides. Dry in oven at 70°C.
2. Dewax in xylene and wash in absolute ethanol.
3. Block endogenous peroxidase activity for 30 min in:
    a. 400 mL analar methanol;
    b. 1.6 mL 1N hydrochloric acid; and
    c. 2.0 mL 30% (v/v) hydrogen peroxide.
    The hydrogen peroxide must be added just before use.
4. Rinse briefly in tap water and then wash in 0.05M TBS, pH 7.6.
5. Trypsinize if necessary. Warm two baths containing 300 mL TBS to 37°C in a water bath. To one, add 300 mg calcium chloride. Place slides in the other bath and add 300 mg trypsin (type II porcine, crude; Sigma) to the bath containing the calcium chloride. After 5 min, transfer the slides to the trypsin/$CaCl_2$ mixture and incubate for 5–30 min (normally 15 min).
6. Wash in running cold water to stop the reaction.
7. Wash in TBS for 3 × 5 min.

8. After wiping around the section to remove excess TBS, pipet on the biotinylated lectin (10 µg/mL in TBS + 1 mM CaCl$_2$). Incubate for 30 min at room temperature in a humidity chamber.
9. Wash off with TBS + CaCl$_2$ and give 3 × 5 min washes in TBS + CaCl$_2$.
10. Wipe off excess buffer and incubate in avidin peroxidase 5 µg/mL in 0.125M TBS, pH 7.6, containing 0.374M NaCl (made by diluting 1 part 0.5M stock TBS with 3 parts distilled water; *see* Note 2) for 1 h at room temperature in a humidity chamber.
11. Wash off with TBS and then wash 3 × 5 min in TBS.
12. Reveal sites of lectin binding in 0.05% (w/v) diaminobenzidine tetrahydrochloride (Aldrich) in TBS with 0.015% (v/v) hydrogen peroxide (diluted from 100 vol) for 5 min at room temperature.
13. Counterstain sections in hematoxylin or methyl green before washing, dehydration, and mounting in XAM (Gurr, BDH).

3.5. Lectin Staining of Semithin, Epoxy Resin Sections (38) (Fig. 1)

1. Cut sections 0.75 µm thick and mount on APES-coated slides.
2. Dry on a hotplate and then at 50°C for 2 d to ensure adhesion.
3. Remove resin in 50% (v/v) saturated sodium ethoxide in ethanol for 15 min at room temperature.
4. Wash in absolute ethanol and gradually rehydrate to distilled water. Wash in TBS 3 × 2 min.
5. Treat with trypsin if necessary: warm two baths containing 300 mL TBS to 37°C in a water bath. To one, add 300 mg calcium chloride. Place slides in the other bath and add either 100 or 300 mg trypsin (type II porcine, crude; Sigma) to the bath containing the calcium chloride. After 5 min, transfer the slides to the trypsin/CaCl$_2$ mixture and incubate for 4 min at 37°C.
6. Wash in running cold water to stop the reaction.
7. Wash in TBS 3 × 5 min.
8. After wiping around the section to remove excess TBS, pipet on the biotinylated lectin (10 µg/mL in TBS + 1 mM CaCl$_2$). Incubate for 45 min at 37°C in a humidity chamber.
9. Wash off with TBS + CaCl$_2$ and give 3 × 5 min washes in TBS + CaCl$_2$.
10. Wipe off excess buffer and incubate in avidin peroxidase 5 µg/mL in 0.125M TBS, pH 7.6, containing 0.374M NaCl (made by diluting 1 part 0.5M stock TBS with 3 parts distilled water; *see* Note 2) for 1 hour at 37°C in a humidity chamber.
11. Wash off with TBS and then wash 3 × 5 min in TBS.
12. Reveal sites of lectin binding in 0.05% (w/v) diaminobenzidine tetrahydrochloride (Aldrich) in TBS with 0.015% (v/v) hydrogen peroxide (diluted from 100 vol) for 5 min at room temperature; then wash in water.
13. Counterstain with hematoxylin if required, wash, then air-dry sections and mount in XAM (Gurr, BDH).

3.6. Lectin Staining of Semithin, Acrylic Resin Sections (Fig. 2)

The method is used as in Section 3.5., but the DAB (step 12) is increased to 10 min and then an intensification method may be used, such as the gold–sulfide–silver technique *(36)*.

1. After treating with DAB for 10 min, wash in distilled water for 6 min.
2. Treat with 0.02% (w/v) gold chloride for 3 min.
3. Wash in distilled water for 6 min.
4. Apply neutralized 0.3% (w/v) sodium sulfide for 3 min.
5. Wash in distilled water for 6 min.
6. Develop in Gallyas' silver solution for about 4 min or until sections appear fully stained; monitor this with a microscope.
7. Fix in 1% (v/v) acetic acid, 3 × 10 min.
8. Dehydrate, clear and mount.

3.7. Lectin Staining of Ultrathin Sections, Acrylic Resin (L.R. White), Using a Peroxidase-Based Revealing System (33)

1. Mount pale gold acrylic sections onto nickel or gold (but not copper) grids. Air-dry.
2. Float sections face down on 10% (v/v) hydrogen peroxide for 10 min to etch the sections and also block endogenous peroxidase activity.
3. Wash in distilled water, then TBS.
4. Incubate in 10 µg/mL biotinylated lectin in TBS with 1 mM $CaCl_2$ for 15 min at room temperature.
5. Jet wash in TBS + $CaCl_2$, then place in drops of 5 µg/mL avidin peroxidase in 0.125M TBS containing 0.374M NaCl for 30 min at room temperature.
6. Jet wash in TBS followed by the DAB-H_2O_2 procedure described in Section 3.5. for 8 min.
7. Intensify the reaction with 0.1% (w/v) gold chloride for 1 min followed by thorough washing. (This step is optional.)
8. Air-dry sections and examine without further staining. Over time, the gold particles may aggregate to form a precipitate over the sections.

3.8. Lectin Staining of Ultrathin Sections Using Colloidal Gold (43) (see Note 3)

1. After appropriate etching pretreatments, immerse grid in drops of TBS + 1% (w/v) BSA for 5 min. Blot edge of grid.
2. TBS + 1% (w/v) gelatin for 10 min. Blot.
3. TBS + 0.2% (w/v) glycine for 3 min. Blot.
4. Place grids in drops of TBS + 1 mM $CaCl_2$ + 0.5% Tween-20, wash in three vials of the same, and place in fresh drops again.
5. Blot and transfer to drops of biotinylated lectin, 10 µg/mL in TBS + 1 mM $CaCl_2$ for 1 h at room temperature.
6. Wash as in step 4. Blot.

7. Place grids in antibiotin gold (10 nm) diluted 1:30 with TBS + 1% (w/v) BSA for 1 h at room temperature.
8. Rinse in drops of distilled water, wash in 5 vials of distilled water and dry.
9. Contrast with 2% (w/v) uranyl acetate and Reynold's lead citrate prior to examination in the electron microscope.

3.9. Pretreatment of Ultrathin Epoxy Resin Sections for Lectin Histochemistry

1. Cut pale gold/gold sections and mount on nickel or gold (but not copper) grids.
2. Etch sections in 1% (v/v) saturated sodium ethoxide in ethanol for 1 min or longer (up to 5 min).
3. Rinse (10 dips) in absolute alcohol, 90% (v/v) alcohol, 70% (v/v) alcohol.
4. Rinse in distilled water.
5. Remove osmium, if necessary, with 0.1 g/mL sodium periodate for 40 min on orbital shaker. This may also etch the resin and may be included with nonosmicated tissue.
6. Distill water washes 3 × 5 min.
7. Proceed with staining protocol.

3.10. Glycosidase Treatment of Sections (42) (see Note 5)

The presence of sialic acid may be confirmed by incubating sections in the enzyme neuraminidase, which cleaves off terminal sialic acid residues. Neuraminidase may be derived from many different sources; the one used in our laboratory is obtained from *Clostridium perfringens* (Type VI, Sigma). Different neuraminidases will vary in their substrate specificity, some being specific for 2–3 or 2–6 linkages.

1. Remove resin, and block endogenous peroxidase activity as in Section 3.4.
2. Treat with trypsin if necessary. Wash in TBS and then in distilled water.
3. Apply neuraminidase solution: 0.1 U/mL in 0.2M acetate buffer, pH 5.5, with 1% (w/v) CaCl$_2$ for 1 h at 37°C in a humidity chamber. Gently remove excess enzyme and replace with fresh solution. Leave for another hour as before. As neuraminidase may be contaminated with proteases, too long of an incubation may result in digestion of the tissue by these enzymes.
4. Rinse off with distilled water, then TBS, and continue staining with the lectins, and so on. Sialic acid may also be removed by mild hydrolysis with 1N sulfuric acid at 60–70°C for 1 h.

3.11. ß-Elimination (35)

Unlike their *N*-linked counterparts, *O*-linked glycans are susceptible to the action of mild alkali, which removes them from the protein backbone (ß-elimination). This technique can be applied to tissue sections to distinguish between *N*- and *O*-linked glycan *in situ*. It is not recommended for plastic sections, or for sections not fixed to the slide by means of an adhesive.

1. Dewax, block endogenous peroxidase, and trypsinize if necessary.
2. Rinse in TBS 3 × 5 min.
3. Rinse briefly in distilled water.
4. Make up the following mixture and allow to warm up to 45°C:
   a. 0.954 g potassium hydroxide; and
   b. 40 mL distilled water.
   Allow to dissolve and add:
   a. 50 mL dimethyl sulfoxide; and
   b. 10 mL Analar ethanol (Potassium hydroxide solution: $0.17M$ final concentration).
5. Incubate sections for 40–60 min at 45°C.
6. Carefully wash twice in 10 mM HCl to neutralize. Sections tend to detach at this stage!
7. Wash in distilled water followed by TBS for 3 × 5 min.
8. Continue staining with lectins as normal.

4. Notes

1. Tris-buffered saline—$0.5M$ stock solution.
   a. Dissolve 60.57 g of Tris (hydroxymethyl) methylamine and 81 g of sodium chloride in 800 mL of distilled water.
   b. Adjust pH to 7.6 using concentrated hydrochloric acid (using correct safety precautions).
   c. Make up to 1 L with distilled water (to give $0.5M$ stock solution).
      This may be stored in a refrigerator in a stoppered bottle until required, when it should be diluted with 9 L of distilled water to make the working concentration of $0.05M$ TBS.
      Warning: Hydrochloric acid is corrosive, very toxic, and an irritant!! Always wear acid resistant gloves and eye protection. Store in acid cupboard and measure out in fume cupboard.
2. The use of a high molarity buffer eliminates nonspecific binding of avidin–peroxidase to tissue sections (44). This is particularly useful when staining highly anionic structures such as mast cells and matrix components (4,27,33).
3. There are other techniques utilizing colloidal gold: the direct staining method using a lectin conjugated directly to gold particles, as well as the method using biotinylated lectins followed by an avidin–gold or streptavidin–gold solution.
4. The lectins chosen for any particular study will depend on the purpose of the study and the amount of information, if any, already available as to the glycans present in the tissues of interest. Tables of lectin-binding characteristics are given in the Appendix and in refs. 26–28, 30, and 31, together with citations of studies defining their specificities.
5. Apart from neuraminidase, other glycosidases (e.g., from Sigma) at a variety of concentrations (normally between 0.1 and 10 U/mL but always in excess) may also be used with appropriate buffers, at various temperatures depending on the source of the enzyme:

a. α-mannosidase in 0.1 M citrate buffer, pH 4.5, at 26°C *(32)*.
b. α-L-fucosidase in 0.1 M phosphate buffer, pH 6.5, at 26°C *(32)*.
c. ß-galactosidase (Grade VIII from *Escherichia coli*, Sigma) in 0.1 M phosphate buffer, pH 7.3, with $1.1 \times 10^{-6} M$ magnesium chloride and a trace of mercaptoethanol at 37°C *(26,32)*.
d. ß-N-acetylglucosaminidase in 0.1 M citrate buffer, pH 5.0, at 26°C *(32)*.

During an incubation period of 1 h, four separate aliquots of enzyme are applied to the sections, each one for 15 min, so that the enzyme is not exhausted. Each aliquot may subsequently be tested by incubation with the appropriate *p*- or *o*-nitrophenyl glycoside, which turns yellow on the addition of a few drops of 10N sodium hydroxide if the enzyme is active. Negative controls, with sections being incubated in the buffer alone, should always be carried out.

References

1. Rademacher, T. W., Parekh, R. B., and Dwek, R. A. (1988) Glycobiology. *Ann. Rev. Biochem.* **57,** 785–838.
2. Cook, G. M. W. and Stoddart, R. W. (1973) *Surface Carbohydrates of the Eukaryotic Cell.* Academic, London, pp. 346.
3. Hughes, R. C. (1975) *Membrane Glycoproteins: A Review of Structure and Function.* Butterworth, London, pp. 416.
4. Kjellén, L. and Lindahl, U. (1991) Proteoglycans: structures and interactions. *Ann. Rev. Biochem.* **60,** 443–475.
5. Garg, H. and Lyon, N. (1993) Structure of collagen-fibril associated small proteoglycans of mammalian origin. *Adv. Carbohydr. Chem. Biochem.* **49,** 239–261.
6. Harding, S. E. (1989) The macrostructure of mucus glycoproteins in solution. *Adv. Carbohydr. Chem. Biochem.* **47,** 345–381.
7. Edelman, G. M. and Crossin, K. L. (1991) Cell adhesion molecules: implications for a molecular histology. *Ann. Rev. Biochem.* **60,** 155–190.
8. Lasky, L. A. (1995). Selectin-carbohydrate interactions and the initiation of the inflammatory response. *Ann. Rev. Biochem.* **64,** 113–139.
9. Jeffrey, G. A. (1992) Hydrogen bonding in carbohydrates and hydrate inclusion compounds. *Adv. Enzymol.* **65,** 217–254.
10. Burgess, W. H. and Macaig, T. (1989) The heparin-binding (fibroblast) growth factor family of proteins. *Ann. Rev. Biochem.* **58,** 575–606.
11. Kornfeld, S. (1992) Structure and function of the mannose-6-phosphate/insulin-like growth factor II receptors. *Ann. Rev. Biochem.* **61,** 307–330.
12. Drickamer, K. and Taylor, M. E. (1993) Biology of animal lectins. *Ann. Rev. Cell Biol.* **9,** 237–264.
13. Turner, M. L. (1992) Cell adhesion molecules: a unifying approach to topographic biology. *Biol. Rev.* **67,** 359–377.
14. Jeanloz, R. W. and Codington, J. F. (1976) The biological role of sialic acid at the surface of the cell, in *Biological Roles of Sialic Acid* (Rosenberg, A. and Schengrund, C. L., eds.), Plenum, New York, pp. 201–238.

15. Thomas, J. (1994) Glycoproteins and glycosylation, in *The Encyclopedia of Molecular Biology* (Kendrew, J., ed.), Blackwell Scientific, Oxford, UK, pp. 439–444.
16. Ashwell, G. and Morell, A. G. (1974) The role of surface carbohydrates in the hepatic recognition and transport of circulating glycoproteins. *Adv. Enzymol.* **41,** 99–128.
17. Rice, K. G. and Lee, Y. C. (1993) Oligosaccharide valency and conformation in determining binding to the asialoglycoprotein receptor of rat hepatocytes. *Adv. Enzymol.* **66,** 41–83.
18. Stoddart, R. W. (1979) Nuclear glycoconjugates and their relation to malignancy. *Biol. Rev.* **54,** 199–235.
19. Hart, G. W., Haltiwanger, R. S., Holt, G. D., and Kelly, W. G. (1989) Glycosylation in the nucleus and cytoplasm. *Ann. Rev. Biochem.* **58,** 841–874.
20. Neufeld, E. F. (1991) Lysosomal storage diseases. *Ann. Rev. Biochem.* **60,** 257–280.
21. Sottocasa, G. L., Sandri, G., Panfili, E., and De Bernard, B. (1971) Glycoprotein located in the inter-membrane space of rat liver mitochondria. *FEBS Lett.* **17,** 100–105.
22. Aota, S.-I. and Yamada, K. M. (1994) Extracellular matrix molecules, in *The Encyclopedia of Molecular Biology* (Kendrew, J., ed.), Blackwell Science, Oxford, UK, pp. 365–369.
23. Stoddart, R. W. (1994) Glycans, in *The Encyclopedia of Molecular Biology* (Kendrew, J., ed.), Blackwell Science, Oxford, UK, pp. 429–436.
24. Stoddart, R. W. and Kiernan, J. A. (1973) Histochemical detection of the α-D-arabinopyranoside configuration using fluorescent-labelled Concanavalin A. *Histochemie* **33,** 87–94.
25. Northcote, D. H. (1994) Lectins, in *The Encyclopedia of Molecular Biology* (Kendrew, J., ed.), Blackwell Science, Oxford, UK., pp. 578–580.
26. Jones, C. J. P., Morrison, C. A. M., and Stoddart, R. W. (1992) Histochemical analysis of rat testicular glycoconjugates II. ß-galactosyl residues in O- and N-linked glycans in seminiferous tubules. *Histochem. J.* **24,** 327–336.
27. Roberts, I. S. D., Jones, C. J. P., and Stoddart, R. W. (1990) Lectin histochemistry of the mast cell: heterogeneity of rodent and human mast cell populations. *Histochem. J.* **22,** 73–80.
28. Bishop, P. N., Boulton, M., McLeod, D., and Stoddart, R. W. (1993) Glycan localization within the human interphotoreceptor matrix and photoreceptor inner and outer segments. *Glycobiology* **3,** 403–412.
29. Jones, C. J. P., Koob, B., Stoddart, R. W., Hoffman, B., and Leiser, R. (1994) Lectin histochemical analysis of glycans in ovine and bovine near term binucleate cells. *Cell Tiss. Res.* **278,** 601–610.
30. McMahon, R. F. T., Panesar, M. J. R., and Stoddart, R. W. (1994) Glycoconjugates of the normal human colorectum: a lectin histochemical study. *Histochem. J.* **26,** 504–518.
31. Chapman, S. A., Bonshek, R. E., Stoddart, R. W., Jones, C. J. P., Mackenzie, K. R., O'Donoghue, E., and McLeod, D. (1995) Glycoconjugates of the human trabecular meshwork: a lectin histochemical study. *Histochem. J.* **27,** 869–881.

32. Robb, J. L. and Stoddart, R. W. (1988) Partial analysis of the UEA-1 binding oligosaccharide of human endothelial cells, in *The Lectins* (Bog-Hansen, T. and Freed, D. L. J., eds.), Sigma, St. Louis, MO, 631–633.
33. Jones, C. J. P., Kirkpatrick, C. J., and Stoddart, R. W. (1988) An ultrastructural study of the morphology and lectin-binding properties of human mast cell granules. *Histochem. J.* **20**, 433–442.
34. Sharon, N. (1975) *Complex Carbohydrates. Their Chemistry, Biosynthesis and Functions.* Addison-Wesley, Reading, MA.
35. Downs, F., Herp, A., Moscher, J., and Pigman, W. (1973) ß-elimination and reduction reactions and some applications of dimethylsulphoxide on submaxillary glycoproteins. *Biochim. Biophys. Acta* **328**, 182–192.
36. Newman, G. R., Jasani, B., and Williams, E. D. (1983) The visualisation of trace amounts of diaminobenzidine (DAB) polymer by a novel gold-sulphide-silver method. *J. Microsc.* **132**, RP1–RP2.
37. Jones, C. J. P. and Stoddart, R. W. (1986) A post-embedding avidin-biotin peroxidase system to demonstrate the light and electron microscopic localization of lectin binding sites in rat kidney tubules. *Histochem. J.* **18**, 371–379.
38. Jones, C. J. P., Dantzer, V., and Stoddart, R. W. (1995) Changes in glycan distribution within the porcine interhaemal barrier during gestation. *Cell Tissue Res.* **279**, 551–564.
39. Hsu, S.-M. and Raine, L. (1982) Versatility of biotin-labeled lectins and avidin-biotin-peroxidase complex for localization of carbohydrate in tissue sections. *J. Histochem. Cytochem.* **30**, 157–161.
40. Sarkar, M., Wu, A. M., and Kabat, E. A. (1981) Immunochemical studies on the carbohydrate specificity of *Maclura pomifera* lectin. *Arch. Biochem. Biophys.* **209**, 204–218.
41. Englund, P. T. (1993) The structure and synthesis of glycosyl phosphatidylinositol protein anchors. *Ann. Rev. Biochem.* **62**, 121–138.
42. Jones, C. J. P., Aplin, J. D., Mulholland, J., and Glasser, S. R. (1993) Patterns of sialylation in differentiating rat decidual cells as revealed by lectin histochemistry. *J. Reprod. Fert.* **99**, 635–645.
43. Chapman, S. A., Bonshek, R. E., Stoddart, R. W., Mackenzie, K., and McLeod, D. (1994) Localisation of $\alpha(2,3)$ and $\alpha(2,6)$ linked terminal sialic acid groups in human trabecular meshwork. *Br. J. Ophthalmol.* **78**, 632–637.
44. Jones, C. J. P., Mosley, S. M., Jeffrey, I. J. M., and Stoddart, R. W. (1986) Elimination of the non-specific binding of avidin to tissue sections. *Histochem. J.* **19**, 264–268.

3

Lectin–Gold Histochemistry on Paraffin and Lowicryl K4M Sections Using Biotin and Digoxigenin-Conjugated Lectins

Jürgen Roth, Christian Zuber, Tetsutaro Sata, and Wei-Ping Li

1. Introduction

A variety of staining reactions for the visualization of cellular and extracellular glycoconjugates at the light microscopic level are available that are based on the detection of carboxyl and sulfate groups or periodic acid reactive configurations *(1,2)*. Starting in the late 1960s, lectins have replaced many of these chemical staining reactions because of their high specificity for defined mono- and oligosaccharidic sequences in both *N*- and *O*-glycosidic-linked oligosaccharide side-chains of glycoproteins and glycolipids. In order to be useful as histochemical reagents, lectins must be tagged with appropriate markers and those employed in immunocytochemistry have been used successfully *(3–9)*. Horisberger and coworkers were the first to prepare lectins labeled with particles of colloidal gold and used them in scanning electron microscopy *(10)*. Subsequently, gold-labeled lectins were applied in transmission electron microscopy to study various aspects of cell surface expression and internalization of lectin-binding sites *(5,8,11,12)*, as well as in postembedding labeling of Lowicryl K4M thin sections *(13)*. Later, it was shown that gold-labeled lectins can be used to stain sections of paraffin-embedded tissues *(14–16)*, as well as semithin sections of Epon *(17)* and Lowicryl K4M-embedded tissues *(18,19)*. However, in order to achieve a visible pink staining, which is the natural color of particles of colloidal gold in transmitted visible light *(20)*, highly concentrated lectin–gold complexes had to be used, thereby allowing the possibility of nonspecific staining. A major improvement resulted through the application of a photochemical silver reaction for signal amplification *(21–24)*, which permitted the use of lectins for light microscopy in concentrations as applied for electron microscopy *(25,26)*.

A variety of direct and indirect lectin labeling techniques have been worked out during the last two decades *(19,27)*. In this chapter, the use of biotin and digoxigenin-conjugated lectins in conjunction with colloidal gold detection systems and silver intensification for light microscopy will be detailed.

1.1. Fixation, Embedding, and Tissue Section Preparation

Tissues, cells grown in suspension or in monolayer culture, and isolated cells can be fixed by buffered aldehydes. In contrast to immunohistochemistry, which often requires specific fixation protocols depending on a particular protein antigen to be detected, aldehyde-fixation-induced alteration of carbohydrate structures recognized by the various lectins is negligible. Formaldehyde and glutaraldehyde either alone or in combination are very suitable fixatives. In our experience, tissues fixed in 10% buffered formalin, embedded in paraffin for diagnostic purposes, and stored for several years are very suitable for retrospective lectin histochemical studies. If at all possible, organs such as liver, kidney, lung, and brain should be fixed by vascular perfusion, whereas others such as stomach, intestine, and bladder can be conveniently fixed by immersion, particularly for studies on the mucosa. Aldehyde fixation should be terminated in a controlled manner by immersion of the cells and tissues in a buffer containing molecules with free amino groups to amidinate reactive aldehyde groups. This is followed by paraffin embedding according to a standard protocol or low temperature dehydration in graded ethanol followed by low temperature Lowicryl K4M embedding as described *(19)* (*see also* instructions for use provided by the manufacturer *[28,29]*). Paraffin or Lowicryl K4M sections mounted on glass slides can be stored at ambient temperature for prolonged periods of time.

1.2. Lectin Histochemistry

1.2.1. Histochemical Staining with Lectin–Digoxigenin Conjugates

The steroid hapten digoxigenin (DIG) has been introduced as a hapten marker for lectins. Digoxigenin provides a major advantage over biotin since it occurs naturally only in plants of the digitalis family and no crossreacting material has been reported to exist in mammalian tissues *(30–32)*. This is in contrast to the natural occurrence of biotin in various tissues and cells, often necessitating section pretreatment to block the endogenous biotin *(33)*.

The lectin–DIG conjugate bound to tissue sections can be visualized using a single anti-DIG antibody gold complex, which is followed by silver intensifi-

Lectin–Gold Histochemistry

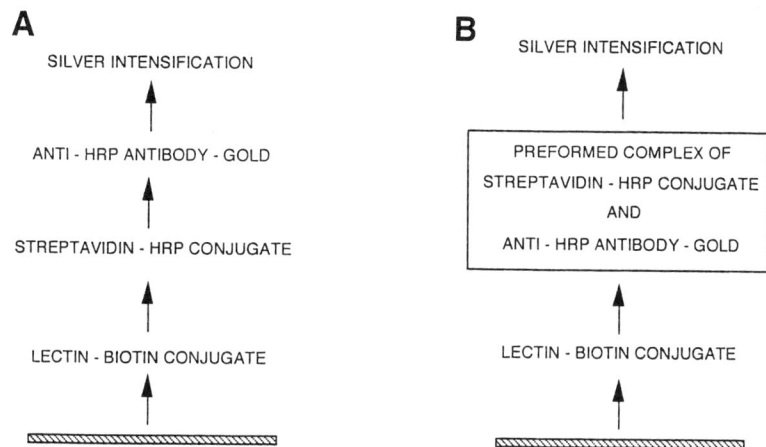

Fig. 1. Schematic presentation of lectin–biotin techniques involving the sequential use of streptavidin–HRP and anti-HRP antibody–gold complex **(A)** or of the preformed complex of streptavidin–HRP and anti-HRP antibody–gold **(B)**.

cation of the gold particles (*see* Section 1.2.3.). As compared to the use of alkaline phosphatase- or horseradish peroxidase-labeled anti-DIG antibodies, the silver-intensified gold labeling provides improved resolution and higher contrast in paraffin sections *(32)*. Furthermore, the enzyme-labeled anti-DIG antibodies cannot be used on Lowicryl K4M sections since the enzyme reaction product does not stick to the sections *(32)*.

1.2.2. Histochemical Staining with Lectin–Biotin Conjugates

The indirect labeling technique using biotin-conjugated lectins described in this chapter is a modification of the well-known (strept)avidin–biotin principle and schematically shown in Fig. 1. The sections are first incubated with a lectin–biotin conjugate followed by incubation with streptavidin–peroxidase conjugate. Then, the histochemical visualization of catalytic activity of horseradish peroxidase (HRP) by the diamino benzidine reaction was replaced by the immunological detection of HRP with a gold-labeled anti-HRP antibody. Consequently, blocking of endogenous peroxidase activity was no longer required. Furthermore, the gold-labeled anti-HRP antibody shows no cross-reactivity with peroxidase molecules present in mammalian tissues. As mentioned in Section 1.2.2., the silver-intensified gold labeling provides better resolution and higher contrast in paraffin sections as compared to the DAB reaction product. Furthermore, the DAB reaction product does not stick to Lowicryl K4M sections *(32)*.

1.2.3. Silver Intensification of Lectin–Gold Labeling

Trace amounts of particles of colloidal gold (and of any other metallic colloid) forming an invisible, latent labeling can be rendered visible for light microscopy by signal amplification using silver intensification. The general process consists of the reduction of ionic silver on particles of colloidal gold, which act as catalysts, in the presence of hydroquinone as reducing agent under acidic conditions. This creates a growing shell of metallic silver around the gold particles and, since metallic silver also catalyzes this reduction, the reaction becomes autocatalytic. Various silver compounds have been used as source for silver ions. The use of the light-sensitive silver nitrate and silver lactate proved not to be very practical the entire procedure had to be performed in the dark. A breakthrough was achieved through the use of silver acetate *(34)*, which permitted the entire silver amplification to be performed under daylight *(32,35–39)*.

Following the silver intensification, the lectin labeling is indicated by the presence of a black staining, which is easily distinguishable from anthracotic and other pigments. The silver-amplified gold particle labeling results in a nondiffusible, permanent staining. Sections can be counterstained with nuclear fast red, with hematoxylin alone, or in combination with eosin and then dehydrated and mounted in xylene-based mountants.

1.2.4. Controls of Specificity

The classical protocol followed to establish the specificity of lectin staining consists in the incubation of the lectin–DIG conjugate with its appropriate binding substrate prior to its use for staining. Most often mono- and oligosaccharides are used for this purpose, although they may not be identical to the lectin-binding sites of cellular glycoconjugates. The closer the inhibitory substrate is to the natural structure recognized by the lectin, the lower the required concentrations are for complete inhibition of lectin binding to tissue sections. Thus, concentrations of 0.1–10 mM can be sufficient in inhibition studies when the respective lectins are used in the range of 1–20 µg/mL. As also mentioned in the chapter by Zuber et al., this type of control also excludes any possible nonsugar-related lectin–protein interactions that sometimes occur. However, abolition of lectin binding by removal of the carbohydrate structure using the appropriate exo- and endoglycosidase provides direct evidence for the specificity of lectin labeling. Such experiments demonstrate not only specificity of lectin-carbohydrate interaction but, at the same time, provide structural information. Furthermore, lectin binding to *N*-glycosidically-linked oligosaccharides or to *O*-glycosidically-linked oligosaccharides can be differentiated by the use of the endoglycosidase PNGase F (endo-F) and the ß-elimination reaction, respectively. Replacement of the lectin-conjugates with the diluting buffer in the incubation sequences serves as a further control.

2. Materials

2.1. Fixation, Embedding, and Preparation of Tissue Sections

1. Paraformaldehyde (*see* Note 1).
2. Glutaraldehyde, 25% aqueous, vacuum distilled, stored under protective gas atmosphere.
3. HBSS: Hanks balanced salt solution, pH 7.4.
4. PBS: 10 mM sodium-phosphate buffer, 150 mM NaCl, pH 7.4.
5. PVP: polyvinyl pyrrolidone, 30,000 mol wt.
6. Sodium nitrite, $NaNO_2$.
7. Fixative for tissue: 3% paraformaldehyde—0.1% glutaraldehyde in HBSS or PBS containing 4% PVP and 70 mM $NaNO_2$ (*see* Note 2). Use freshly prepared fixative or store frozen at –20°C. Alternatively, 10% formaldehyde in 100 mM sodium-phosphate buffer (pH 7.4) may be used. Check pH of the fixative!
8. Fixative for cells: 2% paraformaldehyde—0.1% glutaraldehyde in HBSS or PBS (*see* Note 1). Use freshly prepared fixative or store frozen at –20°C. Check pH of the fixative!
9. Quenching solution: 50 mM NH_4Cl in PBS. Store at ambient temperature and prepare freshly every 4 wk.
10. Reagents and equipment for standard paraffin embedding.
11. Lowicryl K4M kit. Lowicryl K4M monomers may cause severe allergic reactions and are toxic. During embedding, it is strongly recommended that Lowicryl K4M-resistant gloves be used and that inhalation of the resin vapor be avoided by performing all steps in a well-ventilated fume hood.
12. Equipment for low temperature dehydration, resin infiltration, and polymerization from Balzers or equivalent.
13. Paraffin microtome and auxiliary equipment for preparation of paraffin sections.
14. Ultramicrotome and auxiliary equipment for preparation of Lowicryl K4M sections (0.1–1 µm).
15. Protein-coated glass slides for mounting of paraffin sections.
16. Poly-L-lysine, 300,000–500,000 mol wt.
17. Poly-L-lysine-activated glass slides for Lowicryl K4M sections. Clean glass slides are marked on one side with a diamond; this region on the other side is covered with aqueous poly-L-lysine solution (1 mg/mL) for 5 min at ambient temperature. After a quick rinse with distilled water, glass slides are air-dried and stored at ambient temperature until use.
18. Mounting of Lowicryl K4M sections: Place sections with a wire loop onto a water drop in the marked region of poly-L-lysine-activated slides and allow to dry overnight at 40°C. Sections can be stored in an ordinary slide box at ambient temperature for prolonged periods of times.

2.2. Lectin Histochemistry

1. TBPBS: PBS containing 1% bovine serum albumin (BSA) and 0.05% Tween-20 plus 0.02% NaN_3.
2. TTBPBS: PBS containing 1% BSA, 0.05% Tween-20, and 0.01% Triton X-100 plus 0.02% NaN_3.

2.2.1. Lectin–Digoxigenin Technique

1. The various lectin–DIG conjugates were obtained from Boehringer Mannheim (Mannheim, Germany). Working dilutions of lectins were made up with TBPBS. According to the manufacturer's instructions, stock solutions are stored at 4°C, and working dilutions are always freshly prepared before use. Since most lectins are mitogenic and some are quite toxic, due care should be exercised.
2. Anti-DIG antibody–gold complex prepared with 8-nm gold particles *(32)* or purchased from Boehringer Mannheim. Working dilutions were prepared with TTBPBS and stored at 4°C.

2.2.2. Lectin–Biotin Technique

1. The various lectin–biotin conjugates were purchased from different commercial sources (Sigma, St. Louis, MO; Boehringer Mannheim; Vector Laboratories, Burlingame, CA). Working dilutions of lectins were made up with TBPBS. According to the manufacturer's instructions, stock solutions are stored at 4°C, and working dilutions are always freshly prepared before use.
2. Streptavidin–HRP conjugate was obtained from Jackson ImmunoResearch Laboratories (West Chester, PA) and working dilutions were prepared with TBPBS.
3. Affinity-purified goat anti-HRP antibody–gold complex (4-nm gold particles) was obtained from Jackson ImmunoResearch Laboratories.
4. Streptavidin–HRP/anti-HRP antibody–gold complex: Mix equal volumes of streptavidin–HRP conjugate (1.25 µg/mL) and anti-HRP antibody gold ($A_{525\ nm}$ = 0.1) 30–60 min prior to use. Complexes can be stored for up to 2 wk at 4°C.

2.2.3. Silver Intensification

1. Solution A: 0.2% w/v silver acetate (from Fluka, Buchs, Switzerland) in double-distilled water (e.g., 100 mg of silver acetate dissolved in 50 mL double-distilled water). Dissolve with a magnetic stirrer in an Erlenmeyer covered with a sheet of aluminum or sonicate for 5 s in a water bath. Silver acetate dissolves slowly; prepare at least 30 min in advance (*see* Note 3).
2. Solution B: 0.5% w/v hydroquinone in citrate buffer ($0.05M$, pH 3.8) (e.g., 250 mg hydroquinone in 50 mL citrate buffer). Generally make twice as much solution B as solution A.
3. Photographic fixative (i.e., Superfix from Agfa).
4. 1% glutaraldehyde in PBS.
5. Commercially available silver intensification kits.

2.2.4. Specificity Controls

1. Various mono- and oligosaccharides, neoglycoproteins, and purified glycoproteins can be obtained from different commercial sources (Sigma; Glycosystems; Chembiomed, Alberta, Canada).
2. Various exo- and endoglycosidases can be obtained from different commercial sources (Sigma; Boehringer Mannheim; Calbiochem, La Jolla, CA). For preparation of working solutions and conditions of use and storage, consult product information sheets of the respective suppliers.
3. ß-elimination solution: $0.1 N$ NaOH.
4. Solution for acidic methanolysis: $0.1 N$ HCl in methanol.

3. Methods
3.1. Fixation, Embedding, and Tissue Sections

1. Deparaffinize and rehydrate paraffin sections according to standard protocol.
2. Rehydrate Lowicryl K4M sections by covering them with PBS. Resin removal is not required (nor possible owing to the nature of the chemical bonds) for subsequent lectin labeling (or immunostaining).

3.2. Lectin Histochemistry
3.2.1. Lectin–Digoxigenin Technique

1. Condition sections with TBPBS or TTBPBS for 5 min at ambient temperature (*see* Note 4).
2. Drain conditioning solution. Avoid drying of sections.
3. Incubate with diluted lectin–DIG conjugate (*see* Table 1) for 45 min at ambient temperature in a moist chamber.
4. Wash sections in PBS (2 × 5 min).
5. Incubate with 8 nm gold-labeled anti-DIG antibody diluted in TTBPBS ($A_{525\,nm}$ = 0.05) for 1 h at ambient temperature in a moist chamber.
6. Wash sections in PBS (2 × 5 min).
7. Immerse slides with sections in 1% glutaraldehyde in PBS for 20 min (*see* Note 5).
8. Rinse sections quickly with PBS and thoroughly with double-distilled water followed by air drying. The air-dried sections can be stored up to several months or can be immediately processed by silver amplification (*see* Section 3.2.3.).

3.2.2. Lectin–Biotin Technique

1. Condition sections with TBPBS or TTBPBS for 5 min at ambient temperature (*see* Note 4).
2. Drain conditioning solution. Avoid drying of sections.
3. Incubate with diluted lectin–biotin conjugate for 45 min at ambient temperature in a moist chamber.
4. Rinse sections with PBS or TBS (2 × 5 min).
5. Incubate with streptavidin–HRP conjugate. Working dilution (1.25 µg/mL) was prepared with TTBPBS.
6. Rinse sections with PBS (2 × 5 min).

Table 1
A List of Selected Lectins with Indications of Their Use

Lectin source	Name	Specificity	Lectin–DIG concentrations	Inhibitors	Inhibitor concentrations
Amaranthus caudatus	ACA	Gal β1,3 GalNAc, Neu5Ac α2,3 Gal β1,3 GalNAc[a]	25 µg/mL	Synthetic T antigen T antigen neoglycoprotein	10 mM 10 mM
Arachis hypogaea	PNA	Gal β1,3 Gal NAc	25 µg/mL	D-Gal	100 mM
Canavalia ensiformis	Con A	Branched hexoses	5 µg/mL	Methyl-α-mannopyranoside	5 mM
Datura stramonium	DSA	GlcNAc (Gal) β1-4 GlcNAc and oligomers	10 µg/mL	Lactosamin Asialofetuin Asialoovomucoid N-N-chitobiose	200 mM 2 µg/mL 500 µg/mL 500 µg/mL
Glycine max	SBA	GalNAc α1,3 Gal	25 µg/mL	GalNAc	10 mM
Helix pomatia	HPA	GalNAc	1 µg/mL	GalNAc	≤10 mM
Limax flavus	LFL	Sialic acids[b]	0.5 µg/mL	Neu5Ac	10 mM
Maackia amurensis	MAA	Neu5Ac α2,3 Gal β1,4 GlcNAc	25 µg/mL	α2,3 sialyllactose	10 mM
Phaseolus vulgaris	PHA-L	GlcNAc β1,6 Man α1	10 µg/mL	Thyroglobulin	≈9 µM
Sambucus nigra	SNA	Neu5Ac α2,6 Gal (GalNAc)	25 µg/mL	α2,6 sialyllactose	1 mM
Anguilla anguilla	AAA	αFuc			
Dolichos biflorus	DBA	GalNAc α1,3 GalNAc			
Erythrina cristagalli	ECA	Gal β1,4 GlcNAc			
Galanthus nivalis	GNA	Man α1,3 Man			
Lens culinaris	LcH	Branched mannoses with α fucose			
Lotus tetragonolobus	Lotus	Fuc α1,3 Gal β1,4 [Fuc α1-3] GlcNAc			
Phaseolus vulgaris	PHA-E	Gal β1,4 GlcNAc β1,2 Man α1,6→			
Triticum vulgare	WGA	(GlcNAc β1-4 GlcNAc)$_n$			
Ulex europaeus	UEA-1	Fuc α1,2 Gal β1,4 GlcNAc			

[a]To determine whether binding is owing to the presence of sialylated or nonsialylated Gal β1,3 GalNAc, sections were pretreated with galactose oxidase (50 U/mL at 37°C for 4 h or 5 U/mL at 37°C overnight) followed by Schiff reagent (15 min). See refs. 37 and 40.
[b]See ref. 41.

Table 2
Various Sialidases and Their Specificities

Source	Specificity	Relative cleavage rate
Vibrio cholerae	(O-acyl) NeuAcα2-X	$\alpha 2,3 > \alpha 2,6 > \alpha 2,8$
Arthrobacter ureafaciens	NeuAcα2-X or NeuGlcα2-X	$\alpha 2,6 > \alpha 2,3 > \alpha 2,8$
Clostridium perfringens	NeuAcα2-X or NeuGlcα2-X	$\alpha 2,3 > \alpha 2,8 \approx \alpha 2,6$
Newcastle disease virus (Hitchner B1 strain)	(O-acyl) NeuAcα2-3 X (O-acyl) NeuAcα2-8 NeuAc	$\alpha 2,3 > \alpha 2,8$; but not $\alpha 2,6$

7. Incubate with gold-labeled anti-HRP antibody (diluted with TTBPBS to $A_{525\ nm}$ = 0.05) for 1 h at ambient temperature in a moist chamber.
8. Rinse sections with PBS (2 × 5 min).
9. Immerse slides with sections in 1% glutaraldehyde in PBS for 20 min (*see* Note 5).
10. Rinse sections quickly with PBS and thoroughly with double-distilled water followed by air drying. The air-dried sections can be stored up to several months or can be immediately processed by silver amplification.

In this protocol, steps 6–8 can be replaced by a single incubation step with streptavidin–HRP/anti-HRP antibody gold complex (*see* item 4 in Section 2.2.2.).

3.2.3. Silver Intensification

1. Silver intensification: Place slides with the sections in solution B to which an equal volume of distilled water has been added for 2–5 min at ambient temperature. Transfer the slides in the developer consisting of equal volumes of solution A and B and incubate for 18–25 min at 20–22°C (*see* Note 6).
2. Rinse briefly with double-distilled water.
3. Place slides with the sections in photographic fixative for 2–3 min.
4. Rinse thoroughly with tap water.
5. Counterstain as required, and mount.

3.2.4. Specificity Controls

1. Incubate lectin solution with the inhibitory and noninhibitory mono- and oligosaccharides, neoglycoproteins, or purified glycoproteins for 30 min at ambient temperature and use in parallel with standard lectin solution in step 3 of the protocols listed under Sections 3.2.1. and 3.2.2. Correct pH when acidic reagents (sialic acids) are used. *See* Table 1 for working concentrations of inhibitory substances.
2. Replace the lectin–DIG or lectin–biotin conjugate with the diluting buffer.
3. Incubate tissue sections after conditioning with the required exo- and endoglycosidase (*see* Table 2). Incubation conditions according to suppliers' instructions. Run control with enzyme-free buffer under identical conditions.
4. ß-elimination reaction: Float Lowicryl sections on large droplets of 0.1–0.2N NaOH for 12–36 h at 37°C in a moist chamber. Afterward, rinse four times, 5 min each with double-distilled water, mount sections on poly-L-lysine-activated slides, air-dry at 40°C overnight, and proceed with lectin labeling.

4. Notes

1. Formaldehyde is freshly prepared from paraformaldehyde by heating to 70°C. Solution is cooled to room temperature, filtered with filter paper, and pH is adjusted to 7.0–7.4 if only formaldehyde is used for fixation.

    The fixative is usually prepared using Hanks balanced salt solution or PBS, and used either freshly prepared or stored frozen at –20°C. We perform fixation either at 35–37°C (tissues and cell cultures) or at ambient temperature.

2. The PVP is added to prevent osmotically induced changes of the extracellular spaces and $NaNO_2$ to paralyze blood vessel's musculature.
3. Due care should be exercised to use quartz-distilled water or equivalent to prepare the solutions required for signal amplification. Furthermore, the glassware used for the silver reaction needs to be scrupulously cleaned and rinsed with double-distilled water. It is generally recommended that freshly prepared solutions A and B be used. However, we noticed that silver acetate and hydroquinone solution stored in a refrigerator for up to 4 wk produced the same staining results as freshly prepared ones. In this case, due care is needed to be taken to warm the solutions to room temperature before use since the photochemical process is temperature dependent.
4. This step is performed prior to the lectin incubation to prevent or minimize nonspecific interactions between the lectin and constituents contained in the paraffin sections. The blocking step may not be required if sections of Lowicryl K4M-embedded tissues are used.
5. Fixation is required to prevent low pH-induced loss of labeling during silver intensification. Alternatively, sections may be quickly rinsed in water to remove salts and air-dried before silver intensification.
6. Mix solution A and B a few seconds before use. Place sections vertically to prevent fallout of autocatalytically formed silver precipitates. Such precipitated silver cannot be removed by rinsing. We found it useful to siliconize the glassware and mandatory to avoid metal slide holders and metal forceps. The silver intensification is performed under daylight and can be monitored microscopically, the slides being observed upside down. In our hands, a developing time of 18 min between 20 and 22°C results in optimal signal amplification. The deposition of silver proceeds exponentially and reaches optimal levels during the last 3–4 min of the developing time. Toward the end of the developing time, the developer starts to turn dark.

References

1. Spicer, S. S., Baron, D. S., Sato, A., and Schulte, B. A. (1981) Variability of cell surface glycoconjugates. Relation to differences in cell function. *J. Histochem. Cytochem.* **29,** 994.
2. Reid, P. E. and Park, C. M. (1990) *Carbohydrate Histochemistry of Epithelial Glycoproteins.* Gustav Fischer Verlag, Stuttgart, pp. 1–170.
3. Nicolson, G. (1974) The interaction of lectins with animal surfaces. *Int. Rev. Cytol.* **39,** 89–190.

4. Roth, J. (1978) The lectins. Molecular probes in cell biology and membrane research. *Exp. Path. Suppl.* **3**, 1–186.
5. Roth, J. and Binder, M. (1978) Colloidal gold, ferritin and peroxidase as markers for electron microscopic double labeling lectin techniques. *J. Histochem. Cytochem.* **26**, 163–169.
6. Roth, J. and Thoss, K. (1974) Light and electron microscopic demonstration of D-mannose and D-glucose like sites at the cell surface by means of the lectin from the Lens culinaris. *Experientia* **3**, 414.
7. Roth, J. and Thoss, K. (1975) The use of fluorescein isothiocyanate labeled lectins for immunohistochemical demonstration of saccharides. I. Methodical investigations with Concanavalin A, Lens culinaris lectin and Ricinus communis lectin. *Exp. Path.* **10**, 258–267.
8. Roth, J. and Wagner, M. (1977) Peroxidase and gold complexes of lectins for double labeling of cell surface binding sites by electron microscopy. *J. Histochem. Cytochem.* **25**, 1181–1184.
9. Roth, J., Binder, M., and Gerhard, J. (1978) Conjugation of lectins with fluorochromes: an approach to histochemical double labeling of carbohydrate components. *Histochemistry* **56**, 265–273.
10. Horisberger, M., Rosset, J., and Bauer, H. (1975) Colloidal gold granules as markers for cell surface receptors in the scanning electron microscope. *Experientia* **31**, 1147.
11. Horisberger, M. and Rosset, J. (1977) Colloidal gold, a useful marker for transmission and scanning electron microscopy. *J. Histochem. Cytochem.* **25**, 295–305.
12. Wagner, M., Roth, J., and Wagner, B. (1976) Gold-labeled protectin from Helix pomatia for the localization of blood group A antigen from human erythrocytes by immuno freeze etching. *Exp. Path.* **12**, 277–281.
13. Roth, J. (1983) Application of lectin-gold complexes for electron microscopic localization of glycoconjugates on thin sections. *J. Histochem. Cytochem.* **31**, 987–999.
14. Roth, J. (1983) Application of immunocolloids in light microscopy. II. Demonstration of lectin-gold complexes or glycoprotein-gold complexes. *J. Histochem. Cytochem.* **31**, 547–552.
15. Roth, J. (1983) in *Techniques in Immunocytochemistry* (Bullock, G. and Petrusz, P., eds.), Academic, London, pp. 217–284.
16. Lucocq, J. and Roth, J. (1985) in *Techniques in Immunocytochemistry* (Bullock, G. and Petrusz, P., eds.), Academic, London, pp. 203–236.
17. Lucocq, J. and Roth, J. (1984) Applications of immunocolloids in light microscopy. III. Demonstration of antigenic and lectin-binding sites in semithin resin sections. *J. Histochem. Cytochem.* **32**, 1075–1083.
18. Roth, J., Brown, D., and Orci, L. (1983) Regional distribution of N-acetyl-D-galactosamine residues in the glycocalyx of glomerular podocytes. *J. Cell Biol.* **96**, 1189–1196.
19. Roth, J. (1989) in *Vesicular transport. Methods in Cell Biology* (Tartakoff, A., ed.), Academic, San Diego, pp. 513–551.

20. Faraday, M. (1857) Experimental relations of gold (and other metals) to light. *Phil. Trans. R. Soc.* **147,** 145–181.
21. Danscher, G. (1981) Localisation of gold in biological tissue. A photochemical method for light and electron microscopy. *Histochemistry* **71,** 81–88.
22. Danscher, G. and Rytter Nörgaard, J. O. (1983) Light microscopic visualization of colloidal gold on resin-embedded tissue. *J. Histochem. Cytochem.* **31,** 1394–1398.
23. Holgate, C. S., Jackson, P., Cowen, P. N., and Bird, C. C. (1983) Immunogold-silver staining: new method of immunostaining with enhanced sensitivity. *J. Histochem. Cytochem.* **31,** 938–944.
24. Springall, D. R., Hacker, G. W., Grimelius, L., and Polak, J. M. (1984) The potential of the immunogoldsilver staining method for paraffin sections. *Histochemistry* **81,** 603–608.
25. Taatjes, D. J., Schaub, U., and Roth, J. (1987) Light microscopical detection of antigens and lectin binding sites with gold labeled reagents on semithin Lowicryl K4M sections: usefulness of the photochemical silver reaction for signal amplification. *Histochem. J.* **19,** 235–245.
26. Taatjes, D. J., Roth, J., Peumans, W., and Goldstein, I. J. (1988) Elderberry bark lectin-gold techniques for the detection of Neu5Ac(α2,6)Gal/GalNAc sequences: applications and limitations. *Histochem. J.* **20,** 478–490.
27. Taatjes, D. and Roth, J. (1991) Glycosylation in intestinal epithelium. *Int. Rev. Cytol.* **126,** 135–193.
28. Carlemalm, E., Garavito, R. M., and Villiger, W. (1982) Resin development for electron microscopy and an analysis of embedding at low temperature. *J. Microsc. (Oxford)* **126,** 123–143.
29. Roth, J., Bendayan, E., Carlemalm, E., Villiger, W., and Garavito, M. (1981) The enhancement of structural preservation and immunocytochemical staining in low temperature embedded pancreatic tissue. *J. Histochem. Cytochem.* **29,** 663–671.
30. Haselbeck, A., Schickaneder, E., and Hösel, W. (1990) Structural characterization of glycoprotein carbohydrate chains by using digoxigenin-labelled lectins on blots. *Anal. Biochem.* **191,** 25–30.
31. Haselbeck, A., Schikander, E., Schmidt, A., v d Eltz, H., and Hösel, W. (1989) *New Techniques for the identification and characterization of glycoproteins on gels and blots,* Proc. 19th FEBS meeting, Rome.
32. Sata, T., Zuber, C., and Roth, J. (1990) Lectin digoxigenin conjugates: a new hapten system for glycoconjugate cytochemistry. *Histochemistry* **94,** 1–11.
33. Wood, G. and Warnke, R. (1981) Suppression of endogenous avidin-binding activity in tissues and its relevance to biotin-avidin detection systems. *J. Histochem. Cytochem.* **29,** 1196–1204.
34. Skutelsky, E., Goyal, V., and Alroy, J. (1987) The use of avidin–gold complex for light microscopic localization of lectin receptors. *Histochemistry* **86,** 291.
35. Hacker, G. W., Grimelius, L., Danscher, G., Bernatzky, G., Muss, W., Adam, H., and Thurner, J. (1988) Silver acetate autometallography: an alternative enhancement technique for immunogold-silver staining (IGSS) and silver amplification of gold, silver, mercury and zinc in tissues. *J. Histotechnol.* **11,** 213–221.

36. Sata, T., Lackie, P., Taatjes, D., Peumans, W., and Roth, J. (1989) Detection of the NeuAc (α2,3) Gal (ß1,4) GlcNAc sequence with the leukoagglutinin from Maackia amurensis: light and electron microscopic demonstration of differential tissue expression of terminal sialic acid in α2,3 and α2,6-linkage. *J. Histochem. Cytochem.* **37,** 1577–1588.
37. Sata, T., Zuber, C., Rinderle, S. J., Goldstein, I. J. and Roth, J. (1990) Expression patterns of the T antigen and the cryptic T antigen in rat fetuses: detection with the lectin Amaranthin. *J. Histochem. Cytochem.* **38,** 763–774.
38. Sata, T., Roth, R., Zuber, C., Stamm, B., and Heitz, P. (1991) Expression of α2,6-linked sialic acid residues in neoplastic but not in normal colonic mucosa. *Am. J. Pathol.* **139,** 1435–1448.
39. Taatjes, D. and Roth, J. (1990) Selective loss of sialic acid from rat small intestinal epithelial cells during postnatal development: demonstration with lectin-gold techniques. *Eur. J. Cell Biol.* **53,** 255–266.
40. Schulte, B. A. and Spicer, S. S. (1983) Light microscopic histochemical detection of terminal galactose and N-acetylgalactosamine residues in rodent complex carbohydrates using a galactose oxidase-Schiff sequence and the peanut lectin-horseradish peroxidase conjugate. *J. Histochem. Cytochem.* **31,** 19–24.
41. Ravindranath, M. H., Higa, H. H., Cooper, E. L., and Paulson, J. C. (1985) Purification and characterization of an O-acetylsialic acid-specific from a marine crab *Cancer antennarius. J. Biol. Chem.* **260,** 8850–8856.

4

Use of Fluorochrome-Labeled Lectins in Light Microscopy

Udo Schumacher and Barry S. Mitchell

1. Introduction

The use of lectins with different carbohydrate specificities has demonstrated that there is a great diversity of carbohydrate residues, particularly in glycoconjugates of the cell membranes, but also in (sub)terminal carbohydrate residues of mucins and their precursors stored in the mucin granules of goblet and other mucin-producing cells *(1,2)*. The methods for the detection of these carbohydrate residues using fluorescence microscopy have mainly been carried out on cryostat or paraffin wax sections, although the use of fluorescent-labeled lectins on semithin sections is in principle possible, but not advisable owing to the lack of amplification signal. The particular advantages of using fluorochrome-labeled lectins are: first, the methodology is quick and easy to perform; second, quantification of the lectin binding intensity is possible by using fluorescence-activated cell sorters (FACS) or analyzers and digital imaging processing.

Standard schedules for tissue processing may be used, since sugars to which lectins bind are usually fixed by most fixatives and not rendered too labile by standard processing (aside from glycolipids, which are soluble in the organic solvents used to process tissues to paraffin wax). In these cases, fresh or fixed cryostat sections may be the method of choice. Optimal lectin binding has been reported after use of ethanol or Bouin's or Carnoy's fixatives *(3)*. After formalin fixation, however, proteolytic enzymes will often be required to improve lectin binding *(4)*, though use of the latter must be optimized as in immunohistochemistry *(5)*. The precise localization of lectin-binding sites depends, therefore, to some extent on the fixation and tissue-processing regimen utilized *(6)*. In a study of lectin binding in human breast cancer cell lines, considerable variations in the binding of *Helix pomatia* agglutinin (HPA) were observed depending on whether living cells, fixed cells, or fixed and paraffin wax embedded cells were used. In addition, the results were influenced by the technique used (direct with fluorescein isothiocyanate [FITC] versus peroxidase antiperoxidase technique). It was observed that the influence of fixation and processing on lectin binding is complex *(6)*. Although it was generally the case in this latter study that fixation and embedding reduced the amount of lectin binding, there was an increase in lectin binding with some lectins, probably owing to exposure of intracellular lectin-binding sites after sectioning. These results serve to emphasize the need to assess each lectin on an individual basis in order to obtain comparable data.

The methods available for demonstration of lectin-binding sites at light level are mostly similar to those used in immunohistochemistry. Those utilizing FITC, either directly or indirectly, are impermanent. The advantages of using FITC, however, are that it is a small molecule and it is less likely to influence the carbohydrate binding site of a lectin as could larger marker molecules like horseradish peroxidase.

The FITC-labeled lectins may also be used in conjunction with the light microscope to investigate receptor-mediated endocytosis mechanisms *(7)*. Studies in which single cells may be distinguished in flow cytometry are another way in which FITC-labeled lectins can be used to sort cells into those that express certain sugar groups and those that do not. Such methods can be of considerable use in quantitative population studies and cell function assays *(6,8)*. Fluorochrome lectins may also be used in double labeling (e.g., FITC and TRITC, rhodamine), and/or with antibodies in studies that facilitate cell population identification.

The methods described in this paper give short accounts of FITC-lectin labeling in paraffin wax sections, use of such lectins in cytophotometric analyses to quantify lectin uptake, and use of lectins in FACS analysis that enables studies of carbohydrate composition of glycoconjugates in studying cell populations.

2. Materials

1. RPMI-1640 tissue culture medium.
2. Fetal calf serum.
3. Penicillin (5000 U/mL) and streptomycin (5000 µg/mL).
4. FITC or otherwise (TRITC, rhodamine)-labeled lectins (*see* Appendix), obtained from Sigma (Poole, Dorset, UK).
5. $0.1 M$ phosphate-buffered saline, pH 7.4.
6. 4% neutral-buffered formalin.
7. Heat-inactivated fetal calf serum.
8. Glutamine.

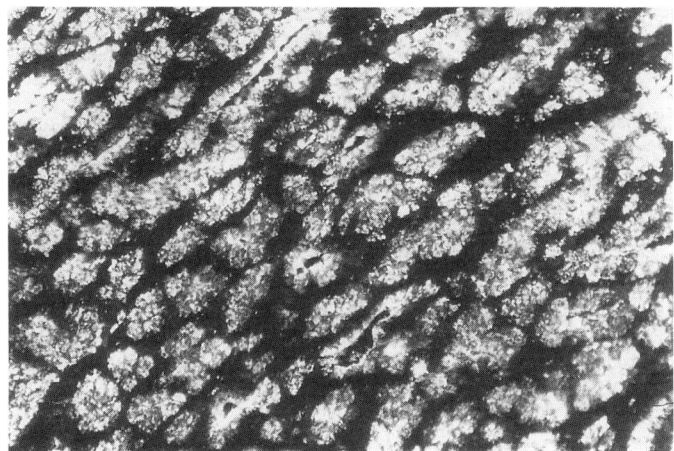

Fig. 1. A paraffin wax section of duodenum, showing Brunner's glands reacted for WGA lectin.

9. 16 μmol/mL Sodium bicarbonate.
10. HEPES buffer.
11. 50 U/mL penicillin and streptomycin.
12. 10% IL-2-containing tissue culture medium supernatant.
13. Shandon Sequenza immunostaining center, Life Sciences International (Basingstoke, Hampshire, UK).
14. Clicks tissue culture medium (Altick, Hudson, WI).
15. Citifluor (University of Kent at Canterbury, UK).

3. Methods

3.1. Method 1: Use of FITC-Labeled Lectins on Sectioned Material

1. Bring paraffin wax sections of duodenum containing Brunner's glands to $0.1M$ Tris-buffered saline (TBS), pH 7.4, containing 2% $CaCl_2$ and incubated for 30 min in a Shandon Sequenza immunostaining center with 100 μL of FITC-labeled lectin (100 μg/mL lectin; see Appendix) at room temperature (see Note 1).
2. Wash sections several times in TBS and mount in Citifluor.
3. Examine the sections on a fluorescence microscope, using exciter and barrier filters appropriate for FITC.
4. Check the specificity of each reaction by using a negative control in which the lectin is incubated with its appropriate binding sugar (see Appendix).
5. Examine and photograph sections (see Note 2 and Fig. 1) on a fluorescence microscope fitted with filters appropriate for FITC. Light microscopic localization of FITC-lectin binding sites is sufficient to visualize labeled membranes as well as intracellular organelles such as the Golgi apparatus (see Note 3).

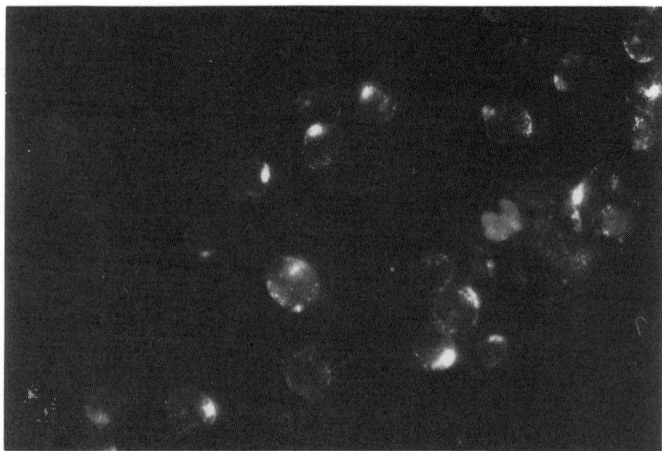

Fig. 2. Cells of the myelomonocytic cell line U937 after incubation with RCA-I lectin. Reactivity can be seen associated with the cell membrane and intracellular granules.

3.2. Method 2: Use of FITC-Labeled Lectins in Cytophotometric Analyses

FITC-lectins have been used in studies of uptake mechanisms in cells. Internalization occurs by endocytosis via both clatharin- and nonclatharin-mediated mechanisms.

1. Culture myelomonocytic cell lines (U 937 obtained from Sundström [9], and HL 60 obtained from the Department of Experimental Haematology, University of Munich) in a tissue culture incubator with 5% carbon dioxide and 95% humidified air maintained at 37°C, using RPMI-1640 medium, supplemented with 10% fetal calf serum and 1% penicillin:streptomycin.
2. Incubate the cell lines (50 µL of 1×10^6 cells) for up to 24 h with a panel of FITC-labeled lectins (50 µg lectin/mL dissolved in $0.15M$ phosphate-buffered saline, pH 7.4).
3. Wash the cells twice after incubation with the lectins in PBS and then fix in 4% neutral-buffered formalin and photograph (*see* Note 7 and Fig. 2).

3.3. Method 3: Use of FITC-Labeled Lectins in Quantitative Internalization Studies

A modification to the methodology described in Section 3.2. is the quantitation of internalization using cytophotometric analysis. A cytophotometric method was used to measure wheatgerm agglutinin (WGA) uptake in activated lymphocytes.

1. Maintain lymphocytes from the cell line CTLL-1 (cytotoxic lymphocytes [10,11]) in Clicks/RPMI-1640 (50/50) tissue culture medium supplemented with

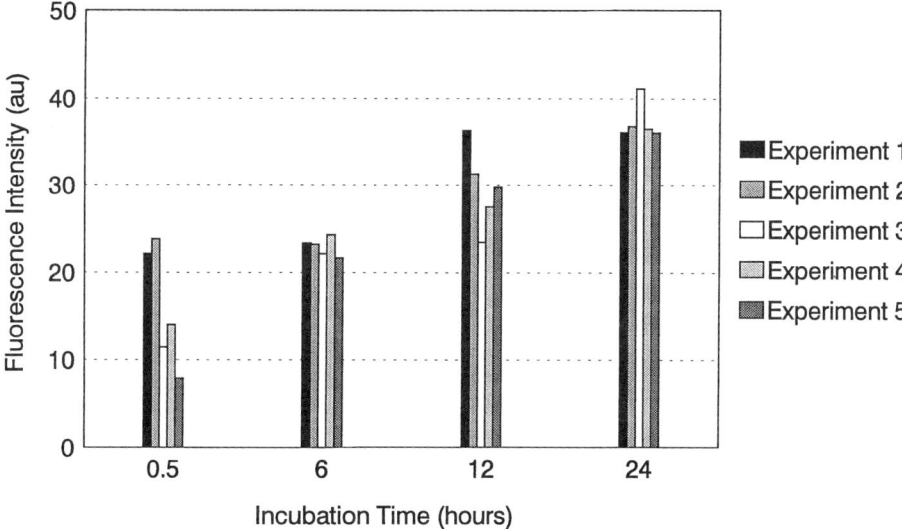

Fig. 3. A graphical representation of the results of an incubation experiment with WGA lectin. The graphs show the effect of extending incubation time on intensity of fluorescence. Experiments 1–5 are replicates.

2% heat inactivated fetal calf serum, 300 μg/mL glutamine, 16 μmol/mL sodium bicarbonate, 25 μmol/mL HEPES buffer, 50 U/mL penicillin and streptomycin, and 10% IL-2 containing supernatant (*see* Note 4).
2. Incubate stimulated cells (500,000/mL) with 4 μg/mL WGA.
3. Place the cells in suspension onto a glass microscope and measure the intensity of fluorescence with a microphotometer (MPV compact, Leitz, Wetzlar, Germany).
4. Examine cells with a ×70 oil immersion lens using a BG 12 excitation filter, and a green analysis filter (515–530 nm) and a 100-W mercury vapor lamp.
5. Measure the background fluorescence in areas free of labeled cells and subtract from the cell-associated fluorescence.
6. Examine 100–120 cells for each experiment to give measures of fluorescence intensity in arbitrary units, which allows a comparison between samples.
7. The results show an increase in WGA uptake, which varied with the length of incubation (*see* Notes 5 and 6, and Fig. 3).

3.4. Method 4: Use of FITC-Labeled Lectins in FACS Analysis

1. Incubate U937 (a myelomonocytic cell line) tumor cells at 4°C for 1 h with one of four lectins (25 μL/10^7): Con-A, UEA-1, PHA-L, and WGA all labeled with FITC.
2. Wash the cells twice with PBS.
3. Pass the cell suspensions (in PBS) through an EPICS(R) Coulter FACS analysis machine.

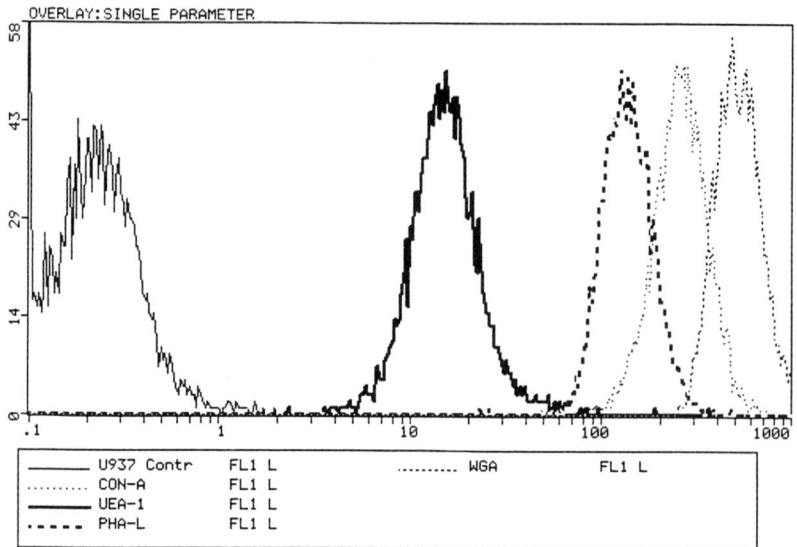

Fig. 4. An overlay of fluorescence intensity measured on a FACS machine for U937 cells incubated with Con A, UEA-1, PHA-L, or WGA lectins. Arbitrary units of relative cell numbers are shown on the *y*-axis, and of relative fluorescence intensity on the *x*-axis.

4. The cells pass through the sample nozzle (which has to be 50–100 μm in diameter depending on the diameter of the cells to be analyzed) in the FACS machine to produce a stream of single cells whose size/fluorescence intensity may be measured.
5. Determine the level of autofluorescence by passage of unlabeled cells through the FACS machine prior to labeled cells. For FITC, a 525-nm exciter filter is used.
6. Pass the cells that have been reacted with the lectin-FITC are then passed through the machine at a rate of 5000 cells/min.
7. Results are represented by Figs. 4–9. In Fig. 4, an overlay of the results is shown. It can be seen that the labeling of this particular population of cells varied accord-

Fig. 5. *(opposite page; top)* The FACS analysis of PHA-L-labeled U937 cells. The black line encircling the scatter distribution in **(A)** represents the cell population that was assessed. On the *y*-axis is FS (forward scatter), which corresponds to cell granularity, and on the *x*-axis is SS (side scatter), which gives a relative measure of cell number. In **(B)** the relative intensity of fluorescence (log scale on the *x*-axis) has been plotted against relative cell number on the *y*-axis.

Fig. 6. *(opposite page; middle)* The FACS analysis of WGA-labeled U937 cells. *See* Fig. 5 for details of axes.

Fig. 7. *(opposite page; bottom)* The FACS analysis of Con A-labeled U937 cells. *See* Fig. 5 for details of axes.

Fluorochrome-Labeled Lectins

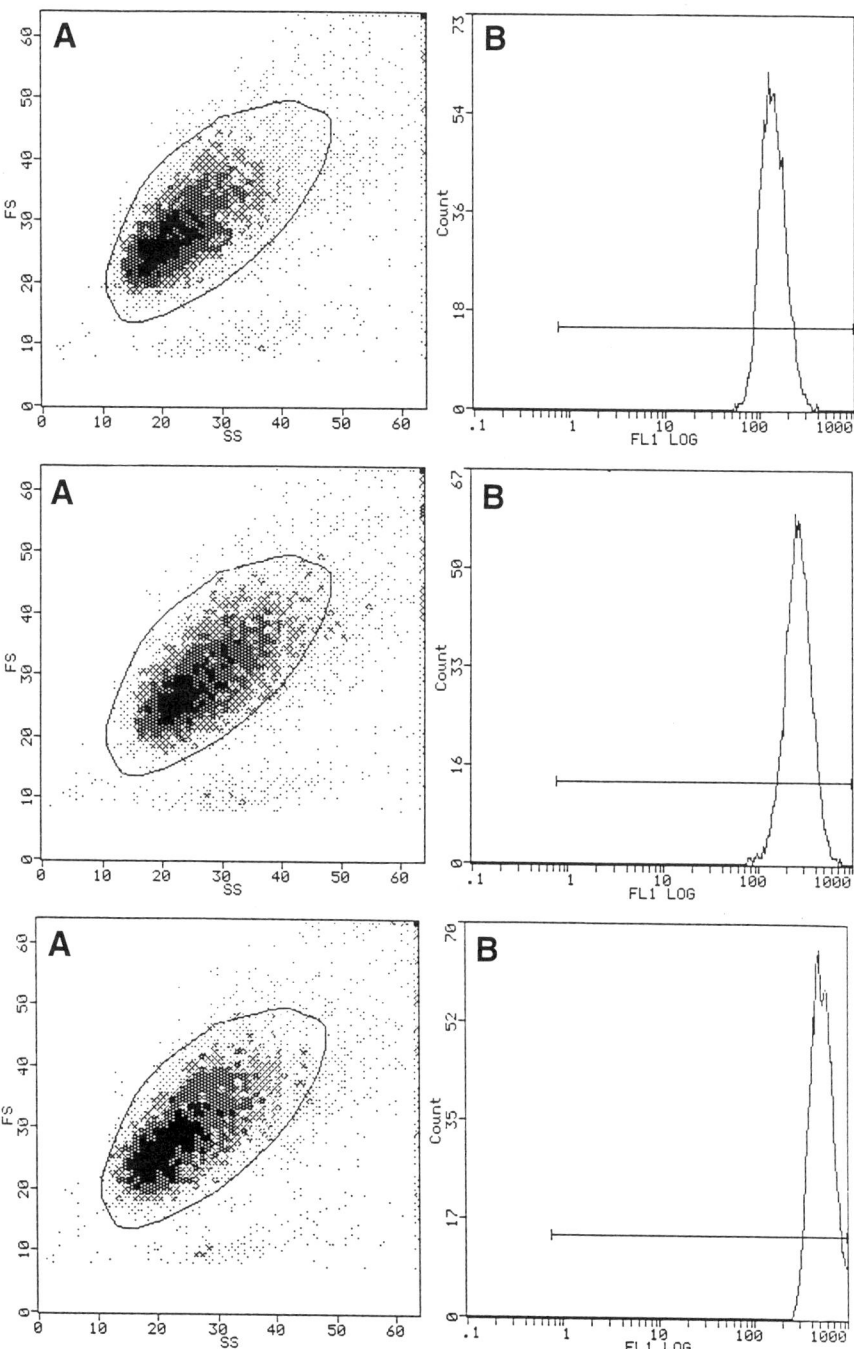

Figs. 5–7.

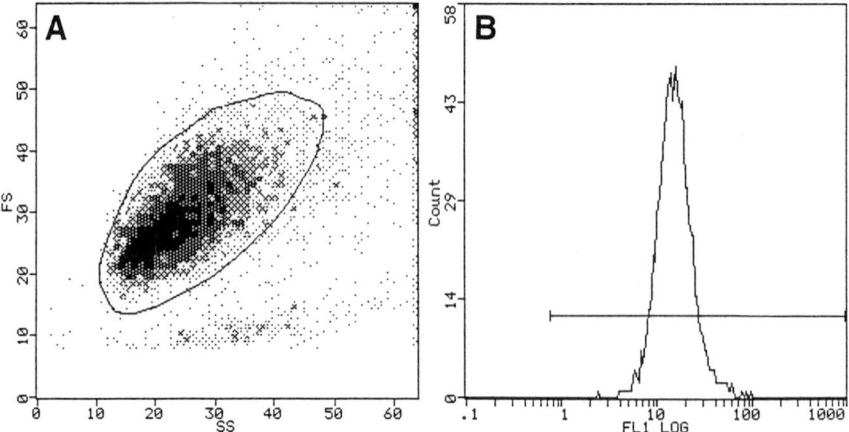

Fig. 8. The FACS analysis of UEA-1-labeled U937 cells. *See* Fig. 5 for details of axes.

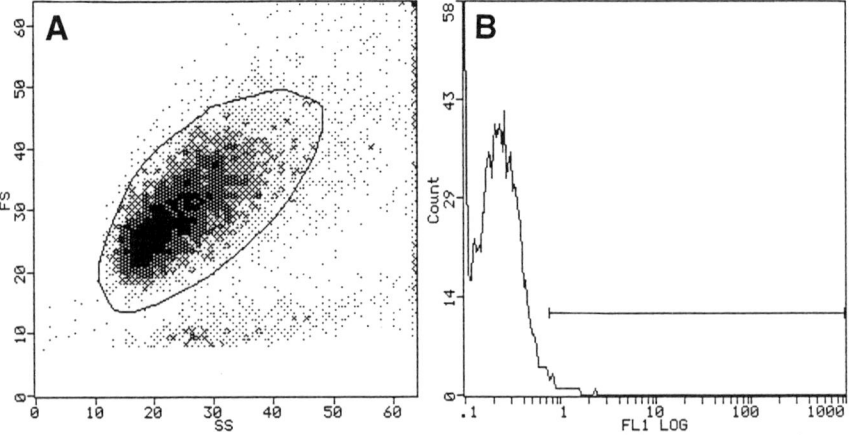

Fig. 9. The FACS analysis of U937 cells without lectin incubation. *See* Fig. 5 for details of axes.

ing to the lectin used; therefore, there is a possibility of sorting the cells into different lectin-positive and lectin-negative populations. Individual representations of FACS analyses for each of the lectins are shown in Figs. 5–9. In Fig. 5A, the field selected for measurement, set arbitrarily, is shown. Figure 5B indicates the fluorescence of cells labeled with WGA. This is the most intensive lectin labeling for U937 cells. In comparison, Fig. 6 shows the results for Con-A, Fig. 7 for PHA-L, and Fig. 8 for UEA-1 of successively decreasing lectin reactivity. The results of the negative control can be seen in Fig. 9, wherein relatively very low levels of fluorescence were detected (*see also* Notes 7–9).

4. Notes

1. Using the Shandon Sequenza immunostaining center, it is possible to ensure that a uniform amount of lectin is delivered to the section in a known volume that is equally applied to all sections. Washing of sections is far more efficient than with manual methods. It is possible to incubate sections with lectins without this apparatus, but its use does give consistent, "clean" results, which are superior to sections incubated on a staining rack. The use of 100 µg lectin/mL is rather high, but it ensures that a reaction takes place if carbohydrate residues specific for that lectin are present. It is recommended that dilutions 50, 25, and 12.5 µg/mL be used to assess background to target ratios. Use of appropriate dilutions will result in considerable savings.
2. In paraffin wax sections, the intensity of fluorescence may be graded semi-quantitatively if required. It is also possible to undertake double labeling experiments wherein two or more different lectin-binding sites are identified using lectins with different markers attached, e.g., FITC and TRITC (rhodamine). In double-labeling studies, it is imperative to ensure that no lectin–lectin interactions occur since this would result in a false positive reaction.
3. Whereas the ability of sections of different tissues varies with respect to their abilities to bind labeled lectins, the latter may be detected in varying intensities. Generally, fluorescence can be detected that is membrane-associated, and also that is located in intracellular organelles, often in endocytotic vesicles, which generally fuse with lysosomes.
4. IL-2-containing supernatant as used in the cytophotometric studies to measure WGA uptake was prepared using 24-h culture medium (RPMI-1640/10% fetal calf serum) from cultures of Con A 2.5 µg/mL stimulated normal DBA/2 mouse spleen cells.
5. The uptake of labeled lectins demonstrates that carbohydrate-containing binding sites are present, and that there are mechanisms to internalize the bound lectin glycoconjugate complexes. Differently labeled lectins could be used to examine couptake and cointernalization. The precise methodology used can result in differences in uptake and binding, particularly in intensity of fluorescence.
6. The estimates resulting from cytophotometric studies allow a quantitative measure to be obtained that may be used in comparative studies. It also allows the monitoring of changes in lectin binding over a period of time, which indicates the changes in expression of cell surface receptors, as in Fig. 3.
7. The cytophotometric approach for quantitative studies of lectin uptake has the advantage over FACS analyses in that in the latter, large cells, including epithelial and tumor cells, sometimes stick to the tubing in the machine and block the flow. If cells are grown on cover slips, no enzymatic pretreatment (trypsinization) for loosening of the cells has to be used. Since enzyme pretreatment can digest away lectin-binding sites under these conditions, it is not recommended (in contrast to fixed material).
8. Adherent cells grown on cover slips can be assessed by both the cytophotometric and FACS methods.
9. Large cell aggregates can be removed by passing the cells through a double layer of lens tissue or forcing the cells through a 25-gage syringe needle.

References

1. Sharon, N. and Lis, H. (1972) Lectins: cell agglutinating and sugar specific molecules. *Science* **177,** 949–955.
2. Lis, H. and Sharon, N. (1986) Lectins as molecules and tools. *Ann. Rev. Biochem.* **55,** 35–67.
3. Rittman, B. R. and MacKenzie, I. C. (1983) Effects of histological processing on lectin binding patterns in oral mucosa and skin. *Histochem. J.* **15,** 467–474.
4. Walker, R. A. (1984) The binding of peroxidase-labelled lectins to human breast epithelium I—normal, hyperplastic and lactating breast. *J. Pathol.* **144,** 279–291.
5. Mepham, B. L., Frater, W., and Mitchell, B. S. (1979) The use of proteolytic enzymes to improve immunoglobulin staining by the PAP technique. *Histochem. J.* **11,** 345–357.
6. Schumacher, U., Adam, E., Brooks, S. A., and Leathem, A. J. (1995) Lectin-binding properties of human breast cancer cell lines and human milk with particular reference to Helix pomatia agglutinin. *J. Histochem. Cytochem.* **43,** 275–281.
7. Horst, H.-A., Schumacher, U., Horny, H.-P., and Lennert, K. (1992) Lectin-binding profile of plasmacytoid monocytes. *Hum. Pathol.* **23,** 1178–1181.
8. Schumacher, U., Wiedmann, B., and Steinmann, G. G. (1989) Wheat germ agglutinin uptake by cytotoxic lymphocytes, a quantitative model of receptor-mediated endocytosis. *Protoplasma* **151,** 124–127.
9. Sundström, C. and Nilsson, K. (1976) Establishment and characterisation of a human histiocytic lymphoma cell line (U937). *Int. J. Cancer.* **17,** 565–577.
10. Gillis, S. and Smith, K. A. (1977) Long term culture of tumour-specific cytotoxic T cells. *Nature* **268,** 154–156.
11. Gillis, S., Ferm, M. M., Ou, W., and Smith, K. A. (1979) T cell growth factor: parameters of production and a quantitative microassay for activity. *J. Immunol.* **120,** 2027–2032.

5

The Use of Lectins in Combination with Enzymatic Digestion for the Study of Glycoconjugates in Cartilage

Sibylle Hoedt-Schmidt

1. Introduction

Cartilage is a highly specialized connective tissue that consists of cells embedded within an abundant extracellular matrix. The extracellular matrix of articular cartilage is composed of several types of collagens, proteoglycans, glycoproteins, noncollagenous proteins, polysaccharides, and water (for review, see refs. *1* and *2*). Besides type II collagen as a major constituent, the predominant proteoglycan of cartilage tissue is the large aggregating proteoglycan (aggrecan) consisting of a core protein to which glycosaminoglycans and *N*- as well as *O*-linked oligosaccharides are attached. This proteoglycan is able to form specific aggregates with hyaluronic acid and link protein.

The integrity of cartilage matrix depends on the equilibrium between anabolic synthetic and catabolic degradative processes. Any perturbation of either the rate of synthesis or rate of catabolism must lead to pathological changes in the structural and functional constitution of the cartilage tissue. Using lectin-binding techniques, alterations in the composition of matrix components (e.g., proteoglycans) in osteoarthritic cartilage can be visualized *(3,4)*. The target structures of lectin-binding patterns in articular cartilage tissue have been evaluated using monoclonal antibodies (MAbs) against a variety of epitopes on proteoglycan molecules *(2)*.

Interactions between carbohydrates and lectins have also been studied for the identification of subcellular compartments containing oligosaccharide moieties *(5–7)*. Furthermore, carbohydrates as components of glycoproteins or glycolipids are important cell-surface molecules involved in cell recognition and growth *(8)*.

1.1. Lectin Histochemistry

The application of lectins for the detection of carbohydrate structures in normal, healthy cartilage tissue results in negative staining patterns for *Ulex europaeus* agglutinin I (UEA I), soybean agglutinin (SBA), and peanut agglutinin (PNA), whereas wheatgerm agglutinin (WGA) and Concanavalin A (Con A) show a distinct positive staining reaction *(3)*. Some lectins detect internal sugar moieties of oligosaccharides or glycosaminoglycans; others bind only to terminal sugar residues. It is thus likely that glycosaminoglycan chain breakdown as a result of osteoarthritic degradation may expose those sugar residues to lectin binding that are not accessible to the lectin in normal healthy tissue. Therefore, positive labeling of tissue with UEA I in osteoarthritic cartilage indicates matrix proteoglycan degradation. It has been shown that UEA I has a high affinity to extracellular matrix in fibrillated human patellar cartilage *(9)*.

In normal healthy cartilage, binding sites for a variety of lectins are masked by steric hindrance, owing to the densely packed proteoglycan aggregates in the matrix. Loss of proteoglycans and bondage breakdown, as can be seen in osteoarthritic cartilage, facilitates the binding of these lectins to specific carbohydrate structures resulting in distinct and defined binding patterns *(3)*. Thus, negative staining in lectin-binding histochemistry does not necessarily imply the absence of lectin-binding sites.

The results of lectin-binding experiments have to be considered with care *(2,6)*. Lectin affinities as established by in vitro analysis may not be consistent with lectin binding *in situ*, in which binding may be affected by charge density, hydrophobic interactions, and accessibility of the potential lectin-binding site *(10)*.

1.2. Use of Fluorescein-Labeled Lectins in Combination with Enzymatic Digestion

To visualize lectin-binding sites in tissue sections, lectins may be conjugated with various labeling substances. The most widely used label is fluorescein isothiocyanate (FITC), a fluorochrome that can be covalently bound to proteins. Other labels that have been applied with lectins include various fluorochromes, ferritin, and horseradish peroxidase. Lectins may also be labeled with ^{3}H-acetyl groups for subsequent detection by autoradiography. Simultaneous double labeling of sugar moieties can be achieved using two lectins with different binding specificity and different labels *(11)*.

Glycosaminoglycan degrading enzymes such as chondroitin ABC lyase (chondroitinase ABC) and endo-β-D-galactosidase (keratanase) can effectively be used to induce a standardized matrix degradation increasing the accessibility of sugar moieties for lectin binding as well as epitopes recognized by MAbs. Chondroitinase ABC removes chondroitin sulfate disaccharides from the proteoglycan monomer, whereas keratanase specifically degrades keratan sul-

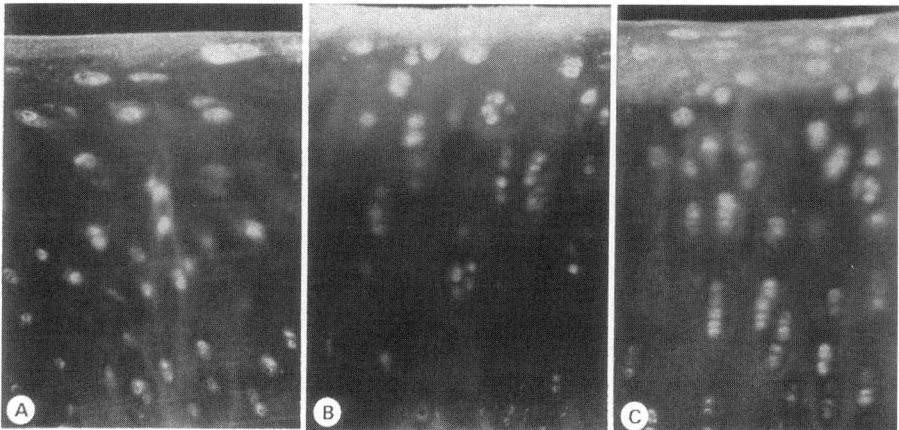

Fig. 1. Lectin-binding histochemistry with FITC-labeled wheatgerm agglutinin (WGA). Reprinted with permission from ref. 2. **(A)** Normal bovine articular cartilage (control). **(B)** Cartilage degradation with chondroitinase ABC prior to fixation. **(C)** Digestion of cartilage tissue with chondroitinase ABC and keratanase.

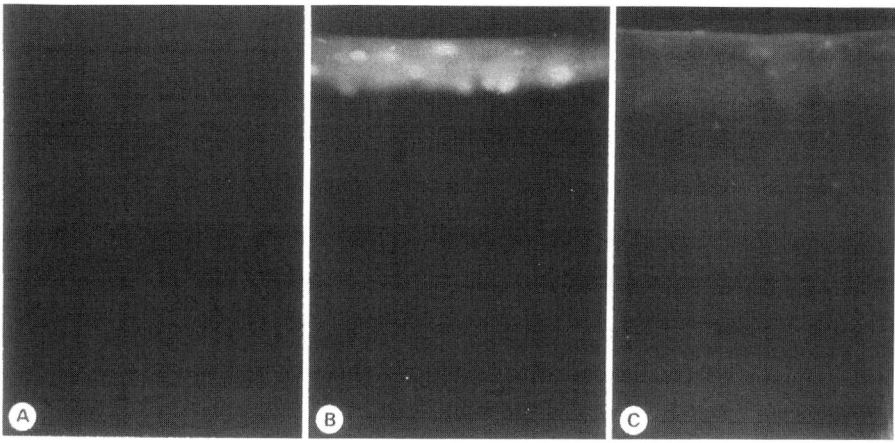

Fig. 2. Lectin-binding histochemistry with FITC-labeled peanut agglutinin (PNA). Reprinted with permission from ref. 2. **(A)** Normal bovine articular cartilage (control). **(B)** Cartilage degradation with chondroitinase ABC prior to fixation. **(C)** Digestion of cartilage tissue with chondroitinase ABC and keratanase.

fate *(12)*. The use of these enzymes results in a defined degradation of cartilage matrix components *(see* Figs. 1 and 2). The changes obtained resemble, at least in part, the alterations in glycosaminoglycan content observed in osteoarthritic

Table 1
Properties of Various Lectins Useful in Cartilage Research and Their Corresponding Carbohydrate Specificity

Lectin	Carbohydrate specificity	Pretreatment
Wheatgerm agglutinin (WGA)	β-D-GlcNAc-(1-4)-β-D-GlcNAc-(1-4)-β-D-GlcNAc > β-D-GlcNAc > Neu5Ac	Not required
Peanut agglutinin (PNA)	β-D-Gal(1-3)-D-GalNAc > D-GalNH$_2$ = α-D-Gal	Neuraminidase, chondroitinase ABC, keratanase
Soybean agglutinin (SBA)	β-D-GalNAc; α-D-GalNAc	Chondroitinase ABC, keratanase
Ulex europaeus agglutinin I (UEA 1)	α-L-Fuc	Chondroitinase ABC, keratanase
Concanavalin A (ConA)	α-D-Man > α-D-Glc > α-D-GlcNAc	Not required

cartilage (2). It may serve as a reproducible protocol to unmask lectin-binding sites in glycosaminoglycan-containing macromolecules.

Table 1 demonstrates the properties of various lectins useful in cartilage research, their different sugar-binding specificities, and the required pretreatment for the detection of lectin-binding sites.

2. Materials

2.1. Lectin Histochemistry

1. 0.01*M* Phosphate-buffered saline (PBS), pH 7.4.
2. Fluorescein isothiocyanate-labeled lectins (various commercial suppliers) are dissolved in PBS buffer at pH 7.4 *(see* Note 1). The binding activities of some lectins (e.g., concanavalin A) depend on cations *(13)*. The PBS buffer can be supplemented with 1 m*M* CaCl$_2$, 1 m*M* MgCl$_2$, and 0.1 m*M* MnCl$_2$ *(14)*. Some of the lectins possess mitogenic properties and others are associated with potent toxins, so they should be handled with care.

2.2. Use of Fluorescein-Labeled Lectins in Combination with Enzymatic Digestion

1. Hanks balanced salt solution containing glucose.
2. Protease-free chondroitinase ABC (Boehringer Mannheim, Mannheim, Germany).
3. Keratanase (Sigma, St. Louis, MO).
4. Neuraminidase (Sigma).
5. 100 m*M* Tris-acetate buffer, pH 7.6.

Lectin Study of Glycoconjugates

6. Proteinase inhibitors: Ethylenediamine tetraacetic acid (EDTA), N-ethylmaleimide, phenylmethanesulfonyl fluoride (PMSF), pepstatin.
7. Incubation buffer for digestion with chondroitinase ABC: 40 mM Tris-HCl, 40 mM sodium acetate, bovine serum albumin (BSA) 0.01% (w/v), pH 7.3.
8. Incubation buffer for digestion with keratanase: 100 mM Tris-HCl, 30 mM sodium acetate, BSA 0.01% (w/v), pH 8.0, including a mixture of proteinase inhibitors consisting of 10 mM EDTA, 10 mM N-ethylmaleimide, 5 mM PMSF, 0.36 mM pepstatin.
9. Incubation buffer for digestion with neuraminidase: 100 mM acetate buffer, 40 mM $CaCl_2$, pH 5.5.

3. Methods

3.1. Lectin Histochemistry of Formaldehyde-Fixed Tissue (see Note 2)

The lectin-binding studies can be performed on 5–8-µm deparaffinated tissue sections. In order to determine the affinity of lectins for tissue components, fluorescein isothiocyanate (FITC)-labeled lectins are reacted with sections at various concentrations. The lectin conjugates should be diluted in 0.01M PBS (pH 7.4) and should be used in concentrations ranging from 3 to 100 µg/mL. The sections are incubated in a 50-µL drop of lectin solution (concentration in stock solution is 1 mg/mL) in a moist chamber at room temperature.

1. Mount 5–8-µm tissue sections on cleaned glass slides.
2. Deparaffinate the sections in xylene and pass them through a decreasing alcohol series into PBS.
3. Incubate with FITC-labeled lectin conjugate for 30–60 min at room temperature in concentrations ranging from 0.003 to 0.1 mg/mL *(see* Note 3).
4. Test the specificity of the obtained staining reaction by preincubating the lectin conjugates in a solution of the appropriate inhibitory sugar *(see* Note 4).
5. Wash off unbound lectin conjugate by 3 × 5-min washes in PBS.
6. Estimate the autofluorescence of the tissue by incubations with PBS buffer alone.
7. Mount the sections using Mowiol *(15)* at pH 8.5.
8. Examine sections using a microscope equipped for epifluorescence with a 450–490-nm excitation filter and a 520–560-nm barrier filter.

3.2. Lectin Histochemistry After Pretreatment with Degrading Enzymes

1. Equilibrate the tissue slices in Hanks balanced salt solution containing glucose for 30–60 min *(see* Note 5).
2. Incubate with protease-free chondroitinase ABC, 0.2 U/mL at pH 7.3 and 37°C for 90 min *(see* Notes 6 and 7).
3. Incubate with keratanase, 0.2 U/mL at pH 8.0 for 90 min at 37°C with proteinase inhibitors *(see* Note 7).

4. Fix the tissue slices in buffered formaldehyde solution *(see* Note 8).
5. See preceding protocol for lectin binding histochemistry.

3.3. Treatment of Tissue Sections with Chondroitinase ABC

1. Treat tissue sections for 90 min at 37°C with chondroitinase ABC at 0.0125 U/50 µL per section in Tris-acetate buffer, pH 7.6, according to Hascall and Heinegård, Poole et al., and Hoedt-Schmidt *(2,16,17).*
2. Perform lectin-binding histochemistry *(see* Section 3.1.).

3.4. Treatment of Tissue Sections with Neuraminidase

1. Incubate tissue sections for 60 min at room temperature with neuraminidase at 1.25 U/mL in acetate buffer, pH 5.5, to expose any sialic acid-covered binding sites *(see* Note 7).
2. *See* Section 3.1. for lectin-binding histochemistry.

4. Notes

1. Fluorochromes other than fluorescein isothiocyanate are also available and can be used in lectin histochemistry.
2. Many lectins (e.g., wheatgerm agglutinin) bind similarly to formaldehyde-fixed and frozen tissue, but there are some exceptions, particularly in terms of sensitivity *(18).* Therefore, it is advisable to perform lectin-binding experiments on frozen and formaldehyde-fixed tissue to compare the staining characteristics.
3. The incubation time may vary between different lectins and should be tested.
4. Some lectins are easier to inhibit than others *(18).* The concentration of inhibitory sugar is usually in the range of 0.2–2M.
5. The equilibration time should be chosen depending on the thickness of the tissue.
6. A mixture of proteinase inhibitors may be included, e.g., ethylenediamine tetra-acetic acid (EDTA), N-ethylmaleimide, phenylmethanesulfonyl fluoride, pepstatin.
7. The duration of the enzymatic digestion and the concentration of the enzyme applied has to be tested and is depending on the tissue of interest.
8. The time of fixation should be tested. Overfixation may influence the detection of lectin-binding sites.

References

1. Heinegård, D. and Oldberg, A. (1989) Structure and biology of cartilage and bone matrix noncollagenous macromolecules. *FASEB J.* **3,** 2042–2051.
2. Hoedt-Schmidt, S., McClure, J., Jasani, M. K., and Kalbhen, D. A. (1993) Immunohistochemical localization of articular cartilage proteoglycans and link protein in situ using monoclonal antibodies and lectin-binding methods. *Histochemistry* **99,** 391–403.
3. Hoedt-Schmidt, S. (1989) Lectin-binding histochemistry of normal and osteoarthritic cartilage tissue. *Clin. Exp. Rheumatol.* **7,** 257–264.

4. Hoedt-Schmidt, S., Scheid, A., and Kalbhen, D. A. (1989) Histomorphological and lectin-histochemical confirmation of the antidegenerative effect of diclofenac in experimental osteoarthrosis. *Drug Res.* **39,** 1212–1219.
5. Roth, J. (1978) *The Lectins. Molecular Probes in Cell Biology and Membrane Research.* Gustav Fischer Verl., Jena.
6. Farnum, C. E. (1985) Binding of lectin-fluorescein conjugates to intracellular compartments of growth-plate chondrocytes in situ. *Am. J. Anat.* **174,** 419–435.
7. Damjanov, I. (1987) Biology of disease: lectin cytochemistry and histochemistry. *Lab. Invest.* **57,** 5–20.
8. Sharon, N. and Lis, H. (1989) Lectins as cell recognition molecules. *Science* **246,** 227–246.
9. Schünke, M., Schumacher, U., and Tillmann, B. (1985) Lectin-binding in normal and fibrillated articular cartilage of human patellae. *Virchows Arch. Pathol. Anat.* **407,** 221–231.
10. Ochoa, J. H. (1981) Consideration of the nature of the lectin-carbohydrate interaction. *J. Chromatogr.* **215,** 351–360.
11. Roth, J., Binder, M., and Gerhard, U. J. (1978) Conjugation of lectins with fluorochromes: an approach to histochemical double labeling of carbohydrate components. *Histochemistry* **56,** 265–273.
12. Oike, Y., Kimata, K., Shinamura, T., and Suzuki, S. (1980) Proteinase activity in chondroitin lyase (chondroitinase) and endo-β-D-galactosidase (keratanase) preparations and a method to abolish their proteolytic effect on proteoglycan. *Biochem. J.* **191,** 203–207.
13. Lönnerdal, B., Borrebaeck, C. A. K., Etzler, M. E., and Ersson, B. (1983) Dependence on cations for the binding activity of lectins as determined by affinity electrophoresis. *Biochem. Biophys. Res. Commun.* **115,** 1069–1974.
14. Farnum, C. E. and Wilsman, N. J. (1984) Lectin-binding histochemistry of non-decalcified growth plate cartilage: a postembedment method for light microscopy of Epon-embedded tissue. *J. Histochem. Cytochem.* **32,** 593–607.
15. Osborn, M. and Weber, K. (1982) Immunofluorescence and immunocytochemical procedures with affinity purified antibodies: tubulin-containing structures, in *Methods in Cell Biology* (Wilson, L., ed.), Academic, New York, pp. 97–132.
16. Hascall, V. C. and Heinegård, D. (1974) Aggregation of cartilage proteoglycans. I. The role of hyaluronic acid. *J. Biol. Chem.* **249,** 4232–4241.
17. Poole, A. R., Pidoux, I., Reiner, A., Tang, L.-H., Choi, H., and Rosenberg, L. (1980) Localization of proteoglycan monomer and link protein in the matrix of bovine articular cartilage: an immunohistochemical study. *J. Histochem. Cytochem.* **28,** 621–635.
18. Rhodes, J. M. and Ching, C. K. (1993) The application of lectins to the study of mucosal glycoproteins, in *Methods in Molecular Biology, vol. 14: Glycoprotein Analysis in Biomedicine* (Hounsell, E. F., ed.), Humana, Totowa, NJ, pp. 247–262.

6

Applications of Lectin Histochemistry and Cytochemistry in Diagnosis and Prognosis

Timothy Richard Helliwell

1. Introduction

The purpose of diagnostic histopathology and cytology is to categorize a disease process and predict how that disease will affect the patient. Pathologists are now combining careful morphological evaluation of the structure of the cell nucleus and cytoplasm with immunohistochemical and lectin histochemical studies to explore the molecular pathophysiology of diseased tissues and provide additional information to determine treatment and prognosis. Many of the antibodies used in diagnostic tumor pathology recognize epitopes that include carbohydrate sequences, e.g., α-fetoprotein, carcinoembryonic antigen, CA125, which may also be identified by lectins. Cancer-associated oligosaccharides are often oncodevelopmental in nature and reflect changes seen during normal development *(1,2)*. Whereas immunohistochemical studies often provide diagnostic information, previous reviews suggest that lectin histochemical data are rarely pathognomonic *(3,4)*. This chapter will summarize more recent data, providing references to recent papers or review articles as an entry to the extensive literature. The lectins quoted, their abbreviations, and nominal specificities are shown in Table 1.

Membrane glycoconjugates are important mediators of growth stimuli, cell–cell recognition, and cell adhesion and cell maturation, disruption of which may become apparent as alterations in response to growth factors or impaired adhesion to substrates. Monitoring of carbohydrate epitopes and their endogenous receptors may detect changes in cellular transformation, biological behavior and prognosis *(1,2)*. Physiologically, changes in cellular carbohydrates will only be relevant if specific endogenous receptors such as lectins

Table 1
Source of Lectins, Specificity, and Abbreviations Used

Lectin origin	Acronym	Specificity	Blood group specificity
Glucose/Mannose			
Concanavalia ensiformis	Con-A	α-D-Glc, α-D-Man	
Lens culinaris	LCA	α-D-Glc, α-D-Man	
Pisum sativum	PSA	α-D-mannose	
Fucose			
Ulex europaeus-1	UEA1	α-L-fucose	H antigen
Tetragonolobus purpurea	LTA	α-L-fucose	H antigen
N-Acetyl Glucosamine			
Triticum vulgaris	WGA	α(1-4)-D-GlcNAc	
Succinyl-WGA	Succ-WGA	GlcNAc	
Bandereirea simplicifolia-2	BS2	GlcNAc	
N-Acetyl Galactosamine/ Galactose			
Arachis hypogaea	PNA	Gal-α-(1-3)-GalNAc	
Dolichos biflorus	DBA	α-D-GalNAc	A1
Bandereirea simplicifolia-1	BS1	α-D-Gal	B
Glycine max	SBA	α-D-GalNAc, α-D-Gal	
Artocarpus integrifolia (Jack fruit)	JFL	D-GalNAc	
Ricinus communis 1	RCA	β-D-Gal	
Sophora japonica	SJA	α-D-GalNAc, α-D-Gal	A,B
Helix pomatia	HPA	GalNAc	A
Vicia villosa	VVA	GalNAc	A1, Tn
Maclura pomifera	MPA	Gal α1-3 GalNAc	
Pokeweed mitogen	PWM	poly-Nacetyllactosamine	

recognize the carbohydrate structures, establishing a modulatory sequence based on specific protein–carbohydrate interactions. The application of labeled, carrier-bound carbohydrates (neoglycoproteins) or natural carbohydrates to tissues, can identify these endogenous lectins, providing a valuable adjunct to more conventional lectin binding studies *(5)*.

1.1. Problems of Methodology

The evaluation of published studies is complicated by the lack of standardization of methods, with potential sources of variation including tissue fixation and processing, lectin concentrations, the methods used to visualize lectin binding, the evaluation of the cellular distribution of lectin binding, and the definition of positive binding.

Lectin binding is best studied using unfixed, frozen sections of tissues, although this is similar to binding by acetone or picric acid-fixed tissues. Formalin fixation generally reduces lectin binding since carbohydrate groups are not protected from either extraction or masking during paraffin embedding. Lipid extraction by chloroform, alcohol, or xylene during processing eliminates lectin binding to glycolipids *(4,6)*, although lipid extraction may reveal lectin-binding sites masked by lipids in frozen sections *(7)*. High lectin concentrations detect both high and low affinity receptors, whereas low concentrations (<20 µg/mL) detect only high affinity receptors and are more specific *(8)*. It is therefore essential that dilutional studies are performed during the initial evaluation of a method.

Visualization methods using lectins directly conjugated to fluorescein or peroxidase are less sensitive than those using multistage amplification procedures, e.g., lectin-antilectin antibody methods and biotinylated lectin–ABC–peroxidase methods, and so may result in different assessments of the value of lectin binding *(4,8,9)*.

For a method to be of use in diagnostic laboratories, it needs to work on formalin-fixed, paraffin wax-embedded tissues (routine processing). Retrospective studies on routinely processed tissues need to be interpreted with caution due to variations in tissue fixation and processing that may occur over a period of time. In research work, the methods of tissue handling are carefully selected, and it is important that these are validated before using data in diagnostic work. Diagnostic laboratories are increasingly receiving biopsy tissues in an unfixed state so that a range of molecular studies can be performed, including lectin histochemistry on frozen tissues.

1.2. The Interpretation of Changes in Lectin Binding to Tissues

Lectins are highly specific for the nature and three-dimensional configuration of the carbohydrate group to which they bind. However, these oligosaccharides may be common to several glycoproteins and/or glycolipids, so that the precise nature of the glycoconjugate is not known. It is useful to select lectins with slight differences in affinity for the same sugar *(10)* and to use lectins as a panel of reagents to identify cell types or functional differences between cells. It is also important to define the normal pattern of endothelial binding of a tissue, because there may be regional variations in glycoconjugate expression, e.g., stomach, colon.

Several lectins have affinity for blood group substances (*see* Table 1), which can be identified by monoclonal antibodies. Whereas the antibodies are more specific than lectins, lectins also identify epitopes on other glycoconjugates *(11)*.

New carbohydrate structures on the surface of cancer cells may represent the accumulation of precursor chains because of decreased activity of synthesizing enzymes, the production of new oligosaccharides because of increased or aberrant glycosylation, a change in density of cell surface carbohydrates, or exposure (unmasking) of chains usually covered by other structures *(1)*. It is important to recognize that changes in lectin binding may reflect increased cell turnover and relative immaturity of the cells rather than changes associated with neoplasia (*see* Section 5.5.2.).

Lectin histochemistry has been most useful for the identification of abnormal storage products in cells and of endothelial cells. Less well-defined uses include the study of cellular changes in inflammatory diseases and neoplasia.

2. Storage Disorders

Lectin histochemistry is a valuable tool for the identification of specific enzyme deficiencies that result in the storage of abnormal glycolipids and glycoproteins *(7)*. Glycoprotein storage disorders, e.g., α-mannosidosis and fucosidosis, can be recognized in routinely-processed tissues, but the identification of storage products in glycolipid storage disorders is limited by lipid extraction in processed tissues. In frozen sections, RCA1 can be used to demonstrate GM1 gangliosidosis (lysosomal β-galactosidase deficiency), and in Fabry's disease, the stored glycolipids show strong binding by BS1 and PNA. In Gaucher's disease (lysosomal glucocerebrosidase deficiency) the lectin binding pattern indicates a wider range of storage products than would be predicted from the enzyme deficiency.

3. Endothelial Cells

3.1. Normal Vessels

The pattern of lectin binding to normal endothelium is influenced by the species of animal and blood group. RCA1 and WGA bind to endothelial cells of all species, UEA1 only binds to human endothelium, and BS1 binds to blood group B human endothelium and is the lectin of choice to identify rodent endothelium *(12,13)*. UEA1 identifies terminal fucose residues of the blood group H antigen and reliably identifies human endothelial cells of lymphatic and blood vessels. For routinely processed tissues, pretreatment of sections with trypsin is essential for optimal lectin binding *(14)*. UEA1 labeling of normal endothelium is used to identify endothelial damage, e.g., in dermatomyositis *(15)*, and to facilitate the identification of vascular invasion by neoplasms

(14,16). As with all labeling techniques, UEA1 binding needs to be combined with careful morphological evaluation, as UEA1 also binds to pancreatic acinar cells and to the epidermis and epithelium of skin appendages.

3.2. Neoplastic Vessels

The binding of UEA1 to endothelial cells is used in the diagnosis of vascular neoplasms. UEA1 binds consistently to the endothelial cells in benign vascular neoplasms and hamartomas, and binds to most malignant vascular neoplasms (angiosarcomas and hemangioendotheliomas) including rare variants such as granular cell angiosarcoma *(17)*. When compared with other vascular endothelial markers, UEA1 is less specific than von Willebrand factor (vWF), but UEA1 is more sensitive and recognizes a higher proportion of malignant vascular tumors *(16,18,19)*. Because of the variation in antigen expression in normal endothelium and in malignant endothelial neoplasms, a panel of markers is used diagnostically, including UEA1, vWF, and CD31 or CD34. At least two markers should be positive before accepting the vascular nature of a tumor, as all markers can crossreact with other tissues *(20)*, and UEA-1 binding is also seen in several carcinomas and in some nonvascular sarcomas *(18,19)*. A panel of reagents is of particular value in the identification of vascular neoplasms in which the endothelial cells assume an epithelioid morphology and can coexpress cytokeratins as well as endothelial markers *(19,21,22)*. In benign lymphangioendothelioma, the cells lining vascular channels are CD34-positive, and less uniformly with UEA1 and vWF *(23)*.

4. Skin

The pattern of lectin binding to routinely processed sections of normal skin is related to cellular maturation *(24)*. All adult epidermal cells bind several lectins, but selective binding occurs with UEA1 (to the upper layers), DBA (to basal cells), and SBA and HPA (to the spinous and granular cell layers). In psoriasis, in which there is increase epidermal proliferation, DBA shows the normal pattern of binding to a single basal layer of cells, whereas both lesional and nonlesional skin show UEA1 binding to all layers of epidermis, implying a general disturbance of epidermal maturation *(25)*.

In skin tumors, ConA binding is absent in benign tumors but present at the cell membranes in all malignant tumors; solar keratoses show increasing binding as dysplasia increases *(26)*. Neoplasms arising from skin appendages show similar lectin binding patterns to normal appendage epithelium *(27)*. PNA binding is seen in normal epithelium and in keratoacanthomas; this may help to distinguish these benign neoplasms from squamous carcinomas *(28)*. Lectin binding may also help in the difficult differential diagnosis between a Spitz nevus and malignant melanoma *(29)*.

5. Alimentary System

5.1. Mouth

The squamous epithelium of the oral mucosa shows regional variations in lectin binding that reflect the degree of keratinization. Normal, nonkeratinizing mucosa shows UEA1 binding to the prickle cells, BS1 to the surface layers, and PNA to the basal layers, with RCA1 receptors in basal and spinous layers *(30)*. BS1 binding is more intense in hyperplasia and mild dysplasia, but binding is lost in severe dysplasia and carcinoma. As dysplasia increases in severity, UEA1 and PNA binding occurs in all layers; whereas PNA and UEA binding is positive in well-differentiated, invasive squamous carcinomas but is lost in poorly differentiated carcinomas. In the inflammatory disease, lichen planus, UEA1 and PNA are positive in all layers *(11)*. Regenerating epithelium and the advancing edge of invasive carcinomas shows less RCA1 binding than normal epithelium *(30)*.

5.2. Salivary Glands

In sections of routinely processed normal salivary glands, serous acinar cells show preferential binding to ConA, whereas mucous cells show more intense binding to WGA and LTA *(31)*. In salivary neoplasms, ConA binds to cells of myoepithelial type in benign and malignant neoplasms, whereas WGA and LTA bind to ductal cells in adenoid cystic carcinomas and to mucous cells of mucoepidermoid carcinomas and cyst lining cells in acinic cell carcinomas *(32)*.

5.3. Stomach

Lectin binding to most cells in the normal gastric mucosa is blood group-independent, however, the expression of carbohydrates in the surface mucous cells is partly dependent on blood group and secretor status *(33,34)*. A comprehensive study of routinely processed gastric neoplasms *(35)* showed that adenomas and intestinal pattern carcinomas have increased cytoplasmic ConA binding and decreased binding by WGA, PNA, and UEA1; these changes are similar to those seen in (premalignant) intestinal metaplasia but differ from those of diffuse-pattern carcinomas and may be owing to increased oligosaccharide sialylation or to loss of glycosyl transferases in the neoplastic cells *(33,35)*.

The expression of HPA binding sites by 59% of gastric carcinomas is not related to tumor differentiation, but is related to clinically aggressive behavior and shorter survival *(36)*. The expression of an endogenous lactose binding lectin (L-31) is greater in carcinoma than in adjacent normal tissue in 50% of cases, and there is higher expression in liver and nodal metastases than in the primary carcinoma in 35% cases. The presence of L-31 is associated with the metastatic phenotype but this association is not sufficiently robust to be used predictively *(37)*.

5.4. Small Intestine

In the small intestine, glycosylation status and lectin binding change according to the site of the cell in the crypt or villus *(38)*. In cystic fibrosis, there is persistent and intense LTA binding indicating abnormal (fetal pattern) fucosylation of mucus *(39)*. Changes in maturation seen in celiac disease are reflected by loss of DBA binding to goblet cells and increased binding by RCA1 and UEA1 *(38)*.

5.5. Large Intestine

5.5.1. Normal

A recent study has highlighted many of the methodological problems of lectin histochemistry *(8)* and demonstrated regional variation in high affinity lectin receptors between the proximal and distal colon, with more diverse glycan structures and greater sialyl content in the distal colon. This may relate to the differing embryological origin and/or activity of the bowel segments. Oligosaccharide chains are modified as the cells move from the base to the surface of the intestinal crypts, probably related to the maturation of glycosyl transferases and the incorporation of mannose residues of N-linked oligosaccharides, and of GalNAc and GlcNAc. Other investigators, using fluorescein-labeled lectins at relatively high concentrations describe a proximal to distal decrease in UEA1 binding to goblet mucin *(40)*.

5.5.2. Inflammatory Bowel Disease

In ulcerative colitis, the more rapid cell turnover and migration accompanying the inflammatory process is reflected by incomplete glycoprotein synthesis and increased PNA binding to enterocytes *(41)*. Compton has reviewed the changes in lectin binding observed in chronic ulcerative colitis and their ability to predict premalignancy. Some areas of dysplastic mucosa show increased PNA binding and reduced DBA binding, but similar changes are also seen in severely inflamed mucosa, and in inactive mucosa without dysplasia *(42)*. The high false positive and false negative rates indicate that lectin histochemistry does not facilitate the diagnosis of premalignancy.

5.5.3. Neoplasia

In colorectal carcinoma, the carbohydrate content of mucin is reduced, owing to fewer oligosaccharide side chains and decreased oligosaccharide chain length. Neutral mucins increase and there is decreased O-acetylated mucin. The normal regional variation of mucins and the constitutional heterogeneity of mucin expression in normal populations make it difficult to define an at-risk phenotype *(43,44)*. Changes in lectin binding are reflected by changes in the expression of blood group antigens, with increased binding of VVA (Tn antigen) accompanied by increased PNA binding (T antigen) and of sialosyl-Tn *(45)*. The HPA binding to the cell surface and cytoplasm is an independent predictor of survival *(46)*.

In patients with the familial adenomatous polyposis syndrome, there appears to be a field change in the intestinal mucosa, which predisposes to the development of neoplasia. This is reflected by abnormal lectin binding to the morphologically normal mucosa with less WGA and more UEA1 and succinyl-WGA binding compared with controls, suggesting that glycoconjugate modifications are early events in the evolution of the neoplastic phenotype *(47)*.

The endogenous lactose-binding lectin, L31, is found in the cytoplasm of normal cells and carcinoma but not in adenomas, with higher levels in more advanced carcinomas *(48)*.

5.6. Pancreas

In routinely processed tissues, human pancreatic acinar cells are uniformly positive for WGA and SBA (islets are negative), UEA1 binds to acinar and ductal tissue and BS2 is specific for ductal epithelium. No consistent, qualitative differences in lectin binding to carcinomas are observed, but PNA and BPA show increased binding to secreted mucins *(49)*.

5.7. Liver

The specialized endothelial cells lining the liver sinusoids undergo transformation to a capillary phenotype in chronic hepatitis and cirrhosis, as well as in hepatocellular carcinomas. This can be demonstrated by a marked increase in UEA1 binding which help the interpretation of small biopsy specimens *(50)*.

6. Upper and Lower Respiratory Tract

There have been few studies of the upper respiratory tract. In the nasopharyngeal epithelium, the basal cells are UEA1- and DBA-positive, whereas all layers of cells bind ConA and WGA *(51)*. Squamous metaplasia leads to the new expression of LTA binding sites and the loss of UEA1 binding *(51)*. PNA binding is reported to be absent in normal laryngeal squamous epithelium but present at the cell membranes in 21/30 squamous carcinomas *(52)*.

The lectin histochemistry of normal and neoplastic pulmonary tissues has been reviewed *(53)*. Type 1 pneumocytes bind to PNA, and type 2 pneumocytes bind to ConA. Bronchial carcinomas show great heterogeneity of labeling both within and between tumors, and there is no cell type-specific labeling, although labeling is generally more intense in adenocarcinomas than in other cell types, and labeling tends to be stronger and more widespread in well-differentiated tumors. The binding of RCA1 and Succ-WGA to most adenocarcinomas may be helpful in distinguishing them from pleural mesotheliomas, which show little binding for these lectins *(54)*. In cytological preparations of cell blocks from serous effusions, UEA1 and SBA bound to 65% of adenocarcinomas and not to reactive mesothelium nor mesotheliomas, suggesting that lectins may be a diagnostically useful adjunct to immunohistochemical markers, e.g., CEA *(55)*.

Blood group A expression on nonsmall cell carcinoma of lung is a favorable prognostic factor related to slower progression and longer survival *(56)*, whereas the observation of a relatively good prognosis for DBA positive, Le^X negative carcinomas is difficult to interpret as multivariate analysis was not performed, and different histological types of carcinoma were combined in the assessment *(57)*.

The binding of neoglycoconjugates to endogenous lectins shows a similar sensitivity and specificity to immunohistochemical methods in the therapeutically important distinctions between small-cell and nonsmall-cell carcinomas (the latter showing greater binding by maltose, fucose, and mannose), and between mesothelioma and metastatic adenocarcinoma (the latter showing binding by GlcNAc) *(59)*. Expression of endogenous receptors for fucose is related to tumor size *(5)*.

Fluorescein-labeling lectin binding has been combined with flow cytometry to study the phenotypic variation of alveolar macrophages obtained by bronchoalveolar lavage. In interstitial lung disease, the increased binding by PNA, UEA1, SJA, and BS1 may reflect the influx of immature blood monocytes or activation of pulmonary macrophages *(58)*.

7. Urinary Tract
7.1. Kidney
7.1.1. Tubules

The application of lectin histochemistry to renal tubular pathology has been reviewed *(60)*. The different segments of the normal nephron show specific lectin binding patterns: LTA and PHA bind to proximal tubules; PNA, DBA, and HPA (with low-mol-wt cytokeratins and epithelial membrane antigen) bind to the distal convoluted tubule and collecting duct; whereas antibodies to Tamm-Horsfall protein label the thick, ascending limb of Henle. These segment-specific glycoproteins are retained in pathological conditions including severe tubular atrophy and acute tubular necrosis, and may be used to define the site of damage. Acute cellular rejection of renal transplants and chronic interstitial nephritis tend mainly to affect the distal nephron and collecting duct *(61)*.

The segmental-specificity of lectins may also define the origins of renal cysts and neoplasms. In autosomal recessive (infantile) polycystic disease, the cysts primarily arise from collecting ducts, whereas in autosomal dominant (adult) polycystic disease, there is more variation, with some cysts showing the phenotype of proximal tubules, some a distal tubular phenotype, and some with no lectin binding *(60)*. In acquired renal cystic disease (dialysis kidney) the cysts originate from both proximal and distal tubules *(62)*. Low-grade renal cell carcinomas are mainly of proximal tubular origin, but high-grade carcinomas show prominent intratumor and intertumor heterogeneity, indicating proximal and distal tubular differentiation. Collecting duct carcinomas bind DBA and SBA, whereas oncocytomas show either a distal or a proximal phenotype *(60)*.

Lectin binding changes during the development of renal tubules, with DBA only binding to 30% of the cells in fetal collecting ducts and RCA binding restricted to the proximal tubules of fetal kidneys *(63)*. This is reflected by the lectin binding pattern in nephroblastoma (Wilms tumor), in which dysplastic tubules show a mainly distal and collecting duct phenotype. Normal fetal blastema and the blastema in Wilms tumors show similar lectin binding profiles and, in particular, the blastema is UEA1-negative. This may help to distinguish nephroblastoma from other small-cell tumors, e.g., neuroblastomas, which are UEA1-positive, and from other childhood renal tumors *(60,63,64)*.

7.1.2. Glomeruli

The glomerular basement membrane is highly sialylated, masking PNA receptors in all components of the glomeruli, whereas SBA and UEA1 bind to glomerular capillary walls *(65)*. Loss of UEA1 binding occurs in areas of glomerulosclerosis in both chronic pyelonephritis and in diffuse proliferative glomerulonephritis *(66)*. Intramembranous electron-dense deposits in type 2 membrane-proliferative glomerulonephritis are uniquely rich in GlcNAc, and show strong binding to succ-WGA *(67)*.

7.2. Bladder

In normal urothelium, cytoplasmic and membranous binding by WGA, GSA, and UEA1 predominate in the luminal cell layers; PNA without neuraminidase pretreatment (PNA-N) is negative, whereas neuraminidases reveal PNA binding sites in the luminal cell layers. In carcinoma *in situ*, there is decreased WGA binding *(68,69)*. In invasive transitional cell carcinomas there is decreased binding by WGA, BS1, and UEA1, and increased PNA-N binding may correlate with invasion and a higher risk of metastasis *(68,70)*. Cytoplasmic binding by neoglycoproteins is greater in the luminal cells than in basal cells of normal urothelium. Nuclear binding by galactose-specific glycoconjugates is characteristic of transitional carcinoma (and corresponds to the increased PNA binding), whereas there is intratumor and grade-dependent variation in the cytoplasmic binding of glycoconjugates, with increased binding to well-differentiated transitional carcinomas, and loss of binding as atypia becomes more severe *(71)*.

7.3. Prostate and Seminal Vesicle

Normal prostatic tissue shows cytoplasmic binding by many lectins, although there is selective binding of Succ-WGA, PNA, and UEA1 to the central zone of glands *(72)*. Seminal vesicles show a similar lectin binding pattern to the central prostatic glands, but uniquely bind BS1. In humans, the number of prostatic cells expressing type 1 oligosaccharides (PNA-binding) decreases with age, and type 1 expression may reflect androgenic drive *(73)*. In prostatic intraepithelial neoplasia, the dysplastic glands show a generalized defect in glycosylation with greatly reduced binding and luminal rather than cytoplasmic positivity *(72)*. Lectin-binding studies of adenocarcinomas give variable and contradictory results with little difference in lectin binding between carcinomas and normal or hyperplastic prostate *(74)*. In metastatic prostatic carcinoma, there is downregulation of type 1 oligosaccharides and masking of type 2 oligosaccharides by sialic acid, and experimental data suggest that the expression of complex branched oligosaccharides is related to tumor invasiveness and metastasis *(73)*.

8. Male and Female Genital Tract and Breast
8.1. Placenta

Placental cells bind many lectins, but some show selective lectin binding of potential diagnostic use. LCA and PSA bind to the trophoblast of first-term but not of third-term placenta. ConA binds preferentially to syncytiotrophoblast *(75)*.

8.2. Uterine Cervix

During the development of cervical intraepithelial neoplasia, the normal ectocervical epithelial binding of ConA and RCA1 becomes more intense as dysplastic changes become more severe, and there is new expression of UEA1 binding sites. Invasive squamous carcinomas show great heterogeneity of lectin binding *(76)*. Incomplete glycoprotein synthesis in squamous carcinomas, reflected by the binding of VVA rather than PNA, correlates with lymphatic permeation, metastasis, and a low 5-yr survival; VVA binding is not a sensitive prognostic marker but it may be useful in conjunction with other markers *(77)*.

In cytological smears, normal ectocervical cells show weak binding by JFL, whereas the intensity of binding and the proportion of positive cells increase as the severity of dysplasia increases *(78)*. This may help diagnostic evaluation, but has not been thoroughly evaluated.

Normal endocervical glands show cytoplasmic binding by WGA, PNA, and UEA1. In intraglandular and invasive neoplasia linear, luminal expression of these lectins is found, together with weaker and more diffuse cytoplasmic labeling *(79)*.

8.3. Uterine Endometrium

Normal endometrial glands show luminal binding by WGA and ConA, with increased sialylation of terminal galactose groups in the late secretory phase *(80)*. DBA binding only occurs in secretory glands and UEA1 binding is generally absent *(81)*. In endometrial adenocarcinomas, lectin binding is very variable, but WGA and ConA binding occurs around the whole of the cell periphery; this loss of polarity may be helpful in the diagnosis of small tissue fragments *(80,81)*.

8.4. Ovaries

PNA binds to the luminal membrane of cells in benign and malignant serous neoplasms, and reacts with the cytoplasm of borderline and malignant mucinous tumors. A combination of PNA binding and immunohistochemical labeling for epithelial membrane antigen may be useful in differential diagnosis *(82)*.

8.5. Breast

Extensive study of carcinomas of the breast has revealed great heterogeneity of lectin binding between different cells in the same tumor and between tumors *(4,10)*. There is some correlation between lectin binding and neoplastic behavior, with increased binding by UEA1 associated with earlier local recurrence and shorter survival *(83)*, and binding by HPA associated with locally aggressive disease, an increased prevalence of nodal metastasis and shorter survival. HPA is not an independent predictor due to its strong association with nodal status *(9)*. Both benign and malignant breast tissue shows increased mannose binding to endogenous lectins in the nucleus and cytoplasm compared with normal tissue. Lactose and GlcNAc receptors are more prevalent in lobular, papillary, or mucinous carcinomas, but there is marked heterogeneity of tumor cells within and between tumors *(84)*.

8.6. Testis

Lectin histochemistry may facilitate the classification of testicular tumors, as seminomas show binding by WGA but not by PNA and UEA1, which tend to bind to embryonal carcinomas and yolk sac tumors *(85)*.

9. Endocrine Organs
9.1. Adrenal

In routinely processed tissues, PNA binds to adrenal medulla and to pheochromocytomas *(86)*, whereas WGA and ConA bind to most cells in the normal cortex, and with increased intensity in hyperplasia. In adenomas and carcinomas, there is more intense granular and reticular binding of WGA and ConA, but lectin binding is not of use in the diagnosis of malignancy *(87)*.

9.2. Thyroid

In normal thyroid, UEA1 positivity discriminates parafollicular from follicular cells. In hyperplastic thyroid follicles, adenomas, and carcinomas there is variable, focal, and inconsistent positivity for PNA and UEA1 of no diagnostic value *(88)*. In medullary carcinoma, lectin binding is not diagnostic, although most cells in most cases show binding for UEA1, ConA, RCA1, Succ-WGA, and SBA; amyloid in the stroma shows weak binding by RCA1 and ConA. Calcitonin (a precursor of which has affinity for ConA) and carcinoembryonic antigen in the neoplastic cells may be important contributors to lectin affinity *(89)*.

10. Musculoskeletal System

10.1. Fibrous Tissue

Many lectins show strong binding to glucose and galactose residues linked to hydroxyproline in collagen, and this may be used to demonstrate tissue architecture and changes in collagen *(90)*.

In a small study of malignant fibrous histiocytoma, RCA1 bound both fibroblastic and histiocytic cells, whereas few cells in fibrosarcomas labeled, and cases of malignant fibrous histiocytoma with foci of benign-appearing histiocytes tended to have a poor prognosis *(91)*.

10.2. Skeletal Muscle

Lectin histochemistry has been studied in muscular dystrophies, and increased binding of many lectins demonstrated in necrotic and regenerating muscle fibers *(92)*. Loss of binding by RCA-1 to the muscle fiber surface in Duchenne muscular dystrophy *(93)* is probably related to the loss of integral membrane glycoproteins accompanying the defects of dystrophin *(94)*. The loss of UEA1 binding to capillaries, in addition to the deposition of complement, can be used to make a diagnosis of dermatomyositis in the absence of degenerative changes in the muscle fibers *(15)*.

10.3. Bone

Little is known about lectin binding to bone and its neoplasms. In nonneoplastic odontogenic cysts of the jaws, most epithelial layers are positive for UEA1 and BS1; a feature that may help to distinguish them from ameloblastoma in small biopsies in which characteristic morphological features may be absent *(95)*. Lectin binding has been studied in chordomas, uncommon neoplasms of the spine that arise from notochordal remnants, but there is no relationship between glycosylation and clinical course *(96)*.

10.4. Joint Pathology

Lectin histochemistry may reveal changes in the function of chondrocytes in normal and degenerating cartilage, with the breakdown of bonds in fibrillated cartilage making sugars accessible, particularly to SBA and PNA.

11. Hemopoietic and Lymphoid Cells

11.1. Bone Marrow

Several lectins bind to normal and neoplastic myeloid cells in routinely processed marrow biopsies, but there is no discrimination between benign and malignant cells *(97)*.

11.2. Macrophages and Histiocytes

PNA binds strongly to the cytoplasm of tissue macrophages and histiocytes *(98)*. The phenotypic variation of macrophages in disease states makes PNA binding a useful adjunct to immunohistochemistry for their identification *(99)*. Interdigitating reticulum cells, and Langerhans cells in histiocytosis-X show strong paranuclear and surface membrane binding with weak cytoplasmic PNA binding *(100)*. Careful morphological interpretation is required since PNA also binds to the Golgi region of plasma cells, and shows membrane binding to neoplastic follicle center cells *(101)*.

11.3. Reed-Sternberg Cells

Reed-Sternberg cells in Hodgkin's lymphoma show a paranuclear dot, with light cytoplasmic staining and strong surface binding by PNA and BPA. Lectin binding identifies Reed-Sternberg cells in more cases and is less affected by formalin fixation than immunohistochemical labeling using Leu-M1 and Ber-H2 *(101,102)*.

11.4. Mast Cells

Human mast cells contain abundant N-linked sequences with few or no O-linked residues *(103)*. The preferential binding of DBA to mast cells has been used to identify these cells in skeletal muscle diseases *(104)*.

12. Central Nervous System

Lectins may be used to identify different types of cell in the central nervous system, and RCA1 has been used to identify microglia, since other glia and neurons are negative *(105,106)*. Lectins may be useful as differentiation markers in gliomas, since well-differentiated astrocytomas and oligodendrogliomas show more intense binding by ConA, RCA, and PNA than more poorly differentiated neoplasms *(106)*. Lectins can be used to identify different cell populations in meningiomas, and PNA binding may represent a differentiation marker, with more intense binding seen in anaplastic and papillary meningiomas *(107)*.

Lectin histochemistry of routinely processed nerves shows that HPA and MPA bind selectively to pathologically altered nerves but are not specific for any type of injury *(108)*.

13. Microbiology

Most fungi will bind to lectins in routinely processed tissues, and PWM and Succ-ConA may be the most useful lectins as they show little binding to normal tissues *(109,110)*. It has been suggested that species-specific lectin binding may occur *(111)*, but this has not been confirmed. Lectin histochemistry is unlikely to supercede established tinctorial methods to detect fungi but may provide biochemical data on changes in the fungal surface in disease.

14. Summary

The widespread diagnostic use of lectin histochemistry is still restricted to the identification of abnormal storage products in cells, of normal and neoplastic endothelial cells, and of fungi. There is increasing evidence that lectins may provide diagnostic information in cytological preparations of effusions, and that more prognostic information can be obtained on the behavior of a wide range of neoplasms. It is difficult to determine the clinical significance of changes in lectin binding owing to the heterogeneity of binding within a group of tumors, between groups of tumors and between laboratories. Most studies have been small and require validation using standardized methodology and multivariate analysis to test whether lectin-binding patterns are independent prognostic variables. Therapeutic decisions for individual patients require a high degree of certainty that the use or withholding of treatment is appropriate, and this is rarely available from lectin histochemical data.

In research, lectins are invaluable for the study of cell-surface interactions and may provide a more reproducible method for the grading of malignancy for many tumors than simple morphological examination. Nevertheless, it is essential that lectin-binding patterns are correlated with careful morphological evaluation of the cells to ensure that, for example, neoplastic rather than reactive cells are being studied.

References

1. Sell, S. (1990) Cancer-associated carbohydrates identified by monoclonal antibodies. *Hum. Pathol.* **21**, 1003–1019.
2. Hakomori, S.-I. (1991) Possible functions of tumor-associated carbohydrate antigens. *Curr. Opin. Immunol.* **3**, 646–653.
3. Caselitz, J. (1987) Lectins and blood group substances as "tumor markers." *Curr. Top. Pathol.* **77**, 245–277.

4. Walker, R. A. (1989) The use of lectins in histopathology. *Path. Res. Pract.* **185**, 826–835.
5. Kayser, K., Bovin, N. V., Korchagina, E. Y., Zeilinger, C., Zeng, F-Y., and Gabius, H. J. (1994) Correlation of expression of binding sites for synthetic blood group A-, B- and H-trisaccharides and for sarcolectin with survival of patients with bronchial carcinoma. *Eur. J. Cancer.* **30**, 653–657.
6. Rittman, B. R. and MacKenzie, I. C. (1983) Effects of histological processing on lectin binding patterns in oral mucosa and skin. *Histochem. J.* **15**, 467–474.
7. Alroy, J., De Gasperi, R, and Warren, C. D. (1991) Application of lectin histochemistry and carbohydrate analysis to the characterization of lysosomal storage diseases. *Carbohydr. Res.* **213**, 229–250.
8. McMahon, R. F. T., Panesar, M. J. R., and Stoddart, R. W. (1994) Glycoconjugates of the normal human colorectum: a lectin histochemical study. *Histochem. J.* **26**, 504–518.
9. Brooks, S. A., Leathem, A. J. C., Camplejohn, R. S., and Gregory, W. (1993) Markers of prognosis in breast cancer—the relationship between binding of the lectin HPA and histological grade, SPF, and ploidy. *Breast Cancer Res. Treat.* **25**, 247–256.
10. Walker, R. A. (1984) The binding of peroxidase-labelled lectins to human breast epithelium 3. altered fucose-binding patterns of breast carcinomas and their significance. *J. Pathol.* **144**, 109–117.
11. Saku, T. and Okabe, H. (1989) Differential lectin-bindings in normal and precancerous epithelium and squamous cell carcinoma of the oral mucosa. *J. Oral Pathol. Med.* **18**, 438–445.
12. Alroy, J., Goyal, V., and Skutelsky, E. (1987) Lectin histochemistry of mammalian endothelium. *Histochemistry* **86**, 603–607.
13. Laitinen, L. (1987) Griffonia simplicifolia lectins bind specifically to endothelial cells and some epithelial cells in mouse tissues. *Histochem. J.* **19**, 225–234.
14. Ordóñez, N. G., Brooks, T., Thompson, S., and Batsakis, J. G. (1987) Use of Ulex europaeus agglutinin I in the identification of lymphatic and blood vessel invasion in previously stained microscopic slides. *Am. J. Surg. Pathol.* **11**, 543–550.
15. Emslie-Smith, A. M. and Engel, A. G. (1990) Microvascular changes in early and advanced dermatomyositis: a quantitative study. *Ann. Neurol.* **27**, 343–356.
16. Ordóñez, N. G. and Batsakis, J. G. (1984) Comparison of Ulex europaeus I lectin and factor VIII-related antigen in vascular lesions. *Arch. Pathol. Lab. Med.* **108**, 129–132.
17. Hitchcock, M. G., Hurt, M. A., and Santa-Cruz, D. J. (1994) Cutaneous granular cell angiosarcoma. *J. Cutan. Pathol.* **21**, 256–262.
18. Leader, M., Collins, M., Patel, J., and Henry, K. (1986) Staining for factor VIII-related antigen and Ulex europaeus agglutinin I (UEA-I) in 230 tumours. An assessment of their specificity for angiosarcoma and Kaposi's sarcoma. *Histopathology* **10**, 1153–1162.
19. Sirgi, K. E., Wick, M. R., and Swanson, P. E. (1993) B72.3 and CD34 immunoreactivity in malignant epithelioid soft tissue tumors: adjuncts in the recognition of endothelial neoplasms. *Am. J. Surg. Pathol.* **17**, 179–185.

20. Kuzu, I., Bicknell, R., Harris, A. L., Jones, M., Gatter, K. C., and Mason, D. Y. Heterogeneity of vascular endothelial cells with relevance to diagnosis of vascular tumours. *J. Clin. Pathol.* **45,** 143–148.
21. Wenig, B. M., Abbondanzo, S. L., and Heffess, C. S. (1994) Epithelioid angiosarcoma of the adrenal glands: a clinicopathologic study of nine cases with a discussion of the implications of finding "epithelial-specific" markers. *Am. J. Surg. Pathol.* **18,** 62–73.
22. Gray, M. H., Rosenberg, A. E., Dickersin, G. R., and Bhan, A. K. (1990) Cytokeratin expression in epithelioid vascular neoplasms. *Hum. Pathol.* **21,** 212–217.
23. Herron, G. S., Rouse, R. V., Kosek, J. C., Smoller, B. R., and Egbert, B. M. (1994) Benign lymphangioendothelioma. *J. Am. Acad. Dermatol.* **31,** 362–368.
24. Virtanen, I., Kariniemi, A.-L., Holthofer, H., and Lehto, V.-P. (1986) Fluorochrome-coupled lectins reveal distinct cellular domains in human epidermis. *J. Histochem. Cytochem.* **34,** 307–315.
25. Kariniemi, A.-L. and Virtanen, I. (1989) Dolichos biflorus agglutinin (DBA) reveals a similar basal cell differentiation in normal and psoriatic epidermis. *Histochemistry* **93,** 129–132.
26. Louis, C. J., Wyllie, R. G., Chou, S. T., and Sztynda, T. (1981) Lectin-binding affinities of human epidermal tumors and related conditions. *Am. J. Clin. Pathol.* **75,** 642–647.
27. Tsubura, A., Fujita, Y., Sasaki, M., and Morii, S. (1992) Lectin-binding profiles for normal skin appendages and their tumors. *J. Cutan. Pathol.* **19,** 483–489.
28. Kannon, G. and Park, H. K. (1990) Utility of peanut agglutinin (PNA) in the diagnosis of squamous cell carcinoma and keratoacanthoma. *Am. J. Dermatopathol.* **12,** 31–36.
29. Kohchiyama, A., Oka, D., and Ueki, H. (1987) Differing lectin-binding patterns of malignant melanoma and nevocellular and Spitz nevi. *Arch. Dermatol. Res.* **279,** 226–231.
30. Dabelsteen, E. and Mackenzie, I. C. (1978) Expression of *Ricinus communis* receptors on epithelial cells in oral carcinomas and oral wounds. *Cancer Res.* **38,** 4676–4680.
31. Tolson, N. D., Daley, T. D., and Wysocki, G. P. (1985) Lectin probes of glycoconjugates in human salivary glands: 1. *J. Oral Pathol.* **14,** 523–530.
32. Daley, T. D., Tolson, N. D., and Wysocki, G. P. (1985) Lectin probes of glycoconjugates in human salivary gland neoplasms: 2. *J. Oral Pathol.* **14,** 531–538.
33. Macartney, J. C. (1986) Lectin histochemistry of galactose and N-acetyl-galactosamine glycoconjugates in normal gastric mucosa and gastric cancer and the relationship with ABO and secretor status. *J. Pathol.* **150,** 135–144.
34. Macartney, J. C. (1987) Fucose-containing antigens in normal and neoplastic human gastric mucosa: a comparative study using lectin histochemistry and blood group immunohistochemistry. *J. Pathol.* **152,** 23–30.
35. Narita, T. and Numao, H. (1992) Lectin binding patterns in normal, metaplastic and neoplastic gastric mucosa. *J. Histochem. Cytochem.* **40,** 681–687.

36. Kakeji, Y., Tsujitani, S., Mori, M., Maehara, Y., and Sugimachi, K. (1991) Helix pomatia agglutinin binding activity is a predictor of survival time for patients with gastric carcinoma. *Cancer* **68**, 2438–2442.
37. Lotan, R., Ito, H., Yasui, W., Yokozaki, H., Lotan, D., and Tahara, E. (1994) Expression of a 31-kDa lactoside-binding lectin in normal human gastric mucosa and in primary and metastatic gastric carcinomas. *Int. J. Cancer* **56**, 474–480.
38. Vecchi, M., Torgano, G., De Franchis, R., Tronconi, S., Agape, D., and Ronchi, G. (1989) Evidence of altered structural and secretory glycoconjugates in the jejunal mucosa of patients with gluten sensitive enteropathy and subtotal villous atrophy. *Gut* **30**, 804–810.
39. Thiru, S., Devereux, G., and King, A. (1990) Abnormal fucosylation of ileal mucus in cystic fibrosis: 1. A histochemical study using peroxidase labelled lectins. *J. Clin. Pathol.* **43**, 1014–1018.
40. Bresalier, R. S., Boland, C. R., and Kim, Y. S. (1985) Regional differences in normal and cancer-associated glycoconjugates of the human colon. *J. Nat. Cancer Inst.* **75**, 249–259.
41. Cooper, H. S., Farano, P., and Coapman, R. A. (1987) Peanut lectin binding sites in colons of patients with ulcerative colitis. *Arch. Pathol. Lab. Med.* **111**, 270–275.
42. Compton, C. C. (1989) Premalignancy in chronic ulcerative colitis. *Hum. Pathol.* **20**, 407–409.
43. O'Brien, M. J., O'Keane, J. C., Zauber, A., Gottlieb, L. S., and Winawer, S. J. (1992) Precursors of colorectal carcinoma. *Cancer* **70**, 1317–1327.
44. Boland, R. C. and Deshmukh, G. D. (1990) The carbohydrate composition of mucin in colonic cancer. *Gastroenterol.* **98**, 1170–1177.
45. Itzkowitz, S. H., Yuan, M., Montgomery, C. K., Kjeldsen, T., Takahashi, H. K., Bigbee, W. L., and Kim, Y. S. (1989) Expression of Tn, sialosyl-Tn and T antigens in human colon cancer. *Cancer Res.* **49**, 197–204.
46. Ikeda, Y., Mori, M., Adachi, Y., Matsushima, T., and Sugimachi, K. (1994) Prognostic value of the histochemical expression of Helix pomatia agglutinin in advanced colorectal cancer. *Dis. Colon Rectum* **37**, 181–184.
47. Sams, J. S., Lynch, H. T., Burt, R. W., Lanspa, S. J., and Boland, C. R. (1990) Abnormalities of lectin histochemistry in familial polyposis coli and hereditary nonpolyposis colorectal cancer. *Cancer* **68**, 502–508.
48. Irimura, T., Matsushita, Y., Sutton, R. C., Carralero, D., Ohannesian, D. W., Cleary, K. R., Ota, D. M., Nicolson, G. L., and Lotan, R. (1991) Increased content of an endogenous lactose-binding lectin in human colorectal carcinoma progressed to metastatic stages. *Cancer Res.* **51**, 387–393.
49. Ching, C. K., Black, R., Helliwell, T., Savage, A., Barr, H., and Rhodes, J. M. (1988) Use of lectin histochemistry in pancreatic cancer. *J. Clin. Pathol.* **41**, 324–328.
50. Dhillon, A. P., Colombari, R., Savage, K., and Scheuer, P. J. (1992) An immunohistochemical study of the blood vessels within primary hepatocellular tumours. *Liver* **12**, 311–318.

51. Gulisano, M., Bryk, S. G., Gheri, G., Sgambati, E., Curreli, A., Masala, W., and Pacini, P. (1994) Histochemical study of human nasopharyngeal epithelium by horseradish peroxidase conjugated lectins. *Epith. Cell Biol.* **3,** 1–6.
52. Feinmesser, R., Freeman, J. L., Noyek, A., and van Nostrand, P. (1989) Lectin binding characteristics of laryngeal cancer. *Otolaryngol. Head Neck Surg.* **100,** 207–209.
53. Alvarez-Fernandez, E. and Carretero-Albinana, L. (1990) Lectin histochemistry of normal bronchopulmonary tissues and common forms of bronchogenic carcinoma. *Arch. Pathol. Lab. Med.* **114,** 475–481.
54. Kawai, T., Greenberg, S. D., Truong, L. D., Mattioli, C. A., Klima, M., and Titus, J. L. (1988) Differences in lectin binding of malignant pleural mesothelioma and adenocarcinoma of the lung. *Am. J. Pathol.* **130,** 401–410.
55. Shield, P. W. (1989) Lectin binding properties of cells from serous effusion and peritoneal washing specimens. *J. Clin. Pathol.* **42,** 1178–1183.
56. Lee, J. S., Ro, J. Y., Sahin, A. S., Hong, W. K., Brown, B. W., Mountain, C. F., and Hittelman, W. N. (1991) Expression of blood-group antigen A—a favorable prognostic factor in non-small-cell lung cancer. *N. Engl. J. Med.* **324,** 1084–1090.
57. Matsumoto, H., Muramatsu, H., Muramatsu, T., and Shimazu, H. (1992) Carbohydrate profiles shown by a lectin and a monoclonal antibody correlate with metastatic potential and prognosis of human lung carcinomas. *Cancer* **69,** 2084–2090.
58. Meyer, K. C., Powers, C., Rosenthal, N., and Auerbach, R. (1993) Alveolar macrophage surface carbohydrate expression is altered in interstitial lung disease as determined by lectin-binding profiles. *Am. Rev. Respir. Dis.* **148,** 1325–1334.
59. Kayser, K., Heil, M., and Gabius, H. J. (1989) Is the profile of binding of a panel of neoglycoproteins useful as a diagnostic marker in human lung cancer? *Path. Res. Pract.* **184,** 621–629.
60. Silva, F. G., Nadasdy, T., and Laszik, Z. (1993) Immunohistochemical and lectin dissection of the human nephron in health and disease. *Arch. Pathol. Lab. Med.* **117,** 1233–1239.
61. Ivanyi, B., Hansen, H. E., and Olsen, S. (1993) Segmental localization and quantitative characteristics of tubulitis in kidney biopsies from patients undergoing acute rejection. *Transplantation* **56,** 581–585.
62. Nadasdy, T., Laszik, Z., Blick, K. E., Johnson, D. L., and Silva, F. D. (1994) Tubular atrophy in the end-stage kidney: a lectin and immunohistochemical study. *Hum. Pathol.* **25,** 22–28.
63. Kumar, S., Carr, T., Marsden, H. B., and Morris-Jones, P. H. (1986) Study of childhood renal tumours using peroxidase conjugated lectins. *J. Clin. Pathol.* **39,** 736–741.
64. Yeger, H., Baumal, R., Harason, P., and Phillips, M. J. (1987) Lectin histochemistry of Wilms' tumor: comparison with normal adult and fetal kidney. *Am. J. Clin. Pathol.* **88,** 278–285.
65. Holthofer, H., Virtanen, I., Pettersson, E., Tornroth, T., Alfthan, O., Linder, E., and Miettinen, A. (1981) Lectins as fluorescence microscopic markers for saccharides in the human kidney. *Lab. Invest.* **45,** 391–399.

66. Yonezawa, S., Irisa, S., Nakamura, T., Uemura, S., Otsuji, Y., Ohi, Y., and Sato, E. (1983) Deposition of α1-antitrypsin and loss of glycoconjugate carrying Ulex europaeus agglutinin-I binding sites in the glomerular sclerotic process. *Nephron* **33,** 38–43.
67. Nevins, T. E. (1985) Lectin binding in membranoproliferative glomerulonephritis: evidence for N-acetylglucosamine in dense intramembranous deposits. *Am. J. Pathol.* **118,** 325–330.
68. Nakanishi, K., Kawai, T., and Suzuki, M. (1993) Lectin binding and expression of blood group-related antigens in carcinoma in-situ and invasive carcinoma of urinary bladder. *Histopathology* **23,** 153–158.
69. Langkilde, N. C., Wolf, H., and Orntoft, T. F. (1989) Lectin histochemistry of human bladder cancer: loss of lectin binding structures in invasive carcinomas. (1989) *APMIS* **97,** 367–373.
70. Limas, C., and Lange, P. (1986) T-antigen in normal and neoplastic urothelium. *Cancer* **58,** 1236–1245.
71. Gabius, H. J., Bahn, H., Holzhausen, H. J., Knolle, J., and Stiller, D. (1992) Neoglycoprotein binding to normal urothelium and grade-dependent changes in bladder lesions. *Anticancer Res.* **12,** 987–992.
72. McNeal, J. E., Leav, I., Alroy, J., and Skutelsky, E. (1988) Differential lectin staining of central and peripheral zones of the prostate and alterations in dysplasia. *Am. J. Clin. Pathol.* **89,** 41–48.
73. Foster, C. S., McLoughlin, J., Bashir, I., and Abel, P. D. (1992) Markers of the metastatic phenotype in prostate cancer. *Hum. Pathol.* **23,** 381–394.
74. Allsbrook, W. C. and Simms, W. W. (1992) Histochemistry of the prostate. *Hum. Pathol.* **23,** 297–305.
75. Lee, M. C. and Damjanov, I. (1984) Lectin histochemistry of human placenta. *Differentiation* **28,** 123–128.
76. Bychkov, V. and Toto, P. D. (1986) Lectin binding to normal, dysplastic, and neoplastic cervical epithelium. *Am. J. Clin. Pathol.* **85,** 542–547.
77. Hirao, T., Sakamoto, Y., Kamada, M., Hamada, S. I., and Aono, T. (1993) Tn antigen, a marker of potential for metastasis of uterine cervix cancer cells. *Cancer* **72,** 154–159.
78. Pillai, K. R., Remani, P., Kannan, S., Mathew, A., Sujathan, K., Vijayakumar, T., and Nair, M. K. (1994) Jack fruit lectin-specific glycoconjugate expression during the progression of cervical intraepithelial neoplasia: a study of exfoliated cells. *Diagn. Cytopathol.* **10,** 342–346.
79. Griffin, N. R. and Wells, M. (1994) Characterisation of complex carbohydrates in cervical glandular intraepithelial neoplasia and invasive adenocarcinoma. *Int. J. Gynaecol. Pathol.* **13,** 319–329.
80. Kluskens, L. F., Kluskens, J. L., and Bibbo, M. (1984) Lectin binding in endometrial adenocarcinoma. *Am. J. Clin. Pathol.* **82,** 259–266.
81. West, K. P. and Cope, J. L. (1989) The binding of peroxidase-labelled lectins to human endometrium in normal cyclical endometrium and endometrial adenocarcinoma. *J. Clin. Pathol.* **42,** 140–147.

82. Ashorn, P., Helle, M., Helin, H., Ashorn, R., and Krohn, K. (1988) Use of immunohistochemical staining panel for characterisation of ovarian neoplasms. *J. Clin. Pathol.* **41,** 12–16.
83. Fenlon, S., Ellis, I. O., Bell, J., Todd, J. H., Elston, C. W., and Blamey, R. W. (1987) Helix pomatia and Ulex europeus lectin binding in human breast carcinoma. *J. Pathol.* **152,** 169–176.
84. Gabius, H. J., Bodanowitz, S., and Schauer, A. (1988) Endogenous sugar-binding proteins in human breast tissue and benign and malignant breast lesions. *Cancer* **61,** 1125–1131.
85. Kosmehl, H., Langbein, L., and Katenkamp, D. (1989) Lectin histochemistry of human testicular germ cell tumours. *Neoplasma* **36,** 29–39.
86. Moorghen, M. and Carpenter, F. (1991) Peanut lectin: a histochemical marker for phaeochromocytomas. *Virchows Archiv. A. Pathol. Anat.* **419,** 203–207.
87. Sasano, H., Nose, M., and Sasano, N. (1989) Lectin histochemistry in adrenocortical hyperplasia and neoplasms with emphasis on carcinoma. *Arch. Pathol. Lab. Med.* **113,** 68–72.
88. Gonzalez-Campora, R., Gallego, F. S., Lacave, I. M., Marin, J. M., Linares, C. M., and Galera-Davidson, H. (1988) Lectin histochemistry of the thyroid gland. *Cancer* **62,** 2354–2362.
89. Martin-Lacave, I., Gonzalez-Campora, R., Fernandez, A. M., Gallego, F. S., Montero, C., and Galera-Davidson, H. (1988) Mucosubstances in medullary carcinoma of the thyroid. *Histopathology* **13,** 55,56.
90. Soderstrom, K. O. (1987) Lectin binding to collagen strands in histologic tissue sections. *Histochemistry* **87,** 557–560.
91. Ueda, T., Aozasa, K., Yamamura, T., Tsujimoto, J., Ono, K., and Matsumoto, K. (1987) Lectin histochemistry of malignant fibrohistiocytic tumors. *Am. J. Surg. Pathol.* **11,** 257–262.
92. Helliwell, T. R., Gunhan, O., and Edwards, R. H. T. (1989) Lectin binding and desmin expression during necrosis, regeneration and neurogenic atrophy of human skeletal muscle. *J. Pathol.* **159,** 43–51.
93. Capaldi, M. J., Dunn, M. J., Sewry, C. A., and Dubowitz, V. (1984) Altered binding of Ricinus communis 1 lectin by muscle membranes in Duchenne muscular dystrophy. *J. Neurol. Sci.* **63,** 129–142.
94. Campbell, K. P. and Kahl, S. D. (1989) Association of dystrophin and an integral membrane glycoprotein. *Nature* **338,** 259–262.
95. Saku, T., Shibata, Y., Koyama, Z., Cheng, J., Okabe, H., and Yeh, Y. (1991) Lectin histochemistry of cystic jaw lesions: an aid for differential diagnosis between cystic ameloblastoma and odontogenic cysts. *J. Oral Pathol. Med.* **20,** 108–113.
96. Kaneko, Y., Iwaki, T., and Fukui, M. (1992) Lectin histochemistry of human fetal notochord, ecchordosis physaliphora, and chordomas. *Arch. Pathol. Lab. Med.* **116,** 60–64.
97. Schumacher, U., Horny, H. P., Welsch, U., and Kaiserling, E. (1991) Lectin histochemistry of human bone marrow: investigation of trephine biopsy specimens in normal and reactive states and neoplastic disorders. *Histochem. J.* **23,** 215–220.

98. Howard, D. R. and Batsakis, J. G. (1980) Peanut agglutinin: a new marker for tissue histiocytes. *Am. J. Clin. Pathol.* **77,** 401–408.
99. Paulli, M., Rosso, R., Kindl, S., Boveri, E., Marocolo, D., Chioda, C., Agostini, C., Magrini, U., and Facchetti, F. (1992) Immunophenotypic characterization of the cell infiltrate in five cases of sinus histiocytosis with massive lymphadenopathy (Rosai-Dorfman disease). *Hum. Pathol.* **23,** 647–654.
100. Ree, H. J. and Kadin, M. E. (1986) Peanut agglutin: a useful marker for histiocytosis X and interdigitating reticulum cells. *Cancer* **57,** 282–287.
101. Ree, H. J., Neiman, R. S., Martin, A. W., Dallenbach, F., and Stein, H. (1989) Paraffin section markers for Reed-Sternberg cells. *Cancer* **63,** 2030–2036.
102. Sarker, A. B., Akagi, T., Jeon, H. J., Miyake, K., Murakami, I., Yoshino, T., Takahashi, K., and Nose, S. (1992) *Bauhinia purpurea*—a new paraffin section marker for Reed-Sternberg cells of Hodgkin's disease. *Am. J. Pathol.* **141,** 19–23.
103. Kirkpatrick, C. J., Jones, C. J. P., and Stoddart, R. W. (1988) Lectin histochemistry of the mast cell: a light microscopical study. *Histochem. J.* **20,** 139–146.
104. Helliwell, T. R., Gunhan, O., and Edwards, R. H. T. (1990) Mast cells in neuromuscular diseases. *J. Neurol. Sci.*, **98,** 267–276.
105. Mannoji, H., Yeger, H., and Becker, L. E. (1986) A specific histochemical marker (lectin *Ricinus communis* agglutinin-1) for normal human microglia, and application to routine histopathology. *Acta. Neuropathol.* **71,** 341–343.
106. Figols, J., Madrid, J. F., and Cervos-Navarro, J. (1991) Lectins as differentiation markers of human gliomas. *Histol. Histopath.* **6,** 79–85.
107. Marafioti, T., Barresi, G., and Batolo, D. (1994) Lectin histochemistry of human meningiomas. *Histol. Histopath.* **9,** 535–540.
108. Estruch, R. and Damjanov, I. (1986) Lectin histochemistry applied to human nerves. *Arch. Pathol. Lab. Med.* **110,** 730–735.
109. Karayannopoulou, G., Weiss, J., and Damjanov, I. (1988) Detection of fungi in tissue sections by lectin histochemistry. *Arch. Pathol. Lab. Med.* **112,** 746–748.
110. Senba, M. (1989) Lectin histochemistry of fungi. *Arch. Pathol. Lab. Med.* **113,** 327,328.
111. Stoddart, R. W. and Herbertson, B. M. (1978) The use of fluorescein-labelled lectins in the detection and identification of fungi pathogenic for man: a preliminary study. *J. Med. Microbiol.* **11,** 315–324.

I

LECTIN HISTOCHEMISTRY AND CYTOCHEMISTRY

B: Electron Microscopy

7

Electron Microscopy

Use of Lectin-Peroxidase Prior to Embedding

Adi Ellinger

1. Introduction

Pre-embedding lectin cytochemistry is one in a range of approaches used for analyzing the subcellular localization of glycoconjugate-classes, and thus enables the determination of the carbohydrate composition of cellular compartments. At the cell's surface, the carbohydrate composition is important in defining cell types; changes in these components are associated with cellular differentiation, maturation, and neoplastic transformation *(1–4)*. Here lectins provide well-defined reagents for the identification of cell types and for the study of these dynamic processes *(5–7)*. Intracellularly, lectins have been used in multiple studies for the analysis of the biosynthetic and endocytic pathway. Such lectin cytochemistry has helped to unravel single steps of the glycosylation process and to elucidate topographical aspects of glycosylation *(8–16)*. Interpreting the lectin binding patterns, it has to be borne in mind that the determinants recognized by a lectin may be common to several different glycoproteins and glycolipids.

1.1. Lectin–Peroxidase Cytochemistry

Since lectins themselves are devoid of enzyme activity and further are not electron-dense, they must be conjugated to marker molecules for visualization of the binding reaction. Besides the use of particulate markers (colloidal gold, ferritin), covalent binding to peroxidase (horseradish peroxidase, HRP) has the most general application. Indirect labeling techniques, which add one or more incubation and wash steps, include the use of biotin- or digoxigenin-labeled (Boehringer Mannheim, Mannheim, Germany) lectins or of antilectin antibodies.

From: *Methods in Molecular Medicine, Vol. 9: Lectin Methods and Protocols*
Edited by: J. M. Rhodes and J. D. Milton Humana Press Inc., Totowa, NJ

The multiple step techniques are derived from immunocytochemical protocols, in which labeling of primary antibodies has several disadvantages. Labeled secondary reagents furthermore are readily available, relatively inexpensive, and usually amplify the primary signal. The pre-embedding labeling procedure used in the author's laboratory is a modification of a technique originally described by Bernhard and Avrameas *(17)*. Variations of this technique have been widely applied for the demonstration of lectin-sugar interactions. In brief, fixed cells or tissue slices are incubated in the lectin-HRP conjugates, the enzyme activity being visualized by means of 3,3'-diaminobenzidine tetrahydrochloride (DAB) conversion.

1.2. Choice of Fixative

In affinity cytochemical procedures, proper fixation is an important aspect for obtaining successful results. Fixation mostly will be a compromise between well-preserved ultrastructure and retention of antigenicity *(18–20)*. Unfortunately, fixatives that best preserve morphology are the most disruptive to antigenic sites. In contrast to the highly fixation-dependent conformational changes in the protein structure, which might result in reduction or even loss of antigenicity, carbohydrate residues have been found to be much less affected. Choice of fixative in lectin-cytochemistry should primarily fulfill requirements of fine structural preservation, prevention of migration, or loss of sugar residues, and should further enable sufficient penetrability of the samples. Fixatives and penetration enhancement procedures applicable in affinity cytochemistry, including the periodate-lysin-paraformaldehyde (FA; *21*), the glutaraldehyde (GA) -borohydride-saponin *(22)* protocol as well as the use of mixtures of the aldehydes—FA/GA have been summarized recently *(23,24)*. Aldehyde mixtures, which are used in our procedure, are based on the idea that the rapidly penetrating monoaldehyde—FA temporarily fixes the specimen until the slower penetrating dialdehyde—GA crosslinks the proteins irreversibly *(19)*; samples fixed in such a way are more easily permeable than as after fixation with GA alone. Large amounts of sugars, like those stored in secretory granules of mucous producing cells, are not always sufficiently stabilized after aldehyde treatment. Extraction in the course of the further processing might result in an artificial loss and false negative results; this loss is less at low temperatures. Lectins, bound in the pre-embedding stage might also exercise a protective role. To increase ultrastructural preservation, samples might be further incubated only after mild fixation and be refixed with higher aldehyde concentrations in a second step after the lectin incubation. Finally, it is difficult to assess the degree of interaction of OsO_4 with carbohydrates; postfixation with OsO_4 is absolutely essential for the formation of electron-dense osmium black precipitates with oxidized DAB.

FA is prepared from paraformaldehyde, a polymerized form that easily dissolves at 60–70°C in the alkaline forms of disodium phosphate or sodium cacodylate buffers (*see* Note 1). Solutions should be prepared immediately before use. FA binds to various groups of proteins, such as amino and amido groups, and may form crossbridges; it is also slightly effective in fixing mucoproteins *(20)*. GA acts by crosslinking proteins and, to a minor degree, also carbohydrates with its two aldehyde groups; the most reactive sites are the ε-amino groups of lysine and polyhydroxyl compounds in carbohydrates. Since impurities may interfere with tissue molecules, dilutions are freshly prepared from purified stock solutions which are packaged under inert gas (*see* Note 2). Free aldehyde groups have to be avoided, since they might cause artifacts, by blocking or by reduction (*see* Note 3).

1.3. Enzyme and Chromogen

Direct labeling with peroxidase is based on the work of Nakane and Kawaoi, who introduced a peroxidase-labeled immunoglobulin for antigen localization *(25)*. The peroxidase enzyme most often used is that found in the horseradish plant, as it is easily obtained. HRP (40-kDa) has an oxidative potency when in conjunction with a source of oxygen; electrons are transferred to an acceptor molecule. From the numerous oxidizable molecules that precipitate as a permanent pigment, the most widely used is 3,3'-diaminobenzidine tetrahydrochloride (DAB; *see* Note 4). DAB precipitates to a brown color when in solution with HRP and H_2O_2, which is further transferred into a fine granular electron-dense precipitate after osmium treatment (osmium black; *see* Note 10). The precipitate is insoluble in alcohol, and therefore, the samples might be dehydrated and embedded by routine electron microscope protocols.

The amount of generated precipitate might be controlled at the lectin incubation (up to 4–8 h) and the DAB (up to ≈30 min) conversion step. Incubation times and concentrations of the reagents are restricted by the generation of diffusion artifacts and background staining. The precipitate formed in the pre-embedding technique is time-dependent, in contrast to the one-to-one binding in the case of particulate markers. Reaction products are formed within the sections or cells, eventually filling membranous compartments or forming smears at the cell's surfaces. The lateral resolution is limited by diffusion of oxidized DAB *(26–28)*. In extracellular spaces, diffusion distances of the reaction product have been estimated to be ≥100 nm and readsorption onto negatively charged cell surfaces has been demonstrated. By reducing the amount of generated DAB, diffusion artifacts are minimized and a resolution of <60 nm is achieved; cytomembranes may form a partial diffusion barrier. However, on plasma membranes a lateral spread of up to 80 nm has been observed and diffusion distances of approx 100 nm or even more have to be expected.

2. Materials

1. Gloves, gowns, and masks.
2. Miscellaneous glass and plastic ware, such as centrifuge tubes, glass cover slips (12 mm ϕ, various standard glass vials (for incubation of sections), Erlenmayer flasks, 35-mm plastic culture dishes—nonsterile for incubations and absorbent paper towels.
3. Adjustable pipets with tips.
4. Glass filtration device, filter papers (Whatman grade 1, prefolded Whatman grade 2).
5. pH test strips.
6. Forceps.
7. Humid chamber for incubations (Petri dish with wet filter paper; *see* Note 5).
8. Beem capsules (polyethylene moulds; Beem 00–1 mm square tip; local electron microscopy supplier).
9. Distilled water.
10. Potassium ferrocyanide ($K_4Fe^{II}[CN]_6$; Merck, Darmstadt, Germany; *see* Note 6).
11. Competitive carbohydrates (various commercial suppliers: Bio Curb, Lund, Sweden; Sigma, St. Louis, MO; Seikagaku, Chuo-ku, Tokyo, Japan; EY Laboratories, San Mateo, CA; Medac, Hamburg, Germany; *see* Note 15).
12. Peroxidase-tagged lectins (various commercial suppliers: Sigma, St. Louis, MO; EY Laboratories, San Mateo, CA; Medac, Hamburg, Germany; Seikagaku Corp., Chuo-ku, Tokyo, Japan; Vector Laboratories, Burlingame, CA; Calbiochem, San Diego, CA; Boehringer Mannheim; Biotrend, Köln, Germany; *see* Note 14).
13. Horseradish peroxidase (grade VI; Sigma).
14. Phosphate-buffered saline (PBS; Sigma).
15. Isopentane (Fluka Chemie AG, Buchs, Switzerland).
16. Bovine serum albumin (BSA; Behring, Marburg, Germany).
17. Saponin (Sigma).
18. Dimethyl sulfoxide (DMSO; Merck, *see* Note 7).
19. Chromogen solution: 3,3'-diaminebenzidine tetrahydrochloride (DAB; Serva, Heidelberg, Germany; *see* Note 4).
20. Hydrogen peroxide 30% (H_2O_2; Merck, *see* Note 4, item d).
21. Tris-HCl buffer (*see* Note 8).
22. Cacodylate buffer (*see* Note 9).
23. Glutaraldehyde (electron microscopy grade, *see* ref. *32*; Serva; *see* Note 2). **All fixatives are toxic. They must always be handled in a fume hood.**
24. Paraformaldehyde (extra pure; Merck; *see* Note 1).
25. Ammonium chloride (Sigma; *see* Note 3).
26. Osmium tetroxide (OsO_4; Merck; local supplier for electron microscopy; *see* Note 10).
27. Ethanol 50, 70, 80, 95, 100%.
28. Propylene oxide (Merck; **propylene oxide is a highly flammable material;** flash point ($\approx 38°C$).
29. Epon 812: 62 mL Epon 812 substitute and 100 mL dodecenyl succinic acid anhydride (DDSA); Epon B: 100 mL EPON 812 substitute and 89 mL nadic methyl anhydride (NMA); mix together 10 mL Epon A and 15 mL Epon B; add

```
Cells attached to coverslips, to culture
dishes, cell pellets or compact tissue
              ⇓
Wash away culture medium with PBS
              ⇓
   Fix in 4% FA + 0.5% GA in PBS
              ⇓
           Wash in PBS
              ⇓
         Compact tissue:
   prepare sections 10 µm-20 µm
              ⇓
          Permeabilize
              ⇓
   Incubate in lectin-HRP solutions
              ⇓
           Wash in PBS
              ⇓
      Transfer to Tris buffer
              ⇓
    Pre-incubate in DAB-Tris buffer
              ⇓
  Incubate in DAB + H₂O₂-Tris buffer
              ⇓
      Wash in distilled water
              ⇓
         Postfix in OsO₄
              ⇓
   Dehydrate in graded ethanol series
              ⇓
          Embed in EPON
```

Fig. 1. The sequence of the steps has to be kept; single steps, however, may have to be adopted to the own system (i.e., fixatives or buffers might be changed). Calculate from at least 2 d up to 1 wk for the whole procedure. Whenever possible, use fresh solutions; all toxic reagents should be handled with care (*see* Sections 2.–4.).

0.5 mL 2,4,6-tri(dimethylaminomethyl)phenol (DMP-30) just prior to use (Serva; *see* Note 11).

3. Methods

The sequence of steps in the case of HRP-labeled lectins is summarized in Fig. 1. Abbreviations as in the text.

3.1. Fixation

1. Rinse the samples (cultured cells, *see* Note 12; small tissue blocks) twice in buffer (PBS or $0.1 M$ sodium cacodylate, pH 7.2), and fix for 30 min to 1 h at 4°C in 4% FA + 0.5% GA in the respective buffers (*see* Note 1).
2. Following fixation, rinse the samples 3–5 times in buffer over a period of 30 min.
3. Quench free aldehyde groups by rinsing the samples in $0.05 M$ ammonium chloride for 2 h at room temperature (*see* Note 3).

4. Rinse for 2–3 d in buffer containing 10% DMSO (*see* Note 7).
5. Tissue blocks: Freeze to –30°C in the presence of isopentane and prepare 10–20 µm cryosections, thawed in PBS (*see* Note 13).

3.2. Lectin Incubations

1. Prepare incubation media, containing the lectin conjugates at a concentration of 30–150 µg/mL in PBS; add 0.1 mg/mL saponin and 1 mg/mL of BSA (*see* Note 14).
2. Incubate the samples for 4–8 h at room temperature. Cells attached to cover slips are placed in a humid chamber (*see* Note 5); cryosections are incubated in Eppendorf tubes and continuously agitated.
3. Rinse three times in PBS over a period of 30 min.

3.3. Control Incubations

The following types of control incubations might be performed. Incubate the samples in pure HRP solution, omitting the lectin-incubation step; perform DAB reaction without lectin-incubation step; add competitive and noncompetitive sugars to the incubation media.

1. Prepare incubation media, containing HRP (grade VI; Sigma) at a concentration of 0.2 mg/mL in PBS; add 0.1 mg/mL saponin and 1 mg/mL BSA.
2. Prepare full incubation media (containing the lectin conjugates at a concentration of 30–150 µg/mL, 0.1 mg/mL saponin, and 1 mg/mL BSA in PBS) and add 0.15–0.6M of the respective sugars (*see* Note 15).

3.4. DAB-Reaction

1. Transfer to 0.05M Tris-HCl buffer, pH 7.6, for 10–30 min.
2. Preincubate in 0.5 mg/mL DAB in 0.05M Tris-HCl buffer, pH 7.6, at room temperature for 10 min (in the dark).
3. Incubate in 0.5 mg/mL DAB and 20 µL/mL 1% H_2O_2 in 0.05M Tris-HCl buffer, pH 7.6, at room temperature for 30 min (in the dark; *see* Note 4).
4. Rinse three times in distilled water over a period of 10 min.
5. Post-fix in 1% osmium ferrocyanide for 30 min (*see* Note 6).
6. Post-fix in 1% veronal-acetate-buffered OsO_4 for 8 h (overnight) at 4°C (*see* Note 10).

3.5. Embedding

1. Dehydrate and infiltrate the samples according to the following schedule:
   a. 50% Ethanol, 15 min.
   b. 70% Ethanol, 15 min.
   c. 80% Ethanol, 15 min.
   d. 95% Ethanol, 15 min.
   e. 100% Ethanol, 15 min.
   f. 100% Ethanol, 15 min; cultured cells on cover slips are directly transferred to the resin (*see* Note 12).

Electron Microscopy

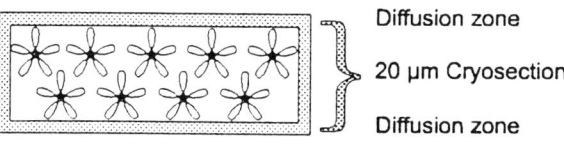

Fig. 2. The margins of cryosections often are marked by diffusion of DAB-reaction product and thus cannot be used for evaluation purposes; also, mechanical damage might affect structural details.

    g. 100% Propylene oxide, 15 min.
    h. 100% Propylene oxide, 15 min.
    i. 1:1 Complete resin:propylene oxide, 1–2 h.
2. Transfer to polyethylene or gelatine-embedding capsules containing the complete resin. Polymerize overnight at 45°C and for 24 h at 60°C *(18,19)*.

3.6. Evaluation of the Sections

The results of the lectin-cytochemical staining are assessed electron microscopically. Crucial questions center on the specificity of the staining and the nature of the sugars or sugar sequences. Different lectins vary considerably in their degree of specificity *(29)*. Use of lectins of overlapping specificity (anomeric specificity; interaction with terminal or internal sugar residues; sequence specificity; high and low affinity binding sites; *29*) or of the sugar competition protocol (*see* Note 15) helps further analysis. Problems of the pre-embedding staining procedure include the diffusion and relocation of detectable molecules and of oxidized DAB. Nonspecific binding of the lectins or DAB might result in false positivity; insufficient penetrability of the samples might result in false negative results. The penetrability of the samples, besides pre- and postfixation treatments also depends on the kind of tissue itself. Soft tissues, like embryonic tissue, are more permeable than compact ones (80–90% of the 10-µm-thick cryosections through rat small intestine show homogeneous precipitate throughout the entire thickness of the sections, almost 100% from embryonic pancreatic tissue, and only 30–40% of sections from the liver *[10–12]*). Strong fixation (e.g., 2% GA 60 min) almost completely hinders penetration of lectins and precipitation within the sections. On the other hand, the surfaces of cryosections are often not sufficiently stabilized and poor structural preservation as well as nonspecific adsorption or diffusion of DAB out of reactive compartments might occur (Fig. 2). The degree of permeability of sections of the sample tissue should be tested by cutting ultrathin cross-sections of the cryosections. Only cryosections that show the reaction product homogeneously distributed through the entire thickness of the sections, and from these only central portions, should be used for evaluation purposes (Figs. 3 and 4). In these regions, ultrastructural preservation is of the quality necessary for precise attri-

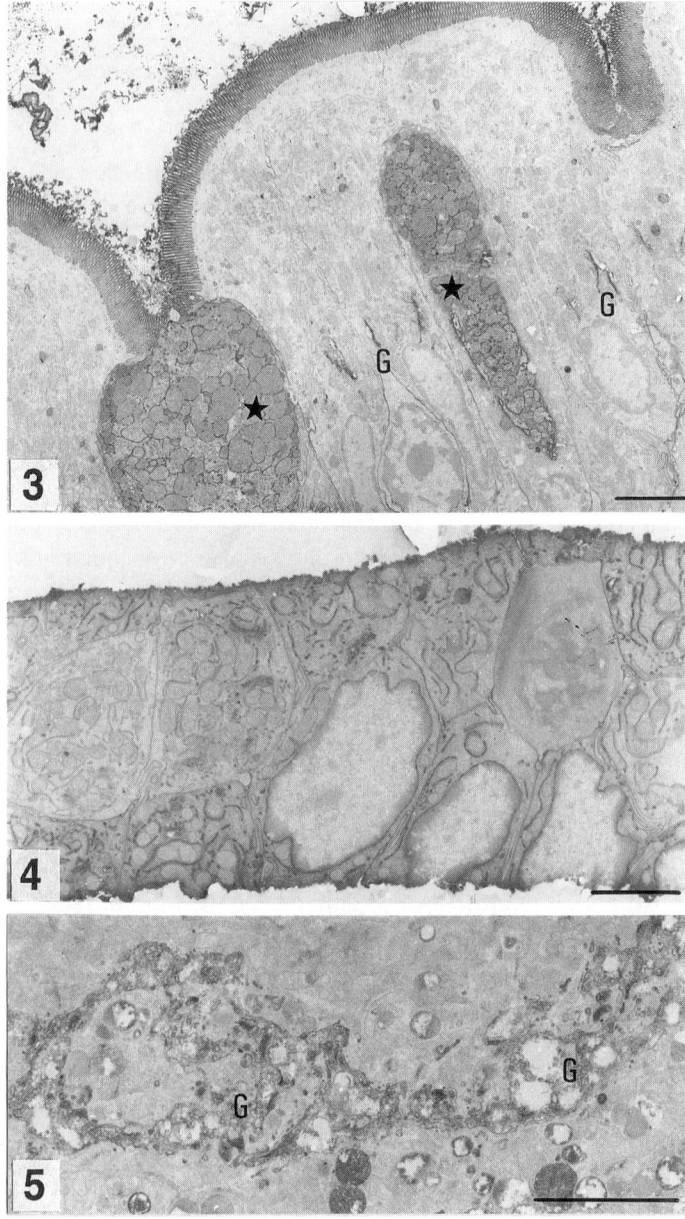

Figs. 3–5.

bution of the reaction products. There are almost no penetration problems in the case of monolayers of cultivated cells, since their thickness, only some few microns, is easily penetrated (Fig. 5).

4. Notes

1. Formaldehyde (FA) is freshly prepared from paraformaldehyde in a fume hood: Add 4 g of paraformaldehyde to 100 mL of sodium cacodylate buffer in a 250-mL flask and heat under constant stirring to 60–70°C. The white powder will dissolve after reaching this temperature; continue with stirring until the milky solution clears. Allow to cool and filter through Whatman No. 1 filter paper. Mix with an equivalent amount of 25% GA (electron microscopy grade) to a final concentration of 4% FA and 0.5% GA (25 mL 4% FA + 0.5 mL 25% GA).
2. Glutaraldehyde tends to polymerize rapidly at pH > 7.5. Polymers together with impurities (cyanide, arsenic) might introduce artifacts. Therefore, fresh dilutions of purified GA stock solutions (25% or above) are to be used. Purified GA remains relatively stable if stored at 4°C or below, at low pH (≈ 5.0) under oxygen-free conditions (inert gas). Diluted, alkaline solutions should be discarded after a few hours at room temperature. They might be stored in the refrigerator for up to a week, but probably indefinitely in a freezer. To avoid undue deterioration of the GA, fixation should take place below 10°C *(32)*.
3. Following aldehyde fixation, the samples generally should be quenched, since incomplete fixation owing to unreacted aldehyde groups should be avoided. This is especially critical after GA fixation, where one free aldehyde group might react with the lectins and nonspecifically bind them. Commonly used quenching agents are ammonium chloride, glycine, or sodium borohydride. Rinse the samples in $0.05M$ ammonium chloride for 2 h (add 27 mg ammonium chloride to 10 mL sodium cacodylate buffer).
4. The chromogen DAB is a suspected carcinogen and should be handled with care.
    a. DAB is purchased in 1-g bottles (Aldrich, Steinheim, Germany; Serva, Heidelberg, Germany; Janssen Chimica, Belgium; Sigma) and stored in the dark.
    b. Work in a fume hood wearing the appropriate gloves, gown, and mask, use a bench coat.
    c. Prepare the working solution directly before use by adding 10 mg DAB to 20 mL $0.05M$ Tris-HCl buffer (or multiple amounts), pH 7.6, and mix gently until dissolved. Filter through two layers of prefolded filter paper and keep covered in the dark. Use one part for preincubation of the probes. Add 200 µL

Fig. 3. *(previous page)* Low-power electron micrograph of a flat section through the center of a cryosection from the small intestinal villus epithelium. *Ricinus communis* I (βGal >Galα>>GalNAc) lectin binding sites mark the brush border, goblet cell mucin granules (★), Golgi stacks (G), and vesicles in enterocytes. Bar = 2 µm.

Fig. 4. *(previous page)* Low-power electron micrograph showing a cross-sectioned cryosection of the villus epithelium from the small intestine. *Lens culinaris* lectin (αMan>αGlc >GlcNAc) binding sites are located throughout the entire thickness of the cryosection. Bar = 2 µm.

Fig. 5. *(previous page)* Localization of *Pisum sativum* lectin (αMan>αGlc=GlcNAc) binding sites in a cultured fibroblast. Reactions mark the extended Golgi apparatus (G), various vesicles, and lysosomes. Bar = 2 µm.

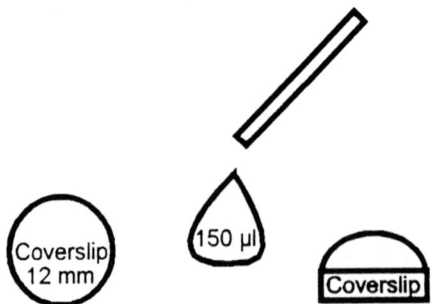

Fig. 6. Cover slips are placed into a wet chamber and incubated with the lectin-conjugates. Place the drops gently onto the surface. Do not touch the margins; the solutes will spread out. At the end of the incubation, withdraw solutions by short suction with filter paper (adsorbed solutes can easily be deposited) and rinse thoroughly. Small volume multiwell plates (96 wells/plate) also proved useful; less than 50 µL of the solutions are sufficient and multiple incubations may be done in parallel. Dehydrate up to 100% ethanol and skip the propylenoxide since it might dissolve the plastic. Finally, the molds are filled with Epon and polymerized.

of 1% H_2O_2 to the second part (10 mL DAB/Tris-HCl solution) for the incubation step. Since oxidation of DAB starts with the addition of H_2O_2 and continues when exposed to the air, it should be prepared directly before use and kept covered.

d. 1% H_2O_2 is diluted from a 30% H_2O_2 stock solution (add 100 µL stock solution to 3 mL distilled water); the stock solution is extremely caustic and should be handled with caution.

e. Contaminated glassware and spills should be thoroughly cleaned up. Wastes from the DAB solution as well as contaminated materials should be collected and disposed as hazardous waste. Preweighed DAB tablets (packages with 50 or 100 U [10 mg/tablet]) as well as 100 mg packages sealed in a 100-mL serum bottles are also commercially available (Sigma).

5. Put two layers of filter paper in a Petri dish and wet them. Put strips of parafilm on the filter paper and place cover slips with the cells upside and incubate with 150–200 µL of the respective solution (*see* Fig. 6).

6. To increase membrane contrast, postfix the samples first for 30 min in 1% osmium ferrocyanide. Prepare a 1:1 mixture of 2% aqueous OsO_4 and 3% aqueous $K_4Fe^{II}[CN]_6$; the yellow color of the clear solution of $K_4Fe^{II}[CN]_6$ changes to brown after mixing with OsO_4 owing to the reduction of osmium.

7. For mild permeabilization, add 10 mL DMSO to 90 mL cacodylate buffer and rinse the samples overnight for up to 2–3 d at 4°C.

8. 0.05M Tris-HCl buffer, pH 7.6 (100 mL): Add 25 mL of 0.2M Tris (hydroxymethyl) aminomethane (2.42 g/100 mL distilled water), 38 mL 0.1N HCl and 37 mL distilled water.

Electron Microscopy

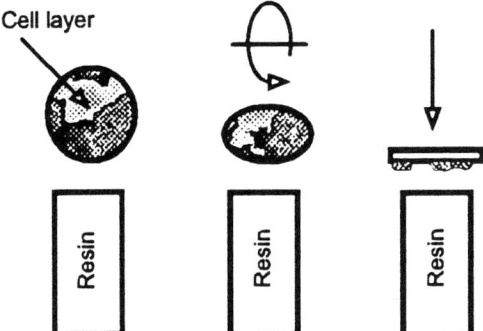

Fig. 7. Take the cover slips, discard the ethanol, and place them on top of resin-filled beem capsules. Slightly overfill the capsules and ensure that the cover slips are completely touched by the resin.

9. $0.05M$ Cacodylate buffer, pH 7.2: 100 mL $0.1M$ sodium cacodylate (21.4 g/L Na $(CH_3)_2 \cdot AsO_2 \cdot 3H_2O$) + 8.3 mL $0.1N$ HCl + 91.7 mL distilled water.
10. Osmium tetroxide ($[OsO_4]$; Merck, Darmstadt, Germany; local supplier for electron microscopy; glass ampule containing OsO_4 crystals): Prepare a 2% stock by adding 1 g OsO_4 to 50 mL distilled water. OsO_4 is toxic because of its high vapor pressure, therefore it should be stored separately and must be handled in a fume hood. Particular care must be taken to avoid breathing the vapor, which is injurious to nose, eyes, and throat. OsO_4 is also supplied as an aqueous solution.
11. Epoxy resins are liquids with irritant, harmful vapors that may cause skin irritation and dermatitis; avoid contact with the skin and use the resin in a fume hood. After curing, polymerized resin blocks are harmless and inert. If cured blocks are sawn, make sure that dust is not inhaled *(18–20)*.
12. Cells grown on cover slips are preferred as they are easily handled in further incubation steps and minimum volumes of media (approx 150 µL/12 mm ϕ cover slip) are sufficient (Fig. 6). Take care that the cells never dry up! At the end of incubation and embedding, the cover slips are placed upside (attached cells) down on resin-filled beem capsules (Fig. 7). The thin cell layer is easily penetrated by the resin; after polymerization of the resin, the cover slips are cut off (mechanically [stick cover slip onto a solid glass plate; Fig. 8] or by repeated warming and cooling [place on top of a heater and on ice or even touch the surface of liquid nitrogen—take care]) and the cells remain top-layer embedded in the resin. Skillful technicians can easily prepare flat sections of the cells; for cutting cross-sections, after removal of the cover slips, the probes should be re-embedded in the resin.
13. The limited penetration of the reagents means that sections should be cut from compact tissue. Tissue chopper, vibratome, or cryocut sections may be used. In our procedure, sections were prepared with a cryomicrotome at $-30°C$. Freezing and thawing of the samples in this procedure might further enhance their penetrability. The specimens are protected against ice crystal damage by use of DMSO (rinse in buffer/DMSO for 1–3 d) and freezing in isopentane.

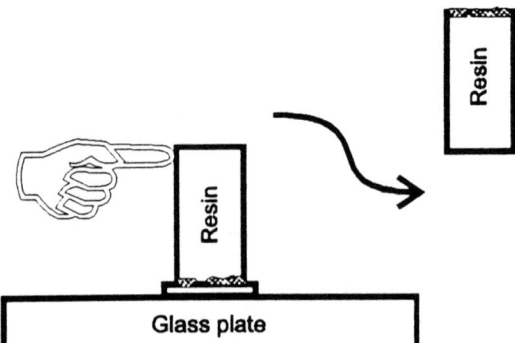

Fig. 8. The fixed beem capsule is removed so that the cover slip remains on the support. Thus, the embedded cells become accessible as outermost layer within the resin. Ensure that no glass splinters remain attached (control under the microscope).

14. To avoid nonspecific binding, the lowest possible concentration of the lectins should be used (≈30–150 μg/mL). High number of sugar residues in the samples might enable further reduction in the lectin concentration. Most lectins can be dissolved in PBS (*see* data sheets from the suppliers); their binding activity often depends on cations, which have to be added to the buffer *(30)*. The specificity and binding characteristics of the various lectins are excellently reviewed by Goldstein and Poretz *(29)*. Some lectins are highly toxic (e.g., *Ricinus communis* I) others are mitogenic (e.g., Concanavalin A), and so should be used with care.
15. The specificity of the binding reaction may be tested with the aid of competitive and noncompetitive sugars. It has to be considered that the binding affinity of the lectins to certain sugar sequences mostly is much higher than to simple monosaccharides, used in inhibition protocols *(29)*. The sugars are added at concentrations of $0.15–0.6M$ to lectin media and allowed to bind with the lectin for 30 min before incubation of the samples. By means of graded series of competitive sugars ($0.001–0.6M$) added to the lectin media, differential labeling is achieved for glycoconjugates that coexist in the tissues and interact with the same lectin; thus high and low affinity binding glycoconjugates can be differentiated in subcellular compartments *(11,12,31)*.

References

1. Edelman, G. M. (1976) Surface modulation in cell recognition and cell growth. *Science* **192**, 218–226.
2. Salik, S. E. and Cook, G. M. W. (1976) Comparison of early embryonic and differentiating cell surfaces. *Biochim. Biophys. Acta* **419**, 119–136.
3. Burridge, K. (1976) Changes in cellular glycoproteins after transformation: identification of specific glycoproteins and antigens in sodium dodecylsulphate gels. *Proc. Natl. Acad. Sci. USA* **73**, 4457–4461.

4. Rapin, A. M. C. and Burger, M. M. (1974) Tumor cell surfaces: general alterations detected by agglutinins. *Adv. Cancer Res.* **20,** 1–94.
5. Simionescu, M., Simionescu, N., and Palade, G. E., (1982) Differentiated microdomains on the luminal surface of capillary endothelium: distribution of lectin receptors. *J. Cell Biol.* **94,** 406–413.
6. Gonatas, N. K. and Avrameas, S. (1973) Detection of plasma membrane carbohydrates with lectin peroxidase conjugates. *J. Cell Biol.* **59,** 436–443.
7. Nicolson, G. L. (1974) The interaction of lectins with animal cell surfaces. *Int. Rev. Cytol.* **39,** 99–190.
8. Hedman, K., Pastan, I., and Willingham, M. C. (1986) The organelles of the trans domain of the cell. Ultrastructural localization of sialoglycoconjugates using *Limax flavus* agglutinin. *J. Histochem. Cytochem.* **34,** 1069–1077.
9. Pavelka, M. and Ellinger, A. (1985) Localization of binding sites for Concanavalin A, *Ricinus communis* I and *Helix pomatia* lectin in the Golgi apparatus of rat small intestinal absorptive cells. *J. Histochem. Cytochem.* **33,** 905–914.
10. Pavelka, M. and Ellinger, A. (1987) The Golgi apparatus in the developing embryonic pancreas: II localization of lectin binding sites. *Am. J. Anat.* **178,** 224–230.
11. Pavelka, M. and Ellinger, A. (1989) Affinity cytochemical differentiation of glycoconjugates of small intestinal absorptive cells using *Pisum sativum* and *Lens culinaris* lectins. *J. Histochem. Cytochem.* **37,** 877–884.
12. Pavelka, M. and Ellinger, A. (1989) Pre-embedding labeling techniques applicable to intracellular binding sites, in *Electron Microscopy of Subcellular Dynamics* (Plattner, H., ed.), CRC, Boca Raton, FL, pp. 199–218.
13. Sato, A. and Spicer, S. S. (1982) Ultrastructural visualization of galactosyl residues in various alimentary epithelial cells with the peanut lectin-horseradish peroxidase procedure. *Histochemistry* **73,** 607–624.
14. Tartakoff, A. M. and Vassalli, P. (1983) Lectin-binding sites as markers of Golgi subcompartments: proximal-to-distal maturation of oligosaccharides. *J. Cell Biol.* **97,** 1243–1248.
15. Yokoyama, M., Nishiyama, F., Kawai, N., and Hirano, H. (1980) The staining of Golgi membranes with *Ricinus communis* agglutinin-horseradish peroxidase conjugate in mice tissue cells. *Exp. Cell Res.* **125,** 47–53.
16. Bretton, R. and Bariety, J. (1976) A comparative ultrastructural localization of Concanavalin A, wheat germ and *Ricinus communis* on glomeruli of normal rat kidney. *J. Histochem. Cytochem.* **24,** 1093–1100.
17. Bernhard, W. and Avrameas, S. (1971) Ultrastructural visualization of cellular carbohydrate components by means of concanavalin A. *Exp. Cell Res.* **64,** 232–236.
18. Glauert, A. M. (1974) Fixation, dehydration and embedding of biological specimens, in *Practical Methods in Electron Microscopy*, vol. 3, Glavert, A. M., ed., North Holland Publishing, Amsterdam, pp. 1–207.
19. Hayat, M. A. (1989) *Principles and Techniques of Electron Microscopy Biological Applications.* 3rd Ed., CRC, Boca Raton, FL.
20. Melan, M. A. (1994) Overview of cell fixation and permeabilization, in *Methods in Molecular Biology, Immunocytochemical Methods and Protocols* (Javois, L. C., ed.), Humana, Totowa, NJ, pp. 55–66.

21. McLean, I. W. and Nakane, P. K. (1974) Periodate-lysin-paraformaldehyde fixative. A new fixative for immunoelectron microscopy. *J. Histochem. Cytochem.* **22,** 1077–1083.
22. Willingham, M. C. (1983) An alternative fixation-processing method for pre-embedding ultrastructural immunocytochemistry of cytoplasmic antigens: the GBS (glutaraldehyde-borohydride-saponin) procedure. *J. Histochem. Cytochem.* **31,** 791–798.
23. Polak, J. M. and Varndell, I. M. (1984) *Immunolabeling for Electron Microscopy*, Elsevier, Amsterdam.
24. Childs, G. V. (1986) Immunocytochemical technology. *Am. J. Anat.* **175,** 127–134.
25. Nakane, P. and Kawaoi, A. (1974) Peroxidase labeled antibody a new method of conjugation. *J. Histochem. Cytochem.* **22,** 1084–1091.
26. Courtoy, P. J., Picton, D. H., and Farquhar, M. G. (1983) Resolution and limitations of the immunoperoxidase procedure in the localization of extracellular matrix antigens. *J. Histochem. Cytochem.* **31,** 945–951.
27. Messing, A., Stieber, A., and Gonatas, N. K. (1985) Resolution of diaminobenzidine for the detection of horseradish peroxidase on surfaces of cultured cells *J. Histochem. Cytochem.* **33,** 837–839.
28. Novikoff, A. B. (1980) DAB cytochemistry: artifact problems in its current uses. *J. Histochem. Cytochem.* **28,** 1036–1038.
29. Goldstein, I. J. and Poretz, R. D. (1986) Isolation, physicochemical characterization, and carbohydrate-binding specificity of lectins, in *The Lectins: Properties, Functions, and Applications in Biology and Medicine* (Liener, I. E., Sharon, N., and Goldstein, I. J., eds.), Academic, London, pp. 33–247.
30. Lonnerdal, B., Borrebaeck, C. A., Etzler, M. E., and Erssoon, B. (1983) Dependence on cations for the binding affinity of lectins as determined by affinity electrophoresis. *Biochem. Biophys. Res. Comm.* **115,** 1069–1074.
31. Ellinger, A. and Pavelka, M. (1992) Subdomains of the rough endoplasmic reticulum in colon goblet cells of the rat: lectin-cytochemical characterization. *J. Histochem. Cytochem.* **40,** 919–930.
32. Prento, P. (1995) Glutaraldehyde for electron microscopy: a practical investigation of commercial glutaraldehydes and glutaraldehyde-storage conditions. *Histochem. J.* **27,** 906–913.

8

Electron Microscopy

Use of Lectin–Gold After Embedding

Rainer Herken and Berti Manshausen

1. Introduction
1.1. Fixation of the Tissue

Fixation of tissues for postembedding gold–lectin histochemistry calls for a fixative that preserves the ultrastructure of the tissue, has no inhibiting effect on lectin binding sites, and binds the glycoconjugates, which carry these lectin binding sites within the tissue so strongly that they cannot be washed out during the dehydration and ensuing embedding steps. There is no "one and only" fixative that has all these properties, and every fixative used for postembedding gold–lectin histochemistry must be viewed, at best, as a compromise. One common fixative, more or less exhibiting the properties required, is a mixture of 4% formaldehyde and 0.5% glutaraldehyde in phosphate buffer at pH 7.35. The use of osmium tetroxide as a fixative must, in general, be avoided for postembedding lectin histochemistry since it disturbs the reactivity of lectin binding sites.

1.2. Embedding of the Tissue

Generally for postembedding lectin histochemistry, hydrophilic resins should be used. Embedding in hydrophobic resins such as Epon is possible, but can hardly be recommended. Postembedding lectin histochemistry on tissue embedded in hydrophobic resins is much more complicated than on tissues embedded in hydrophilic resins because the resin must be removed from the tissue or the surface of the section etched with chemicals. These procedures increase the danger of errors due to nonspecific binding of lectins as well as disturbances of lectin-binding sites.

From: *Methods in Molecular Medicine, Vol. 9: Lectin Methods and Protocols*
Edited by: J. M. Rhodes and J. D. Milton Humana Press Inc., Totowa, NJ

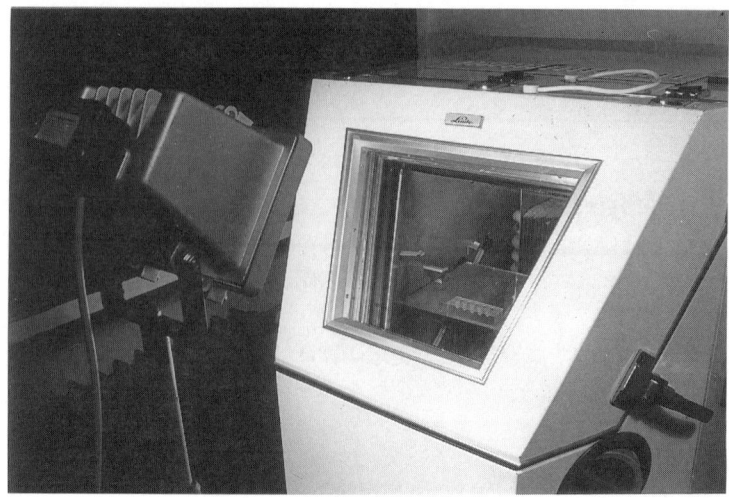

Fig. 1. Cryotome equipped with capsule holder for hardening LR-Gold at −20°C.

For postembedding gold–lectin histochemistry, we prefer embedding in the hydrophilic resin LR–Gold, which can be hardened at −25°C with the help of a halogen lamp *(1)*. Another hydrophilic resin widely used for lectin histochemistry is Lowicryl K4M *(2)*. A clear advantage of the use of LR–Gold is the fact that tissue embedded in this resin not only allows localization of numerous lectin binding sites but also postembedding immunohistochemistry for a greater number of antigens than does Lowicryl. One disadvantage of LR–Gold compared with Lowicryl K4*M* is the fact that the hardening of this resin is less reliable than the hardening of Lowicryl, probably owing to variations in the chemical composition of the resin. Moreover, the cutting of tissue embedded in LR–Gold requires a great deal of patience.

1.3. Hardening of the Resin

For hardening of the resin in the capsules, some form of closed cooling equipment that has a temperature of −25 to −30°C and that also allows the hardening of the LR–Gold with the light of a halogen lamp with maximal brightness is necessary.

Such a piece of equipment can be built, for example, from an old cryotome (Fig. 1). For this purpose, the microtome of the cryotome is replaced by a capsule holder, which contains the capsules with the LR–Gold solution to be hardened.

The low temperature of approx −25°C is not important for the hardening of the resin, but rather for the preservation of the lectin binding sites. The most important factor for a sufficient hardening of the resin is the amount of light that reaches the LR–Gold solution. We are faced with the problem that a 20-W halogen lamp (which can be placed inside the cryotome without affecting the temperature) does not give enough light for a sufficient hardening of the resin. On the other hand, brighter lamps produce so much heat that the temperature inside the cryotome housing becomes too high. Therefore, the light to harden the resin should come from a source placed outside the cryotome shining through the glass screen of the cryotome housing. A 500-W halogen lamp mounted on a stand, a model now available in any garden or hobby center is suitable as a light source.

1.4. Preparation of Sections

General principles of cutting sections for electron microscopy are described in any textbook dealing with methods of electron microscopy *(3)*. In this chapter, we will only deal with the details of cutting of tissue embedded in LR–Gold, which are different from those of cutting, for example, Epon-embedded tissue.

Generally speaking, it is possible to cut ultrathin sections of LR–Gold-embedded tissue with glass knives. However, it must be remembered that most investigators in the field use Lowicryl K4M rather than LR–Gold just to avoid the difficulties involved in cutting LR–Gold. The use of diamond knives makes the work with LR–Gold much easier.

Another problem in cutting LR–Gold for electron microscopy is the fact that the determination of the thickness of the sections on the basis of their interference color is, to say the least, difficult. The fact that not small crumbs, but rather a total section can be cut from the resin block is the marker for the right thickness of the section. These sections, which are suitable for postembedding lectin histochemistry, have a yellowish color and a thickness of approx 60 nm.

1.5. Lectins

Numerous lectins coupled to colloidal gold particles of different sizes are available. The gold particles should not be too small. Colloidal gold with a diameter of <10 nm is difficult to recognize under the electron microscope. If there are no special reasons for using such small gold particles as, for example, detection of lectin binding sites in structures smaller than 10 nm, lectins coupled to larger gold particles (15–20 nm) are preferable.

For some lectins, it is easy to localize their binding sites at the ultrastructural level, whereas, for others more experience in this field is needed. For "beginners" we recommend the ultrastructural localization of WGA or RCA I binding sites, because these lectin-binding sites are so strong that they can easily "forgive" many of the mistakes made with the histochemical reactions.

The most important factor for the success of the localization of lectin binding sites is the use of the correct buffer for each lectin. For LM lectin histochemistry and for postembedding gold–lectin histochemistry, the basis of lectin reactions is physiological saline, i.e., 0.9% NaCl. Depending on the range of the pH, which leads to optimal lectin binding to the lectin binding sites, i.e., to the specific carbohydrates, this saline must be buffered with either phosphate or Tris buffer. Moreover, different lectins need different electrolytes for their binding reactions to the carbohydrates. Therefore, the buffered physiological saline must be supplemented with different electrolytes.

A wide range of lectin reactions can be performed with the PBS solution described below. Therefore, this buffer is our standard buffer for all lectins for which no special buffer systems for postembedding gold–lectin histochemistry have been described.

For the control of the specificity of lectin binding within the tissue, the binding sites of the gold-coupled lectin used are blocked by incubation of the lectin with its specific carbohydrate before the reactions are run.

The tissue section is incubated with the blocked gold-coupled lectin, i.e., with a lectin that now has no available carbohydrate-binding sites. No staining should be observed, indicating the specificity of the staining.

2. Materials

2.1. Fixation

1. Commercially available formaldehyde contains formic acid and methanol, which is unsuitable for the preservation of the ultrastructure. Therefore, the formaldehyde solution must be prepared from paraformaldehyde. Details of the method of preparation of phosphate-buffered formaldehyde solution from paraformaldehyde can be found in every textbook dealing with methods of electron microscopy *(1)*.
2. As a final step, glutaraldehyde is added to the solution. Commercially available glutaraldehyde solution is usually an aqueous solution of 25% glutaraldehyde, which is added to the phosphate-buffered formaldehyde to a final concentration of 0.5% glutaraldehyde.

2.2. Dehydration

Ammonium chloride: 10 mM in PBS. To prepare PBS: Add 8 g NaCl, 0.2 g KCl, 0.1 g $CaCl_2 \cdot H_2O$, 0.1 g $MgCl_2$, 1.15 g Na_2HPO_4, 0.2 g KH_2PO_4 to 1 L distilled water. Adjust the pH of the solution to 7.35 (*see* Note 7).

2.3. Hardening

1. 500-W halogen lamp.

2.4. Sections

1. Diamond knife.
2. Formvar-coated nickel grids.

2.5. Lectins

1. Upon delivery, the gold-coupled lectins (EY, San Mateo, CA) should be portioned into aliquots of 10 µL in Eppendorf tubes and stored at 4°C. It is important to avoid freezing of the gold-coupled lectins because this disturbs the gold coupling. Generally, the shelf life of gold-coupled lectins given is 1 yr. However, our experience has shown that under good storage conditions these lectins are usable even after 2 or 3 yr. For the lectins WGA, RCA I, SBA and Lotus use PBS.
2. For Con A and LFA, use the following buffers:
   a. Tris buffer: To 100 mL distilled water, add 6.04 g trihydroxymethylaminomethane ($0.5M$) and adjust the pH of the solution to 7.0 with $1N$ HCl (*see* Note 8).
   b. Add to this solution 876 mg NaCl (final concentration $0.15M$ NaCl), 14.7 mg $CaCl_2 \cdot 2H_2O$ ($0.001M$ $CaCl_2$), 20.3 mg $MgCl_2 \cdot 6H_2O$ ($0.001M$ $MgCl_2$).
   c. Add 19.8 mg $MnCl_2 \cdot 4H_2O$ ($0.001M$ $MnCl_2$) (*see* Note 9).

3. Methods

3.1. Fixation

1. For fixation in ultrastructural lectin histochemistry, the size of the tissue sample should be kept as small as possible (~1 mm^3). The duration of the fixation depends on the lectin binding site to be localized. We recommend a fixation time of not more than 1 h at 4°C.
2. Following fixation, any free aldehyde groups remaining in the tissue, which would influence the histochemical reaction, must be blocked by incubation of the tissue for 45 min in 10 mM ammonium chloride in PBS.

3.2. Dehydration of the Tissue

1. Rinse the tissue 3 × 5 min in $0.1M$ PBS.
2. Dehydrate in an ascending ethanol series at 4°C (*see* Note 1).
   a. 15 min in 30% ethanol.
   b. 15 min in 50% ethanol.
   c. 30 min in 70% ethanol.
   Note: If necessary, the tissue can be stored in 70% ethanol for several days or even weeks at 4°C (*see* Note 2).
3. Dehydrate further in 90% ethanol for 30 min (*see* Note 3).

3.3. Procedure for Embedding in LR–Gold

1. Introduce the dehydrated tissue into beem capsules (which can be hermetically sealed) (Fig. 2), fill with LR–Gold, and store for 1 h in the dark at –20°C.
2. Transfer the tissue to a capsule containing LR–Gold, supplemented with 0.8% of the accelerator benzil.
3. Store the capsule overnight, i.e., for 12–18 h, at –20°C in the dark.
4. Embed the tissue in a fresh capsule containing LR–Gold with 0.8% benzil (*see* Note 4).

Fig. 2. Beem capsule suitable for embedding tissue samples in LR-Gold.

3.4. Hardening of the Resin

1. Place the 500-W halogen lamp at a minimum distance of 1 m away from the glass screen of the cryotome. It is important that the light reaches all capsules in the cryotome in order to render the hardening process as effective as possible.
2. After 24–36 h, when the LR–Gold should be hard, remove the capsules from the cryotome (*see* Note 5).

3.5. Preparation of Sections

LR–Gold-embedded tissue, especially tissue dehydrated only up to 70% ethanol, is extremely hydrophilic. The sections must, therefore, be removed from the water surface of the knife as quickly as possible to avoid absorbance of water. Otherwise, they can neither be picked up with an eyelash nor with a grid (*see* Note 6).

3.6. Lectin Histochemistry

3.6.1. Incubation of Sections with Gold-Coupled Lectins

1. Line a Petri dish with wet filter paper to create a damp chamber.
2. Place a strip of paraplast (plastic film) on the filter paper and pipet drops of the lectin–gold solution onto the film.
3. Gently place the nickel grids containing the sections face downward on top of the drops with the aid of fine tweezers.
4. Cover the Petri dish and leave at room temperature for the necessary incubation time.

3.6.2. Detailed Example of the Procedure for the Localization of WGA Binding Sites (Figs. 3 and 4)

1. Incubate the grids for 5 min with PBS.
2. Incubate for 60 min with gold-coupled WGA (EY Laboratories) diluted 1:30 with PBS.
3. Rinse with distilled water.
4. Postfix with 2.5% glutaraldehyde for 10 min.
5. Postfix and stain with 1% osmium tetroxide for 30 min.

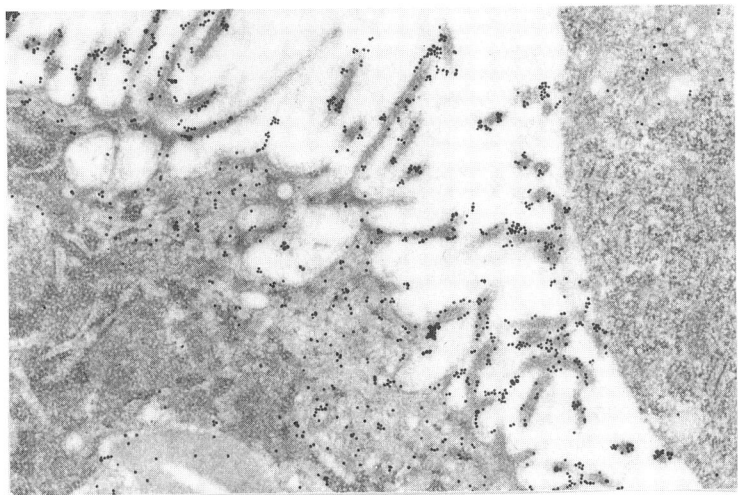

Fig. 3. Staining for WGA-binding sites on an endodermal cell of a mouse embryo.

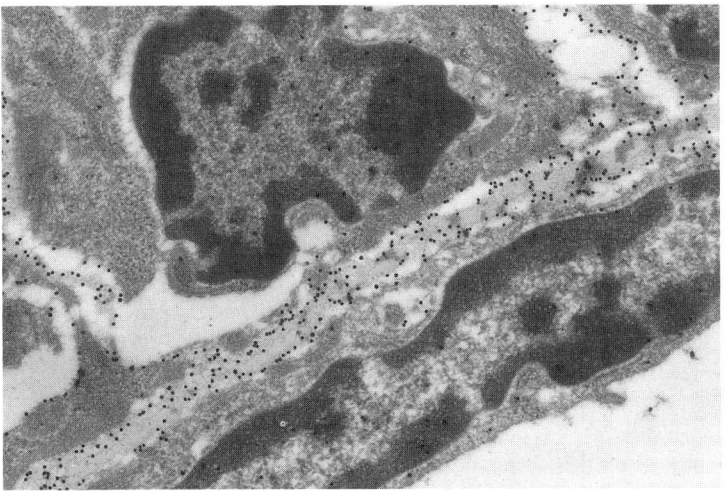

Fig. 4. Artery in mouse kidney stained for SBA-binding sites.

6. Rinse with distilled water.
7. Stain with 1% uranyl acetate for 10 min.
8. Rinse with distilled water.
9. Stain with 1% lead citrate for 10 min.
10. Rinse with distilled water.
11. Air-dry the nickel grids (*see* Note 10).

3.7. Control Experiments

Experimental procedure for the control experiment for localization of WGA binding sites:

1. Dilution of gold-coupled WGA (EY Laboratories) with PBS solution.
2. Addition of the inhibiting sugars N-N'-diacetylchitobiose (0.1 mM), N,N',N''-triacetylchitotriose (0.1 mM), and N-acetyl-D-glucosamine (0.5 mM) to the solution.
3. The solution is allowed to stand for 30 min at room temperature
4. Performance of the lectin reaction with this solution instead of the unblocked gold coupled WGA.

4. Notes

1. Some of the glycoproteins may be washed out of the tissue during the dehydration procedure. This problem can be diminished by carrying out the procedure in the cold, i.e., at 4°C.
2. Dehydration only up to 70% ethanol leads to an optimal preservation of the reactivity of lectin binding sites *(2)*. However, the preservation of the ultrastructure is limited. If the preservation of the ultrastructure is not satisfactory, dehydration can be extended further up in 90% ethanol.
3. It is important to note that the tissue should not be stored in 90% ethanol, but in 70% ethanol.
4. For the hardening of the resin, it is vital that the capsule is sealed air tight and that no air bubbles remain once the tissue has been transferred.
5. A good polymerization of the resin can be recognized by the fact that the resin is both hard and clear. "Milkiness" of the resin is an indication of insufficient polymerization, which renders the resin more or less useless for the ultrathin sections necessary for electron microscopy. Insufficient polymerization is mainly owing to two factors:
   a. The capsules containing the LR–Gold were not sealed air tight or there were still air bubbles in the capsule.
   b. The amount of light reaching the capsules was too low.
6. In our experience, investigators who have been cutting conventional EM sections (for example, Epon sections) for many years have much more trouble cutting LR–Gold than those cutting ultrathin sections for the first time and who are beginning their cutting activities with LR–Gold-embedded tissue.
7. This solution has been tested in our laboratory and been shown to be suitable for lectin histochemistry for WGA, RCA I, SBA, Lotus, LFA, and Con A binding sites *(4–6)*. If the PBS solution described in Section 2.2. does not lead to satisfactory localization of lectin binding sites, we recommend the use of another buffer system that (for example) allows the localization of binding sites for Con A and LFA.
8. This solution must be clear. If it becomes brown upon addition of the following substances, the pH is incorrect.
9. It is important that $MnCl_2$ be added last, otherwise it precipitates out of the solution.

10. For the localization of RCA I binding sites, for example, we increase the incubation time (step 2) from 60 to 90 min. Incubation of sections with gold-coupled Con A gives better results if the sections are incubated at 4°C overnight instead of 60 or 90 min at room temperature.

References

1. Herken, R., Fussek, M., and Thies, M. (1988) Light and electron microscopical postembedding lectin histochemistry for WGA-binding sites in the renal cortex of the mouse embedded in polyhydroxy aromatic resins LR–White and LR–Gold. *Histochemistry* **89,** 277–282.
2. Roth, J. (1983) Application of lectin–gold complexes for electron microscopic localization of glycoconjugates on thin sections. *J. Histochem. Cytochem.* **31,** 987–999.
3. Robinson, D. G., Ehlers, U., Herken, R., Herrmann, B., Mayer, F., and Schürmann, F.-W. (1987) *Methods of Preparation for Electron Microscopy,* Springer-Verlag, Berlin.
4. Herken, R., Sander, B., and Hofmann, M. (1990) Ultrastructural localization of WGA, RCA I, LFA and SBA binding sites in the seven-day-old mouse embryo. *Histochemistry* **94,** 525–530.
5. Salamat, M., Götz, W., Horster, A., Janotte, B., and Herken, R. (1993) Ultrastructural localization of carbohydrates in Reichert's membrane of the mouse. *Cell. Tiss. Res.* **272,** 375–381.
6. Salamat, M., Götz, W., Werner, J., and Herken, R. (1993) Ultrastructural localization of lectin-binding sites in different basement membranes. *Histochem. J.* **25,** 464–468.

9

Amplification of Lectin–Gold Histochemistry

Juan F. Madrid, Francisco Hernández, and José Ballesta

1. Introduction

Colloidal gold was developed as a probe for microscopy in the late 1970s. However, Roberts *(1)* had previously injected gold salts in living animals and detected the tissue–gold interactions microscopically by a photochemical method. With this method, metallic silver was deposited around gold particles. Roberts based his method on the idea of Liesegang and Rieder *(2)*, who visualized silver compounds injected into animals by a photographic process.

In 1981, Danscher developed a silver-enhancement reaction for the localization of gold in the tissues *(3,4)*. Two years later, Holgate et al. *(5)* combined the immunogold staining of Faulk and Taylor *(6)* with the photochemical method of Danscher *(3,4)* for the localization of antigenic sites.

Silver-enhancement procedures have been widely used for the identification of gold-labeled antibody binding sites at the light microscopical level *(5,7–10)*. Gold-labeled lectins have also been localized with silver-enhancement procedures with the light microscope *(9,11,12)*. The use of silver-enhancement procedures shows some advantages with respect to other procedures. These advantages have been mainly described for immunocytochemistry *(7–9)*: the silver-enhancement procedure may give positive immunostaining when other methods fail; the antibodies can be considerably more diluted, which might help to reduce costs; and it is more sensitive than other methods (e.g., the silver-enhanced immunogold procedure was found to have a four- to eightfold higher sensitivity than the standard PAP procedure *[13]*). Some authors have found similar advantages with lectin–gold-labeled sections *(9)*.

1.1. Usefulness of Silver-Enhancement

Colloidal gold has a red color when visualized with the light microscope, but the intensity is somewhat low. Thus, sensitive visualization is provided by the gold-induced silver reduction in which the silver precipitate appears black. At the electron microscopic level, the very high electron density of gold particles makes them an excellent tracer. This is probably the reason why silver-enhancement procedures have scarcely been used at electron microscopic level.

The diameter of the gold particles varies depending on the method used for the synthesis of colloidal gold. The gold diameter is very important. Kehle and Herzog *(14)* have observed that the binding efficiency of 6-nm IgG-Au to *Staphylococcus aureus* bacterium is 40%, whereas the 20-nm IgG-Au complexes only reach a binding efficiency of 6.7%; the binding efficiency of 6-nm gold particles is six-fold higher than that of 20-nm particles. When small gold particles are used, a higher intensity of labeling (measured in gold particles/μm^2) is obtained *(14,15)*. This has also been observed with lectin–gold procedures *(16,17;* personal observation). On the contrary, 20-nm IgG-Au complexes bind with higher affinity to *S. aureus* surface than the smaller 6-nm IgG-Au complexes *(14)*. This could be originated by the high number of IgG molecules bound to 20-nm gold particles (about 110), whereas 6-nm gold particles bind only one or two IgG molecules. In summary, with receptors showing low binding affinity and relatively high receptor density, the use of large gold particles is recommended. On the other hand, if the receptor density is very low, small gold particles should be selected.

The sections observed with the transmission electron microscope are usually 50 nm in thickness. Moreover, the fixation, as well as the embedding procedures, and the resin can mask the receptor. For these reasons, the number of identifiable receptors can be very low. In many cases, the use of small gold particles would be recommendable.

In the past years ultrasmall gold particles have been obtained (≤ 1 nm). They are used in transmission electron microscopy to obtain a higher number of particles binding to the ultrathin sections *(18)*. However, these gold particles are so small that they can be confused with some cell structures making their visualization difficult (Fig. 1A). On the other hand, the ultrasmall size of the gold particles enforces the photography at high magnification; thus decreasing the number of different cell structures in the photographs.

The silver-enhancement procedures with electron microscopy allow the use of small or ultrasmall gold particles as protein-markers (lectins, antibodies, and so on). During enhancement, the gold particles increase the size to the desired diameter (Fig. 1). The final size depends on the procedure used and the time. If the time is too long, the closely adjacent particles will fuse together (Fig. 2) generating large conglomerates. Scopsi et al. *(19)* enlarged 5–6 nm gold particles to ~85 nm after 60 min. Lackie et al. *(20)* found that after 7 min, the volume of silver deposited around the 5-nm gold particles was ~5000 times the original particle. This higher speed is due to the absence of protective col-

Amplification of Lectin–Gold Histochemistry

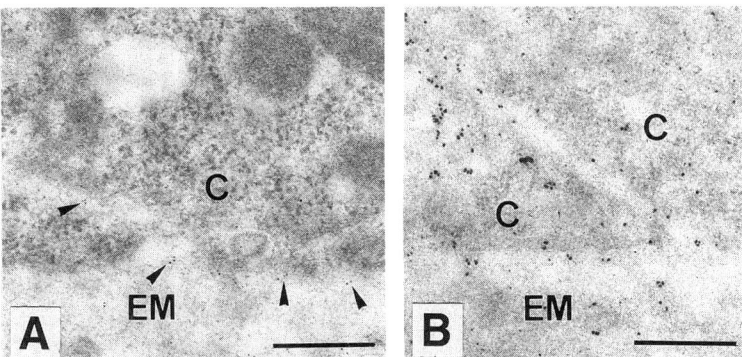

Fig. 1. *Aleuria aurantia* lectin, labeled with 5-nm gold particles. Human large intestine. Basal region of the enterocytes. **(A)** Gold granules are hardly visualized (arrowheads). **(B)** After 10 min of silver-enhancement, the gold granules are clearly observed. The plasmatic membrane and the extracellular matrix (EM) are labeled. C, cell cytoplasm. Bar = 0.5 µm.

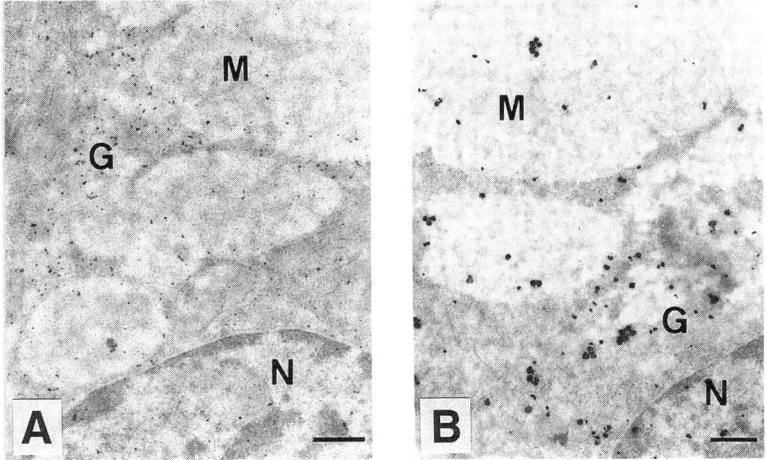

Fig. 2. *Aleuria aurantia* lectin, labeled with 5-nm gold particles. Human large intestine. Goblet cells. **(A)** The labeling in the Golgi apparatus (G) is clearly observed after 10 min of silver-enhancement. **(B)** After 20 min of silver-enhancement, some granules are fused together. N, nucleus; M, mucous granules. Bar = 0.5 µm.

loid in the developer. The use of small or ultrasmall gold particles is recommended for pre-embedding labeling. Van den Pol *(21)* observed that 20–40-nm gold particles did not penetrate 30-µm vibratome sections, whereas positive immunostaining was achieved with 5–10-nm gold particles. Thus, for pre-embedding lectin–gold methods, the use of small gold particles followed by silver-enhancement, is the method of choice.

Silver-enhancement procedures can also be used with scanning electron microscopy. However, large particles (≥20 nm) are usually employed for conventional scanning electron microscopy; smaller particles are hardly visualized and may be covered by the metal coating. Improvements in detection have made it possible to use back-scattered electron imaging for detection of particles as small as 5 nm *(22)*. In both conventional scanning and back-scattering electron microscopy, the use of smaller gold particles would increase the density of labeling. Smaller gold particles can be used if silver-enhancement procedures are also performed *(23)*. Namork and Heier *(24)*, using back-scattering electron microscopy, have described a size increase of 1.5–2.0 nm/min in 15- and 40-nm gold particles.

Double labeling in combination with silver enhancement can also be performed in transmission and scanning electron microscopy. Gold particles of different sizes bound to different molecules (one lectin and one antibody, two different lectins, and so on) must be used, but care must be taken. The growth rate for 15- and 40-nm gold particles was close to linear by time, and it is the same for both sizes (1.5–2.0 nm/min) *(24)*. In agreement with the data of Namork and Heier *(24)*, if 15- and 40-nm gold particles are used, the 15-nm gold particles are initially 37.5% of the diameter of the 40-nm gold particles. After 10 min of silver enhancement, the size is approx 35.0 ± 4.5 and 56.0 ± 4.5 nm, respectively (now the smallest represents 62.5% of the diameter of the largest). In summary, for double labeling with protein–gold and silver enhancement, gold particles should have a large difference in diameter, i.e., if 15- and 20-nm gold particles are used, after 10 min they may have a similar diameter and they cannot be discriminated.

1.2. Lugol Iodine Pretreatment

The silver-enhancement procedures were first applied in immunohistochemical studies. They include a pretreatment with lugol iodine, before the immunocytochemical method. This pretreatment is necessary to obtain a positive result *(5,7)*. Other authors have claimed that this step is unnecessary and that this treatment itself or the use of thiosulfate for decolorizing the sections could alter the receptors *(11,13,20,25)*. Electron microscopic studies have not considered the use of this pretreatment *(20,26,27)*.

1.3. Silver Ions

For decades, silver nitrate has represented the source of silver ions in many techniques (e.g., Grimelius stain for endocrine cells). The original method also used silver nitrate *(28)*. The two other components of the developer are a reducing agent (usually hydroquinone) and a buffer (usually citrate buffer). Introduction of silver lactate instead of silver nitrate without altering the hydroquinone concentration improves the specificity of the technique *(3,4)*. The major disadvantage of silver nitrate is the unspecific argyrophilic reaction, whereas this is required with the Grimelius silver impregnation technique *(29)*. Silver lactate has a low dissociation coefficient, which allows a more controllable reduction. Scopsi *(25)* indicated that in tissue sections, the sensitivity of silver nitrate and lactate is similar, but with silver nitrate the reaction progresses more quickly.

For light microscopy, the reaction can be controlled in the light microscope. To avoid excessive background staining, silver lactate requires development in darkness. The degree of background staining is lower when silver acetate is used *(30)*; it allows an easier control with the microscope. Scopsi *(25)* recommended that both silver lactate and acetate require darkness for a better reaction. For electron microscopy, control of the reaction is not possible; therefore, the reaction should be developed in darkness to avoid background induced by light. The sensitivity of the reaction is similar when silver lactate or acetate are used *(30)*. Since silver acetate is cheaper, it is preferred.

1.4. Reducing Agent

Most studies use hydroquinone (1-4-dihydroxybenzene) as reducing agent. Danscher *(3)* also tested metol and ascorbic acid, and concluded that 0.85% hydroquinone was the method of choice. Fujimori and Nakamura *(31)* have also used bromohydroquinone (0.33%). Recently, hydroquinone was substituted for N-propyl gallate for the amplification of small gold probes (1–1.4 nm) in electron microscopy and a more uniform development was obtained *(26)*.

1.5. Protective Colloid

Gum arabic has been widely used as a protective colloid. It is not essential, but serves to inhibit the otherwise inevitable autocatalytic interaction between the first two components, hydroquinone and silver salts. Gum arabic also controls the speed of the amplifying process. The first articles of Danscher *(3,4)* used a 30% final concentration of gum arabic. This concentration has been widely used. Scopsi *(25)* employed a 15% final concentration for both light and electron microscopy and lesser amounts were not recommended for electron microscopy. Burry et al. *(26)* have found that at least a 5% final concentration is needed. They used a final concentration of 25%. Some alternatives to gum arabic have also been tested: gelatin, bovine serum albumin, dextran, polyethylene glycol, and so on. However, these alternative reagents increased the background staining and the particles grew asymmetrically, with formation of elongated or otherwise irregular shapes *(3,25,32)*. Some studies have avoided the protective colloid *(26,33,34)*. Burry et al. *(26)* have described that in the absence of gum arabic, silver lactate and hydroquinone reacted completely in a matter of seconds, leaving none to react with the gold particles in the ultrathin section.

1.6. Buffer

The optimal reduction conditions need a low pH. The low pH buffering systems have included citrate (pH 3.5), acetate (pH 4.5), and lactate (pH 3.5) buffers *(25,35,36)*. However, these low pH buffers impair the ultrastructural preservation of the tissue samples. The pH of the silver-enhancement solution must be close to that of physiological pH to ensure a good preservation for electron microscopic studies. Lah et al. *(37)* used 10 mM HEPES buffer (pH 6.8), obtaining a good preservation. More recently, Burry et al. *(26)* by using 200 mM HEPES buffer (pH 6.8) have obtained a silver-enhancement solution with a pH of 5.8; the tissue was well preserved when observed with the electron microscope.

1.7. Water

Deionized distilled water should be used. Even minimal impurities (e.g., chloride) will induce a whitish precipitate and render the enhancement solution useless. The use of deionized water minimizes the contamination with chloride ions. For the same reason, it is also important to use very clean glassware and to avoid contact between metallic objects and the silver solutions. Silver ions and the final developer must be protected against the light. Should the developer be rendered isotonic, sucrose can be added without affecting the reaction.

1.8. Incubation

The duration of the incubation is determined by the physical developer and the temperature. It usually is about 5–30 min. The rate of development generally increases with increasing temperature. To ensure the reproducibility, a constant temperature should be employed. Danscher *(38)* recommended 26°C, whereas Hacker *(30)* used 20°C. Higher temperatures shorten the intensification time, but increase the risk of nonspecific silver precipitation. If the temperature is reduced from 26 to 4°C, the enhancing time increases from 8–12 to 15–30 min *(39)*.

1.9. Fixer

Photographic fixer (e.g., Amphix, Ilfospeed, or Agefix [Ilford, Agfa]) can be used after developer, diluted 1:4–1:10 for 1–30 min *(8,20,25,33)*. Alternatively, a previous step, a stop acid bath (e.g., 1% acetic acid) can be introduced

(40). To improve the morphology of the tissues at electron microscopical level, a neutral fixer has also been employed *(25,26)*.

1.10. Osmium Tetroxide

Osmium tetroxide is a deleterious agent for silver-enhancement methods. If the samples are postfixed with osmium tetroxide before the silver-enhancement, osmium will be able to catalyze the reduction of silver ions. Thus, to avoid obscuring the specific staining, a pretreatment with periodic acid or sodium metaperiodate have been recommended for immunocytochemical methods *(25)*. This recommendation must be used with caution for lectin–gold methods; if possible, it should be omitted. Periodic acid and sodium metaperiodate modify the carbohydrates (oxidizing vic-glycol groups of carbohydrates to aldehyde). If these methods have to be used, the treatment should be performed after the lectin labeling and before the silver-enhancement.

Another possibility is the postfixation with osmium tetroxide, after silver-enhancement. Osmium tetroxide treatment partially reduces the diameter of silver-enhanced gold particles. Burry et al. *(26)* have found that 1% OsO_4 for 30 min decreases the mean diameter of silver-enhanced particles from 40–80 nm to 40–60 nm. Treatment with 1% OsO_4 for 60 min generated a large number of particles below 20 nm. They recommended 0.1% OsO_4 for 30 min, which did not show a large shift in particle diameter.

1.11. Glutaraldehyde Postfixation

Postfixation with 2% glutaraldehyde could be necessary, especially if a low pH buffer is selected for the developer. The low pH environment could release lectins from the binding sites.

2. Materials

1. Buffer: Traditionally, citrate buffer (0.5M) has been used. It is obtained by dissolving 25.5 g of citric acid × 1H_2O and 23.5 g of sodium citrate·2H_2O in 100 mL of deionized distilled water. The pH is adjusted to 3.8 with citric acid. This can be kept at 4°C for at least 2–3 wk.

    Alternatively, HEPES buffer can be used. The following solutions should be prepared: 50 mM HEPES, 200 mM sucrose, pH 5.8 (1.19 g HEPES, 6.85 g sucrose, 100 mL deionized distilled water); 1M HEPES, pH 6.8 (23.83 g HEPES, 100 mL deionized distilled water). The pH is adjusted with NaOH. Chloride ions should be avoided.
2. Gum arabic stock solution: 50 g of gum arabic in 100 mL of deionized distilled water are agitated by magnetic stirring over several days. The solution is filtered through six single layers of coarse gauze and stored at –20°C in small tubes. This solution can be stored at least 6 mo. Thawed tubes should be used in the day. Alternatively, the gum stock solution can be kept at 4°C after addition of thymol crystals to prevent bacterial growth; however, freezing is preferred.

3. Reducing agent: Hydroquinone usually has been used as reducing agent. 0.85 g of hydroquinone is dissolved in 15 mL of deionized distilled water. Alternatively, N-propyl-Gallate (NPG) can be employed: 10 mg of NPG dissolved in 250 µL of 100% ethanol is brought to 5 mL with deionized distilled water. This solution should be mixed on the day of use.
4. Silver solution: 36 mg of silver lactate dissolved in 5 mL of deionized distilled water is prepared just before use and stored in the dark.
5. Silver-enhancement solution:
   a. Solution A: The mixing of the different solutions is made at several steps, starting 30 min before use: 5 mL of the gum arabic stock and 2 mL of 1M HEPES buffer are mixed with agitation; about 3 min before use, 1.5 mL of the NPG stock is added with agitation; 1 min before use, 1.5 mL of silver lactate stock is added. The total volume will be 10 mL.
   b. Solution B: If citrate buffer and hydroquinone are preferred, the mixing will be: 6 mL of gum arabic stock solution, 1 mL of citrate buffer, 1.5 mL of hydroquinone solution, and then 1.5 mL of silver lactate solution.
   Solution A shows a less acidic pH; therefore, it is preferred.
6. Fixer: Photographic fixer (e.g., Ilfospeed, Amphix, or Agefix) diluted 1:4–1:10 may be used. A neutral fixer solution composed of 250 mM sodium thiosulfate and 20 mM HEPES (3.95 g sodium thiosulfate, 0.48 g HEPES, 100 mL deionized distilled water) at pH 7.4 can also be used.
7. 1.6% glutaraldehyde in PBS.
8. 0.1% OsO_4 in PBS.

3. Methods

The reagents used in this method have been precisely described in Section 2.

1. Place 500 Å ultrathin sections on formvar-carbon coated nickel grids. Stain the grids with a lectin-gold method (*see* Note 1).
2. Fix in 1.6% glutaraldehyde in PBS for 15 min (*see* Notes 2 and 3).
3. Rinse in HEPES buffer rinse (50 mM HEPES with 20 mM sucrose, pH 5.8) four times at room temperature, 7 min each. The buffer is prepared with 1.19 g HEPES, 6.85 g sucrose, and 100 mL deionized distilled water. The pH is adjusted to 5.8 with NaOH (*see* Note 2).
4. Incubate with silver-enhancement solution: 5–20 min at 20°C in the dark. Thirty minutes before use, mix 5 mL of gum arabic stock solution (50 g of gum arabic in 100 mL of deionized distilled water agitated over several days and then filtered through six single layers of coarse gauze) with 2 mL of 1M HEPES buffer, pH 6.8 (23.83 g HEPES, 100 mL deionized distilled water, and the pH adjusted with NaOH). About 3 min before use, this solution is mixed with 1.5 mL of NPG solution (10 mg of N-propyl-Gallate is dissolved in 250 µL of 100 % ethanol and brought to 5 mL with deionized distilled water). One minute before use, 1.5 mL of silver lactate stock solution (36 mg of silver lactate is dissolved in 5 mL of deionized distilled water) is added (*see* Notes 2 and 4–10).

5. Rinse three times in neutral fixer for 5 min each at room temperature. The fixer is composed of 3.95 g sodium thiosulfate, 0.48 g HEPES, and 100 mL deionized distilled water at pH 7.4 (*see* Note 2).
6. Stain with 0.1% OsO_4 for 30 min at room temperature (*see* Notes 2 and 11–13).
7. Counterstain with uranyl acetate and lead citrate in the usual manner.

4. Notes

1. The lectin–gold method used does not affect the silver-enhancement procedure. With the small or ultrasmall gold particles, a more intense labeling will be observed. With pre-embedding procedures, the use of small or ultrasmall gold particles is very important to improve the tissue penetration of gold particles.
2. The grids are floated face down on a drop of the different solutions.
3. Glutaraldehyde fixation after lectin–gold methods is an optional step that fixes the lectin to the binding sites. The subsequent treatments will not release the lectin.
4. For silver enhancement, solution A is preferred. Its low acidic pH improves the tissue preservation for ultrastructural studies. This is the solution described in step 4 of Section 3.
5. All the components, except the protective colloid, should be made fresh, a few minutes before use.
6. If the protective colloid is omitted in the silver-enhancement solution, an equivalent volume of deionized distilled water must replace it.
7. The most adequate time for silver-enhancement has to be tested. The adequate time will depend on both the size of the probes and the magnification used in the electron microscope to study the tissue. Burry et al. *(26)* have described that tissue sections required enhancement 3–4 times higher than those needed for cell samples obtained from cultures.
8. The selection of a constant temperature is important for the reproducibility of the technique.
9. Although silver-enhancement solution can be used in light, darkness is recommended to avoid unnecessary background.
10. Contact of metals should be avoided with silver solutions.
11. Postfixation with osmium tetroxide is an optional step. Osmium postfixation will improve the quality of the images in the electron microscope. However, osmium precipitates could appear on the specimen.
12. The time and concentration of osmium postfixation must be low to avoid a deleterious effect on silver deposits.
13. The use of osmium tetroxide before silver enhancement should be avoided. Osmium could act as a nucleating agent for silver ions.
14. For postembedding methods, nickel grids have been used with satisfactory results.
15. Deionized distilled water should be used. Impurities could act as nucleating agent.
16. Glassware should be very clean (at least rinsed carefully with deionized distilled water) to avoid impurities.

Acknowledgments

Grant support: This work was partially supported by a grant from DGICYT of Spain (PB 93/1123). We are greatly indebted to María D. López-López, Joaquón Moya, María C. Gónzalez, and Cristina Otamendi for excellent technical assistance. We wish to thank María T. Hernández-Alfaro for typing the manuscript.

References

1. Roberts, W. J. (1935) A new method for the detection of gold in animal tissues: physical development. *Proc. R. Acad. Sci. Amsterdam* **38,** 540–544.
2. Liesegang, R. E. and Rieder, W. (1921) Versuche mit einer «Keim-methode» zum Nachweis von Silber in Gewebsschnitten. *Z. Wissench. Mikrosk.* **38,** 334–338.
3. Danscher, G. (1981) Histochemical demonstration of heavy metals. A revised version of the sulphide silver method suitable for light and electronmicroscopy. *Histochemistry* **71,** 1–16.
4. Danscher, G. (1981) Localization of gold in biological tissue. A photochemical method for light and electron microscopy. *Histochemistry* **71,** 81–88.
5. Holgate, C. S., Jackson, P., Cowen, P. N., and Bird, C. C. (1983) Immunogold-silver staining: new method of immunostaining with enhanced sensitivity. *J. Histochem. Cytochem.* **31,** 938–944.
6. Faulk, W. P. and Taylor, G. M. (1971) An immunocolloidal method for the electron microscopy. *Immunochemistry* **8,** 1081–1083.
7. Springall, D. R., Hacker, G. W., Grimelius, L., and Polak, J. M. (1984) The potential of the immunogold silver staining for paraffin sections. *Histochemistry* **81,** 603–608.
8. Hacker, G. W., Grimelius, L., Danscher, G., Bernatzky, G., Muss, W., Adam, H., and Thurner, J. (1988) Silver acetate autometallography: an alternative enhancement technique for immunogold–silver staining (IGSS) and silver amplification of gold, silver, mercury and zinc in tissues. *J. Histotechnol.* **11,** 213–221.
9. Taatjes, D. J., Schaub, U., and Roth, J. (1987) Light microscopical detection of antigens and lectin binding sites with gold-labelled reagents on semithin Lowicryl K4M sections: usefulness of the photochemical silver reaction for signal amplification. *Histochem. J.* **19,** 235–245.
10. Krenácks, T., Lászik, Z., and Dobó, E. (1989) Application of immunogold-silver staining and immunoenzymatic methods in multiple labelling of human pancreatic Langerhans islet cell. *Acta Histochem.* **85,** 79–85.
11. Skutelsky, E., Goyal, V., and Alroy, J. (1987) The use of avidin-gold complex for light microscopic localization of lectin receptors. *Histochemistry* **86,** 291–295.
12. Madrid, J. F., Castells, M. T., Martínez-Menárguez, J. A., Avilés, M., Hernández, F., and Ballesta, J. (1994) Subcellular characterization of glycoproteins in the principal cells of human gallbladder. A lectin cytochemical study. *Histochemistry* **101,** 195–204.

13. Scopsi, L. and Larsson, L.-I. (1985) Increased sensitivity in immunocytochemistry. Effects of double application of antibodies and of silver intensification on immunogold and peroxidase-antiperoxidase staining techniques. *Histochemistry* **82**, 321–329.
14. Kehle, T. and Herzog, V. (1990) Quantification of colloidal gold by electron microscopy, in *Colloidal Golds: Principles, Method, and Application*, vol. 3. (Hayat, M. A., ed.), Academic. San Diego, CA, pp. 117–137.
15. Yokata, S. (1988) Effect of particle size on labeling density for catalase in protein A-gold immunocytochemistry. *J. Histochem. Cytochem.* **36**, 107–119.
16. Roth, J., Lucocq, J. M., and Charest, P. M. (1984) Light and electron microscopic demonstration of sialic acid residues with the lectin from *Limax flavus*: a cytochemical affinity technique with the use of fetuin-gold complexes. *J. Histochem. Cytochem.* **32**, 1167–1177.
17. Roth, J. (1986) Post-embedding cytochemistry with gold labeled reagent: a review. *J. Microsc.* **143**, 125–137.
18. Sibon, O. C. M., Humbel, B., De Graaf, A., Verkleij, A. J., and Cremers, F. F. M. (1994) Ultrastructural localization of epidermal growth factor (EGF)-receptor transcripts in the cell nucleus using pre-embedding in situ hybridization in combination with ultra-small gold probes and silver enhancement. *Histochemistry* **101**, 223–232.
19. Scopsi, L., Larsson, L.-I., Bastholm, L., and Nielsen, M. H. (1986) Silver-enhanced colloidal gold probes as markers for scanning electron microscopy. *Histochemistry* **86**, 35–41.
20. Lackie, P. M., Hennessy, R. J., Hacker, G. W., and Polak, J. M. (1985). Investigation of immunogold-silver staining by electron microscopy. *Histochemistry* **83**, 545–550.
21. van den Pol, A. N. (1986) Tyrosine hydroxylase immunoreactive neurons throughout the hypothalamus receive glutamate decarboxylase immunoreactive synapses: a double preembedding immunocytochemical study with particulate silver and HRP. *J. Neurosci.* **6**, 877–897.
22. Walther, P. and Müller, M. (1985) Detection of small (5–15 nm) gold-labelled surface antigens using backscattered electrons, in *The Science of Biological Specimen Preparation for Microscopy and Microanalysis* (Müller, M., Becker, R. P., Bodye, A., and Wolosweick, J. J., eds.), SEM, Chicago, pp. 195–201.
23. Herter, P., Laube, G., Gronczewski, J., and Minuth, W. W. (1993) Silver-enhanced colloidal-gold of rabbit kidney collecting-duct cell surfaces imaged by scanning electron microscopy. *J. Microsc.* **171**, 107–115.
24. Namork, E. and Heier, H. E. (1989) Silver enhancement of gold probes (5–40 nm) single and double labeling of antigenic sites on cell surface imaged with backscattered electrons. *J. Electr. Microsc. Tech.* **11**, 102–108.
25. Scopsi, L. (1989) Silver-enhanced colloidal gold method, in *Colloidal Golds: Principles, Method, and Application*, vol. 1. (Hayat, M. A., ed.), Academic, San Diego, pp. 251–295.
26. Burry, R. W., Vandré, D. D., and Hayes, D. M. (1992) Silver enhancement of gold antibody probes in pre-embedding electron microscopic immunocytochemistry. *J. Histochem. Cytochem.* **40**, 1849–1856.

27. Shimizu, H., Masunaga, T., Ishiko, K., Hashimoto, T., Garrod, D. R., Shida, H., and Nishikawa, T. (1994) Demonstration of desmosomal antigen by electron microscopy using cryofixed and cryosubstituted skin with silver-enhanced gold probe. *J. Histochem. Cytochem.* **42,** 687–692.
28. Liesegang, R. (1928) Histologische Versilberung. *Z. Wiss. Mikr.* 45, 273–279.
29. Grimelius, L. (1968) A silver nitrate stain for A_2 cells of human pancreatic islets. *Acta Soc. Med. Ups.* **73,** 243.
30. Hacker, G. W. (1989) Silver-enhanced colloidal gold for light microscopy, in *Colloidal Golds: Principles, Method, and Application*, vol. 1. (Hayat, M. A., ed.), Academic, San Diego, pp. 297–321.
31. Fujimori, O. and Nakamura, M. (1985) Protein A gold-silver staining method for light microscopic immunohistochemistry. *Arch. Histol. Jpn.* **48,** 449–459.
32. DiFiglia, C. J. and Fields, K. L. (1987) A comparison of indirect immunofluorescence and immunogold-silver staining for surface antigens of PNS and CNS cells. *J. Neuroimmunol.* **16,** 43–49.
33. Geuze, H. S., Slot, J. W., Yanagibashi, K., McCraken, J. A., Schwartz, A. L., and Hall, P. F. (1987) Immunogold cytochemistry of cytochrome P-450 in porcine adrenal cortex. *Histochemistry* **86,** 551–561.
34. Goode, D. and Maugel, T. K. (1987) Backscattered electron imaging of immunogold-labelled and silver-enhanced microtubules in mammalian cells. *J. Electron Microsc. Tech.* **5,** 263–273.
35. Danscher, G. (1983) A silver method for counterstaining plastic embedded tissue. *Stain Technol.* **58,** 365–372.
36. Lageman, A. and Buccheim, P. (1985) Die Gold-Silber-Technik—ein neues immunohistochemisches Verfahren zum Nachweis von Zellmembranantigenen. *Acta Histochem.* **76,** 113–123.
37. Lah, J. J., Hayes, D. M., and Burry, R. W. (1990) A neutral pH silver development method for the visualization of 1-nanometer gold particles in pre-embedding electron microscopic immunocytochemistry. *J. Histochem. Cytochem.* **38,** 503–513.
38. Danscher, G. (1982) Exogenous selenium in the brain. A histochemical technique for light and electron microscopical localization of catalytic selenium bonds. *Histochemistry* **76,** 281–293.
39. De Waele, M. (1989) Silver-enhanced colloidal gold for the detection of leukocyte cell surface antigens in dark-field and epipolarization microscopy, in *Colloidal Golds: Principles, Method, and Application*, vol. 2. (Hayat, M. A., ed.), Academic, San Diego, pp. 443–467.
40. Manigley, C. and Roth, J. (1985). Applications of immunocolloids in light microscopy. IV. Use of photochemical silver staining in a simple and efficient double-staining technique. *J. Histochem. Cytochem.* **33,** 1247–1251.

10

Electron Microscopic Methods for the Demonstration of Lectin-Binding Sites in Cancer Cell Lines

Barry S. Mitchell and Udo Schumacher

1. Introduction

The use of lectins *(1)* in electron microscopical histochemistry enables specific questions to be answered about the distribution of carbohydrates in cellular and extracellular components. For example, the occurrence of defined carbohydrate residues in particular stacks of the Golgi apparatus *(2)*, or in particular domains of the rough endoplasmic reticulum *(3)* as demonstrated by lectins indicate the presence of carbohydrate processing enzymes, and the lectin binding can therefore serve as an indicator of cellular function. Because of these functional implications, lectins are used as markers of different cellular populations in studies of development and regenerative processes *(4–10)*, in the analysis of the molecular mechanisms of metastasis *(11)*, and in pathological processes such as the carbohydrate disorder globoid cell leukodystrophy (a lysosomal storage disease) *(12)*.

The advantage of using lectin binding methods with the electron microscope compared to light microscopy is that a more precise localization of the binding sites is possible. The methods depend on the presence of an electron-dense product in much the same way as for immunoelectron microscope methods. The principles of the technique are similar, nevertheless, to those of any lectin histochemical method at light microscopical level. Localization of lectin binding sites in paraffin sections is now a matter of routine *(13)*, and direct or indirect methods are commonly used to demonstrate lectin binding sites at the ultrastructural level as well, though there are limits to this method due to the nature of the processing of the tissue for electron microscopy. It is well known from light microscopical studies that lectin localization can depend critically on fixation and embedding *(14)*. Since more limitations on fixation and processing apply to electron microscope than to light microscope methods, it might be difficult to correlate the results of both methods. For example, unfixed frozen sections can readily be used in light microscopy, whereas not all electron microscopic units are set up to process this kind of preparation. Before starting any electron microscopy with lectins, it is therefore wise to decide whether a particular carbohydrate residue has to be localized by electron microscopy and whether the effort is worthwhile.

A number of methodologies are available for use in conjunction with ultrastructural localization: pre-embedding and postembedding methods, both of which have advantages and disadvantages. There are a few recent reports on the use of pre-embedding methods, which, as the name implies, indicates that the tissues are exposed to the lectin before embedding and sectioning. In contrast, in a postembedding method, the lectins are applied to the ultrathin sections of tissues, that is, *after* the tissues have been embedded and sectioned.

Electron microscope histochemistry has fundamentally changed in recent years. Whereas previously the electron-dense marker of choice was horseradish peroxidase, colloidal gold particles are now commonly used either directly conjugated to lectins or used indirectly with biotinylated lectins in conjunction with the use of (strept)avidin gold complexes for visualization. This has the advantage that there is now no need for peroxidase reactions to take place *in situ*, thus avoiding the problem of nonspecific peroxidase activity. Partly as a consequence of this, the pre-embedding methods such as the ones used by Pavelka and Ellinger *(2)* and Sorobin and Hoyt *(15)* are less popular than the postembedding methods. The pre-embedding methods are furthermore restricted in their suitability mainly because of the difficulties of penetration of lectins and lectin-marker complexes into cells. This is because the molecular size of lectins is relatively large, especially when conjugated to a marker such as horseradish peroxidase (HRP), or colloidal gold particles *(16)*. For this reason, a permeabilization step may be incorporated. The pre-embedding methods may be utilized in either a direct or indirect method.

Postembedding methods have become the method of choice for electron microscopic demonstration of lectin binding sites, despite the fact that the tissue processing necessary for electron microscopy may occasionally destroy carbohydrate integrity and thus reactivity for lectins. With this in mind, it is thus obligatory to ascertain whether the lectin ligand in question requires a particular type of fixation by trying a range of different fixatives to determine which gives the optimum lectin reactivity. One advantage lectin histochemistry has over antibody localization methods is that often stronger glutaraldehyde fixation than used in immunohistochemistry may be employed in lectin histochemistry, which although it might denature protein epitopes, does not harm lectin-binding sites. Modern embedding media such as LR–White or

LR–Gold *(17)*, which are hydrophilic, are preferred to Araldite, which is hydrophobic, and do not have the same problems of penetration because of less extensive crosslinking as seen with epoxy resins. The use of hydrophilic resins such as Lowicryl K4M obviates the need for etching in postembedding methods *(18)*, though use of Araldite results in better structural preservation. Postembedding techniques are either one-, two-, or three-step methods. In other words, a direct method in which the lectin is directly bound to a marker such as colloidal gold; an indirect method in which there is a second stage binding to the lectin, for example biotinylated lectins being detected by (strept)avidin conjugated to HRP; or a three-step method in which, for example, the lectin is linked to digoxigenin, which is detected by an antidigoxigenin, which itself is detected by a second gold-conjugated antibody. The advantage of the latter methods is the amplification of the signal. The postembedding methods are also popular because a range of lectins can be applied to serial sections, and this is not possible with pre-embedding methods.

Of the electron-dense markers used for demonstration of lectin binding, such as HRP, ferritin, and colloidal gold-labeled lectins, the latter is becoming the marker of choice *(19)*. Often it is more convenient, however, to use a biotinylated lectin in conjunction with a (strept)avidin gold complex for visualization; e.g., *see* Zhou et al. *(20)*. Indeed, one-, two-, or three-step procedures have been applied successfully in a variety of situations *(21,22)*.

Our more recent work has focused on the development of an electron microscope method for the demonstration of lectin-binding sites in human breast cancer cell lines utilizing gold-conjugated HPA, previously demonstrated to be a good predictor of metastasis formation in breast cancer *(23)*. Since cell-to-cell and cell-to-matrix interactions play a crucial role in this process, we concentrated on lectin binding sites at the surface of cell membranes.

2. Materials

1. Fixation is carried out in 4% paraformaldehyde and 0.5% glutaraldehyde in $0.1M$ phosphate-buffered saline (PBS), pH 7.4. Making up the paraformaldehyde solution is a critical step. The paraformaldehyde solution needs to be heated to about 65–70°C using a heated magnetic stirrer, and is then dissolved by dropwise addition of $1N$ sodium hydroxide before making up to volume with the buffer to attain pH 7.4. Fixatives should preferably be used ice-cold. Although aldehyde fixation is preferred, numerous permutations of the combination of paraformaldehyde and glutaraldehyde have been used (Table 1), the optimal permutation must be determined by experimentation.
2. HPA gold particles can be obtained from Sigma, Poole, Dorset, UK (as indeed can many other lectin–gold conjugates).
3. Acetone and methanol, both made up in 50, 75, 95, and 100% v/v (in distilled water) concentrations.

Table 1
Permutations of Glutaraldehyde and Paraformaldehyde Fixatives Used in Electron Microscope Lectin Studies

Applications	Pf, %	Glut, %	Ref.
Con A, DSA, LFA, HPA, WGA binding sites in human bronchial glands	—	1	24
DSA, LFA, RCA-I binding sites in torpedo electric organ	—	1	25
Influence of SO$_4$ groups on PNA binding sites in rat intestine	—	1	19
Con A, LPA, PNA, RCA, WFA, WGA binding sites in nematode cuticles	—	1	26
DBA, PNA, SBA, UEA-I, WGA binding sites in human gastric mucosal cells	—	1	27
GSA-I binding sites on rat macrophages in embryonic CNS, liver and lung	—	1	28
AAA, Con A, DSA, GNA, HPA, LFA, LTA, MAA, PNA, RCA, UEA-I, WGA binding sites in principal cells of human gall bladder	—	1	22
AAA, Con A, DSA, LFA, MAA, PNA, RCA-I, SBA, WGA binding sites in rat zona pellucida	—	2	21
Con A, PNA, RCA-I, WGA binding sites in globoid cell leukodystrophy (Krabbe's disease)	—	2	12
GSA-II, MPA, SBA, UEA-I binding sites in Golgi apparatus of mouse Brunner's glands	—	2.5	29
HPA binding sites in papillae of rat tongue	—	2.5	30
DSA, LFA, RCA-I binding sites in electric organ of torpedo	3	—	25
BPA binding sites in human tonsil and peripheral blood	—	3	31
LCA binding sites in cat oocyte cortical granules	3.7	—	32
DSA, LFA, RCA-I binding sites in electric organ of torpedo	2	0.1	25
UEA-I binding sites in nuclei of root primordium in green peas *Pisum sativum*	2	2	33
LEA, MPA binding sites in rat lung comparing LR–Gold and LR–White resins	4	0.05	17
Con A, GSA-I, GSA-II, PHA-P, PNA, RCA-I, SBA, UEA-I, WGA binding sites in guinea pig prostate	4	0.25	18
Con A, GSA-I, GSA-II, PNA, RCA, SBA, UEA-I, WGA binding sites in developing mouse Leydig cells	2	2.5	20
Con A, RCA-I, SBA, WGA binding sites on different basement membranes of mice	4	0.5	34
RCA-I, WGA binding sites in rat secretory cells (intestine, parotid, pancreas)	4	0.5	35

Pf, paraformaldehyde; Glut, glutaraldehyde.

AAA, *Anguilla anguilla*; BPA, *Bauhinia purpurea*; Con A, *Canavalia ensiformis*; DSA, *Datura stramonium*; GSA-I, *Griffonia simplicifolia*; GSA-II, *Griffonia simplicifolia*; HPA, *Helix pomatia*; LEA, *Lycopersicon esculentum*; LFA, *Limax flavus*; LPA, *Limax polyphemus*; MAA, *Maackia amurensis*; MPA, *Maclura pomifera*; PHA-P, *Phaseolus vulgaris*; PHA-L, *Phaseolus vulgaris*; PNA, *Arachis hypogaea*; RCA-I, *Ricinus communis*; RCA-II, *Riciuns communis*; SBA, *Glycine max*; UEA(-I), *Ulex europeaus*; WGA, *Triticum vulgaris*.

4. Araldite, LR–White or LR–Gold resins. There are adequate manufacturer's instructions for making up these embedding media, and for the processing of the tissues prior to embedding.
5. 0.5% Osmium tetroxide and 0.5% potassium ferricyanide.
6. 1% Noble agar (Difco, E. Molesley, Surrey, UK).
7. Nickel grids for electron microscopy.
8. Normal goat serum, 0.1% bovine serum albumin, 0.1% gelatin, and 50 mM glycine to make the blocking buffer. This solution blocks potential free aldehyde groups, which may act as sites of nonspecific binding of the lectin.
9. A rocking table that ensures, as evenly as possible, mixing of the cells and the incubation solutions.
10. Phosphate-buffered saline (PBS): 0.1M phosphate buffer, pH 7.4, 0.15M sodium chloride.
11. 0.1% Saponin.
12. Human breast cancer cell lines (from the European Animal Cell Culture Collection or the American Type Culture Collection, Porton Down, Salisbury, UK), including MCF7, HBL100, T47D, BT549, and BT20.

3. Methods
3.1. Cell Maintenance

1. Maintain cell lines in media as recommended by the manufacturer in a humidified atmosphere in 5% carbon dioxide and 95% air at 37°C.
2. Change media every 3 d; once confluent, harvest cultures with a rubber policeman and pellet by gentle centrifugation.

3.2. Pre-Embedding Technique

1. Fix cell pellets by immersion in 4% paraformaldehyde and 0.5% glutaraldehyde in 0.1M PBS, pH 7.4, for 15 min (*see* Note 3).
2. After a brief wash in PBS, react some cell samples with 0.05% sodium borohydride to reduce background reactivity (*see* Note 9).
3. Wash pellets in PBS before treating with saponin (0.1%) for 30 min to permeabilize the cell membranes prior to incubation with the lectin (*see* Notes 6–8).
4. Incubate with lectin on a rocking table using HPA conjugated to gold particles (10-nm diameter) diluted 1:100 for 4 h at room temperature.
5. Optimal dilutions and incubation times are previously established by use of a chessboard titration. For this, incubate three dilutions of the lectin (1:10, 1:100, and 1:200) at each of three times (2, 4, and 6 h).
6. Wash pellets briefly in PBS.
7. Carry out postfixation in 1% glutaraldehyde in PBS for 15 min, followed by washes in PBS.
8. Fix the cell pellets further in 0.5% osmium tetroxide and 0.5% potassium ferricyanide before block staining in 1% uranyl acetate for 30 min.
9. After a brief wash in distilled water, embed the cell pellets in 1% agar (prepared by dissolving 1 g/100 mL PBS and heating until the agar is dissolved). After

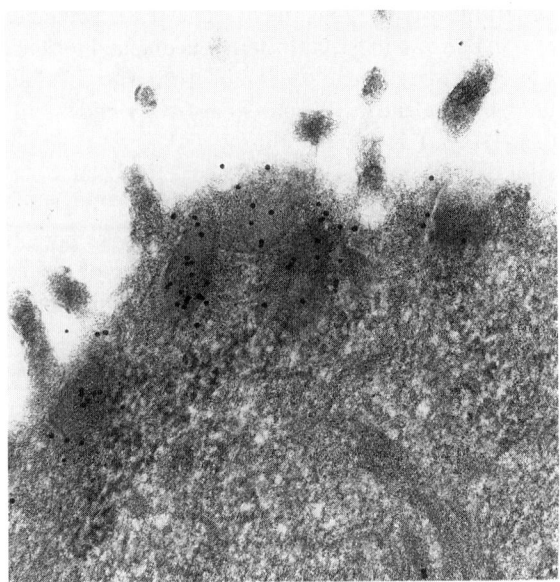

Fig. 1. An electron micrograph of MCF7 breast cancer cells reacted with HPA in a pre-embedding method. Note the reactivity over the submembranous particles characteristic of this cell line.

cooling to a temperature just above the setting point of the agar, the cell pellets become embedded in it, allow them to set and cut into small pieces for processing in Araldite.
10. Dehydrate pellets in 50% acetone for 15 min, 75% for 15 min, 95% for 20 min, and finally 100% acetone, three changes of 20 min duration.
11. Infiltrate the pellets in a 1:1 mixture of acetone and Araldite for 18 h, and then in neat Araldite for 4 h and then for 18 h before polymerizing in Araldite in gelatin capsules for 18 h at 60°C.
12. Use of this method enables visualization of lectin binding sites on the cell surface (Fig. 1). Note the HPA reactivity of the submembranous granules. This reactivity pattern was not observed after use of the postembedding technique, possibly because of the lipid nature of the granules, which may have been washed out by the solvents used in the processing of the tissues (*see also* Note 2).

3.3. Postembedding Technique

1. Fix cell pellets in the same way as for pre-embedding; embed in 1% agar and, for LR–White resins, dehydrate in 50% ethanol for 15 min, 75% ethanol for 15 min, 95% ethanol for 20 min, and finally three changes of 20 min duration in 100% ethanol (*see* Notes 4 and 10).
2. Infiltrate pellets in a 1:1 mixture of 100% acetone and LR–White resin for two 1-h changes, then a 2:1 mixture of LR–White resin and acetone for two

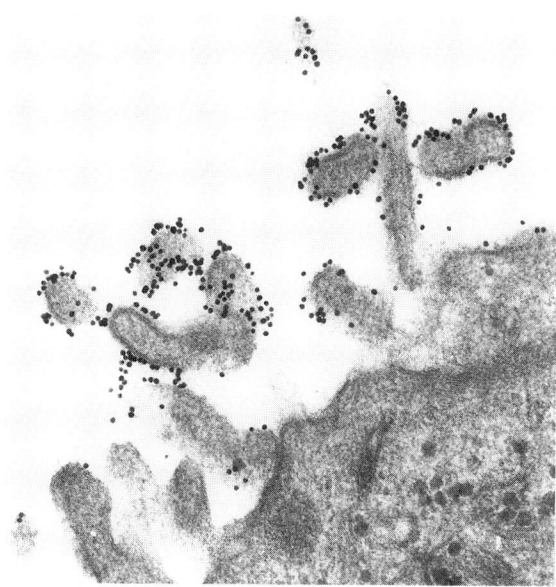

Fig. 2. An electron micrograph of MCF7 breast cancer cells reacted with HPA in a postembedding method. Notice the reactivity associated with cell surface projections.

changes of 1-h duration, and finally LR–White monomer for changes of 1- and 18-h duration, followed by two further changes of 1-h duration before polymerization in gelatin capsules in LR–White resin for 18 h at 60°C.
3. For LR–Gold, dehydrate pellets in 50% methanol for 15 min, then 75% methanol for 15 min at –20°C, then in 90% methanol at –20°C, and finally three changes of 20 min duration in 100% methanol at –20°C.
4. Infiltrate with LR–Gold resin at –20°C, first exposing pellets to the LR–Gold monomer for 2 h, then monomer for 18 h, then a 1:1 mixture of monomer and 0.5% benzoin methyl ether for 4 h, then for 18 h.
5. Allow polymerization to take place in gelatin capsules with a 1:1 mixture of fresh LR–Gold monomer and 0.5% benzoin methyl ether, and place the capsules 9 cm above UV light source for 24 h at –20°C. Place the gelatin capsules in the wells of a clear, plastic 96-multiwell assay plate on a stand above the light source (a 20-cm fluorescent strip, 6 W).
6. Cut ultrathin sections and mount on nickel grids *(24)*, treat with blocking buffer containing 99 mL PBS, 1 mL normal goat serum, 1 mg/mL bovine serum albumin, 0.1% gelatin, and 50 mM glycine (*see* Note 5).
7. Incubate sections on a rocking table with HPA-gold conjugate (10-nm diameter) diluted 1:100 in blocking buffer and as determined by previous experiments for 4 h.
8. Use of this method results in specific lectin labeling on the cell surface (Fig. 2).

4. Notes

1. In both pre- and postembedding methods, the HPA stage was omitted as a negative control. A sugar control was included whereby the lectin was preincubated with its appropriate ligand (1:50 lectin mixed in an equal volume of $0.6M$ sugar, which gives a final dilution of 1:100 lectin in $0.3M$ sugar) and then applied in the histochemical procedure. This is the equivalent to an absorption control in immunohistochemistry.
2. Use of pre-embedding methods may result in uneven distribution of binding because of the inequities of the penetration of the lectin into different cells.
3. It is often the case that higher concentrations of aldehyde fixation are permitted with lectin binding compared to antibody localization work. There is inevitably a compromise between achieving structural integrity and lectin binding intensity.
4. Successful use of the low temperature resin Lowicryl K4M has been reported *(17,21)*.
5. Ultrathin sections are normally mounted on Formvar-coated nickel grids, though in our experience it is not necessary to coat grids.
6. In pre-embedding methods, relatively thick sections can be cut on a Vibratome and reacted with lectins, after which they can be embedded. This has an advantage in that lectin binding can occur before any possible deleterious effects of dehydration or embedding have taken place.
7. Lectin penetration problems may be overcome by use of detergents or proteolytic enzymes to permeabilize cell membranes. Such procedures will, of course, damage the proteins that carry the carbohydrate residues and may be inappropriate depending on the expected location of the lectin ligand.
8. To facilitate entry of lectins, permeabilization steps may be included with Saponin and/or other detergents *(37)*.
9. Fixation with aldehydes may produce high background labeling; some authorities claim that treatment of sections with 0.05% sodium borohydride decreases background labeling in antibody localization methods *(38,39)* (in our experience, it increases lectin binding *[11]*). It has been suggested that this treatment reduces the number of Schiff bases that arise during the fixation process.
10. Some workers include treatment with $0.5M$ ammonium chloride in PBS for 1 h in postembedding methods to amidimate free aldehyde groups to reduce background reaction *(24)*.

References

1. Sharon, N. and Lis, H. (1972) Lectins: cell agglutinating and sugar specific molecules. *Science* **177,** 949–955.
2. Pavelka, M. and Ellinger, A. (1993) Early and late transformations occurring at organelles of the Golgi area under the influence of Brefeldin-A—an ultrastructural study and lectin cytochemical study. *J. Histochem. Cytochem.* **41,** 1031–1042.
3. Ellinger, A. and Pavelka, M. (1992) Subdomains of the rough endoplasmic-reticulum in colon goblet cells of the rat—lectin cytochemical characterization. *J. Histochem. Cytochem.* **40,** 919–930.

4. Taatjes, D. J., Barcomb, L. A.., Leslie, K. O., and Low, R. B. (1990) Lectin binding patterns to terminal sugars of rat lung alveolar epithelial sugars. *J. Histochem. Cytochem.* **38,** 233–244.
5. Shimizu, T., Nettesheim, P., Mahler, J. F., and Randell, S. H. (1991) Cell type-specific lectin staining of the tracheobronchial epithelium of the rat: Quantitative studies with *Griffonia simplicifolia* I isolectin B4. *J. Histochem. Cytochem.* **39,** 7–14.
6. Gheri, G., Gheri-Bryk, S., and Petrelli, V. (1990) Histochemical detection of sugar residues in the chick embryo mesonephros with lectin-horseradish peroxidase conjugates. *Histochemistry* **95,** 63–71.
7. Gordon, S. R. and Marchand, J. (1990) Lectin binding to injured corneal epithelium mimics patterns observed during development. *Histochemistry* **94,** 455–462.
8. Wang, J. J., Yin, C. S., and Tang, C. S. (1990) Lectin bindings and diethylstilbestrol effects on the recognition of Mullerian inhibiting substance (MIS) on chick Mullerian ducts by MIS-antiserum. *Histochemistry* **95,** 55–61.
9. Malmi, R., Fröjdmann, K., and Söderström, K. O. (1990) Differentiation-related changes in the distribution of glycoconjugates in rat testis. *J. Histochem. Cytochem.* **94,** 387–395.
10. Kalina, M. and Socher, R. (1990) Internalization of pulmonary surfactant into lamellar bodies of cultured rat pulmonary type II cells. *J. Histochem. Cytochem.* **38,** 483–492.
11. Mitchell, B. S., Vernon, K., and Schumacher, U. (1995) Ultrastructural localization of *Helix pomatia* agglutinin (HPA)-binding sites in human breast cancer cell lines and characterization of HPA-binding glycoproteins by Western blotting. *Ultrastruct. Pathol.* **19,** 51–59.
12. Figols, J., Zimmer, C., Warzok, R., and Cervos-Navarro, J. (1992) Immuno-lectin histochemistry and ultrastructure in two cases of globoid cell leukodystrophy (Krabbe's disease). *Clin. Neuropathol.* **11,** 312–317.
13. Leathem, A. (1986) Lectin histochemistry, in *Immunocytochemistry*, 2nd ed. (Polak, J. and Van Noorden, S., eds.), Wright, Bristol, UK, pp. 167–187.
14. Schumacher, U., A.dam, E., Brooks, S. A., and Leathem, A. J. (1995) Lectin-binding properties of human breast cancer cell lines and human milk with particular reference to *Helix pomatia* agglutinin. *J. Histochem. Cytochem.* **43,** 275–281.
15. Sorobin, S. P. and Hoyt, R. F. (1992) Macrophage development. I. Rationale for using *Griffonia simplicifolia* isolectin B4 as a marker for the line. *Anat. Rec.* **232,** 520–526.
16. Roth, J. (1983) Application of lectin-gold complexes for electron microscopic localization of glycoconjugates on thin sections. *J. Histochem. Cytochem.* **31,** 987–999.
17. Kasper, M. and Migheli, A. (1993) LR–Gold and LR–White embedding of lung tissue for immunoelectron microscopy. *Acta Histochemica.* **95,** 221–227.
18. Chau, L. and Wong, Y. C. (1992) Localization of prostatic glycoconjugates by the lectin-gold method. *Acta. Anatomica.* **143,** 27–40.
19. Martinéz-Menánguez, J. A.., Ballesta, J., Avilés, M., Madrid, J. F., and Castells, M. T. (1992) Influence of sulphate groups in the binding of peanut agglutinin.

Histochemical demonstration with light- and electron-microscopy. *Histochem. J.* **24**, 207–216.
20. Zhou, X. H., Kawakami, H., and Hirano, H. (1992) Changes in lectin binding patterns of Leydig cells during fetal and post-natal development. *Histochem. J.* **24**, 354–360.
21. Avilés, M., Martinéz-Menárguez, J. A., Castells, M. T., Madrid, J. F., and Ballesta, J. (1992) Cytochemical characterisation of oligosaccharide side chains of the glycoproteins of rat zona pellucida: an ultrastructural study. *Anat. Rec.* **239**, 137–149.
22. Madrid, J. F., Castells, M. T., Martinéz-Menárguez, J. A., Avilés, M., Hernández, F., and Ballesta, J. (1994) Sub-cellular characterisation of glycoproteins in the principal cells of human gall bladder. *Histochemistry* **101**, 195–204.
23. Brooks S. A. and Leathem, A. J. C. (1991) Prediction of lymph node involvement in breast cancer by detection of altered glycosylation in the primary tumour. *Lancet* **338**, 71–74.
24. Castells, M. T., Balesta, J., Madrid, J. F., Martinéz-Menárguez, J. A., and Avilés, M. (1992) Ultrastructural localization of glycoconjugates in human bronchial glands: the subcellular organisation of N- and O-linked oligosaccharide chains. *J. Histochem. Cytochem.* **40**, 265–274.
25. Egea, G. and Marsal, J. (1992) Carbohydrate patterns of the pure cholinergic synapse of torpedo electric organ: a cytochemical and immunocytochemical electron microscopic approach. *J. Histochem. Cytochem.* **40**, 513–521.
26. Peioxto, C. A. and De Souza, W. (1992) Cytochemical characterization of the cuticle of *Caenorhabditiselegans* (Nematoda, Rhabditoidea). *J. Submicrosc. Cytol. Pathol.* **24**, 425–435.
27. Ríos-Martin, J. J., Díaz-Cano, S. J., and Rivera-Huetro, F. (1993) Ultrastructural distribution of lectin-binding sites on gastric superficial mucus-secreting epithelial cells. *Histochemistry* **99**, 181–189.
28. Sorokin, S. P., Hoyt, R. F., Jr., Blunt, D. G., and McNelly, N. A. (1992) Macrophage development: II. Early ontogeny of macrophage populations in brain, liver and lungs of rat embryos as revealed by a lectin marker. *Anat. Rec.* **232**, 527–550.
29. Suzaki, E. and Kataoka, K. (1992) Lectin cytochemistry in the gastrointestinal tract with special reference to glycosylation in the Golgi apparatus of Brunner's gland cells. *J. Histochem. Cytochem.* **40**, 379–385.
30. Witt, M. and Miller, I. J., Jr. (1992) Comparative lectin histochemistry on taste buds in foliate, circumvallate and fungiform papillae of the rabbit tongue. *Histochemistry* **98**, 173–182.
31. Sarker, A. B., Akagi, T., Yoshino, T., Fujiwara, K., and Murakami, I. (1993) *Bauhinia purpurea* lectin in hyperplastic human tonsil and peripheral blood: immunohistochemical, immunoelectron microscopic, and flow cytometric analysis. *J. Histochem. Cytochem.* **41**, 811–817.
32. Byers, A. P., Barone, M. A., Donoghue, A. M., and Wildt, D. E. (1992) Mature domestic cat oocyte does not express a cortical granule-free domain. *Biol. Reprod.* **47**, 709–715.

33. Chamberland, H. and Lafontaine, J. G. (1992) Ultrastructural localization of L-fucose residues in nuclei of root primordia of the green pea *Pisum sativum*. *Histochem. J.* **24,** 1–8.
34. Salamat, M., Götz, W., Werner, J., and Herben, R. (1993) Ultrastructural localization of lectin binding sites in different basement membranes. *Histochem. J.* **25,** 464–468.
35. Tamaki, H. and Yamashina, S. (1994) Improved method for postembedding cytochemistry using reduced osmium and LR white resin. *J. Histochem. Cytochem.* **42,** 1285–1293.
36. Varndell, I. M. and Polak, J. M. (1984) Double immunostaining procedures and applications, in *Immunolabelling for Electron Microscopy* (Polak, J. M. and Varndell, I. M., eds.), Elsevier, Amsterdam, pp. 157–177.
37. Yamawaki, M., Zurbriggen, A., Richard, A., and Vandevelde, M. (1993) Saponin treatment for in situ hybridisation maintains good morphological preservation. *J. Histochem. Cytochem.* **41,** 105–109.
38. Weber, K., Rathke, P. C., and Osborn, M. (1978) Cytoplasmic microtubular images in glutaraldehyde-fixed tissue culture cells by electron microscopy. *Proc. Natl. Acad. Sci. USA* **75,** 1820–1824.
39. Willingham, M. C. (1983) A.n alternative fixation-processing method for cytoplasmic antigens. *J. Histochem. Cytochem.* **31,** 791–798.

II

USE OF LECTINS FOR STRUCTURAL ANALYSIS OF OLIGOSACCHARIDE CHAINS

11

Application of Sequential Smith Degradation to Lectin Blots

Chi Kong Ching

1. Introduction

Glycoconjugates on cell surfaces are known to have important physiological and biological functions. The oligosaccharide components of the cell surface glycoconjugates are recognized as important mediators of cellular interaction and interaction of cells with ligands such as hormones and growth factors (1,2). In normal tissues, such molecules (receptors) form part of the transmembrane structures linking extracellular signals with intracellular transducers of the received information (3). Disruption of these receptors results in loss of transmembrane communication and hence cellular dysfunction. Oligosaccharide domains of cellular glycoproteins and glycolipids are synthesized by a series of hierarchically organized glycosyltransferase enzymes within the Golgi apparatus. Structural modifications to oligosaccharide domains of these glycoconjugates occur during ontogeny and oncogenesis (4,5). It is increasingly recognized that these modified oligosaccharide domains (antigens) on cancer cells may represent the accumulation of precursor chains with absence of more complex structures because of decreased activity of synthesizing enzymes for terminal components of the carbohydrate chains. In addition, neosynthesis of new saccharide structures may occur because of enhanced synthesis of new components, often because of aberrant fucosylation or sialylation or a combination of both. It has been postulated that altered glycoconjugates interact in vivo with lectins associated with the membrane of corresponding cells, as is thought to be the case in the interaction between tumor cells and endothelium in the course of seeding of metastases (6). Increasing awareness of the major significance of cell surface glycoconjugates has led to a rapid expansion in techniques for their analysis.

From: *Methods in Molecular Medicine, Vol. 9: Lectin Methods and Protocols*
Edited by: J. M. Rhodes and J. D. Milton Humana Press Inc., Totowa, NJ

Altered cellular, structural, or secretory glycoconjugate oligosaccharide components of cancer cells can be detected by antitumor antibodies or lectins. Characterization of these altered structures is important for rationalizing the use of antitumor antibodies and/or lectins in tumor biology and clinical oncology. The composition of the purified oligosaccharides is most accurately achieved by a combination of gas chromatography and fast atom bombardment mass spectrometry. The sequence of monosaccharides, their linkage type and anomeric configuration (alpha or beta) is deduced by nuclear magnetic resonance scanning *(7)*. Alternatively, considerable information about oligosaccharide structure could be deduced more simply by a combination of lectin binding activities and sequential chemical degradation processes, a technique that was pioneered by Irimura and Nicholson *(8)*. This method was then shown to be applicable to electroblotted glycoproteins *(9)*, which have the advantage of being more durable to repeat chemical treatment and manual handling than polyacrylamide gels. A panel of enzyme-conjugated (peroxidase-tagged) lectins is selected for their differing carbohydrate epitope specificities and is used to evaluate carbohydrate expression by the electroblotted glycoprotein of interest (Fig. 1). Electroblotted glycoprotein is treated on the "replica" blots by mild acid hydrolysis, which cleaves off terminal sialic acid and fucose, and is then re-evaluated by the same panel of enzyme-conjugated lectins (Fig. 2). Further analysis of the subterminal carbohydrate structures is then achieved by one, two, or three Smith degradation cycles and lectin binding after each Smith degradation process (Figs. 3–5). The latter consists of sequential oxidation by periodate, reduction by borohydride, and mild acid hydrolysis with sulfuric acid. Smith degradation, which depends on the availability of vicinal diols for periodate oxidation, destroys all monosaccharides at the nonreducing terminal of side-chains and, in addition, will destroy any nonterminal monosubstituted saccharides unless they are substituted at the C3 position for hexopyranoses and C3 or C4 for 2-acetamido hexopyranoses *(10)*. Since the majority of glycoproteins have more than one oligosaccharide side-chain, this method gives an overall picture of the degree of variability of the side-chain structures rather than full sequential information.

2. Materials

1. Polyacrylamide gels (2–16% gradient gels used for mucin studies; *see* Note 1).
2. Sample buffer: $0.01M$ Tris-HCl, $0.001M$ EDTA, 2% SDS, and 5% 2-mercaptoethanol, pH 8.0.
3. 40% Glycerol, 0.0001% bromophenol blue.
4. Electrophoresis buffer: $0.09M$ Tris, $0.08M$ boric acid, $0.003M$ disodium ethylenediaminetetraacetic acid, 0.1% sodium dodecyl sulfate, pH 9.0.

Sequential Smith Degradation

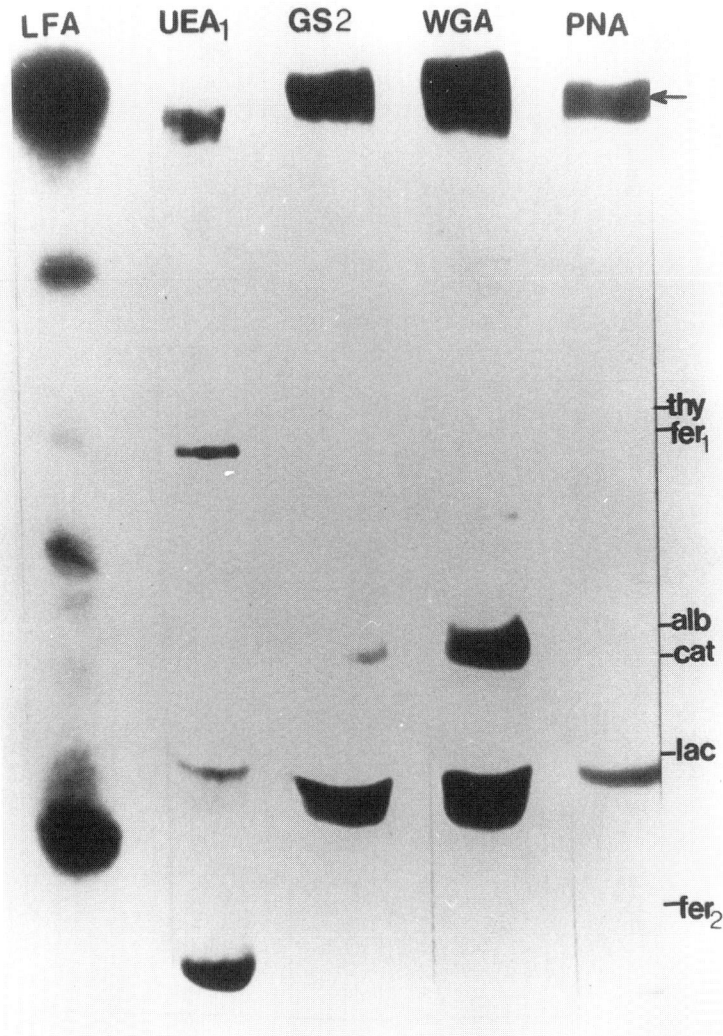

Fig. 1. Lectin blotting of five aliquots of a pancreatic cancer serum following SDS-PAGE (2–16%). the turnover-related mucin is arrowed. LFA, *Limax flavus*; UEA, *Ulex eurapeaus*; GS2, *Griffonia simplicfolia*; WGA, wheat-germ agglutinin; PNA, peanut agglutinin. Reproduced with permission from ref. 9.

5. Whatman No. 1 filter papers.
6. Nitrocellulose membrane (slightly larger than the gels).
7. Electrophoretic transfer buffer (Tris/glycine buffer): 1.25 mM Tris and 96 mM glycine, pH 8.3 (*see* Note 2).

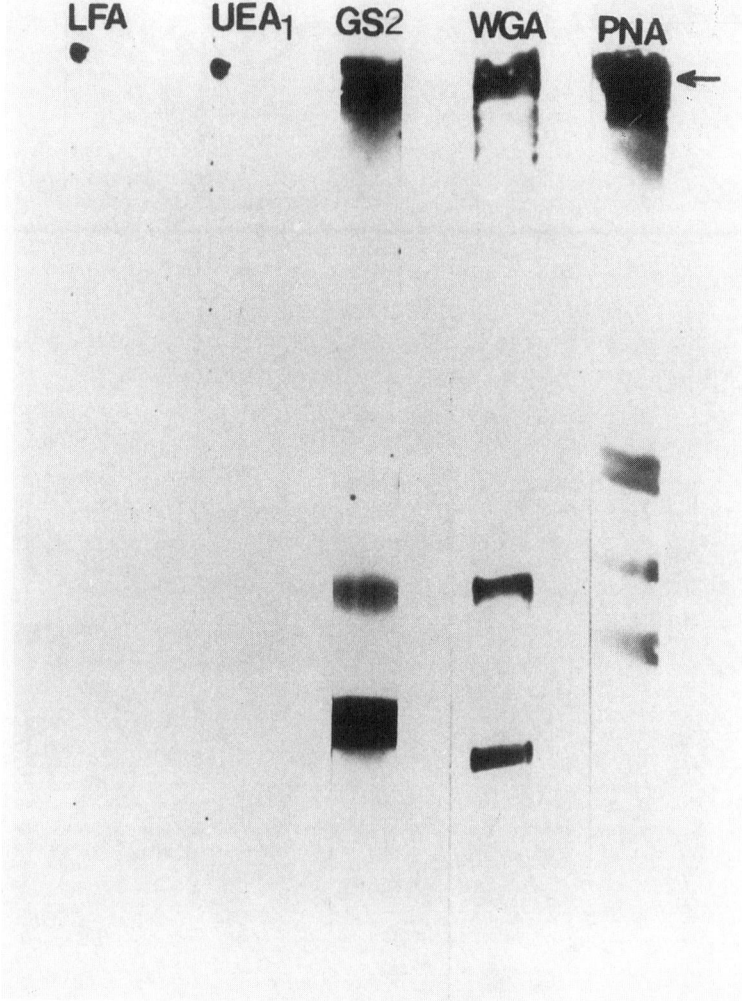

Fig. 2. Lectin blotting after mild acid hydrolysis of the blot showing the absence of LFA (sialic acid) and UEA[1] (fucose) bindings. Reproduced with permission from ref. 9.

8. 0.1% Tween-20 (0.1%) in 0.01M phosphate-buffered saline (PBS), pH 7.2, prepared fresh.
9. 0.2% Ponceau S in distilled water.
10. 50 and 25 mM Sulfuric acid.
11. 75 mM Sodium periodate in 50 mM sodium acetate, pH 4.0—made up fresh avoiding strong sunlight.
12. 0.1% 1,2-ethanediol in deionized, distilled water.
13. 0.1M sodium borohydride in 0.1M sodium borate buffer, pH 8.0.

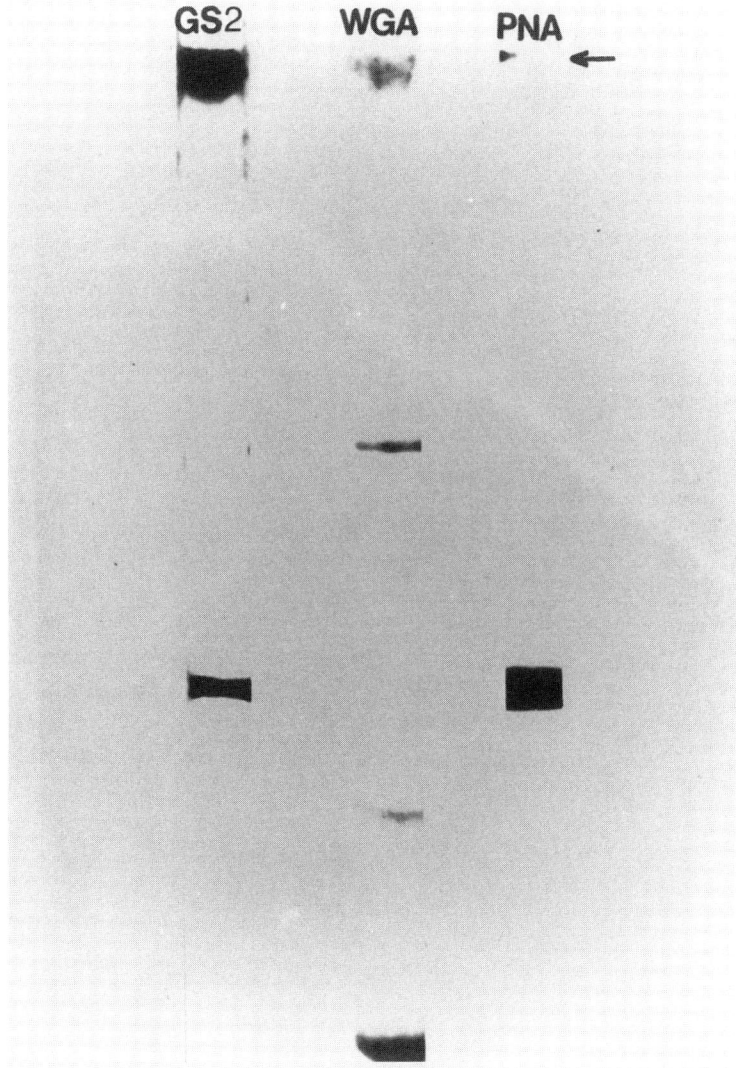

Fig. 3. Lectin blotting after mild acid hydrolysis followed by one Smith degradation showing disappearance of PNA binding and persistence of GS2 binding. Reproduced with permission from ref. 9.

14. For the lectin solutions, prepare PBS buffer (pH 7.2) containing 10 mM Ca^{2+}, Mg^{2+}, and Mn^{2+}. Separately prepare PBS buffer, pH 6.8, containing the same amount of Ca^{2+}, Mg^{2+}, and Mn^{2+} ions by titrating with 1N HCl. Prepare a stock solution (at 10X the final lectin concentration) for each lectin in the appropriate

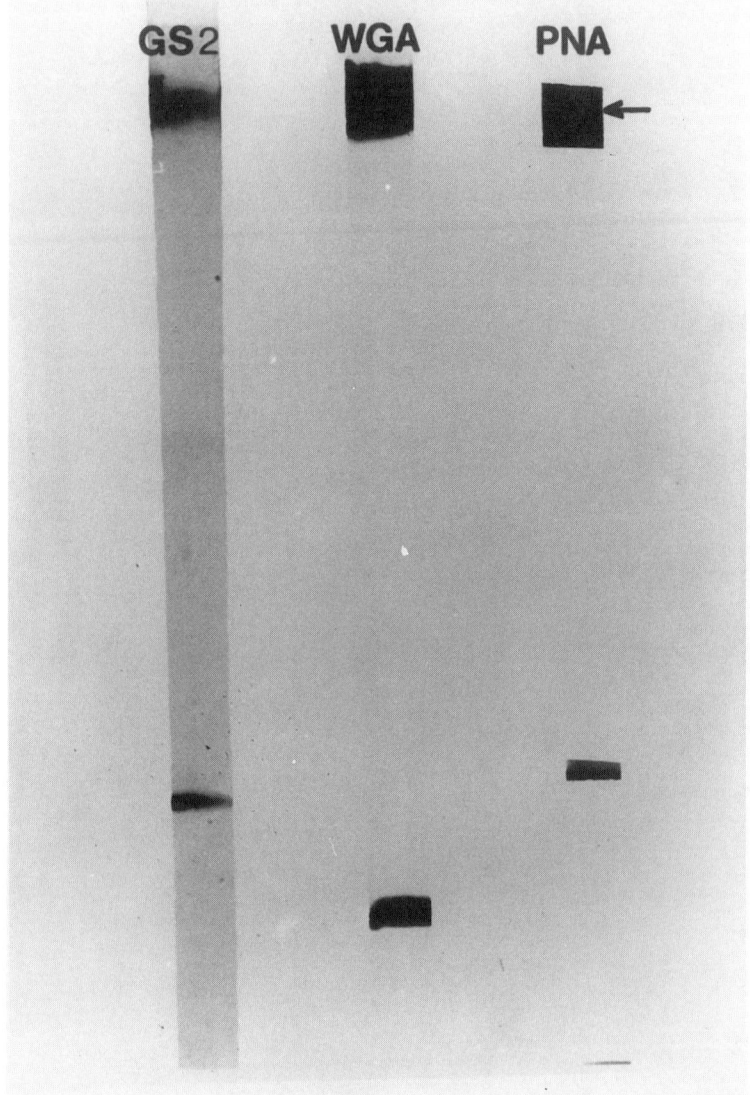

Fig. 4. Lectin blotting after mild acid hydrolysis and two Smith degradation cycles showing reappearance of PNA binding. Reproduced with permission from ref. *9*.

buffer (*peanut agglutinin* [PNA] and *concanavalin A* [ConA] in PBS, pH 6.8; UEA I, WGA, *Limax flavus* [LFA], and *Griffonia simplicifolia* [GS2] in PBS, pH 7.2, for example). Store the lectin solutions in aliquots at –20°C until used. Dilute 10 times in the appropriate buffer prior to use. Suggested lectin-peroxidase concentrations are: PNA 6.25 µg/mL, LFA 6.25 µg/mL, WGA 6.25 µg/mL, UEA I 25 µg/mL, GS2 25 µg/mL (*see* Note 3).

Sequential Smith Degradation

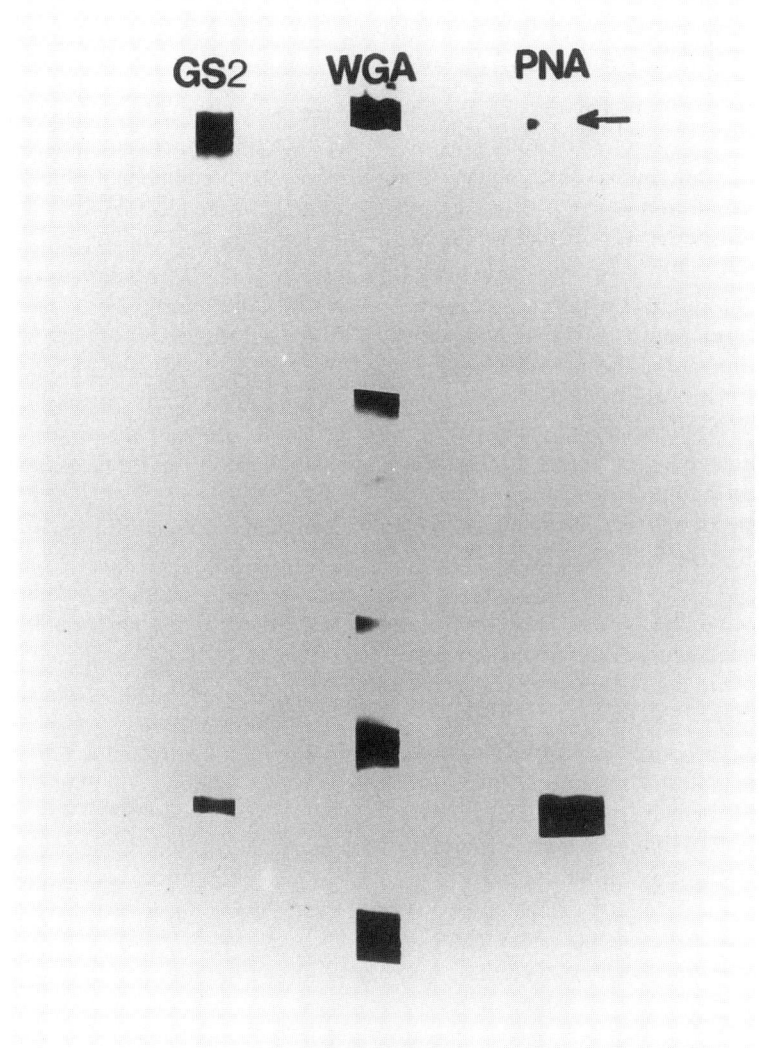

Fig. 5. Lectin blotting after mild acid hydrolysis and three Smith degradation cycles showing loss of PNA binding. Reproduced with permission from ref. *9*.

15. 4-Chloro-1-naphthol in hydrogen peroxide prepared fresh just before use by mixing 1 part 3 mg/mL 4 chloro-1-naphthol in methanol with 4 parts H_2O_2 in PBS (50 μL 30% H_2O_2 in 40 mL PBS, pH 7.2). **Caution:** 4-Chloro-1-naphthol is a carcinogen and should be handled with care.
16. Methanol.

3. Methods
3.1. Sodium Dodecyl Sulfate Polyacrylamide Gel Electrophoresis (SDS-PAGE) and Western Blotting

1. Mix glycoprotein under investigation with the sample buffer and heat at 100°C in a water bath for 5–10 min to dissociate the glycoprotein into subunits.
2. Add glycerol to a final concentration of 40% and bromophenol blue to a final concentration of 0.0001% to the treated solution before electrophoresis.
3. Equilibrate the polyacrylamide gel in the electrophoresis buffer for 15 min at 125 V followed by sample loading (into individual lane or one broad lane) and a 5-min run in at 300 V.
4. Electrophoresis is then conducted at 150 V for a further 60–180 min depending on the size of the molecule studied (*see* Note 4). Keep the electrophoresis buffer constantly below 10°C by using a supercooling coil containing ice-chilled water (*see* Note 5). Enough lanes will have to be run to allow a panel of at least five lectins to be tested after each of four degradation steps, i.e., 20 lanes, not allowing for duplicates.
5. Equilibrate the electrophoresed gel in electrophoretic transfer buffer for 30–60 min at room temperature.
6. Assemble the equilibrated gel next to the presoaked nitrocellulose membrane and then sandwich these between Whatman No. 1 filter papers. This should be performed while everything is immersed in the transfer buffer. Care should be taken to avoid poor contact or air bubbles being trapped between the gel and the nitrocellulose membrane, which will result in inefficient transfer.
7. Carry out high intensity transfer at 100 V for 2 h in a transblot cell with the temperature constantly kept below 10°C by a supercooling coil containing ice-chilled water (*see* Note 5).
8. After high intensity transfer, carefully separate the nitrocellulose membrane (blot) from the polyacrylamide gel.
9. Quench the blot by incubation in PBS/Tween-20 for 1 h with two changes of buffer at room temperature.
10. Reveal the position of the glycoprotein bands on the nitrocellulose blot using Ponceau S as a temporary stain.
11. Cut replica lanes from the blot, creating a series of "mini blots" (or 0.25-cm wide strips if one broad lane has been run), and mark the site of the glycoprotein under investigation on each blot with a pencil. Remove the Ponceau S stain by repeated washings in PBS/Tween-20 until clear.

3.2. Mild Acid Hydrolysis, Smith Degradation, and Lectin Blotting

1. Incubate the miniblots in 50 mM H_2SO_4 at 80°C for 3 h (*see* Notes 6 and 7).
2. Wash (5 × 3 min) with PBS/Tween-20 buffer.
3. Store at a sufficient number of miniblots for analysis by the complete range of lectins at –20°C. The remaining miniblots are used for Smith degradation (*see* Note 8).

4. Incubate the miniblots with sodium periodate/sodium acetate in the dark at 4°C for 48 h (see Note 6).
5. Discard the periodate solution, and wash the blots with PBS/Tween-20 buffer (3 × 3 min).
6. Incubate the blots in 0.1% ethanediol solution at room temperature for 1 h to stop the oxidation process.
7. Repeat step 2.
8. Incubate the blots in $0.1 M$ sodium borohydride in $0.1 M$ sodium borate buffer, pH 8.0, at room temperature for 4 h.
9. Repeat the washing step.
10. Add 25 mM H_2SO_4 to the miniblots and incubate at 80°C for 1 h.
11. Repeat the washing step, which completes the first Smith degradation cycle. Further cycles of Smith degradation can be carried out on the same blots by repeating steps 4–10.
12. Quench blots by incubation in PBS/Tween-20 for 1 h at room temperature.
13. Incubate the blots in the appropriate lectin solutions at 4°C for 16 h on a rotating platform (see Notes 3 and 9).
14. Discard the lectin solutions, and wash the blots three times with PBS/Tween-buffer to remove unbound lectin.
15. Identify bound peroxidase-tagged lectin by incubation for 5–10 min until satisfactory color density is achieved in 4-chloro-1-naphthol and H_2O_2 solution (see Note 10).

4. Notes

1. 2–16% polyacrylamide gradient gels were used in our experiments because mucins, the molecule of our research interest, have high molecular weight and only enter the gel at low polyacrylamide concentration. The 2% end of the gel is rather soft and becomes adherent to the nitrocellulose membrane after high intensity transfer. The adherent gel can be carefully rubbed off with a presoaked tissue while immersed in the Tris/glycine buffer. Every attempt should be made to remove the gel completely; otherwise, it will cause unwanted artifacts. For lower molecular weight glycoproteins, gradient gels of 5–10% or homogeneous gels of appropriate percentage acrylamide could be used.
2. Fresh buffer is required for every high-intensity transfer procedure. The use of old buffer results in rapid temperature rise and swelling of the gel, which will result in distortion of the glycoprotein bands. The addition of SDS (0.01%) to the transfer buffer has been claimed to enhance transfer, but it causes excessive foaming during the high intensity transfer and should be avoided. Methanol is commonly included in transfer buffers for Western blotting, but we have avoided it because it reduces the pore size of the gels and reduces the speed and efficiency of the electrophoretic transfer.
3. Lectin-peroxidase conjugates are available from numerous commercial sources. The signal of each product may be different depending on the conjugation process. Therefore, titration studies are advisable to determine the optimum lectin-

enzyme conjugate concentrations. If radiolabeled lectin is used together with autoradiography as the detection device, a much lower concentration of lectin is required. Alternatively, the two steps, unconjugated lectin and then followed by enzyme- or heavy metal-conjugated antilectin antibody, technique can be used.
4. Conventionally, the electrophoresis is stopped about 30 min after the tracking dye (bromophenol blue) has been electrophoresed out of the bottom of the gel. If the size of the molecule is small, it is advisable to stop at this point; otherwise, the molecule of study may be lost. For analyzing molecules of larger size that are present among other smaller molecules, a slightly prolonged electrophoresis may help to clear the gel of these unwanted lower molecular weight molecules.
5. Keeping the electrophoresis buffer and transfer buffer at temperature below 10°C will prevent gel swelling with heat which will cause distortion of the transferred glycoprotein bands.
6. The duration of mild acid hydrolysis and each Smith degradation step may be shortened for some glycoproteins.
7. Exo-glycosidases, e.g., α-fucosidase and neuraminidase, can be used instead of mild acid hydrolysis to remove fucose and sialic acid.
8. Miniblots should be stored between hard filter papers to prevent damage. Thaw stored miniblots at room temperature prior to study and reimmerse them in PBS/Tween-20 buffer for 30 min before further study.
9. The lowest possible concentration of each lectin solution should be used to avoid nonspecific binding.
10. Developed blots can be stored between sheets of hard filter paper at −20°C for future photography. The blots should be thawed slowly, reimmersed in deionized, distilled water, and kept at 4°C prior to photography.

Acknowledgment

I thank Jonathan M. Rhodes for his most helpful supervision and invaluable advice during this piece of research work.

References

1. Carpenter, G. (1987) Receptors for epidermal growth factor and other polypeptide mitogens. *Ann. Rev. Biochem.* **56,** 881–914.
2. Rauvala, H., Carter, W. G., and Hakomori, S. I. (1981) Studies on cell adhesion and recognition. I. Extent and specificity of cell adhesion triggered by carbohydrate-reactive proteins (glycosidases and lectins) and by fibronectin. *J. Cell Biol.* **88,** 127–137.
3. Norris, W. E. (1989) Evidence for a second class of membrane glycoprotein involved in cell adhesion. *J. Cell. Sci.* **93,** 631–640.
4. Johnson, L. V. and Calarco, P. G. (1981) Mammalian preimplantation development. The cell surface. *Anat. Rec. 1981* **196,** 201–219.
5. Warren, L., Buck, C. A., and Tuszynski, G. P. (1978) Glycopeptide changes and malignant transformation. A possible role for carbohydrate in malignant behaviour. *Biochim. Biophys. Acta.* **516,** 97–127.

6. Dennis, J. W. and Laferte, S. (1987) Tumour cell surface carbohydrates and the metastatic phenotype. *Cancer Metastasis Rev.* **5,** 185–204.
7. Hounsell, E. F. (1987) Structural and conformational characterization of carbohydrate differentiation antigens. *Chem. Soc. Rev.* **16,** 161–185.
8. Irimura, T. and Nicolson, G. L. (1983) Carbohydrate chain analysis by lectin-binding to mixtures of glycoproteins, separated by polyacrylamide slab-gel electrophoresis, with *in situ* chemical modifications. *Carbohydrate Res.* **115,** 209–220.
9. Ching, C. K. and Rhodes, J. M. (1990) Purification and characterization of a peanut-agglutinin-binding pancreatic cancer-related scrum mucus glycoprotein. *Int. J. Cancer* **45,** 1022–1027.
10. Beeley, J. G. (1985) Structural analysis, in *Glycoprotein and Proteoglycan Techniques* (Burdon, R. H. and van Knippenberg, R. H., eds.), Elsevier, Amsterdam, p. 285.

12

Blot Analysis with Lectins for the Evaluation of Glycoproteins in Cultured Cells and Tissues

Christian Zuber, Wei-Ping Li, and Jürgen Roth

1. Introduction

The carbohydrate-binding properties of plant and animal lectins have found many diverse applications in biomedical research that are comprehensively treated in the various chapters of this book. In our studies, lectins are often applied for the *in situ* detection of their respective binding sites in specific cells either grown in culture or present in various tissues (*see* Chapter 3). In order to complement these *in situ* localization studies, blots are analyzed using lectins to identify glycosylation patterns of glycoproteins that contain accessible lectin-binding sites in their oligosaccharide side-chains. The isolation and characterization of lectins which exhibit a narrow spectrum of reactivity with oligosaccharide structures has provided new possibilities for such kind of analyses. In comparison to Concanavalin A, which recognizes glucose/mannose residues in terminal nonreducing position as well as internal linkages *(1)*, lectins such as the *Bowringia milbraedii* lectin *(2)*, the *Narcissus pseudonarcissus* lectin *(3)*, and *Galanthus nivalis* lectin *(4)* react with well defined processing intermediates of asparagine-linked oligosaccharides. Another example for such a high degree of specificity is provided by sialic acid-specific lectins such as the *Sambucus nigra* lectin I *(5)*, the *Maackia amurensis* lectin *(6,7)*, and the *Amaranthus caudatus* lectin *(8–10)*. These lectins, in contrast to the general sialic acid binding *Limulus polyphemus (11)* and *Limax flavus (12,13)* lectins, discriminate terminal sialic acids present in different ketosidic linkages. Thus, the narrow specificity of these lectins is based on their binding to defined oligosaccharidic sequences in which the sugar moieties are present in specific linkages.

1.1. Blot Analysis by Lectins

The same principle as used in Western (immuno-) blotting is applied for the analysis of glycoproteins by lectins. For this purpose, proteins from tissue and cell homogenates are resolved by conventional SDS-PAGE which is followed by their transfer and immobilization onto suitable membranes. Variations to this basic protocol need to be established depending on the type of material (cell cultures and isolated cells vs different kinds of tissues and organs), the nature of the glycoproteins under study (membrane proteins, mucins, and so on) and the type of sugar moieties (sialic acids vs hexoses and hexosamines).

For the visualization of lectin binding to the immobilized glycoproteins, lectins have been tagged with various markers such as enzymes, ^{125}I, fluorochromes, and biotin. The introduction of the steroid hapten digoxigenin (DIG) as a marker for lectins has provided a number of advantages *(14,15)*. All the different lectin-DIG conjugates can be visualized using a single alkaline phosphatase-conjugated anti-DIG antibody. DIG occurs naturally only in plants of the digitalis family and no crossreactive material has been reported to exist in mammalian tissues *(14–16)*. Furthermore, and in contrast to histochemical applications, endogenous alkaline phosphatase activity poses no problems when blots are analyzed. In a comparative study, we found that the DIG system as compared to biotinylated lectin visualized with alkaline phosphatase-conjugated streptavidin was superior with respect to specificity *(17)*.

1.2. Specificity Controls and Structural Analysis

Glycoproteins in homogenates or immobilized onto membranes can be subjected to various kind of treatments to obtain information about both the specificity of the lectin interaction and the nature of the sugar moieties detected. A classical type of control for specificity of lectin binding involves the incubation of a lectin with its appropriate binding substrate. Since the latter may not be available, mono- or oligosaccharides shown to be most potent inhibitors in lectin-mediated hemagglutination and glycoprotein precipitation assays are usually applied. When used in the millimolar range, they will result in abolition of lectin binding *(10,16)*. Less suitable inhibitory sugars, for example *N*-acetylglucosamine in conjunction with wheat germ agglutinin, which reacts best with oligo/poly-*N*-acetylglucosamine, need to be applied in the molar range. However, this type of control actually excludes only nonsugar related lectin–protein interactions which may sometimes occur. However, removal of the carbohydrate structure as defined by the specificity of the lectin by the appropriate exo- and endoglycosidase provides direct evidence for the specificity of lectin labeling. Such experiments demonstrate not only specificity of

Blot Analysis with Lectins

lectin–carbohydrate interaction, but at the same time provide structural information. The sequential use of a variety of exoglycosidases or sequences of Smith degradations will supply more detailed structural information. The distinction between lectin binding sites being present either on asparagine, N-glycosidically-linked oligosaccharides, or serine/threonine, O-glycosidically-linked oligosaccharides can be made by the use of the endoglycosidase PNGase F (endo-F) and the ß-elimination reaction, respectively.

2. Materials
2.1. Blotting

1. Polyacrylamide gels (3–10 or 3–15% gradient gels).
2. Regular filter papers (slightly smaller than the gels).
3. Nitrocellulose and PVDF membranes (same size as filter papers).
4. Semidry blotting equipment (*see* Note 1).
5. Ponceau S (working solutions prepared according to instructions of supplier, i.e., Sigma, St. Louis, MO).
6. PBS: 10 mM Na-phosphate buffer, 150 mM NaCl, pH 7.4.
7. TBS: 100 mM Tris-HCl, pH 7.5, 150 mM NaCl.
8. TPBS: PBS containing 0.05% Tween-20, pH 7.4.
9. TTBS: TBS containing 0.05% Tween-20.
10. MTTBS: TTBS containing 1% (v/w) defatted milk powder.
11. MTPBS: TPBS containing 1% (v/w) defatted milk powder.
12. Blocking reagent from Boehringer Mannheim (Mannheim, Germany) for DIG system.
13. Homogenization buffer: PBS containing 1% Triton X-114 and protease inhibitors (*see* Note 2).
14. Transfer buffer: 0.02M Tris-HCl, 0.15M glycine, pH 8.5, and methanol (0–40%) (*see* Note 3).
15. The various lectin–DIG conjugates and alkaline phosphatase-conjugated sheep anti-DIG Fab were obtained from Boehringer Mannheim. Working dilutions of lectins were made up with TPBS (*see* Note 4) and of the anti-DIG antibody with MTTBS. According to the manufacturer's instructions, stock solutions are stored at 4°C, and working dilutions were always freshly prepared before use. Since most lectins are mitogenic and some quite toxic, due care should be exercised.
16. NBT stock solution: 75 mg nitroblue tetrazolium dissolved in 700 µL dimethyl formamide (anhydrous) and mixed with 300 µL distilled water. Stored at –20°C. NBT is toxic.
17. BCIP stock solution: 50 mg BCIP (bromo-chloro-indoyl-phosphate) dissolved in 1 mL dimethyl formamide. Stored at –20°C.
18. Color reaction solution: Add 22.5 µL of NBT stock solution and 17.5 µL of BCIP stock solution to 10 mL of TBS, pH 9.4. Use always freshly prepared solution. Keep protected from light.

2.2. Controls

1. Various mono- and oligosaccharides, neoglycoproteins, and purified glycoproteins can be obtained from different commercial sources (Boehringer Mannheim; Chembiomed, Alberta, Canada; Glycosystems, Oxford; Sigma).
2. Various exo- and endoglycosidases can be obtained from different commercial sources (Boehringer Mannheim, Calbiochem, Sigma). For preparation of working solutions and conditions of use and storage, consult product information sheets of the respective suppliers.
3. ß-elimination solution: $0.1N$ NaOH.
4. Solution for acidic methanolysis: $0.1N$ HCl in methanol.

3. Methods

3.1. Blot Analysis

1. Prepare homogenates of the samples (*see* Notes 5 and 6).
2. Determine protein concentration of samples by Bradford assay or equivalent.
3. Apply samples to the gradient gels and run gels. Stop the gel when the methylene blue front reached the bottom of the gel or run gradient gel to equilibrium when studying high molecular mass glycoproteins.
4. Wash nitrocellulose membranes in distilled water and PVDF membranes in 100% methanol followed by two short rinses with distilled water and keep membranes wet. Soak filter paper and membranes in transfer buffer for 30 min. Use them to form a sandwich with the gel (filter papers, membrane, gel, filter papers in transfer buffer).
5. Electroblot the proteins to the membrane using the semidry blotting equipment (*see* Note 1) at a constant current not exceeding 3 mA/cm^2 gel (graphite electrodes) for one to several hours (*see* Note 7).
6. Separate membrane from gel with a flush of distilled water to prevent any sticking of the upper part (low percentage) of the gel to the membrane.
7. Wash membrane in distilled water.
8. Stain membrane with Ponceau S for 1 min, mark the position of the lanes and the molecular mass standards with a pencil (or make a hard copy), place membranes in distilled water (containing approx 5% of transfer buffer) for destaining and transfer afterwards into PBS.
9. Block membranes in blocking solution (*see* Note 8) for 1 h at ambient temperature.
10. Incubate with lectin solution for 1 h at ambient temperature.
11. Wash twice in TPBS for 10 min each and once in TTBS for 5 min.
12. Incubate with alkaline phosphatase-conjugated anti-DIG antibody (5000-fold diluted in MTTBS) for 1 h at ambient temperature or 30 min at 37°C.
13. Wash five times in TTBS for 10 min each at ambient temperature, one time in TBS for 5 min followed by two rinses in TBS, pH 9.4, for 2 min.
14. Incubate membranes in color reaction solution protected from light.
15. Stop color reaction by rinsing with water and air-dry.

3.2. Specificity Controls and Structural Analysis

1. Incubate lectin solution with the inhibitory and noninhibitory mono- and oligosaccharides, neoglycoproteins, or purified glycoproteins for 30 min at ambient temperature and use in parallel with standard lectin solution in step 10 of Section 3.1. Correct pH when acidic reagents (sialic acids) are used. *See* Table 1 in Chapter 3 for working concentrations of inhibitory substances.
2. Incubate glycoproteins immobilized on membranes after blocking (step 9 of Section 3.1.) with the required exo- and endoglycosidase (*see* Tables 1 and 2 in Chapter 3). Incubation conditions according to suppliers' instructions. Run control with enzyme-free buffer under identical conditions.

   Homogenate samples may be subjected to enzyme digestion prior to SDS-PAGE (*see* Note 9).
3. ß-elimination reaction: Incubate glycoproteins immobilized on PVDF membranes before blocking (step 9 of Section 3.1.) in $0.1N$ NaOH at 37°C for 30 min to 1 h (*see* Note 10). Wash membranes 5 times for 2 min each in PBS to neutralize alkaline. ß-elimination of homogenate prior to SDS-PAGE is not recommended (*see* Note 10).
4. Acid methanolysis of sialic acids is performed in $0.1N$ HCl in methanol at 37°C for 1 h up to overnight, or at 65°C for 10 min to 2 h (*see* Note 11).

4. Notes

1. Semidry transfer can be performed at ambient temperature making the use of a cooling system unnecessary.
2. We use routinely 1 m*M* PMSF (or AEBSF) and 1% aprotinin (Sigma). To evaluate which is the appropriate detergent, the sample is homogenized in PBS containing protease inhibitors and then either boiled in SDS sample buffer or membrane proteins are extracted by using various detergents (Triton X-100, Triton X-114, NP-40, deoxycholate, and so on), centrifuged, and the supernatant boiled in SDS sample buffer.
3. The methanol content of the transfer buffer is varied according to the molecular mass of the studied glycoproteins, i.e., >200 kDa, ≤20% methanol; <200 kDa, ≥20% methanol.
4. In our experience, the addition of trace amounts of particular cations such as Mn^{2+}, Co^{2+}, Mg^{2+} to the dilution buffer is not required.
5. Cell cultures: Harvest cells and keep on ice during the entire homogenization procedure. Centrifuge at $100g$, wash the pelleted cells with PBS, centrifuge again, dissolve the resulting cell pellet in homogenization buffer (4 vol buffer/1 vol of cell pellet) for 30 min, remove nuclei, and cell debris by centrifugation at $100–400g$ for 10 min and use for SDS-PAGE.

   Tissues: Use freshly obtained samples or samples which were shock-frozen in liquid nitrogen and stored at –70°C. All steps are performed on ice. Soft tissues (for example normal liver, kidney, brain, and so on) are directly homogenized in homogenization buffer (5–10 mL/g tissue) using a Dounce homogenizer. Other tissues such as skeletal muscle, cartilage, or pathological materials rich in connective tissue are homogenized with an appropriate mixer in homogenization buffer not containing detergent. Afterwards detergent is added to the homogenate. Homogenates of the different tissues in detergent-containing buffer are now kept for 30 min on ice. Remove nuclei and tissue debris by centrifugation at 100–400g for 10 min and use for SDS-PAGE.

6. Shock freeze homogenate supernatant in liquid nitrogen and store at –70°C.
7. Transfer time depends on the nature of the glycoproteins under study and the type of gel used. Gradient gels provide the advantage that high molecular mass glycoproteins present in the low percentage region of the gels are eluted equally efficiently as the low-mol-wt glycoproteins in the higher percentage regions of the gels. In homogenous gels, elution efficiency for high and low molecular mass glycoproteins will differ significantly.
8. In addition to commercially available blocking solution, others such as MTPBS, TPBS, or TPBS containing up to 10% bovine serum albumin can be used. Exclude reactivity of the lectin with blocking compounds. At this point, it should be mentioned that depending on the lectin applied, the nature and composition of the diluting buffer may be of importance with regard to background. Thus, for each lectin, the use of either phosphate or Tris buffer, its appropriate pH value (usually pH 7.4) and the specific additives such as Tween-20 alone or in combination with Triton X-100, defatted milk, and bovine serum albumin has to be determined empirically.
9. Enzyme digestion prior to SDS-PAGE may give doubtful results owing to proteolytic degradation, which may occur during the incubation even in the presence of protease inhibitors, and owing to the presence of endogenous glycosidases. Further, some endoglycosidases such as PNGase F require denatured substrates for full deglycosylation to occur. Although acetate buffer is usually recommended for work with sialidases, this will result in partial protein precipitation when the enzyme treatment is performed in solution. This can be overcome by the use of citrate buffer of the same molarity and pH.
10. Longer incubation times are not suitable since protein loss from membrane may occur under alkaline conditions. To monitor the protein amount which remains immobilized in the membrane, an immunoblot with antibody directed against the protein backbone is recommended. Following alkaline hydrolysis at temperatures higher than 37°C, more stringent blocking conditions (higher concentrations of blocking compounds, longer blocking time) may be required to prevent nonspecific protein–protein interactions. Note, concentrated NaOH (>1N) solutions should not be stored in glass bottles.

    Denaturation and precipitation of proteins and detergent may occur at alkaline pH applied for the ß-elimination reaction. However, if ß-elimination needs to be performed prior to SDS-PAGE, the pH in the reaction mixture must be neutralized afterwards since the SDS of the sample buffer will precipitate at alkaline pH.
11. Methanol has a boiling point at 65°C. To prevent evaporation and incidents, the incubation has to be performed in sealed plastic bags. As a control experiment, membranes should be incubated under the same conditions in methanol only.

References

1. Goldstein, I. and Poretz, R. (1986) Isolation, physicochemical characterization, and carbohydrate-binding specificity of lectins, in *The Lectins. Properties, Functions and Applications in Biology and Medicine*, vol. 35 (Liener, I., Sharon, N., and Goldstein, I., eds.), Academic, Orlando, FL.
2. Animashaun, T. and Hughes, R. (1989) *Bowringia milbraedii* agglutinin. Specificity of binding to early processing intermediates of asparagine-linked oligosaccharide and use as a marker of endoplasmic reticulum glycoproteins. *J. Biol. Chem.* **264**, 4657–4663.
3. Schulte, B. A. and Spicer, S. S. (1983) Light microscopic histochemical detection of terminal galactose and N-acetylgalactosamine residues in rodent complex carbohydrates using a galactose oxidase-Schiff sequence and the peanut lectin-horseradish peroxidase conjugate. *J. Histochem. Cytochem.* **31**, 19–24.
4. Shibuya, N., Goldstein, I., E. J. M., v. Damme, and Peumans, W. (1988) Binding properties of a mannose-specific lectin from snowdrop (*Galanthus nivalis*) bulb. *J. Biol. Chem.* **263**, 728–734.
5. Shibuya, N., Goldstein, I., Broekaert, W., Nsimba-Lubaki, M., Peeters, B., and Peumans, W. (1987) The elderberry (*Sambucus nigra L.*) bark lectin recognizes the Neu5Ac(a2,6)Gal/GalNAc sequence. *J. Biol. Chem.* **262**, 1596–1601.
6. Wang, W. and Cummings, R. (1988) The immobilized leukoagglutinin from the seeds of *Maackia amurensis* binds with high affinity to complex-type asn-linked oligosaccharides containing terminal sialic acid α2,3 to penultimate galactose residues. *J. Biol. Chem.* **263**, 4576–4585.
7. Knibbs, R. N., Goldstein, I. J., Ratcliffe, R. M., and Shibuya, N. (1991) Characterization of the carbohydrate binding specificity of the leukoagglutinating lectin from Maackia amurensis: comparison with other sialic acid-specific lectins. *J. Biol. Chem.* **266**, 83–88.
8. Rinderle, S., Goldstein, I., and Remsken, E. (1990) Physicochemical properties of amaranthin, the lectin from *Amaranthus caudatus* seeds. *Biochemistry* **29**, 10,555–10,561.
9. Rinderle, S., Goldstein, I., Matta, K., and Ratcliffe, R. (1989) Isolation and characterization of amaranthin, a lectin present in the seeds of *Amaranthus caudatus*, which recognizes the T-(or cryptic T) antigen. *J. Biol. Chem.* **264**, 16,123–16,131.
10. Sata, T., Zuber, C., Rinderle, S. J., Goldstein, I. J., and Roth, J. (1990) Expression patterns of the T antigen and the cryptic T antigen in rat fetuses: detection with the lectin Amaranthin. *J. Histochem. Cytochem.* **38**, 763–774.
11. Marchalonis, J. and Edelman, G. (1968) Isolation and characterization of a hemagglutinin from Limulus polyphemus. *J. Mol Biol.* **32**, 453–465.
12. Miller, R., Collawan, J., and Fish, W. (1982) Purification and molecular properties of a sialic acid-specific lectin from the slug *Limax flavus*. *J. Biol. Chem.* **257**, 7574–7580.
13. Ravindranath, M. H., Higa, H. H., Cooper, E. L., and Paulson, J. C. (1985) Purification and characterization of an O-acetylsialic acid-specific from a marine crab Cancer antennarius. *J. Biol. Chem.* **260**, 8850–8856.

14. Haselbeck, A., Schickaneder, E., and Hösel, W. (1990) Structural characterization of glycoprotein carbohydrate chains by using digoxigenin-labelled lectins on blots. *Anal. Biochem.* **191**, 25–30.
15. Haselbeck, A., Schikander, E., Schmidt, A., v d Eltz, H., and Hösel, W. (1989) *New Techniques for the Identification and Characterization of Glycoproteins on Gels and Blots.* Proc. 19th FEBS meeting, Rome.
16. Sata, T., Zuber, C., and Roth, J. (1990) Lectin digoxigenin conjugates: a new hapten system for glycoconjugate cytochemistry. *Histochemistry* **94**, 1–11.
17. Li, W., Zuber, C., and Roth, J. (1993) Use of *Phaseolus vulgaris* leukoagglutinating lectin in histochemical and blotting techniques: a comparison of digoxigenin- and biotin-labeled lectins. *Histochemistry* **100**, 347–356.

13

Characterization of HIV gp120 Envelope Glycoprotein by Lectin Analysis

Gregers J. Gram and John-Erik Stig Hansen

1. Introduction

The envelope glycoprotein gp120 of the human immunodeficiency virus (HIV), the causative agent of AIDS, contains approx 24 potential sites for N-glycosylation (Asn-X-Ser/Thr; X≠Pro) *(1)*, all of which are utilized and constitute about 50% of the molecular mass *(2)*. Binding of gp120 to the CD4 molecule is an initial step in viral infection of cells bearing the CD4 molecule on the surface, and fusion of HIV-infected cells with uninfected CD4-positive cells *(3)*.

Using recombinant gp120 produced in H9 cells or Chinese hamster ovary cells (CHO), 29 different N-glycosylated glycans of all three main types, high mannose, hybrid, and complex, have been found *(4–6)*, and a number of studies have indicated that the glycans of gp120 play a role in gp120/CD4 binding. Thus, gp120 deglycosylated by chemical methods, has a much reduced *(7)* or slightly reduced *(8)* affinity for the CD4 receptor, and inhibition of Golgi-mediated glycosylation in infected cells destroys their ability to form multinucleated cells (syncytia) and reduces the infectivity of the viral particles produced *(9)*. Furthermore, the lectins concanavalin A (ConA) and *Lens culinaris* agglutinin (LCA), which both bind N-glycosylated biantennary oligosaccharides, block syncytium formation when added to infected cells before coculture with uninfected, CD4 bearing cells *(10)*.

To evaluate the significance of glycans in creating vaccines against HIV or other blocking agents of CD4-gp120 binding, it is important to identify accessible glycan structures of gp120 and their influence on CD4 binding. To this end, we have examined whether lectins with different glycan specificities bound gp120 from different HIV isolates using *in situ* staining on electrophoretically separated and electroblotted HIV antigens.

1.1. HIV Glycoprotein Preparation

The HIV-1 envelope glycoprotein gp120 is produced in various amounts in different cells. As retroviruses can change cell-specific glycosylation patterns, it is important that preparations of gp120 used to identify oligosaccharide structures are produced in nonvirally transformed cell lines. A problem with this is the small or moderate amount of gp120 production in these cell lines (e.g., H9, CEM, or 6D5), which necessitates concentration of gp120 containing culture supernatants.

Cell-free supernatants from the CD4-positive cell line 6D4, which was chronically infected with the Centers for Disease Control (CDC) HIV-1 isolate 451, was used as a source of gp120. HIV virions in supernatant from infected cells were collected by ultracentrifugation, and the pelleted material was inactivated by incubation with Triton X-100 before use.

1.2. Electrophoretic Polyacrylamide Gel Separation and Electroblotting

Electrophoretic separation of proteins according to molecular mass by polyacrylamide gel electrophoresis (PAGE) and subsequent electroblotting is widely used for identification and characterization of various proteins. The method is easy to perform and it is suitable for simple and rapid identification purposes. We performed the separations in 10% homogeneous polyacrylamide gels and subsequently the separated proteins were transferred by semidry electroblotting onto 0.45-µm nitrocellulose membranes.

1.3. Lectin and Antibody Staining

Lectins are carbohydrate binding proteins, and when interacting with glycoproteins, the lectins bind to oligosaccharides and glycans linked to the peptide backbone via asparagine (*N*-linked) or via serine/threonine (*O*-linked, mucin type), with association constants in the order of $10^{-6} M$ *(11)*.

Binding of lectins to immobilized, electroblotted glycoproteins is fairly sensitive but only semiquantitative at best. Still, it is very easily done and suitable for simple and rapid identification of oligosaccharide structures present on glycoproteins, particularly as it allows oligosaccharide identification without prior separation and purification from the protein backbone.

We performed *in situ* staining of gp120 with horseradish peroxidase-conjugated lectins, antibodies, and pooled patient serum on commercially available electroblotted nitrocellulose strips from Du Pont containing HTLV-IIIB antigens and electroblotted strips containing purified HIV antigens from the HIV-1 isolate CDC 451. Antibody staining was included for identification purposes. Binding conditions for antibodies or lectins to electroblotted glycoproteins are not identical; glycoproteins bind antibodies better at low-salt/high-pH and lectins better at high-salt/neutral pH.

1.4. Glycosidase Degradation of Oligosaccharide Structures

Glycosidases are molecules that cleave oligosaccharide structures and enzymes with various specificities and are now commercially available. Because glycoproteins usually carry several different oligosaccharide structures, glycosidases can be very useful in characterizing multispecific lectin binding patterns. We concentrated our attention on neuraminidase, which is known to cleave sialic acids on N- and O-linked glycans. Glycosidase treatment of glycoproteins can be performed both on immobilized electroblotted material or glycoproteins in solution. Glycosidase treatment on immobilized glycoproteins is performed on the blotted strip just before binding of lectins, whereas glycosidase treatment in solution can be performed at any step before SDS-PAGE separation of the sample. The latter treatment may influence the mobility of the glycoprotein in SDS-PAGE.

2. Materials
2.1. HIV Glycoproteins Preparation

1. 6D5 cells are cultured at 37°C, 5% CO_2, using RPMI-1640 (Gibco-BRL, Gaithersburg, MD) supplemented with 10% fetal calf serum, 100 IU/mL penicillin, 20 µg/mL Gentamicin, 100 IU/mL streptomycin, and 2.3 µg/mL amphotericin B. Cells are maintained at a concentration of 2–8 × 10^5/mL, and medium is changed twice weekly.
2. Phosphate-buffered saline (PBS), pH 7.4, containing 1% Triton X-100 (TX-100).

2.2. Electrophoretic Polyacrylamide Gel Separation and Electroblotting

1. Prepare a 16 × 11 cm, 1-mm-thick 10% homogeneous polyacrylamide separation gel (T = 10%; C = 2%) with a 4% stacking gel (T = 4%; C = 2%) on top.
2. Sample buffer: 2% w/v SDS, 0.05 g dithiotreitol, 0.2% w/v tetramethylenediamine, 0.01% w/v pyronin Y and 10% v/v glycerol in a 0.4M Tris/H_2SO_4 buffer, pH 7.2.
3. Tris/Taurine electrophoresis buffer: 70 mM Taurine, 70 mM Tris, 0.1% w/v SDS (sample buffer as blot marker).
4. 0.45-µm nitrocellulose filters.
5. Filter paper.
6. Transfer buffer: 48 mM glycine, 32 mM Tris, 1.3 mM SDS, and 20% v/v ethanol.

2.3. In Situ Lectin and Antibody Staining

1. Antibody (Ab)-incubation buffer: 50 mM Tris-HCl, pH 10.2, 150 mM NaCl, 0.05% v/v Tween-20, and 5 mM NaN$_3$. Lectin (Lec)-incubation buffer: 10 mM Tris-HCl, pH 7.2, 1M NaCl, 0.05% v/v Tween-20, and 5 mM NaN$_3$.
2. Ab-blocking buffer: 2% v/v Tween-20 in Ab-incubation buffer. Lec-blocking buffer: 2% v/v Tween-20 in Lec-incubation buffer.
3. 30 µL/mL pooled HIV positive serum or 6 µg/mL anti-gp120 antibody (Sheep-anti gp120; Biochrome KG, Berlin, Germany). Use peroxidase-conjugated lectins at a subunit concentration of $10^{-6}M$ (1–20 µg/mL) (Sigma, St. Louis, MO; Kem-En-Tec, Copenhagen, Denmark).
4. 3 µg/mL peroxidase-conjugated rabbit antibody against sheep IgG (DAKO, Copenhagen, Denmark). 0.05M sodium acetate containing 4% v/v 0.1% 3-amino, 9-ethyl carbozole in acetone and 0.1 % v/v 30% hydrogen peroxide adjusted pH to 5.5 with acetic acid. Prepare just before use *(see* Note 1).
5. 50 mM sodium disulfite (Na$_2$S$_2$O$_5$).

2.4. Glycosidase Degradation of Oligosaccharide Structures

1. 0.5 U Neuraminidase from *Clostridium perfringens* (Boehringer Mannheim, Mannheim, Germany) in 200 µL 50 mM sodium-acetate buffer, pH 5.5.

3. Method
3.1. HIV Glycoproteins Preparation

1. Harvest culture supernatants from 6D5 cells chronically infected with CDC-451.
2. Clear supernatants by centrifugation for 15 min at 4°C, 18,000g.
3. Pellet virus particles by centrifugation for 2.5 h at 4°C, 150,000g.
4. Resuspend pellet in 3.6 mL PBS, pH 7.4, containing 1% v/v TX-100 *(see* Note 2).
5. Incubate for 1 h at 4°C and store at –80°C until analysis.
6. As a negative control, repeat steps 1–5 with an uninfected 6D5 culture supernatant.

3.2. Electrophoretic Polyacrylamide Gel Separation and Electroblotting

1. Prepare the SDS-PAGE gel with several small lanes.
2. Dilute samples 1:1 in 50 µL sample buffer and boil for 3 min *(see* Note 3). Load the HIV antigen samples into lanes with molecular protein marker in the outer lanes.
3. Perform electrophoresis until the electrophoresis marker, pyronin Y, has migrated 11 cm or just above the saltbridge *(see* Note 4). Before stopping the electrophoresis run, add sample buffer to each lane and electrophorese until the pyronine Y marker has migrated into the resolving gel *(see* Note 5).
4. Put the resolving gel on a 0.45-µm nitrocellulose paper and place between several pieces of filter paper presoaked in transfer buffer *(see* Note 6).

5. Place in blotting apparatus in the following order: cathode/filters/separation gel/nitrocellulose paper/filters/anode. Transfer for 1 h at 0.8 mA/cm^2 gel.
6. Cut the nitrocellulose paper in strips according to the pyronin Y marker and number each.

3.3. In Situ *Lectin and Antibody Staining*

1. Wash strips once for 10 min in antibody- or lectin-incubation buffer (Ab/Lec-incubation buffer).
2. Block for 2 min in antibody- or lectin-blocking buffer (Ab/Lec-blocking buffer).
3. Wash strips for 10 min in Ab/Lec-incubation buffer.
4. Incubate strips overnight on a rocking table in 2 mL Ab/Lec-incubation buffer with an appropriate concentration of antibody (6 µg/mL) or peroxidase-conjugated lectin (1–20 µg/mL).
5. Wash strips 3 × 10 min in Ab/Lec-incubation buffer. For binding of secondary peroxidase-conjugated antibody, incubate strips with 3 µg/mL peroxidase-conjugated antibody for 1 h on a rocking table and repeat the washing step.
6. Stain the strips in 3-amino,9-ethyl carbozole and hydrogen peroxide for 2–10 min *(see* Note 7).
7. Wash strips in distilled water.
8. Fix the strips in sodium disulfite and photograph *(see* Figs. 1 and 2).

3.4. Glycosidase Degradation of Oligosaccharide Structures

1. Before separation on SDS-PAGE gel, 100 µL of the semipurified material is incubated with 0.5 U Neuraminidase in 200 µL 50 mM sodium acetate-buffer, pH 5.5, for 3 h at room temperature.
2. Proceed with electrophoresis, but include an antibody detection on a replica strip as described for antibody staining.

4. Notes

1. Immediately before use, prepare a mixture of the following three stock solutions: 2.5 mL (1% 3-amino,9-ethyl carbozole in acetone), 40 mL (sodium acetate buffer, pH 5.0), and 30 µL (30% hydrogen peroxide).
2. Incubation for 1 h in 1% TX-100 at temperatures above 3°C inactivates HIV. Aliquots can be stored at –80°C until analysis.
3. Use Eppendorf tubes and make a hole in the lid before boiling to avoid tube-pops and evaporation of the sample.
4. Electrophoresis can be performed at 5 or 30 mA for a duration of approx 24 or 3 h, respectively.
5. The pyronin Y blot marker buffer should not contain sugar for lectin blots.
6. It is very important to avoid any air bubbles between separation gel and nitrocellulose paper.
7. Be aware that staining will not stop immediately after soaking in water. Soak in water just before perfect staining appears.

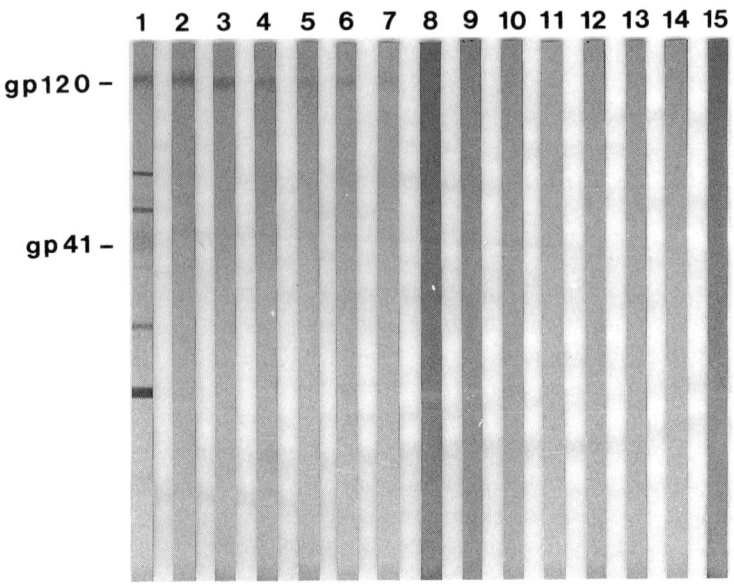

Fig. 1. Electroblotted strips from Du Pont containing HTLV-IIIB antigens incubated with:

1. 30 µg/mL pooled HIV positive serum.
2. 5 µg/mL Sheep anti-gp120.
3. 2 µg/mL concannavalin A (Con A).
4. 8 µg/mL *Lens culinaris* agglutinin (LCA).
5. 6 µg/mL *Vicia faba* agglutinin (VFA).
6. 15 µg/mL *Pisum sativum* agglutinin (PSA).
7. 15 µg/mL phytohaem (erythro) agglutinin (PHA-E).
8. 4 µg/mL *Tricium vulgaris* (wheat-germ) agglutinin (WGA).
9. 15 µg/mL soybean agglutinin.
10. 15 µg/mL peanut agglutinin.
11. 15 µg/mL *ulex europaeus 1* agglutinin (UEA-1).
12. 15 µg/mL *Griffonia simplicifolia 2* agglutinin (GSA-2).
13. 15 µg/mL phytohaem (leuko) agglutinin (PHA-L).
14. 15 µg/mL *Dolichos biflorus* agglutinin.
15. 15 µg/mL *Vicia villosa* agglutinin.

All lectins were conjugated to horseradish peroxidase and incubations were performed in 2 mL incubation buffer. Indicated is the envelope glycoproteins of HIV: gp120 and gp41. gp120 was stained with the following lectins: Con A (3), LCA (4), VFA (5), PSA (6), PHA-E (7).

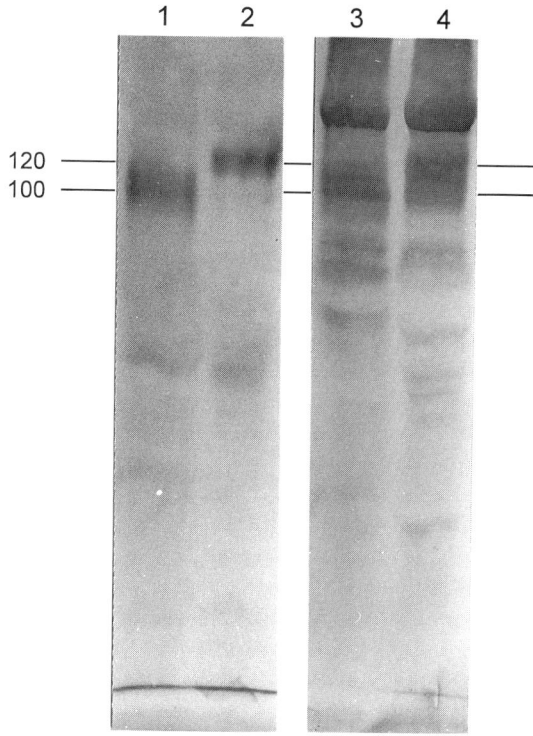

Fig. 2. Neuraminidase treatment of a gp120 preparation (CDC isolate 451) resulted in a shift from approx 120,000 to 100,000 M_r in staining with sheep anti-gp120 and Con A. Lanes 1 and 3: 100 µL purified HIV-1 supernatant incubated with 0.5 U neuraminidase from *Clostridium perfringens* (Boehringer Mannheim) in 200 µL 50 mM sodium acetate-buffer, pH 5.5, for 3 h at room temperature. Lanes 2 and 4: Mock treated supernatant. Lanes 1 and 2 were incubated with 6 µg/mL sheep anti-gp120. Lanes 3 and 4 were incubated with 2 µg/mL Con A. Note that the faint band at 100,000 in lane 2 is relatively stronger in lane 4. All four lanes are nitrocellulose strips blotted from the same polyacrylamide gel electrophoresis (PAGE).

Acknowledgment

The HIV isolate, CDC isolate 451, was obtained from Paul Feorino, Centers for Disease Control, Atlanta, GA.

References

1. Kornfeld, R. and Kornfeld, S. (1985) Assembly of asparagine-linked oligosaccharides. *Ann. Rev. Biochem.* **54,** 631–664.
2. Leonard, C. K., Spellman, M. W., Riddle, L., Harris, R. J., Thomas, J. N., and Gregory, T. J. (1990) Assignment of intrachain disulfide bonds and characteriza-

tion of potential glycosylation sites of the type 1 recombinant human immunodeficiency virus envelope glycoprotein (gp120) expressed in Chinese hamster ovary cells. *J. Biol. Chem.* **265,** 10,373–10,382.
3. Dalgleish, A. G., Beverley, P. C. L., Clapham, P. R., Crawford, D. H., Greaves, M. F., and Weiss, R. A. (1984) The CD4 (T4) antigen is an essential component of the receptor for the AIDS retrovirus. *Nature* **312,** 763–767.
4. Mizuochi, T., Spellmann, M. V., Larkin, M., Soloman, J., Basa, L. J., and Feizi, T. (1988) Carbohydrate structures of the human-immunodeficiency-virus (HIV) recombinant envelope glycoprotein gp120 produced in Chinese hamster ovary cells. *Biochem. J.* **263,** 599–603.
5. Mizuochi, T., Matthews, T. J., Kato, M., Hamako, J., Titani, K., Solomon, J., and Feizi, T. (1990) Diversity of oligosaccharide structures on the envelope glycoprotein gp120 of human immunodeficiency virus 1 from the lymphoblastoid cell line H9. Presence of complex-type oligosaccharides with bisecting N-acetylglucosamine residues. *J. Biol. Chem.* **265,** 8519–8524.
6. Geyer, H., Holsbach, C., Hunsmann, G., and Scheider, J. (1988) Carbohydrate of human immunodeficiency virus. *J. Biol. Chem.* **263,** 11,760–11,767.
7. Mathews, T. J., Weinhold, K. J., Lyerly, H. K., Langlois, A. J., Wigzell, H., and Bolognesi, D. P. (1987) Interaction between the human T-cell lymphotropic virus type IIIB envelope glycoprotein gp120 and the surface antigen CD4. Role of carbohydrate in binding and cell fusion. *Proc. Natl. Acad. Sci. USA* **84,** 5424–5428.
8. Fenouillet, E., Clerget-Raslain, B., Gluckmann, J. C., Guetard, D., Montagnier, L., and Bahraoui, E. (1989) Role of N-linked glycans in the interaction between the envelope glycoprotein of human immunodeficiency virus and its CD4 cellular receptor. *J. Exp. Med.* **169,** 807–822.
9. Gruters, R. A., Neefjes, J. J., Tersmette, M., de Goede, R. E. Y., Tulp, A., Huismann, H. G., Miedema, F., and Ploegh, H. L. (1987) Interference with HIV-induced syncytium formation and viral infectivity by inhibitors of trimming glycosidases. *Nature* **330,** 74–77.
10. Lifson, J., Coutre, S., Huang, E., and Englemann, E. (1986) Role of envelope glycoprotein carbohydrate in human immunodeficiency virus (HIV) infectivity and virus-induced cell fusion. *J. Exp. Med.* **164,** 2101–2106.
11. Osawa, T. and Tsuji, T. (1987) Fractionation and structural assessment of oligosaccharides and glycopeptides by use of immobilized lectins. *Ann. Rev. Biochem.* **56,** 21–42.

14

Use of Lectins for Characterization of O-Linked Glycans of Herpes Simplex Virus Glycoproteins

Sigvard Olofsson and Anders Bolmstedt

1. Introduction

The existence of *O*-linked glycans in viral glycoproteins was described in the early 1980s for enveloped viruses such as herpes simplex virus type 1 (HSV-1), vaccinia virus, and mouse hepatitis virus *(1–4)*. Glycoprotein C of HSV-1 (designated gC-1) was demonstrated to contain domains, in which numerous *O*-linked glycans were concentrated to pronase-resistant clusters *(5–7)*, thereby resembling the organization of mucins *(8)*. This glycoprotein, containing nine sites for *N*-linked glycosylation in addition to the *O*-linked glycans, is responsible for several important biological activities, including virus receptor binding *(9)* and binding of factor C3b of the complement system *(10)*. The function of the *O*-linked glycans in these activities remains unclear, but it is conceivable that their clustered appearance may cause gC-1 to adopt an extended fibrous conformation *(11)*, as originally demonstrated for the *O*-linked glycans of mucins *(8)*. Use of lectins facilitates a structural analysis of clustered *O*-linked glycans of gC-1 and it is possible that the methodology presented here may be of more general use, as similar arrangements of clustered *O*-linked glycans are present in an increasing number known glycoproteins of other enveloped viruses including herpes simplex virus type 2 *(12,13)*, Epstein-Barr virus *(14)*, and respiratory syncytial virus *(15)*.

1.1. HSV Glycoproteins

HSV-1 is a large, enveloped DNA virus with a double-stranded genome of more than 10^6 bp, encoding more than 75 proteins. Several open reading frames with features characteristic of membrane-associated glycoproteins have been identified and today, nine major glycoprotein species, designated from gB-1 to

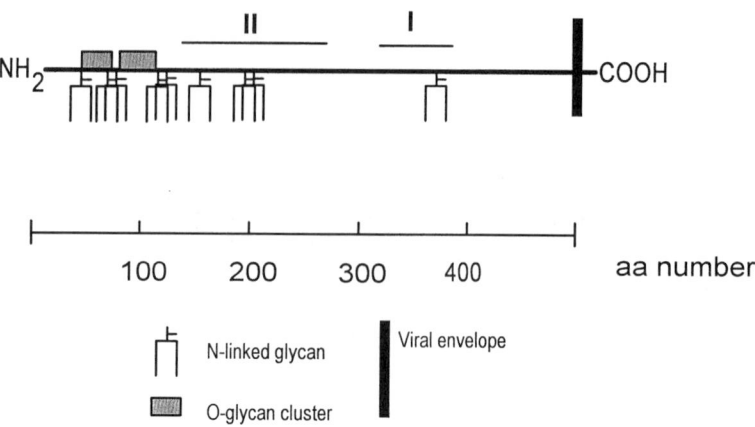

Fig. 1. Molecular organization of gC-1, specified by HSV-1. Sequence data taken from ref. *28*. The position of the clustered *O*-linked glycan units *(29)* is marked in gray. These stretches are pronase-resistant and constitute the glycopeptides isolated and analyzed after Step C (Fig. 3). The receptor-binding region (B) of gC-1 as identified by Trybala and coworkers *(30)* is indicated.

gL-1 (with gA-1 and gF-1 excluded because of initial ambiguity as to their identity) have been classified. The glycobiology of these glycoproteins was recently reviewed *(11)*. Although other HSV glycoproteins, such as gD-1 may contain *O*-linked oligosaccharides *(16)*, only gC-1 is discussed further in the present chapter.

The molecular organization of gC-1, which is a membrane-associated glycoprotein with a short cytoplasmic tail, is presented in Fig. 1. All *O*-linked glycans and most *N*-linked glycans are situated in the *N*-terminal part of gC-1. The receptor binding domain of gC-1, binding to cell-associated heparan sulfate, is also located in this region.

O-linked glycosylation of the mucin-type employs serine and threonine residues as acceptors for addition of an initial *N*-acetylgalactosamine residue (GalNAc) but the signal requirements of acquisition of *O*-linked glycans are not as well understood as for addition of *N*-linked glycans. One explanation for this is probably that the GalNac-transferase, carrying out the first step in *O*-linked glycosylation, operates in the Golgi region when the glycoprotein substrate has adopted its final three-dimensional conformation. It is, therefore, possible only to score a probability of a given serine or threonine residue to function as an acceptor for *O*-glycosylation *(17)*. A preliminary analysis, using the Elhammer algorithm for prediction of *O*-glycosylation sites *(17)*, indicated about 20 serine or threonine residues within the shaded area of gC-1 (Fig. 1) to be highly probable targets for *O*-glycosylation.

Owing to the high number of clustered O-linked glycans in the N-terminal portion, gC-1 may be purified by subjecting extracts from HSV-1-infected cells to lectin affinity chromatography, resulting in separation from not only the other HSV glycoproteins, but also most of the contaminating proteins, expressed by the host cell *(1)*. The efficacy of this method is illustrated by the fact that one single lectin chromatography step is sufficient to produce an excellent antigen for use in serological diagnosis of HSV infection *(18)*.

Host cell-specific protein synthesis is very efficiently shut down by HSV-1 infection and it is therefore possible to use radiolabeled carbohydrate metabolites for selective labeling of HSV-glycoproteins *(2)*. This ensures that lectin-based analysis of HSV-1 glycoproteins is not obscured by presence of host-cell derived glycoproteins.

1.2. Lectin-Aided Analysis of gC-1

Owing to its high content of O-linked and N-linked glycans, gC-1 is characterized by a high degree of heterogeneity (Fig. 2). Precursor gC is an intracellular intermediate with unprocessed N-linked glycans and a set of identical high mannose oligosaccharides. As soon as this precursor has entered the Golgi region, processing of N-linked glycans and addition of O-linked glycans take place. There are differences in lectin binding properties between the 130- and the 115-kDa part of the broad gC-band, reflecting differences mainly in O-linked glycosylation between the extreme parts of the band. For details of variations in O-glycosylation of viral glycoproteins, the reader is referred to refs. *11* and *19*.

Thus, the gC-1 molecules of the 115-kDa part of the band contain mainly short O-glycans, in which most of these units in fact constitute single N-acetyl-D-galactosamine (GalNAc) units attached to the polypeptide backbone. This population of gC-1 molecules will bind mainly to *Vicia villosa* isolectin B_4 (VVA B_4) or *Helix pomatia* lectin (HPA), with affinity for clustered GalNAc units of a polypeptide. The major O-linked glycan of the middle region is the disaccharide Galβ(1–3)GalNAc, and this part of the glycoprotein band will only bind to Peanut *(Arachis hypogaea)* lectin (PNA). Lastly, the 130-kDa part of gC contains a high proportion sialylated O-linked glycans, and these gC-1 molecules therefore bind to sialic acid-binding lectins such as wheat-germ *(Triticum vulgaris)* lectin (WGA).

It is important to note that whereas WGA and many other lectins bind to a wide variety of viral and host cell species, it is easy to adjust the experimental conditions for HPA and PNA to achieve a pronounced selectivity for the clustered O-linked glycans of gC-1 *(19)*. HPA and PNA chromatography therefore constitute powerful chromatographic tools not only for isolation of O-linked glycans, but also for purification of gC-1 from other HSV and cellular glyco-

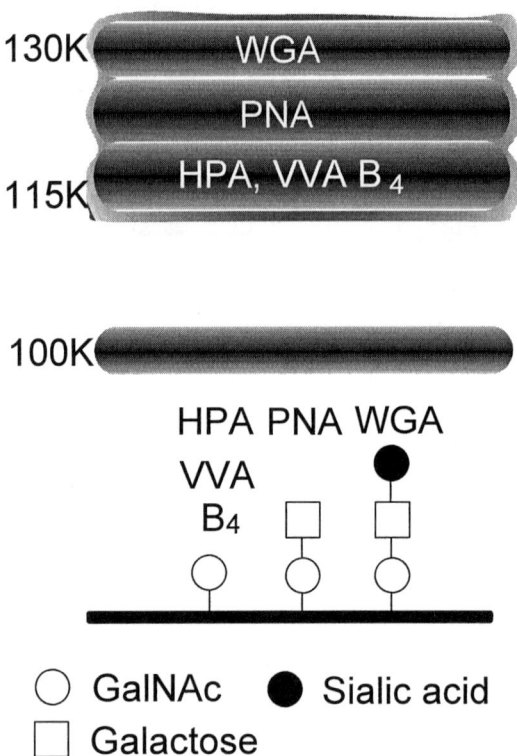

Fig. 2. Schematic representation showing the electrophoretic complexity of gC-1, which appears as one broad band ranging in apparent molecular weight from about 115–130 kDa, and one precursor band with an apparent molecular weight of 100 kDa. The broad band, referred to as gc, contains O-linked glycans and complete N-linked glycans, whereas the narrow band is referred to as precursor-gC (pgC), containing no O-linked glycans and only high mannose N-linked glycans. The three sub-bands binding with preference for WGA, PNA, and HPA, respectively are indicated. At the bottom of the figure are given structures of representative O-linked glycans with affinity for each lectin. Note that boundaries of the different lectin-binding areas are diffuse with partial overlapping specificities.

proteins. The procedures for use of lectins as tools for isolation and characterization of O-linked glycans of gC-1 are outlined in Fig. 3.

It is possible to use either lectin (HPA or PNA) affinity chromatography or gC-specific immunosorbent to obtain sufficiently pure [^{3}H]-GlcNAc-labeled gC-1 (Step A in Fig. 3) *(1,20,21)*. However, the gC populations isolated may vary in their properties, dependent on the choice of purification procedure. Thus, HPA chromatography will select for gC-1 variants with an apparent

Lectin Characterization of HSV Glycoproteins

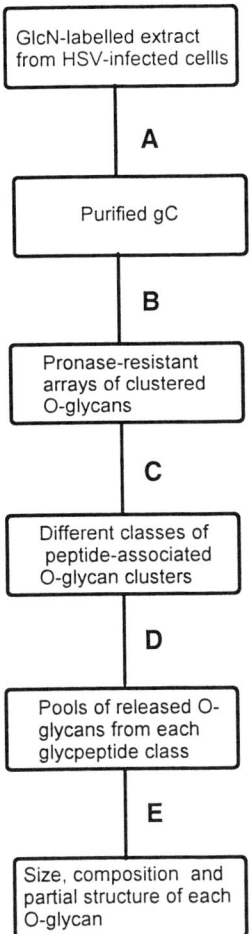

Fig. 3. Outline of the experimental procedures of the present protocol. For details, *see* Section 1.2.

molecular weight of 115 kDa enriched in *O*-glycosidic GalNAc units (Fig. 2) *(22)*, whereas PNA chromatography will select for larger gC-1 molecules, dependent on their content of larger *O*-linked glycans. Both these preparations are devoid of most contaminating HSV non-gC-1 glycoproteins. On the other hand, if a representative picture of all glycoforms of gC-1 is preferred, immunosorbent purification is the method of choice.

Pronase treatment (Step B) will digest most of gC-1 with exception for the peptide stretches containing clustered *O*-linked glycans *(20,23)*. These pronase-resistant glycopeptides will be of different sizes dependent on the length of the pronase-resistant peptide chain and the complexity of the O-glycans associated with the glycopeptide. The size distribution of the O-glycan glycopeptides may be determined by gel filtration, using Sephadex G50 or corresponding resins from other manufacturers. Usually, the pronase digest is too heterogeneous to permit a more detailed analysis.

The structural characterization of the O-linked glycans is also facilitated by separation of the glycopeptides according to their lectin affinities (Step C). The different classes of pronase-resistant glycopeptides may be separated dependent on O-glycan composition by lectin chromatography, using HPA, VVA B_4, or PNA. HPA-chromatography will enrich glycopeptides preferentially containing O-linked GalNAc residues and, in addition, a few larger O-linked glycans. Gel-bound PNA will retain glycopeptides with a high content of Gal β(1–3)-GalNAc, and WGA is useful for glycopeptides of the 130K region, containing mature gC-1 with a large proportion of sialylated O-glycans. Gel filtration on Sephadex G50 will give important information as to the size of the different classes of peptides with O-linked glycans. Identically sized HPA-binding glycopeptides are obtained from gC-1, isolated from tunicamycin-treated, HSV-infected cells, indicating that there are no N-linked glycans in those particular glycopeptides *(23)*.

The separated glycopeptides are useful starting material for release of individual O-glycans (Step D). For a long time, the only reliable method for release of O-linked glycans was to use treatment with weak alkali in the presence of a molar concentration of $NaBH_4$. This causes a β-elimination of the O-linked glycan, in which the innermost GalNAc is converted to a N-acetylgalactosaminol, which is easy to detect as a chemical proof for a successful β-elimination. This treatment is relatively simple to perform. The borohydride does not take any active part in the β-elimination; its function is merely to stop "peeling" of the oligosaccharide, i.e., stripping off monosaccharides one by one from the reducing end. In our hands, the methods of Carlson *(24)* and Spiro *(25)* work well. After the reaction, it is essential to eliminate the high concentration of $NaBH_4$, which may be converted to volatile methylboronates by several cycles of addition of methanol and subsequent evaporation.

It is also possible to release O-linked glycans with the specific endo-glycosidase O-glycanase. However, this enzyme has a very narrow specificity for only certain O-glycans *(26)*.

Hydrazinolysis has been described as a reproducible method for quantitative release of O-glycans *(26,27)*. However, owing to the extreme health hazards associated with hydrazine, this technique is not easily available in most laboratories for cell biology work. Recently, a commercial kit for hydrazinolysis of O-linked glycans has become available.

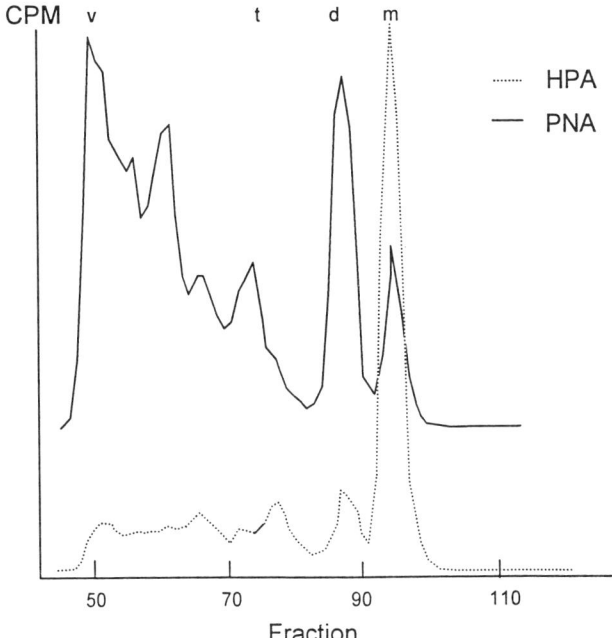

Fig. 4. Typical Bio-Gel P2 gel filtration of O-linked glycans released by alkaline borohydride treatment of HPA-binding or PNA-binding pronase-resistant glycopeptides of gC-1. The positions of the void volume (v), a trisaccharide (t), a disaccharide (d), and a monosaccharide (m) are indicated.

The released O-glycans are relatively easily separated by gel filtration on Bio-Gel P2, where up to octasaccharides are readily separated as distinct peaks, and each glycan may be structurally determined (Step E). A typical chromatogram is shown in Fig. 4. The fractions representing each peak are pooled and may be used for further structural determination by thin layer chromatography (TLC) after acid hydrolysis. This latter treatment, carried out at 100°C in $3M$ HCl cleaves the glycosidic bonds between the individual monosaccharides. Two side reactions should be considered:

1. The acetamide sugars GalNAc and GlcNAc are converted to corresponding hexosamines.
2. Sialic acids are destroyed, leaving radiolabeled breakdown products, which are detectable in TLC. We previously used a system based on plates coated with cellulose *(2)*, but later we found a system based on silica-coated plates described by Dall'Olio et al. *(7)* more convenient. This system is described in detail in Section 3.7., step 7. A typical TLC chromatogram and its interpretation in structural terms is given in Fig. 5.

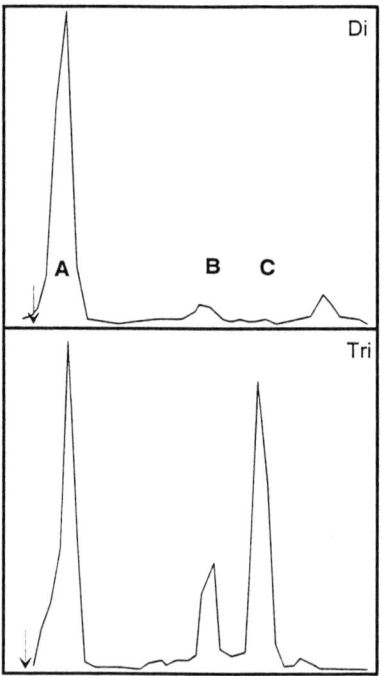

Fig. 5. Thin layer chromatography on silica of the disaccharide and trisaccharide peaks of Fig. 4. The positions of origin (arrow), GalNAc-ol (A), GalNAc (B), and GlcNAc (C) (as detection of markers after ninhydrin staining) are given. The disaccharide contains essentially one radiolabeled monosaccharide, because galactose does not carry any label during the present conditions. The trisaccharide contains both GalNAc and GlcNAc label in addition to the GalNAc-ol label. Since essentially no GalNAc or GlcNAc label is present in the disaccharide profile of Fig. 4, it is reasonable that the trisaccharides are GlcNAc-Gal-GalNAc-Ser/Thr and GalNAc-Gal-GalNAc-Ser/Thr. The relative amounts of GalNAc and GlcNAc suggest that the latter O-linked glycan is the far most abundant one.

In conclusion, gC-1 constitutes a heterogeneous glycoprotein with respect to its content of O-linked glycans, in which different populations of gC-1 contains different sets of O-linked glycans. However, use of lectins for two purposes, i.e., for selection of differently O-glycosylated gC-1 populations, and, after pronase digestion, for selection of different O-glycan-containing peptides from a given gC-1 population, makes it possible to achieve surprisingly exact structural and distributional analysis of the O-glycans of gC-1 by applying relatively simple biochemical and chromatographic techniques.

Table 1
Eluting Conditions for Some Lectins in the Procedures of the Present Chapter

Lectin	Eluting sugar	%-(w/v)	Comments
HPA	GalNAc	0.1	
SBA	GalNAc	1.0	Bind and wash at 4°C
PNA	GalN	1.0	Bind and wash at 4°C
WGA	GlcNAc	1.0	
RCA	Lactose	1.0	Bind and wash at 4°C
SJA	GalNAc	0.1	
DBA	GalNAc	0.1	
WA B$_4$	GalNAc	0.1	
BSA	GlcNAc	1.0	

SBA, Soybean *(Glycine max)* lectin; RCA, *Ricinus communis* lectin; SJA, *Sophora japonica* lectin; DBA, *Dolichos biflorus* lectin; BSA, *Bandeiraea simplicifolia* lectin; GalN, Galactosamine; GlcNAc, *N*-acetyl-D-glucosamine.

2. Materials
2.1. Infection and Radiolabeling of HSV-Infected Cells

1. Various Green monkey kidney cell lines are susceptible to herpes simplex virus and they produce relative large amounts of virus-specified glycoproteins. Vero cells can be obtained from American Type Culture Collection (ATCC; Rockville, MD) (*see* Note 1).
2. Plastic culture utensils: 75-cm^2 flasks for propagation of cells; 50-mm Petri dishes for virus infection of Vero cells.
3. Radiochemicals: [^{3}H]-*N*-acetylglucosamine ([^{3}H]-GlcN) and [^{3}H]-galactose ([^{3}H]-Gal).
4. Eagle's MEM (CO$_2$/bicarbonate buffer system), containing 10% fetal calf serum for propagation of cells; without serum during virus infection and radiolabeling (*see* Note 2).
5. Established laboratory HSV-1 strains (F; KOS [Consult ATCC catalog]) or plaque-purified field isolates may be used.
6. Standard rubber policeman for cell culture harvest.

2.2. Purification of HSV-1 gC-1

1. Agarose- or sepharose-bound HPA or PNA (various commercial suppliers).
2. Protein G-Sepharose or Protein A-Sepharose (Pharmacia, Uppsala, Sweden).
3. Polyclonal or monoclonal antibodies to gC-1 (*see* Note 3).
4. Disposable columns for small-scale lectin chromatography (e.g., Bio-Rad, Hercules, CA) (*see* Note 4).
5. Tris-buffered saline (TBS), i.e., 0.15*M* NaCl, 0.02*M* Tris-HCl, pH 7.4.
6. Eluting sugars for lectin chromatography (*see* Table 1).

7. Eluting buffer for immunosorbent: 0.1M Glycine-HCl, pH 2.5–3.0.
8. Neutralizing buffer 1M Tris base.

2.3. Pronase Digestion

Pronase from *Streptomyces griseus* (various commercial suppliers).

2.4. Lectin Affinity Chromatography of Pronase Digests

1. Agarose- or sepharose-bound lectins (various commercial suppliers), including VVA B_4, HPA, PNA, and WGA.
2. Eluting sugars including GalNAc (WA B_4 and HPA), GalN (PNA), and GlcNAc (WGA). More information is given in Table 1.
3. Disposable columns for small-scale lectin chromatography (e.g., Bio-Rad).

2.5. Chemical Release of O-Linked Glycans

1. Standard hydrolysis glass (Pyrex or similar) tubes with Teflon-coated screw caps (various suppliers).
2. $NaBH_4$ (The p.a. quality from Fluka, Buchs, Switzerland, proved to be sufficient in our laboratory).
3. Methanol, p.a.
4. Acetic acid, p.a.
5. *O*-glycan Recovery Kit, based on hydrazinolysis (Oxford Glycosystems, Oxford, UK).

2.6. Gel Filtration of Pronase-Resistant and O-Linked Glycans

1. Chromatography columns (120 cm; $\phi = 0.8$ cm; e.g., Bio-Rad econo columns).
2. Pyridine-acetate buffer (0.1M; pH 7.8).
3. Bio-Gel P2; 200–400 mesh (Bio-Rad).
4. Sephadex G-50, Superfine (Pharmacia-LKB, Stockholm, Sweden).

2.7. Determination of Monosaccharide Composition

1. 3M Hydrochloric acid.
2. Standard hydrolysis glass (Pyrex or similar) tubes with teflon-coated screw caps (various suppliers).
3. Thin layer chromatography (TLC) tank of glass, suitable for 20 × 20 cm plates.
4. Hamilton syringe, 100–200 µL.
5. Disposable silica gel TLC plates (DC Alufolien, Merck, Darmstadt, Germany).
6. Glucosamine and galactosamine solutions in water (5 mg/mL).
7. Ethanol/pyridine/1-butanol/acetic acid/water (100:10:10:3:30); freshly prepared.
8. Standard liquid scintillation cocktail, compatible with water (various suppliers).
9. Ready-to-use Ninhydrin spray (Merck, Darmstadt, Germany).

3. Methods

3.1. Infection and Radiolabeling of HSV-Infected Cells

1. Remove medium from Petri dish cultures of Vero cells and wash the cells once with 1 mL Eagle's MEM without serum.

Lectin Characterization of HSV Glycoproteins

2. Inoculate the cells by adding 0.5 mL HSV-1 suspension (10–100 PFU/cell). The plate should be agitated gently to spread the suspension over the entire surface. Incubate at 37°C (CO_2-incubator under standard conditions) for 30 min to ensure attachment of virus.
3. Aspirate virus suspension from Petri dish and add 2 mL Eagle's MEM. Incubate at 37°C (CO_2-incubator under standard conditions) for 3.5 h.
4. At 4 h postinfection (p.i.), remove medium, and add 1.5 mL Eagle's MEM, containing 50 µCi/mL of [^{3}H]-GlcN or [^{3}H]-Gal (*see* Note 5). Incubate at 37°C (CO_2-incubator under standard conditions) until harvest.
5. Harvest cells at 18 h p.i. by scraping with a standard rubber policeman, transfer the cell suspension to a centrifuge tube, and spin down the cells by slow speed centrifugation.
6. Discard radiolabeled medium (*see* Note 6) and resuspend the cells in 1 mL of TBS, containing 1% Triton X-100. Sonicate in an ice-cold water bath for 3 × 2 min. Keep the samples in the cold at least 1 min between the sonication periods.
7. Centrifuge at 100,000g for 1 h. Collect the supernatant, which defines a "soluble protein fraction" (*see* Note 7). This material can be kept frozen at –70°C prior to further experimentation.

3.2. Purification of HSV-1 gC-1

3.2.1. Lectin Affinity Chromatography

1. Prepare a lectin bed, consisting of approx 500 µL (hydrated volume) HPA-Sepharose or PNA-Sepharose, in a disposable plastic chromatography column (*see* Note 8).
2. Wash the column with at least five gel volumes of TBS, containing 1% Triton X-100.
3. Apply the 200 µL of the soluble fraction to the column (*see* Note 9).
4. Allow to adsorb for 10 min at room temperature and apply another 200 µL of the soluble fraction, if necessary. These steps may be repeated several times.
5. Wash by adding 10 vol each of 500 µL TBS, containing 1% Triton X-100. Collect the pass-through fractions and count 10 µL of each in a β-counter to ensure that the last wash fractions has reached a background level.
6. Elute with 5 vol (500 µL) of 10 mM GalNAc (HPA-Sepharose) or 100 mM galactosamine (PNA-Sepharose). Identify the peak fractions after counting of 10 µL each of the eluted fractions. These fractions are pooled and stored frozen (–20°C would be sufficient) prior to further experimentation.

3.2.2. Immunosorbent Chromatography

1. Prepare an immunosorbent using either mouse monoclonal antibodies (MAb) or rabbit hyperimmune serum to gC-1 (*see* Note 10).
2. Apply the 200 µL of the soluble fraction to a disposable plastic chromatography column, containing approx 500 µL immunosorbent.
3. Allow to adsorb for 10 min at room temperature and apply another 200 µL of the soluble fraction. These steps may be repeated several times.

4. Wash by adding 10 vol each of 500 µL TBS, containing 1% Triton X-100. Collect the pass-through fractions and count 10 µL of each in a β-counter to ensure that the last wash fractions has reached a background level.
5. Elute bound glycoprotein with 5 vol (500 µL) 0.1 mM glycine-HCl, pH 2.5–3.0, and neutralize as quickly as possible with 1M Tris base (*see* Note 11).
6. Identify the peak fractions after counting of 10 µL each of the eluted fractions. These fractions are pooled and stored frozen prior to further experimentation.

3.3. Pronase Digestion

1. Predigest Pronase (10 mg/mL in TBS) by incubation for 2 h at 37°C.
2. Remove eluents (Formate or GalNAc) from the purified gC-1 fraction using a short, prefabricated Sephadex G25 column (Pharmacia PD10 or its equivalent). It is recommended not to apply more than 2.0 mL purified protein solution to PD10 (*see* Note 12).
3. Wash by adding 0.5 mL vol of TBS and collect 20 fractions.
4. Determine the position of the radiolabeled glycoprotein by counting 10 µL of each fraction in a β-counter (*see* Note 13).
5. Mix 0.5 mL of the pooled glycoprotein fraction with 5 µL 0.15M CaCl$_2$ and 10 µL of the predigested pronase solution in a Teflon-capped glass tube.
6. Add one drop of toluene to prevent microbial growth and incubate at 37°C for 6 d. Replace autodigested pronase by addition of another 10 µL of the predigested pronase solution after 3 d.
7. The reaction is terminated by boiling for 3 min and the resulting suspension is clarified by low speed centrifugation (*see* Note 14).
8. The resulting pronasate may be stored indefinitely at –70°C.

3.4. Lectin Affinity Chromatography of Pronase Digests

1. Lectin chromatography of pronase digests for isolation of pronase-resistant peptides with clustered *N*-linked glycans is performed essentially as described above in Section 3.2.1., for lectin affinity chromatography of detergent-solubilized gC-1. One important exception is that detergent should be omitted from all solutions used in experimentation.
2. Wash the column with at least 5 gel vol of TBS, *without* Triton X-100.
3. Apply the 200 µL of the pronase digest to a disposable plastic chromatography column, containing approx 500 µL gel-bound lectin. Consult Table 1 for choice of suitable lectins and eluting conditions.
4. Allow to adsorb for 10 min at room temperature and apply another 200 µL of the soluble fraction. These steps may be repeated several times.
5. Wash by adding 10 vol each of 500 µL TBS without detergent.
6. Collect the pass-through fractions and count 10 µL of each in a β-counter to ensure that the last wash fractions have reached a background level.
7. Elute with 5 vol (500 µL) of sugar solution recommended in Table 1. Identify the peak fractions after counting of 10 µL each of the eluted fractions. These fractions are pooled and stored frozen prior to further experimentation.

3.5. Release of O-Linked Glycans

3.5.1. Alkaline Borohydride Treatment (β-Elimination)

1. Prepare a fresh stock solution, consisting of $2M$ NaBH$_4$ in $0.1M$ NaOH.
2. Desalt Pronase-resistant glycopeptides (obtained as described in Section 3.4., step 7) by gel filtration in a Pharmacia PD-10 G25 column (see Note 15). Elute stepwise by adding 500 µL vol of water. Identify the peak fractions after counting of 10 µL each of the eluted fractions. The glycopeptides should elute with the void vol of approx 2.2–4.5 mL.
3. Add freshly prepared $2M$ NaBH$_4$ in $0.1M$ NaOH to the desalted glycopeptides in a 8–10 mL hydrolysis tube (Teflon cap). Avoid larger volumes than 2 mL.
4. Deaerate by streaming nitrogen through the solution and cap the tube cautiously to avoid oxygen contamination.
5. Incubate at 48°C for 16 h.
6. Add a drop of 0.01% phenol red solution and neutralize by adding $1M$ acetic acid dropwise (see Note 16). The solution is considered neutral when a light red to orange color is achieved.
7. Evaporate to dryness under a stream of nitrogen.
8. Add 5 mL methanol and a drop of acetic acid and evaporate to dryness under a stream of nitrogen. Repeat this step at least 10 times to remove excess borate.

3.5.2. O-Glycanase and N-Acetylgalactosaminidase Digestion

1. This approach is not suitable for quantitative recovery of all possible O-glycan structures and has been used only occasionally in the author's laboratory. For a recent laboratory manual, the reader is referred to ref. 26.

3.5.3. Hydrazinolysis Kit for Release of O-Linked Glycans

1. A commercial kit for release of O-linked glycans, based on hydrazinolysis, is available from Oxford Glycosystems. Hydrazinolysis gives reproducible, almost quantitative, yields of O-linked glycans, independent of their structure, and the product is claimed to be relatively easy to handle. The product has not yet been evaluated in the authors' laboratory. Procedures for hydrazinolyisis of O-linked glycans were recently given in a laboratory manual (27).

3.6. Gel Filtration of O-Linked Glycans

1. Swell Bio-Gel P2 (200–400 mesh) in $0.1M$ pyridine/acetate, pH 7.8. Consult the manufacturer's instructions for details regarding swelling and degassing of the slurry. Equilibrate the slurry at 4°C (see Note 17).
2. Pack the gel in a long chromatography column ($\phi = 0.9$ cm) at 4°C (see Note 18). The height of the gel should be 120 cm or more. Use a plastic funnel with wide outlet, fixed to the top of the column as shown in Fig. 6 as a packing aid, pour the slurry gently into the funnel, avoiding generation of air bubbles, and allow the beads to pack, using hydrostatic pressure of 1–1.5 m (see Fig. 6).

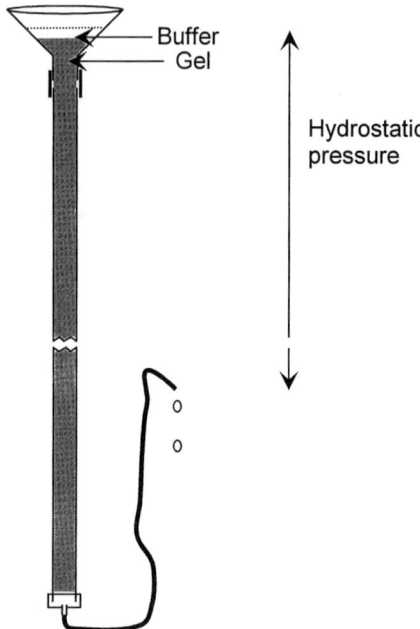

Fig. 6. Funnel arrangement for easy packing of long columns. A funnel is connected with the top of the column via a few centimeters of silicone tubing. Be sure that the connection is leakproof. Add 1–2 cm of water at the bottom of the column by using a syringe, attached to outlet tubing. Thereafter, the slurry, diluted to approx 1.5 times the volume of the swollen gel in pyridine acetate buffer, is carefully poured into the funnel. The distance between the buffer surface and the tubing outlet is adjusted to give an appropriate hydrostatic pressure and flow rate as indicated in the text.

3. After packing, attach a peristaltic pump and start washing the column. Consult the manufacturer's instructions for details as to the flow rate (*see* Note 19).
4. Dissolve the released *O*-linked glycans in $0.1M$ pyridine/acetate, pH 7.8, and apply to the column in a vol not exceeding 1 mL. Use the same eluting conditions as for washing of the column as described above. Collect 0.5 mL fractions and count about 10% of each in a liquid scintillation counter.
5. Determine the positions of each peak and pool appropriate fractions. For further analysis, *see* Note 20.

3.7. Determination of Monosaccharide Composition

1. Dry the pooled peaks of separated *O*-linked glycans under a stream of nitrogen.
2. Dissolve the material in $3M$ HCl, using a stream of nitrogen to remove oxygen.
3. Incubate at 100°C for 3 h and dry under a stream of nitrogen. Use Teflon-capped hydrolysis tubes.

4. Dissolve the samples in the smallest possible volume of water for further analysis on thin layer chromatography.
5. Draw a light pencil line on the white side of a precoated silica plate (Merck, Alufolien), defining the position for the samples to be applied. This line should be situated at approx 1 cm from the edge of the plate.
6. Aspirate 50–100 µL of the samples, using a Hamilton syringe, and apply small vol along the pencil line, forming a 0.5–1.0 cm broad and 2–3 cm long spot. Use an electric hair dryer if necessary to restrict the width of the spot. Each sample spot should be separated by at least 0.5 cm.
7. Pour the developing solution (ethanol/pyridine/1-butanol/acetic acid/water [100:10:10:3:30]) into a standard chromatography tank. The level of developer should be lower than the distance between the edge of the plate and the pencil line, serving as a chromatography origin. Put the lid on top of the tank and allow the developer to equilibrate with the gas phase for a few minutes.
8. Carefully place the plate in the tank and develop until the solvent front has moved to about 1 cm from the opposite edge.
9. Dry the plate and identify the position of aminosugar markers after staining with a ninhydrin spray.
10. Divide (pencil) each lane into 5-mm strips by using a pair of scissors. Define the positions of the separated monosaccharides by liquid scintillation counting (*see* Note 21) and determine the amount of radioactivity associated with the stained monosaccharide markers.

4. Notes

1. The ATCC catalog is easily accessed via internet (World Wide Web: gopher://culture.atcc.org:70/1).
2. It is advisable to add antibiotics such as penicillin (10.000 IU/mL) and streptomycin (100 µg/mL).
3. Small quantities of mouse monoclonal antibodies to gC-1 may be available from the authors' laboratory. Please inquire for further details.
4. Such columns can be reused repeatedly after careful washing.
5. The efficiency of labeling is proportional to the specific radioactivity (µCi/mL) of the medium. It is economical with respect to radioactivity costs to use a small volume of medium for incubation, but it is essential not to let the cells dry out during the overnight incubation. In our hands, 1.5 mL is the smallest practical volume.
6. Up to 100 µCi may be discarded directly in the waste in most countries. Consult your local authorities.
7. For screening of different variants of gC-1, it could be more convenient to centrifuge for 10 min at 11,000g in an Eppendorf centrifuge. This will in most cases result in elimination of the majority of the particulate material.
8. The PNA column should be kept at 4°C during binding and washing to achieve optimal binding affinity.
9. Be careful not to contaminate the inner wall of the chromatography column with radiolabeled sample. This may cause unnecessary tailing.

10. The quantities of MAbs to gC-1 available from established laboratories are restricted and not sufficient for extensive experimentation. If immunosorbent-purified gC-1 is available, a polyclonal gC-1 rabbit serum should be prepared by immunizing rabbits with either lectin-purified or monoclonal-purified gC-1.
11. Calculate the amount of $1M$ Tris base needed to neutralize the volume of an eluted fraction and add to each fraction tube prior to elusion.
12. Equilibrate the PD-10 column with 3 bed vol of TBS prior to addition of the sample.
13. If the sample volume is 2 mL the glycoprotein is elated with the void volume (approx fraction 5–9) and the low molecular weight contaminants elute at fraction 13–20.
14. Three minutes of boiling is sufficient for destroying most of the pronase activity. However, if subsequent experimentation includes long incubations with sensitive proteins (i.e., lectins) at room temperatures or higher, longer treatment at 100°C is recommended. It should be noted, however, that both sialic acid and fucose could be destroyed by prolonged harsh conditions.
15. Most glycopeptides will elute with the void volume, but smaller units may be partly included.
16. The color may shift from red to yellow after addition of one single drop. If no additional drop was added after shift to yellow, there is no need for increasing the pH. Note that it may be necessary to add more phenol red to replace indicator being destroyed by alkaline BH_4.
17. Warm slurry to 50–60°C and deaerate by applying tap vacuum for 10 min.
18. Higher temperatures such as 55–60°C will give higher resolution, but keeping the column in a refrigerated room is probably the most easy way to store the column at a constant temperature. The conditions described give satisfactory resolution of N-linked glycans at least up to nonasaccharide.
19. Usually it is advantageous not to apply the highest flow rate possible but to adjust the rate so one gel filtration is completed conveniently overnight (16 h). If signs of gel compression are noted, it is advisable to reverse the flow and apply the sample to the bottom of the column.
20. It is important not to lower concentration of pyridine-acetate in the eluting buffer since this may cause unwanted retention of the glycans on the polyacrylamide matrix of Bio-Gel P2.
21. Care should be taken to protect the operator from breathing the tritium-containing dust when fractionating the TLC strips. Whereas tritium-labeled metabolites in solution are more or less harmless, the risks associated with internal radiation of inhaled tritium should not be neglected. The use of a fume hood is strongly recommended.

Acknowledgments

This work was supported by grants from The Swedish Medical Research Council (Grants 9083), The National Swedish Board for Technical Development (Project 87 0256P), The Medical Faculty, University of Göteborg.

References

1. Olofsson, S., Jeansson, S., and Lycke, E. (1981) Unusual lectin-binding properties of a herpes simplex virus type 1-specific glycoprotein. *J. Virol.* **38**, 564–570.
2. Olofsson, S., Blomberg, J. and Lycke, E. (1981) O-glycosidic carbohydrate-peptide linkages of herpes simplex virus glycoproteins. *Arch. Virol.* **70**, 321–329.
3. Shida, M. and Dales, S. (1981) Biogenesis of vaccinia virus: carbohydrate of the hemagglutinin molecule. *Virology* **111**, 56–72.
4. Niemann, H. and Klenk, H. D. (1981) Coronavirus glycoprotein E1, a new type of viral glycoprotein. *J. Mol. Biol.* **153**, 993–1010.
5. Olofsson, S., Sjoblom, I., Lundström, M., Jeansson, S., and Lycke, E. (1983) Glycoprotein C of herpes simplex virus type 1: characterization of O-linked oligosaccharides. *J. Gen. Virol.* **64**, 2735–2747.
6. Lundström, M., Olofsson, S., Jeansson, S., Lycke, E., Datema, R., and Mansson, J. E. (1987) Host cell-induced differences in O-glycosylation of herpes simplex virus gC-1.1. Structures of nonsialylated HPA- and PNA-binding carbohydrates. *Virology* **161**, 385–394.
7. Dall'Olio, F., Malagolini, N., Speziali, V., Campadelli-Fiume, G., and Serafini-Cessi, F. (1985) Sialylated oligosaccharides O-glycosidically linked to glycoprotein C from herpes simplex virus type 1. *J. Virol.* **56**, 127–134.
8. Gottschalk, A. (1960) Correlation between composition, structure, shape and function of a salivary mucoprotein. *Nature* **186**, 949–951.
9. WuDunn, D. and Spear, P. G. (1989) Initial interaction of herpes simplex virus with cells is binding to heparan sulfate. *J. Virol.* **63**, 52–58.
10. Cines, D. B., Lyss, A. P., Bina, M., Corkey, R., Kefalides, N. A., and Friedman, H. M. (1982) Fc and C3b receptors induced by herpes simplex virus on cultured endothelial cells. *J. Clin. Invest.* **69**, 123–128.
11. Olofsson, S. (1992) Carbohydrates in herpesvirus infections. *APMIS* **100(suppl. 27)**, 84–95.
12. Serafini-Cessi, F., Malagolini, N., Dall'Olio, F., Pereira, L., and Campadelli-Fiume, G. (1985) Oligosaccharide chains of herpes simplex virus type 2 glycoprotein G.2. *Arch. Biochem. Biophys.* **24**, 866–876.
13. Olofsson, S., Lundström, M., Marsden, H., Jeansson, S., and Vahlne, A. (1986) Characterization of a herpes simplex virus type 2-specified glycoprotein with affinity for N-acetylgalactosamine-specific lectins and its identification as g92K or gG. *J. Gen. Virol.* **67**, 737–744.
14. Thorley-Lawson, D. A. and Poodry, C. A. (1982) Identification and isolation of the main component (gp350-gp220) of Epstein-Barr virus responsible for generating neutralizing antibodies in vivo. *J. Virol.* **43**, 730–736.
15. Wertz, G. W., Collins, P. L., Huang, Y., Gruber, C., Levine, S., and Ball, L. A. (1985) Nucleotide sequence of the G protein gene of human respiratory syncytial virus reveals an unusual type of viral membrane protein. *Proc. Natl. Acad. Sci. USA* **82**, 4075–4079.
16. Serafini-Cessi, F., Dall'Olio, F., Malagolini, N., Pereira, L., and Campadelli-Fiume, G. (1988) Comparative study on O-linked oligosaccharides of glycoprotein D of herpes simplex virus types 1 and 2. *J. Gen. Virol.* **69**, 869–877.

17. Elhammer, A. P., Poorman, R. A., Brown, E., Maggiora, L. L., Hoogerheide, J. G., and Kezdy, F. J. (1993) The specificity of UDP-GalNAc:polypeptide N-acetylgalactosaminyltransferase as inferred from a database of in vivo substrates and from the in vitro glycosylation of proteins and peptides. *J. Biol. Chem.* **268,** 10,029–10,038.
18. Svennerholm, B., Olofsson, S., Jeansson, S., and Vahlne, A. (1984) Herpes simplex virus type-selective enzyme-linked immunosorbent assay with *Helix pomatia* lectin-purified antigens. *J. Clin. Microbiol.* **19,** 235–239.
19. Olofsson, S., Jeansson, S., and Hansen, J.-E. S. (1994) Use of lectins in general and diagnostic virology, in *Lectin-Microrganism Interactions* (Doyle, R. J. and Slifkin, M., eds.), Marcel Dekker, New York, pp. 67–109.
20. Lundström, M., Olofsson, S., Jeansson, S., Lycke, E., Datema, R., and Mansson, J.-E. (1987) Host cell induced differences in O-glycosylation of the herpes simplex virus gC-1.1. Structures of non-sialylated HPA- and PNA-binding carbohydrates. *Virology* **161,** 385–394.
21. Olofsson, S., Lundström, M., Jeansson, S., and Lycke, E. (1985) Different populations of herpes simplex virus glycoprotein C discriminated by the carbohydrate-binding characteristics of N-acetylgalactosamine specific lectins (soybean and *Helix pomatia*). *Arch. Virol.* **86,** 121–128.
22. Olofsson, S., Norrild, B., Andersen, A., Pereira, L., Jeansson, S., and Lycke, E. (1983) Populations of herpes simplex virus glycoprotein with and without affinity for the N-acetylgalactose-specific lectin of *Helix pomatia*. *Arch. Virol.* **76,** 25–38.
23. Olofsson, S., Sjoblom, I., Lundström, M., Jeansson, S., and Lycke, E. (1983) Glycoprotein C of herpes simplex virus: characterization of O-linked oligosaccharides. *J. Gen. Virol.* **64,** 2735–2747.
24. Carlson, D. M. (1968) Structure and immunochemical properties of oligosaccharides isolated from pig submaxillary mucins. *J. Biol. Chem.* **243,** 616–626.
25. Spiro, R. G. (1966) Characterization of carbohydrate units of glycoproteins. *Meth. Enzymol.* **8,** 26–52.
26. Piller, F. and Piller, V. (1993) Structural characterization of mucin-type O-linked oligosaccharides, in *Glycobiology. A Practical Approach* (Fukuda, M. and Kobata, A., eds.), IRL Press at Oxford University Press, Oxford, UK, pp. 291–328.
27. Patel, T. P. and Parekh, R. B. (1994) Release of oligosaccharides from glycoproteins by hydrazinolysis, in *Methods in Enzymology. Guide to Techniques in Glycobiology* Lennarz, W. J. and Hart, G. W., eds.), Academic, New York, pp. 57–66.
28. Frink, R. J., Eisenberg, R., Cohen, G., and Wagner, E. K. (1983) Detailed analysis of the portion of herpes simplex virus genome encoding glycoprotein C. *J. Virol.* **45,** 634–647.
29. Sjöblom, I., Sjögren-Jansson, E., Glorioso, J. C., and Olofsson, S. (1992) Antigenic structure of the herpes simplex virus type 1 glycoprotein C: demonstration of a linear epitope, situated in an environment of highly conformation-dependent epitopes. *APMIS* **100,** 229–236.
30. Trybala, E., Bergström, T., Svennerholm, B., Jeansson, S., Glorioso, J. C., and Olofsson, S. (1994) Localization of a functional site of HSV-1 gC-1 involved in binding to cell surface heparan sulfate. *J. Gen. Virol.* **75,** 743–752.

III

LECTINS FOR DETECTION OF ALTERED GLYCOSYLATION OF CIRCULATING GLYCOPROTEINS

15

Use of Lectin for Detection of Agalactosyl IgG

Naoyuki Tsuchiya, Tamao Endo, Naohisa Kochibe, Koji Ito, and Akira Kobata

1. Introduction
1.1. Agalactosyl IgG

The immunoglobulin G (IgG) molecule contains two biantennary complex-type oligosaccharide chains, each linked to the heavy chain at asparagine 297 within the CH2 domain *(1)* (*see* Note 1). X-ray crystallographic analysis suggested that the sugar chains of IgG play a role in maintaining the 3D structure of its Fc portion by bridging the two CH2 domains *(2–4)*. Although these sugar chains can possess the complete structure shown in Fig. 1, normally only 25% of the sugar chains are sialylated, which is unusual because the sugar chains of other serum glycoproteins are highly sialylated. Also characteristic is the extremely high microheterogeneity resulting from the presence or absence of the two galactose (Gal), the bisecting N-acetylglucosamine (GlcNAc), and the fucose (Fuc) residues *(1)*.

It was originally reported by Mullinax that the galactose content of IgG is significantly decreased in the patients with rheumatoid arthritis (RA) *(5)*. Much interest has been focused on the IgG, which contains two oligosaccharide chains lacking galactose and terminates in N-acetylglucosamine—various names were given to this molecule: agalactosyl IgG, agalacto IgG, Gal (0), and G(0)—since quantitative structural analysis using hydrazinolysis clearly demonstrated that such IgG is markedly increased in the patients with RA *(4)*.

Subsequent studies using lectin binding assays demonstrated that the level of agalactosyl IgG correlated with the disease activity of RA *(6–8)*. Age-related increase in agalactosyl IgG is noted in healthy individuals *(9)*. Patients with

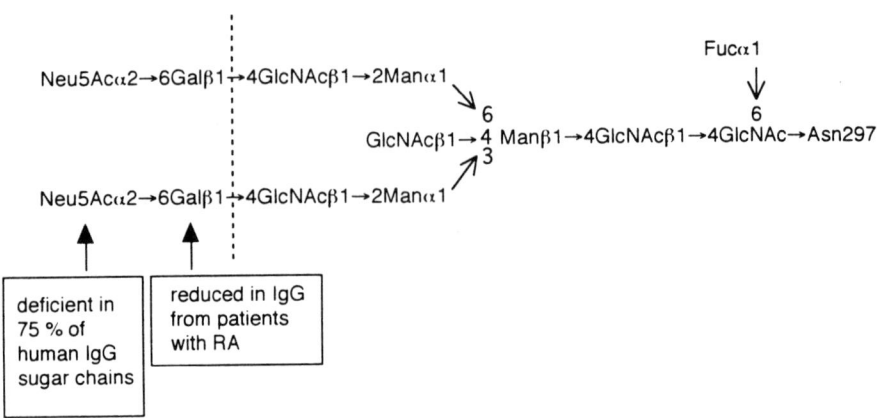

Fig. 1. Structure of asparagine-linked sugar chain of human IgG.

tuberculosis, juvenile rheumatoid arthritis, or Crohn's disease are also shown to have increased circulating agalactosyl IgG *(6,8,10,11)*. Increased agalactosyl IgG has also been reported in rodent models of arthritis *(12–14)*.

The decrease of galactose seems to be caused by insufficient galactosylation during the biosynthesis of the sugar moieties, rather than by postsecretory modification *(15,16)*. Significant reduction of IgG-specific galactosyl-transferase (GalTase) is reported in B lymphocytes from patients with RA. Kinetic studies demonstrated that the reduction is caused by the decreased affinity of GalTase for UDP-galactose, the donor of galactose residues *(17)*. Another line of evidence suggests the role of interleukin-6 for the increase of agalactosyl IgG *(14,18–20)*. Clear explanation has not been given of how the cytokine overproduction is associated with the alteration of the enzyme kinetics. Another report suggested the role of serum antibodies to GalTase *(21)*. The significance of this observation remains unclear.

The biological functions of IgG such as complement activation and binding to Fcγ receptors are dependent on the Fc portion of IgG. In addition, the epitope for rheumatoid factors, autoantibodies to IgG predominantly found in patients with RA, is located within the Fc portion. It had previously been shown that total removal of sugar chains from IgG resulted in the decrease of its affinity for C1q or Fcγ receptors *(22,23)*. Enzymatically degalactosylated IgG also showed reduced affinity for C1q or human Fcγ receptor type I *(24)*. However, binding to IgM-rheumatoid factor or protein A, which are dependent on the CH2–CH3 interface region of IgG, remained intact *(10,24,25)*. These results seem to indicate that the removal of galactose has a significant effect on the structure of CH2 domain, but not on that of CH2–

CH3 interface region. Another report suggested that the absence of galactose may result in only slight reduction in the affinity of IgG4 myeloma protein for Fcγ receptor type I *(26)*. Because IgG4 represents a small fraction of total serum IgG, further studies using other IgG subclasses are anticipated.

In summary, serum agalactosyl IgG is increased in the serum from patients with some of the chronic inflammatory diseases, most prominently in RA. The production of agalactosyl IgG is associated with a change in the activity of GalTase. Thus far, evidence suggesting that agalactosyl IgG is the autoantigen for the rheumatoid factor has not been reported, although there may be a defect in the clearance of immune complexes formed by agalactosyl IgG.

1.2. Detection of Serum Agalactosyl IgG

Attempts have been made to utilize the measurement of agalactosyl IgG for the diagnosis or clinical monitoring of RA *(6–8,27)*. Although structural analysis affords the most precise determination of the galactose content *(4,28)*, it is virtually impossible for average clinical laboratories to use it for the purpose of screening a large number of samples. For this reason, investigators made an effort to establish a simple method to estimate the relative level of agalactosyl IgG. Sumar et al. used a dot-blot assay of purified IgG, which was subsequently heat-denatured to expose the sugar moiety. Then the blot was probed with either *Griffonia simplicifolia* (*Bandeiraea simplicifolia*) II (BS-II) or a monoclonal antibody against N-acetylglucosamine *(29)*. Parkkinen used an ELISA-based assay, in which the microtiter plates coated with BS-II were incubated with the serum and bound IgG was detected by antibody to human IgG-Fc *(30)*.

It has been recently reported that *Psathyrella velutina* lectin (PVL) *(31)* reacts specifically with N-acetylglucosamine. Whereas other GlcNAc-specific lectins from higher plants preferentially interact with the GlcNAcβ1→4 residues, PVL reacts preferentially with GlcNAcβ1→2 residues *(32)*, exposed at the termini of agalactosyl IgG. We therefore took advantage of this and used PVL for the detection of agalactosyl IgG in serum and synovial fluid. Agalactosyl IgG could easily be detected by lectin blotting of purified IgG (Fig. 2). Moreover, relative estimation was possible by ELISA-based immunoassay, without heat-denaturation of IgG (Fig. 3B) (*see* Note 2). Using this system, we were able to confirm that agalactosyl IgG was markedly increased in patients with RA (Fig. 4) *(8)*.

In the following sections, we describe the procedures for the detection of agalactosyl IgG using PVL.

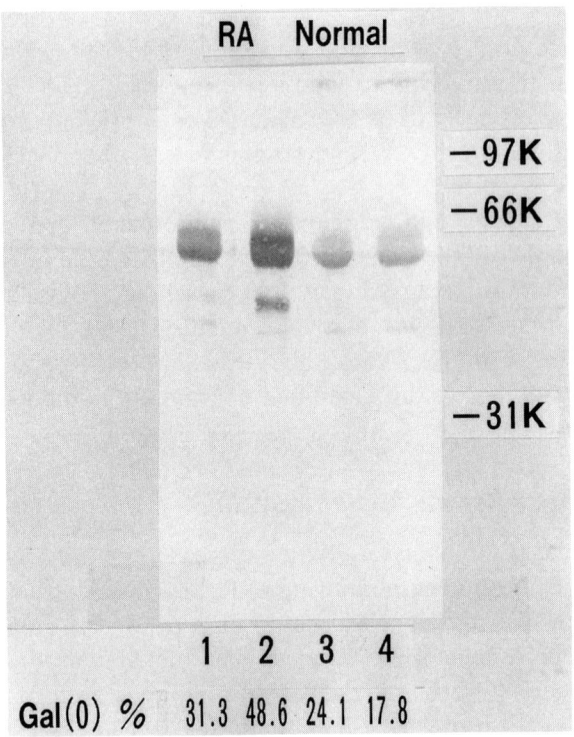

Fig. 2.

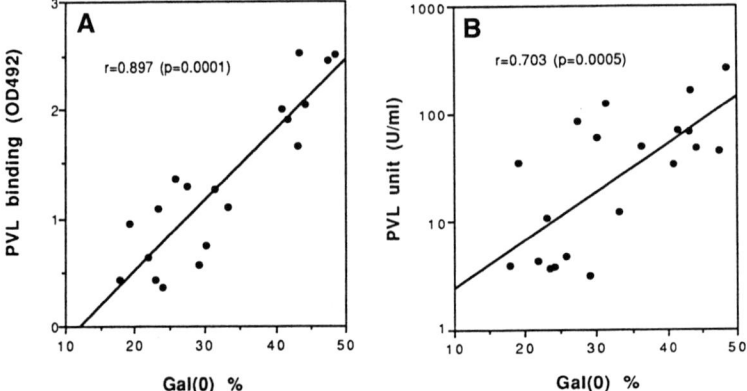

Fig. 3.

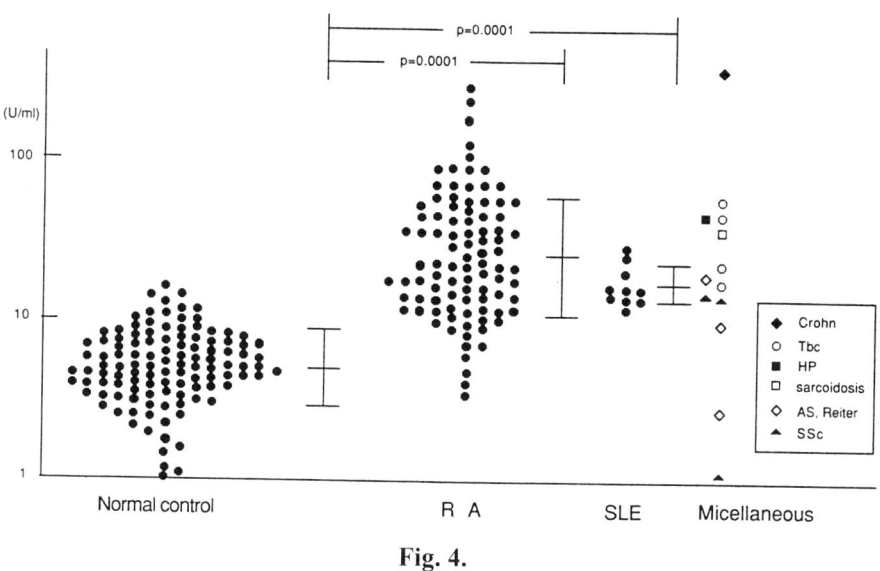

Fig. 4.

Fig. 2. *(previous page)* Lectin-blotting of purified serum IgG. Protein G-purified serum IgG was separated by SDS-PAGE (12.5%) and was transferred to PVDF membrane. The blot was probed with biotinylated PVL, and was visualized by avidin-biotin-peroxidase complex. Lanes 1 and 2, patients with RA; lanes 3 and 4, normal controls. 31K, 66K, 97K: molecular weight markers. Gal (0)% determined from the structural analysis is shown under each lane. Reprinted with permission from ref. *8*. Copyright 1993, *Journal of Immunology*.

Fig. 3. *(previous page)* Correlation of PVL binding assay and chemically-determined Gal (0)%. Gal (0)% obtained from the structural analysis of serum IgG purified from 14 patients with RA and 6 normal controls are compared with the PVL binding assays. **(A)** The serum was reduced and alkylated, and assayed for the reactivity with PVL at the final concentration of 1:10,000. **(B)** The serum was assayed without reduction at the final concentration of 1:200. PVL binding is shown by the actual absorbance. Reprinted with permission from ref. *8*. Copyright 1993, *Journal of Immunology*.

Fig. 4. *(this page)* PVL binding of inflammatory diseases and normal controls. In order to standardize interassay variability, incremental concentration of a reference serum was included in each assay. All sample sera were assayed at 1:200 dilution. The bars indicate mean ± SD values. SLE: systemic lupus erythematosus, Tbc: tuberculosis, HP: hypersensitivity pneumonitis, AS: ankylosing spondylitis, SSc: systemic sclerosis. Reprinted with permission from ref. *8*. Copyright 1993, *Journal of Immunology*.

2. Materials
2.1. Biotinylation of PVL
1. PVL is purified from fruiting bodies of *P. velutina* as described in ref. *31*. It is now commercially available through Wako Chemicals (Osaka, Japan).
2. Immunopure NHS-LC biotinylation kit (Pierce, Rockford, IL).
3. 5 mM Phosphate buffer, 150 mM NaCl, 10% glycerol, pH 7.2.
4. *N*-acetylglucosamine.

2.2. Purification of Serum IgG
1. Protein G-coupled agarose (*see* Note 3).
2. PBS: 15 mM phosphate buffer, 150 mM NaCl, pH 7.4.
3. 0.1M Glycine-HCl, pH 2.5.
4. 1M Tris-HCl, pH 8.0.
5. Centriprep-30 (Amicon, Beverly, MA).

2.3. Lectin-Blotting of Purified IgG
1. Polyacrylamide gels (12.5%) (acrylamide:bis = 37.5:1).
2. Laemmli sample buffer (2% SDS, 10% glycerol, 5% 2-mercaptoethanol, 0.06M Tris, 0.025% bromphenol blue).
3. PVDF membrane (Immobilon P; Millipore, Marlborough, MA).
4. PBS (pH 7.4) containing 1% nonfat dry milk (*see* Note 4).
5. PBS containing 1% nonfat dry milk and 0.05% Tween-20.
6. Avidin–biotin–peroxidase complex (Vectastain ABC Elite kit; Vector, Burlingame, CA).
7. Substrate solution: PBS containing 0.4 mg/mL of diaminobenzidine and 0.001% hydrogen peroxide (freshly prepared).

2.4. Enzyme-Linked Lectin Binding Assay
1. ELISA plates (e.g., Immulon 4; Dynatech, Chantilly, VA).
2. Recombinant protein G (*see* Note 5).
3. PBS.
4. PBS containing 0.05% Tween-20 (PBS–Tween).
5. PBS containing 0.05% Tween-20 and 1% bovine serum albumin (BSA; RIA grade such as Sigma A7030; *see* Note 6; PBS–BSA–Tween).
6. Biotinylated PVL.
7. Avidin–biotin–peroxidase complex (Vectastain ABC Elite kit; Vector).
8. Substrate solution: *o*-phenylenediamine (0.4 mg/mL) in phosphate-citrate buffer, pH 5.0, containing 0.001% hydrogen peroxide (freshly prepared).
9. 2.5N H_2SO_4.

3. Methods
3.1. Biotinylation of PVL
1. Prepare 1 mL of PBS containing 2.5 mg PVL, 10 mM *N*-acetylglucosamine and 10% glycerol (*see* Note 7).

2. Add 2 mg of NHS-LC biotin to 100 µL of distilled water. Add 35 µL of biotin solution to the PVL solution.
3. Incubate at room temperature for 3 h.
4. Dialyze the biotinylated PVL solution extensively against PBS containing 10% glycerol at 4°C.
5. Purify the biotinylated PVL by passing through a 10 mL desalting column equilibrated with PBS-10% glycerol.
6. Store frozen in PBS containing 10% glycerol and 0.1% sodium azide.

3.2. Lectin Blotting of Purified IgG

1. Dilute serum (0.5–1.0 mL) with an equal volume of PBS.
2. Pass through a column containing protein G-coupled Sepharose.
3. Wash the column with at least 10 vol of PBS.
4. Elute IgG with $0.1M$ glycine-HCl (pH 2.5). Immediately neutralize the eluate with 1/10 vol of $1M$ Tris-HCl buffer, pH 8.0.
5. Concentrate the eluate by Centriprep-30. Exchange the buffer by the repetitive addition of PBS and centrifugation.
6. Determine the concentration of IgG by measuring the absorbance at 280 nm.
7. Mix PBS solution containing 15 µg of IgG with an equal volume of Laemmli sample buffer.
8. Heat at 60°C for 30 min.
9. Centrifuge at $10,000g$ for 1 min.
10. Run the supernatant on a SDS-PAGE gel (12.5%).
11. Soak PVDF membrane in methanol for 20 s, and then in deionized water for 5 min with shaking.
12. Electroblot at 100 mA for 1.5 h using a semidry blotter.
13. Carefully separate the membrane from the gel and soak in the blocking solution (PBS containing 1% nonfat dry milk) at least for 1 h. If the membrane is blocked for more than overnight, addition of 0.02% sodium azide to the blocking solution is recommended to inhibit bacterial growth.
14. Wash the membrane with PBS containing 1% nonfat dry milk and 0.05% Tween-20 for 5 min.
15. Incubate the membrane with biotinylated PVL (3 µg/mL of the same buffer) for 1.5 h.
16. Prepare the avidin–biotin–peroxidase solution (Vectastain ABC Elite kit, Vector) according to the supplier's instruction.
17. Wash the membrane three times with the same buffer.
18. Put the membrane in avidin–biotin–peroxidase solution and incubate for 20 min.
19. Wash the membrane four times with the same buffer.
20. Add PBS containing 0.4 mg/mL of diaminobenzidine and 0.001% hydrogen peroxide.
21. After the bound lectin is visualized, stop the reaction by washing the membrane with distilled water.

3.3. Enzyme-Linked PVL Binding Assay (see Note 8)

1. Coat ELISA plates with 75 µL/well of recombinant protein G solution (5 µg/mL of PBS).
2. Incubate overnight at 4°C.
3. Wash plates three times with PBS-0.05% Tween-20.
4. Block nonspecific protein binding sites with 300 µL/well of PBS–BSA–Tween-20 for 1 h at room temperature.
5. Wash plates with PBS–Tween.
6. Dilute serum 1:200 with PBS–BSA–Tween. Add 75 µL/well of diluted serum to the duplicate wells. Incubate for 1 h at room temperature.
7. Wash the plates four times with PBS–Tween.
8. Add 75 µL/well of biotinylated PVL (5 µg/mL of PBS–BSA–Tween). Incubate for 1 h at room temperature.
9. Prepare the avidin–biotin–peroxidase solution (Vectastain ABC Elite kit, Vector) according to the supplier's instruction.
10. Wash the plates with PBS–Tween four times.
11. Add 75 µL/well of avidin–biotin–peroxidase solution and incubate for 20 min.
12. Wash the plates with PBS–Tween four times.
13. Add 75 µL/well of substrate buffer and incubate for 20 min.
14. Stop the reaction by addition of 50 µL/well of $2.5N\ H_2SO_4$.
15. Read the absorbance at 490 nm by using an ELISA reader.

4. Notes

1. It should be noted that the sugar chains linked to asparagine 297 do not account for all of the oligosaccharide chains of IgG. IgGs from human sera contain 2.8 mol of sugar chains on the average, among which 2.0 mol are located at Asn 297 and the remainder are linked at the variable region (1). The procedures described in this chapter detect not only Asn 297-linked sugar chains, but also those in the variable regions.
2. Under normal conditions, IgG oligosaccharide chains are directed inside the Fc portion and not exposed. Some procedures such as reduction and alkylation or heat aggregation have been shown substantially to increase the sensitivity of the detection of agalactosyl IgG by lectins, perhaps by exposing the sugar chains. The protocol for reduction and alkylation is described below.
   a. Dilute serum 1:100 in $0.1M$ Tris-HCl buffer, pH 8.0, containing $0.12M$ 2-mercaptoethanol.
   b. Incubate at 37°C for 1 h and on ice for 15 min.
   c. Dilute the reaction mixture with 100 vol of PBS containing 1% BSA and 0.05% Tween-20.
   d. Add 75 µL/well of the reaction mixture to the ELISA plates coated with protein G and quenched with BSA, incubate for 1 h and follow the protocol for PVL binding assay from step 7 of Section 3.3.

   Figure 3A shows the correlation between ELISA absorbance obtained by this protocol and the percentage of agalactosyl IgG determined by the structural analysis.

3. FPLC-compatible prepacked columns such as HiTrap protein G (Pharmacia, Uppsala, Sweden) can also be used.
4. The use of nonfat dry milk in the buffer may substantially diminish the reaction of the lectins such as RCA, and possibly PVL as well. In such a case, the use of bovine serum albumin is recommended.
5. Protein A is also compatible with the system. However, recombinant protein G is preferred because protein A does not react with human IgG3 and reacts with a part of other Ig isotypes *(33,34)*.
6. It is essential to use BSA with the least contamination of bovine IgG, because bovine IgG binds to protein G and may inhibit the reaction between human IgG and protein G.
7. PVL tends to aggregate in a buffered saline, which is prevented by adding 10% (v/v) glycerol to the solution. *N*-acetylglucosamine was added during biotinylation to prevent the loss of reactivity by the biotinylation of *N*-acetylglucosamine binding site on PVL. *N*-acetylglucosamine can be removed from the binding site by dialysis.
8. It is essential to include reference sera in each ELISA plates for interassay standardization. The ideal references should be IgG samples with known percentage of agalactosyl IgG, determined by the structural analysis. Alternatives include myeloma IgGs lacking galactose, or polyclonal IgG from which galactose residues were artificially removed. For the relative comparison between different ELISA plates, simply placing 3–5 reference sera with variable reactivity with PVL or incremental concentration of a single reference serum may be acceptable.

References

1. Kobata, A. (1990) Function and pathology of the sugar chains of human immunoglobulin G. *Glycobiology* **1**, 5–8.
2. Deisenhofer, J. (1981) Crystallographic refinement and atomic models of a human Fc fragment and its complex with fragment B of protein A from *Staphylococcus aureus* at 2.9- and 2.8-Å resolution. *Biochemistry* **20**, 2361–2370.
3. Sutton, B. J. and Phillips, D. C. (1983) The three-dimensional structure of the carbohydrate within the Fc fragment of immunoglobulin G. *Biochem. Soc. Trans.* **11**, 130–132.
4. Parekh, R. B., Dwek, R. A, Sutton, B. J., Fernandes, D. J., Leung, A., Stanworth, D., Rademacher, T. W., Mizuochi, T., Taniguchi, T., Matsuta, K., Takeuchi, F., Nagano, Y., Miyamoto, T., and Kobata, A. (1985) Association of rheumatoid arthritis and primary osteoarthritis with changes in the glycosylation pattern of total serum IgG. *Nature* **316**, 452–457.
5. Mullinax, F. (1975) Abnormality of IgG structure in rheumatoid arthritis and systemic lupus erythematosus. *Arthritis Rheum.* **18**, 417.
6. Parekh, R. B., Roitt, I. M., Isenberg, D. A., Dwek, R. A., Anzell, B. M., and Rademacher, T. W. (1988) Galactosylation of IgG associated oligosaccharides: reduction in patients with adult and juvenile onset rheumatoid arthritis and relation to disease activity. *Lancet* **ii**, 966–969.

7. Rook, G. A. W., Steele, J., Brealey, R., Whyte, A., Isenberg, D., Sumar, N., Nelson, J. L., Bodman, K. B., Young, A., Roitt, I. M., Williams, P., Scragg, I., Edge, C. J., Arkwright, P. D., Ashford, D., Wormald, M., Rudd, P., Redman, C. W. G., Dwek, R. A., and Rademacher, T. W. (1991) Changes in IgG glycoform levels are associated with remission of arthritis during pregnancy. *J. Autoimmunity* **4**, 779–794.
8. Tsuchiya, N., Endo, T., Matsuta, K., Yoshinoya, S., Takeuchi, F., Nagano, Y., Shiota, M., Furukawa, K., Kochibe, N., Ito, K., and Kobata, A. (1993) Detection of glycosylation abnormality in rheumatoid IgG using *N*-acetylglucosamine-specific *Psathyrella velutina* lectin. *J. Immunol.* **151**, 1137–1146.
9. Parekh, R., Roitt, I., Isenberg, D., Dwek, R., and Rademacher, T. (1988) Age-related galactosylation of the N-linked oligosaccharides of human serum IgG. *J. Exp. Med.* **167**, 1731–1736.
10. Tomana, M., Schrohenloher, R. E., Koopman, W. J., Alarcon, G. S., and Paul, W. A. (1988) Abnormal glycosylation of serum IgG from patients with chronic inflammatory diseases. *Arthritis Rheum.* **31**, 333–338.
11. Dubé, R., Rook, G. A. W., Steele, J., Brealey, R., Dwek, R., Rademacher, T., and Lennard-Jones, J. (1990) Agalactosyl IgG in inflammatory bowel disease: correlation with C-reactive protein. *Gut* **31**, 431–434.
12. Bond, A., Cooke, A., and Hay, F. C. (1990) Glycosylation of IgG, immune complexes and IgG subclasses in the MRL-*lpr/lpr* mouse model of rheumatoid arthritis. *Eur. J. Immunol.* **20**, 2229–2233.
13. Mizuochi, T., Hamako, J., Nose, M., and Titani, K. (1990) Structural changes in the oligosaccharide chains of IgG in autoimmune MRL/Mp-*lpr/lpr* mice. *J. Immunol.* **145**, 1794–1798.
14. Thompson, S. J., Hitsumoto, Y., Zhang, Y. W., Rook, G. A. W., and Elson, C. J. (1992) Agalactosyl IgG in pristane-induced arthritis. Pregnancy affects the incidence and severity of arthritis and the glycosylation status of IgG. *Clin. Exp. Immunol.* **89**, 434–438.
15. Furukawa, K. and Kobata, A. (1991) IgG galactosylation—its biological significance and pathology. *Mol. Immunol.* **28**, 1333–1340.
16. Bodman, K. B., Sumar, N., MacKenzie, L. E., Isenberg, D. A., Hay, F. C., Roitt, I. M., and Lydyard, P. M. (1992) Lymphocytes from patients with rheumatoid arthritis produce agalactosylated IgG *in vitro. Clin. Exp. Immunol.* **88**, 420–423.
17. Furukawa, K., Matsuta, K., Takeuchi, F., Kosuge, E., Miyamoto, T., and Kobata, A. (1990) Kinetic study of a galactosyltransferase in the B cells of patients with rheumatoid arthritis. *Int. Immunol.* **2**, 105–112.
18. Hirano, T., Matsueda, T., Turner, M., Miyasaka, N., Buchan, G., Tang, B., Sato, K., Shimizu, M., Maini, R., Feldmann, M., and Kishimoto, T. (1988) Excessive production of interleukin 6/B cell stimulatory factor-2 in rheumatoid arthritis. *Eur. J. Immunol.* **18**, 1797–1801.
19. Nakao, H., Nishikawa, A., Nishiura, T., Kanayama, Y., Tarui, S., and Taniguchi, N. (1991) Hypogalactosylation of immunoglobulin G sugar chains and elevated serum interleukin 6 in Castleman's disease. *Clin. Chim. Acta* **197**, 221–228.

20. Hitsumoto, Y., Thompson, S. J., Zhang, Y. W., Rook, G. A. W., and Elson, C. J. (1992) Relationship between interleukin 6, agalactosyl IgG and pristane-induced arthritis. *Autoimmunity* **11**, 247–254.
21. Axford, J. S., Sumar, N., Alavi, A., Isenberg, D. A., Young, A., Bodman, K. B., and Roitt, I. M. (1992) Changes in normal glycosylation mechanisms in autoimmune rheumatic disease. *J. Clin. Invest.* **89**, 1021–1031.
22. Nose, M. and Wigzell, H. (1983) Biological significance of carbohydrate chains on monoclonal antibodies. *Proc. Natl. Acad. Sci. USA* **80**, 6632–6636.
23. Leatherbarrow, R. J., Rademacher, T. W., Dwek, R. A., Woof, J. M., Clark, A., and Burton, D. (1985) Effecter functions of a monoclonal aglycosylated mouse IgG2a: binding and activation of complement component C1 and interaction with human monocyte Fc receptor. *Mol. Immunol.* **22**, 407–415.
24. Tsuchiya, N., Endo, T., Matsuta, K., Yoshinoya, S., Aikawa, T., Kosuge, E., Takeuchi, F., Miyamoto, T., and Kobata, A. (1989) Effects of galactose depletion from oligosaccharide chains on immunological activities of human IgG. *J. Rheumatol.* **16**, 285–290.
25. Newkirk, M. M., Lemmo, A., and Rauch, J. (1990) Importance of the IgG isotype, not the state of glycosylation, in determining human rheumatoid factor binding. *Arthritis Rheum.* **33**, 800–809.
26. Lund, J., Takahashi, N., Pound, J. D., Goodall, M., Nakagawa, H., and Jefferis, R. (1995) Oligosaccharide-protein interaction in IgG can modulate recognition by Fcγ receptors. *FASEB J.* **9**, 115–119.
27. Young, A., Sumar, N., Bodman, K., Goyal, S., Sinclair, H. Roitt, I., and Isenberg, D. (1991) Agalactosyl IgG: an aid to differential diagnosis in early synovitis. *Arthritis Rheum.* **34**, 1425–1429.
28. Mizuochi, T., Taniguchi, T., Shimizu, A., and Kobata, A. (1982) Structural and numerical variations of the carbohydrate moiety of immunoglobulin G. *J. Immunol.* **129**, 2016–2019.
29. Sumar, N., Bodman, K. B., Rademacher, T. W., Dwek, R. A., Williams, P., Parekh, R. B., Edge, J., Rook, G. A. W., Isenberg, D. A., Hay, F. C., and Roitt, I. M. (1990) Analysis of glycosylation changes in IgG using lectins. *J. Immunol. Methods* **131**, 127–136.
30. Parkkinen, J. (1989) Aberrant lectin-binding activity of immunoglobulin G in serum from rheumatoid arthritis patients. *Clin. Chem.* **35**, 1638–1643.
31. Kochibe, N. and Matta, K. L. (1989) Purification and properties of an *N*-acetylglucosamine-specific lectin from *Psathyrella velutina* mushroom. *J. Biol. Chem.* **264**, 173–177.
32. Endo, T., Ohbayashi, H., Kanazawa, K., Kochibe, N., and Kobata, A. (1992) Carbohydrate binding specificity of immobilized *Psathyrella velutina* lectin. *J. Biol. Chem.* **267**, 707–713.
33. Sasso, E. H., Silverman, G. J., and Mannik, M. (1989) Human IgM molecules that bind staphylococcal protein A contain $V_H III$ H chains. *J. Immunol.* **142**, 2778–2783.
34. Sasso, E. H., Silverman, G. J., and Mannik, M. (1991) Human IgA and IgG F(ab')$_2$ that bind to staphylococcal protein A belong to the $V_H III$ subgroup. *J. Immunol.* **147**, 1877–1883.

16

Lectins for Detection of Altered Glycosylation of Circulating Glycoproteins

α-1-Antitrypsin

Yutaka Aoyagi and Hitoshi Asakura

1. Introduction

Human α-1-antitrypsin (α-1-anti-T), one of the most important serum protease inhibitors, is a glycoprotein of mol wt 51,000 *(1–3)*. α-1-anti-T neutralizes the activity of enzymes such as elastase, trypsin, and chymotrypsin. The molecule consists of a single polypepide chain of 394 amino acids, with three sugar chains that are N-linked to asparagine residue at positions 46, 83, and 247 *(2–4)*. The major chemical structures of the sugar chains have been reported to be biantennary and triantennary oligosaccharides *(4,5)*. Genetic polymorphism of α-1-anti-T has been elucidated by many investigators. About 40 genetic variants have been recognized, implying considerable heterogeneity and polymorphism. These complexities include variations in the degree of sialylation on the sugar chains and complex formation with low molecular weight thiols by disulfide bridges of the reactive single cysteine *(6)*. Hereditary deficiency of α-1-anti-T predisposes to degenerative lung disease and liver disease *(7,8)*. Nucleotide sequence studies of α-1-anti-T have shown single or two-base substitution or dinucleotide deletion with subsequent single amino acid substitutions or deletions. In the common S and Z variants, A to T (264 Glu to Val) and G to A (342 Glu to Lys) mutations were identified *(9)*.

Our previous work has revealed the presence of two species of α-1-anti-T: the *Lens culinaris* agglutinin (LCA)-reactive and LCA-nonreactive species determined by crossed immunoaffinoelectrophoresis (CIAE) with an increase in the percentage of LCA-reactive species in hepatocellular carcinoma (HCC) compared to nonneoplastic liver diseases and normal controls *(10)*.

From: *Methods in Molecular Medicine, Vol. 9: Lectin Methods and Protocols*
Edited by: J. M. Rhodes and J. D. Milton Humana Press Inc., Totowa, NJ

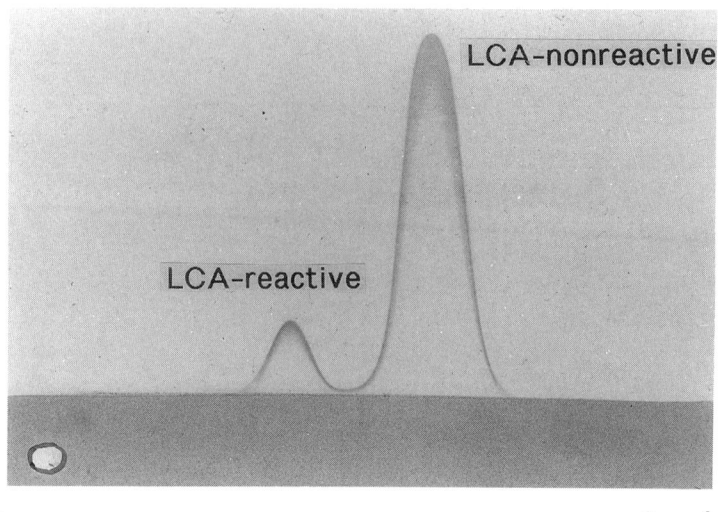

Fig. 1. Crossed immunoaffinoelectrophoretic (CIAE) pattern of α-1-anti-T from a patient with hepatocellular carcinoma.

1.1. Outline of Procedure

The presence of altered glycosylation of α-1-anti-T was determined by CIAE in the presence of LCA essentially according to the procedure by Bøg-Hansen *(11)*. CIAE was carried out in 1% agarose in 0.02M barbital buffer, pH 8.6, containing 0.5 mg/mL of soluble LCA. Serial dilution of the serum samples was subjected to CIAE with 10 mM phosphate buffer containing 0.15M NaCl from 1 part in 16 to 1 in 64 depending on the serum concentrations of α-1-anti-T. In the first dimension, 10 µL of a diluted serum sample was run at 10 V/cm for 3 h. Then, electrophoresis in the second dimension was performed at 2 V/cm for 20 h in an antibody containing gel. Two species of α-1-anti-T were detected on CIAE (Fig. 1). Migration of one of the two α-1-anti-T species was retarded (LCA-reactive species) and that of the other remained unchanged (LCA-nonreactive species). The area under the peak of immunoprecipitation after the second dimension run was measured. When CIAE was performed in gels containing 0.1, 0.3, 0.5, or 1.0 mg/mL of soluble LCA, the best resolution was obtained in the gel containing 0.5 mg/mL of soluble LCA *(10)*.

1.2. Increase in LCA-Reactive Species of α-1-Anti-T in HCC

An increase in LCA-reactive species of α-1-anti-T in HCC was observed as compared with that in liver cirrhosis. As shown in Fig. 2, the percentages of

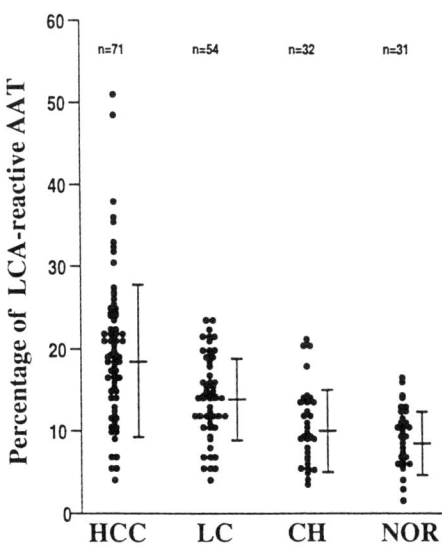

Fig. 2. The percentage of LCA-reactive species of α-1-anti-T in patients with hepatocellular carcinoma (HCC), liver cirrhosis (LC), chronic hepatitis (CH), and normal control (NOR). The percentages of LCA-reactive species in HCC (19 ± 9%, mean ± standard deviation) were significantly higher than those in liver cirrhosis (13 ± 5%), in acute hepatitis (15 ± 5), chronic hepatitis (10 ± 5), and normal control (8 ± 4). Vertical lines in each column indicate mean ± standard.

LCA-reactive species in HCC (19 ± 9%, mean ± standard deviation) were significantly higher than those in liver cirrhosis (13 ± 5%), acute viral hepatitis (15 ± 5), chronic hepatitis (10 ± 5) and normal control (8 ± 4). There was no correlation between the percentage of LCA-reactive species of α-1-anti-T and serum α-1-anti-T concentration in any group *(10)*.

When we compared LCA-reactive species of α-1-anti-T in the acute stage of acute viral hepatitis with that in the convalescent stage, there was no significant difference. However, serum α-1-anti-T concentration in the acute stage was significantly higher than that in convalescence *(10)*.

Next, we compared LCA-reactive species of α-1-anti-T in 15 pairs of serum samples from patients with a long history of chronic liver diseases before and after the subsequent development of HCC. The percentage of LCA-reactive species of α-1-anti-T after HCC development (21 ± 9%) was significantly higher than before (11 ± 5%), although there was no significant difference in serum concentrations of α-1-anti-T before and afterwards *(10)*.

1.3. Carbohydrate Structure of α-1-Anti-T Assessed by Lectin Reactivity

Chemical structures of the sugar chains of human α-1-anti-T have been proposed by Hodges et al. *(5)* and Mega et al. *(4)*. These authors have shown that human serum α-1-anti-T from normal individual contains two types of sugar chains, biantennary sugar chain, Galβ1-4GlcNAcβ1-2Manα1-6(Galβ1-4GlcNAcβ1-2Manα1-3)Manβ1-4GlcNAcβ1-4GlcNAc, and triantennary sugar chain, Galβ1-4GlcNAcβ1-4(Galβ1-4GlcNAcβ1-2)Manα1-3(Galβ1-4GlcNAcβ1-2Manα1-6)Manβ1-4GlcNAcβ1-4GlcNAc. They reported that the ratio of the biantennary sugar chain to the triantennary sugar chain was about 2:1 or 3:1. Bayard et al. *(12)* have shown the presence of another type of sugar chain, a biantennary sugar chain with a bisecting *N*-acetylglucosamine residue, and that the percentages of biantennary, biantennary with a bisecting *N*-acetylglucosamine residue, and triantennary sugar chains were 80, 14, and 6%, respectively.

Kornfeld et al. indicated that LCA binds biantennary sugar chains with a fucose residue at the innermost *N*-acetylglucosamine at the trimannosyl core (fucosylated biantennary sugar chain); and with both a fucose and a bisecting *N*-acetylglucosamine residue (fucosylated and *N*-acetylglucosaminylated biantennary sugar chain) *(13)*. On the other hand, Con A which binds the biantennary and the fucosylated biantennary sugar chains, will not bind if the biantennary sugar chain structure has undergone bisecting-glucosaminylation and/or further branching leading to the formation of triantennary and tetraantennary structures *(14)*.

Recently, a very sensitive and convenient method has been developed to study the fine detail of carbohydrate structures by a combination of derivatization into fluorescent oligosaccharides and separation by HPLC *(15–17)*. Since a number of PA-oligosaccharides, the structures of which have been established by ^{1}H-nuclear magnetic resonance, have become commercially available, we applied this method in combination with exoglycosidase digestion to study the fine structures of various α-1-anti-T species *(18)*. Eleven α-1-anti-T specimens were prepared from serum samples of six patients with HCC and five healthy controls by immunoaffinity chromatography. Subsequent affinity chromatography was performed with LCA-Sepharose to prepare the LCA-reactive and LCA-nonreactive species of α-1-anti-T. The LCA-reactive species of α-1-anti-T from patients with HCC carried both the biantennary sugar chain with a fucose residue at innermost *N*-acetylglucosamine residue, Galβ1-4GlcNAcβ1-2Manα1-6(Galβ1-4GlcNAcβ1-2Manα1-3)Manβ1-4GlcNAcβ1-4(Fucα1-6)GlcNAc and the biantennary one without a fucose residue, Galβ1-4GlcNAcβ1-2Manα1-6(Galβ1-4GlcNAcβ1-2Manα1-3)Manβ1-4GlcNAcβ1-4GlcNAc at a ratio about 1:0.6. The LCA-nonreactive species of α-1-anti-T contained the biantennary sugar chain, Galβ1-4GlcNAcβ1-2Manα1-6(Galβ1-4GlcNAcβ1-2Manα1-3)Manβ1-4GlcNAcβ1-4GlcNAc, as a major component. These results indicate that a characteristic feature of the carbohydrate chains of α-1-anti-T from patients with HCC is an increment in core fucosylation.

2. Materials

1. 1% Agarose (type with low gelling temperature) in $0.02M$ barbital buffer, pH 8.6.
2. 2% Agarose in $0.02M$ barbital buffer, pH 8.6.
3. Salt-free lyophilized LCA (L-5880 from Sigma, St. Louis, MO).
4. Monospecific goat antihuman α-1-anti-T serum.
5. Staining solution: 5 g Coomassie brilliant blue, 100 mL acetic acid, 450 mL ethanol, and 450 mL distilled water.
6. Destaining solution: Same as staining solution without Coomassie brilliant blue.
7. Phosphate-buffered saline: 10 mM phosphate buffer, pH 7.0, $0.15M$ NaCl.
8. Glass plates: 2.6 × 8.5 cm and 8.5 × 10 cm.
9. Gel puncher.
10. Immunoviewer, an apparatus by which immunoprecipitation lines can be visualized directly before staining.

3. Methods
3.1. First Dimension of Crossed Immunoaffinoelectrophoresis

1. Prepare 2% agarose in barbital buffer in a boiling water bath and then allow to cool to 56°C.
2. Dissolve 1.5 mg LCA in 1.5 mL $0.02M$ barbital buffer, pH 8.6, at room temperature and then warm to 56°C (*see* Note 3).
3. Mix 1.5 mL 2% agarose (step 1) and 1.5 mL LCA solution (step 2) at 56°C, and layer onto the 2.6 × 8.5 cm glass plate on the levelling table (*see* Note 2).
4. Punch two 2-mm wells at cathodal end of the plate with gel puncher (commercially available), and then suck out agar plugs with a Pasteur pipet connected to a water vacuum pump.
5. Dilute serum samples with PBS from 1 part in 16 to 1 in 64 depending on the serum concentrations of α-1-anti-T. Two samples, 10 μL of each, can be loaded on one 2.6 × 8.5 cm glass plate.
6. Electrophorese at 10 V/cm for 3 h (*see* Note 1).
7. Separate the first dimension gel into two pieces of gel strip by cutting the center of the gel along the major axis of the glass plate.

3.2. Second Dimension of Crossed Immunoaffinoelectrophoresis

1. Melt the 1% agarose in $0.02M$ barbital buffer, pH 8.6, in a boiling water bath and transfer to 56°C.
2. Add an appropriate amount of anti-α-1-anti-T antibody to 12 mL of agar at 56°C, and mix well, and layer onto the 10 × 8.5 cm glass plate on the levelling table (*see* Note 4).

3. Cut and remove a 1.3 × 8.5 cm longitudinal strip, the same size as a half of the first dimension gel, from one end of the gel, and replace with the longitudinal strip of the first dimension gel (*see* Note 5).
4. Place the plate in the electrophoresis apparatus. The cathode must be at the end of the plate where the longitudinal strip of the first dimension gel was placed.
5. Electrophorese at 2 V/cm overnight.

3.3. Staining

1. Wash the glass plate in several changes of PBS to remove free protein from second dimension gel for 12–24 h. Next, wash the plate in distilled water to remove salt from the gel (*see* Note 7).
2. Dry overnight.
3. Stain the precipitation line with Coomassie brilliant blue solution for 1–2 min (*see* Note 8), and then destain (*see* Notes 9–11).

4. Notes

1. An albumin and bromophenol blue complex can be used as a marker in the first dimension electrophoresis since this will migrate very rapidly. However, if excess dye has been added, a band of free dye will run in front of the albumin towards the anode.
2. The prepared gels should stored at 4°C in a humid chamber for 30 min before use. This procedure hardens the gel, and makes it easier to punch wells.
3. It is not always easy to dissolve LCA powder directly in 1% agarose. Mixing equal volumes of 2% agarose and LCA solution prevents foaming of the gel.
4. Antibody concentrations in the second dimension gel:
   a. α-1-anti-T concentration in diluted sample sera: 10–20 mg/dL.
   b. IgG concentration: 0.5–1 mg/mL.
   c. Anti-α-1-anti-T antibody/12 mL gel: 20–40 µL.
5. When the longitudinal strip of the first dimension gel is placed on the second dimension plate, avoid bubbles between the first and second dimension gels as this will disturb the formation of precipitation lines.
6. Precoating with 0.5% agar solution (0.5 g of agar in 100 mL of distilled water) holds the second dimension gel in place during the washing procedure *(18)*.
7. The presence of an immunoprecipitation line can be assessed by holding the gel over a black background and illuminating it from the side. There is a commercially available apparatus called an "Immunoviewer" (Jookoo, Tokyo, Japan).
8. It is enough for a few minutes to stain a second dimension gel.
9. The area under the peak of immunoprecipitation after the second dimension run can be easily quantitated by cutting out the peak from photocopy and weighing the paper, or by image scanner.
10. The staining and destaining solution can be stored for several months in a stoppered bottle at room temperature.
11. The used destaining solution can be regenerated by passing through powdered charcoal.

References

1. Chan, S. K., Luby, J., and Wu, Y. C. (1973) Purification and chemical compositions of human α-1-antitrypsin of the MM type. *FEBS Lett.* **35,** 79–81.
2. Carrell, R. W., Jeppsson, J. O., Vaughan, L., Brennan, S. O., Owen, M. C., and Boswell, D. R. (1981) Human α-1-antitrypsin: carbohydrate attachment and sequence homology. *FEBS Lett.* **135,** 301–303.
3. Carrell, R. W., Jeppsson, J. O., Laurell, C. B., Brennan, S. O., Owen, M. C., Vaughan, L., and Boswell, D. R. (1982) Structure and variation of human α-1-antitrypsin. *Nature* **298,** 329–334.
4. Mega, T., Lujuan, E., and Yoshida, A. (1980) Studies on the oligosaccharide chains of human α-1-protease inhibitor. *J. Biol. Chem.* **255,** 4053–4061.
5. Hodges, L. C., Laine, R., and Chan, S. K. (1979) Structure of the oligosaccharide chains in human α-1-protease inhibitor. *J. Biol. Chem.* **254,** 8208–8212.
6. Putnam, F. W. (1984) Alpha, beta, gamma, omega. The structure of the plasma proteins, VI α-glycoproteins, C protease inhibitors in *The Plasma Proteins vol. IV* (Putnam, F. W., ed.), Academic, Orlando, FL, pp 76–86.
7. Eriksson, S. (1964) Pulmonary emphysema and α-1-antitrypsin deficiency. *Acta Med. Scand.* **175,** 197–205.
8. Sharp, H. L., Bridges, R. A. Krivit, W., and Freier, E. F. (1969) Cirrhosis associated with α-1-antitrypsin deficiency. *J. Lab. Clin. Med.* **73,** 934–939.
9. Carrell, R. W., Jeppsson J.-O., Laurell C.-B., Brennan, S. O., Owen, M. C., Vaughan, L., and Boswell, R. (1982) Structure and variation of human α-1-antitrypsin. *Nature* **298,** 329–334.
10. Sekine, C., Aoyagi, Y., Suzuki, Y., and Ichida, F. (1987) The reactivity of α-1-antitrypsin with *Lens culinaris* agglutinin and its usefulness in the diagnosis of neoplastic diseases of the liver. *Br. J. Cancer* **56,** 371–375.
11. Bøg-Hansen, T. C. (1973) Crossed immuno-affinoelectrophoresis an analytical methods to predict the result of affinity chromatography. *Anal. Biochem.* **56,** 480–488.
12. Bayard, B., Kerckaert, J. P., Laine, A., and Hayem, A. (1982) Uniformity of glycans within molecular variants of alpha-protease inhibitor with distinct affinity for concanavalin A. *Eur. J. Biochem.* **124,** 371–376.
13. Kornfeld, K., Reitman, M. L., and Kornfeld, R. (1981) The carbohydrate-binding specificity of pea and lentil lectins. *J. Biol. Chem.* **256,** 6633–6640.
14. Baenziger, J. U. and Fiete, D. (1979) Structural determinations of concanavalin A specificity for oligosaccharides. *J. Biol. Chem.* **254,** 2400–2407.
15. Hase, S., Ibuki, T., and Ikenaka, T. (1984) Reexamination of the pyridylamination used for fluorescence labeling of oligosaccharides and its application to glycoproteins. *J. Biochem.* **95,** 197–203.
16. Yamamoto, S., Hase, S., Fukuda, S., Sano, O., and Ikenaka, T. (1989) Structures of the sugar chain of interferon-γ produced by human myelomonocyte cell line HBL-38. *J. Biochem.* **105,** 547–555.

17. Tomiya, N., Awaya, J., Kurono, M., Endo, S., Arata, T., and Takahashi, N. (1988) Analysis of N-linked oligosaccharides using a two dimensional mapping technique. *Anal. Biochem.* **171,** 73–90.
18. Saitoh, A., Aoyagi, Y., and Asakura, H. (1993) Structural analysis on the sugar chains of human α-1-antitrypsin: presence of fucosylated biantennary glycan in hepatocellular carcinoma. *Arch. Biochem. Biophys.* **303,** 281–287.
19. Hudson, L. and Hay, F. R. (1976) Antibody interaction with antigen in *Practical Immunology* Blackwell, Oxford, UK, pp. 88–133.

17

Detection of Altered Glycosylation of α-Fetoprotein Using Lectin-Affinity Electrophoresis

Kazuhisa Taketa, Miao Liu, and Hiroko Taga

1. Introduction

α-Fetoprotein (AFP) is an oncofetal glycoprotein with a molecular mass of 70,000 Dalton and a carbohydrate content of approx 4%, i.e., one carbohydrate chain per molecule. High concentrations of AFP are found in the fetal serum as well as in amniotic fluid. Serum concentration of AFP in cord blood at birth range from 10,000–100,000 ng/mL, and it becomes suppressed to levels below 10 ng/mL by 300 d after birth (1). AFP is re-expressed frequently in hepatocellular carcinomas and yolk sac tumors, and much less frequently in gastrointestinal carcinomas.

Carbohydrate structures of AFP purified from cord serum at birth (2), from ascites of a patient with hepatocellular carcinoma (3), and from serum of mice transplanted with a human yolk sac tumor (4) are principally biantennary complex-type oligosaccharides linked to asparagine. AFP sugar chains of hepatocellular carcinomas and yolk sac tumors resemble those of fetal livers (2) and amniotic fluids, respectively, in early gestation stages (5,6). Sugar chain heterogeneity of AFP was first demonstrated by Smith and Kelleher (7) by means of affinity chromatography with concanavalin A (Con A) bound to agarose gel.

1.1. Lectin-Affinity Electrophoresis

Lectin-affinity electrophoresis was introduced by Bøg-Hansen et al. (8) in 1975 and applied to separation of AFP variants. Br^{e}borwicz et al. (5) and Miyazaki et al. (9) separated AFP variants of patients with hepatocellular carcinomas by crossed-immunoaffinoelectrophoresis and demonstrated a potential diagnostic use. The sensitivity of this method for detection of separated AFPs was low, even if immunoenzymatic amplification was employed (10). Taketa et al. (11) circumvented this difficulty by introducing antibody-affinity blotting followed by immunoenzymatic amplification. Our new technique allowed us not only to increase the sensitivity in detecting separated AFP bands, but also to make a direct comparison of band mobilities by parallel running of samples. It also made it possible to run two-dimensional (2D) lectin-affinity electrophoresis *(12)*. This method can detect a 20 pg AFP band with a coefficient of variation <10% when 4 µL of 100 ng/mL AFP is applied.

An isotacophoretic and electrofocusing modification of lectin-affinity electrophoresis of AFP has been reported by Schranz et al. *(13)*.

1.1.1. One-Dimensional (1D) Lectin-Affinity Electrophoresis

Serum samples are applied to troughs in agarose gel plates, which contain lectin uniformly throughout the gel, and electrophoresed. AFP isoforms with different sugar chain structures separate into bands depending on the affinity of the isoforms for the lectin employed, AFPs with high affinities for a lectin are retarded relative to those without affinity. Resolution of AFP isoforms is not necessarily complete with single lectins; so, the resulting single AFP bands may or may not represent single AFP isoforms. There are different systems of nomenclature for the identification of separated AFP bands. The system proposed by Taketa et al. *(14)* can be applied uniformly for all the lectins used independent of the number of separated AFP bands for one lectin. Briefly, the major AFP bands are numbered consecutively from the anode so that the most anodic is band 1, and the band numbers are suffixed to the capitalized initial letters of lectins used: for example, AFP-C1 and -C2 for Con A; AFP-L1, -L2, and -L3 for *Lens culinaris* agglutinin-A (LCA-A); AFP-P1, -P2, -P3, -P4, and -P5 for erythroagglutinating phytohemagglutinin of *Phaseolus vulgaris* (E-PHA); AFP-A1, -A2, and -A3 for *Allomyrina dichotoma* lectin (allo A); AFP-R1, -R2, and -R3 for *Ricinus communis* agglutinin-120 (RCA-120); AFP-D1, -D2, -D3, -D4 and -D5 for *Datura stramonium* agglutinin (DSA), and AFP-AA1, -AA2, -AA3 and -AA4 for *Aleuria aurantia* lectin (AAL). Unresolved bands of AFP-L2 and AFP-L3 are expressed as AFP-L2-3, slow- and fast-migrating bands of AFP-P3 as AFP-P3s and -P3f, respectively, and so on. Representative patterns of AFP bands separated by 1D affinity electrophoresis with different lectins are shown in Fig. 1 for different pathological conditions. This may be used as a reference for identification of AFP bands.

1.1.2. 2D Lectin-Affinity Electrophoresis

The 2D lectin-affinity electrophoresis is a combination of two 1D lectin-affinity electrophoreses with different lectins for the first and second dimension electrophoreses. Identification of resolved AFP spots is also based on the combination of the nomenclature of each 1D affinity electrophoresis *(12)*. An example of 2D lectin-affinity electrophoresis is shown in Fig. 2. AFP isoforms

Detection of α-Fetoprotein

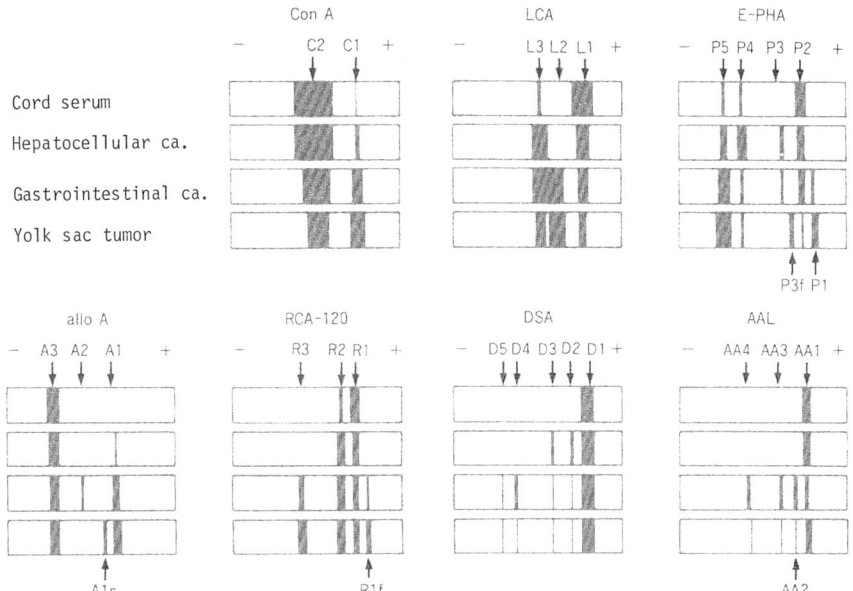

Fig. 1. Schematical illustration of AFP bands separated by 1D affinity electrophoresis with different lectins and their nomenclature. The lower panels correspond each to the upper panels with respect to pathological conditions. (Modified from Fig. 1 in ref. *15* with permission.)

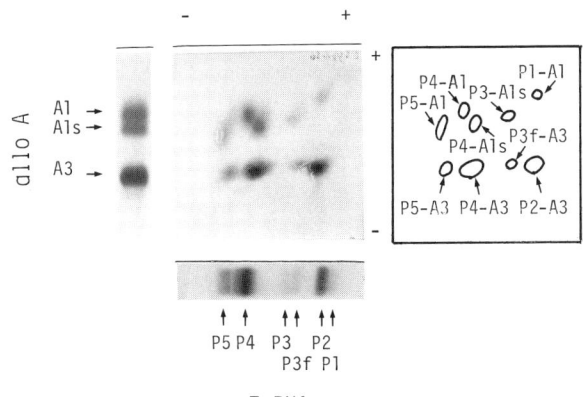

Fig. 2. Correspondence of 1D lectin-affinity electrophoretic AFP bands to 2D lectin-affinity electrophoretic AFP spots and their nomenclature for an unusual case (TY) of hepatocellular carcinoma with a yolk sac tumor-type sugar chin structure of AFP.

migrating faster in the first dimension lectin gel migrate into the second lectin gel a little faster, resulting in a slightly slanted front line with lesser migration distances toward the origin of the first dimension electrophoresis *(12)*.

1.2. Antibody–Affinity Blotting

Because of the presence of an excessively high concentration of lectin protein in the gel as compared with the applied amount of AFP, the lectin blocks the nitrocellulose (NC) membrane during conventional blotting, preventing the binding of AFP to the membrane. Therefore, we precoat NC membranes with horse (or goat) antihuman AFP antibodies (*see* Note 1) as a capture antibody in order to specifically transfer AFP to NC membrane *(11)*.

1.3. Immunoenzymatic Amplification and Visualization

The AFP specifically transferred to NC membrane can then be treated with rabbit (or mouse) antihuman AFP antibodies (*see* Note 2), followed by horseradish peroxidase-labeled goat antibodies against rabbit (or mouse) IgG. Horseradish peroxidase reaction can then be visualized by the tetrazolium method of Taketa et al. *(16)* employing phenol, Nitro Blue tetrazolium (NBT), NADH, and H_2O_2. Intensities of the resulting formazan bands can be quantified by densitometric scanning after immersing the developed and dried NC membranes in decalin.

2. Materials
2.1. Lectin-Affinity Electrophoresis

1. Dissolve lectins at twice the concentrations indicated below in the electrode buffer, barbital/barbital-Na buffer, pH 8.6, ionic strength (I) = 0.05 (*see* Note 3). The optimum concentrations of lectins in agarose gels are as follows: Con A, 1.0 mg/mL; LCA-A or LCA (*see* Note 4), 0.2 mg/mL; E-PHA (*see* Note 5), 0.3–0.5 mg/mL (*see* Note 6); allo A, 0.3 mg/mL; DSA, 0.4 mg/mL; and AAL, 0.2 mg/mL (Con A from Pharmacia, Uppsala, Sweden; allo A from Cosmo Bio Co., Tokyo, Japan; LCA, E_4-PHA, DSA, and AAL from Seikagaku, Tokyo, Japan).
2. Aqueous solution of agarose: 2% agarose with an electroendosmosis (-M_r) of 0.13 and a gelling temperature of 36 or 42°C.
3. Molds for preparation of agarose gels for 1D affinity electrophoresis, made of acryl resin board, having a gel-space size of 80 × 75 × 1.0 mm with 7 combs, 0.8 × 5.0 mm, projecting 1.0 mm into the gel space and 4 mm distances between the adjacent projections, as illustrated in Fig. 3, and a mold for preparation of an agarose gel for 2D affinity electrophoresis, made of acryl resin board, having a gel-space size of 90 × 90 × 1.0 mm with blank space-forming arms as illustrated in Fig. 4. In 2D affinity electrophoresis, the lectin gels used for the first-dimension electrophoresis are the same as for 1D affinity electrophoresis except that wells of 1.6 mm diameter are punched out in place of the troughs for sample application.

Detection of α-Fetoprotein

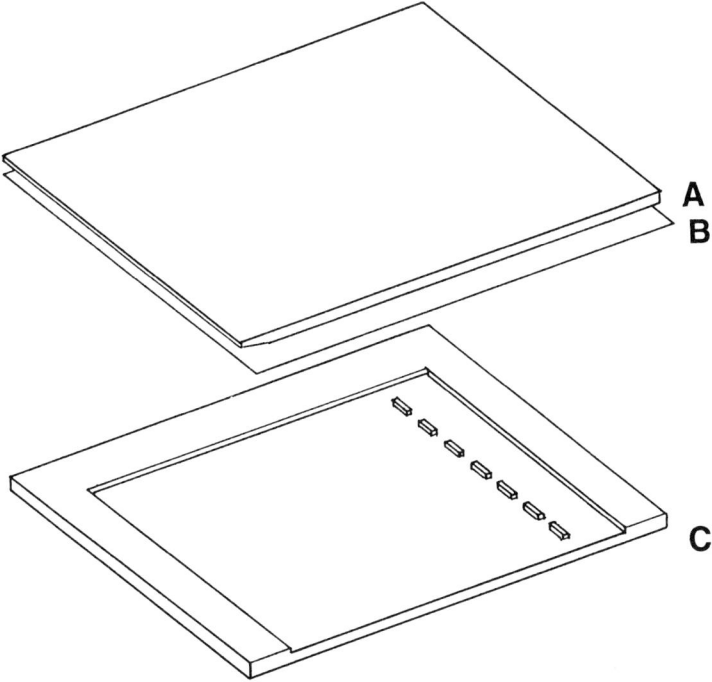

Fig. 3. An illustrated design of mold for 1D lectin-affinity electrophoresis before assembling. **(A)** Covering plate, **(B)** GelBond film, and **(C)** Gel former.

4. Five milligrams/milliliter bromophenol blue (BPB) in the electrode buffer (*see* Note 7).
5. Frozen body fluids or purified AFP dissolved in 0.3% bovine serum albumin in phosphate-buffered saline (0.01M, pH 7.2). Frozen cord serum obtained at full-term delivery has uniform and stable patterns of AFP bands for the lectins used and may serve as a reference.

2.2. Antibody–Affinity Blotting

1. NC membrane (50 × 92 mm or any desired size of this surface area. Schleicher and Schuell, Dassel, Germany).
2. Tris-buffered saline (TBS): 20 mM Tris-HCl, 500 mM NaCl, pH 7.5.
3. Affinity-purified polyclonal horse (or goat) antibodies to human AFP: 100 µg/mL TBS (*see* Note 8).
4. Glutaraldehyde (25%).
5. NaBH$_4$: 0.2 mg/mL TBS.
6. Tween-20: 0.5% in TBS.
7. Thick filter paper pads.
8. A weight of approx 500 g and a glass plate of approx 150 × 150 × 3 mm.
9. Washing buffer: 0.05% Tween-20 in TBS (T-TBS).

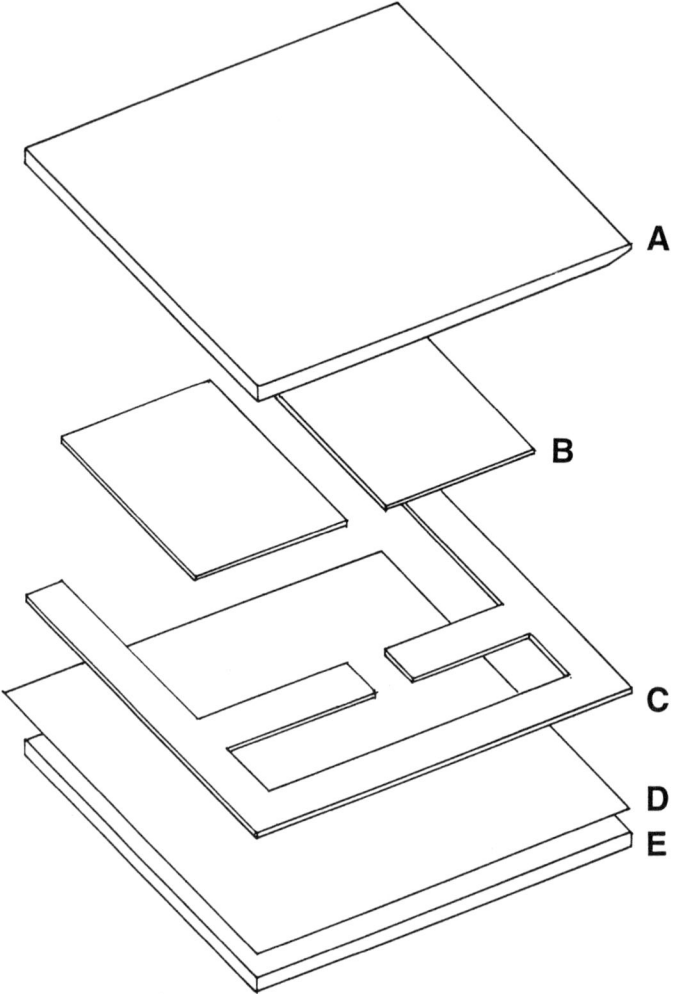

Fig. 4. An illustrated design of mold for 2D lectin-affinity electrophoresis before assembling. **(A)** Covering plate, **(B)** Blocking piece, **(C)** Blank-space former, **(D)** GelBond film, and **(E)** Supporting plate.

10. α-Methyl-D-mannoside: $0.2M$ in T-TBS.
11. α-Lactose: $0.2M$ in T-TBS.

2.3. Enzyme Immunodetection

1. Gelatin: 1.0% in TBS (G-TBS) (Bio-Rad Laboratories, CA).
2. IgG fraction of polyclonal rabbit or mouse antibodies to human AFP, diluted 1000-fold in G-TBS depending on the titer of the antibody preparation commercially available (*see* Note 9).

Detection of α-Fetoprotein

3. Affinity-purified goat antirabbit (or mouse) IgG (H+L)-horseradish peroxidase conjugate (Dako A/S, Denmark), diluted 1000-fold in G-TBS.
4. Staining solution consisting of 0.3 mg/mL NBT, 2.0 mg/mL NADH, 0.4 mg/mL phenol, and 0.02% H_2O_2 in 50 mM phosphate buffer, pH 7.0 (*see* Note 10).
5. Decalin.

3. Methods

3.1. Preparation of NC Membranes Precoated with Anti-AFP Antibodies

1. Immerse NC membranes into 10 mL of 100 µg/mL horse (or goat) antihuman AFP antibodies for 30 min at 25°C.
2. Take out, blot, and hang the membranes in a closed chamber of approx 200 × 200 × 200 mm, containing approx 1 mL of glutaraldehyde solution sprinkled onto filter paper pads placed inside.
3. Take out the membrane after 30 min and reduce the unreacted glutaraldehyde with 0.02% $NaBH_4$ in TBS for 2 min.
4. Wash the membranes with TBS and block them with 0.5% Tween-20 in TBS for 30 min.
5. Wash the blocked membranes with TBS, dry, and store at 4°C in a desiccator under a reduced pressure.

3.2. Lectin-Affinity Electrophoresis

One- and two-dimensional lectin-affinity electrophoreses share common materials and methods. Accordingly, full methods are described for 1D lectin-affinity electrophoresis and their modifications for 2D lectin-affinity electrophoresis.

3.2.1. 1D Lectin Affinity Electrophoresis

1. Take 4 mL of molten 2% agarose solution at 60°C.
2. Take 4 mL of a twice-concentrated lectin solution in the electrode buffer at room temperature.
3. Assemble the mold by placing a GelBond film between the gel former and covering plate, with the hydrophobic side facing the covering plate and the hydrophilic side toward the space for agarose gel (*see* Note 11).
4. Quickly mix the molten agarose and the lectin solution and cast the lectin containing agarose into the mold (*see* Note 12). Stand the gel-containing mold upright until the gel becomes solidified (10 min at room temperature and then cool for 10 min at 4°C).
5. Prepare AFP solutions by diluting sera with barbital/barbital-Na buffer (I = 0.025), pH 8.6 (made by two-fold dilution of electrode buffer as in Section 2.1.1.), to give an AFP concentration of 100 ng/mL and a BPB concentration of 0.5 mg/mL.
6. Disassemble the mold and take out the solidified agarose gel with GelBond. Set the gel plate on a cooled ceramic plate (10°C) of an electrophoretic chamber

(Denshi-Reikyaku-Eidoso, Model TC-3; Kayagaki-Irika-Kogyo, Tokyo, Japan). Place the sample trough-side of the gel to the cathodic side.
7. Apply 4-µL samples (*see* Note 13) to troughs in the gel and connect the gel and the electrode buffer with wicks moistened with the electrode buffer (*see* Note 14). The width of the wick is identical with that of the gel. Adjust the overlapping edges of gel and wicks to 5 mm in both side of the gel.
8. Run electrophoresis at 10°C using a constant voltage of 1 5 V/cm until unbound BPB has migrated 55 mm from the origin.
9. Take out the gel plates from the electrophoretic chamber and overlay the gels with antibody-precoated NC membranes (*see* Note 15), followed by thick filter paper pads, a glass plate, and a weight.
10. After 20 min at 25°C, remove the filter paper pads and wash the NC membranes with 20 mL of 0.05% T-TBS containing $0.2M$ α-methyl-D-mannoside for Con A and LCA, or $0.2M$ α-lactose for allo A, followed by a similar washing with T-TBS alone.

3.2.2. 2D Lectin-Affinity Electrophoresis

1. Apply 2 µL of AFP preparations (200–400 ng/mL AFP) to wells of the 1D lectin-affinity electrophoresis.
2. Run electrophoresis as for the 1D affinity electrophoresis.
3. Assemble the molds for 2D affinity electrophoresis in Fig. 4 by making an inverted T-shape space. Cast 2.1 mL of molten 1% plain agarose in barbital/barbital-Na buffer (1 = 0.025, pH 8.6) (dilute 2% agarose with an equal volume of the electrode buffer) to fill up the inverted T-shape space. After gels are formed, remove two pieces of plate (Fig. 4B), making two separate spaces.
4. Prepare lectin-containing agarose for the 2D affinity electrophoresis, and cast it into the two separate spaces where two plates (Fig. 4B) were present. The same lectin gel (6 mL) or two different lectin gels (each 3 mL) can be made for one mold.
5. When unbound BPB has migrated 55 mm from the origin in the first dimension electrophoresis, take out the gel and cut out 8 × 40 mm gel strip together with GelBond and migrated AFP spots. Make two of them either from the same lectin gel or from different ones.
6. Remove frame **C** in Fig. 4 by disassembling the mold. Place the cut out 8 × 40 mm gel strips gel side (not GelBond side) down to the 10 × 40 mm blank spaces. Seal the junctional gaps with the molten 1% plain agarose solution to connect the first dimension gels to the second dimension gels.
7. A small drop of BPB solution can be placed on the gel along the starting line (a line connecting the two sample wells).
8. Run the 2D affinity electrophoresis similarly until unbound BPB migrates 55 mm from the connected lectin-containing gel border.

3.3. Enzyme Immunodetection

1. Place the washed AFP transfers in 10 mL of diluted rabbit (or mouse) antihuman AFP antibodies in G-TBS for 60 min at 25°C.

Detection of α-Fetoprotein

2. Wash the membranes twice each for 10 min with T-TBS. Transfer the membranes to 10 mL of affinity-purified goat antirabbit (or mouse) IgG (H+L)-horseradish peroxidase conjugate diluted 1000-fold with G-TBS for 60 min at 25°C.
3. Wash the membranes twice with T-TBS for 10 min.

3.4. Peroxidase Staining

1. Transfer the antibody-treated membranes into 10 mL of freshly prepared staining solution and allow to react for 30 min at 25°C.
2. Wash the stained membranes with water and dry.
3. Treat the dried membranes with decalin for densitometric scanning.
4. Photographs of the stained and dried membranes can be taken with monochrome films using a yellow filter with a cutoff level of 480 nm.
5. The stained AFP bands can be identified using our system of nomenclature in the light of standard patterns of AFP bands for each lectin used (Fig. 1). Reference values of AFP band intensities are available elsewhere *(14)*.
6. Correspondence between the major AFP isoforms and the sugar chain structures is shown in Fig. 5. This is based on the results obtained by comparison of paired 2D affinity electrophoreses, each having one sharing lectin.

4. Notes

1. When polyclonal horse (or goat) antibodies are used, the antibodies should be affinity purified. Mouse MAb can be used after isolation of the IgG fraction.
2. Mouse MAb can be used when the capture antibodies are horse (or goat) antibodies.
3. Although some of the lectins used require divalent cations for efficient binding of sugars, no cations were added to the dissolving buffer in order to avoid unduly high electric current through the lectin-containing agarose gels. When dissolved E-PHA (5 mg/mL) is to be kept longer than one month, it should be dissolved in 5 mM Tris-HCl, pH 7.5, containing 0.1 mM $MgCl_2$, 0.1 mM $MnCl_2$ and 0.1 mM $CaCl_2$, and further diluted with the electrode buffer immediately before use. Other lectin solutions can be kept at 4°C for a couple of months without deterioration.
4. LCA consists of two isolectins, LCA-A and LCA-B. Originally LCA-A was used. Some preparations of LCA can be used if they contain LCA-A alone and have higher affinities than purified LCA-A for AFP sugar chain.
5. The tetramer of E-PHA, E_4-PHA, without contamination of leukoagglutinating PHA (L-PHA) should be used.
6. Some preparations of E_4-PHA have lower or higher affinities for AFP isoforms. Accordingly, a suitable concentration should be chosen for each preparation of E_4-PHA to give an (AFP-P4-P5)/(AFP-P2-P5) ratio of approx 0.2 for AFP of cord serum at full term. A new batch of other lectins may be checked similarly for their affinities by comparing with known isoforms as references (Fig. 1).
7. BPB was not added for DSA affinity electrophoresis, because it interfered with the DSA-AFP interaction.
8. The polyclonal antibodies can be replaced by a mouse MAb to human AFP.

AFP isoforms	Sugar chain structures

AFP-C2-L1-P2-A3-R1

```
        NeuSAcα2→6Galβ1→4GlcNAcβ1→2Manα1
                                           \6
                                            Manβ1→4GlcNAcβ1→4GlcNAc→Asn
                                           /3
        NeuSAcα2→6Galβ1→4GlcNAcβ1→2Manα1
```

AFP-C2-L3-P2-A3-R1

```
                                                    Fucα1
                                                      ↓
        NeuSAcα2→6Galβ1→4GlcNAcβ1→2Manα1              6
                                           \6
                                            Manβ1→4GlcNAcβ1→4GlcNAc→Asn
                                           /3
        NeuSAcα2→6Galβ1→4GlcNAcβ1→2Manα1
```

AFP-C2-P4-A3-R2

```
                                                   ±Fucα1
                                                      ↓
        Galβ1→4GlcNAcβ1→2Manα1                        6
                                \6
                                 Manβ1→4GlcNAcβ1→4GlcNAc→Asn
                                /3
        NeuScα2→6Galβ1→4GlcNAcβ1→2Manα1
```

AFP-C2-P3-A3

```
                                                   ±Fucα1
                                                      ↓
        GlcNAcβ1→2Manα1                               6
                        \6
                         Manβ1→4GlcNAcβ1→4GlcNAc→Asn
                        /3
        NeuScα2→6Galβ1→4GlcNAcβ1→2Manα1
```

AFP-C2-P4-A1s

```
                                                   ±Fucα1
                                                      ↓
        Galβ1→4GlcNAcβ1→2Manα1                        6
                                \6
                                 Manβ1→4GlcNAcβ1→4GlcNAc→Asn
                                /3
        Galβ1→4GlcNAcβ1→2Manα1
```

AFP-C2-P5-A3-R1

```
                                                   ±Fucα1
                                                      ↓
        NeuSAcα2→3Galβ1→4GlcNAcβ1→2Manα1              6
                                          \6
                                           Manβ1→4GlcNAcβ1→4GlcNAc→Asn
                                          /3
        NeuSAcα2→6Galβ1→4GlcNAcβ1→2Manα1
```

AFP-C1-P5-A3-R3

```
                                       GlcNAcβ1      ±Fucα1
                                           ↓            ↓
        Galβ1→4GlcNAcβ1→2Manα1             4            6
                                \6
                                 Manβ1→4GlcNAcβ1→4GlcNAc→Asn
                                /3
        NeuSAcα2→6Galβ1→4GlcNAcβ1→2Manα1
```

Fig. 5. Isoforms of human serum AFP and their corresponding sugar chain structures.

9. If nonspecific bands appear, the use of F(ab')$_2$ fragment (12 µg/mL) is recommended in place of IgG fraction.
10. A 10 mL staining solution was made by dissolving 20 mg of NADH in 5 mL of 0.1 M phosphate buffer, pH 7.0, and adding to it 4 mL of 1.0 mg/mL phenol, 1 mL of 3.0 mg/mL NBT, and finally 20 µL of 10% H$_2$O$_2$ immediately before starting the reaction.
11. Clip the gel former and the covering plate with GelBond securely to prevent leakage of molten agarose.
12. Pour the lectin-agarose mixture against the GelBond film sticking out slightly from the edge of gel former. This allows quick pouring of molten agarose without solidification or trapping air bubbles in the gel.
13. Apply seven samples for one gel plate. Two gel plates (14 samples) can be run simultaneously.
14. Use urethan foam sponge or 3–5 layers of thick filter paper.
15. Moisten anti-AFP antibody-precoated NC membranes with TBS and remove extra buffer by gently pressing with filter paper. When overlaying the membrane on the gel surface, caution should be taken to avoid trapping air bubbles.

References

1. Tsuchida, Y., Endo, Y., Saito, S., Kaneko, M., Shiraki, K., and Ohmi, K. (1978) Evaluation of alpha-fetoprotein in early infancy. *J. Pediatr. Surg.* **13,** 155,156.
2. Yamashita, K., Taketa, K., Nishi, S., Fukushima, K., and Ohkura, T. (1993) Sugar chains of human cord serum α-fetoprotein: characteristics of *N*-linked sugar chains of glycoproteins produced in human liver and hepatocellular carcinomas. *Cancer Res.* **53,** 2970–2975.
3. Yoshima, H., Mizuochi, T., Ishii, M., and Kobata, A. (1980) Structure of the asparagine-linked sugar chains of α-fetoprotein purified from human ascites fluid. *Cancer Res.* **40,** 4276–4281.
4. Yamashita, K., Hitoi, A., Tsuchida, Y., Nishi, S., and Kobata, A. (1983) Sugar chain of α-fetoprotein produced in human yolk sac tumor. *Cancer Res.* **43,** 4691–4695.
5. Bręborowicz, J., Mackiewicz, A., and Bręborowicz, D. (1981) Microheterogeneity of α-fetoprotein in patient serum as demonstrated by lectin affino-electrophoresis. *Scand. J. Immunol.* **14,** 15–20.
6. Kerckaert, J. P., Bayard, B., and Biserte, G. (1979) Microheterogeneity of rat, mouse and human α$_1$-fetoprotein as revealed by polyacrylamide gel electrophoresis and by crossed immuno-affino-electrophoresis with different lectins. *Biochim. Biophys. Acta* **576,** 99–108.
7. Smith, C. J. and Kelleher, P. C. (1973) α$_1$-Fetoprotein: separation of two molecular variants by affinity chromatography with concanavalin A-agarose. *Biochim. Biophys. Acta* **317,** 231–235.
8. Bøg-Hansen, T. C., Bjerrum, O. J., and Ramlau, J. (1975) Detection of biospecific interaction during the first dimension electrophoresis in crossed immunoelectrophoresis. *Scand. J. Immunol.* **4(Suppl.),** 141–147.

9. Miyazaki, J., Endo, Y., and Oda, T. (1981) Lectin affinities of alpha-fetoprotein in liver cirrhosis, hepatocellular carcinoma and metastatic liver tumor. *Acta Hepatol. Jpn.* **22,** 1559–1568.
10. Taketa, K. (1983) Enzymatic amplification of electroimmunoprecipitates in agarose gels with horseradish peroxidase-labeled protein A. *Electrophoresis* **4,** 371–373.
11. Taketa, K., Ichikawa, E., Taga, H., and Hirai, H. (1985) Antibody-affinity blotting, a sensitive technique for the detection of α-fetoprotein separated by lectin affinity electrophoresis in agarose gels. *Electrophoresis* **6,** 492–497.
12. Taketa, K., Ichikawa, E., Sato, J., Taga, H., and Hirai, H. (1989) Two-dimensional lectin affinity electrophoresis of α-fetoprotein: characterization of erythroagglutinating phytohemagglutinin-dependent microheterogeneity forms. *Electrophoresis* **10,** 825–829.
13. Schranz, D., Morkowski, S., Karamova, E. R., and Abelev, G. I. (1991) Counterflow affinity isotachophoresis on cellulose acetate membranes. *Electrophoresis* **12,** 414–419.
14. Taketa, K., Sekiya, C., Namiki, M., Akamatsu, K., Ohta, Y., Endo, Y., and Kosaka, K. (1990) Lectin-reactive profiles of α-fetoprotein characterizing hepatocellular carcinoma and related conditions. *Gastroenterology* **99,** 508–518.
15. Taketa, K. (1995) Structures of α-fetoprotein sugar chain. *J. Med. Technol.* **39,** 66–70 (Japanese).
16. Taketa, K. (1987) A tetrazolium method for peroxidase staining: application to the antibody-affinity blotting of α-fetoprotein separated by lectin affinity electrophoresis. *Electrophoresis* **8,** 409–414.

18

Use of Lectin-Affinity Electrophoresis for Quantification and Characterization of Glycoforms of α-1 Acid Glycoprotein

Thorkild C. Bøg-Hansen

1. Introduction

Quantification and characterization of glycoproteins may be achieved by lectin-affinity electrophoresis. Here we show the identification and quantification of three naturally occurring glycoforms of human serum α-1 acid glycoprotein (AGP) or orosomucoid. The method described here is two-dimensional (2D); a combination of lectin-affinity electrophoresis in the first dimension, and quantitative immunoelectrophoresis with specific antibodies against the glycoprotein in the second dimension. This type of affinity electrophoresis is carried out in agarose gels and was first termed crossed immunoaffinoelectrophoresis because it is a variant of quantitative crossed immunoelectrophoresis *(1)*. For general descriptions, applications, and techniques of crossed immunoelectrophoresis the reader is referred to other manuals *(2,3)*.

With this method it may be possible to obtain the following kinds of information about a glycoprotein:

1. Identification of a protein as a glycoprotein.
2. Carbohydrate structure.
3. Number of glycosylation sites.
4. Number of glycoforms (degree of microheterogeneity).
5. Dissociation constant for binding to lectin.
6. Quantification of each individual glycoform.
7. Prediction of the result of a preparative separation on immobilized lectin.

Free lectin is incorporated into the first dimension (1D) gel at a uniform concentration. A typical experiment would use 25 mg Con A in 18 mL agarose

From: *Methods in Molecular Medicine, Vol. 9: Lectin Methods and Protocols*
Edited by: J. M. Rhodes and J. D. Milton Humana Press Inc., Totowa, NJ

gel, giving a lectin concentration of approx $10^{-5}M$. During the 1D electrophoresis, the glycoprotein(s) in the sample will interact with the lectin under equilibrium conditions. The reversible interaction between a glycoprotein and the lectin will result in a change of the net mobility of the glycoprotein dependent on the strength of the interaction. The large pores of the 1% agarose gel allow free movement of lectin glycoprotein complexes. However, glycoproteins with two or more binding sites may form a precipitate. Thus the first dimension net mobility of each glycoform is dependent primarily on the affinity between the glycoform and the lectin. Other factors determining the 1D net mobility are the concentration and electrophoretic mobility of the lectin and the electrophoretic mobility of the glycoform.

Polyclonal antibodies against the glycoprotein(s) in question are incorporated into the 2D gel, as in crossed immunoelectrophoresis. An inhibitor of the lectin glycoprotein complex may be included in this gel, resulting in liberation of glycoforms bound in a precipitate in the 1D gel. The 2D electrophoresis is performed perpendicularly to the 1D electrophoresis and the separated glycoproteins (and other proteins) are carried into the antibody containing gel, being precipitated with their specific antibodies giving rise to rocket-shaped or arch-shaped precipitates. The size of the area covered by a precipitate is a quantitative measure of the amount of a protein, in our case of a glycoform.

The method has been used to study the variation of glycoforms in health, during diseases and during pregnancy. The following reviews contain references to α-fetoprotein, α-1 acid glycoprotein (orosomucoid), α-1 antitrypsin, α-1 antichymotrypsin, ceruloplasmin, transferrin, haptoglobin, serum ferritin, blood clotting factors, glycophorin, thyroglobulin, mucins, alkaline phosphatase, acid phosphorylase, and other glycoproteins *(4–9).*

2. Materials

1. Electrophoresis apparatus with a cooling surface as the original Holm-Nielsen apparatus supplied by Bie and Berntsen, Copenhagen, Denmark, or as the Multiphor from Pharmacia, Uppsala, Sweden. The cooling surface may be thermostatted at around 18°C.
2. Power pack capable of delivering 300–500 V DC at 200 mAmp and voltmeter.
3. Water bath at 55°C for keeping the agarose melted.
4. Level table for working with the agarose gels.
5. Various utensils for working with the agarose gels: scalpel, well cutter, cutting blades, and plate holder. Glass plates or GelBond film (FMC) for agarose support. Filter paper No. 1 from Whatman as wicks for connection. Test tubes, pipets, and parafilm.
6. Buffer for electrode vessels and gels: Make 5 L Tris/Veronal stock solution, pH 8.6, ionic strength 0.1 with 221.5 g Tris (Sigma, St. Louis, MO) and 112 g 5,5-diethylbarbituric acid (veronal or barbital) in distilled water to 5000 mL.

Dilute 1 + 4 before use to obtain an ionic strength of $0.02M$. The diluted buffer can be used three times for electrophoresis before it is renewed.
7. Agarose gel, 1% w/v agarose: Agarose (HSA, Litex) 1 g in 100 mL Tris/barbital buffer, ionic strength 0.02, pH 8.6. The gel is boiled for 5 min and can be kept in solution in a water bath at 55°C. The gel can be stored at 4°C and is then ready for use again after a short boiling. An agarose with an electroendosmosis (EEO) around –0.13 is suitable for this type of experiments, as for instance the Litex agarose HSA 1000 protein grade (FMC, Vallensbaek Strand, Denmark) or agarose type I, IA or VIA (Sigma).
8. Electrophoresis marker, serum mixed with bromophenol blue. Albumin will react with the dye and will appear strongly stained in the gel. Free bromophenol blue will migrate anodically to albumin at the described conditions. Note that with nonionic detergents the free bromophenol blue will migrate slower than albumin because it is taken up into micelles.
9. Staining solution: Dissolve 5 g Coomassie brilliant blue R 250 in 450 mL 96% ethanol, 100 mL glacial acetic acid, and 450 mL water. After mixing the ingredients, the solution is left overnight with gentle mixing. The following day the solution is filtered.
10. Destaining solution: 450 mL 96% ethanol, 100 mL glacial acetic acid, and 450 mL water.
11. Serum samples may be stored under normal conditions and diluted appropriately with water.
12. Antibodies are available from DAKO Immunoglobulins (Ejby, Denmark) as rabbit antibodies against orosomucoid, immunoglobulin fraction, cat. no. A0011.
13. Concanavalin A (Con A) is available from Kem-En-Tec (Copenhagen, Denmark) as a freeze-dried powder.

3. Methods

1. Label and rinse a 10 × 10-cm glass plate, the first dimension plate. Boil the agarose solution and equilibrate in the water bath at 55°C.
2. Weigh 25 mg Con A into a test tube and dissolve in two drops of buffer and equilibrate for half a minute in the water bath at 55°C, then pour 18 mL agarose solution (see Section 2., item 7) into the test tube and mix, then pour onto the glass plate.
3. Punch wells according to plan (Fig. 1), and apply the serum samples (see Note 1; 2 µL human serum as for instance 10 µL diluted serum) and the electrophoresis marker.
4. Place the plate on the electrophoresis tank. Connect the gel to the buffer reservoirs with eight layers of filter paper. Perform electrophoresis at 10 V/cm until the marker has migrated 6 cm (see Notes 2–5).
5. Prepare five 10 × 7-cm glass plates, the 2D plates: The plates may be marked with a diamond or a glass drill, then wash with detergent and rinse in alcohol.
6. Remove the first dimension plate from the electrophoresis tank and cut the agarose gel according to Fig. 1, then transfer the gel strips to the 10 × 7-cm glass plates, as in Fig. 2 with a long razorblade (cutting blades).

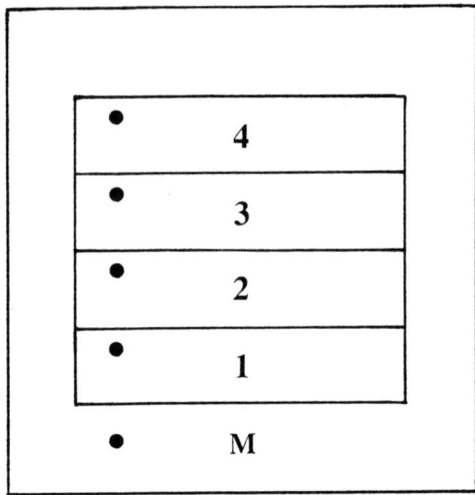

Fig. 1. Plan of the 1D agarose gel with Con A. The agarose gel is cast on a 10 × 10-cm glass plate. Samples to be applied in lanes 1–4 and marker (bromophenol blue-stained albumin) to be applied in lane M. After the 1D electrophoresis (anode right), each lane will be cut along the guidelines and transferred to the 2D glass plate (*see* Fig. 2).

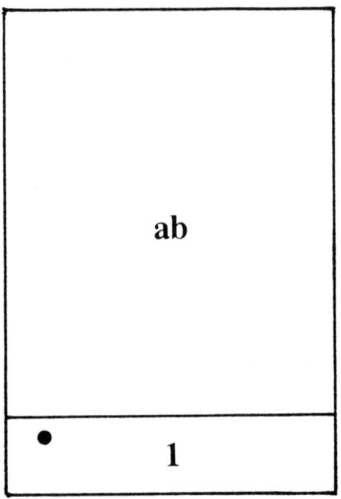

Fig. 2. Plan of the 2D agarose gels on a 7 × 10-cm glass plate. One lane from the 1D gel (1) is transferred to the 2D glass plate. Then the 2D gel (ab) with inhibitor (α-methyl-D-glucopyranoside) is mixed with the antibody and cast on the upper part of the plate (anode at the top).

α-1 Acid Glycoprotein Glycoforms

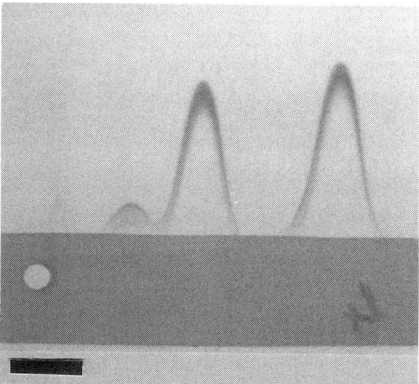

Fig. 3. Crossed immunoelectrophoresis of orosomucoid (AGP, α-1 acid glycoprotein) with Con A in the 1D gel. Three glycoforms can be identified and quantified; from right to left: glycoform 1 (48%), glycoform 2 (43%), glycoform 3 (9%). Glycoform 1 represents the position of orosomucoid without Con A in the 1D gel. Two microliter normal human serum was applied at the origin. The 1D agarose gel contained 2.3 mg Con A (Kem-En-Tec, Copenhagen; freeze-dried concanavalin A, lot 20 626). The 1D electrophoresis was run for 1.5 h at 10 V/cm, giving bromophenol blue-stained albumin a migration of 6.0 cm (anode right). The 2D gel contained specific antibodies against orosomucoid (DAKO Immunoglobulins, Eiby, Denmark; catalog number A0011, 0.58 µL/cm^2). The 2D electrophoresis was run overnight at 2 V/cm (anode at the top). Stained for protein with Coomassie. Bar = 1 cm.

7. Prepare five test tubes in the water bath each with 10 mL agarose with 3% inhibitor (α-methyl glucopyranoside). Pipet antibody (50 µL anti-AGP) into the agarose gel and mix carefully. Without delay, the mixture is poured onto the free upper part of the glass plate.
8. The 2D electrophoresis is performed at 1–2 V/cm overnight. Use five layers of filter paper wicks (*see* Note 6).
9. Remove the plates from the electrophoresis tank and inspect the plates (*see* Note 7).
10. Press the plates under absorbent paper until flat, wash in water for 10 min, and press again under absorbent paper until flat.
11. Dry the plates in a stream of (hot) air until clear as glass.
12. Stain for 10 min, destain for 2 min, and dry. Inspect the plates or photograph (Fig. 3; *see* Notes 8–10).

4. Notes

1. A control experiment is always performed without Con A. Identification of a protein as a glycoprotein is based on a shift of the net mobility in the first dimension electrophoresis indicating binding to Con A.

2. A microheterogenous glycoprotein may appear as an immunoprecipitate with multiple peaks, with each peak representing one glycoform.
3. The dissociation constant, K_d, may be calculated from a set of several experiments with a different concentration of Con A in the first dimension gel. A concentration interval from 0.1×10^{-5} to $2 \times 10^{-5} M$ Con A may be chosen for this set of experiments. The data may be plotted as:

$$(R_{mo} - R_{mi})^{-1} = K_d \cdot (R_{mo} - R_{mc})^{-1} \cdot c^{-1} + (R_{mo} - R_{mc})^{-1}$$

where K_d is the dissociation constant of the Con A-glycoprotein complex; c is the concentration of Con A expressed in moles of binding sites per L; R_{mo} is the mobility of the glycoprotein without Con A; R_{mi} is the mobility of the glycoprotein with Con A; and R_{mc} is the mobility of the Con A-glycoprotein complex *(10)*.
4. A glycoprotein with several glycosylation sites may form a precipitate in the 1D gel. This precipitate may be dissolved by inclusion of an inhibitor in the 2D gel, here for instance 1% α-methyl-D-mannopyranoside or 3% α-methyl-D-glucopyranoside.
5. Con A is known to bind to the mannose core of biantennary carbohydrates. Other lectins will react with other carbohydrate structures, thus identifying other elements of the carbohydrate of a glycoprotein.
6. Affinity immunoelectrophoresis may be performed in presence of nonionic detergents. Therefore membrane glycoproteins may be analyzed and characterized by this method.
7. The area enclosed by an immunoprecipitate is proportional to the amount of antigen. Therefore the method allows a quantitation of the relative amount of the glycoforms. From a normal donor, serum orosomucoid is often found to have the following distribution: 45% of glycoform 1, 45% of glycoform 2, and 10% of glycoform 3 (Fig. 3).
8. The resolving power of this analysis is high, making it possible to quantify a large number of proteins with a multispecific antibody preparation in one set of experiments. The precision is 2–18% depending on the antigen-antibody system, and the technique is very versatile and numerous modifications exists *(2)*. The method may be generally useful for biomolecular characterization of antigens without having them in a pure state *(1)*.
9. A prerequisite for this type of electrophoresis is the availability of polyclonal antibodies specific for the glycoprotein under study. Other types of affinity electrophoresis have been described with or without the need for specific antibodies (*see* Chapters 16 and 17).
10. Con A and other lectins may be supplied as freeze-dried powder or as a solution with or without salts. For electrophoresis purposes, it is highly recommended to avoid salts. However, if a salt is present, the proper control would consist of an experiment with the same concentration of salt. When working with freeze dried lectin, care should be taken to choose a preparation that is fully soluble. Other lectins may be used similarly *(11)*.

References

1. Bøg-Hansen, T. C. (1973) Crossed immuno-affinoelectrophoresis. An analytical method to predict the result of affinity chromatography. *Anal. Biochem.* **56,** 480–488.
2. Axelsen, N. H. (ed.) (1983) *Handbook of Immunoprecipitation-in-Gel Techniques.* Blackwell, Oxford, UK alias *Scand. J. Immunol.* **17(Suppl. 10),** 1–169.
3. Bøg-Hansen, T. C. (1990) Immunoelectrophoresis, in *Gel Electrophoresis. A Practical Approach*, 2nd ed., (Hames, B. D. and Rickwood, D., eds.), IRL, Oxford, pp. 273–300.
4. Takeo, K. (1987) Affinity electrophoresis, in *Advances in Electrophoresis*, vol. 1 (Chrambach, A., Dunn, M. J., and Radola, B. J., eds.), VCH, Weinheim, Germany, pp. 229–279.
5. Bøg-Hansen, T. C. (1983) Affinity electrophoresis of glycoproteins, in *Solid Phase Biochemistry. Analytical and Synthetic Aspects* (Scouten, W. H., ed.), Wiley, New York, pp. 223–251.
6. Bøg-Hansen, T. C. (1983) Affinity electrophoresis with lectins for study of glycoproteins, in *Handbook of Immunoprecipitation-in-Gel Techniques* (Axelsen, N. H., ed.), *Scand. J. Immunol.* **17(Suppl. 10),** 243–253.
7. Bøg-Hansen, T. C. (1983) Affinity electrophoresis of glycoproteins, in *Solid Phase Biochemistry: Analytical and Synthetic Aspects* (Scouten, W. H., ed.), Wiley, New York, pp. 223–251.
8. Heegaard, N. H. H. and Bøg-Hansen, T. C. (1990) Affinity electrophoresis in agarose gels. Theory and some applications. *Appl. Theor. Electrophoresis* **1,** 249–259.
9. Heegaard, P. M. H., Heegaard, N. H. H., and Bøg-Hansen, T. C. (1992) Affinity electrophoresis is for the characterization of glycoproteins. The use of lectins in combination with immunoelectrophoresis, in *Affinity Electrophoresis* (Mackiewicz, A. and Breborowicz, J., eds.), CRC, Boca Raton, FL, pp. 3–21.
10. Bøg-Hansen, T. C. and Takeo, K. (1980) Determination of dissociation constants by affinity electrophoresis: complexes between human serum proteins and concanavalin A. *Electrophoresis* **1,** 67–71.
11. Bøg-Hansen, T. C. (1996) Lectin links internet page http://plab.ku.dk/tcbh/lectin-links.htm.

19

ABO(H) Blood Group Expression on Circulating Glycoproteins

Taei Matsui and Koiti Titani

1. Introduction

ABO(H) blood group antigens are typical carbohydrate antigens widely expressed on erythrocytes, digestive tissue, respiratory tissue, and in secreted body fluids such as milk, urine, and saliva. These antigens are mostly distributed as glycolipids or glycoproteins *(1,2)*. Recently, human plasma von Willebrand factor (vWF) *(3,4)*, coagulation factor VIII (FVIII) *(5)*, and a portion of α_2-macroglobulin (α_2M) *(6)* have been found to possess covalently associated ABO(H) blood group antigens. There is no information about the physiological significance of the blood group antigens on these plasma glycoproteins, but some interesting relationship between concentration of vWF and blood groups has been reported *(7,8)*. Moreover, a number of blood group-related antigens have been found as clinical markers for carcinogenesis *(9)*.

1.1. Blood Group-Specific Lectins

Since *Phaseolus lunatus* lectin (LBA) has been found selectively to agglutinate blood group A erythrocytes, a variety of lectins have been used to study ABO(H) blood group specificities by hemagglutination assay *(10)*. Lectins from *Dolichos biflorus* (DBA), *Helix pomatia* (HPA) *(11)*, *Glycine max* (SBA), and *Griffonia simplicifolia* lectin-I (GS-I [B_4]) agglutinate blood group A and B erythrocytes, respectively *(10)*. Lectins from *Ulex europaeus* (UEA-I) *(12)* and *Lotus tetragonolobus* (LTA) recognize blood group O(H) substance on erythrocytes. DBA and UEA-I have been used clinically for the determination of blood subgroup A_1 and secretor status, respectively. These lectins have been used for the detection of blood group substances on glycoproteins and glycolipids as well as on erythrocytes and tissues *(13,14)*.

Table 1
Reactivity of ABO(H) Blood Group Specific Lectins Examined

Lectins	Blood group specificity (hemagglutination)	Binding to vWF and α_2M	
		Blotting[a]	ELISA[b]
DBA	A_1	No binding	Weakly bound to A
HPA	A	Bound to A and AB	Bound to A and AB[c]
LBA	A	No binding	Weakly bound to A
SBA	A	No binding	ND[d]
SJA	B, A	No binding	No binding
EEA	B, O	No binding	ND
GS-I (B_4)	B	No binding	No binding
LTA	O	No binding	ND
UEA-I	O	Bound better to O	Bound better to O[c]

[a]See Section 3.3.
[b]See Section 3.4.
[c]Very weak reactivity to α_2M (see Note 16).
[d]Not determined.

1.2. Application of Lectins to Proteins Immobilized on Membranes or in ELISA Systems

Highly purified lectins conjugated with enzymes, biotin, gold particles, fluorescein, or gel matrix are commercially available and these labeled lectins are helpful for the detection or the separation of oligosaccharides and glycoconjugates *(15,16)*. Lectin binding analysis of proteins adsorbed to membranes *(3,17–19)* or in ELISA systems *(6,19,20)* is also simple and highly sensitive for the survey of carbohydrate species on the glycoproteins. However, the interaction of a lectin with sugar chains seems to vary according to the state of the sugar chains at the molecular or cellular level. It is important to note that the affinity of lectins to sugar chains is affected by temperature, density of sugar chains, ionic requirements, steric hindrance of protein (accessibility), and the structure of internal sugar moiety in addition to nonreducing terminal sugars recognized *(4,15,16,21)*. We have found that not all the blood group-specific lectins bind to blood group substances on glycoproteins immobilized on membranes or plates (see Table 1 and Note 11), and that the recognition by the blood group-specific lectins shows reduced specificity and sensitivity compared with the group-specific antibodies (see Note 16). In this chapter, we describe methods for the detection of blood group substances (A and H) on plasma glycoproteins by lectin blotting and the use of lectins in ELISA systems in comparison with group-specific antibodies.

2. Materials
2.1. Buffers

1. Tris-buffered saline (TBS): 150 mM NaCl, 10 mM Tris-HCl, pH 7.5.
2. TBS containing Tween-20 (Tw/TBS): 150 mM NaCl, 10 mM Tris-HCl, pH 7.5, 0.05% Tween-20.
3. Blocking buffer: TBS containing 1% BSA (see Note 1).
4. Phosphate-buffered saline (PBS): 150 mM NaCl, 10 mM sodium phosphate, pH 7.2.
5. Coating buffer: 100 mM Na_2CO_3, 100 mM $NaHCO_3$, pH 9.5.
6. SDS-buffer: 2% SDS, 5% 2-mercaptoethanol (2-ME), 10% glycerol, 62.5 mM Tris-HCl, pH 6.8 (see Note 2).
7. Transfer buffers: 10 mM CAPS-NaOH buffer, pH 11.0, containing 10% methanol for polyvinylidene difluoride (PVDF) membrane (see Note 3).
8. 3,3-Diaminobenzidine-HCl (DAB) solution: 0.2 mg/mL of DAB in 50 mM Tris-HCl buffer, pH 7.5, containing 100 mM NaCl and 0.006% H_2O_2 (see Note 4).
9. o-phenylenediamine-HCl (OPD) solution: 0.5 mg/mL of OPD in 50 mM Tris-HCl buffer, pH 7.5, containing 200 mM NaCl and 0.03% H_2O_2.
10. Protein stain for PVDF membrane: 0.1% Fast green dissolved in 50% methanol and 10% acetic acid. The solution can be repeatedly used.
11. Protease inhibitor cocktail: 200 mM EDTA, 200 mM N-ethylmaleimide and 5000 kallikrein inhibitor U/mL of aprotinin.

2.2. Biotin-Conjugated Lectins

1. Biotinylated derivatives of HPA and UEA-I are available from several companies such as EY Laboratories (San Mateo, CA), Seikagaku Kogyo (Tokyo, Japan), Vector (Burlingame, CA), and Sigma (St. Louis, MO).
2. Dissolve each lyophilized lectin at a concentration of 1 mg/mL in PBS or TBS containing 0.05% sodium azide and store at –80°C in aliquots until use (avoid repeated freezing and thawing).

2.3. Antibodies

1. Anti-blood group A and B monoclonal antibodies (MAbs) are available from clinical reagent companies such as Ortho Diagnostic Systems (Raritan, NJ) (see Note 5). Polyclonal goat antibodies against human vWF and α_2M are available from several companies such as Medical and Biological Laboratories (MBL; Nagoya, Japan).
2. Horseradish peroxidase (HRP)-conjugated streptavidin, anti-mouse IgM, and anti-goat IgG are available from Vector, Zymed (San Francisco, CA) and MBL, respectively.

3. Methods
3.1. Preparation of Plasma

1. Centrifuge (1000g, 20 min at 25°C) the citrated whole blood from a healthy donor with known blood group to remove cells and mix the supernatant with 1/50 vol of protease inhibitor cocktail.

2. Divide it into several Eppendorf tubes and centrifuge again (13,000g, 15 min at 25°C) to remove remaining platelets. Store the plasma at –80°C until use.

3.2. Immunoprecipitation of vWF and α_2M (see Note 6)

1. Mix each plasma (1 mL) with 100 µL of anti-vWF or anti-α_2M antibody in an Eppendorf tube and incubate for 24 h at 4°C.
2. Collect the immunoprecipitate by centrifugation (13,000g, at 4°C for 30 min) and carefully wash the precipitate with TBS by gentle pipeting. Repeat centrifugation and washing two more times.
3. Dissolve the washed precipitate in 100 µL of SDS-buffer containing 0.8M urea at 95°C for 5 min.

3.3. Lectin-Blotting Analysis

1. Dilute glycoprotein, immunoprecipitate, or plasma appropriately with SDS-buffer and incubate at 95°C for 5 min (see Note 7).
2. Apply aliquots of a sample (1–5 µL/lane containing 1–5 µg of glycoprotein, 1/40–1/300 vol of immunoprecipitate or 0.2–0.4 µL of plasma) and perform SDS-PAGE (see Note 8) as described by Laemmli (22).
3. Soak a PVDF membrane (8.5 × 10 cm) and Whatman No. 3 filter papers (10 × 10 cm, 2 pieces/membrane) in 50% methanol for more than 1 min followed by CAPS buffer before SDS-PAGE (step 2) is finished (see Note 9).
4. After SDS-PAGE, soak the gel in the blotting buffer for 5 min, and put the gel and the PVDF membrane between two pieces of filter paper soaked with CAPS buffer (step 3), taking care not to form bubbles. Perform electrotransfer at 4°C for 3 h under constant voltage of 4.4 V/cm.
5. Mark lane numbers on the membrane with a ball-point pen after removing the gel carefully and soak the membrane at 4°C for 18 h in Tw/TBS (see Note 10).
6. Cut the membrane into appropriate pieces if necessary and incubate them separately with either biotinylated HPA, UEA-I (2 µg/mL in Tw/TBS) or any antibody (diluted with Tw/TBS) at 25°C for 1.5 h with gentle agitation (see Note 11).
7. Wash the membrane with Tw/TBS for 5 min, 5 times.
8. Incubate the membrane at 25°C for 1 h with HRP-conjugated avidin or the second antibody diluted to 1:1000.
9. Wash the membrane with Tw/TBS as above and incubate at 25°C for 1–5 min with DAB solution for HRP-reaction (see Note 12).
10. Wash the membrane with distilled water several times and photograph with instant camera immediately after dried (see Note 13).
11. Proteins on the membrane can be stained with Fast green solution for 5 min followed by several washings with destaining solution (50% methanol, 10% acetic acid). Figures 1–4 indicate the reactivity of HPA and UEA-I to vWF, α_2M and the whole plasma proteins separated by SDS-PAGE and transferred to a PVDF membrane.

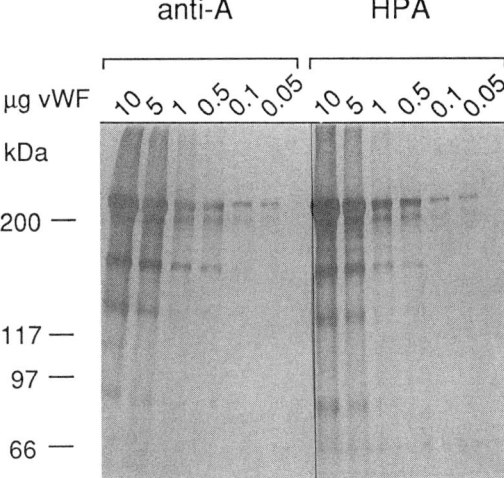

Fig. 1. Reactivity of HPA and anti-A antibody to human vWF blotted on PVDF membrane. Purified human vWF (0.05–10 μg) was applied to SDS-PAGE followed by electrotransfer to a PVDF membrane. The membrane was treated with anti-A MAb or HPA-biotin followed by the detection with HRP-conjugated anti-mouse IgM and streptavidin, respectively. vWF shows a major 270-kDa subunit band with some minor subbands. HPA detects A-substance even on as low as 0.05 μg of vWF.

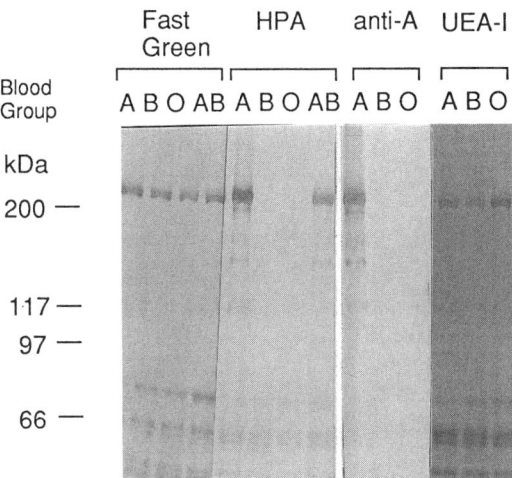

Fig. 2. Reactivity of HPA and UEA-I to vWF prepared from each blood group plasma. 2% of the immunoprecipitate obtained with anti-vWF antibody from 1 mL of each blood group plasma was blotted to a PVDF membrane after SDS-PAGE and treated with anti-A MAb, HPA-biotin, UEA-I-biotin, and HPA-biotin in the presence of 50 mM GalNAc, respectively, or protein-stained by Fast green. vWF from blood group A and AB reacted with HPA. vWF from group O-reacted better to UEA-I. Several minor bands (fibrinogen) weakly reacted with HPA and UEA-I in a manner independent on the blood group.

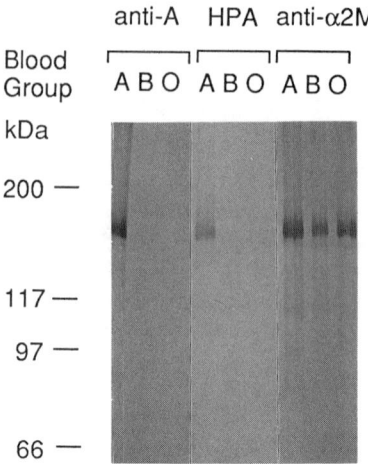

Fig. 3. Reactivity of HPA to α_2M prepared from each blood group plasma. 0.4% of α_2M immunoprecipitated with anti-α_2M antibody from 1 mL of each blood group plasma was treated with HPA-biotin, anti-A, and anti-α_2M antibodies, respectively. α_2M (180-kDa band) from blood group A plasma reacted with HPA as well as anti-A.

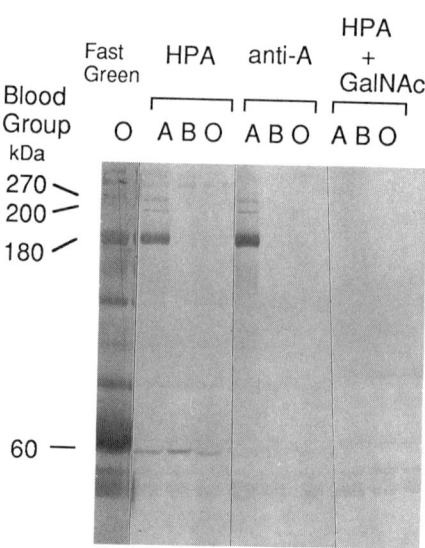

Fig. 4. Reactivity of HPA to plasma glycoproteins. The whole plasma (equivalent to 0.2 µL) from each blood group was treated with HPA-biotin, anti-A antibody and HPA in the presence of 50 mM GalNAc, respectively. HPA reacted to 270-kDa (vWF), 200-kDa (unknown), and 180-kDa (α_2M) components in group A plasma being consistent with the results by anti-A antibody. HPA also reacted with a 60-kDa component in all blood group plasma. The reaction disappeared in the presence of GalNAc at incubation with lectins.

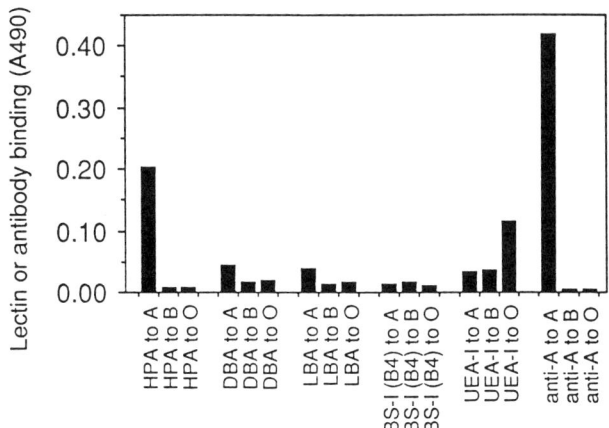

Fig. 5. Reactivity of blood group-specific lectins to plasma vWF immobilized on ELISA plates. vWFs in blood group A, B, and O plasma (1/10 diluted) were immobilized on the plate coated with anti-vWF IgG. A- and H-substances on vWF were detected by HPA and UEA-I, respectively. DBA and LBA reacted weakly with vWF derived from A plasma.

3.4. ELISA

1. Coat an ELISA plate (Maxisorp, Nunc InterMed, Denmark) with 50 µL of anti-vWF or anti-α_2M antibody dissolved at 20 µg/mL in coating buffer at 4°C for 24 h.
2. Wash the plate with 200 µL of TBS twice and incubate at 4°C for 24 h with 200 µL of blocking buffer.
3. Wash the plate with 200 µL of TBS twice (*see* Note 14).
4. Prepare serially diluted plasma (1/10 to 1/500) with Tw/TBS and incubate the plate with 50 µL of each diluted plasma at 25°C for 1 h.
5. Wash the plate four times with 200 µL of Tw/TBS and incubate at 25°C for 1 h with 50 µL of Tw/TBS containing a biotinylated lectin (2–5 µg/mL) or an antibody.
6. Wash the plate with Tw/TBS as above and incubate at 25°C for 1 h with 50 µL of HRP-conjugated streptavidin or the second antibody (diluted to 1:1000 with Tw/TBS).
7. After washing with 200 µL of Tw/TBS for 4 times, incubate the plate with 100 µL of OPD solution for 10–30 min in the dark.
8. Add 100 µL of 9N sulfuric acid to stop HRP-reaction and put in the dark for 10 more min (*see* Note 15).
9. Measure the absorbance of the reaction products at 490 nm (or 492 nm) with a microplate reader using the buffer solution as a blank (*see* Note 16). Figures 5 and 6 shows the detection of A and H substances on vWF immobilized on ELISA plate by HPA and UEA-I.

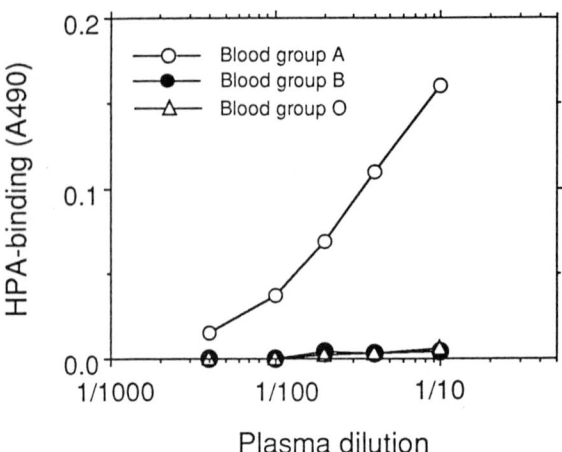

Fig. 6. Reactivity of HPA-biotin with plasma vWF on ELISA plate. HPA specifically detected A-substance on plasma vWF from blood group A immobilized on the plate with anti-vWF IgG. The binding was inhibited in the presence of 50 mM GalNAc (not shown).

4. Notes

1. After dissolving BSA in TBS, filter the solution with a disposable cartridge filter (0.2- or 0.45-µm pore-size) and store at −20°C after dividing into several tubes. Avoid using the solution kept more than 1 wk at 4°C.
2. It is recommended to prepare two-fold concentrated SDS-buffer without 2-ME and store at below −20°C. This buffer is mixed with equal volume of a sample solution (or distilled water) followed by addition of 2-ME before use.
3. It is recommended to prepare five-fold concentrated stock buffer of 50 mM CAPS containing 50% methanol (pH is adjusted to 11.0 with 10N NaOH before addition of methanol) and freshly dilute it with chilled water before use. Alternatively, use 25 mM Tris containing 192 mM glycine for nitrocellulose (NC) membrane.
4. DAB is carcinogenic, and should be handled with care. DAB and OPD solutions should be prepared just before use.
5. These antiblood group MAbs for diagnostics are usually a mixture of two to three clones of mouse IgM (rarely IgG). Because these are prediluted antibodies (approx 1–3 µg IgM/mL), use them in dilution at 1:5 or 1:10. Other MAbs from a single clone are also available from Dakopatts (Glostrup, Denmark) and Seikagaku Kogyo, and others.
6. Synthetic neoglycoproteins such as BSA or HSA with covalently linked blood group sugar chains are useful as standard marker proteins. They are very expensive, but it is fairly difficult to synthesize them. vWF and α_2M are excellent marker proteins for studying blood group antigens on plasma glycoproteins. They can be prepared from the whole blood easier than FVIII.

7. We prepare five-fold diluted plasma with distilled water and mix with equal volume of two-fold concentrated SDS-buffer. Undiluted plasma often results in bad resolution by mini-slab SDS-PAGE for its high concentration of proteins.
8. It is recommended that mini-slab gels (7.5 × 8.5 cm, 1-mm thick) of 7.5 and 5% acrylamide should be used for separating and stacking gel, respectively.
9. Use disposable gloves or forceps for handling blotting membranes because fingerprints sometimes cause contamination of blood group antigens. In the case of NC membrane, soak it with Tris/Glycine buffer without methanol treatment.
10. Keep wet and avoid drying the membrane during marking. Staining with 0.2% Ponceau red followed by washing with distilled water makes trace identification easier. Repeated washing with Tw/TBS destains the membrane.
11. We have also examined the reactivities of vWF and α_2M with other blood group-specific lectins such as DBA, LBA, SBA, *Sophora japonica* (SJA), *Euonymus europaeus* (EEA), GS-I (B_4) and LTA, but no other lectin than HPA and UEA-I specifically detected A, B, and H substances on these glycoproteins (Table 1). No suitable lectin has yet been found for detecting B-substance on these glycoproteins. It is important to examine the nonspecific binding of lectins. We usually incubate the blotted membrane and/or the plasma-incubated plate with lectins in the presence of specific sugars (50 mM GalNAc for HPA and 100 mM L-Fuc for UEA-I) to see the nonspecific binding. It is also necessary to check the specificity of a lectin to a sample individually prepared from each blood group donor by anti-A or anti-B antibody. HPA reacted with a 60-kDa protein in plasma irrelevant to blood groups (Fig. 4). In the case of UEA-I, it binds to several glycoproteins in plasma suggesting that UEA-I also reacts with α-fucosyl groups other than type 2 H-structure if they exist abundantly on a protein despite the weaker affinity.
12. Gently agitate the solution for reducing the nonspecific deposition of HRP products. Among many HRP-reaction reagents commercially available, 4-chloro-1-naphthol is less sensitive and shows low contrast and rapid decolorization of the product. Care should be taken to reduce nonspecific binding when using a highly sensitive reagent for HRP reaction.
13. The membrane negatively-reacted to a lectin can be used for another lectin (or antibody) experiment even after drying, if it is immersed in the Tw/TBS again.
14. Wash the plate with TBS containing 0.05% sodium azide and store at 4°C after wrapping in a plastic sheet to avoid drying. The plates thus stored should be used within 2 mo.
15. Avoid forming microbubbles in the well when 9N sulfuric acid is added. It is recommended to use a dispenser pipet for immediate mixing.
16. When α_2M is immobilized on the plate coated with anti-α_2M IgG by incubating with plasma, the binding of HPA and UEA-I to the plate is very low probably because the amount of α_2M containing blood group antigens in plasma is comparatively low *(6)*. However, anti-A and anti-B antibodies clearly bind to α_2M immobilized on the plate, indicating a higher sensitivity of the antibodies compare to the lectins.

References

1. Kabat, E. A. (1982) Philip Levine award lecture. Contribution of quantitative immunochemistry to knowledge of blood group A, B, H, Le, I and i antigens. *Am. J. Clin. Pathol.* **78,** 281–292.
2. Clausen, H. and Hakomori, S. (1989) ABH and related histo-blood group antigens; immunochemical differences in carrier isotypes and their distribution. *Vox Sang.* **56,** 1–20.
3. Matsui, T., Kihara, C., Fujimura, Y., Mizuochi, T., and Koiti, T. (1991) Carbohydrate analysis of human von Willebrand factor with horseradish peroxidase-conjugated lectins. *Biochem. Biophys. Res. Commun.* **178,** 1253–1259.
4. Matsui, T., Titani, K., and Mizuochi, T. (1992) Structures of the asparagine-linked oligosaccharide chains of human von Willebrand factor; occurrence of blood group A, B, and H(O) structures. *J. Biol. Chem.* **267,** 8723–8731.
5. Hironaka, T., Furukawa, K., Esmon, P. C., Fournel, M. A., Sawada, S., Kato, M., Minaga, T., and Kobata, A. (1992) Comparative study of the sugar chains of factor VIII purified from human plasma and from the culture media of recombinant baby hamster kidney cells. *J. Biol. Chem.* **267,** 8012–8020.
6. Matsui, T., Fujimura, Y., Nishida, S., and Titani, K. (1993) Human plasma α_2-macroglobulin and von Willebrand factor possess covalently linked ABO(H) blood group antigens in subjects with corresponding ABO phenotype. *Blood* **82,** 663–668.
7. Gill, J. C., Endres-Brooks, J., Bauer, P. J., Marks, W. J., and Mongomery, R. R. (1987) The effect of ABO group on the diagnosis of von Willebrand disease. *Blood* **69,** 1691–1695.
8. Shima, M., Fujimura, Y., Nishiyama, T., Tsujiuchi, T., Narita, N., Matsui, T., Titani, K., Katayama, M., Yamamoto, F., and Yoshioka, A. (1995) ABO blood group genotype and plasma von Willebrand factor in normal individuals. *Vox Sang.* **68,** 236–240.
9. Hakomori, S. (1989) Aberrant glycosylation in tumors and tumor-associated carbohydrate antigens. *Adv. Cancer Res.* **52,** 257–331.
10. Goldstein, I. J. and Poretz, R. D. (1986) Isolation, physicochemical characterization, and carbohydrate-binding specificity of lectins, in *The Lectins; Properties, Functions, and Applications in Biology and Medicine* (Liener, I. E., Sharon, N., and Goldstein, I. J., eds.), Academic, London, pp. 35–247.
11. Hammarstrom, S. and Kabat, E. A. (1969) Purification and characterization of a blood-group A reactive hemagglutinin from the snail *Helix pomatia* and a study of its combining site. *Biochemistry* **8,** 2696–2705.
12. Pereira, M. E. A., Kisailus, E. C., Gruezo, F., and Kabat, E. A. (1978) Immunochemical studies on the combining site of the blood group H-specific lectin 1 from *Ulex europeus* seeds. *Arch. Biochem. Biophys.* **185,** 108–115.
13. Smith, D. F. and Torres, B. V. (1989) Lectin affinity chromatography of glycolipids and glycolipid-derived oligosaccharides. *Methods Enzymol.* **179,** 30–45.
14. Baker, D. A, Sugii, S., Kabat, E. A., Ratcliffe, R. M., Hermentin, P., and Lemieux, R. U. (1983) Immunochemical studies on the combining sites of forssman hapten

reactive hemagglutinins from *Dolichos biflorus*, *Helix pomatia*, and *Wistaria floribunda*. *Biochemistry* **22,** 2741–2750.
15. Cummings, R. D. and Kornfeld, S. (1982) Fractionation of asparagine-linked oligosaccharides by serial lectin-agarose affinity chromatography; a rapid, sensitive, and specific technique. *J. Biol. Chem.* **257,** 11,235–11,240.
16. Osawa, T. and Tsuji, T. (1987) Fractionation and structural assessment of oligosaccharides and glycopeptides by use of immobilized lectins. *Ann. Rev. Biochem.* **56,** 21–42.
17. Kijimoto-Ochiai, S., Katagiri, Y. U., and Ochiai, H. (1985) Analysis of N-linked oligosaccharide chains of glycoproteins on nitrocellulose sheets using lectin-peroxidase reagents. *Anal. Biochem.* **147,** 222–229.
18. Kijimoto-Ochiai, S., Katagiri, Y. U., Hatae, T., and Okuyama, H. (1989) Type analysis of the oligosaccharide chains on microheterogenous components of bovine pancreatic DNAase by the lectin-nitrocellulose sheet method. *Biochem. J.* **257,** 43–49.
19. Rhodes, J. M. and Ching, C. K. (1993) The application of lectins to the study of mucosal glycoproteins. *Methods Mol. Biol.* **14,** 247–262.
20. McCoy, J. P., Varani, J., and Goldstein, I. J. (1983) Enzyme-linked lectin assay (ELLA): use of alkaline phosphatase-conjugated *Griffonia simplicifolia* B4 isolectin for the detection of α-D-galactopyranosyl end groups. *Anal. Biochem.* **130,** 437–444.
21. Yamashita, K., Totani, K., Ohkura, T., Takaski, S., Goldstein, I. J., and Kobata, A. (1987) Carbohydrate binding properties of complex-type oligosaccharides on immobilized *Datura stramonium* lectin. *J. Biol. Chem.* **262,** 1602–1607.
22. Laemmli, U. K. (1970) Cleavage of structural proteins during the assembly of the head of bacteriophage T4. *Nature* **227,** 680–685.

IV

USE OF LECTINS IN QUANTIFICATION OF SOLUBLE GLYCOPROTEINS

20

Lectin/Antibody "Sandwich" ELISA for Quantification of Circulating Mucin as a Diagnostic Test for Pancreatic Cancer

Neil Parker

1. Introduction
1.1. Use of Lectin Capture ELISA System for Serological Diagnosis of Pancreatic Cancer (Lectin/Antibody ELISA)

Pancreatic cancer in the nonjaundiced patient can be difficult to diagnose, even with the aid of modern scanning techniques. Studies have shown that mucus glycoproteins (mucins) are frequently detectable in the serum of patients with pancreatic cancer *(1,2)*, and these mucins may express a wide range of different carbohydrate structures. The detection and quantification of mucin can be used in serological diagnosis of the disease.

A novel monoclonal antibody (MAb) called CAM 17.1, which has been used in the detection of purified mucin *(3)* and has been shown to detect mucin in the serum of patients with pancreatic cancer *(4)* should prove useful as an aid to the diagnosis of the disease. CAM 17.1 is an immunoglobulin M antibody which probably binds the sialylated blood group I antigen *(4)*, and has been shown immunohistochemically to be highly specific for intestinal mucus, particularly in the colon, small intestine, biliary tract, and pancreas *(5)*.

The assay employed for detection of glycoprotein uses a lectin capture system, in which wheatgerm agglutinin (WGA) which binds N-acetylglucosamine and sialic acid (N-acetylneuraminic acid) is bound to an ELISA plate. The WGA will bind to any mucin present in the serum in addition to other sialylated glycoproteins. The mucin is then detected by enzyme labeled CAM 17.1.

A retrospective study using this WGA/CAM 17.1 assay *(4)* showed a sensitivity of 78% for pancreatic cancer with a specificity of 76%. The assay has since been commercialized with similar results obtained for retrospective studies, e.g., sensitivity of 82% and a specificity of 83% *(6)*; however, in a prospective study, even better sensitivity and specificities have been obtained (84 and 92% respectively), indicating that the assay probably performs better on fresh samples *(7)*.

1.2. Wheatgerm Agglutinin as a Nonspecific Mucin Capture System

Initially, the CAM 17.1 assay was intended to be developed as a conventional sandwich assay with CAM 17.1 bound to the plate, followed by sample incubation, then detection by conjugated CAM 17.1. However, attempts to use biotinylated CAM 17.1 resulted in loss of antibody binding activity and, so, interest then turned to the use of a lectin as the initial capture system. If this were successful it would then enable an enzyme-labeled anti-IgM to be used as the final detection system for bound CAM 17.1.

Gastrointestinal mucins contain up to 85% carbohydrate, consisting mainly of *O*-linked oligosaccharide sidechains linked to the protein backbone *(8)*. The carbohydrate part of these chains consists of fucose, galactose, *N*-acetyl-galactosamine, *N*-acetyl-glucosamine and sialic acid *(9)*. The specific carbohydrate binding capabilities of several lectins were considered. It is known that the lectin wheatgerm agglutinin binds to *N*-acetyl-glucosamine, and sialic acid, which two carbohydrates make up approx 33% of the total carbohydrate of pure mucin *(9)*. WGA might therefore bind all mucosal glycoproteins. For this reason, this lectin was tested as a replacement for CAM 17.1 as the initial mucin capture agent. Initial titering experiments were set up for the WGA coating, using diluents that had been shown to work in enzyme-linked lectin binding assays (ELLA) *(10)*. Serum from known pancreatic cancer patients and normals were run in the assay followed by limited CAM 17.1 and peroxidase-labeled anti-IgM checkerboard titering. Once a working assay was established, i.e., the sera from pancreatic cancer patients produced substantially more color response than the normal sera, further WGA, CAM 17.1, and anti-IgM titering was carried out to optimize the assay.

This lectin capture technique is likely to work well with other antibodies and may save the inconveniences of conjugation, repurification, and repeat testing of antibodies.

2. Materials

1. Cobalt-irradiated ELISA plates (e.g., M 124B Dynatech, Chantilly, VA).
2. Lectin WGA (Sigma, St. Louis, MO) dissolved in carbonate coating buffer containing 50 mM Na$_2$CO$_3$, 50 mM NaHCO$_3$, pH 9.6.
3. Tween-20 (Sigma) (0.1%) in 0.01M phosphate-buffered saline (PBS) pH 7.2 (PBS/Tween).

4. CAM 17.1 culture supernatant (Euro DPC, Llanberis, UK).
5. Peroxidase-conjugated rabbit antimouse immunoglobulin (Sigma).
6. Freshly made up ortho phenylenediamine (Sigma) 5 mg in 12.5 mL 0.2M phosphate/0.1M citrate buffer, pH 5.0, with 12.5 µL 30% hydrogen peroxide (OPD substrate) (*see* Note 1).
7. Sulfuric acid (4M).

3. Methods

1. Add 100 µL of WGA (2.5 mg/mL) diluted 1/600 in carbonate coating buffer to each well of an ELISA plate and incubated for 16 h at 4°C (*see* Note 2).
2. Remove unbound WGA by washing three times with PBS/Tween.
3. Block by incubation in PBS/Tween for 1 h at room temperature to lower nonspecific binding (*see* Note 3).
4. Incubate 100 µL of prediluted serum (1/20 in PBS/Tween; *see* Note 4) in duplicate for 2 h at 37°C (*see* Notes 3 and 5). Several wells on each plate should be incubated with buffer alone as blanks.
5. Tap out wells, and wash three times with PBS/Tween to remove unbound serum.
6. Add 100 µL of CAM 17.1 supernatant diluted 1/10 in PBS/Tween to the plate and incubate at 37°C for 2 h (*see* Note 6).
7. Tap out wells, and wash three times with PBS/Tween to remove unbound CAM 17.1.
8. Add 100 µL of peroxidase-conjugated rabbit antimouse immunoglobulin diluted 1/600 in PBS/Tween to each well and incubate for 2 h at 37°C (*see* Note 6).
9. Tap out wells, and wash three times with PBS/Tween to remove unbound rabbit immunoglobulins.
10. Add 100 µL OPD substrate to each well and incubate for 10 min.
11. Stop the reaction by addition of 100 µL of 4M H_2SO_4 (*see* Note 7).
12. Read the plate at 492 nm in a suitable ELISA plate reader (*see* Note 8).
13. Results can be expressed in arbitrary U/mL, with one unit being equivalent to the optical density of a constant positive control sample that was run in duplicate on each ELISA plate.

4. Notes

1. Ortho phenylenediamine is carcinogenic and very unstable, especially in light and should be used within 15 min of making it up. Owing to this light sensitivity, the OPD solution should be made up in a dark container or one wrapped in silver foil.
2. It is best to prepare as many plates as possible at one time with the WGA coating to eliminate any coating to coating variation. Prepared plates can then be stored at 4°C in individual resealable laminated foil bags with desiccant. Adding antibacterial agents to the coating buffer should reduce the risk of bacterial contamination. Bags should be labeled with a specific lot number, e.g., date of manufacture and initial. The plates should remain stable for several months if treated in this manner.
3. Shaking a plate on a plate shaker will shorten incubation times if required.

4. In Section 3., step 4, it states that the serum should be diluted 1/20 in PBS/Tween. This is because when the assay was developed, sample conservation was a high priority. With a relatively large sample of 100 µL being used, it was decided to predilute the sample. The step of predilution is unnecessary unless it is required to protect the sample stocks, 100 µL of neat sample being used instead. If neat sample is used, further titering may have to be carried out to lower the color produced in the assay.
5. If shaking is combined with a controlled temperature environment, it should greatly reduce assay-to-assay variation of color development. This happens because enzyme conjugates are extremely temperature dependent.
6. The CAM 17.1 can be combined with the peroxidase conjugated rabbit antimouse immunoglobulin to form one solution. This would shorten the assay by one incubation, the incubation time would be as normal, i.e., 2 h at 37°C. The CAM 17.1 and the peroxidase-conjugated rabbit antimouse immunoglobulins would however have to be titered first to obtain optimal NSB, maximum binding (highest color intensity), and good method comparison with patients with known CAM 17.1 values. It is best to leave the different titers of CAM 17.1/anti-IgM overnight to equilibrate before testing.
7. An alternative to end-point reading is kinetic analysis, in which the rate of change of the development of color is measured. This technique integrates a series of readings obtained over several minutes of enzyme/substrate interaction. The rate of increase in optical density can be measured precisely even in wells which obtain a final color of 3.0 or 4.0 OD U, this gives a broader active range and much better discrimination with improved signal at the low end of the assay. In this case more color production is required than for an end point assay so the system would require retitering, probably at the CAM 17.1 and/or peroxidase-conjugated rabbit antimouse immunoglobulin stages.
8. The assay can be better calibrated by the use of a patient sample that gives a very high concentration of CAM 17.1. activity and of which there is a large sample pool available. The sample can then be diluted serially in a range between the NSB of the assay and the maximum binding (i.e., highest color) to form a calibration curve. A suitable diluent would have to be tested to give linear dilutability, e.g., pooled normal human serum that has been shown to be negative for CAM 17.1 or bovine cadet serum or fetal calf serum could be tested, neat diluent should be used as the "zero," nonspecific background (NSB) calibrator. Providing there is enough of each "calibrator" made, then the calibrators can be aliquoted and frozen after assigning them a CAM 17.1 value.

If calibrators are produced as described and run in the assay, it is advised to produce a log–log graph for the "percent-bound" values of the nonzero calibrators. This is carried out by firstly correcting the assay response (OD) for each nonzero calibrator by subtracting the average OD of the NSB wells (zero calibrator). The percent bound is then calculated for each well as a percent of the maximum binding, with the average NSB-corrected response of the calibrator with the highest CAM 17.1 value taken as 100%. A plot of percent bound vs concentration for each of the nonzero calibrators is made, with a curve being drawn through the path of the points. CAM 17.1. values for unknowns are then determined from the curve by reading off the concentration from the relevant percent bound.

Controls could be made by taking other patient samples with CAM 17.1 results which fall in the calibration curve and have large enough sample pools. These should then be aliquoted and frozen; after several assays have been performed they can be assigned control ranges of ±2SD of the mean. Samples which give OD readings greater than the maximum detectable can be diluted into the measurable range.

References

1. Herlyn, M., Sears, H. F., Steplewski, Z., and Koprowski, H. (1982) Monoclonal antibody detection of a circulating tumor-associated antigen: 1. Presence of antigen in the sera of patients with colorectal, gasric and pancreatic carcinoma. *J. Clin. Immunol.* **2**, 135–140.
2. Magnani, J. L, Steplewski, Z., Koprowski, H., and Ginsberg, V. (1983) Identification of the gastrointestinal and pancreatic cancer-associated antigen detected by monoclonal antibody 19-9 in the sera of patients as a mucin. *Cancer Res.* **43**, 5489–5492.
3. Parker, N., Finnie, I. A., Raouf, A. H., Ryder, S. D., Campbell, B. J., Tsai, H. H., Iddon, D., Milton, J. D., and Rhodes, J. M. (1993) High performance gel filtration using monodisperse highly cross-linked agarose as a one-step system for mucin purification. *Biomed. Chromatogr.* **7**, 68–74.
4. Parker, N., Makin, C. A., Ching, C.-K., Eccleston, D., Taylor, O. M., Milton, J. D., and Rhodes, J. M. (1992) A new enzyme-linked lectin/mucin antibody sandwich assay (CAM 17.1/WGA) assessed in combination with CA19-9 and peanut lectin binding assay for the diagnosis of pancreatic cancer. *Cancer* **70**, 1062–1068.
5. Makin, C. A. (1985) Monoclonal antibodies for the study of colorectal and other carcinomas (PhD thesis). Faculty of Medicine, University of London, pp. 67–70.
6. Ching, C.-K., Parker, N., Hand, C., Charney, R., and Long, R. G. (1993) CAM 17.1/WGA assay for exocrine pancreatic cancer. *Gut* **34**, S28 (abstract).
7. Yiannakou, J. Y., Calder, F., Newland, P., Kingsnorth, A. N. and Rhodes, J. M. (1995) Prospective clinical appraisal of CAM 17.1: an effective serological pancreatic tumour marker. *Gastroenterology* **108**, A556 (abstract).
8. Laboisse, C. L. (1986) Structure of gastrointestinal mucins: searching for the Rosetta stone. *Biochemie* **68**, 611–617.
9. Podolsky, D. K. and Isselbacher, K. J. (1983) Composition of human colonic mucin. Selective alteration in inflammatory bowel disease. *J Clin. Invest.* **72**, 142–153.
10. Rhodes, J. M. and Ching, C. K. (1993) The application of lectins to the study of mucosal glycoproteins, in *Methods in Molecular Biology*, vol 14. *Glycoprotein Analysis in Biomedicine* (Hounsell, E. F., ed.), Humana, Totowa, NJ, pp. 247–262.

21

Quantification of Intestinal Mucins

Jeremy D. Milton and Jonathan M. Rhodes

1. Introduction

1.1. Nature of Intestinal Mucins

Intestinal mucins are glycoproteins containing a very high proportion of carbohydrate, up to 90%, and they can occur either in a membrane-bound or a soluble form *(1)*. The general structure of these molecules includes discrete *C*-terminal and *N*-terminal regions, which are relatively poorly glycosylated, at either end of a long repeating structure, up to 50 repeats, which is heavily *O*-glycosylated on serine and threonine residues the first sugar in each oligosaccharide always being *N*-acetyl-galactosamine *(2)*. These oligosaccharide chains can vary between 1 and about 15 saccharides. The *C*- and *N*-terminal regions are likely to be of importance in determining the membranous or secreted nature of the molecule *(2)*.

1.2. Lectins vs Antibodies for Mucin Quantitation

Both lectins and antibodies can be used to characterize and quantify soluble glycoproteins, but lectins are particularly appropriate for analysis of heavily glycosylated substances such as mucins. Lectins generally have rather broader specificity than antibodies, and of course only react with carbohydrate groups. Most, but by no means all, antibodies raised against mucins tend to be against the carbohydrate components *(3,4)*, though antibodies against the protein components have been produced and are of great value in identifying the core protein structures of deglycosylated mucin *(5)*.

Lectins of different carbohydrate specificity can be used not only for quantification but also to obtain information about the nature of mucin glycosylation. Enzyme-linked lectin binding assays are highly sensitive and can be extremely useful for quantification of mucins after column fractionation when concentrations are often beyond the limits of sensitivity of conventional protein or carbohydrate assays.

From: *Methods in Molecular Medicine, Vol. 9: Lectin Methods and Protocols*
Edited by: J. M. Rhodes and J. D. Milton Humana Press Inc., Totowa, NJ

1.3. ELISA vs Slot Blot

Two commonly used techniques for measuring individual proteins, which require immobilization of the protein, are the ELISA method using polystyrene plates and dot/slot blots, which use nitrocellulose. Though the ELISA technique is much more sensitive and can be more accurately quantitated than slot blots, the latter are reasonably quantitative and have the considerable advantage that bound protein or glycoprotein material can be treated with chemicals and enzymes when immobilized on nitrocellulose; thereafter changes in its antibody *(6)* or lectin reactivity can be measured. We will describe both procedures.

1.4. Quantification of Purified Glycoprotein by Estimation of Wheatgerm Agglutinin Binding

When a purified glycoprotein is known to contain sialic acid or N-acetylglucosamine, wheatgerm agglutinin is a very useful ligand. Provided the glycoprotein is known to be pure, a standard curve can be established in comparison with protein or carbohydrate assay and a binding assay with conjugated wheatgerm agglutinin will then prove a very sensitive and robust system for quantification. We have used this in particular for analysis of mucins after separation by column chromatography *(7)*. Other lectins can readily be used to determine individual sugar availability of different mucin preparations, such as sialic acid using limax lectin or Gal-GalNAc determinants using mushroom lectin or peanut agglutinin *(8)*.

2. Materials

1. ELISA plates: Immulon 4 flat-bottomed, 96-well ELISA plates (Dynatech, Billingshurst, UK).
2. Carbonate coating buffer pH 9.5: 2.92 g $NaHCO_3$, 1.6 g Na_2CO_3 made up to 1 L in distilled water.
3. Peroxidase-labeled lectins: These can be obtained from Sigma or Vector or can be prepared by the two-stage glutaraldehyde conjugation procedure *(8)*, using unlabeled lectin, from Sigma or Vector and peroxidase from Sigma. When using this method, we have not found it necessary to purify the conjugate from unbound peroxidase, though this can be achieved simply using a Biogel P100 column.
4. Phosphate-buffered saline (PBS): Dulbecco A (Oxoid, Basingstoke, UK) or made up as follows: 8 g NaCl, 0.2 g KCl, 1.15 g Na_2HPO_4, 0.2 g KH_2PO_4 made up to 1 L in distilled water.
5. PBS/Tween: 0.1% Tween-20 (Sigma) in PBS (1 mL/L).
6. OPD substrate: Orthophenylenediamine (OPD) (Sigma), 10-mg tablet dissolved in 25 mL phosphate citrate buffer, pH 5.0, to which is added 10 µL 30% H_2O_2.

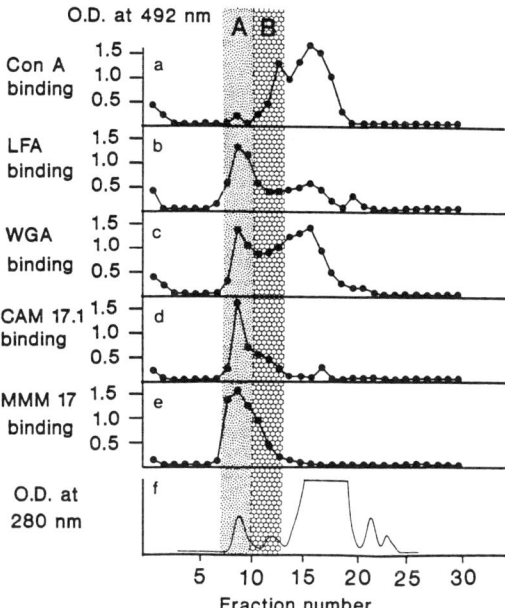

Fig. 1. Example of Con A, *Limax flavus* agglutinin (LFA), and wheatgerm agglutinin (WGA) used in ELLA to quantify Superose gel filtration fractions of human colonic mucin. CAM 17.1 and MMM 17 are antimucin MAbs (Reproduced with permission from ref. 7).

   Phosphate citrate buffer is prepared by dissolving 9.15 g $Na_2HPO_4·2H_2O$ and 5.11 g citric acid monohydrate in 500 mL and is diluted 1:1 for use.
 7. ELISA plate reader: Any microplate reader is satisfactory; we use a CLS 962 microplate reader (Cambridge Life Sciences, Cambridge, UK).
 8. Nitrocellulose membrane: Cellulose nitrate 0.45-µm pore size, (Schleicher & Schuell Dassel, Germany; Sigma).
 9. Chloronaphthol substrate: 20 mL PBS, 5 mL chloronaphthol solution (3 mg/mL in methanol), 25 µL 30% hydrogen peroxide.
10. Mixing machine: A machine that both rocks and rolls samples such as the Multimix Major (Luckham, Burgess Hill, UK).
11. Slot blot apparatus (Hoeffer Scientific Instruments, San Francisco, CA).
12. Vacuum pump: Any reasonably powerful pump.

3. Methods
3.1. ELISA

A simple single layer ELISA in which the mucin containing material is attached directly to the solid phase and detected directly by labeled lectin is very effective with partially purified mucin (Fig. 1; *see* Note 1).

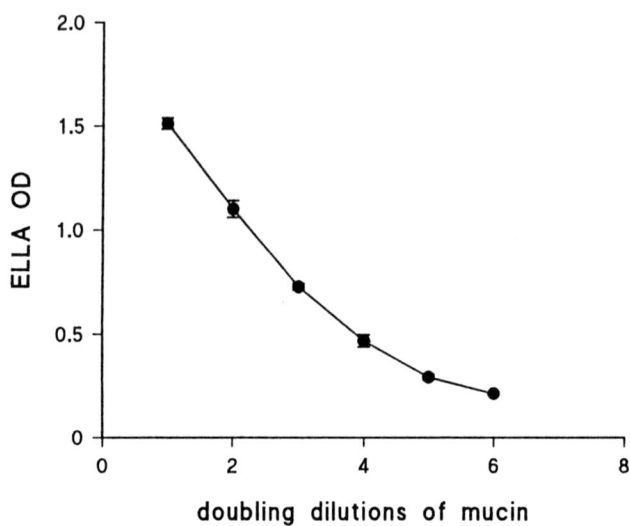

Fig. 2. An ELLA curve of OD 492 against mucin concentration. Human colonic mucin and wheatgerm agglutinin.

1. Coat ELISA plate wells with 100 μL sample in coating buffer (*see* Note 2).
2. Incubate at 4°C for 16 h.
3. Wash twice with PBS.
4. Add 120 μL blocking buffer, PBS/Tween (*see* Note 3).
5. Incubate at 37°C for 2 h.
6. Wash three times with PBS.
7. Add 100 μL peroxidase-lectin conjugate in PBS/Tween (*see* Note 4).
8. Incubate at 37°C for 4 h (*see* Note 5).
9. Wash five times with PBS/Tween.
10. Add 100 μL OPD substrate.
11. Incubate at 20°C for 6 min (*see* Note 6).
12. Add 100 μL 4M sulfuric acid.
13. Read absorbance at 492 nm (Fig. 2; *see* Note 7).

3.2. Slot Blots

1. Put PBS-soaked, nitrocellulose in Hoeffer slot blot apparatus.
2. Connect vacuum line to apparatus and apply vacuum for a few minutes, then clamp line with vacuum held.
3. Add 100 μL sample and allow to run through (*see* Note 8).
4. Add 400 μL PBS and allow to run through.
5. Reconnect vacuum, dissemble apparatus, and remove nitrocellulose membrane (*see* Note 9).
6. Place strips in plastic, 30-mL universal containers.

Quantification of Intestinal Mucins

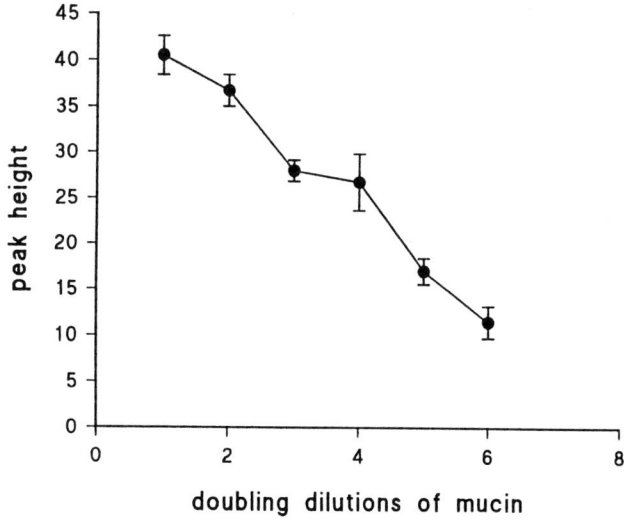

Fig. 3. A slot blot assay of peak height of absorption against mucin concentration. Bovine submaxillary mucin and wheatgerm agglutinin.

7. Block with 2 mL 0.5% BSA/PBS or 0.5% Tween/0.5% BSA/PBS for 1–2 h on a rocking and rolling machine (Multimix Major, Luckham, UK) at room temperature. All incubations and washes were done on the Multimix (*see* Note 10).
8. Wash 3 × 3 min with 5 mL PBS.
9. Add lectin in BSA/PBS or Tween/BSA/PBS and incubate for 2–4 h at room temperature (*see* Note 11).
10. Wash 3 × 3 min with 5 mL PBS.
11. Transfer strip to a Petri dish and incubate in chloronaphthol solution until color develops (*see* Note 12).
12. Rinse thoroughly in several rinses of PBS (30 s, then 3 × 5 min). Blots may be left in PBS in the cold and the dark for several days before measuring color intensity.
13. Dry blots between filter papers and read on a densitometer (Hoeffer or Shimadzu, Kyoto, Japan) measuring the peak heights. These have been shown to give a good measure of mucin added (Fig. 3).

4. Notes

1. We have frequently used this assay system for quantitation of mucin eluted from high performance gel filtration columns or from ion exchange columns. If mucus glycoprotein is to be quantified without prior purification then an antibody-lectin sandwich technique might be more appropriate (*see* Chapter XX).
2. Whereas an alkaline coating buffer is generally used for ELISAs, the nature of the coating buffer is not critical and we have found that mucin column eluate fractions that are in Tris-HCl, pH 8.0, diluted with either the same Tris-HCl or with PBS produce the same reaction as diluting with carbonate coating buffer.

3. A common blocking buffer for ELISAs is BSA/Tween/PBS, but we have found that the presence of BSA reduces the sensitivity considerably, with no reduction in background.
4. A lectin concentration of approx 2 µg/mL was usually satisfactory, but obviously this needs to be determined for any individual lectin preparation, with a balance between speed of reaction and background.
5. Incubation for 2–4 h at 37°C or 16 h at 4°C is equally acceptable—the longer period gives higher readings, but a somewhat higher background makes it no more sensitive (10).
6. Of course this is an arbitrary time and 10 min would be equally acceptable; if no color appears within this time leaving it longer is unlikely to be much use.
7. Plates can be left for 30 min or so at room temperature or overnight at 4°C before reading.
8. 100 µL is a convenient size of sample, but not critical, similarly the 400 µL washing volume is convenient and effective.
9. Deviations from the manufacturers instructions are permissible depending on the availability of vacuum pumps and pressure gages.
10. Blocking with Tween as well as BSA seems to be essential to keep the background to acceptable levels, though this does vary a little from lectin to lectin.
11. However, we have also found that if Tween is present in the lectin incubation, one can do without it in the blocking buffer. We have found no disadvantage in having Tween present at both stages.
12. Color usually starts to develop within 1 min and by 5 min is likely to be optimal as the background will usually start to creep up after such a time.

References

1. Rose, M. C. (1992) Mucins: structure, function, and role in pulmonary diseases. *Am. J. Physiol.* **263**, L413–429.
2. Gum, J. R., Hicks, J. W., Toribara, N. W., Siddiki, B., and Kim, Y. S. (1994) Molecular cloning of human intestinal mucin (MUC2) cDNA. Identification of the amino terminus and overall sequence similarity to prepro-von Willebrand factor. *J. Biol. Chem.* **269**, 2440–2446.
3. Magnani, J. L., Steplewski, Z., Koprowski, H., and Ginsburg V. (1982) Identification of the gastrointestinal and pancreatic cancer-associated antigen detected by the monoclonal antibody 19-9 in the sera of patients as a mucin. *J. Biol. Chem.* **257**, 14,365–14,369.
4. Lan, M. X., Khorrami, A., Kaufman, B., and Metzgar, R. S. (1987) Molecular characterisation of a mucin-like antigen associated with human pancreatic cancer. The DU-PAN-2 antigen. *J. Biol. Chem.* **262**, 12,863–12,870.
5. Siddiqui, J., Abe, M., Hayes, D., Shani, E., Yunis, E., and Kufe, D. (1988) Isolation and sequencing of a cDNA coding for the human DF3 breast carcinoma-associated antigen. *Proc. Natl. Acad. Sci.* **85**, 2320–2323.
6. Milton, J. D., Eccleston, D., Parker, N., Raouf, A., Cubbin, C., Hoffman, J., Hart, C. A., and Rhodes, J. M. (1993) Distribution of O-acetylated sialomucin in the

normal and diseased gastrointestinal tract shown by a new monoclonal antibody. *J. Clin. Pathol.* **46,** 323–329.
7. Raouf, A., Parker, N., Iddon, D., Ryder, S., Langdon-Brown, B., Milton, J. D., Walker, R., and Rhodes, J. M. (1991) Ion-exchange chromatography of purified colonic mucus glycoproteins in inflammatory bowel disease: absence of a selective subclass defect. *Gut* **32,** 1139–1146.
8. Yu, L., Fernig, D. G., Smith, J. A., Milton, J. D., and Rhodes, J. M. (1993) Reversible inhibition of proliferation of epithelial cell lines by *Agaricus bisporus* (edible mushroom) lectin. *Cancer Res.* **53,** 4627–4632.
9. O'Sullivan, M. J. and Marks, V. (1981) Methods for the preparation of enzyme-antibody conjugates for use in enzyme immunoassay. *Meth. Enzymol.* **73,** 147–166.
10. Rhodes, J. M., Milton, J. D., Parker, N., and Mullen, P. J. (1993) Use of lectins in quantitation and characterization of soluble glycoproteins in *Lectins and Glycobiology* (Gabius, H.-J. and Gabius, S., eds.), Springer-Verlag, Berlin, pp. 150–157.

V

LECTINS IN AFFINITY PURIFICATION OF SOLUBLE GLYCOPROTEINS

22

Purification and Characterization of Human Serum and Secretory IgA1 and IgA2 Using Jacalin

Michael A. Kerr, Lesley M. Loomes, Brian C. Bonner, Amy B. Hutchings, and Bernard W. Senior

1. Introduction

Human immunoglobulin A (IgA) is found in serum and secretions in several different forms *(1)*. Serum IgA is produced by the bone marrow and is predominately monomeric (M_r = 160 kDa). Most of the IgA in secretions (sIgA) is polymeric, predominantly dimeric, comprising two monomer subunits linked together by intersubunit disulfide bonds and a cysteine-rich polypeptide termed J-chain (M_r = 16 kDa). Polymeric IgA in secretions, but not in serum, contains an additional heavily glycosylated protein called secretory component (SC) (M_r = 70 kDa).

Serum or secretory IgA exists as two isotypic forms, IgA1 and IgA2, distinguished by antigenic, biochemical, and biological properties. The two isotypes have characteristic distributions in both serum and secretions; 89% of serum IgA is IgA1, whereas up to 65% of the IgA in external secretions is IgA2. The major structural difference between the two isotypes lies in the amino acid sequence of the hinge region of their heavy chains. The 13 amino acid hinge region of the α_1-chain is composed of an unusual repeating sequence rich in proline, serine, and threonine, the serines carrying *O*-linked oligosaccharides. The α_2-chain lacks this hinge region segment and hence the *O*-linked sugars, which has important biological consequences for binding to some IgA receptors *(2)* and in rendering it resistant to cleavage by proteinases produced by several pathogenic bacteria *(3)*.

The purification of IgA was revolutionized by the discovery that jacalin, a lectin isolated from the jackfruit, selectively binds IgA1 *(4,5)*. The lectin has since been used in both small and large scale purification of IgA1 from human serum and secretions and for the separation of IgA1 from IgA2 in more complicated procedures.

1.1. History of the Use of Jacalin in the Purification of IgA

Following the initial demonstration of the binding of IgA by jacalin, Gregory and coworkers *(6)* showed that jacalin binds specifically to terminal Galβ1-3GalNAc moieties of *O*-linked oligosaccharides in the hinge region of IgA1. Thus affinity chromatography using jacalin appeared to provide a simple procedure, not only for the isolation of IgA in high yields but also for the resolution of the two IgA isotypes. However, a subsequent report *(7)* contradicted these findings suggesting that jacalin also bound IgA2, albeit with lower affinity than for IgA1. Other observations suggested considerable heterogeneity in jacalin preparations from different sources, which might explain this discrepancy *(8,9)*. A large number of subsequent publications *(10,11)* showed the marked specificity of jacalin for IgA1 and its usefulness in purifying the immunoglobulin.

Commercial preparations of jacalin now available have high specificity for IgA1 compared with IgA2. It should, however, be stressed that jacalin is not specific for IgA1 in serum and has been shown in a number of studies to have strong affinity for IgD, C1-inhibitor, and at least five other serum proteins (possibly including albumin), which would therefore be present as contaminants *(12-15)*.

The methods described were developed from those of Loomes et al. *(16)* who showed that milligram quantities of IgA1 and IgA2 could be purified from serum in concentrations which reflect their relative proportions using jacalin-agarose to resolve the isotypes. The same conditions could be used to separate IgA1 and IgA2 from colostrum and from other secretions (unpublished results). These methods have been used extensively in our laboratory over many years to produce IgA for studies on FcαR binding and IgA proteinase studies *(17–23)*.

1.2. Properties of Jacalin

Jacalin is a tetrameric lectin purified from jackfruit seeds. It has an apparent M_r of 40,000–50,000 and is comprised of two distinct types of polypeptide chain *(10,24)*. The lectin has specificity for galactose residues such as those on the hinge region of IgA1 *(9)*. Jacalin does not bind to IgA from mouse, rat, pig, goat, horse, cow, or dog IgA indicating their structural similarity to the human IgA2 isotype *(25)*. Jacalin does bind to human secretory IgA1 (but not sIgA2) showing that binding is not inhibited by secretory component. The interaction with IgA is unaffected by bivalent metal ions or by many detergents, although it is inhibited by denaturing agents *(9)*.

1.3. Binding of Jacalin to IgA1

IgA contains 2–5 N-linked glycosylation sites depending on the isotype. The N-glycosylation sites on serum IgA1 have been defined in detail *(26)*. The O-linked sugars found on the five serine residues of the hinge of serum IgA1 were originally reported to comprise one site having only N-acetyl galactosamine and four others having galactosyl β1-3 N-acetylgalactosamine (GalNAc) residues *(27)*. The O-linked sugars of serum IgA1 have recently been defined in more detail The results suggest a high degree of sialylation of the O-linked chains *(26)*. The O-linked sugars isolated from milk sIgA have been suggested to be even more complex and heterogeneous containing sialic acid, fucose, and N-acetylglucosamine *(28,29)*.

O-linked Galβ1-3GalNAc and sialylated derivatives are the preferred ligands for jacalin *(9)*. The number of binding sites on IgA1 for jacalin has not been accurately determined, but it must be greater than 1 since the lectin is a potent precipitator of monomeric IgA1. The number of IgA1 binding sites on the lectin is believed to be 3 or 4 *(10)*. Jacalin precipitates IgA1 and the fragments F(abc)$_2$, F(ab')$_2$, and Fc in agar gel and in solution. It does not precipitate F(ab) to any significance, only binding to GalNac linked to Ser 224 *(30)*. The fragments F(abc')$_2$, lacking CH3; (Fab')$_2$, lacking CH2 and CH3; and Fc appear to have a higher jacalin-binding capacity than the intact IgA1 molecule, suggesting that some of the binding sites are inaccessible in the whole molecule *(30)*. Differences in the jacalin binding to IgA1 from patients with IgA nephropathy have been reported in a number of publications *(31,32)*.

2. Materials
2.1. Preparation of a Crude Globulin Fraction of Serum

1. Glass wool (Fisher, Loughborough, UK) in the neck of a glass filter funnel.
2. Ammonium sulfate: Low in heavy metals (Fisher).

2.2. Partial Purification of IgA by Anion Exchange Chromatography

1. Chromatography column: 100 × 5 cm (e.g., Pharmacia, Uppsala, Sweden, XK 50/100).
2. Sepharose 6B (Pharmacia).
3. 50 mM Tris-HCl, pH 8.0.
4. Plates for radial immunodiffusion made using affinity-purified anti-IgA (α-chain specific) antibody (Sigma, St. Louis, MO): 100 µL of antibody are added to 10 mL, 1% (w/v) agarose (Sigma A4679) dissolved in PBS, pH 7.4, at 56°C, and then the plates are allowed to cool. For assays, 5-µL samples are added to wells cut in the agarose.

5. Chromatography column: 40 × 2.6 cm (e.g., Pharmacia XK 26/40).
6. Q-sepharose FF (Pharmacia).
7. Reagents for SDS-PAGE carried out as described by Laemmli (34). Gels should be 5–15% acrylamide gradient gels. Standard slab gels (1.5 mm thick) are run at 60 mA for approx 5 h.

2.3. Purification of IgA1 by Jacalin-Agarose Affinity Chromatography

1. Chromatography column: 40 × 2.6 cm (e.g., Pharmacia XK 26/40).
2. Jacalin-agarose: Approx 170 mL (4 mg lectin/mL agarose; Vector, Burlingame, CA).
3. 50 mM Tris-HCl, pH 8.0, 0.15M NaCl.
4. 50 mM Tris-HCl, pH 8.0, 0.15M NaCl containing 0.8M D-galactose or 0.1M melibiose (Sigma).

2.4. Purification of IgA2 by Sepharose Fastflow S Cation Exchange Chromatography

1. Column of protein G sepharose (15 × 1.6 cm in Pharmacia XK16/20).
2. 50 mM Sodium acetate, pH 4.5.
3. Chromatography column: 40 × 2.6 cm (e.g., Pharmacia XK26/40).
4. SP-sepharose FF (Pharmacia).

2.5. Characterization of Serum IgA1 and IgA2 Isotypes

1. Reagents for dot blotting or Western blotting; Use nitrocellulose 0.22-µm pore size. After direct application of sample (dot blotting) or Western blotting, using transfer buffer comprising 25 mM Tris, 192 mM glycine, 20% methanol, 0.1% SDS, pH 8.3, the nitrocellulose is blocked with 5% milk powder (Marvel) in PBS for 1 h before addition of anti-IgA antibody.
2. Human isotype-specific mouse monoclonal antibodies (MAbs) 2D7 (anti-A1 and -A2), NIF2 (anti-A1), and 2E2 (anti-A2) are used to detect the IgA of the two subclasses. These antibodies are widely available from commercial sources (e.g., Binding Site, Birmingham UK). The antibodies are diluted in 5% milk powder (Marvel) in PBS and are incubated with blocked nitrocellulose membranes for 1 h at room temperature. After washing the membranes in PBS, the bound antibodies are detected using goat anti-mouse IgG-alkaline phosphatase conjugate (Sigma) diluted in milk powder/PBS. The washed membranes are developed using NBT/BCIP solution made up from 200 µL nitroblue tetrazolium salt solution (50 mg/mL), 100 µL 5-bromo-4-chloro-3-indolyl phosphate (50 mg/mL) in 30 mL Tris buffer, pH 9.5 (12.12 g Tris, 5.84 g NaCl, 1.02 g $MgCl_2 \cdot 6H_2O$).

2.6. Preparation of a Crude Protein Extract from Colostrum

1. 4N HCl.
2. 2M Tris.
3. 0.22-µm Filters (Millipore, Bedford, MA).

2.7. Purification of sIgA from the Crude Colostrum Protein Extract Using Gel Filtration and Ion-Exchange Chromatography

1. Materials for gel filtration and ion exchange chromatography and for analysis of IgA are as in Section 2.2.

2.8. Fractionation of the sIgA1 and sIgA2 Isotypes

1. Materials for lectin affinity chromatography and for analysis of IgA are as in Sections 2.3. and 2.5.

3. Methods
3.1. Purification and Separation of Human Serum IgA1 and IgA2 (16)
3.1.1. Preparation of a Crude Globulin Fraction of Serum

1. Clarify 100 mL of fresh, pooled human serum by filtration through glass wool followed by centrifugation at 2500g for 1 h at 4°C.
2. Remove the supernatant and add 29.1 g ammonium sulfate to give 50% saturation and stir for 1 h at 4°C before centrifugation at 23,000g for 1 h. After centrifugation, carefully remove the supernatant (see Note 1) and redissolve the pellet in 50 mL distilled water.

3.1.2. Partial Purification of IgA by Anion Exchange Chromatography

1. Apply the redissolved ammonium sulfate pellet to a column, 90 × 5 cm, of Sepharose 6B equilibrated in 20 mM Tris-HCl, pH 8.0. Wash the column with the same buffer until all proteins have been eluted. Identify those fractions containing IgA by radial immunodiffusion (33) using a good quality affinity-purified anti-IgA (α-chain specific) antibody. Pool the main IgA containing fractions.
2. Apply the pool to a column (40 × 2.6 cm) of Q-sepharose anion exchange column equilibrated in 20 mM Tris-HCl, pH 8.0. If required, collect the column runthrough, which contains highly purified IgG. Wash the column with 200 mL 20 mM Tris-HCl, pH 8.0, then elute the bound proteins with a linear salt gradient of 0–0.5M NaCl (500 mL of each) in 20 mM Tris-HCl, pH 8.0.
3. Identify the fractions containing IgA by radial immunodiffusion and/or by SDS-PAGE (34). The IgA should elute in the 0.1–0.2M region of the salt gradient. Pool the main IgA containing fractions.

3.1.3. Purification of IgA1 by Jacalin-Agarose Affinity Chromatography (see Note 2)

1. Apply this pool to a column (32 × 2.6 cm; 170 mL) of jacalin-agarose (4 mg lectin/mL agarose; Vector) equilibrated in 20 mM Tris-HCl, pH 8.0, 0.15M NaCl. After loading, wash the column with the same buffer until the OD$_{280}$ of the effluent falls to zero. Elute the bound IgA1 with the same buffer containing 0.8M D-galactose (Fig. 1). Confirm the purity of the IgA1 by SDS-PAGE (see Note 3).

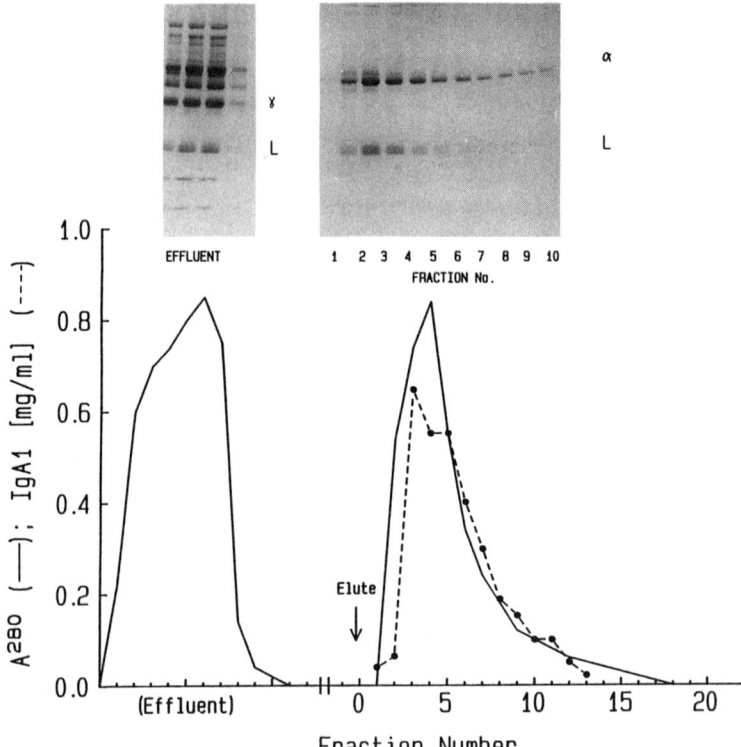

Fig. 1. Chromatography of the IgA pool from the Q-sepharose column on a jacalin-agarose. Fraction volume collected was 7.5 mL. Protein elution profile (—) and IgA1 concentration, mg/mL (– –), are shown. The disparity between the A^{280} and IgA1 protein elution profiles arises from the D-galactose in the elution buffer which gives an elevated A^{280} reading. The photograph shows Coomassie-blue stained SDS-PAGE gels of the column effluent and eluate. Fractions collected for the eluate are numbered across the bottom of the gel. (Reproduced with permission from ref. *16*.)

3.1.4. Purification of IgA2 by Sepharose Fastflow S Cation Exchange Chromatography

1. The run-through from the jacalin-agarose column contains IgA2 and a number of contaminants including some IgG. The IgG is removed by passage of the pool through a column of protein G sepharose (15 × 1.6 cm). To purify the IgA2, dialyze the pooled effluent from the protein G column against 50 m*M* sodium acetate, pH 4.5., then apply to a column (40 × 2.6 cm) of SP-sepharose FF (Pharmacia) equilibrated in the same buffer. Wash with one column vol of the same buffer and elute the bound proteins with a linear gradient of 0–0.5*M* NaCl in the same buffer. The IgA2 is the first protein eluted (Fig. 2). Detect the IgA2 by radial immunodiffusion or by SDS-PAGE (*see* Note 4).

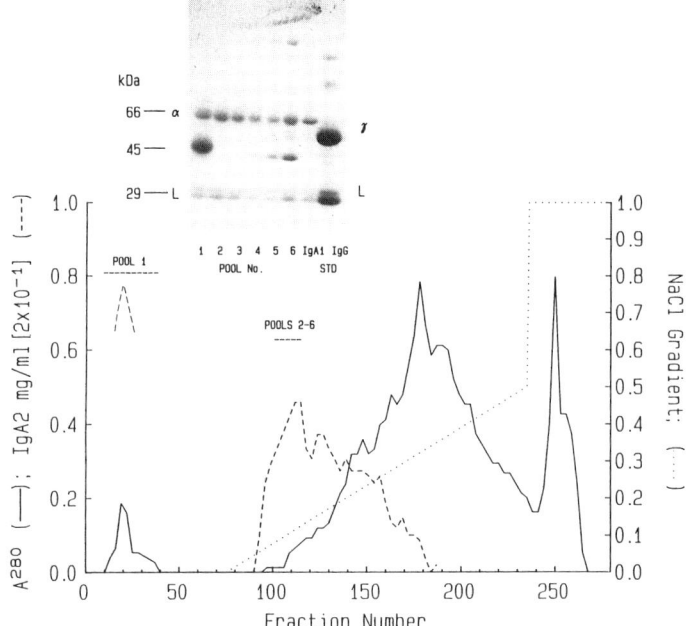

Fig. 2. Typical profile of cation exchange chromatography of the IgG depleted pooled effluents from the jacalin-agarose column on a sepharose Fastflow S equilibrated in 50 mM sodium acetate (pH 4.5). The fraction volume collected was 7 mL. The protein elution profile (—), IgA concentration (– –), and NaCl concentration (.....) are shown. The photograph shows a Coomassie blue-stained reduced SDS-PAGE gel of six pools made across the IgA2 elution profile: lane 1, Pool 1 unbound material; lanes 2–6, Pools 2–6 as indicated; lanes 7 and 8 show reference preparations of IgA1 and IgG standards, respectively. (Reproduced with permission from ref. *16*.)

3.1.5. Characterization of Serum IgA1 and IgA2 Isotypes

1. The purity of serum IgA1 and IgA2 preparations can be confirmed by immunodot blotting or Western blotting (unreduced) with human isotype-specific mouse MAbs 2D7 (anti-A1 and-A2), NIF2 (anti-A1), and 2E2 (anti-A2). These antibodies are widely available from commercial sources (e.g., Binding Site, Birmingham, UK).

3.2. Purification and Separation of Secretory IgA1 and IgA2 from Colostrum

IgA is the predominant immunoglobulin in almost all secretions. In humans, colostrum and early milk are the body fluids richest in secretory IgA. The values for colostral IgA reported in the literature range from 1.6 to 85.9 mg/mL, with an average of (11.4 ± 5.5 mg/mL) in pooled samples of colostrum *(35)*.

The concentration of sIgA falls rapidly from the very high concentrations found in days 1–3, IgG levels increase during this time (36). In colostrum collected between 1 and 3 d, the ratio for IgA1:IgA2 is around 53:47.

3.2.1. Preparation of a Crude Protein Extract from Colostrum

1. Mix 50 mL colostrum or early milk with 25 mL of isotonic saline. Centrifuge at 100,000g for 1 h at 4°C and then remove the clarified colostrum from between the surface layer of fat and the pellet of cell debris.
2. Lower the pH of the colostrum to 4.0 with HCl to precipitate casein and after centrifugation at 30,000g for 30 min at 4°C, remove the supernatant and neutralize to pH 7.0 with 2M Tris. Recentrifuge and then pass through a 0.22-μm filter (Millipore, Watford, UK).

3.2.2. Purification of sIgA from the Crude Colostrum Protein Extract Using Gel Filtration and Ion-Exchange Chromatography

1. Load the defatted and decaseinated colostrum onto a column (90 × 5 cm) of sepharose 6B equilibrated in 50 mM Tris-HCl, pH 8.0. Wash with 2 column vol and analyze fractions by radial immunodiffusion and SDS-PAGE (Fig. 3; see Note 5).
2. Pool the sIgA-containing fractions and load onto a column 40 × 2.6 cm) of Q-sepharose equilibrated in 50 mM Tris-HCl, pH 8.0. Wash with a column volume of the same buffer and then elute with a linear gradient of 0–0.5M NaCl in the same buffer (500 mL of each). Pool the IgA containing fractions (Fig. 4; see Note 6).

3.2.3. Fractionation the sIgA1 and sIgA2 Isotypes

1. Load the pool onto a column (32 × 2.6 cm) of jacalin-agarose (Vector) equilibrated in 50 mM Tris-HCl, 0.15M NaCl, pH 8.0. After loading, wash the column with the same buffer until the OD_{280} of the effluent falls to zero. The run-through contains sIgA2. Elute the bound sIgA1 with three column volumes of 50 mM Tris-HCl, 0.15M NaCl, pH 8.0, containing 0.8M D-galactose (Fig. 5).
2. Confirm the purity of the IgA1 and IgA2 by SDS-PAGE and by dot blotting using specific MAbs (see Note 7).

4. Notes

1. Care should be taken to remove all of the supernatant which contains most of the albumin, which, if retained, limits the capacity of subsequent ion exchange columns.
2. The capacity of the commercial jacalin-agarose column (4 mg lectin/mL agarose) is around 4 mg IgA/mL agarose. We have used columns of jacalin-agarose repeatedly for many preparations over a number of years without appreciable loss of capacity. Monomeric serum IgA1 elutes from jacalin-sepharose as a relatively sharp peak. The trailing edge of the peak tends to be enriched for dimeric IgA. In contrast to its serum counterpart, the sIgA1 from colostrum elutes as a broad peak of relatively low protein concentration (0.1–0.2 mg/mL).

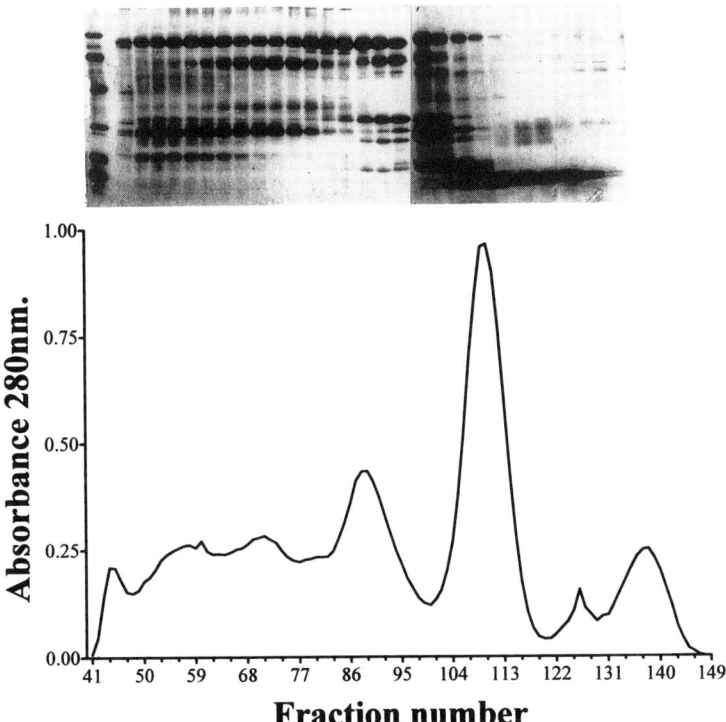

Fig. 3. Gel filtration of a decaseinated crude protein extract from colostrum on a sepharose 6B column equilibrated in 50 mM Tris-HCl. Fraction volume was 10 mL. Protein elution profile A^{280} (—); the photograph shows a Coomassie blue-stained SDS-PAGE gel. Lane 1, mol-wt markers (97.4, 66.2, 42.7, 37.0, 21.5, 14.4 kDa). Lanes 2–28, every third column fraction from 44 to 134.

3. The yield of pure IgA1 should be around 90% of that in the original serum. The pure IgA1 can be dialyzed to remove the galactose and concentrated by ultrafiltration to give a maximum concentration of 5 mg/mL. The major contaminant is likely to be C1-inhibitor (M_r = 100 kDa), which stains poorly on SDS-PAGE. This can be removed by gel filtration of the pool on appropriate resin, e.g., Sephacryl S300.
4. This scheme will yield some fractions containing only IgA2. Together these fractions constitute 10–20% of the original serum content of IgA2. For higher yield, the IgA2 containing fractions can be pooled, concentrated by ultrafiltration then gel filtered on a column of Sephadex G200 or equivalent resin. The first protein peak eluting from these columns is apparently pure IgA2. The yield of serum IgA2 from a typical preparation is around 50%. On SDS-PAGE run under nonreduced conditions, the IgA2 runs as bands of apparent M_r = 200 kDa corre-

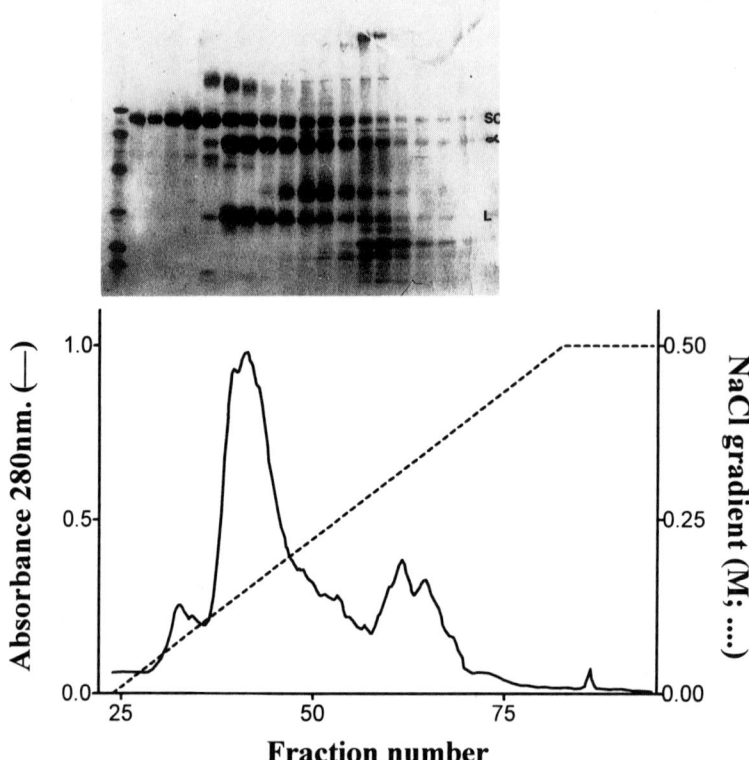

Fig. 4. Anion exchange chromatography of IgA containing fractions from the Sepharose 6B column on a Q-Sepharose column equilibrated in 50 mM Tris-hydrochloride (pH 8.0). Fraction volume collected was 5.0 mL. Protein elution profile, A^{280} (—); linear gradient (– –) of 0–0.5M NaCl. The photograph shows a Coomassie blue-stained SDS-PAGE gel; Lane 1, molecular weight markers (97.4, 66.2, 42.7, 37.0, 21.5, 14.4 kDa). Lanes 2–20, column fractions 10, 20, 30 then every third fraction to 78. α, IgA heavy chain; L, light chain; SC, secretory component.

sponding to intact IgA, 140 kDa corresponding to H-chain dimer and 50 kDa, light-chain dimer (Fig. 6).

5. SDS-PAGE of fractions across the gel filtration column show that sIgA (M_r = 385 kDa) is eluted mainly in the second minor broad peak in the protein elution profile. The first peak contains IgM (M_r = 971 kDa). The sIgA peak corresponds mainly to dimer. Higher molecular weight forms of sIgA elute earlier A small amount of IgG (M_r = 150 kDa) and free secretory component (SC) (M_r = 80–90 kDa) are eluted together in the third peak of the profile. The fourth and major peak of the protein profile contain predominantly lactoferrin (M_r = 90 kDa) and α-lactalbumin (M_r = 12.5 kDa), respectively.

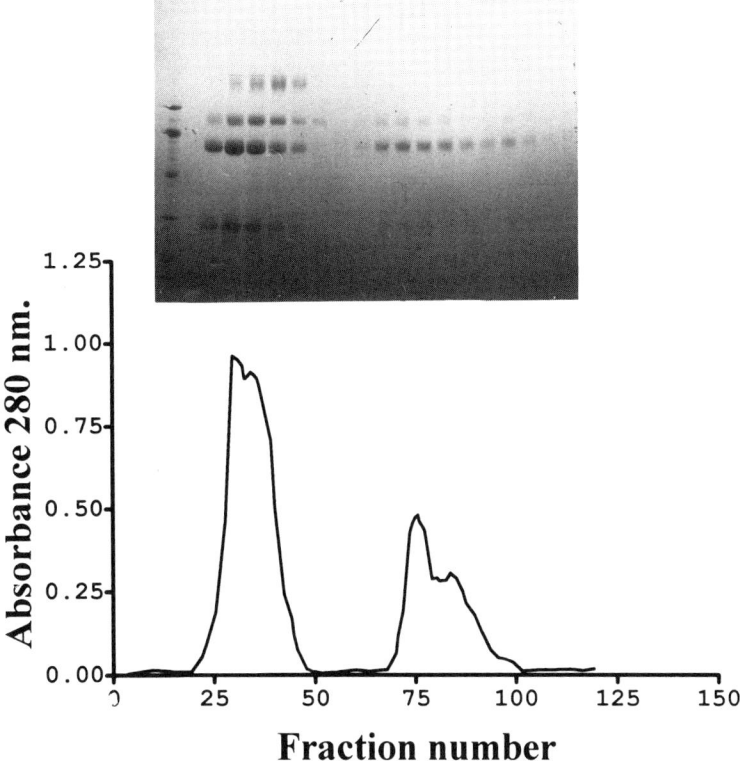

Fig. 5. Affinity chromatography of the sIgA pool from the Q-sepharose column on a jacalin-agarose column. The fraction volume was 2 mL. The jacalin column effluent contained functionally pure sIgA2. sIgA1 bound to the column was eluted with buffer containing 0.8M D-galactose. Protein elution profile, A^{280} (—); the photograph shows a Coomassie blue-stained gel; lane 1, mol-wt markers (97.4, 66.2, 42.7, 37.0, 21.5, 14.4 kDa); lanes 2–20, every fifth column fraction from 25–110; α, IgA heavy chain; L, light chain; SC, secretory component.

6. The major contaminant at this stage, lactoferrin, is eluted from the column ahead of sIgA in the 0.05–0.1M region of a linear gradient of 0–0.5M NaCl; sIgA is eluted in the 0.125–0.375M region of the salt gradient. The yield of pure sIgA is around 50% of the original sIgA content of colostrum.
7. The final yield of sIgA after resolution of its isotypes on the jacalin-agarose is approx 30%; 33% of the sIgA1 and around 24% of sIgA2. A protein characterized by its broad band on SDS-PAGE can contaminate the sIgA2. In some preparations, lactoferrin, which often forms complexes with polymeric IgA, can be a contaminant. Lactoferrin can be removed by a heparin affinity column *(38)*.

8. The elution of IgA from jacalin can also be effected using 0.1M melibiose. Melibiose (galactosyl-α-D glucose) has around seven-fold higher affinity for jacalin than galactose. It has been reported to be more effective for eluting bound proteins from jacalin. Used at 0.1M, it is no more expensive than galactose.

Acknowledgments

This work was supported by research grants from the Arthritis and Rheumatism Council and Medical Research Council. We are grateful to Helen Cowper and Stewart MacPherson for assistance in preparation of the manuscript.

References

1. Kerr, M. A. (1990) The structure and function of human IgA. *Biochem. J.* **271**, 285–296.
2. Rudd, P. M., Fortune, F., Patel, T., Parekh, R. B., Dwek, R. A., and Lehner, T. (1994) A human T-cell receptor recognises *O*-linked sugars from the hinge region of human IgA1 and IgD. *Immunology* **83**, 99–106.
3. Senior, B. W., Loomes, L. M., and Kerr, M. A. (1991) Microbial IgA proteases and virulence. *Revs. Med. Microbiol.* **2**, 200–207.
4. Roque-Barreira, M. C. and Campos-Neto, A. (1985) Jacalin: an IgA- binding lectin. *J. Immunol.* **134**, 1740–1743.
5. Kondoh, H., Kobayashi, K., Hagiwara, K., and Kajii, T. (1986) Jacalin, a jackfruit lectin, precipitates IgA1 but not IgA2 subclass on gel diffusion reaction. *J. Immunol.* **88**, 171–173.
6. Gregory, R. L., Rundegren, J., and Arnold, R. R. (1987) Separation of human IgA1 and IgA2 using jacalin-agarose chromatography. *J. Immunol. Methods* **99**, 101–106.
7. Aucouturier, P., Duarte, F., Mihaesco, E., Pineau, N., and Preud'homme, J.-L. (1988) Jacalin, the human IgA1 and IgD precipitating lectin, also binds IgA2 of both allotypes. *J. Immunol. Meth.* **113**, 185–191.
8. Kondoh, H., Kobayashi, K., and Hagiwara, K. (1987) A simple procedure for the isolation of human secretory IgA of IgA1 and IgA2 subclass by a jackfruit lectin, jacalin, affinity chromatography. *Mol. Immunol.* **24**, 1219–1222.
9. Hashim, O. H., Ng, C. L., Gendeh, S., and Jaafar, M. I. N. (1991) IgA binding lectins isolated from distinct *Artocarpus* species demonstrate differential specificity. *Mol. Immunol.* **28**, 393–398.
10. Hagiwara, K., Collet-Cassart, D., Kobayashi, K., and Vaeman, J. (1988) Jacalin: isolation, characterization and influence of various factors on its interaction with human IgA1 as assessed by precipitation and latex agglutination. *Mol. Immunol.* **25**, 69–83.
11. Skea, D. L., Christopoulous, P., Plaut, A. G., and Underdown, B. J. (1988) Studies on the specificity of the IgA-binding lectin, Jacalin. *Mol. Immunol.* **25**, 1–6.
12. Aucouturier, P., Mihaesco, E., Mihaesco, C., and Preud'homme, J.-L. (1987) Characterisation of Jacalin, the human IgA and IgD binding lectin from jackfruit. *Mol. Immunol.* **24**, 503–511.

13. Zehr, B. D. and Litwin, S. D. (1987) Human IgD and IgA1 compete for D-galactose-related binding sites on the lectin jacalin. *Scand. J. Immunol.* **26,** 229–236.
14. Hiemstra, P. S., Gorter, A., Stuurman, M. E., Van Es, L. A., and Daha, M. R. (1987) The IgA-binding lectin jacalin induces complement activation by inhibition of C1-inactivator function. *Scand. J. Immunol.* **26,** 111–117.
15. Biewenga, J., Steneker, I., and Hameleers, D. M. H. (1988) Effect of serum albumin on the recovery of human IgA1 from immobilized jacalin. *J. Immunol. Methods* **115,** 199–207.
16. Loomes, L. M., Stewart, W. W., Mazengera, R. L., Senior, B. W., and Kerr, M. A. (1991) Purification and characterisation of human immunoglobulin IgA1 and IgA2 isotypes from serum. *J. Immunol. Methods.* **141,** 209–218.
17. Kerr, M. A., Stewart, W. W., Bonner, B. C., Greer, M. R., MacKenzie, S. J., and Steele, M. G. (1994) The diversity of leucocyte IgA receptors. *Contrib. Nephrol.* **111,** 60–65.
18. Mazengera, R. L. and Kerr, M. A. (1990) The specificity of the IgA receptor purified from human neutrophils. *Biochem. J.* **272,** 159–165.
19. Stewart, W. W. and Kerr, M. A. (1990) The specificity of the human IgA receptor (FcαR) determined by measurement of chemiluminescence induced by serum or secretory IgA1 or IgA2. *Immunology* **71(3),** 328–334.
20. Stewart, W. W. and Kerr, M. A. (1991) The measurement of respiratory burst induced in polymorphonuclear neutrophils by IgA and IgG anti-gliadin antibodies isolated from coeliac serum. *Immunology* **73,** 491–497.
21. Stewart, W. W. and Kerr, M. A. (1994) Unaggregated serum IgA binds to neutrophil FcαR at physiological concentrations and is endocytosed but crosslinking is necessary to elicit a respiratory burst. *J. Leuk. Biol.* **56,** 481–487.
22. Senior, B. W., Loomes, L. M., and Kerr, M. A. (1991) Microbial IgA proteases and virulence. *Revs. Med. Microbiol.* **2,** 200–207.
23. Loomes, L. M., Senior, B. W., and Kerr, M. A. (1992) Proteinases of Proteus spp.: purification, properties and detection in urine of infected patients. *Infect. Immun.* **60,** 2267–2273.
24. Mahanta, S. K., Sastry, M. V. K., and Surolia, A. (1990) Topography of the combining region of a Thomsen-Friedenrich-antigen-specific lectin jacalin (*Artocarpus integrifolia agglutinin*). *Biochem. J.* **265,** 831–840.
25. Wilkinson, R. and Neville, S. (1988) Jacalin: its binding reactivity with immunoglobulin A from various mammalian species. *Vet. Immunol. Immunopathol.* **18,** 195–198.
26. Field, M. C., Amatayakul-Chantler, S., Rademacher, T. W., Rudd, P. M., and Dwek, R. A. (1994) Structural analysis of the N-glycans from human immunoglobulin A1: comparison of normal human serum immunoglobulin A1 with that isolated from patients with rheumatoid arthritis. *Biochem. J.* **299,** 261–275.
27. Pierce-Cretel, A., Debray, H., Montreuil, J., Spik, G., Van Halbeek, H., Mutsaers, J. H. G. M., and Vliegenthart, J. F. G. (1984) Primary structure of N-glycosidically linked asialoglycans of secretory immunoglobulins A from human milk. *Eur. J. Biochem.* **139,** 337–349.

28. Pierce-Cretel, A., Plamblanco, M., Strecker, G., Montreuil, J., and Spik, G. (1981) Heterogeneity of the glycans O-glycosidically linked to the hinge region of secretory immunoglobulin A from human milk. *Eur. J. Biochem.* **114,** 169–178.
29. Pierce-Cretel, A., Plamblanco, M., Strecker, G., Montreuil, J., Spik, G., van Halbreek H., and Vliegenthart, J. F. G. (1982) Primary structure of N-glycosidically linked asialoglycans of secretory immunoglobulins A from human milk. *Eur. J. Biochem.* **125,** 383–388.
30. Biewenga, J., Hiemstra, P. S., Stenkler, I., and Daha, M. R. (1989) Binding of human IgA1 and IgA1 fragments to jacalin. *Mol. Immunol.* **26,** 275–281.
31. Andre, P. M., Le Pogamp, P., and Chevet, D. (1990) Impairment of jacalin binding to serum IgA in IgA nephropathy. *J. Clin. Lab. Anal.* **4,** 115–119.
32. Allen, A. C., Harper, S. J., and Feehally, J. (1995) Galactosylation of N- and O-linked carbohydrate moieties of IgA1 and IgG in IgA nephropathy. *Clin. Exp. Immunol.* **100,** 470–474
33. Mancini, G., Vaerman, J. P., Carbonara, A. O., and Heremans, J. F. (1965) Immunochemical quantitation of antigens by single radial immunodiffusion. *Immunochem.* **2,** 235–243.
34. Laemmli, U. K. (1970) Cleavage of structural proteins during the assembly of the head of bacteriophage T4. *Nature* **227,** 680–685.
35. Delacroix, D. L., Dive, C., Rambaud, J. C., and Vaerman, J. P. (1982) IgA subclasses in various secretions and in serum. *Immunology* **47,** 383–385.
36. Goldblum R. M. and Goldman, A. S. (1994) Immunologic components of Milk: formation and function, in *Handbook of Mucosal Immunology* (Ogra, P. L. et al., eds.), Academic, San Diego, CA, pp. 643-652.
37. Ladjeva, I., Peterman, J. H., and Mestecky, J. (1989) IgA subclasses of human colostral antibodies specific for microbial and food antigens. *Clin. Exp. Immunol.* **78,** 85–89.

23

Use of Lectins in Affinity Purification of HIV and SIV Envelope Glycoproteins

Gustav Gilljam

1. Introduction

The human immunodeficiency viruses (HIV-1 and HIV-2) are the etiologic agents of the acquired immunodeficiency syndrome (AIDS) and related disorders *(1–3)*. Simian immunodeficiency virus (SIV) is the corresponding virus for nonhuman primates. SIVmac has been isolated from rhesus monkeys *(Macaca mulatta)* with immunosuppression and malignant lymphomas *(4)*.

These are retroviruses with a surface that is characteristically made up of knobs (HIV-1 has 72 knobs) containing trimers or tetramers of the major envelope glycoproteins *(5–8)*. Glycoprotein gp160 is a precursor protein produced in HIV- and SIV-infected cells. This protein is further cleaved by cellular enzymes into two glycoproteins: an external surface envelope glycoprotein (gp120) and a transmembrane protein (gp41) *(9)*. Glycoprotein 120 contains the binding site for the cellular receptor, the CD4 molecule *(10–13)*, and also the major neutralizing domains *(14–17)*. This makes gp120 and gp160 potential vaccine candidates. The primary sequence of gp120 has 24 potential N-linked glycosylation sites *(18)*. The carbohydrate portion of the gp120 molecule has been calculated to be about the same molecular weight as the peptide backbone, which was estimated to be about 60 kDa after deglycosylation using endoglycosidase-F *(19–21)*. This carbohydrate portion has been reported to be important for the gp120–CD4 interaction *(20–22)*.

1.1. Purification of Outer Envelope Glycoproteins Using Galanthus Nivalis Agglutinin

Several lectins have been shown to bind to gp120 *(23,24)*, and it has been shown that the α-*(1-3)*- and α-(1-6)-D-mannose-specific plant lectins can inhibit the virus from binding to and infecting CD4 positive T-cells *(25)*. Lectins

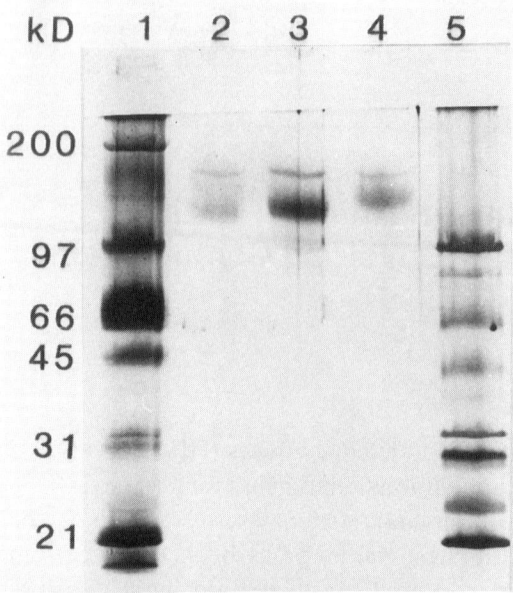

Fig. 1. GNA-purified envelope glycoproteins separated on an 8–25% SDS-polyacrylamide gel and silver stained. Lane 1, high-mol-wt marker; lane 2, gp120 from HIV-1(IIIB) (20 µg/mL); lane 3, gp125 from HIV-2(SBL6669) (40 µg/mL); lane 4, gp148 from SIV_{mac251}(18 µg/mL); lane 5, low-mol-wt marker. Reprinted from ref. *27* with permission.

have also been used in purification schemes for the gp120 molecule and especially for recombinant gp160 for use in vaccine trials. The lectin used for this purpose is Lentil lectin (LCA), which has a specificity for mannose and glucose. This lectin also binds to several other glycoproteins in the viral suspension, which makes it necessary to include other purification steps. Other lectins have been used in combinations to purify the outer envelope glycoprotein of SIV *(26)*. The lectin from *Galanthus nivalis* (snow drop) bulbs (GNA), shows an inhibitory effect on syncytium formation in HIV-1 infected cells *(25)*, and it binds to the outer envelope glycoprotein from HIV-1, HIV-2, and SIV *(27)*.

GNA has a specificity for terminal high mannose residues especially those possessing Man(α1-3)Man. Internal mannosyl residues, D-glucose and N-acetyl-D-glucosamine do not bind to GNA *(28,29)*. Glycoprotein 120 from HIV-1 possesses 7–9 of these mannose residues *(18)* and they are conserved between different isolates *(30,31)*. This type of carbohydrate residue is rarely found in mammalian glycoproteins and only a few glycoproteins that possess such mannose groups have so far been recognized). Using GNA coupled to a matrix it has been possible to purify the outer major glycoprotein from HIV-1, HIV-2, and SIV to a high purity *(26,27)* (Figs. 1–3). We have been able to purify gp120

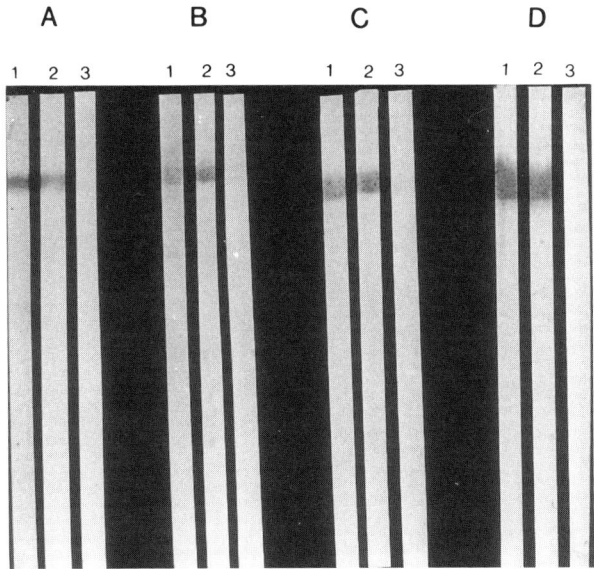

Fig. 2. The GNA-purified envelope glycoproteins were separated on an SDS-polyacrylamine gel and transferred to a nitrocellulose filter. The strips were analyzed with polyclonal sera and monoclonal antibodies. (A) gp120 HIV-1(IIIB): (1) polyclonal human anti-HIV1 serum; (2) mouse monoclonal, T9 anti-gp 120; (3) human monoclonal 2F5 anti-gp41. (B) gp120 HIV-1(SF2): (1) polyclonal anti-HIV-1 serum; (2) mouse monoclonal, T9 anti-gp120; (3) human monoclonal 2F5 anti-gp41. (C) gp125 HIV-2(SBL6669): (1) human polyclonal anti-HIV-2 serum; (2) mouse monoclonal, KK12 antigp125; (3) mouse monoclonal, KK41 anti-gp36. (D) gp148 SIV_{mav251}(1) macaque polyclonal anti-SIV serum; (2) mouse monoclonal KK8 anti-gp148; (3) mouse monoclonal KK20 anti-gp32. Reprinted from ref. 27 with permission.

from all HIV-1, SIV, and HIV-2 strains that we have tried and the choice of cell line (H9, C8166, U937-2, or Jurkat tat) for virus production does not appear to be of consequence. This is because the glycosylation that leads to high terminal mannose groups is very conserved (30,31). The outer envelope glycoprotein from different HIV-1 isolates grown in PBMC also binds strongly to GNA.

1.2. Detection and Quantification of the Outer Envelope Glycoprotein Using GNA in an ELISA

The GNA molecule can also be used in an ELISA, to detect and quantify the amount of envelope glycoprotein from HIV-1, HIV-2, and SIV, present in the viral suspension or in the purified products (Fig. 4). The advantage of using

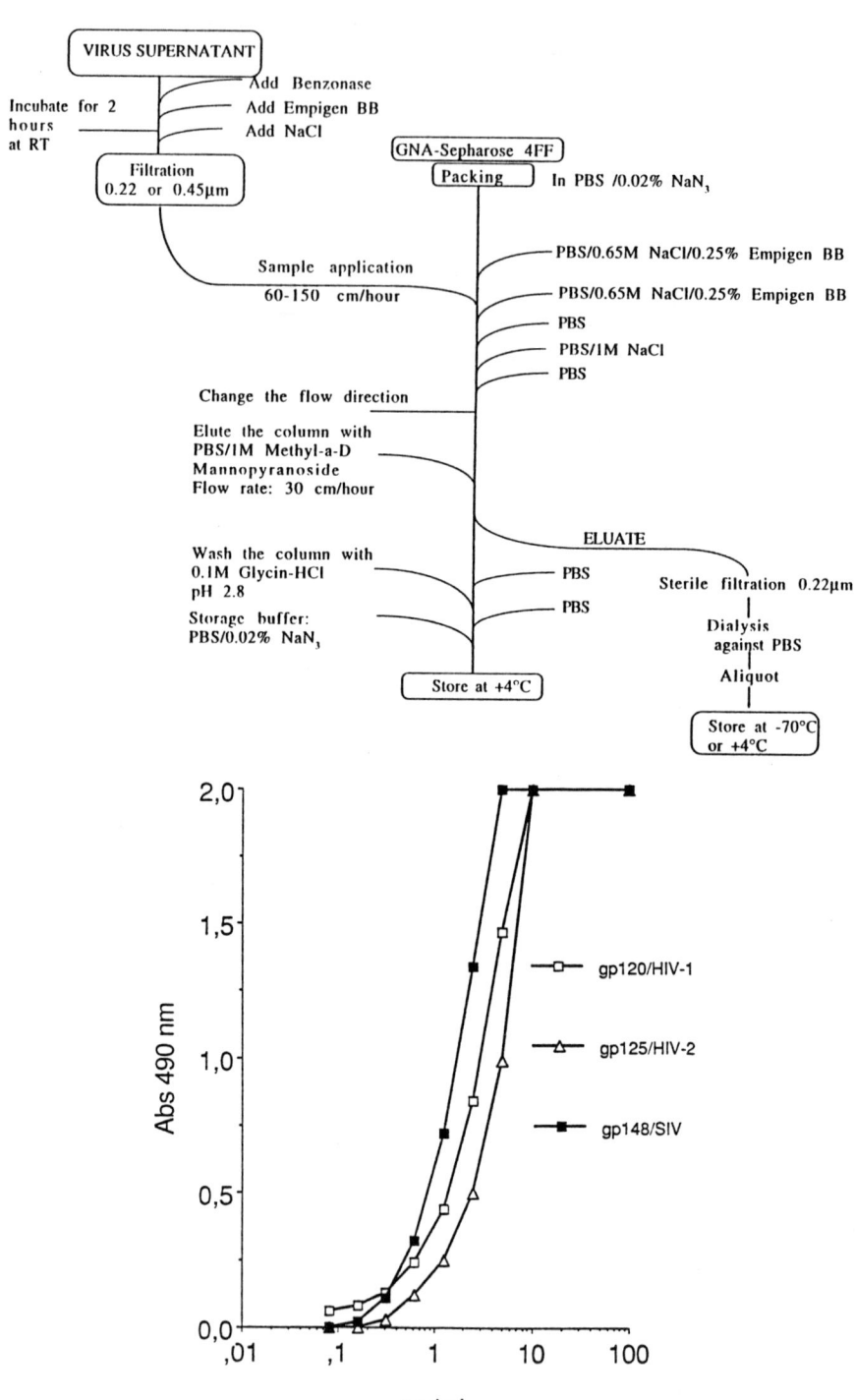

Figs. 3–4.

HIV and SIV Envelope Glycoproteins

GNA is that it does not bind any glycoproteins from the fetal calf serum present in the culture medium, which could block the gp120 binding and give lower sensitivity. The standard used in the GNA ELISA should be detected equally well as the purified gp120 by the detecting antibody so the amount of gp120 in the purified product can be properly quantified. This method is suitable for calculating the recovery of the product.

2. Materials
2.1. Coupling GNA to CNBr-Activated Sepharose 4 Fast Flow

1. CNBr activated Sepharose Fast Flow (Pharmacia, Uppsala, Sweden; *see* Note 1).
2. *Galanthus nivalis* agglutinin (GNA) is bought as a powder (Boehringer, Mannheim, Germany) and can be dissolved in PBS, in which it forms a tetramer. This lectin is not dependent on cations for its binding capacity, has a pH range of 2–12 (stable at pH 6.0–12), and can also withstand heating at 70°C for 10 min, but sodium chloride concentrations above $2M$ lead to irreversible inactivation.
3. 1 mM HCl (should be used cold, 4°C).
4. Coupling buffer: 0.25M NaHCO$_3$/0.2M methyl-α-D-mannopyranoside, pH 8.3 (*see* Note 2).
5. 1M ethanolamine, pH 8.0 (use conc. HCl to adjust the pH).
6. Buffer I: 50 mM Tris-HCl, 1M NaCl, pH 8.0.
7. Buffer II: 50 mM glycine, 1M NaCl, pH 3.5.
8. 0.01M Phosphate-buffered saline without Ca^{2+} or Mg^{2+}, pH 7.3: PBS (this type of PBS is used in all assays).
9. PBS with 0.01% merthiolate.

2.2. Sample Preparation

1. Benzonase, which possesses a DNase and RNase activity in the presence of the zwitterion detergent (Empigen BB, Calbiochem, LaJolla, CA). This enzyme has optimum activity between, pH 8.0-9.2 and in the presence of 1-2 mM Mg^{2+}; it is active against all DNA and RNA.
2. A zwitterion detergent: Empigen BB (stock solution is 30%, *see* Note 3).
3. Sodium chloride: solid and pro analysis quality.

2.3. GNA Chromatography

1. GNA coupled to Sepharose 4 Fast Flow.
2. Phosphate-buffer saline: PBS (*see* Note 4).
3. PBS, 0.25% Empigen BB, 0.65M NaCl, pH 7.3.

Fig. 3. *(previous page; top)* A description of the purification process.

Fig. 4. *(previous page; bottom)* Sensitivity analysis of purified envelope glycoproteins on a GNA ELISA. gp120 HIV-1(IIIB); gp125 HIV-2(SBL6669); gp148 SIV$_{mav251}$. Reprinted from ref. *27* with permission.

4. PBS, 1M NaCl, pH 7.3.
5. 1M Methyl-α-D-mannopyranoside in PBS, pH 7.3.
6. 0.1M Glycine-HCl, pH 2.8.

2.4. GNA ELISA

1. 96-Well microplates (e.g., polystyrene, Nunc Immunoplate MaxiSorp, Nunc, Aahus, Denmark).
2. Coating buffer: PBS, pH 7.3 with 0.02% NaN_3 (*see* Note 5).
3. Washing buffer: 0.9% NaCl, 0.05% Tween-20.
4. Dilution buffer A: PBS, pH 7.3, 0.25% Empigen BB, 0.5% bovine serum albumin (BSA).
5. Dilution buffer B: PBS, pH 7.3, 0.5% BSA, 0.05% Tween-20.
6. Blocking buffer: PBS, pH 7.3, 15% fetal calf serum (FCS).
7. Polyclonal sera or MAb directed to the gp120 that will be assayed (*see* Note 6).
8. Antibodies conjugated with peroxidase, and directed to the species of the detecting antibody used (e.g., goat antimouse IgG-HRP-labeled conjugate).
9. Substrate: Orthophenyl diamine (OPD) 10 mg dissolved in 10 mL substrate buffer (0.0667M sodium phosphate/0.0347M citric acid and pH adjusted to 5.5 using NaOH) activated by 3–4 μL H_2O_2 (35%) (*see* Note 7).
10. 2.5M Sulfuric acid.

3. Methods
3.1. Coupling GNA to CNBr Activated Sepharose 4 Fast Flow

1. Suspend the CNBr-activated gel in 1 mM HCl for 30 min to allow the gel to swell.
2. Wash the gel with 15 gel volumes of cold 1 mM HCl in a sinterred glass funnel.
3. Wash the gel in coupling buffer. Check that the pH of the flow through is 8.3 (*see* Note 8).
4. Dissolve the GNA in the coupling buffer to a concentration of around 1.5 mg/mL, adjust the pH to 8.3. Use glass bottles or glass beakers.
5. Add the washed gel to the ligand solution. The ratio between liquid and gel should be around 1. Incubate the solution at 4°C with gentle rocking overnight (*see* Note 9).
6. Wash the gel in 1M ethanolamine, pH 8.0 in a sinterred glass funnel. Resuspend the gel in a large volume (3 times the gel volume) of 1M ethanolamine pH 8.0 for 2–4 h at room temperature to block the unused activated sites on the gel.
7. Wash the gel eight times with 4 gel volumes each time, shifting 4 times between buffers I and II.
8. Wash the gel with 10 gel volumes of PBS.
9. Store the coupled gel in PBS with 0.01% Merthiolate at 4°C.

3.2. Sample Preparation

Sample preparation prior to purification of the gp120 on the lectin column will be presented as two alternatives depending on the sample volume.

If the sample volume is such that cells can be removed by centrifugation or sedimentation, then the procedure is as follows.

1. Remove the cells.
2. Add Benzonase to the supernatant, 2500 U/L, adjust the pH to 8.0 (using 1M NaOH), and add Mg^{2+} to a final concentration of 2 mM (see Note 10).
3. Add the zwitterion detergent (Empigen BB) to a final concentration of 0.25%, and incubate the solution for 2 h at room temperature.
4. Add NaCl to a final concentration of 0.65M (the medium itself contains 0.15M; see Note 11).
5. Filter the solution through a 0.2 or 0.45-µm filter (see Note 12).

For larger volumes of tissue culture supernatant, the following procedure is used.

Ultrafiltration units are used both to remove the cells and to concentrate the viral particles. Follow the guidelines in the manual when using the ultrafiltration equipment.

1. Allow the suspension to pass through a filter unit with a pore size of 1.2 µm, which removes the cells.
2. The material that passed through the filter unit contains the viral particles and is concentrated using a filter unit with a 300-kDa cutoff. The volume is concentrated to approx 2 L.
3. Clarify the sample by centrifugation at 7000g for 30 min at 4°C.

The method above is then followed from step 2 in Section 3.2.1. (adding the Benzonase, and so forth).

3.3. GNA Chromatography

1. Equilibrate the GNA-Sepharose 4 Fast Flow with PBS containing 0.25% Empigen BB and 0.65M NaCl (see Note 13).
2. Apply the sample to the column with a flow rate of 60–150 cm/h (this is equal to 2–5 mL/min using a column with a diameter of 1.6 cm; see Note 14).
3. Wash the gel with PBS/0.25% Empigen BB/0.65M NaCl/pH 7.3 until the absorbance at 280 nm reaches the background.
4. Wash the gel with PBS to remove the detergent.
5. Wash the gel with PBS containing 1M NaCl to remove material bound nonspecifically.
6. Wash the gel with PBS to remove the salt which can be measured using a conductivity meter.
7. Elute the gp120 from the GNA column using 1M methyl-α-D-mannopyranoside at a flow rate of 30 cm/h (1 mL/min using a column with a diameter of 1.6 cm; see Note 15).
8. The eluted material is sterile-filtered through a hydrophilic durapore (PVDF) 0.22-µm filter (e.g., a Millex GV filter, Millipore, Bedford, MA).

9. After the sterile filtration, the sugar in the sample is removed by dialysis against PBS for at least 20 h, using dialysis tubes (regenerated cellulose, cutoff 12–14,000) washed in distilled water followed by boiling in distilled water for 30 min. The PBS has to be exchanged 3–4 times (*see* Note 16).
10. Aliquot the products in suitable volumes and freeze at –70°C (*see* Note 17).
11. Clean the column using glycine-HCl.
12. Neutralize the gel immediately thereafter with PBS.

3.4. GNA ELISA

1. Dissolve the GNA in PBS with 0.02% NaN_3 to a concentration of 10 μg/mL.
2. Add 100 μL to each well and incubate the plate overnight at room temperature and at least 1 d at 4°C.
3. Wash the plate three times.
4. Add the blocking solution and incubate the plate for 30 min at 37°C.
5. Wash the plate 1 time.
6. Add the sample diluted in buffer A to the well and incubate overnight at +4°C.
7. Wash the plate three times.
8. Add a MAb or polyclonal antibody directed to the gp120 diluted in buffer B and incubate the plate for 90 min at 37°C.
9. Wash the plate three times.
10. Add the conjugate, diluted in buffer B, and incubate the plate for 90 min at 37°C.
11. Wash the plate three times.
12. Add the substrate, 100 μL/well and incubate at room temperature for 30 min.
13. Stop the reaction by adding 100 μL of $2.5M$ H_2SO_4.
14. The absorbance is then read at 490 nm.

4. Notes

1. GNA can be coupled to other matrices, e.g., pore glass, which is preformed by Bio Processing, Durham, UK. These matrices have been chosen by us, to make it possible to use higher flow rates without reaching high backpressure.
2. Use sterile filtered solutions and check that the sugar lot used does not give a high absorbance at 280 nm. A_{280} = 0.06 is normal for $1M$ methyl-α-D-mannopyranoside. This is also important when the sugar is used for eluting the material. Make this buffer prior to use.
3. Empigen BB has been shown to be very effective in solubilizing the gp120 from the virus particle and in retaining the biological activity of the protein *(27)*. Another advantage in using Empigen BB is that it is easy to remove.
4. Use sterile or sterile-filtered buffers through the whole chromatographic procedure.
5. The 0.02% NaN_3 is used to prevent microbial growth, which can lead to degradation of the GNA molecule.
6. Mouse IgM is not recomended, because it binds to the GNA. We have also had some problems using rabbit sera. This will also be important for the conjugate used.
7. This solution should be freshly made and the activation by H_2O_2 should be done just prior to use. The, pH adjustment to 5.5 is done to prevent a color change of

the substrate buffer over time. This substrate buffer can be stored for several months at 4°C.
8. The sugar is included in the coupling buffer in order to protect the binding sites on GNA, during the coupling.
9. This coupling can also be performed at room temperature, and the coupling time is then 3–4 h.
10. It is important that the Benzonase is added before the detergent and that it is evenly dispersed in the solution. Otherwise the DNA/RNA can aggregate and the enzyme can have difficulties in fully degrading it.
11. The NaCl is added to the sample in order to prevent ionic binding to the GNA coupled gel, during the chromatographic step, which can lead to loss of material. A more specific binding to the GNA molecule may also result. The outer envelope glycoproteins of HIV-1, HIV-2, and SIV bind very well to GNA in this salt concentration.
12. The sterile filtration and the Benzonase treatment of the solution is important for the life of the column, and also for the recovery of the glycoproteins. These impurities can stick to the matrix, which can result in trapping the bound glycoproteins.
13. The binding of the outer envelope glycoproteins to the GNA molecule can be performed both at room temperature and at 4°C. The method described here can be used in both cases. All solutions used should have the same temperature as the gel, to avoid air bubbles.
14. We have seen that a flow rate of 300 cm/h is also possible during this affinity step. But this can depend on how high the GNA concentration is per mL matrix (we have used 0.5–1.5 mg GNA/mL matrix), and the glycoprotein concentration in the media.
15. Before the material is eluted, change the flow direction. This can increase the concentration of the eluted material.
16. The dialysis of the eluted products gives rise to at least a doubling of the volume. This must be taken acount of when preparing the dialysis tubes.
17. If the purified glycoproteins are to be used in a few days, keep them sterile at 4°C. It is important not to freeze and thaw the glycoproteins several times, as this gives a much lower recognition by monoclonal antibodies.

References

1. Barre-Sinoussi, F., Chermann, J.-C., Rey, F., Nugeyre, M. T., Chamaret, S., Gruest, J., Dauguet, C., Axler-Blin, C., Vezinet-Brun, F., Rouzioux, C., Rozenbaum, W., and Montagnier, L. (1983) Isolation of a T-lymphotropic retrovirus from a patient at risk for acquired immune deficiency syndrome (AIDS). *Science* **220,** 868–871.
2. Gallo, R. C., Salahuddin, S. Z.., Popovic, M., Shearer, G. M., Kaplan, M., Haynes, B. F., Palker, T. J., Redfield, R., Oleske, J., and Safai, B. (1984) Frequent detection and isolation of cytopathic retroviruses (HTLV-III) from patient with AIDS and at risk for AIDS. *Science* **224,** 500–503.

3. Clavel, F., Guetard, D., Brun-Vezinet, F., Chamaret, S., Rey, M.-A., Santos-Ferreira, M. O., Laurent, A. G., Dauguet, C., Katlama, C., Rouzioux, C., Klatzmann, D., Champalimaud, J. L., and Montagnier, L. (1986) Isolation of a new human retrovirus from West African patients with AIDS. *Science* **233**, 343–346.
4. Daniel, M. D., Letvin, N. L., King, N. W., Kannagi, M., Sehgal, P. K., Hunt, R. D., Kanki, P. J., Essex, M., and Desrosiers, R. C. (1985) Isolation of a T-cell tropic HTLV III-like retrovirus from macaques. *Science* **228**, 1201–1204.
5. Gelderblom, H. R., Özel, M., Hansmann, E. H. S., Winkel, T., Pauli, G., and Kock, M. A. (1988) Fine structure of human immunodeficiency virus (HIV), immunolocalization of structural proteins and virus-cell relation. *Micron Microscop.* **19**, 41–60.
6. Özel, M., Pall, G., and Gelderblam, H. R. (1988) The organization of the envelope projections on the surface of HIV. *Arch. Virol.* **100**, 255–266.
7. Earl, P. L., Doms, R. W., and Moss, B. (1990) Oligomeric structure of the human immunodeficiency virus type 1 envelope glycoprotein. *Proc. Natl. Acad. Sci.* USA **87**, 648–652.
8. Weiss, C. D., Levy, J. A., and White, J. M. (1990) Oligomeric organization of gp120 on infectious human immunodeficiency virus type 1 particles. *J. Virol.* **64**, 5674–5677.
9. McCune, J. M., Rabin, L. B., Feinberg, M. B., Lieberman, M., Kosek, J. C., Reyes, G. R., and Weisman, I. L. (1988) Endoproteolytic cleavage of gp160 is required for the activation of the human immunodeficiency virus. *Cell* **53**, 55–67.
10. Lasky, L. A., Nakamura, G., Smith, D. H., Fennie, C., Shimasaki, C., Patzer, E., Berman, P., Gregory, T., and Capon, D. J. (1987) Delineation of a region of the human immunodeficiency virus type 1 gp120 glycoprotein critical for interaction with the CD4 receptor. *Cell* **50**, 975–985.
11. Nygren, A., Bergman, T., Matthews, T., Jornvall, H.pand Wigzell, H. (1988) 95- and 25-kDa fragment of the human immunodeficiency virus envelope glycoprotein gp120 bind to the CD4 receptor. *Proc. Natl. Acad. Sci. USA* **85**, 6543–6546.
12. Sattentau, Q. J. and Weiss, R. A. (1988) The CD4 antigen: physiological ligand and HIV receptor. *Cell* **52**, 631–633.
13. Capon, D. J. and Ward, R. H. (1991) The CD4-gp120 interaction and AIDS pathogenesis. *Annu. Rev. Immunol.* **9**, 649–678.
14. Matthews, T. J., Langlois, A. J., Robey, W. G., Chang, N. T., Gallo, R. C., Fischinger, P. J., and Bolognesi, D. P. (1986) Restricted neutralization of divergent human T-Lymphotropic virus type III isolates by antibodies to the major envelope glycoprotein. *Proc. Natl. Acad. Sci USA* **83**, 9709–9713.
15. Ho, D. D., Kaplan, J. C., Rackanskas, I. E., and Gurney, M. E. (1988) Second conserved domain of gp120 is important for HIV-infectivity and antibody neutralization. *Science* **239**, 1021–1023.
16. Haigwood, N. S., Shuster, J. R., Moore, G. K., Lee, H., Skiles, P. V., Higgins, K. W., Barr, P. J., Georg-Nascimento C., and Steimer. K. S. (1990) Importance of hypervariable regions of HIV-1 gp120 in the generation of virus-neutralizing antibodies. *AIDS Res. Hum. Retroviruses* **6**, 855–869.

17. Broliden, P.-A., von Gegerfellt, A., Clapham, P., Rosen, J., Fenyo, E.-M., Wharen, B., and Broliden, K. (1992) Identification of human neutralization-inducing regions of the human immunodeficiency virus type 1 envelope glycoproteins. *Proc. Natl. Acad. Sci. USA* **89**, 461–465.
18. Leonard, C. K., Spellman, M. W., Riddle, L., Harris, R. J., Thomas, J. Gland Gregory, T. J. (1990) Assignment of intrachain disulfide bonds and characterization of potential glycosylation sites of the type 1 recombinant human immunodeficiency virus envelope glycoprotein (gp120) expressed in Chinese hamster ovary cells. *J. Biol. Chem.* **260**, 10,373–10,382.
19. Allan, J. S., Coligan, J. E., Barin, F., McLane, M. F., Sodrowski, J. G., Rosen, C. A., Haseltine, W. A., Lee, T. H., and Essex, M. (1985) Major glycoprotein antigens that induce antibodies in AIDS patients are encoded by HTLV-III. *Science* **228**, 1091–1094.
20. Putney, S. D., Matthews, T. J., Robey, W. G., Lynn, D. L., Robert-Guroff, M., Mueller, W. T., Langlois, A. J., Ghrayeb, J., Petteway, S. R., Weinhold, K. J., Fischinger, P. J., Wong-Staal, F., Gallo, R. C., and Bolognesi, D. P. (1986) HTLV-III/LAV-neutralizing antibodies to an *E. coli*-produced fragment of the virus envelope. *Science* **234**, 1392–1395.
21. Matthews, T. J., Weinhold, K. J., Lyerly, H. K., Langlois, A. J., Wigzell, H., and Bolognesi, D. P. (1987) Interaction between the human T-cell lymphotropic virus type IIIB envelope glycoprotein gp120 and the surface antigen CD4: role of carbohydrate in binding and cell fusion. *Proc. Natl. Acad. Sci. USA* **84**, 5424–5428.
22. Steimer, K. S., Klasse, P. J., and McKeating, J. A. (1991) HIV-1 neutralization directed to epitopes other than linear V3 determinants. *AIDS* **5**, S135–S143.
23. Eriksson, S., Bhikhabhai, R., and Hammar, L. Analysis of glycoprotein. Lectin binding of HIV glycoproteins. *PhastSystem™*, Application file No 301, Pharmacis LKB Biotechnology, Uppsala, Sweden.
24. Hammar, L., Eriksson, S., and Morein, B. (1989) Human immunodeficiency virus glycoproteins: lectin binding properties. *AIDS Res.* **5**, 495–506.
25. Balzarini, J., Schols, D., Neyts, J., Van Damme, E., Peumans, W., and De Clercq, E. (1991) α-(1-3) and α-(1-6)-D mannose specific plant lectins are markedly inhibitory to human immunodeficiency virus and cytomegalovirus infection in vitro. *Antimicrob. Agents Chemotherapy* **35**, 410–416.
26. Gilljam, G., Siridewa, K., and Hammar, L. (1994) Purification of simian immunodeficiency virus, SIVmac251, and of its external envelope glycoprotein, gp148. *J. Chromatogr. A.* **675**, 89–100.
27. Gilljam, G. (1993) Envelope glycoproteins of HIV-1, HIV-2, and SIV purified with *Galanthus nivalis* agglutinin induce strong immune responses. *AIDS Res. Human Retrovir.* **9**, 431–438.
28. Van Damme, E. J. M., Allen, A. K., and Peumans, W. J. (1987) Isolation and characterization of a lectin with exclusive specificity towards mannose from snowdrop *(Galanthus nivalis)* bulb. *FEBS Lett.* **215**, 140–144.
29. Shibuya, N., Goldstein, I. J., Van Damme, E. J. M., and Peumans, W. J. (1988) Binding properties of a mannose-specific lectin from the snowdrop *(Galanthus nivalis)* bulb. *J. Biol. Chem.* **263**, 728–734.

30. Willey, R. L., Rutledge, R. A., Dias, S., Folks, T., Theodore, T., Buckler, C. E., and Martin, M. A. (1986) Identification of conserved and divergent domains within the envelope gene of the acquired immunodeficiency syndrome retrovirus. *Proc. Natl. Acad. Sci. USA* **83,** 5038–5042.
31. Modrow, S., Hahn, B. H., Shaw, G. M., Gallo, R. C., Wong-Staal, F., and Wolf, H. (1987) Computer-assisted analysis of envelope protein sequence of seven human immunodeficiency virus isolates: prediction of antigenic epitopes in conserved and variable regions. *J. Virol.* **61,** 570–578.

24

T-Cell Receptor Purification

Kelly P. Kearse

1. Introduction

The antigen receptor expressed on most T-lymphocytes (T-cell antigen receptor [TCR]) is a multisubunit complex consisting of at least six different polypeptides (α, β, γ, δ, ε, and ζ), several of which are modified by addition of N-linked oligosaccharide chains. Oligosaccharide side chains on TCR glycoproteins undergo well-characterized processing events within the endoplasmic reticulum (ER) and Golgi systems, including removal of mannose residues and subsequent addition of galactose and sialic acid oligosaccharides. In the current chapter, a method is described for the isolation of immature and mature TCR complexes from murine T-lymphocytes using lectin affinity chromatography. This technique is rapid, sensitive, and does not disrupt the integrity of assembled TCR complexes.

1.1. Assembly, Intracellular Transport, and Posttranslational Modification of Murine T-Cell Receptor Proteins

The TCR complex is composed of three different families of proteins: clonotypic $\alpha\beta$ polypeptides, invariant CD3-$\gamma,\delta,\varepsilon$ subunits, and invariant ζ chains *(1)*. Assembly of the multisubunit TCR complex takes place within the ER and is initiated by formation of noncovalently associated pairs of CD3$\delta\varepsilon$ and CD3$\gamma\varepsilon$ proteins *(2–5)*. Nondisulfide-linked α and β polypeptides then assemble with CD3 chains to form intermediate TCR complexes consisting of $\alpha\delta\varepsilon$ and $\beta\gamma\varepsilon$ subunits, which, in turn, assemble to form incomplete $\alpha\beta\delta\varepsilon\gamma\varepsilon$ TCR complexes *(6,7)*; clonotypic $\alpha\beta$ proteins are subsequently disulfide-linked to each other, and ζ proteins are added to yield complete $\alpha\beta\delta\varepsilon\gamma\varepsilon\zeta\zeta$ TCR complexes *(6,7)*. The intracellular transport of TCR proteins is directly related to their assembly status. Unassembled individual TCR proteins and partial complexes of CD3 components are retained within the ER and, depending on the

Table 1
Glycosylation of Murine T-Cell Receptor Proteins

TCR proteins	Oligosaccharide side chains		
	High mannose-type[a]	Complex-type[b]	Total
α	0	3–4	3–4
β	1	2–3	3–4
γ	0	1	1
δ	2	1	3
ε	0	0	0
ζ	0	0	0

[a]Number of N-linked oligosaccharide side chains on mature protein susceptible to digestion with Endoglycosidase H.

[b]Number of N-linked oligosaccharide side chains on mature protein resistant to digestion with Endoglycosidase H, containing complex-type sugars.

particular protein, degraded (8,9). Incomplete αβδεγ TCR complexes and complete αβδεγεζζ complexes egress the ER and transit through the Golgi apparatus (8,9). Incomplete TCR complexes are targeted to lysosomes for degradation; only complete TCR complexes are effectively transported to the plasma membrane (10–12).

Several members of the TCR complex are posttranslationally modified by the addition of N-linked oligosaccharide side chains: clonotypic α,β proteins and invariant CD3-γ,δ chains (Table 1). On egress from the ER and transit through Golgi system, some, but not all, N-linked sugar chains of TCR glycoproteins are converted from high mannose-type oligosaccharides to complex-type sugar chains. Indeed, mature TCRα and CD3γ proteins contain exclusively complex-type N-linked oligosaccharide chains, whereas mature TCRβ and CD3δ chains contain both high mannose and complex-type oligiosaccharide chains (Table 1). Because of their unique binding specificity for oligosaccharides, lectins have been widely used in the purification of glycoproteins (13). Regarding the TCR complex, lectins have been utilized in the purification of TCR proteins from immature CD4$^+$CD8$^+$ thymocytes and splenic T-lymphocytes (14,15), murine T hybridoma cells (16), and human T-lymphocytes (17). In the current chapter, a method is described for the isolation of immature and mature TCR complexes from murine T-lymphocytes using immobilized lectins specific for oligosaccharides added in the trans Golgi compartment of the cell.

2. Materials
2.1. Lectin Affinity Chromatography
1. The principle behind using lectin affinity chromatography for the separation of immature and mature TCR complexes is shown in Fig. 1. Immobilized lectins are

T-Cell Receptor Purification

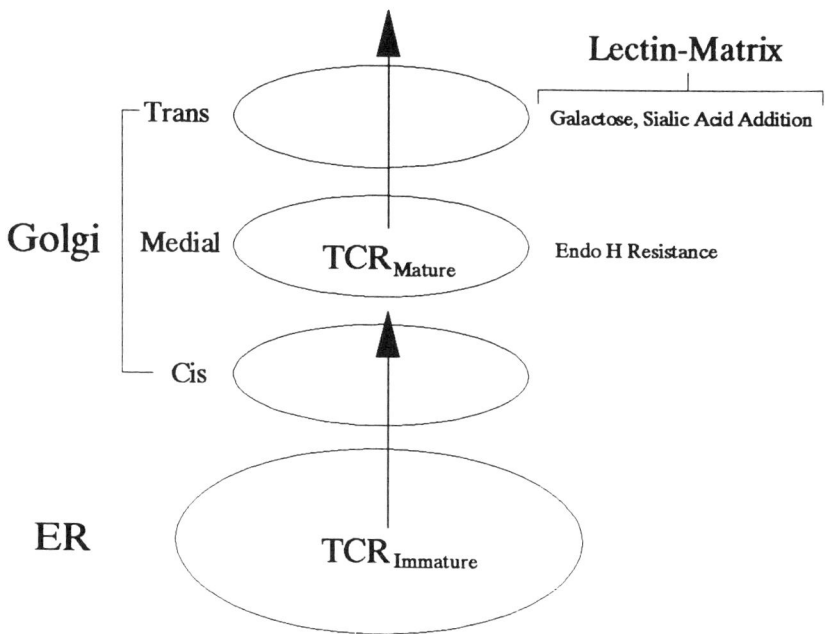

Fig. 1. Separation of immature and mature TCR complexes by lectin affinity chromatography: TCR complexes localized within the ER contain immature high mannose oligosaccharide chains and are, therefore, not bound by lectin matrices specific for complex-type sugars (galactose, sialic acid). TCR complexes which effectively egress the ER and transit to the Golgi system undergo processing by Golgi glycosidases and glycosyltransferases (e.g., galactosyltransferase, sialyltransferase) and are therefore effectively retained on lectin matrices specific for complex-type sugars.

available from numerous commercial suppliers. Wheatgerm agglutinin (WGA) and *Ricin communis* agglutin (RCA) conjugated to agarose matrices (EY Laboratories, San Mateo, CA) were utilized in our studies *(14–16)*.
2. Tris-buffered saline (TBS): 20 mM Tris-HCl, 150 mM NaCl, pH 7.2, with 10 mM iodoacetamide, 20 µg/mL leupeptin, and 40 µg/mL aprotinin.
3. Digitonin solution: Digitonin (Wako, Kyoto, Japan) is prepared as a 2% stock solution (w/v) and boiled for 10 min. Solution is allowed to cool at room temperature at least 30 min before use.
4. Lectin wash buffer: TBS containing 0.5% digitonin.
5. Precipitation wash buffer: TBS containing 0.2% digitonin.
6. Elution buffer: precipitation wash buffer containing appropriate competing oligosaccharide, typically at concentrations of 100–500 mM. For WGA, affinity chromatography, 500 mM N-acetylglucosamine (GlcNAc) is used; for RCA, 150 mM β-lactose is used. Oligosaccharides were purchased from either Sigma (St. Louis, MO) or EY.

7. PBS, pH 7.2.
8. 20% Bleach solution is required in experiments using RCA (*see* Note 1).

2.2. Immunoprecipitation, Glycosidase Digestion, and Gel Electrophoresis

1. Cell lysis buffer: TBS containing 1% digitonin.
2. TCR specific antibodies conjugated to protein A-sepharose (Pharmacia, Uppsala, Sweden).
3. Precipitation wash buffer: TBS containing 0.2% digitonin.
4. PBS, pH 7.2
5. 1% SDS solution.
6. Glycosidase digestion buffer: 75 mM sodium phosphate, pH 6.1, 75 mM EDTA, 0.1% NP-40.
7. Endoglycosidase H (Genzyme, Cambridge, MA).
8. For analysis of TCR proteins, 13% polyacrylamide gels are typically used. Standard gel electrophoresis reagents and gel equipment are also needed.

3. Methods

3.1. Lectin Affinity Chromatography

1. A flow diagram of this procedure is presented in Fig. 2. Resuspend cells (approx 1 × 10^8 cells for murine thymocytes or splenic T-lymphocytes; 5 × 10^7 cells for murine T-cell hybridomas) in 800–1000 µL cell lysis buffer, and incubate at 4°C for 20 min.
2. Remove insoluble material by centrifugation and transfer lysate (Sn) to a new tube.
3. Mix lysate with an equal volume of TBS to yield a final detergent concentration of 0.5%. Place half of the sample at 4°C until further analysis (unfractionated material); mix the other half with approx 200–400 µL of immobilized lectin slurry that was previously washed three times with 1 mL PBS and once with 1 mL lectin wash buffer.
4. Incubate material for 3–4 h at 4°C with rocking.
5. Centrifuge sample, remove Sn (unbound material), and transfer to a new tube. Recentrifuge material, remove supernatant, and transfer to a new tube. Repeat this process twice more (four times total) to ensure that no carryover of lectin beads has occurred. Place samples at 4°C until further analysis.
6. Wash lectin beads containing bound material five times in 1 mL precipitation wash buffer, resuspend in 1 mL elution buffer containing appropriate oligosaccharide, and incubate at 4°C for 60 min.
7. Centrifuge lectin beads, remove supernatant (containing previously bound material) transfer to a new tube. Repeat this process three more times to ensure that no carryover of beads has occurred. Place samples at 4°C until further analysis.
8. Note: All materials that come in contact with RCA must be decontaminated with 20% bleach solution, as this material is extremely toxic. Also, proper ventilation and protective clothing must be used to avoid inhalation or contact with this material (*see* Note 1).

T-Cell Receptor Purification

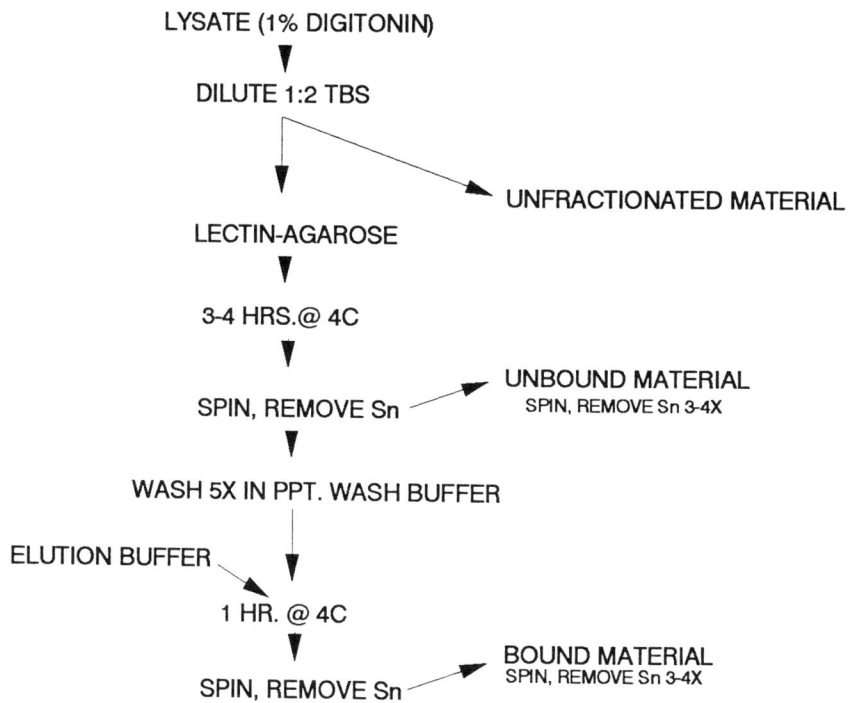

Fig. 2. Isolation of immature and mature TCR complexes by lectin affinity chromatography: Flow diagram of isolation of immature and mature TCR complexes by lectin–agarose matrices. The reader is referred to Section 3.1. for details.

3.2. Immunoprecipitation, Glycosidase Digestion, and Gel Electrophoresis

1. Mix unfractionated material, unbound material, and bound material with TCR-specific MAb preabsorbed to protein A sepharose (approx 60 µL of slurry), and incubate for 3 h at 4°C.
2. Wash immunoprecipitates three times with 1 mL precipitation and once with 1 mL PBS. Prior to the last wash with PBS divide precipitates in half for glycosidase digestion.
3. Add 5 µL of 1% SDS to washed beads, and boil samples for 5–10 min.
4. Allow samples to cool for 5 min at room temperature and then add 35 µL of glycosidase digestion buffer.
5. Add 5 µL (10 mU) of Endo H (Genzyme, Cambridge, MA) to Endo H digested groups; mock treated samples receive 5 µL of glycosidase digestion buffer.
6. Samples are incubated overnight at 37°C.
7. Stop glycosidase digestion by addition of 60 µL of 3X gel electrophoresis sample buffer.

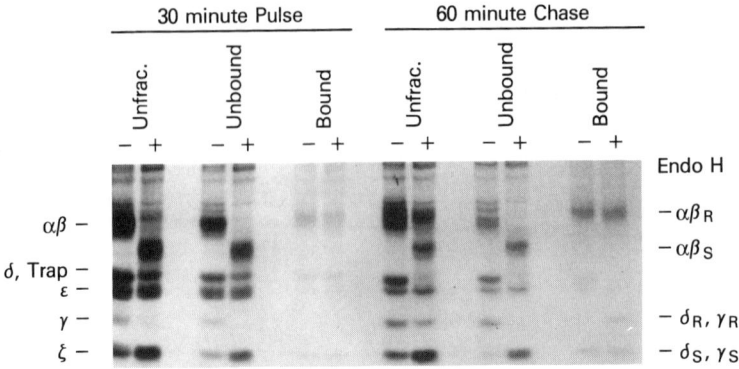

Fig. 3. Separation of immature and mature TCR complexes from splenic T lymphocytes by RCA matrices: Radiolabeled lysates of splenic T-cells were either unfractionated or incubated with *Ricin communis* agglutin (RCA) conjugated to agarose beads. Unfractionated and RCA-fractionated material was precipitated with MAb to CD3ε; precipitates were either mock treated or digested with Endo H and analyzed by one dimensional SDS-PAGE under reducing conditions. The positions of α,β,γ,δ, TRAP (TCR associated protein), ε, and ζ chains are indicated. The mobilities of TCR glycoproteins containing immature, Endo H sensitive oligosaccharide chains ($\alpha\beta_S$, δ_S, and γ_S) and mature, Endo H resistant oligosaccharide chains ($\alpha\beta_R$, δ_R, and γ_R) are indicated.

8. Run precipitates on SDS-PAGE gels using standard electrophoretic techniques and analyze by appropriate biochemical methods: immunoblotting, autoradiography, and so on.
9. The separation of immature and mature TCR complexes of splenic T-lymphocytes by RCA matrices is shown in Fig. 3. In this experiment, purified splenic T-cells were metabolically labeled with [^{35}S]methionine for 30 min and chased in medium containing excess unlabeled methionine for 60 min to allow movement of metabolically labeled proteins from the ER to the Golgi apparatus. Cells were solubilized in 1% digitonin and lysates were either unfractionated or exposed to RCA matrices. Material was separated into RCA-unbound and RCA-bound fractions. TCR complexes present in unfractionated material, RCA-unbound, and RCA-bound fractions were immunoprecipitated with anti-CD3ε specific MAb and precipitates digested with Endo H. Samples were analyzed on 13% SDS-PAGE gels under reducing conditions, and gels processed for autoradiography. It can be seen that RCA effectively separated immature and mature TCR complexes of splenic T-cells in these studies as Endo H-sensitive TCR glycoproteins were only present in RCA-unbound fractions and Endo H-resistant TCR chains existed exclusively in RCA-bound fractions (Fig. 3).

4. Notes

1. **Caution:** *Ricin communis* agglutin is *extremely* toxic. Therefore, appropriate safety measures must be used when handling this material. It is recommended

that commercial suppliers and local supervisors be consulted prior to use of this reagent for specific safety guidelines.
2. Digestion of isolated material with Endo H glycosidase is a useful means for determining the efficiency of separation of immature and mature TCR complexes using WGA and RCA lectins; TCR glycoproteins containing immature (Endo H sensitive) oligosaccharides are localized within the unbound fraction, whereas mature (Endo H resistant) species are restricted to the bound fraction (Fig. 3). Alternatively, separation of immature and mature TCR complexes by lectin matrices may be assessed by measuring addition of sialic acid residues to oligosaccharide chains using neuraminidase digestion.
3. Although lectin binding is compatible with many detergent solutions, the use of digitonin is recommended in these studies to maintain the integrity of assembled TCR complexes.
4. The binding activity of many lectins (e.g., Concanavalin A) is cation-dependent and thus requires addition of Ca^{2+}, Mg^{2+}, Mn^{2+}, to buffer solutions. However, the lectins WGA and RCA do not require cations for their binding activity and, thus, can be used with most standard laboratory buffer solutions.
5. It should be noted that the lectin of choice for isolation of TCR complexes is dependent on the cell type being studied. For example, TCR complexes expressed on murine 2B4 T-cell hybridomas are efficiently bound by WGA matrices, whereas TCR complexes expressed on immature thymocytes and mature splenic T-cells are not. The reason for this difference is unclear, but may reflect differential processing of oligosaccharide chains of TCR complexes upon T-cell activation. In contrast to these findings, TCR complexes expressed on all of the abovementioned T-cell types are efficiently retained on RCA matrices. Thus, the use of RCA seems best suited for isolation of TCR complexes of any cell type, although it is conceivable that other, less hazardous, lectins may also be used, for example *Sambucus nigra* (SNA) or *Limax flavus* (LFA), specific for sialic acid containing oligosaccharides.

References

1. Klausner, R. D., Lippincott-Schwartz, J., and Bonifacino, J. S. (1990) The T-cell antigen receptor: insights into organelle biology. *Ann. Rev. Cell. Biol.* **6,** 403–431.
2. Ohashi, P. S., Mak, T. W., van den Elsen, P., Wangai, Y., Yoshikai, Y., Calman, A. F., Terhorst, C., Stobo, J. D., and Weiss, A. (1985) Reconstitution of active surface T3/T-cell antigen receptor by DNA transfer. *Nature* **316,** 606–609.
3. Saito, T., Weiss, A., Guner, K. C., Shevach, E. M., and Germain, R. N. (1987) Cell surface T3 expression requires the presence of both α-β-chains of the T-cell receptor. *J. Immunol.* **139,** 625–628.
4. Alarcon, B., Berkhout, B., Breitmeyer, J., and Terhorst, C. (1988) Assembly of the human T-cell receptor-CD3 complex takes place in the endoplasmic reticulum and involves intermediary complexes between the CD3-gamma, delta, and epsilon core and single T-cell receptor alpha or beta chains. *J. Biol. Chem.* **263,** 2953–2961.

5. Bonifacino, J. S, Chen, C., Lippincott-Schwartz, J., Ashwell, J. D., Klausner, R. D. (1988) Subunit interactions within the T-cell receptor: clues from the study of partial complexes. *Proc. Natl. Acad. Sci. USA* **35,** 6929–6923.
6. Kearse, K. P., Roberts, J. L., and Singer, A. (1995) TCRα-CD3δε association is the initial step in αβ dimer formation in murine T-cells and is limiting in immature CD4$^+$CD8$^+$ thymocytes. *Immunity* **2,** 391–399.
7. Minami, Y., Weissman, A. M., Samelson, L. E., and Klausner, R. D. (1987) Building a multichain receptor: synthesis, degradation, and assembly of the T-cell antigen receptor. *Proc. Natl. Acad. Sci. USA* **84,** 2688–2692.
8. Chen, C., Bonifacino, J. S., Yuan, L., and Klausner, R. D. (1988) Selective degradation of T-cell antigen receptor chains retained in a pre-Golgi compartment. *J. Cell Biol.* **107,** 2149–2161.
9. Lippincott-Schwartz, J., Bonifacino, J. S., Yuan, L., and Klausner, R. D. (1989) Degradation from the endoplasmic reticulum: disposing of newly synthesized proteins. *Cell* **54,** 209–220.
10. Wileman, T., Carson, G. R., Concino, J., Ahmed, A., and Terhorst, C. (1990). The γ and ε subunits of the CD3 complex inhibit pre-Golgi degradation of newly synthesized T-cell receptors. *J. Cell Biol.* **110,** 973–986.
11. Sussman, J. J., Bonifacino, J. S., Lippincott-Schwartz, J., Weissman, A. M., Saito, T., Klausner, R. D., and Ashwell, J. D. (1988) Failure to synthesize the T-cell ζ chain: structure and function of a partial T-cell receptor complex. *Cell* **52,** 85–95.
12. Hall, C. Berkhot, B., Alarcon, B., Sancho, J., Wileman, T., and Terhorst, C. (1991) Requirements for cell surface expression of the human TCR/CD3 complex in non-T-cells. *Int. Immunol.* **3,** 359–368.
13. Lotan, R., Beattie, G., Hubbell, W., and Nicolson, G. L. (1977) Activities of lectins and their immobilized derivatives in detergent solutions. Implications on the use of lectin affinity chromatography for the purfication of membrane glycoproteins. *Biochemistry* **16,** 1787–1794.
14. Kearse, K. P., Wiest, D. L., and Singer, A. (1993) Subcellular localization of T-cell receptor complexes containing tyrosine-phosphorylated ζ proteins in immature CD4$^+$CD8$^+$ thymocytes. *Proc. Natl. Acad. Sci. USA* **90,** 2438–2442.
15. Kearse, K. P., Roberts, J. L., Munitz, T., Wiest, D. L., Nakayama, T., and Singer, A. (1994) Developmental regulation of αβ T-cell antigen receptor expression results from differential stability of nascent TCRα proteins within the endoplasmic reticulum of immature and mature T-cells. *EMBO J.* **13,** 4504–4514.
16. Kearse, K. P. and Singer, A. (1994) Isolation of immature and mature T-cell receptor complexes by lectin affinity chromatography. *J. Immunol. Methods* **167,** 75–81.
17. Chilson, O. P. and Kelly-Chilson, A. E. (1989) Mitogenic lectins bind to the antigen receptor on human lymphocytes. *Eur. J. Immunol.* **19,** 389–396.

VI

LECTINS IN FLOW CYTOMETRY

25

Use of Monomeric, Monovalent Lectin Derivatives for Flow Cytometric Analysis of Cell Surface Glycoconjugates

Hanae Kaku and Naoto Shibuya

1. Introduction

Because of their unique carbohydrate binding characteristics, lectins serve as invaluable tools in biological and medical research; separation and characterization of glycoproteins and glycopeptides, histochemistry of cells and tissues, and the study of cell differentiation *(1)*. Direct analysis of individual cells carrying glycoconjugate on the cell surface by flow cytometry/cell sorting should give invaluable information on the distribution, dynamics, and biological role of these glycoconjugates. However, native lectins with multiple binding sites often suffer from their property to agglutinate cells in these applications. Single cell suspensions are required for these applications to avoid the stacking problems and also for accurate measurement. To attain this, lectins should be used in the concentration range, in which the agglutination does not occur *(2)*. This has been one of the major problems for the application of lectins in this field.

We have solved this problem by establishing monomeric, monovalent derivatives of lectins which have lost their cell agglutinating property while maintaining their carbohydrate binding activity *(3,4)*. In this chapter, we describe the preparation of a stable monomeric subunit of Japanese elderberry *(Sambucus sieboldiana)* bark lectin (SSA) and *Maackia arnurensis* leukoagglutinin (MAL), by selective reduction and alkylation of the disulfide bond between the subunits and demonstrate its usefulness in the flow cytometric analysis of cell surface glycoconjugates. These lectins are specific to Neu5Acα2 6GalGalNAc and NeuSAcα23Galβ1 4Glc/GlcNAc sequences, respectively *(5–8)*, and have been shown to be quite useful for the analysis of various sialylated glycoconjugates, which are widely distributed in animal cell surface and have been reported to play important roles in biological recognition systems *(9,10)*. Combined use of these new monomeric, monovalent derivatives (MSSA and MMAL) for flow cytometric analysis could show the different pattern of the expression of α2-3- and α2-6-linked sialylated oligosaccharides on the cell surface of various cell lines.

2. Materials

2.1. Preparation of Lectins

1. Bark lectin (SSA) is purified from the extract of the twigs of Japanese elderberry by affinity chomatography on fetuin-agarose (Sigma, St. Louis, MO). The twigs are chopped into small pieces, homogenized, and extracted with 10 vol of PBS. After the extract is centrifuged at 27,000g for 20 min, the supernatant is applied to a fetuin-agarose column (0.7 × 10 cm). The column is washed with PBS until the absorbance at 280 nm falls below 0.05 and is then eluted with PBS containing 0.5M lactose. The lactose eluate is dialyzed against H_2O at 4°C and lyophilized *(11)*.
2. *Maackia amurensis* leukoagglutinin (MAL) is purified from the PBS extract of the seeds of *Maackia arnurensis* by affinity chomatography on immobilized murine laminin (EY Laboratories, San Mateo, CA) as described in *(8)*.

   These lectins are also commercially available from several sources (EY; Seikagaku Kogyo, Tokyo, Japan; Wako, Osaka, Japan).

2.2. Buffers and Reagents

1. Phosphate-buffered saline (PBS): 150 mM NaCl, 10 mM Na_2HPO_4–NaH_2PO_4, pH 7.2.
2. EMA buffer: 0.4M N-ethylmorpholine acetate buffer, pH 8.3.
3. DTT: dithiothreitol, store at 4°C in the dark.
4. 4-Vinylpyridine: store at 4°C in the dark.
5. Sephadex G-25 column: Pharmacia PD 10 column packed with Sephadex G-25M (1.6 × 5 cm).
6. FITC: fluorescein isothiocyanate. store at –20°C in the dark.
7. CB: 0.5M carbonate buffer, pH 9.5.

2.3. Cell Culture

1. Chinese hamster ovary (CHO) cells are maintained with Ham's nutrient mixture F12 (ICN Biomedicals, High Wycombe, UK) containing 10% fetal bovine serum (Gibco-BRL, Gaithersburg, MD) and gentamicin (50 μg/mL) in culture plates. Cells are incubated at 37°C under 5% CO_2 *(12; see* Note 1).
2. Human histiocytic lymphoma U937 (U 937) cells are maintained with Dulbecco's modifed eagle medium (Gibco-BRL) containing 10% fetal calf

serum and 1% penicillin-streptomycin *(13)*. Culture conditions are the same as CHO cells.

3. Methods

3.1. Preparation of Monomeric, Monovalent Lectin Derivative (MMLD)

Monomeric, monovalent derivatives of SSA and MAL (MSSA and MMAL) are prepared by the selective reduction of the disulfide linkage between subunits and the stabilization by S-β-4-pyridylethylation *(3,4)*.

3.1.1. Monovelent SSA (MSSA)

1. Incubate 5 mg SSA in 1 mL of EMA buffer containing 1% DTT and $0.25M$ lactose (to protect the carbohydrate binding site) at 20°C for 3 h under N_2 *(see* Note 2).
2. Add 4-vinylpyridine (8 µL/mL) to the reaction mixture using a syringe and react at 20°C for 15 min under N_2. Add 1 mL of H_2O to the reaction mixture at the end of the reaction.
3. Separate the stabilized monomeric subunit of SSA (MSSA) from the excess reagents using a Sephadex G-25 column equilibrated with PBS.
4. Collect the void fraction and store at 4°C *(see* Notes 3 and 4).

3.1.2. Monovalent MAL

1. Incubate 2 mg MAL in 1 mL PBS containing 0.35% DTT and $0.4M$ lactose at 4°C for 4 h under N_2.
2. Add 4-vinylpyridine (8 µL/mL) to the reaction mixture, and allow to react for 15 min under N_2 at 4°C. Proceed as described in Section 3.1.1.

3.2. Labeling of Monovalent Lectin with FITC

1. Incubate monomeric, monovalent lectin derivative (MSSA or MMAL, 1 mg) in CB buffer containing $0.2M$ lactose with FITC solution (50 µL, 2 mg/mL CB buffer) in the dark at 4°C for 4 h.
2. Apply the reaction mixture to the Sephadex G-25 equilibrated with PBS containing 0.01% NaN_3 and elute with the same buffer. Collect the void fraction and keep in the dark at 4°C *(see* Note 5).

3.3. Flow Cytometry

1. Wash the cells three times with PBS prior to use. Incubate approx $1–2 \times 10^5$ cells with the FITC-labeled monovalent lectin (5–25 µg/mL) in PBS containing 0.1% NaN_3 for 30 min in the dark at room temperature.
2. Wash the cells 2 times with PBS ($50g$, 5 min), and keep on ice in the dark until use. Analyze an aliquot of the cells with a flow cytometer (EPICS PROFILE II, Coulter, Hialeah, FL, *see* Fig. 1).

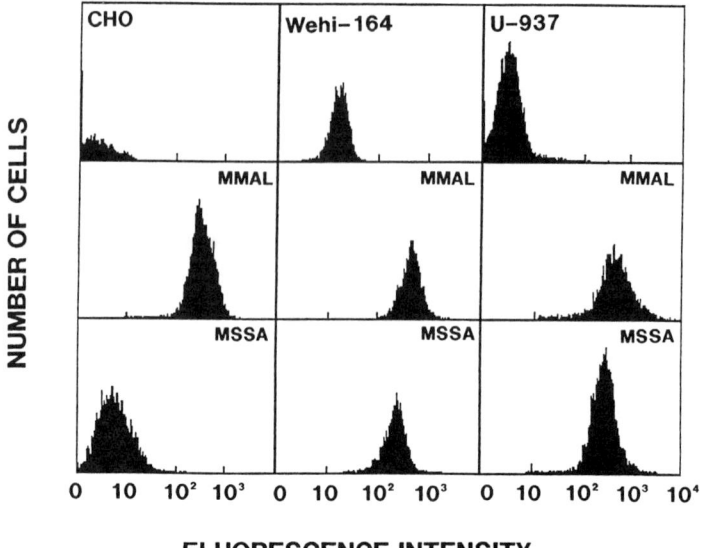

Fig. 1. Flow cytometric analyses of three different kinds of cell lines stained with FITC-Monomeric MAL or FITC-Monomeric SSA *(4)*. For each calculation, 5000 cells were used. Top row: CHO, Wehi-164 (mouse fibrosarcoma clone 28-4), and U-937 cell lines without lectins; middle row: CHO, Wehi-164, and U-937 cell lines stained with FITC-MMAL (25 µg/mL); bottom row: CHO, Wehi-164, and U 937 cell lines stained with FITC-MSSA (5 µg/mL).

4. Notes

1. CHO cells grow on the plastic surface of the culture plate. To obtain single cell suspension for flow cytometric analysis or cell sorting, the plate (2×10^6 cell) is washed twice with PBS, then treated with trypsin (125 µg/mL PBS) for 1 min at room temperature (or, until the cells start to release from the plate). Culture medium (2 mL) is added and the released cell suspension is passed though a Pasteur pipet that has been packed with glass wool and autoclaved. Do not overdigest with trypsin.
2. A small reactivial with mininert valve (Pierce, Rockford, IL) is recommended as the reaction vessel.
3. Formation of monovelent lectin can be followed by HPLC with a gel filtration column (for example, TSK-GEL G3000sw column) or by SDS PAGE in the absence of β-mercaptoethanol.
4. Carbohydrate binding activities of monovelent lectin can be analyzed by enzyme linked immunosorbent assay (ELISA) *(3,4)*. On the other hand, quantitative precipitation and hemagglutination assay (the most popular methods for the study of lectins) are not applicable, because these methods require the presence of multiple binding sites in the lectin molecule. The inability of

the agglutination or precipitation, however, is a good indication of the monovalency of the preparation, if the binding activity itself is confirmed by the ELISA method.
5. The molar ratio of fluorescein bound to MMLD is calculated from the absorbance at 495 and 280 nm *(14)*. Recommendable molar ratio of fluorescein:protein is 0.2:1.0.

Acknowledgment

This research is supported in part by a Grant in Aid (Bio Media Program) from the Ministry of Agriculture, Forestry, and Fisheries (BMP 95-V-4-1) and a special coordination fund from the Science and Technology Agency of the Japanese Government.

References

1. Goldstein, I. J. and Poretz, R. D. (1986) in *The Lectins* (Liener, I. E., Sharon, N., and Goldstein, I. J., eds.), Academic, Orlando, FL, pp. 35–357.
2. Reimann, J., Ehman, D., and Miller, R. G. (1984) Differential binding of lectins to lymphopoietic and myelopoietic cells in murine marrow as revealed by flow cytometry. *Cytometry* **5,** 194–203.
3. Kaku, H. and Shibuya, N. (1992) Preparation of a stable subunit of Japanese elderberry *(Sambucus sieboldiana)* bark lectin and its application for the study of cell surface carbohydrates by flow cytometry. *FEBS Lett.* **306,** 176–180.
4. Kaku, H., Mori, Y., Goldstein, I. J., and Shibuya, N. (1993) Monomeric, monovalent derivative of *Maackia amurensis* leukoagglutinin: preparation and application to the study of cell surface glycoconjugates by flow cytometry. *J. Biol. Chem.* **268,** 13,237–13,241.
5. Shibuya, N., Goldstein, I. J., Broekaert, W. F., Nsimba Lubaki, M., Peeters, B., and Peumans, W. J. (1987) The elderberry *(Sambucus nigra* L.) bark lectin recognizes the Neu5Ac(α2-6)Gal/GalNAc sequence. *J. Biol. Chem.* **262,** 1596–1601.
6. Shibuya, N., Tazaki, K., Song, Z., Tarr, G. E., Goldstein, I. J., and Peumans, W. J. (1989) A comparative study of bark lectins from three elderberry (*Sambucus)* species. *J. Biochem.* **106,** 1098–1103.
7. Wang, W. C. and Cummings, R. D. (1988) The immobilized leukoagglutinin from the seeds of *Maackia amurensis* binds with high affinity to complex-type Asn linked oligosaccharides containing terminal sialic acid linked α 2,3 to penultimate galactose residues. *J. Biol. Chem.* **263,** 4576–4585.
8. Knibbs, R. N., Goldstein, I. J., Ratcliffe, R. M., and Shibuya, N. (1991) Characterization of the carbohydrate binding specificity of the leukoagglutinating lectin from *Maackia amurensis. J. Biol. Chem.* **266,** 83–88.
9. Schauer, R. (1982) Chemistry, metabolism, and biological functions of sialic acids. *Adv. Carbohydr. Chem. Biochem.* **40,** 131–234.
10. Chavin, S. I. and Weidner, S. M. (1984) Blood clotting factor IX. *J. Biol. Chem.* **259,** 3387–3390.

11. Tazaki. K. and Shibuya. N. (1989) Purification and partial characterization of a lectin from the bark of Japanese elderberry *(Sambucus sieboldiana)*. *Plant Cell Physiol.* **30,** 899–903.
12. Kao, F., Chasin, L., and Puck, T. T. (1969) Genetics of somatic mammalian cells, X. Comp lamentation analysis of glycine-requiring mutants. *Proc. Natl. Acad. Sci. USA* **64,** 1284–1291.
13. Sundstrom, C. and Nilsson, K. (1976) Establishment and characterization of a human histiocytic lymphoma cell line (U-937). *Int. J. Cancer* **17,** 565–577.
14. Chistensen, J. and Leslie, R. G. Q. (1990) Quantitative measurement of Fc receptor activity on human peripheral blood monocytes and the monocyte like cell line, U937, by laser flow cytometry. *J. Immunol. Methods* **132,** 211–219.

26

Analysis of Subcellular Components by Fluorescent-Lectin Binding and Flow Cytometry

Rosa M. Guasch and José-Enrique O'Connor

1. Introduction

Because of their extensive availability and the wide spectrum of carbohydrates that may be specifically bound, lectins have become essential reagents for detection and quantitation of glycoconjugates in solution and in cell surfaces, identification and separation of cells, and functional studies based on membrane properties *(1,2)*.

Flow cytometry is an analytical process in which several physical and/or chemical properties of individual cells or particles may be measured at a high velocity. Because of these features, flow cytometry is used extensively in basic and clinical research *(3)*. In the context of glycoconjugate analysis, the application of flow cytometry and fluorescent lectins started in the early 1970s *(4)*. Although the advent of MAb decreased the use of lectins as tools for cell identification, there are still a number of interesting flow cytometric applications which involve fluorescent lectin binding, especially in multiparametric combination with other structural and functional cellular features *(5,6)*.

1.1. Application of Fluorescent Lectins to Subcellular Analysis by Flow Cytometry

Intracellular membrane systems, such as endoplasmic reticulum, Golgi complex, mitochondria, glyoxysomes and chloroplasts contain accessible carbohydrates bound to lipids or proteins *(7,8)*. Such glycoconjugates are thought to play important roles in maintaining the structure and function of organelle membranes. Thus, glycosylated proteins may provide conformational stability, surface charge, water-binding capacity, and resistance to proteases *(9)*, whereas glycolipids are involved in the regulation of membrane fluidity and shape, ion binding, and membrane interactions *(10)*.

From: *Methods in Molecular Medicine, Vol. 9: Lectin Methods and Protocols*
Edited by: J. M. Rhodes and J. D. Milton Humana Press Inc., Totowa, NJ

In spite of such interesting properties, there have been only limited attempts to apply fluorescent lectins and flow cytometry to the analysis of intracellular glycoconjugates. This lack of information reflects in part the methodological diffculties in the flow cytometric analysis of some intracellular structures. Thus, subcellular analysis in whole cells may require careful fixation and permeabilization procedures when relatively large (e.g., conjugated lectins or antibodies) or nonpermeable fluorescent molecules are used as reagents *(11)*. On the other hand, the use of isolated subcellular particles demands proper instrumental sensitivity and resolution to establish functional or structural criteria to distinguish single subcellular elements from background noise, as will be pointed out later in this chapter. Nevertheless, flow cytometric analysis of subcellular compartments may be performed satisfactorily with most of the currently available instruments *(7,12)*.

Thus, Petit and coworkers *(7,13)* have been pioneers in the application of fluorescent lectins to the flow cytometric analysis of isolated subcellular elements, namely mitochondria, chloroplasts, and thylakoids from different plant tissues. More recently, we have applied flow cytometry to the detection and quantitation of the binding of several glycoconjugate specific lectins to isolated *cis* and *trans* Golgi fractions from rat liver *(14,15)*.

1.2. Application of Fluorescent Lectins to the Analysis of Isolated Golgi Elements

The maturation and distribution of membrane and secretory glycoproteins involve the transit of newly synthesized proteins through the Golgi complex *(16)*, which is organized into three functionally distinct regions, the cis Golgi network, the medial Golgi, and the *trans* Golgi network *(17)*. The *cis* and *trans* Golgi networks constitute the entry and exit faces of the stack, respectively, and have morphological and functional differences because each one includes distinct processing enzymes *(17)*.

A substantial part of the current knowledge of the functions of the Golgi complex and the heterogeneity of glycosylation events within its subcompartments has arisen from cytochemical and biochemical analysis of lectin binding to specific carbohydrates *(18,19)*. In view of the capabilities of flow cytometry, we have extended these studies to the detection and quantitation of the binding of several fluorescein conjugated lectins, namely concanavalin A (Con A), wheatgerm agglutinin (WGA), *Ulex europaeus* agglutinin (UEA) and *Arachis hypogaea* agglutinin (PNA) to isolated *cis* and *trans* Golgi fractions from the liver of normal and ethanol treated rats *(14,15)*. Our results showed that flow cytometric analysis, using low-power, air cooled laser, is adequate to study carbohydrates present in isolated Golgi membranes. Thus, the use of specific lectins allows the determination of: the intensity of

Fluorescent-Lectin Binding

specific binding of different lectins to each Golgi subcompartment; the percentage of elements that bind different lectins specifically within the same subcompartment; and the intensity and percentage of lectin binding when comparing *cis* and *trans* fractions. All these parameters were affected to different extents by chronic exposure to ethanol *(14,15)*, thus indicating the feasibility of applying the technique described here to detect alterations in the pathways of glycoconjugate metabolism.

2. Materials
2.1. Purification and Isolation of Golgi Subfractions

1. Isolation buffer 1: 0.25M sucrose, 10 mM Tris-HCl, pH 7.4.
2. Isolation buffer 2: 1.35M sucrose.
3. Discontinuous sucrose density gradient: 2.00, 1.35, 1.15, 0.86, and 0.25M sucrose. 7.5 mL of each concentration are used for each single tube in step 7 of Section 3.1.
4. Potter Elvehjem homogenizer with Teflon® pestle.
5. Beckman L5-65 ultracentrifuge with Beckman SW 27 rotor (Fullerton, CA).

2.2. Lectin Preparation

1. Fluorescent lectins, conjugated to fluorescein isothiocyanate (FITC, green fluorescing) or to tetramethylrhodamine isothiocyanate (TRITC, red fluorescing) (Sigma, St. Louis, MO).
2. Lectin binding inhibitory sugars: specific to the lectins selected for the study (Sigma) (*see* Note 1 and 2).

2.3. Flow Cytometry

1. Dulbecco phosphate-buffered saline (DPBS) with calcium and magnesium.
2. Any laser or mercury lamp based flow cytometer, able to provide 488 nm excitation light (blue) and to detect forward angle light scatter, right angle light scatter, and green or red fluorescence.

3. Methods
3.1. Isolation of Golgi Cis and Trans Fractions

1. Golgi *cis* and *trans* subcompartments may be isolated from the liver of rats (150–200 g) following a modification *(20)* of the procedure of Bretz et al. *(21)*, as follows.
2. Sacrifice the animals by cervical dislocation, remove their livers, wash them briefly in ice-cold isolation buffer 1, and weigh them.
3. Homogenize the livers in ice-cold isolation buffer 1.
4. Dilute the homogenate to 20% (w/v) concentration with isolation buffer 1 and centrifuge for 10 min at 10,000g at 4°C.
5. Separate the resulting supernatant and centrifuge for 90 min at 105,000g at 4°C to obtain a microsomal pellet.

6. Resuspend the pellet in 7.5 mL of isolation buffer 2.
7. Layer the sample under a discontinuous sucrose density gradient (7.5 mL each of 2.00, 1.35, 1.15, 0.86, and 0.25M sucrose) prepared in polyallomer tubes and centrifuge it at 82,000g for 3 h at 4°C with a Beckman SW 27 rotor.
8. Collect separately the material floating to the 0.25/0.86M interface *(trans* Golgi fraction) and to the 0.86/1.5M interface (*cis* Golgi fraction). Keep stored at 4°C and analyze before 24 h.
9. Dilute Golgi fractions in DPBS to an equivalent to 100 µg protein/mL for flow cytometric analysis (*see* Notes 3–6).

3.2. Flow Cytometry

3.2.1. Instrumental Settings and Parameter Selection

1. Excite the FITC- or TRITC-conjugated lectins with argon-ion laser or mercury lamp, both tuned at 488 nm.
2. Use a filter setting homologous to the one suggested here for the photomultiplier (PMT) arrangement of the Coulter EPICS Elite Cell Sorter (it may vary depending on the instrument): 488 nm dichroic + 488 band pass (in PMT1, right angle light scatter) + 488 nm blocking + 550 nm dichroic + 525 nm band pass (in PMT2, green fluorescence) + 575 nm dichroic + 610 long pass (in PMT3, red fluorescence) (*see* Note 7).
3. Estimate particle morphology by logarithmic amplification of forward angle light scatter (proportional to particle size) and right angle light scatter (dependent on particle texture). Quantitate lectin binding by logarithmic amplification of green (FITC) or red (TRITC) fluorescence.

3.2.2. Discrimination of Golgi Elements from Background Noise

1. Specific selection of Golgi particles based on light scatter is necessary, prior to the analysis of fluorescence. Because of their small size, these particles are not easily discriminated from background noise.
2. Incubate with a fluorescent lectin (e.g., FITC-Con A, 200 µg/mL, for 1 h at 37°C in DPBS).
3. Perform the backgating procedure as shown in Fig. 1.

3.2.3. Titration of Lectin-Specific Binding

1. To establish the optimal conditions for lectin binding, incubate Golgi fractions (100 µg protein/mL) for 1 h at room temperature with increasing concentrations of each conjugated lectin. Following incubation with the lectin, centrifuge each sample for 10 min at 15,000g and resuspend in 1 mL of DPBS, before flow cytometric analysis.
2. Nonspecific binding is estimated in aliquots preincubated for 1 h with excess of adequate inhibitory sugars.
3. Specific lectin binding is calculated as the difference between the mean fluorescence intensities of total (samples incubated with fluorescent lectin in the absence

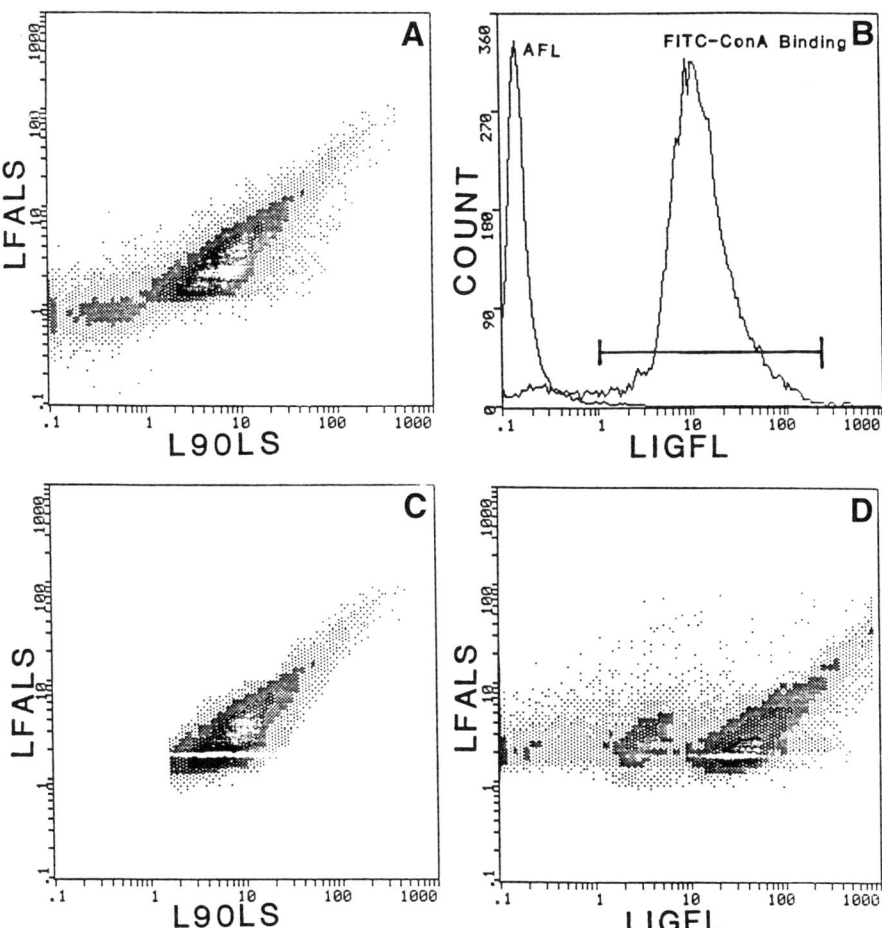

Fig. 1. Flow cytometric detection of isolated Golgi elements. **(A)** Biparametric histogram of forward angle light scatter (LFALS) vs right angle light scatter (L9OLS) showing the population of Golgi elements and background noise; **(B)** superimposition of histograms of green (LIGFL) autofluorescence (AFL) and FITC Con A fluorescence of Golgi fractions, showing the lectin binding population considered for backgating, included within a linear cursor; **(C)** biparametric histogram of LFALS vs L9OLS gated on LIGFL showing the cytometric morphological properties only of the elements included within the cursor in (B), which are able to bind FITC Con A and, thus, selected for fluorescence analysis; and **(D)** biparametric histogram of LFALS vs LIGFL showing that the morphological region selected indudes positive and negative Con A binding fractions in the Golgi population separated from noise. Thus, the morphological region shown in (C) is selected within an electronic bitmap and all the fluorescence measurements restricted to these elements. The described procedure should be performed for each individual preparation.

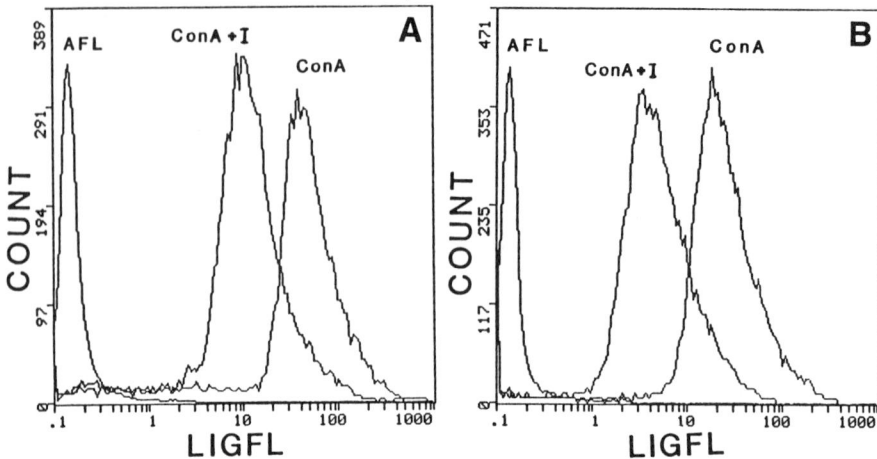

Fig. 2. Representative histograms of the binding of Con A to isolated *cis* (**A**) and *trans* (**B**) Golgi fractions: Each panel is a superimposition of three histograms: autofluorescence (AFL) of unlabeled fractions, nonspecific binding (Con A +I) of the lectin in samples preincubated with the presence of $0.01M$ α-D-methylmannose, and total binding (Con A) of the lectin without preincubation with the inhibitor. Con A binding shows an homogeneous population that shifts in the presence of the inhibitor.

of inhibitor) and nonspecific binding (samples incubated with fluorescent lectin in the presence of inhibitor) in monoparametric histograms as shown in Fig. 2. When no difference in the mean fluorescence of total and nonspecific binding is observed, Overton substraction of histograms must be applied *(22)*.

4. For performing flow cytometric analysis with lectins, select the concentration giving the maximal degree of specific binding in the titration curve, as exemplified in Fig. 3. As a guide, Table 1 shows the concentrations of lectins and competitive sugars that we found optimal in our experiments with isolated Golgi fractions.

4. Notes

1. Handle lectins with care, since some of them may be mitogenic or toxic.
2. Perform all the steps involving fluorescent lectins in the dark or under subdued light. Keep the stock solutions protected with aluminium foil.
3. The purity of the Golgi preparations may be assessed by biochemical analysis in the homogenate and in the fractions of the following enzyme markers: 5'-nucleotidase (plasma membrane), glucose 6 phosphatase (endoplasmic reticulum), acid phosphatase (lysosomes), alcohol dehydrogenase (cytosol), glutamate dehydrogenase (mitochondria) *(20)*.
4. The relative enrichment in *cis* and *trans* Golgi in the fractions may be quantified by assaying galactosyl transferase, a specific marker of *trans* Golgi *(20)*.

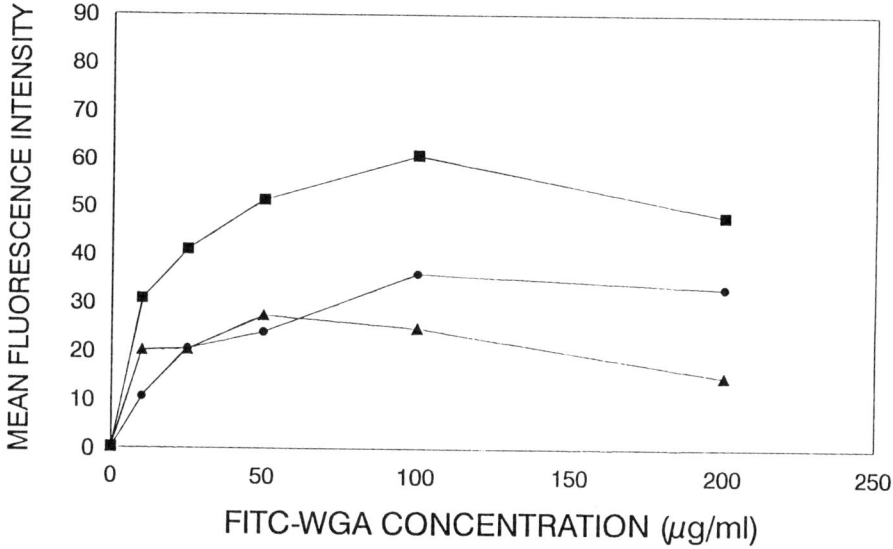

Fig. 3. Characterization of the binding of FITC WGA to isolated Golgi elements: Total and nonspecific binding were determined in the absence and presence of the sugar inhibitor, respectively, as shown in Fig. 2. Specific binding was calculated by substracting the mean value of nonspecific binding from that of total binding. Each point represents the mean of four experiments. In this example, the optimal lectin concentration was shown to be 50 μg/mL. ■, total binding; ▲, specific binding; ●, nonspecific binding.

Table 1
Optimal Concentrations of Lectins and Inhibitors

Lectin	Inhibitor	Lectin concentration, μg/mL	Inhibitor concentration
Con A	α-D-methylmannose	200	$0.01M$
WGA	N-acetylglucosamine	50	$0.1M$
UEA	L-fucose	25	$0.1M$
PNA	D-galactose	75	$0.5M$

5. The cytochemical detection of thiamine pyrophosphatase, a *trans* Golgi marker, may be used to assess the purity of fractions by electron microscopy *(20)*.
6. Avoid freezing and thawing the preparations of Golgi subfractions, since this may disrupt the isolated vesicles and increase very much the background noise.
7. The flow cytometer settings recommended here have been tested and shown adequate for the range of Coulter EPICS instruments using low power (15 mW), air cooled argon laser and quartz flow chambers (EPICS Elite, Profile II and XL). Petit

et al. *(7,13)* have performed their flow cytometric analysis of lectin binding to isolated plant mitochondria, chloroplasts, and thylakoids using an EPICS V with open flow chamber, which demands a higher laser power (400 mW; ref. *13*).

8. To provide an adequate data rate (400–500 events/s), while avoiding particle coincidence during the flow cytometric analysis, a dilution of subcellular fractions equivalent to about 100 µg protein/mL is recommended.
9. When possible, store flow cytometric data in the form of list mode files, since this allows one to reanalyze data off line as desired, thus permitting gate relocation, different parameter combination, and so on.
10. All current flow cytometers have more than adequate software for performing on-line and off-line data analysis and there are several commercial sources for off-line analysis on MS-DOS, Windows, and Macintosh systems. However, several excellent freeware or shareware software for offline data analysis can be obtained through the internet. WinMDI is a list mode data analysis and display software for Windows 3.1, developed by Joseph Trotter at the Salk Institute. It is available by anonymous FTP to *flosun.sal.edu*, in the directory */pub/pc/*. MFI is a program that calculates median fluorescence intensities from list mode data and has been developed by Eric Martz, at the Department of Microbiology, University of Massachusetts: *emartz@microbio.umass.edu*. AutoGate is an analysis and classification software for Macintosh, developed by Gary Salzman at the Los Alamos National Laboratory and available by e mail at *gs@lanl.gov*.

References

1. Goldstein, I. J. and Poretz, R. D. (1986) Isolation, physicochemical characterization, and carbohydrate binding specificity of lectins, in *The Lectins: Properties, Functions and Applications in Biology and Medicine* (Liener, I. E., Sharon, N., and Goldstein, I. J., eds.), Academic, New York, pp. 22–247.
2. Damianov, I. (1987) Biology of disease. Lectin cytochemistry and histochemistry. *Lab. Invest.* **57**, 5–20.
3. Shapiro, H. M. (1995) *Practical Flow Cytometry,* 3rd. ed. Liss, New York.
4. Kraemer, P. M., Tobey, R. A., Van Dilla, M. A. (1973) Flow microfluorimetric studies of lectin binding to surfaces of living cells. I. General features. *J. Cell Physiol.* **81**, 305–314.
5. McCoy, M. P., Jr. (1987) The application of lectins to the characterization and isolation of mammalian cell populations. *Cancer Metastasis Rev.* **6**, 595–613.
6. Shapiro, H. M. (1995) *Practical Flow Cytometry,* 3rd. ed. Liss, New York, pp. 298,299.
7. Mellor, R. B., Krusuius, T., and Lord, J. M. (1980) Analysis of glycoconjugate saccharides in organelles isolated from castor bean endosperm. *Plant Physiol.* **67**, 470–473.
8. Petit, P. X., Diolez, P., Muller, P., and Brown, S. C. (1986) Binding of concanavalin A to the outer membrane of potato tuber mitochondria detected by flow cytometry. *FEBS Lett.* **196**, 65–70.

9. Paulson, J. C. (1989) Glycoproteins: what are the sugar chains for? *Trends Biochem. Sci.* **14,** 272–276.
10. Curatolo, W. (1987) Glycolipid function. *Biochim. Biophys. Acta* **906,** 137–160.
11. Robinson, J. P. (ed.) (1993) *Handbook of Flow Cytometry Methods,* Wiley-Liss, New York.
12. O'Connor, J. E., Vargas, J. L., Kimler, B. F., Hernandez Yago, J., and Grisolia, S. (1988) Use of rhodamine 123 to investigate alterations in mitochondrial activity in isolated mouse liver mitochondria. *Biochem. Biophys. Res. Commun.* **151,** 568–573.
13. Schroder, W. P. and Petit P. X. (1992) Flow cytometry of spinach chloroplasts: determination of intactness and lectin binding properties of the envelope and the thylakoid membranes. *Plant Physiol.* **100,** 1092–1102.
14. Guasch, R. M., Guerri, C., and O'Connor, J. E. (1993) Flow cytometric analysis of concanavalin A binding to isolated Golgi fractions from rat liver. *Exp. Cell. Res.* **207,** 136–141.
15. Guasch, R. M., Guerri, C., and O'Connor, J. E. (1995) Study of surface carbohydrates on isolated Golgi subfractions by fluorescent lectin binding and flow cytometry. *Cytometry* **19,** 112–118.
16. Rothman, J. E. and Orci, L. (1992) Molecular dissection of the secretory pathway. *Nature* **355,** 409–415.
17. Mellman, I. and Simmons, K. (1992) The Golgi complex: in vitro veritas? *Cell* **68,** 829–840.
18. Lucocq, J. M., Berger, E. G., and Roth, J. (1987) Detection of terminal N-linked N-acetylglucosamine residues in the Golgi apparatus using galactosyltransferase and endoglucosaminidase F/peptide N glycosidase F: adaptation of a biochemical approach to electron microscopy. *J. Histochem. Cytochem.* **35,** 67–74.
19. Tartakoff, A. M. and Vassalli, P. (1983) Lectin binding sites as markers of Golgi subcompartments: proximal to distal maturation of oligosaccharides. *J. Cell Biol.* **97,** 1243–1248.
20. Guasch, R. M., Renau Piqueras, J., and Guerri, C. (1992) Chronic ethanol consumption induces accumulation of proteins in the liver Golgi apparatus and decreases galactosyltransferase activity. *Alcohol Clin. Exp. Res.* **16,** 942-948.
21. Brew, R., Brew, H., and Palade, G. E. (1980) Distribution of terminal glycosyltransferases in hepatic Golgi fractions. J. *Cell Biol.* **84,** 87–101.
22. Overton, W. R. (1988) Modified histogram technique for analysis of flow cytometry data. *Cytometry* **9,** 619–626.

VII

Lectins as Tools for Cell Purification/Purging

27

Lectins as Tools for the Purification of Liver Endothelial Cells

Daniel E. Gomez and Unnur P. Thorgeirsson

1. Introduction

Successful procedures for the isolation and culture of large-vessel endothelial cells (EC), were first reported in the early seventies (1,2). Since then, microvascular EC have been isolated from various organs, such as adrenal gland (3), brain (4), skin (5), retina (6), and myocardium (7). The initial steps of the conventional methods for EC isolation involve mechanical and/or enzymatic dissociation of the tissues, followed by filtration and pelleting of cells. A number of special techniques have been developed to eliminate contaminating cell types and enrich endothelial cells in mixed cell populations. These include: manual removal of nonendothelial cell types; use of selective media; plating cells on gelatin or fibronectin-coated dishes, and Percoll gradient centrifugation (8,9). The main problem with the conventional methods is that they are labor intensive and often do not produce pure EC populations. A more advanced approach is to use fluorescent-activated cell sorting (FACS), which allows sorting based on specific surface antigens or metabolic differences. Auerbach et al. used FACS for EC isolation, using an antibody against angiotensin converting enzyme (10). Later, Voyta et al. sorted EC, based on their uptake of acetylated low-density lipoprotein (11). Cell separation techniques using magnetic affinity are based on similar principles as the FACS, but do not involve expensive equipment. In this chapter, we describe liver endothelial cell isolation, using lectin-coated magnetic beads (12).

1.1. Lectins as Markers of Endothelial Cells

Vascular EC surface carbohydrates act as ligands for mammalian lectins. Human vascular EC possess L-fucose groups on their surface; these groups

From: *Methods in Molecular Medicine, Vol. 9: Lectin Methods and Protocols*
Edited by: J. M. Rhodes and J. D. Milton Humana Press Inc., Totowa, NJ

bind to *Ulex europaeus* I (UEA-I) lectin *(13)*, which is selective for primate EC *(14,15)*. In contrast, *Griffonia simplicifolia* (GSL) lectin and the isolectin GSL-B4 bind to murine EC and most other species other than humans *(16)*. *Lycopersicon esculentum* exclusively recognizes endothelium of small vessels in rats *(17)* and *Evonymus europaeus* agglutinin (EEA) binds to EC of monkey, swine, ox, sheep, dog, and human *(14)*.

It is of interest to note that different galactose-binding lectins have been isolated from mammalian tissues *(18)*, and that sequence homology has been found between C-type of animal lectins and the endothelial adhesion molecules, ELAM-1 and GMP-140 *(19,20)*.

1.2. Magnetic Affinity Cell Sorting

Magnetic cell sorting offers a rapid and inexpensive method for isolation of a single cell population from a mixture of cells *(21)*. For EC isolation, we used 4.5 μ magnetic polystyrene-coated microspheres (M-450 Dynabeads). Dynabeads, which were developed by Ugelstad and coworkers *(22)*, have hydroxyl groups on their surface, allowing covalent binding of antibodies or antibody binding molecules *(23)*. They are commercially available from Dynal (Oslo, Norway).

1.3. Isolation of Liver Endothelial Cells

Several methods have been devised to purify EC from liver after enzymatic digestion with collagenase, pronase, or both. Liver EC have been isolated by elutriation *(24)* or velocity sedimentation after pronase disaggregation *(25)*. Since pronase destroys certain cell surface receptors, differential centrifugation methods after collagenase perfusion were developed *(26)*, including Nycodenz centrifugation *(27)*, sedimentation above Metrizamide *(28)*, and Percoll separation *(29)*, combined in some cases with special culture conditions *(30)*. All of these methods led to EC enrichment in the range of 80–95%. However, even a small number of contaminating cell types, such as fibroblasts, will rapidly overgrow the EC in culture. When employing the magnetic cell sorting technique described in this chapter as a final purification step, it was possible to attain 100% purity of liver endothelial cells.

1.4. Outline of the Procedure

Briefly, the major procedures involved in this protocol are the following (Fig. 1):

1. Evaluation of the lectin-binding capacity of the target cells.
2. Preparation of lectin-coated magnetic beads.
3. Preparation of target-cell enriched fraction.
4. Immunomagnetic separation.
5. Removal of magnetic particles.
6. Evaluation of purity.

Purification of Liver Endothelial Cells

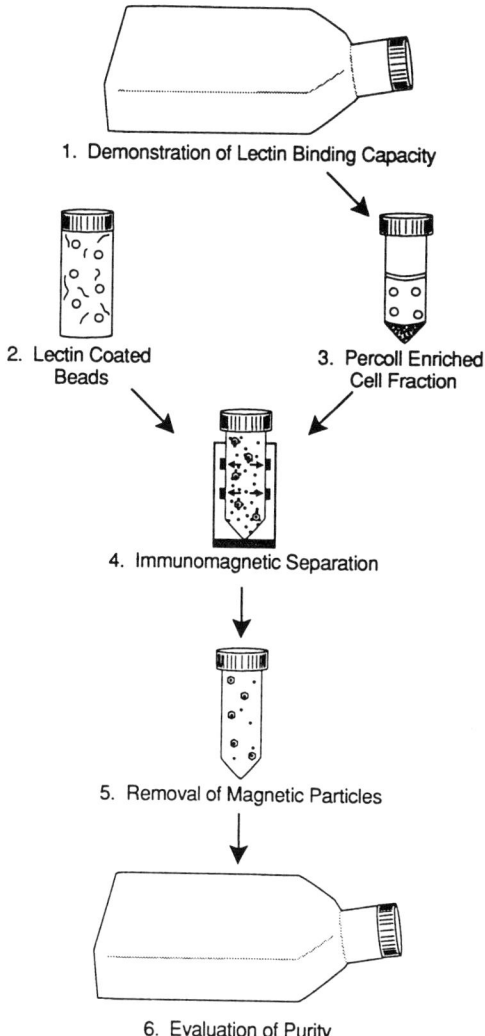

Fig. 1. Schematic outline of liver endothelial cell purification using lectin-coated magnetic beads.

2. Materials
2.1. Lectin Immunofluorescence
1. Fluorescein-labeled EEA lectin (Sigma, St. Louis, MO).
2. Ham F-12 K medium (Biofluids, Rockville, MD) with 10% FCS.
3. 1X PBS.
4. 4% Paraformaldehyde in PBS.

5. EC growth medium: 350 mL Ham's F12 K media, 150 mL FBS, 10 mL glutamine, 2 mL bovine brain extract (Clonetics, San Diego, CA) containing approx 3 mg/mL of protein, 0.5 mL 10 µg/mL epidermal growth factor (Clonetics), 0.5 mL hydrocortisone, 25 mg Gentamicin sulfate, and 25 µg Amphotericin-B.

2.2. Preparation of Lectin-Coated Magnetic Beads

1. EEA lectin: 0.1 mg of the lectin is dissolved in 0.5 mL of $0.5 M$ borate solution at pH 9.5.
2. Tosylactivated Dynabeads (*see* Note 1). An average diameter of 4.5 µ is ideal for cell separation. The beads are supplied in a solution of phosphate-buffered saline (PBS) and 0.1% bovine serum albumin (BSA) at a concentration of 4×10^8 beads/mL.
3. $0.1 M$ PBS containing 0.1% BSA.
4. Hanks' balanced salt solution (HBSS) plus 5% fetal bovine serum (FBS).

2.3. Preparation of Liver EC Enriched Fraction

1. HBSS with $0.01 M$ HEPES, pH 7.4.
2. 300 mL Williams E medium containing 200 mg collagenase type IV. Prepare fresh each time.
3. Nylon gauze (mesh width, 50 µm).
4. 1X Dulbecco PBS (DPBS).
5. 10X DPBS.
6. Percoll.
7. 1X PBS.
8. HBSS, 5% FBS.

2.4. Immunomagnetic Separation

1. Magnetic particle concentrator (MPC).
2. HBSS, 5% FBS.

2.5. Removal of Magnetic Particles

1. Magnetic particle concentrator (MPC).
2. HBSS, 5% FBS containing $0.1 M$ L-fucose.
3. HBSS, 5% FBS.
4. EC growth medium.

2.6. Evaluation of Purity

1. Acetylated low-density lipoprotein (Biomedical Technologies, Stoughton, CA) labeled with 1,1'-dioctadecyl-3,3,3',3'-tetramethylindocarbocyanine perchlorate (DiI-Ac-LDL).
2. EC growth medium.

3. Methods
3.1. Lectin Immunofluorescence

1. Take 100 μL of enriched cell fraction and dilute it with 900 μL of endothelial cell growth medium. Seed 500 μL per well in a 24-well plate.
2. Allow the cells to attach overnight.
3. Discard the medium and wash twice with PBS.
4. Incubate the cells with 25 μg of fluorescein-labeled EEA in Ham F-12 K medium with 10% FCS for 20 h.
5. Wash three times with PBS.
6. Fix the cells with 4% paraformaldehyde in PBS for 15 min.
7. Observe under fluorescence microscope (*see* Note 2).

3.2. Preparation of Lectin-Coated Magnetic Beads

1. Add 0.5 mL of tosylactivated beads (2×10^8 beads) to 0.5 mL of EEA solution (*see* Note 3).
2. Rotate the mix slowly end-over-end at 22°C for 24 h.
3. Collect the beads with the MPC. Allow the beads to attach to the magnet for 1–2 min. Discard the supernatant without disturbing the beads.
4. Add 5 mL of $0.01M$ PBS containing 0.1% BSA. Wash with end-over-end rotation for 20 min at room temperature.
5. Repeat steps three times.
6. Incubate the magnetic beads with $0.01M$ PBS containing 0.1% BSA. Wash with end-over-end rotation at 4°C overnight.
7. Resuspend the washed beads in Hanks' balanced salt solution (HBSS) containing 5% FBS at a final concentration of 4×10^8 beads/mL (*see* Note 4).

3.3. Preparation of Liver EC Enriched Fraction

1. Mince 100 g of freshly harvested liver tissue under sterile conditions.
2. Wash with 50 mL of HBSS and $0.01M$ HEPES, pH 7.4. Centrifuge at $100g$ for 5 min. Repeat three times.
3. Incubate the cells in 300 mL of Williams E medium containing 200 mg collagenase type IV at 37°C for 30 min.
4. After disruption of liver fragments, filter the suspension through nylon gauze (mesh width, 50 μm) to remove debris and cell aggregates.
5. Wash three times with 1X DPBS.
6. Remove the hepatocyte fraction by centrifugation at 50 g for 2 min at 4°C. Save the supernatant and discard the pellet.
7. Prepare a stock isotonic Percoll (SIP) by mixing 5 mL of 10-fold concentrated DPBS and 45 mL of Percoll.
8. Furnish four 50-mL Falcon tubes with a two-step Percoll gradient. The bottom cushion should consist of 7.5 mL SIP and 7.5 mL PBS. The upper cushion should consist of 5 mL SIP and 15 mL PBS (*see* Note 5).

9. Adjust the supernatant from step 7 to a total volume of 40 mL with 1X DPBS. Layer 10 mL on top of the preformed Percoll gradients.
10. Centrifuge the tubes at 800g at 4°C for 15 min.
11. Remove and discard the upper 15 mL of the top Percoll cushion in each tube. Collect the next 5 mL from each tube. This fraction is enriched in EC.
12. Dilute the recovered fraction with an equal volume of PBS, and centrifuge at 800g for 10 min.
13. Resuspend the pellet in HBSS, 5% FBS (6×10^5 cells/mL).

3.4. Immunomagnetic Separation

1. Take 100 µL of the enriched cell suspension and mix it with 30 µL of EEA-coated beads (*see* Note 6).
2. Incubate with end-over-end rotation for 15 min at 4°C (*see* Note 7).
3. Collect the cells attached to the EEA-coated beads with the MPC for 1 min. Discard the supernatant (*see* Note 8).
4. Add 5 mL of HBSS, 5% FBS. Rotate for 1 min. Apply magnetic force for 1 min. Discard supernatant. Repeat seven times (*see* Note 9).
5. Resuspend the cells with 1 mL of HBSS, 5% FBS.

3.5. Removal of Magnetic Particles

1. *See* Note 10, and Fig. 2.
2. Apply magnetic force to the washed cells for 1 min. Discard the supernatant.
3. Resuspend the cells in 1 mL of HBSS, 5% FBS with 0.1M fucose. Rotate end-over-end for 10 min at room temperature.
4. Wash five times with 1 mL of HBSS, 5% FBS and collect with the MPC. Collect the supernatant from each wash, pool it, and centrifuge it at 100g for 5 min.
5. Resuspend the cells in EC growth media. Incubate at 37°C, 5% CO_2.

3.6. Evaluation of Purity

1. Aseptically dilute the DiI-Ac-LDL to 10 µg/mL in EC growth medium (*see* Note 11).
2. Remove the medium from the selected cells and add the diluted DiI-Ac-LDL. Incubate for 4 h at 37°C.
3. Remove DiI-Ac-LDL and wash five times with EC growth medium.
4. Observe using standard rhodamine excitation-emission filters. EC should display an intense uniform cytoplasmic staining (Fig. 3).

4. Notes

1. Tosylactivated particles are uncoated and activated magnetic beads for the covalent binding of secondary antibodies and other proteins (i.e., lectins).
2. Positive cells should present a peripherial ring of fluorescence with the nuclei being consistently negative.
3. Make a uniform suspension of the magnetic particles before using. Vortex briefly if necessary.

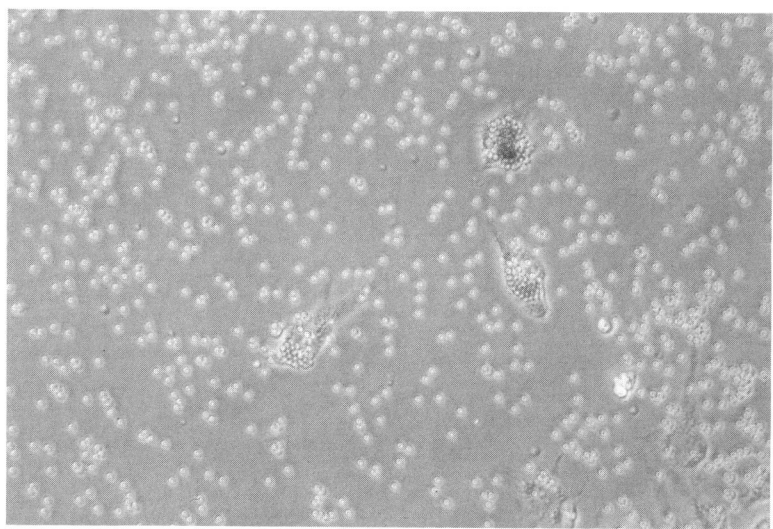

Fig. 2. Liver endothelial cells in culture 30 h after selection with EEA-coated beads. Note that a few beads still remain attached to the cell.

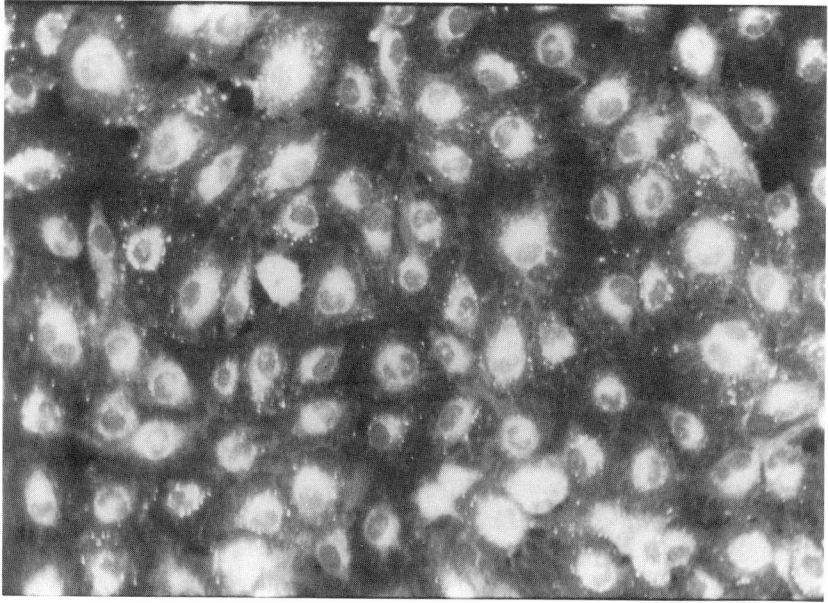

Fig. 3. Characterization of endothelial cells selected by EEA-coated beads using acetylated LDL.

4. Lectin-coated magnetic particles may be stored at 4°C for at least 6 mo. Sodium azide (0.02%) (final concentration) may be used as a preservative. Before use, the beads should be washed to eliminate the preservatives.
5. Caution should be exerted not to mix the two Percoll cushions. The upper cushion should be added dropwise along the wall of the tube.
6. For positive selection, a relatively low bead-to-cell ratio is to be used (3:1). However, negative selection requires higher bead to cell ratio (10:1).
7. Short incubation time at a low temperature (4°C) is essential to avoid phagocytic response, especially when dealing with EC.
8. When collecting the supernatant, avoid disturbing the particles attaching to the flask wall.
9. Examine the recovered beads under the microscope. If there is a substantial number of cells not attached to magnetic particles, the number of washes should be increased.
10. An incubation at 37°C overnight will result in spontaneous release of the magnetic beads from the selected cells, except for a few beads that will remain, and in our experience they do not alter the expected in vitro behavior of the cells (Fig. 2). However, if remaining beads are unacceptable, release of beads from the cells should be carried out before culture.
11. LDL products tend to aggregate. These aggregates can interfere with the visualization of endothelial cells. For this reason it is convenient to spin in a microfuge for 2 min the diluted DiI-Ac-LDL.

References

1. Jaffe, E. A., Nachman, R. L., Becker, C. G., and Minick, C. R. (1973) Culture of human endothelial cells derived from umbilical veins. Identification by morphologic and immunologic criteria. *J. Clin. Invest.* **52,** 2745–2756.
2. Gimbrone, M. A., Jr. (1975) Culture of vascular endothelium, in *Progress in Hemostasis and Thrombosis,* vol. 3 (Spaet, T., ed.) Grune Stratton, Orlando, FL, pp. 1–28.
3. Del Vecchio P. J., Ryan, U. S., and Ryan, J. W. (1977) Isolation of capillary segments from rat adrenal gland. *J. Cell Biol.* **75,** 73a, abstract no. CU041.
4. Brendel, K., Meezan, E., and Carlson, E. C. (1974) Isolated brain microvessels: a purified, metabolically active preparation from bovine cerebral cortex. *Science* **185,** 953–955.
5. Folkman, J., Haudenschild, C. C., and Zetter, B. R. (1979) Long-term culture of capillary endothelial cells. *Proc. Natl. Acad. Sci. USA* **76,** 5217–5221.
6. Buzney, S. M. and Massicotte, S. J. (1979) Retinal vessels: proliferation of endothelium in vitro. *Invest. Opthalmol. Vis. Sci.* **18,** 1191–1195.
7. Gerritsen, M. E. and Cheli, C. D. (1983) Arachidonic acid and prostaglandin endoperoxide metabolism in isolated rabbit and coronary microvessels and isolated and cultivated coronary microvessel endothelial cells. *J. Clin. Invest.* **72,** 1658–1671.
8. Balconi, G. and Dejana, E. (1986) Cultivation of endothelial cells: limitations and perspectives (review). *Medical Biol.* **64,** 231–245.

9. Gerritsen, M. E. (1988) Microvascular endothelial cells: isolation, identification, and cultivation. *Adv. Cell Cult.* **6**, 35–67.
10. Auerbach, R. L., Alby, J., Grieves, J., Joseph, C., Lindgren, L. W., Morrissey, Y. A., Sidky, M., Tu, Z., and Watt, S. L. (1982) Monoclonal antibody against angiotensin-converting enzyme: its use as a marker for murine, bovine and human endothelial cells. *Proc. Natl. Acad. Sci. USA* **79**, 7891–7895.
11. Voyta, J. C., Via, D. P., Butterfield, C. E., and Zetter, B. R. (1984) Identification and isolation of endothelial cells based on their increased uptake of acetylated low-density lipoprotein. *J. Cell Biol.* **99**, 2034–2040.
12. Gomez, D. E., Hartzler, J. L., Corbitt, R. H., Nason, A. M., and Thorgeirsson, U. P. (1993) Immunomagnetic separation as a final purification step of liver endothelial cells. *In Vitro Cell Dev. Biol.* **29A**, 451–455.
13. Holthofer, H., Virtanen, I., Kariniemi, A., Hormia, M., Linder, E., and Miettinen, A. (1982) *Ulex Europaeus* I lectin as a marker for vascular endothelium in human tissues. *Lab. Invest.* **47**, 60–66.
14. Roussel, F. and Dalion, J. (1988) Lectins as markers of endothelial cells; comparative study between human and animal cells. *Lab. Anim.* **22**, 135–140.
15. Jackson, C. J., Garbett, P. K., Nissen, B., and Schrieber, L. (1990) Binding of human endothelium to *Ulex europaeus* I coated Dynabeads: application to the isolation of microvascular endothelium. *J. Cell. Sci.* **96**, 257–262.
16. Laitinen, L. (1987) *Griffonia simplicifolia* lectins bind specifically to endothelial cells and some epithelial cells in mouse tissues. *Histochem. J.* **19**, 225–234.
17. Porter, G. A., Palade, G. E., and Milici, A. J. (1990) Differential binding of the lectins Griffonia simplicifolia I and Lycopersicon esculentum to microvascular endothelium: organ-specific localization and partial glycoprotein characterization. *Eur. J. Cell. Biol.* **51**, 85–95.
18. Kataoka, M. and Tavassoli, M. (1985) Identification of lectin-like substances recognizing galactosyl residues of glycoconjugates on plasma membrane of marrow sinus endothelium. *Blood* **65**, 1163–1171.
19. Bevilacqua, M. P., Stengelin, S., Gimbrone, M. Y., and Seed, B. (1989) Endothelial leukocyte adhesion molecule 1: an inducible receptor for neutrophils related to complement regulatory proteins and lectins. *Science* **243**, 1160–1164.
20. Johnston, G., Cook, R., and MacEver, R. P. (1989) Cloning of GMP-140, a granule membrane protein of platelets and endothelium: sequence similarity to proteins involved in cell adhesion and inflammation. *Cell* **56**, 1033–1044.
21. Padmanabhan, R., Corsico, C., Holter, W., Howard, T., and Howard, B. (1989) Purification of transiently transfected cells by magnetic-affinity cell sorting. *J. Immunogenet.* **16**, 91–102.
22. Kemshead, J. T. and Ugelstad, J. (1975) Magnetic separation techniques: their application to medicine. *Mol. Cell. Biochem.* **67**, 11–16.
23. Kemshead, J. T., Heath, L., Gibson, F. M., Katz, F., Richmond, F., Treleaven, J., and Ugelstad, J. (1986) Magnetic microspheres and monoclonal antibodies for the depletion of neuroblastoma cells from bone marrow: experiences, improvements, and observations. *Br. J. Cancer* **54**, 771–778.

24. Knook, D. L., Blanjaar, N., and Sleyter, C. H. (1977) Isolation and characterization of Kupffer and endothelial cells from the rat liver. *Exp. Cell. Res.* **109**, 317–329.
25. Morin, O., Patry, P., and Lafleur, L. (1984) Heterogeneity of endothelial cells of adult rat liver as resolved by sedimentation velocity and flow cytometry. *J. Cell. Physiol.* **119**, 327–334.
26. Carpena, T. J. and Garvey, J. S. (1982) Antigen handling in aging. II. The role of Kupffer and endothelial cells in antigen processing in Fischer 344 rats. *Mech. Ageing Dev.* **20**, 205–221.
27. Blomhoff, R., Smedsrod, B., and Eskild, W. (1984) Preparation of isolated liver endothelial cells and Kupffer cells in high yield by means of an enterotoxin. *Exp. Cell. Res.* **150**, 194–204.
28. Nagelkerke, J. F., Barto, K. P., and VanBerkel, T. J. C. (1983) *In vivo* and *in vitro* uptake and degradation of acetylated low-density lipoprotein by rat liver endothelial, Kupffer, and parenchymal cells. *J. Biol. Chem.* **258**, 12,221–12,227.
29. Nnalue, N. A., Shnyra, A., Hultenby, K., and Lindberg, A. A. (1992) *Salmonella choleraesuis* and *salmonella typhimurium* associated with liver cells after intravenous inoculation of rats are localized mainly in Kupffer cells and multiply intracellularly. *Infect. Immun.* **60**, 2758–2768.
30. Smedsrod, B. and Pertoft, H. (1985) Preparation of pure hepatocytes and reticuloendothelial cells in high yield from a single rat liver by means of Percoll centrifugation and selective adherence. *J. Leukocyte Biol.* **38**, 213–230.

28

The Use of Soybean Agglutinin (SBA) for Bone Marrow (BM) Purging and Hematopoietic Progenitor Cell Enrichment in Clinical Bone Marrow Transplantation

Arnon Nagler, Shoshana Morecki, and Shimon Slavin

1. Introduction

Soybean agglutinin (SBA) is a plant lectin, a glycoprotein of nonimmune origin that binds specifically to cell surface carbohydrates through noncovalent combinations and thus provokes agglutination of the bound cells. SBA has been used, therefore, to fractionate a variety of cell types. SBA binds approx 60–90% of bone marrow mononuclear cells, including mature myeloid, erythroid, and lymphoid cells, but has very low binding affinity and no toxic effect to the human hematopoietic progenitor cells. In addition, it binds very effectively to certain tumor cells, including neuroblastoma, breast cancer, and Burkitt's lymphoma cells. Based on these characteristics, SBA has multiple potential applications for clinical bone marrow, transplantation (BMT).

SBA has been used clinically to enrich bone marrow aspirates for early progenitor cells as a first step toward depletion of mature T-cells from the bone marrow graft in order to present graft-vs-host disease (GVHD). Effective T-cell depletion achieved by depletion of E-rosette forming cells in stem cell fractions enriched by SBA, enables the successful performance of bone marrow transplant across major histocompatibility complex (MHC) barriers, including successful outcome with no GVHD in recipients of haploidentically mismatched donors. In addition, elimination of bone marrow-derived fibroblasts, stromal cells, red blood cells, as well as mature myeloid and lymphoid cells by SBA enables use of SBA agglutination or binding as an effective separation step resulting in significant enrichment of the CD34$^+$ hematopoietic progenitor cells from both human bone marrow and umbilical cord blood. Positive $CD34^+$ selection is an attractive way of achieving an effective T-cell depletion of bone marrow graft and thus preventing GVHD and enabling haploidentical transplants. On the other hand, in the setting of autologous stem cell transplantation, SBA treatment of the graft can be used successfully for purging the autologous graft from contaminating tumor cells binding to SBA, including neuroblastoma breast cancer, and Burkitt's lymphoma in conjunction with enrichment of the $CD34^+$ progenitor cells.

SBA, which is a harmless and inexpensive compound, may thus be considered as an important molecule in experimental and clinical BMT with multiple potential applications for positive and negative selections. In our chapter, we will describe in detail the various practical applications of SBA in preclinical and clinical hematopoeitic cell transplantation in hemato-oncological malignant diseases.

The plant lectin, soybean agglutinin (SBA) is a tetrameric glycoprotein that, in its native form, exists primarily as β pleated sheets (1). Each of its four subunits are 30 kDa with an abundance of acidic and hydroxylic amino acid residues. One SBA molecule contains four carbohydrate binding sites which have the highest affinity for N-acetylgalactosamine and its derivatives (1). SBA has been used for a variety of purposes, e.g., to purify glycoproteins, to separate heterogeneous cell populations, and to purge tumor cells. In mice, SBA has been used to fractionate splenocytes into B-cell-enriched (SBA+) and T-cell-enriched (SBA–) fractions (2). Moreover, murine hematopoietic stem cells bind both SBA and peanut agglutinin (PEA), and, when isolated, will engraft allogeneic recipients without occurrence of graft vs host disease (3,4).

1.1. SBA Purging for Autologous Stem Cell Transplantation

In the last decade, autologous bone marrow transplantation (ABMT) has increasingly become the alternative treatment modality for patients with hemato-oncological malignancies, such as lymphomas and solid tumors. This therapy involves the use of high dosage chemoradiotherapy in conjunction with autologous marrow or peripheral blood stem cell rescue (5–13). Unfortunately, high relapse rates following autologous stem cell transplantation are unavoidable, partially owing to lack of graft vs leukemia/tumor (GVL) effect, and also potentially due to residual tumor cells in the graft (14,15).

Under certain conditions, heavy graft involvement with the tumor cells may become a major obstacle in improving relapse-free survival following autologous stem cell transplantation. However, in view of major limitations in eradicating tumor cells in the host, it is still unknown whether a small number of residual tumor cells reinfused with the autologous graft contribute to the relapse. It is worth nothing that the majority of relapses following high dose therapy and autologous stem cell transplantation occur in sites of prior bulky disease and relatively few new sites of relapse are usually seen. Most relapses may thus be caused by failure of the ablative regimen to eradicate residual disease in the patient. If reinfused tumor cells would have contributed to relapse, then new sites of relapse should be more frequently observed. Alternatively, reinfused tumor cells may preferentially lodge and proliferate in prior sites of bulk disease; these sites may provide the optimal microenvironment to support neoplastic cell growth. Purging of the bone marrow or peripheral blood graft from the contaminating tumor cells may reduce the relapse rate postautologous stem cell transplantation (ASCT), as was recently demonstrated unequivocally by Brenner et al. *(16)*. Contamination of the graft may contribute to the relapse post-ABMT, thus explaining why ex vivo purging of the graft resulted in significant improvement of the disease-free survival, at least in B-cell lymphoma patients *(17)*.

Lectins, including SBA are glycoproteins of nonimmune origin that can bind specifically to cell surface carbohydrates and, thus, provoke agglutination of these cells. This agglutination is reversible in the presence of an excess of the corresponding sugar. Lectins may therefore be used for positive or negative selection of contaminating tumor cells. Since SBA binds to cells transformed by viral or chemical agents, this compound that binds preferentially to tumor cells sparing the hematopoietic progenitor cells, may be used safely for effective purging.

1.1.1. SBA Purging for Neuroblastoma Stage IV

Neuroblastoma (NB) is the second most commonly diagnosed solid tumor in children *(18)*. Cure of advanced stage IV NB in children over 1 yr old remains a major challenge. Despite the increasing number of patients who enter complete remission, only <10% are long-term survivors *(19)*. The high relapse rate, even in patients who were successfully induced into complete remission (CR) by frontline therapy, indicates that conventional chemotherapy cannot eradicate stage IV disease and that a significant number of undetectable tumor cells remain after conventional chemotherapy.

Several reports have already described the use of high-dose chemotherapy followed by ABMT with response rates of 20–40%, and delayed relapse in comparison with conventional chemotherapy *(20–22)*. Assuming more successful eradication of tumor cells in vivo, elimination of undetectable minimal residual disease in the autograft may be important to improve disease-free survival time.

Neuroblastoma cells have been previously demonstrated to bind differentially to SBA *(23,24)*. Reisner et al. *(23)* have mixed in vitro human bone marrow cells with radioactively labeled neuroblastoma tumor cell lines ($n = 9$) and the cell mixtures were separated by differential agglutination with SBA. SBA treatment resulted in effective purging of NB tumor cells with varying efficiencies depending on the expression of SBA receptors (detected by flow cytofluorimetry with fluoresceinated SBA).

The agglutination step alone removes 64–76% of the radiolabeled NB cells and 85–98% of the clonogenic cells from the tumor/bone marrow cell mixtures *(24)*. Passage of the unagglutinated radiolabeled cells through SBA sepharose columns results in further purging of 28–53% of the NB cells. More than 98% of the cells of all the cell lines tested, specifically bound to the lectin, whereas no specific binding could be detected in the stem cell-enriched bone marrow cell fractions *(24)*.

Based on this background information we have auto transplanted eight children with stage IV neuroblastoma with SBA-purged marrow *(25)*. Out of the eight patients, five were males and three females, with median age of 3 yr (range, 2–8 yr). The interval between diagnosis and ABMT was 6–18 mo (mean, 10 mo) and 1–6 mo (mean, 3 mo) between cryopreservation of autologous bone marrow and the actual transplant. All children were treated with a combination of conventional chemotherapy, surgery, and local radiotherapy to eliminate local residual disease and to induce remission pre-ABMT. The chemoradiotherapy regimen given for induction of remission included vincristine, cyclophosphamide, cisplatin VM26, adriamycin, and VP16, in various combinations. Involved field radiation of 1200–1500 cGy was added on completion of the cytotoxic regimen in three patients. Bone marrow was harvested while the children were in CR, at which time an extensive staging procedure showed no evidence of disease and no marrow involvement. Marrow cells were purged with SBA and cryopreserved in 10% dimethyl sulfoxide and 10% heat-inactivated AB serum at −194°C in liquid nitrogen. High dose chemotherapy used for ABMT conditioning in two patients included vincristine (1.5 mg/m^2 bolus) followed by continuous infusion (0.5 mg/m^2/d for 5 d), melphalan (140 mg/m^2), and total body irradiation (200 cGy twice daily for a total of 1200 cGy). Six patients were conditioned with a newly designed polychemotherapy regimen (BETCAM) consisting of BCNU (300 mg/m^2 once), etoposide (200 mg/m^2/d for 4 d), thiotepa (0.6–1.2 mg/kg/d for 4 d), cytarabine (200 mg/m^2/d for 4 d), cyclophosphamide (60 mg/kg once), and melphalan (60 mg/m^2 twice).

All patients were transplanted in CR with no macroscopic evidence of disease in the bone marrow at the time of admission for transplant. At the time of transplant, 0.68×10^8 to 2×10^8 SBA-purged cells/kg (mean, 1.16×10^8 cells) were infused iv. Three of the eight children received granulocyte-macrophage colony stimulating factor (GM-CSF) (5 µg/kg) by continuous infusion from d 1 to d 12 post-BMT. Engraftment following purging with SBA was satisfactory with white blood cell counts (WBC) of $>1 \times 10^9$/L at d 10–20 (mean, 14), neutrophils (ANC) of $>0.5 \times 10^9$/L at d 10–23 (mean, 15), and platelets $>20 \times 10^9$/L at d 16–42 (mean, 19). Three years posttransplant, one of the patients developed carcinoma of the thyroid and 1 yr after that, papilloma was detected on the vocal cords. Two years later, he developed high-grade lymphoma and died. Two additional patients relapsed 8 and 11 mo post-BMT, respectively. The remaining patients are alive with no evidence of disease 38 mo to 9 yr (median, 2 yr) posttransplant *(25)*. These encouraging results with a relatively low rate of relapse suggests that purging of NB cells with SBA is feasible and safe. However, larger numbers of patients and longer follow-up periods are needed to assess the role of purging NB cells in improving disease-free survival. It seems unlikely that any obvious benefits will be demonstrable, unless better tumor eradication can be accomplished prior to ABMT/ASCT.

1.1.2. SEA Purging for Breast Cancer

ASCT following high dose chemotherapy has gained an increasing role in the treatment of high-risk breast cancer patients. A recent randomized study reported a survival advantage to those treated in this approach *(26)*. A retrospective long-term follow-up study in early stage breast cancer patients has shown a higher response rate and a better survival rate for those patients treated with high-dose chemotherapy (HDC) and ASCT compared to those treated with conventional doses of chemotherapy *(27)*. Although the results are encouraging, relapse rate is still worrisome in HDC+ASCT-treated patients *(26–28)*. The contribution of malignant breast cancer cells infused with the hematopoietic blood cells to the relapse rate is still unknown and will be difficult, if at all possible to assess. Several studies reported detection of tumor cells in otherwise normal BM *(29,30)* or peripheral blood (PB) collection *(31–33)* that ranged from 46–70% and 19–85%, respectively. The clinical importance of this marrow and peripheral blood tumor cells contamination is still questionable. Landys et al. *(34)* reported that the median overall survival was significantly shorter in tumor cell-positive patients being 1.9 yr compared to 11.7 yr in patients with no BM involvement. Mansi et al. *(35)* followed 82 women with localized disease by repeated BM aspirates and concluded that micrometastases in the bone marrow have questionable clinical importance. Sharp et al. *(36)* claimed that breast cancer patients undergoing peripheral blood stem cell transplantation in which the graft was free of tumor would have a better prognosis than those without tumor purging. In contrast, Ybanez et al. *(37)* found no difference in disease-free survival in 41 metastatic breast cancer patients undergoing HDC+ASCT in which residual tumor cells were detected in the BM or PB.

Several approaches for breast cancer purging have been reported, including pharmacological methods by using various chemotherapeutic agents, immunological methods (MAb), and physical methods, such as counterflow elutriation *(38–45)*. We have studied in an experimental model the ability of SBA to remove breast cancer cells from normal bone marrow cells. For this purpose 40×10^6 normal marrow cells were mixed with breast cancer tumor cells line (T-47D) (5, 10, and 20%) in order to mimic bone marrow aspirates of metastatic breast cancer patients *(46)*. We were able to show that T-47D tumor cells bound to SBA in a specific fashion that could be blocked by D-galactose (the specific binding was reduced from 92 to 5% in the presence of $0.9M$ D-galactose [determined by FACS analysis]). Tumor cells were effectively purged (2 orders of magnitude) by both SBA agglutination and depletion of cells bound to magnetic beads linked to SBA. A depletion of 3–4 orders of magnitude of tumor cells was consistently accomplished by combining one step of agglutination followed by one cycle of SBA-magnetic beads depletion, as 99.5% of the radioactive tumor-labeled cells were depleted by agglutinating SBA-positive cells *(46)*. There were no marked differences in degree of tumor cell depletion whether soluble SBA agglutination or SBA-bound to magnetic beads were used for purging. Neither procedure affects stem cell recovery assessed by CFU-GM colony growth which was evaluated before and after purging.

In order to evaluate the effect of our purging techniques on residual SBA-negative clonogenic tumor cells, we established the conditions for a tumor colony growth assay in agar *(47)*. Clonogenic efficiency was mostly above 65%. Contamination as low as 0.005% of tumor cells seeded in 2×10^5 BM cells could be detected by our clonogenic assay. This provided us with a more precise, biologically meaningful evaluation of residual tumor cells than could be obtained from the radiolabeling assay. The degree of tumor cell depletion after two cycles of SBA agglutination was 3.5 logs, as determined by immediate radiolabeling, whereas the clonogenic assay indicated a depletion of 4.3 logs. The clonogenic assay is of practical relevance, since it can provide indication of the growth potential of residual tumor cells following marrow purging in clinical autologous transplantation.

The use of SBA-magnetic beads has several advantages over direct agglutination by soluble SBA: Separation of the SBA-negative fraction is technically simple and faster. This aspect is of particular importance when large volumes are involved, as is the case in clinical situations in preparation for autologous BMT. Also, the yield of total nucleated cells is better than with soluble SBA. As with the case with direct agglutination by soluble SBA, two cycles of interaction with SBA-magnetic beads improved elimination of tumor cells (>4 logs) when compared with one cycle of separation (2–3.5 logs) and spared the GM-CFU hematopoietic cells *(47)*.

We have also tried to improve our breast cancer purging efficiency by using SBA in conjunction with chemocytotoxic agents like VP-16 and 4HC and MC-540 (a dye that mediates photosensitization) *(48)*. Treatment with high concentration of etoposide (VP-16) (10–80 μg/mL) resulted in a maximal depletion of 1.5 log of T-47D breast cancer tumor cells, whereas more efficient tumor cell eradication (2.5 log) was achieved by 30 min incubation with 100 μg/mL 4HC at 37°C. Photosensitization by exposure for 90 min to daylight in the presence of MC-540 could remove only 1 log of T-47D cells. Artificial mixtures containing 10–14% T-47D cells in fresh normal BM cells were subjected to SBA magnetic beads, and the SBA-negative fraction was further treated with 4HC. The combination of SBA and 4HC resulted in a consistent tumor cell depletion of >4.4 logs *(48)*.

Based on our extensive in vitro studies, we can conclude that SBA can serve as a useful and relatively inexpensive tool for selective depletion of breast cancer cells without compromising hematopoeitic reconstitution following ABMT. We subsequently used the SBA purging technique to treat the BM graft of 13 consecutive breast cancer patients (adjuvant stage II-III $n = 8$; metastatic $n = 5$) who underwent HD chemotherapy following by ABMT.

The mean interval from diagnosis to ABMT was 7.8 ± 2.4 mo in patients with stage II-III and 38.1 ± 26.3 mo in patients who had metastatic disease. Mean age at transplant was 40.6 ± 1.3 and 32.0 ± 2.7 yr for adjuvant and metastatic patients, respectively. For the adjuvant patients status at diagnosis was (T1-1, T2-4, T3-2, and Tx-1) with 13.4 ± 2.8 positive lymph nodes. They received five courses of CAF (cyclophosphamide, adriamycin and 5-fluoro-uracil) at diagnosis. Out of the five metastatic breast cancer patients, two were in complete remission at time of transplant, whereas three achieved only partial response.

The pretransplant conditioning regimen consisted of etoposide 200 mg/m^2 as 2 h infusion (on d-7, -5, and -3), thiotepa 10 mg/m^2 as 30 min infusion (d-7, -5, and -3), carboplatin 500 mg/m^2 as a 2 h infusion (d-6, -4, and -2), carmustine 300 mg/m^2 as 30 min infusion (d-4), and melphalan 60 mg/m^2 as 30 min infusion (d-4, -3) (BCTEM protocol). The patients were transplanted with $0.7 \pm 0.4 \times 10^8$ SBA purged bone marrow cells/kg. Granulocyte-macrophage-colony stimulating factor (GM-CSF) (5 μg/kg) was administered to all patients as a single daily SC injection from d 1 until neutrophil count exceeded 0.5×10^9/L for three consecutive days. Engraftment was normal with WBC $> 1 \times 10^9$/L on d 9.3 ± 1.0, ANC 0.5×10^9/L on d 10.4 ± 1.5, and platelets $>25 \times 10^9$/L on d 22.3 ± 10 posttransplant. No tumor recurrence was observed in the eight patients that underwent ABMT with SBA-purged marrow in the adjuvant setting, during a median follow-up period of 30 mo. Kaplan Meier survival and disease-free survival were 73% at 30 mo (Figs. 1 and 2). Out of the patients with metastatic disease, one died due to transplant-related toxicity and four

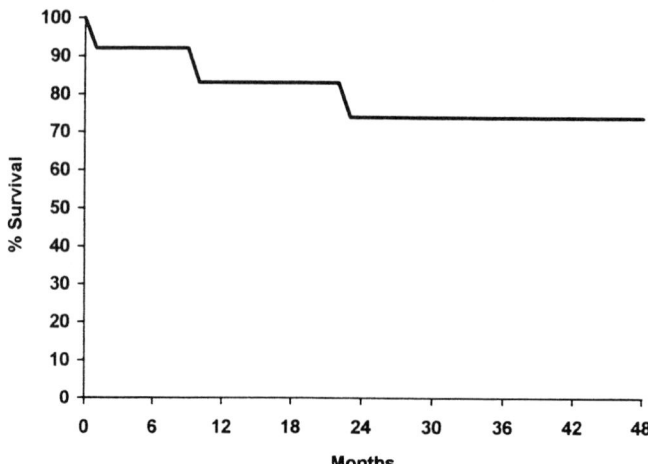

Fig. 1. Kaplan Meier survival of adjuvant cancer patients transplanted with SBA purged autologous BM.

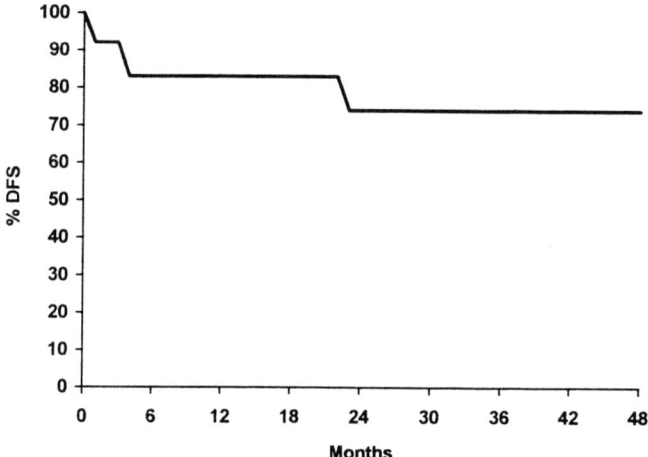

Fig. 2. Kaplan Meier disease-free survival (DFS) of adjuvant breast cancer patients transplanted with SBA purged autologous BM.

developed progressive disease and subsequently died with a median follow-up of 23 mo.

We conclude that breast cancer tumor cell purging by SBA is feasible without major side-effects, however, its effect on relapse and overall survival is going to be difficult if at all possible to document.

1.1.3. SBA Purging for Burkitt's Lymphoma (BL)

High-dose cytotoxic regimens followed by autologous stem cell transplantation have been successfully used to induce long-term remission in patients with advanced or relapsed Burkitt's lymphoma, resistant to conventional doses of chemotherapy *(49)*. Cryopreserved autologous BM cells may contain undetectable clonogenic tumor cells that might cause treatment failure due to relapse after ABMT. Ideally, BM should be harvested in complete remission and then purging may not be required. In reality, contamination of Burkitt's lymphoma (BL) in a remission marrow (defined by cytology) was found in 40% of the patients *(50)* when tumor cells were detected by a liquid-culture system *(49)*. Because BM may be harvested during partial remission or relapse, in patients with or without prior BM involvement, marrow purging may offer a reduced chance of reinfusion of BL cells following successful eradication of clonogenic cells by high-dose chemoradiotherapy given in the course of ABMT. Recently Gribben et al. *(17)* purged the bone marrow (by a cocktail of anti-B cell MAb) of 114 patients with B-cell non-Hodgkin's lymphoma in which *bcl*-2 translocations were detectable by PCR. A strong association between the ability to purge the bone marrow of residual lymphoma cells and disease-free survival after ABMT was observed. Out of the 114 patients, no lymphoma cells could be detected in the marrow of 57 patients after purging. Disease-free survival was increased in these 57 patients as compared with those whose marrow contained detectable residual lymphoma ($p < 0.00001$). The ability to purge residual lymphoma cells was not associated with the degree of BM involvement ($p = 0.45$) or the previous response to therapy ($p = 0.12$) *(17)*.

We have tested the potential use of SBA to purge BL cells from human bone marrow aspirate *(51)*. Four BL tumor cell tines (Raji, Daudi, BL69, and BL103) were mixed with normal human BM (10%) prior to addition of SBA for agglutination. The efficiency of BL tumor cell elimination was assessed by flow cytofluorometry and radioactive labeling. We achieved a 2-log depletion of BL tumor cells by a two-step purging procedure with SBA agglutination. Using SBA-coated magnetic beads, a 4 logs depletion was observed and the number of residual BL was below four cells in 10^5 BM cells, concomitantly with enrichment in the hematopoietic progenitor cell compartment *(51)*. This purging procedure for BL cells is yet to be examined in the clinic.

1.2. The Use of SBA for T-Cell Depletion in Allogeneic Bone Marrow Transplantation

1.2.1. Sibling Transplants

The prevention of GVHD is a major goal in allogeneic BMT *(14,52)*. The high incidence of GVHD in allogeneic BMT is a significant cause of morbidity and mortality, despite pharmacological prophylactic regimens. T-cell depletion is the only way for complete prevention of GVHD *(14,52)*. SBA has been previously demonstrated to effectively bind to helper T-cells and to deplete bone marrow allografts of T-cells *(53–56)*. Effective depletion of T-cells was achieved by combining the SBA agglutination with neuraminidase-treated sheep erythrocyte (SRBC) rosetting (E-rosette), eliminating 99% of the T-cells *(57,58)*. Using this technique, only <10% of the patients developed usually very mild (grade I-II) acute GVHD *(57)*. The NK and lymphokine-activated killer cell activity was found to be normal as early as 3 wk following SBA/SRBC T-cell depletion of allogeneic BMT *(59,60)*. However, the immunologic recovery of proliferative responses to IL-2 and the appearance of cells regulating in vivo activation of T-cells, appeared to be more delayed in patients receiving SBA/SRBC T-cell depleted BMT *(61)*.

1.2.2. Haploidentical Transplant

The application of allogeneic bone marrow transplantation (BMT) for the treatment of patients with hematological malignancies and other diseases is hampered by the lack of availability of suitable major histocompatibility complex MHC matched donors. In contrast, nearly all patients have an HLA-haploidentical relative (parent, child, sibling) who could serve as a donor. A successful T-cell depletion by SBA and SRBC enables transplantation of haplotypes mismatched marrow cells into patients with severe combined immunodeficiency (SCID) without apparent GVHD within the major limitation of mismatched BMT *(62–65)*. Cowan et al. reported on such patients *(65)*: Seven of the nine patients (78%) had documented T-lymphocyte engraftment based on HLA typing and/or chromosomal analysis. Six patients showed evidence of B-cell immunity on the basis of increased immunoglobulin levels, and/or HLA-DR typing of non-T-cells. Neutrophil engraftment occurred between 9 and 17 d posttransplantation, erythrocytes engrafted within 3–4 wk of transplantation, and platelet recovery was seen between d 17 and 49 following the transplants. Three patients had no GVHD, two had transient rash and/or fever, and two developed mild focal (stage 1) chronic cutaneous GVHD.

However, the use of T-cell-depleted haploidentical grafts in leukemia patients had poor success. Recently Aversa et al. have reported the successful engraftment of T-cell depleted haploidentical "three-loci incompatible" transplants in leukemia patients by addition of recombinant human granulocyte colony stimulating factor-mobilized peripheral blood progenitor cells to bone marrow inoculum *(66)*. The sole graft-vs-host disease (GVHD) prophylaxis consisted or T-cell depletion or the graft by the soybean agglutination and E-rosetting technique. Out of 17 patients (mean age, 23.2 yr; range. 6–51 yr) with end-stage chemoresistant leukemia, only one patient, who received a much greater quantity of lymphocytes than any other patient, died from grade IV acute GVHD There were no other cases of GVHD ≥ grade II. One patient rejected the graft and the other 16 had early and sustained full donor type engraftment. However, only six patients were alive and event free at a median follow-up of 230 d. The authors conclude that a highly immunosuppressive and myeloablative conditioning followed by transplantation of stem cells depleted of T-lymphocytes by soybean agglutination and E-rosetting technique, has made transplantation of three-HLA-antigen disparate grafts possible, with only rare cases of GVHD.

1.3. The Use of SBA for CD34⁺ Hematopoietic Progenitor Cell Enrichment

1.3.1. Bone Marrow

CD34⁺ hematopoietic progenitor cells (HPC) have been previously demonstrated to produce colony-forming units in vitro, which allow establishment of long-term bone marrow cultures, and more importantly restore hematopoietic function following myeloablative therapy *(67)*. These cells are therefore very attractive candidates for both transplantation and gene therapy. Various attempts have been made over the years to isolate the human CD34⁺ HPC. It was previously shown by Reisner et al. and others both in the murine and human systems that SBA can be used for enrichment of hematopoietic progenitor cells *(68,69)*. Recently SBA was covalently attached to derivatized polystyrene tissue culture devices in order to achieve rapid, large-scale BM processing *(41)*. In preclinical experiments, $2-10 \times 10^9$ mononuclear bone marrow cells (BMMC) were loaded into the sterilized flasks and were allowed to settle onto the SBA-coated surface during a 1-h room temperature incubation. After incubation, the nonadherent (SBA⁻) fraction was removed and analyzed by FACS. An average of 26.2% of the BMMC were recovered in the nonbound SBA fraction *(41)*. Following SBA treatment a 2.3–5.8-fold enrichment was observed in CD16⁺ natural killer (NK) cells, and 1.9–3.3-fold enrichment in the HLA-DR⁺ cells in the nonbound SBA cell fraction, which suggest progenitor cell enrichment. Most importantly, CD34⁺ were enriched by 2.5-fold in the SBA(–) cell fraction, and the mean CD34⁺ cell recovery was 65.8% (SD 47.7) *(41)*. The SBA separation resulted in 5.0-fold enrichment in BFU-E (erythroid) colony growth and 4.9-fold enrichment in CFU-GEMM (multipotential) colony growth.

The SBA-negative BM fraction was further used for purification of the CD34⁺ HPC using polystyrene devices that were covalently immobilized with anti-CD34⁺ MAb. Following this final enrichment step it was demonstrated that 1.2–4.2% of the SBA(–) cells were CD34⁺ HPC (70–90% purity) *(41)*.

The potential utility of SBA in the selection of human stem cells is suggested by the fact that SBA-negative fraction is enriched in $CD34^+$ cells. Because $CD34^+$ cells only comprise 0.5–3% of BMMC *(70–72)*, the pre-enrichment of these cells using SBA can greatly facilitate the isolation and collection of a pure $CD34^+$ cell population.

1.3.2. Human Umbilical Cord Blood (HUCB)

Human umbilical cord blood represents a unique source of HPC, which is an attractive candidate for both transplantation and gene therapy *(73)*. Because of the relatively small number of nucleated cells and the fact that several reports *(74,75)* suggest that manipulation of cord blood samples resulted in a substantial loss of HPC, unseparated cord blood has been used for most of the reported human umbilical cord blood transplants. Development of efficient cell processing methods for processing HUCB is perhaps the most critical issue in determining the ultimate utility of HUCB for transplantation. Volume reduction of the cord blood for the purpose of cell banking is one of a number of benefits to be derived from HUCB manipulation *(76)*. Also, it would be desirable to remove the possibility of blood type (ABO) incompatibility reactions in the recipient of an ABO mismatched HUCB by adding a process for red blood cell (RBC) depletion, assuming the method will not result in unacceptably high loss of progenitor cells. Additionally, a process for partial T-cell depletion might be beneficial in some HUCB transplant settings where GVHD is a concern.

Based on the satisfactory enrichment of $CD34^+$ HPC from BM using SBA, we evaluated the use of SBA to maximize removal of RBC, achieving a partial T-cell depletion in an attempt to enrich $CD34^+$ HPC from HUCB *(77)*. As a first stage, lectin binding to cord blood RBC was assessed using six different lectins that cover a broad spectrum of different monosaccharide and oligosaccharide structures: elderberry bark lectin, jacalin, peanut agglutinin, *Maackia amurensis* I, *Maackia amurensis* II, and SBA, which recognize specific carbohydrate moieties (Table 1). It was found that SBA had the highest affinity for binding cord blood RBC, followed by *Maackia amurensis* II and *Maackia amurensis* I, whereas jacalin and elderberry bark lectin had very low affinity for binding cord blood RBC. Peanut agglutinin did not bind cord blood RBC at all (Table 1). Cord blood was sequentially processed by 3% gelatin followed by SBA covalently attached to polystyrene tissue culture flasks and evaluated for cell yield, viability, and phenotypic profile. The cell recovery (mean ± SEM) for the SBA depletion step was 48.4 ± 7.4%. Combined with the 72.4 ± 7.3% recovery of nucleated cells following 3% gelatin, the total precept recovery of nucleated cells after sequential treatment with 3% gelatin followed by SBA was 35.1%. The viability of the cells following sequential processing was (mean ± SEM) 93.0 ± 4.2%. FACS analysis showed a trend of increasing

Table 1
Lectin Affinities for Red Blood Cells (RBC) Present in Human Umbilical Cord Blood

Lectin	Carbohydrate recognized	RBC affinity
Elderberry bark lectin	Sialic acid linked to terminal galactose	+
Jacalin	Galactosyl (1,3) N-acetylgalactosamine	+
Peanut agglutinin	Galactosyl (1,3) N-acetylgalactosamine	0
Maackia amurensis I	Galactosyl (1,3) N-acetylglucosamine	++
Maackia amurensis II	Sialic acid in alpha 2,3 linkage	++/++
Soybean agglutinin	N-acetylgalactosamine and galactose	++

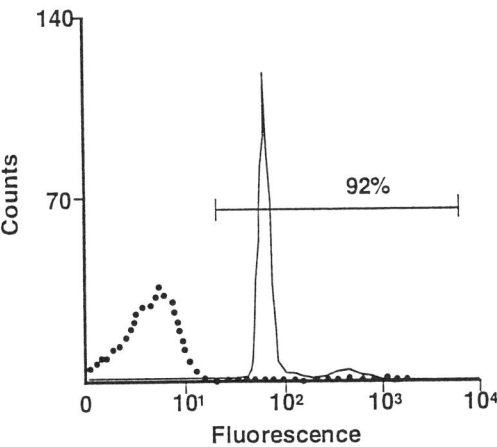

Fig. 3. FACS analysis of CD34$^+$ HPC separated from human umbilical cord blood.

percentages of CD38- and CD45-positive cells. There was no apparent loss of CD34$^+$ cells by the combined 3% gelatin and SBA cell processing steps. The 35% nucleated cell recovery after the two process steps resulted in a threefold enrichment of the percentage of CD34$^+$ cells in the SBA-negative fraction: 3.4 ± 1.3% post-SBA compared to 1.1 ± 0.2% in the unseparated cord blood.

We next tried to purify the CD34$^+$ HPC from the SBA(–) fraction using the same technique that we described above. Using CD34$^+$ MAb covalently attached to polystyrene tissue culture flasks, we were able to purify the HUCB CD34$^+$ HPC (Fig. 3). However, a wide range of purity was observed with different cord blood samples, ranging from 10–92% with a median of 19.8% in the final CD34$^+$ adherent fraction. The recovered CD34$^+$ cell fraction represented 0.92 ± 0.28% of the cord blood mononuclear cells and had a viability of

Table 2
Progenitor Assays[a] for Human Umbilical Cord Blood During CD34+ Enrichment Procedure, Reported as Colonies per 10^5 Cells Seeded

Fraction	BFU-E/10^5	CFU-GM/10^5	CFU-GEMM/10^5
Post ficoll	257 ± 110	81 ± 29	79 ± 40
Post SBA	557 ± 135	178 ± 5	272 ± 115
CD34+	3710 ± 430	740 ± 60	390 ± 210

[a]14-d methylcellulose cultures in the presence of IL-3, GM-CSF, and SCF.
Data are given as mean ± SEM.

84 ± 5% *(77)*. Plating these cells in clonogenic assays yielded 740 ± 60, 3710 ± 430, and 390 ± 210 CFU-GM, BFU-E, and CFU-GEMM/10^5 CD34+ cells, respectively (Table 2). This represented a 9-, 14-, and 5-fold enrichment for CFU-GM, BFU-E, and CFU-GEMM, respectively, in the CD34+ fraction relative to the cord blood mononuclear cell fraction. We conclude that SBA can be combined as an ex vivo processing technique to reduce the volume of the HUCB product by depleting RBC and T-cells while still maintaining a high recovery of HPC, which may enhance the utility of HUCB for storage and transplantation. Moreover, separation of highly enriched CD34+ cells from HUCB is achievable and opens up the possibility of using CD34+ cells from HUCB for ex vivo progenitor cell expansion for transplantation, transfusion, and gene therapy.

2. Materials

1. Phosphate-buffered saline (PBS) containing Ca^{2+}, Mg^{2+} (cPBS), pH 7.2 (*see* Note 1).
2. Soybean agglutinin (SBA) kindly supplied by Y. Reisner, Department of Biophysics, Weizmann Institute and Bio Yeda, Rehovot, Israel. Prepare aliquots of 2 mg/mL. Keep suspensions in –70°C (*see* Note 2).
3. 0.2*M* D-galactose serum.
4. Bovine serum albumin (BSA) (fraction V, Sigma, St. Louis, MO) 5% in cPBS.
5. Polystyrene magnetic beads (Dynabeads M-450, Dynal, Norway).
6. 4-Hydroperoxycylclophosphamide (4HC) is dissolved in PBS immediately before use. All solutions are sterilized by 0.2-μm disposable tissue culture filters (Nalgene, Nalge, Rochester, New York).

3. Methods

3.1. SBA Agglutination for Tumor Cell Depletion in Autologous BMT or for T-Cell Depletion in Allogeneic BMT (Originally established and described by Y. Reisner [23])

1. Mix heparinized BM cells with Hetastarch (Hespan) at a final concentration of 0.66%. Allow to settle until the erythrocytes have sedimented to half volume.

2. Collect the leukocyte-rich fraction, wash with PBS by centrifugation for 10 min at 300g.
3. Resuspend in cPBS containing 1% BSA at a cell concentration of 400×10^6/mL.
4. Mix 1 mL SBA (9 mg/mL) with 1 mL of cell suspension in 5-mL test tubes (*see* Note 3). Leave for 3 min for agglutination. Then, collect every three test tubes and load contents on top of 30 mL of 5% BSA solution in 50-mL test tubes (Greiner Labortechnik, Germany). Agglutinated cells sediment to the bottom within 5–10 min, while nonagglutinated cells remained at the top.
5. After sedimentation of agglutinated cells, remove the top fraction carefully with Pasteur pipets, and wash with $0.2M$ D-galactose (top fraction consists of hematopoietic cells, whereas bottom fraction includes mature human T-cells, or if present, tumor cells like neuroblastoma or breast cancer cells). Wash twice with PBS and than resuspend cells in saline solution for infusion.

3.2. Experimental Models

3.2.1. Binding of SBA to Magnetic Beads

1. Mix 105 mg polystyrene (PS) magnetic beads with 5 mg of SBA in a total volume of 6.6 mL of PBS for 16 h at 4°C by end-over-end rotation.
2. Collect Dynabeads by placing the suspension in a cobalt samarium magnetic particle concentrator (MPC, Dynal) and wash twice immediately with PBS and then twice, 2 h each, by end-over-end rotation at 4°C. Fifth wash is overnight in PBS containing 0.1% BSA.
3. Resuspend the SBA-PS-magnetic beads to a concentration of 30 mg/mL with 0.1% BSA/PBS and keep until use. The SBA-PS beads are stable at 4°C for at least 2 mo.

3.2.2. Depletion of Tumor Cells by SBA-Magnetic Beads

1. Mix 12.3 mg of SBA-PS with 20×10^6 BM cells containing 10% tumor cells in a total volume of 0.8 mL 0.1% BSA/PBS for 15 min at room temperature.
2. Remove magnetic beads by MPC, collect unbound cells, and wash once with 0.1% BSA/PBS.
3. For a second cycle of separation: Mix 0.9 mg of SBA-PS beads with 2×10^6 cells for 15 min at room temperature. Collect the unbound cells and wash twice with 0.1% BSA/PBS.

4. Notes

1. PBS solution: Ca^{2+} and Mg^{2+} are added to a final concentration of 0.19 µg/mL and 0.1 mg/mL, respectively.
2. Preparation of SBA stock solution should be adjusted to a final concentration of 2 mg/mL according to optical density (OD) reading at 280 nm (1 mg/mL SBA = 1.3 OD).
3. Try first in a small scale agglutination with your SBA stock solution (2 mg/mL) and the cell suspension (400×10^6/mL): take 50 µL of SBA stock solution and 50 µL of cell suspension. If agglutination is too fast (quicker than 2 min) and all

the cells are agglutinated immediately, you should decrease your SBA concentration. If however, there is a weak agglutination within 3 min, you should increase SBA concentration. Define optimal conditions of agglutination in small scale, before application to large volumes.

5. Summary

In this chapter, we described the possible uses of SBA for fractionation of a variety of cell types in the field of BMT. Then studies are based on extensive preclinical research. SBA is a molecule that has multiple possible applications for both allogeneic and autologous stem cell transplantation. SBA binds very effectively to most of the accessory cells in the BM without binding the human stem cells, and therefore it is a simple and elegant way to enrich for the $CD34^+$ hematopoietic progenitor cells.

References

1. Georgiou, G. M., Roberton, D. M., Ellis, Wnm, Shen, B. J., Ekert, H., and Hosking, C. S. (1983) CFU-C enrichment from human bone marrow using a discontinous percoll gradient and soybean agglutinin in comparison with ficoll-hypaque. *Clin. Exp. Immunol.* **53,** 491–496.
2. Reisner, Y., Ravid, A., and Sharon, N. (1976) Use of soybean agglutinin for the separation of mouse B and T lymphocytes. *Biochem. Biophys. Res. Commun.* **72,** 1585–1591.
3. Reisner, Y., Itzicovitch, L., Meshorer, A., and Sharon, N. (1978) Hematopoietic stem cell transplantation using mouse bone marrow and spleen cells fractionated by lectins. *Proc. Natl. Acad. Sci. USA* **75,** 2933–2936.
4. Reisner, Y., Ikehara, S., Hodes, M. Z., and Good, R. A (1980) Allogeneic hematopoietic stem cell transplantation using mouse spleen cells fractionated by lectins: in vitro study of cell fractions. *Proc. Natl. Acad. Sci. USA* **77,** 1164–1168.
5. McMillan, A. K. and Goldstone, A. H. (1991) Autologous bone marrow transplantation of non Hodgkin's lymphoma. *Eur. J. Haematol.* **46,** 129–135.
6. Vose, J. M. and Armitage, J. O. (1993) Role of ABMT in non-Hodgkin's lymphoma. Autologous bone marrow transplantation. *Hematol. Oncol. Clin. North Am.* **7,** 577–590.
7. Freedman, A. S., Takvorian, T., Neuberg, D., Mauch, P., Rabinowe, S. N., Anderson, K. C., Soiffer, R. J., Spector, N., Grossbard, M., and Robertson, M. J. (1993) Autologous BMT in poor-prognosis intermediate-grade and high-grade B-cell non-Hodgkin's lymphoma in first remission: a pilot study. *J. Clin. Oncol.* **11,** 931–936.
8. Freedman, A. and Nadler, L. M. (1992) BMT in low grade non-Hodgkin's lymphoma. *Marrow Transplant Rev.* **2,** 33,34.
9. Philip, T., Armitage, J. O., Spitzer, G., Chauvin, F., Jagannath, S., Cahn, J. Y., Colombat, P., Goldstone, A. H., Gorin, N. C., Flesh, M., et al. (1987) High dose therapy and ABMT after failure of conventional chemotherapy in adults with

intermediate grade or high grade non-Hodgkin's lymphoma. *N. Engl. J. Med.* **316**, 1493–1499.
10. Chopra, R., McMillan, A. K., Linch, D. C., Yuklea, S., Taghipour, G., Pearce, R., Patterson, K. G., and Goldstone, A. H. (1993) The place of high dose BEAM therapy and ABMT in poor risk Hodgkin's disease: a single center eight year study of 155 patients. *Blood* **81**, 1137–1145.
11. Vose, J. M., Bierman, P. J., and Armitage, J. O. (1990) Hodgkin's disease: the role of bone marrow transplantation. *Semin. Oncol.* **17**, 749–757.
12. Peters, W. P., Ross, M., Vredenburgh, J., Meisenberg, B., Rosner, G., Marks, L., Mathias, B., Henderson, C., Hurd, D., Budman, D., Norton, L., and Weiss, R. B. (1992) High dose alkylating agents and ABMT for stage II/III breast cancer involving 10 or more axillary lymph nodes. *Proc. Am. Soc. Clin. Oncol.* **11**, 58–65.
13. Meyers, S. E. and Williams, S. F. (1993) Role of high dose chemotherapy and autologous stem cell support in treatment of breast cancer. *Hematol. Oncol. Clin. North Am.* **7**, 631–645.
14. Slavin, S. and Nagler, A. (1991) New developments in bone marrow transplantation. *Curr. Opinion Oncol.* **3**, 254–271.
15. Varadi, G. and Nagler, A. (1994) Conditioning regimens in autologous bone marrow transplantation. *Clin. Immunother.* **2**, 342–351.
16. Brenner, M. K., Rill, D. R., Moen, R. C. Krance, R. A., Jr., J. M., Anderson, W. F., and Ihle, J. N. (1993) Gene marking to trace origin of relapse after autologous bone marrow transplantation. *Lancet* **341**, 85,86.
17. Gribben, J. G., Freedman, A. S., Neuberg, D., Roy, D. C., Blake, K. W., Woo, S. D., Grossbard, M. L., Rabinowe, S. N., Coral, F., Freeman, G. J., et al. (1991) Immunologic purging of marrow assessed by PCR before autologous bone marrow transplantation for B-cell lymphoma. *N. Engl. J. Med.* **325**, 1525–1533.
18. Young, J. L. and Miller, R. W. (1975) Incidence of malignant tumors in US children. *J. Pediatr.* **86**, 254–258.
19. Evans, A. K., D'Angio, G. J., Propert, K., Anderson, J., and Hann, H. L. (1987) Prognostic factors in neuroblastoma. *Cancer* **59**, 1853–1857.
20. Philip, T., Bernard, J. L., Zucker, J. M., Pinkerton, R., Lutz, P., Bordigoni, P., Plouvier, E., Robert, A., Carton, R., Philippe, N., et al. (1987) High dose chemotherapy with bone marrow transplantation as consolidation treatment in neuroblastoma: an unselected group of stage IV patients over 1 yr of age. *J. Clin. Oncol.* **5**, 266–271.
21. Hartmann, O., Benhamou, E., Beaujeane, F., Kalifa, C., Lejars, O., Patte, C., Behard, C., Flamant, F., Thyss, A., Deville, A., et al. (1987) Repeated high dose chemotherapy followed by purged autologous bone marrow transplantation as consolidation therapy in metastatic neuroblastoma. *J. Clin. Oncol.* **5**, 1205–1211.
22. Dini, G., Philip, T., Hartmann, O., Pinkerton, R., Chauvin, F., Garaventa, A., Lanino, E., and Dallorso, S. (1989) Bone marrow transplantation for neuroblastoma: a review of 509 cases. EBMT group. *BAT 4* (Suppl. 4), 42–46.
23. Reisner, Y. (1983) Differential agglutination by soybean agglutinin of human leukemia and neuroblastoma cell lines: potential application to autologous bone marrow transplantation. *Proc. Natl. Acad. Sci. USA* **80**, 6657–6661.

24. Reisner, Y. and Gan, J. (1985) Differential binding of soybean agglutinin to human neuroblastoma cell lines: potential application to autologous bone marrow transplantation. *Cancer Res.* **45,** 4026–4031.
25. Kapelushnik, J., Nagler, A., Or, R., Naparstek, Cividali, G., Aker, M., Mehta, J., Mumcuoglu, M., and Slavin, S. (1993) Autologous bone marrow transplantation for stage IV neuroblastoma: the role of soybean agglutinin purging. *Trans. Proceed.* 25, 2375,2376.
26. Bezwoda, W., Seymour, L., and Dansey, R. (1995) High dose chemotherapy with hematopoietic rescue as primary treatment for metastatic breast cancer: a randomized trial. *J. Clin. Oncol.* **13,** 2483–2489.
27. Peters, W. P., Berry, D., Vredenburgh, J. J., Hussein, A., Rubin, P., Elkordy, M., Ross, M., Henderson, I. C., Budman, D., Norton, L., Weiss, R., and Hurd, D. (1995) Five year follow-up of high dose combination alkylating agents with ABMT as consolidation after standard dose CAF for primary breast cancer involving >10 axillary lymph nodes (Duke CALGB 8782). *ASCO Proceedings* **933** (abstract).
28. Ayash, L., Wheeler, C., Fairclough, D., Schwartz, G., Reich, E., Warren, D., Schnipper, L., Antman, K., Frei, E., III, and Elias, A. (1995) Prognostic factors for prolonged progression free survival with high dose chemotherapy with autologous stem cell support for advanced breast cancer. *J. Clin. Oncol.* **13,** 2043–2049.
29. Diel, J., Kaufmann, M., Goerner, R., Costa, S. D., Kaul, S., and Bastert, G. (1992) Detection of tumor cells in bone marrow of patients with primary breast cancer: a prognostic factor for distant metastasis. *J. Clin. Oncol.* **10,** 1534–1539.
30. Cote, R. J., Rosen, P. P., Lesser, M. L. Old, L. J., Osborne, M. P. (1991) Prediction of early relapse in patients with operable breast cancer by detection of occult bone marrow micrometastases. *J. Clin. Oncol.* **9,** 1749–1756.
31. Ross, A. A., Cooper, B. W., Lazarus, H. M., Mackay, W., Moss, T. J., Ciobanu, N., Tallman, M. S., Kennedy, M. J., Davidson, N. E., Sweet, D., et al. (1993) Detection and viability of tumor cells in peripheral blood stem cell collections from breast cancer patients using immunocytochemical and clonogenic assay techniques. *Blood* **82,** 2605–2610.
32. Fields, K., Elfenbein, G., Moscinski, L., Trudeau, W., and Janssen, W. (1995) PCR detection of breast cancer cells in peripheral blood stem cell collections: incidence and correlations. *Blood* **86,** 2474 (abstract).
33. Schulze, R., Schulze, M., Wischnik, A., Ehnle, S., Doukas, K., Behr, W., and Schlimok, G. (1995) Tumor cell contamination of peripheral blood stem cell transplants (PBSCT) and/or bone marrow (BM) in high risk breast cancer patients (N≥10). *Blood* **86,** 923 (abstract).
34. Landys, K., Persson, S., Kovarik, J.. Hultborn, R., and Holmberg, E. (1995) Prognostic value of bone marrow biopsy in operable breast cancer patients at the time of initial diagnosis: results of a 19 year median follow-up. *Eur. J. Cancer* **31A(Suppl. 5),** 647 (abstract).
35. Mansi, J. L., Berger, U., McDonnell, T., Pople, A., Rayter, Z., Gazet, J. C., and Coombes, R. C. (1989) The fate of bone marrow micrometastases in patients with primary breast cancer. *J. Clin. Oncol.* **7,** 445–449.

36. Sharp, J., Kessinger, A., and Vaughan, W. (1992) Detection and clinical significance of minimal tumor contamination of peripheral blood stem cell harvests. *J. Cell. Cloning* **10(Suppl. 1),** 92–94.
37. Ybanez, J., Lazarus, H., Ross, A., and Cooper, B. (1995) Impact of occult tumor (OT) reinfusion on outcome after high dose chemotherapy (HDC) in patients with advanced breast cancer. *Blood* **86,** 1532 (abstract).
38. Shppall, E. J., Cagnoni, P., Gehling, U., Hami, L. S., and Hogan, C. J. (1995) Bone marrow purging, in *High Dose Cancer Therapy,* vol. 2 (Armitage, J. and Antman, K., eds.), Williams & Wilkens, Baltimore, MD, pp. 289–318.
39. Shppall, E., Jones, R., and Bast, R. (1991) 4-HC purging of breast cancer from the mononuclear cell fraction of bone marrow in patients receiving high dose chemotherapy and autologous marrow support: a phase I trial. *J. Clin. Oncol.* **9,** 85–93.
40. Coombes, R. and Buckman, R. A. F. (1986) In vitro and in vivo effects of monoclonal antibody toxin conjugate for patients with breast cancer. *Cancer Res.* **50,** 1170–1175.
41. Lebkowski, J. S., Schain, L. R., Okrongly, D., Levinsky, R., Harvey, J. M. J., and Okarma, T. B. (1992) Rapid isolation of human CD34 hematopoietic stem cells purging of human tumor cells. *Transplantation* **53,** 1011–1019.
42. Mapara, M. Y., Korner, I. J., Hildebrandt, M., Bargou, R., and Dorken, B. (1995) Immunomagnetic enrichment of CFD34+ peripheral blood progenitor cells in breast cancer patients: monitoring of purging efficiency by cytokeratin 19 and EGF-R based RT-PCR. *Blood* **86,** 922 (abstract).
43. Thomas, T. E., Zant, G. V., Phillips, G. L., and Lansdorp, P. M. (1995) Simultaneous direct purging of breast carcinoma cells and enrichment of CD 34+ cells using one step high gradient magnetic cell depletion. *Blood* **86,** 2485 (abstract).
44. Passos-Coelho Ross, A. A., Davis, J. M., Huelskamp, A. M., Clarke, B., Noga, S. J., Davidson, N. E., and Kennedy, M. J. (1994) Bone marrow micrometastases in chemotherapy responsive advanced breast cancer: effect of ex vivo purging with 4-hydroperoxycylclophosphamide. *Cancer Res.* **54,** 7366–7371.
45. Mklebust, A. T., Godal, A., Juell, S., Pharo, A., and Fodstad, O. (1994) Comparison of two antibody based methods for elimination of breast cancer cells from human bone marrow. *Cancer Res.* **54,** 909–214.
46. Morecki, S., Pavlotzky, F., Margel, S., and Slavin, S. (1987) Purging breast cancer cells in preparation for autologous bone marrow transplantation. *Bone Marrow Transplant.* **1,** 357–363.
47. Morecki, S., Margel, S., and Slavin, S. (1988) Removal of breast cancer cells by soybean agglutinin in an experimental model for purging human marrow. *Cancer Res.* **48,** 4573–1577.
48. Morecki, S. and Slavin, S. (1988) Combination of magnetic and chemocytotoxic cancer cell depletion for autologous bone marrow transplantation. *Israel J. Med. Sci.* **24,** 488–493.
49. Philip, I., Favort, M. C., and Philip, T (1985) Detection of Burkitt's cells in remission marrow by a cell culture monitoring system implications for autologous bone marrow transplantation, in *Autologous Bone Marrow Transplant: Proceedings of*

the *First International Symposium* (Dicke, K. A., Spitzer, G., and Zander, A. R., eds.), University of Texas M.D. Anderson Hospital and Tumor Institute, Houston, TX, p. 341.
50. Favrot, M. C., Philip, I., Philip, T., Protoukalian, J., Dore, J. F., and Lenoir, G. M. (1984) Distinct reactivity of Burkitt's lymphoma cell lines with eight monoclonal antibodies correlated with the ethnic origin. *J. Natl. Cancer Inst.* **73,** 841–847.
51. Momcuoglu, M., Favort, M., and Slavin, S. (1990) Lectin binding properties of Burkitt's lymphoma cell lines: application to bone marrow purging. *Exp. Hematol.* **18,** 55–60.
52. Marmont, A. M., Gale, R. P., Butturini, A., Goldman, J. M.: Martelli, M. F., Prentice, H. G., Slavin, S., Storb, R., Truitt, R. L., and Van Bekkum, D. W. (1989) T cell depletion in allogeneic bone marrow transplantation progress and problems. *Hematologia* **74,** 235–248.
53. Reisner, Y., Kapoor, N., Kirkpatrick, D., Pollack, M. S., Cunningham-Rundles, S., Dupont, B., Hodes, M. Z., Good, R. A., and O'Reilly, R. J. (1983) Transplantation for severe combined immunodeficiency with HLA, A, B, DR incompatible parent marrow cells fractionated by soybean agglutinin and sheep red blood cells. *Blood* **61,** 341–348.
54. Reisner, Y., Kapoor, N., Kirkpatrick, D., Pollack, M. S., Dupont, B., Good, R. A., and O'Reilly, R. J. (1981) Transplantation for acute leukemia with HLA A and B nonidentical parent marrow cells fractionated with soybean agglutinin and sheep red blood cells. *Lancet* **2,** 327–331.
55. Reisner, Y., Pahwa, S., Chiao, J. W., Sharon, N., Evans, R. L., and Good, R. A. (1980) Separation of antibody helper and antibody suppressor human T-cells by using soybean agglutinin. *Proc. Natl. Acad. Sci. USA* **77,** 6778–6782.
56. Morecki, S., Weigensberg, M., and Slavin, S. (1985) Lectin separation of nonlymphoid suppressor cells induced by lymphoid irradiation. *Eur. J. Immunol.* **15,** 138–148.
57. Slocombe, G. W., Newland, A. C., Yeatman, N. W., Nacey, M., Jones, H. M., and Knott, L. (1986) Allogeneic bone marrow transplantation for adult leukaemia with soybean lectin fractionated marrow. *Bone Marrow Transplant.* **1,** 31–39.
58. Frame, J. N., Collins, N. H., Cartagena, T., Waldmann, H., O'Reilly, R. J., Dupont, B., and Kernan, N. A. (1989) T-cell depletion of human bone marrow. Comparison of Campath-1 plus complement, Anti-T-cell ricin, A chain immunotoxin, and soybean agglutinin alone or in combination with sheep erythrocytes or immunomagnetic beads. *Transplantation* **47,** 984–988.
59. Keever, C. A., Welte, K., Sullivan, M., and O'Reilly, R. J. (1987) Phenotype and functional characterization of NK and LAK cells following T depleted bone marrow transplantation. *Prog. Clin. Biol. Res.* **244,** 423–432.
60. Keever, C. A., Welte, K., Small, T., Levick, J., Sullivan, M., Hauch, M., Evans, R. L., and O'Reilly, R. J. (1987) Interleukin 2 activated killer cells in patients following transplants of soybean lectin separated and E-rosette depleted bone marrow. *Blood* **70,** 1893–1903.

61. Welte, K., Keever, C. A., Levick, J., Bonilla, M. A., Merluzzi, V. J., Mertelsmann, R., Evans, R., and O'Reilly, R. J. (1987) Interleukin 2 production and response to interleukin-2 by peripheral blood mononuclear cells from patients after bone marrow transplantation: II. Patients receiving soybean lectin separated and T-cell depleted bone marrow. *Blood* **70,** 1595–1603.
62. Friedrich, W., Goldmann, S. F., Vette, U., Fliedner, T. M., Heymer, B., Peter, H. H., Reisner, Y., and Kleihauer, E. (1984) Immunoreconstitution in severe combined immunodeficiency after transplantation of HLA-haploidentical, T-cell depleted bone marrow. *Lancet* **1(8380),** 761–764.
63. Minegishi, M., Tsuchiya, S., Imaizumi, M., Yamaguchi, Y., Goto, Y., Tamura, M., Konno, T., and Tada, K. (1985) Successful transplantation of soybean agglutinin fractionated, histoincompatible, maternal marrow in a patient with severe combined immunodeficiency and BCG infection. *Eur. J. Pediatr.* **143,** 291–294.
64. Nespoli, L., Porta, F., Locatelli, F., Aversa, F., Carotti, A., Lanfranchi, A., Gibardi, A., Marchesi, M. E., Abate, L., Martelli, M. F., et al. (1990) Successful lectin separated bone marrow transplantation in adenosine deaminase deficiency related severe immunodeficiency. *Hematologica* **75,** 546–550.
65. Cowan, M. J., Wara, D. W., Weintrub, P. S., Pabst, H., Ammann, A. J. (1985) Haploidentical bone marrow transplantation for severe combined immunodeficiency disease using soybean agglutinin-negative, T depleted marrow cells. *Clin. Immunol.* **5,** 370–376.
66. Aversa, F., Tabilio, A., Terenzi, A., Velardi, A., Falzetti, F., Giannoni, C., Iacucci, R., Zei, T., Martelli, M. P., Gambelunghe, C., et al. (1994) Successful engraftment of T-cell depleted haploidentical "three loci" incompatible transplants in leukemia patients by addition of recombinant human granulocyte colony stimulating factor mobilized peripheral blood progenitor cells to bone marrow inoculum. *Blood* **84,** 3394–3955.
67. Berenson, R. J., Andrews, R. G., Bensinger, W. I., Kalamasz, D., and Knitter, G. (1988) Antigen CD34+ marrow cells engraft lethally irradiated baboons. *J. Clin. Invest.* **81,** 951–955.
68. Reisner, Y., Kapoor, N., Hodes, M. Z., O'Reilly, R. J., and Good, R. A. (1982) Enrichment for CFU-C from murine and human marrow using soybean agglutination. *Blood* **59,** 360–363.
69. Ebell, W., Castro-Malaspina, H., Moore, M., and O'Reilly, R. J. (1985) Depletion of stromal elements in human marrow grafts separated by soybean agglutinin. *Blood* **65,** 1105–1111.
70. Andrews, R. G., Singer, J. W., Bernstein, I. D. (1986) Monoclonal antibody 12-8 recognizes a 115 Kd molecule present on both unipotent and multipotent hematopoietic differentiation antigens. *Blood* **67,** 842–845.
71. Civin, C. I., Strauss, L. C., Brovall, C., Fackler, M. J. Schwartz, J. F., and Shaper, J. H. (1984) A hematopoietic progenitor cell surface antigen defined by a monoclonal antibody raised against KG-la cells. *J. Immunol.* **133,** 157–165.
72. Watt, S. M., Karhi, K., Gatter, K., Furley, A. J. W., Katz, F. E., Healy, L. E., Atlass, L. J., Bradley, N. J., Sutherland, D. R., Levinsky, R., and Greaves, M. F.

(1987) Distribution and epitope analysis of the cell membrane glyco-protein (HPCA-I) associated with human hematopoietic progenitor cells. *Leukemia* **1,** 417–426.
73. Varadi, G., Elchalal, U., Brautbar, C., and Nagler, A. (1995) Human umbilical cord blood for hematopoietic progenitor cells transplantation. *Leukemia and Lymphoma* **20,** 51–58.
74. Broxmeyer, H. E., Douglas, G. W., Hangoc, G., Cooper, S., Brad, J., English, D., Arny, M., Thomas, L., and Boyse, E. A. (1989) Human umbilical cord blood as a potential source of transplantable hematopoietic stem/progenitor cells. *Proc. Natl. Acad. Sci. USA* **86,** 3828–3832.
75. Thierry, D., Traineau, R., Adam, M., Delachaux, V., Brossard, Y., Richard, P., Gerotta, A., Devergie, A., Benbuman, M., and Gluckman, E. (1990) Hematopoietic stem cell potential from umbilical cord blood. *Nouv. Rev. Fr. Hematol.* **32,** 439,440.
76. Wagner, J. E. (1993) Umbilical cord blood stem cell transplantation. *Am J. Ped. Hematol/Oncol.* **15,** 169–174.
77. Nagler, A., Peacock, M., Tantoco, M., Lamons, D., Okarma, T. B., and Okrongly, D. A. (1994) Red blood cell depletion and enrichment of CD34+ hematopoietic progenitor cells from human umbilical cord blood using soybean agglutinin and CD34+ immunoselection. *Exp. Hematol.* **22,** 1134–1140.

29

Combined Lectin/Monoclonal Antibody Purging of Bone Marrow for Use in Conjunction with Autologous Bone Marrow Transplantation in the Treatment of Multiple Myeloma

Elizabeth G. H. Rhodes

1. Introduction
1.1. Myeloma: The Reasons for Using Bone-Marrow Purging

What was thought at the time to be a real breakthrough in myeloma therapy came in 1983 when McElwain reported that complete remission could be gained by means of a single, very high, dose of melphalan (HDM). Between 20 and 30% of patients attain complete remission with HDM (1). These observations led directly to exploration of the role of autologous bone marrow rescue following high-dose therapy, since the morbidity and mortality associated with prolonged bone marrow aplasia caused by these doses are unacceptably high (2). Most patients with myeloma are relatively old for allogeneic bone marrow transplantation, since graft-vs-host reactions tend to be poorly tolerated in people over 40. Autologous transplantation, however, carries the obvious disadvantage of potential reinfusion of malignant cells. Efforts have therefore been made to develop systems for in vitro purging to remove potentially malignant cells prior to reinfusion of the marrow (3,4). Although the nature of the initial malignant cell in myeloma is unknown, it is assumed that purging should entail at least the removal of plasma cells and B-lymphocytes.

In unpurged autologous bone marrow transplant (ABMT) only the Royal Marsden group reported complete remission rates as high as 50%; other investigators found that 20–30% achieved complete remission. In all series the relapse-free survival did not exceed 18–24 mo, but overall survival at best was 80% at 4 yr, which represents a modest improvement over chemotherapy alone.

1.2. The Reasons for Using Lectins as Purging Agents

In the search for a surface reacting agent for the selective removal of plasma cells, it was soon clear that there are few monoclonal antibodies (MAbs) that bind all plasma cells with sufficiently high affinity, owing to the relative paucity of cell surface antigens on these cells. Most of the surface antigens that distinguish hemopoietic cell types from each other are predominantly carbohydrate, and for this reason it was thought worthwhile examining lectins, which have specificity for carbohydrates, for their selective reactivity with plasma cells.

After initial lectin cytochemical studies showing good selectivity of peanut lectin for plasma cells *(5,6)*, pilot experiments to use peanut agglutinin (PNA) to remove plasma cells selectively from myeloma bone marrow explored: simple agglutination as in the soybean agglutinin techniques described by Reisner et al. *(7,8)* to remove T-lymphocytes from allogeneic bone marrow donor cells; erythrocyte rosetting techniques using either PNA-agglutinable rabbit erythrocytes or autologous desialylated erythrocytes, which are also PNA-agglutinable; and bone marrow cells preincubated with PNA were run down a column of sepharose conjugated with anti-PNA antibody and eluted cells were assessed for depletion of target cells.

The method utilizing desialylated autologous erythrocytes proved to be a highly effective means of separating PNA-positive cells from other bone marrow cells and some preliminary purging experiments were performed on a larger scale using peripheral blood buffy coat cells mixed with a PNA-positive tumor cell line. However, that method entails prior incubation of bone marrow cells with relatively high concentrations of lectin, and although no toxicity of PNA for hemopoietic colony assays could be demonstrated, it was deemed desirable to look for alternative techniques for purging, which had been proven to be safe for clinical use, hence the series of experiments using PNA linked to magnetized microspheres.

In addition to removing plasma cells from myeloma bone marrow by means of PNA-coated magnetic beads, evidence implicating earlier cells in B-cell development as being part of the myeloma clone, including our own myeloma colony work, suggested that effective purging of myelomatous marrow should include anti-B-cell elements as well. The CD19 antigen is expressed on B-lymphocytes from the early pre-B stage right up to, but not including, plasma cells; the IgM anti-CD19 MAb has been chemically bound to Dynabeads (Dynal, UK) and successfully used in several purging systems, such as lymphoma and acute lymphoblastic leukemia. This preparation was chosen to combine with PNA-coated magnetic beads in an attempt to remove all plasma cells and $CD19^+$ B-lymphocytes from myeloma bone marrow. Model systems designed to ascertain optimal magnetic bead:target cell ratios were comprised of normal peripheral blood or normal bone marrow mononuclear cells (MNC) mixed with a small proportion of tumor cell lines known to express receptors in high number for CD19 or PNA.

Toxicity experiments were performed on normal or myeloma bone marrow MNC by incubating them with increasing concentrations of PNA and assessing their ability to generate normal hemopoietic colonies of various lineages.

Finally, the effect on normal hemopoietic precursors of purging normal and myeloma bone marrow with PNA and anti-CD19 (each one separately then in combination) was assessed by cell numbers, phenotypic profile, and normal hemopoietic colony numbers at various stages of purging *(9,10)*.

Note: Clearly it is most important that similar studies are repeated by any group wishing to use such purging techniques in treatment of patients.

1.3. Development of Large-Scale Lectin Purging Techniques

In order to adapt small-scale in vitro manipulation of bone marrow to the large volumes and cell numbers required for clinical application in autologous bone marrow transplantation (ABMT) certain criteria had to be met:

1. An adequate supply of normal hemopoietic stem cells is necessary to ensure predictable bone marrow reconstitution following marrow ablative therapy. We had already shown that PNA does not bind to $CD34^+$ putative progenitors, that the lectin was not directly toxic to committed hemopoietic progenitors, and that removal of all detectable PNA^+ cells from small samples of bone marrow did not specifically remove cells capable of generating in vitro colonies. It was now necessary to repeat these experiments on a large scale. Although there is by now a sufficiently large body of experience in ABMT for guidelines to be published concerning the minimum number of nucleated bone marrow cells required for reliable reconstitution *(11)*, the data for minimum CFU-GM numbers necessary is not so clear-cut *(11,12)*, and some authorities consider them to be a poor guide to predicting hemopoiesis following ABMT *(13)*. However, when dealing with purged marrow, the final nucleated cell count is inevitably going to be far below the usual marrow preparation used for ABMT. For this reason, in vitro colony formation had to be the major criterion by which to assess the viability of the processed marrow, together with close inspection of both the morphology and colony-forming characteristics of the cells removed from the marrow.

2. A suitable preparation for efficient purging is one depleted as far as possible of mature cells, such as erythrocytes and granulocytes that might interfere mechanically with optimal target cell/magnetic bead interaction. This entailed a great deal of preparative bench work to find cell separation procedures that would achieve a balance between desired depletion of unwanted cells and minimum nonspecific loss of stem cells, at the same time as maintaining a strictly sterile preparation. The use of automated cell separators instead of simple centrifugation allows for preferential concentration of progenitor cells with increased elimi-

nation of other hemic cells. Either intermittent-flow or continuous-flow cell separators have been described, with elimination of granulocytes and erythrocytes by the addition of density separation media, such as Ficoll-Hypaque or sedimenting agents, such as hydroxyethyl starch *(14)*.
3. The entire bone marrow processing time must be kept to a minimum, both to preserve cell viability and to reduce the risk of bacterial contamination.
4. The final method chosen to process and purge myeloma marrow that attempted to meet the above criteria was that described by Baker et al. *(15)*. Initial concentration of mononuclear cells was achieved with the lymphocyte collection program on a Hemonetics V50 cell separator (Hemonetics, Braintree, Essex, UK). Further depletion of erythrocytes and granulocytes was effected by a density gradient centrifugation step using the Ficoll program of the Hemonetics V50. The processed marrow was then transferred in a standard sterile blood pack to a Class 1 sterile work station for magnetic bead separation.

Although the technique described here has been successfully applied to a considerable number of bone marrow samples in vitro, only two patients with myeloma have so far undergone autologous bone marrow transplantation, so this approach must be considered as highly experimental. Both patients, however, remained well and in clinical remission 6 yr after their course of radiochemotherapy and purged bone marrow transplantation *(16)*.

2. Materials
2.1. Bone Marrow Harvest

1. 2000-mL blood collection bag containing 150 mL acid/citrate/dextrose (Formula A) (ACD-A) for bone marrow harvest.
2. Hemonetics (Leeds, UK) V50 cell separator for preparation of a leukocyte-enriched preparation.
3. Lymphoprep (150 mL) (Pharma, Oslo, Norway) for density gradient centrifugation to concentrate mononuclear cells.
4. Wash solutions: 500 mL sterile physiological saline; 500 mL 4.5% human albumin solution (BPL, Elstree, UK); and 500 mL IMDM (Iscove's Modified Dulbecco's Medium) containing 15% (v/v) ACD-A and 20% (v/v) 4.5% human albumin solution.

2.2. Preparation of Magnetic Beads

1. Dynabeads (M-450 uncoated beads and anti-CD19-coated immunomagnetic beads) from Dynal, Wirral, UK.
2. 20 mg Peanut lectin agglutinin (PNA) (Serotec, UK).
3. Phosphate-buffered saline: pH 7.0 for washing beads, containing 0.1% human albumin (BPL) for the final wash.
4. Limulus test (Marine Biologicals, NJ, via Lysate and Radio Pharmaceutical Consultancy, Bradford upon Avon, UK) for exclusion of endotoxin in bead supernatant.

2.3. Purging with PNA- and CD19-Coated Magnetic Beads

1. IMDM: 100 mL for suspension of bead/bone marrow preparation.
2. Dynal magnetic separation unit, separated from the magnet by a removable iron plate, for separation of cell/magnetic bead complexes.
3. Peristaltic pump (Ismatec SA, Zurich, Germany) for retrieval of free cells.
4. Small magnetic trap (Dynal) to catch residual bead/cell complexes.
5. Transfer pack (Fenwal, 400 mL) (Baxter, Newbury, Berks, UK) for purged marrow collection.

2.4. Cryopreservation

1. 20% (v/v) Dimethyl sulfoxide (DMSO, Sigma, Poole, Dorset, UK).
2. Two Gambro hemofreeze bags (Gambro Dialysatoren, Hechingen, Germany).
3. Planar Kryo 10 controlled rate freezer (Planar, Sunbury, UK).

2.5. Normal Progenitor Cell Assay

1. IMDM containing 0.3% (w/v) agar, 20% fetal calf serum and 10% medium conditioned by C 5637 bladder carcinoma cell line, as source of GCSF (Sloan-Kettering Institute, New York) for CFU-GM assay.
2. 0.8% (w/v) Methylcellulose in IMDM supplemented with 30% FCS, 10% C 5637 CM, 2.5 µ/mL erythropoietin for CFU-GEMM assay.

3. Methods
3.1. Bone Marrow Harvest

1. Aspirate bone marrow in multiple 3–5 mL vol (to a maximum of 1 L) from the posterior iliac crests and transfer to a 2000-mL blood collection bag containing 150 mL acid/citrate/dextrose (Formula A) (ACD-A) (*see* Note 1).
2. Prepare a leukocyte-enriched product using a Hemonetics V50 cell separator, collecting 40 mL buffy coat at each pass using the lymphocytapheresis two-arm procedure program. Collect 240 mL total in six passes (*see* Note 2).
3. Further concentrate bone marrow mononuclear cells (MNC) by a density centrifugation step using Lymphoprep (Pharma) (*see* Note 3). Bone marrow buffy coat cells are pumped into the centrifuge bowl and, after addition of 50 mL Lymphoprep, increase the centrifuge speed from 400–600g in 50g increments with at least 15 s in between. Restart the pump to add the remaining 100 mL of Lymphoprep and commence product collection.
4. Wash the MNC on the Hemonetics V50 employing the washing protocol as follows. Transfer the cells into the centrifuge bowl spinning at 4000 rpm and wash to remove the Lymphoprep, firstly by the addition of sterile physiological saline, then by 500 mL of 4.5% human albumin solution (BPL, Elstree, UK) and finally in 500 mL IMDM containing 15% (v/v) ACD-A and 20% (v/v) 4.5% human albumin solution. The final volume of washed MNC should be 200 mL.

Fig. 1. The setup used for large-scale purging showing the magnet supported on a rocking deck with a peristaltic pump for removal of nonadherent cells.

3.2. PNA Coating of Magnetic Beads

1. Wash 90 mg uncoated Dynabeads twice in phosphate-buffered saline (PBS) and resuspend in 6.6 mL PBS containing 5 mg PNA for 16 h at 4°C with end-over-end rotation.
2. Wash again in PBS, and incubate again for 16 h at 4°C in PBS containing 0.1% human albumin.
3. Resuspend the beads at 28.5 mg/mL in PBS/0.1% albumin (*see* Note 4).
4. Assess lectin coating efficiency by adding 10 µL of washed PNA-coated beads to 40 µL of a 2% suspension of neuraminidase-treated red cells. Incubate the mixture for 30 min at 22°C with gentle agitation, and then observe under a microscope to ensure that all the PNA-coated beads bind treated red cells.
5. Test supernatant from the lectin-coated magnetic beads for the presence of endotoxin by means of the Limulus test (Marine Biologicals).

3.3. Purging with PNA- and CD19-Coated Magnetic Beads

1. Wash the coated magnetic beads three times in sterile saline and suspend in 100 mL IMDM.
2. Add PNA- and CD19-coated beads to the bone marrow MNC in the blood collection bag at a bead to target cell ratio of 40:1 for each bead type and incubate the mixture at 22°C for 1 h with constant gentle agitation (*see* Note 5).
3. Load the bag containing bone marrow cells and magnetic beads onto the Dynal magnetic separation unit, with the bag separated from the magnet by a removable iron plate (Fig. 1).

4. After removing the plate, lower the platform in 4 stages, 5 min for stages 1 and 2, 3 min each for stages 3 and 4.
5. As soon as the bag has reached the lowest stage, pump the free cells out of the bag using the peristaltic pump (Ismatec) at 10 mL/min.
6. Pass through the small magnetic trap (Dynal) and collect the unbound cells into a 400-mL Fenwal transfer pack.
7. Take samples of purged bone marrow for cytochemical analysis, bone marrow culture, and microbiological culture.
8. Centrifuge the remaining bone marrow sample at 400g for 5 min at 4°C. Discard the supernatant and add 100 mL autologous plasma to the bone marrow cells (*see* Notes six and seven for efficiency and yield of purging).

3.4. Cryopreservation

1. Operating in a Class 1 sterile work station, support the bone marrow/plasma mixture on ice, and gradually add 100 mL plasma containing 20% (v/v) dimethyl sulfoxide (DMSO) at 10 mL/min with thorough mixing.
2. Transfer the final purged bone marrow/plasma/DMSO mixture into two Gambro hemofreeze bags (Gambro) and seal the entry ports.
3. Freeze the bags using the Planar Kryo 10 controlled rate freezer, and then transfer to liquid nitrogen *(17)*.

3.5. Normal Progenitor Assays

3.5.1. Granulocyte-Macrophage Colony Forming Units (CFU-GM)

Assay CFU-GM in IMDM containing 0.3% (w/v) agar, 20% FCS, and 10% medium conditioned by C 5637 bladder carcinoma cell line.

3.5.2. Granulocyte-Erythrocyte-Macrophage-Megakaryocyte Colony Forming Units (CFU-GEMM)

Suspend 1×10^5 cells in 0.8% (w/v) methylcellulose in IMDM and supplement with 30% FCS, 10% C 5637 CM, 2.5 µ/mL erythropoietin. The full method is described in ref. *18*. To obtain the data provided in Note 8, 10% bovine serum albumin (previously batch tested) was substituted for the serum of a patient with severe aplastic anaemia, and PHA-LCM omitted.

4. Notes

1. Written informed consent is given by each patient to undergo bone marrow harvest under general anaesthetic.
2. Plasma, 180 mL, is collected at this stage for addition to the cells prior to cryopreservation.
3. The density gradient stage is desirable first because the efficiency of purging is reduced in the presence of large numbers of erythrocytes and second PNA is known to bind some granulocytes. After density gradient separation, fewer PNA-

Table 1
Characteristics of Bone Marrow Harvest

Volume harvested (l)	WBC[a] in harvested BM	WBC in concd BM	WBC postficoll	WBC postpurge
1.47 ± 0.31	12.35 ± 4.76	9.09 ± 3.63	3.18 ± 1.03	1.78 ± 0.69

[a]Absolute white blood cell count ($\times 10^9$/L), as measured on a Coulter StkS automated machine. Results shown are mean ± standard deviation.

coated magnetic beads are required, thus reducing the cost and also improving efficiency.

4. Plasma is omitted since previous studies had shown that its inclusion at this stage greatly reduced purging efficiency, presumably because of the presence of PNA binding glycoproteins in normal plasma (18).
5. Initial work involving small-scale experiments on normal peripheral blood and bone marrow spiked with tumor cell lines, and on myeloma marrow, had shown that efficient removal of plasma cells and CD19$^+$ lymphocytes could be achieved at a magnetic bead:target cell ratio of 40:1 without serious loss of hemopoietic progenitors. In the present technique for purging large volumes of bone marrow, the numbers of magnetic beads required to achieve a 40:1 ratio are calculated on bone marrow aspirates taken from the patients just prior to the main purged bone marrow harvest. No detectable PNA$^+$ or CD19$^+$ cells were found in the final product in the first four cases. On this basis, the magnetic bead:target cell ratio of 40:1 as well as the incubation times and temperature have not been altered in subsequent procedures.
6. Cell recoveries during processing and purging. The mean volume of bone marrow aspirated in the first four patients studied was 1.47 ± 0.31 L (SD) (see Table 1) with a mean absolute white blood cell count (WBC) of 12.35 ± 4.76 × 10^9/L. After the initial leukocyte concentration stage using the Hemonetics V50 there remained 9.09 (mean) ± 3.63 × 10^9/L WBC, and after further processing with Lymphoprep the final absolute WBC prior to purging was 3.18 ± 1.03 × 10. After incubation with combined PNA- and CD19-coated magnetic beads and separation on a magnet, the mean absolute WBC was 1.78 ± 0.69 × 10^9.

Percentage recovery of bone marrow nucleated cells at each stage of the processing is shown in Table 2. The final purged product yielded 14.4% of the original nucleated cell preparation.

There was a 26.7% cell loss during the initial leukocyte concentration stage and a 64.9% cell loss of the leukocyte concentrate after the Lymphoprep processing. Examination of cytocentrifuge preparations (Romanowski stain) showed mainly mononuclear cells with very few granulocytes. The purging procedure itself resulted in a 44% cell loss.

7. Efficiency of removal of target cells from bone marrow harvests. In the four patients whose data are shown here, cytocentrifuge preparations were made of the bone marrow at each stage and stained for cells reactive with PNA, CD19

Table 2
Numbers of Cells Recovered at Each Stage of Processing

	After concentration	After ficoll separation	After purging
Nucleated cell recovery, mean ± SD%	72 ± 9.1	28 ± 4.2	20 ± 3.7
Range	60–80%	25–34%	15–23%

Table 3
Efficiency of Removal of Target Cells from Bone Marrow Harvests, Phenotyping Studies

	PNA^+, mean ± SD%	$CD19^+$, mean ± SD%
Before purging	8.7 ± 10.0	5.2 ± 3.0
After purging	None detected	None detected

Table 4
Recovery of Normal Hemopoietic Progenitors After Purging

A

	CFU-GM per 2×10^5 cells plated (mean ± SD)	CFU-GEMM per 10^5 cells plated (mean ± SD)	Total nucleated cells ($\times 10^9$/L) (mean ± SD)
Before purging	42.5 ± 20.3	11.0 ± 4.9	3.3 ± 1.1
After purging	87.5 ± 56.4	15.5 ± 6.0	2.0 ± 0.6

B

	CFU-GM	CFU-GEMM	Nucleated cells
Percent recovery	78.2	79.0	61.0

MAb, and cytoplasmic kappa and lambda. The proportion of cells reacting with PNA prior to processing the bone marrow was 18.4% ± 9.5 (mean ± SD; range 12–35%). After processing and purging the bone marrows, PNA^+ cells were not detected by immunoalkaline phosphatase (*see* Table 3). Also, cytoplasmic immunoglobulin was absent in the bone marrow samples after purging. $CD19^+$ cells were 5.0% ± 3.0 (mean ± SD; range 2–10%) in samples prior to processing, but were undetectable in the final product after purging (*see* Table 3).

8. Recovery of normal hemopoietic progenitors after purging the bone marrow harvests in six patients is shown in Table 4. In each case, numbers of CFU-GM and

CFU-GEMM were estimated from aliquots of bone marrow withdrawn after the density gradient separation stage (prepurge) and after the purging procedure (postpurge). A mean of 42.5 ± 20.3 CFU-GM per 2×10^5 cells plated was obtained before purging, and 87.5 ± 56.4 per 2×10^5 after purging. When adjusted for cell loss there was (78.2%) recovery of CFU-GM following purging. A mean of 11.0 ± 4.9 ($\pm$ 1SD) CFU-GEMM per 1×10^5 plated cells was obtained before purging, and 15.5 ± 6.0 per 1×10^5 cells plated after purging. When adjusted for nucleated cell loss there was a 79% CFU-GEMM recovery following purging. Calculations based on patient's body weight predicted a mean CFU-GM dose of $0.86 \pm 0.32 \times 10^4$/kg recipient (range $0.5–1.3 \times 10^4$/kg).

Hemopoietic progenitor cells (as assessed by CFU-GM and CFU-GEMM) were concentrated in the final MNC product, whereas colony assays of cells bound to magnetic beads yielded less than 3% of total CFU-GM (data not shown). The hemopoietic potential of the final product calculated as CFU-GM dose per kilogram body weight was greater than 0.75×10^4/kg in four out of the six cases. The one patient who was harvested twice yielded 0.5×10^4/kg CFU-GM on both occasions.

Acknowledgments

This work was supported by a grant from the Leukemia Research Fund and was carried out in the University Department of Haematology, Liverpool, UK.

References

1. Gore, M. E., Selby, P. J., Viner, C., Clark, P. I., Meldrum, M., Millar, B., Bell, J., Maitland, J. A., and Milan, S., et al. (1989) Intensive treatment of multiple myeloma and criteria for complete remission. *Lancet* **2,** 879–882.
2. Barlogie, B. and Gahrton, G. (1991) Bone marrow transplantation in multiple myeloma. *Bone Marrow Transplant* **7,** 71–79.
3. Anderson, K. C., Barut, B. A., Ritz, J., Freedman, A. S., Takvorian, T., Rabinowe, S. N., Soiffer, R., Heflin, L., Coral, F., Dear, K., Mauch, P., and Nadler, L. M. (1991) Monoclonal antibody-purged autologous bone marrow transplantation therapy for multiple myeloma. *Blood* **77,** 712–720.
4. Gobbi, M., Cavo, M., Tazzari, P. L., et al. (1989) Autologous bone marrow transplantation with immunotoxin purged marrow for advanced multiple myeloma. *Eur. J. Haematol.* **43(Suppl 51),** 176–181.
5. Rhodes, E. G. H. and Flynn, M. P. (1989) PNA shows specificity for bone marrow plasma cells. *Br. J. Haematol.* **71,** 183–187.
6. Slupsky, J. R., Duggan-Keen, M., Booth, L. A., Karpas, A., Rhodes, E. G. H., Cawley, J. C., and Zuzel, M. (1993) The peanut-agglutinin (PNA)-binding surface components of malignant plasma cells. *Br. J. Haematol.* **83,** 567–573.
7. Reisner, Y., Itzicovitch, L., Meshorer, A., and Sharon, N. (1978) Hemopoietic stem cell transplantation using mouse bone marrow and spleen cells fractionated by lectins. *Proc. Natl. Acad. Sci. USA* **75,** 2933–2936.

8. Reisner, Y., Kapoor, N., Hodes, M. Z., O'Reilly, R. J., Good, R. A. (1982) Enrichment for CFU-C from murine and human bone marrow using soybean agglutinin. *Blood* **59**, 360–363.
9. Rhodes, E. G. H., Baker, P. K., Rhodes, J. M., Davies, J. M., and Cawley, J. C. (1991) Peanut agglutinin in combination with CD19 MAb has potential as a purging agent in myeloma. *Exp. Hematol.* **19**, 833–837.
10. Rhodes, E. G. H., Baker, P. K., Duguid, J. K. M., Davies, J. M., and Cawley, J. C. (1992) A method for clinical purging of myeloma bone marrow using peanut agglutinin as an anti-plasma cell agent, in combination with CD19 MAb. *Bone Marrow Transplantation* **10**, 485–489.
11. Gorin, N. C., David, R., Stachowick, J., et al. (1981) High-dose chemotherapy and autologous bone marrow transplantation in adult leukemias, malignant lymphomas and solid tumors. A study of 23 patients. *Eur. J. Cancer* **17**, 557–568.
12. To, L. B., Haylock, D. N., Kimber, R. J., and Juttner, C. A. (1984) High levels of circulating hemopoietic stem cells in very early remission from acute non-lymphoblastic leukemia and their collection and cryopreservation. *Br. J. Haematol.* **58**, 399–410.
13. Kaiser, H., Stuart, R. K., Brookmeyer, R., Beschorner, W. E., Braine, H. G., et al. (1985) Autologous bone marrow transplantation in acute leukemia: a phase I study of in vitro treatment of marrow with 4-hydroperoxycyclophosphamide to purge tumor cells. *Blood* **65**, 1504–1510.
14. Baker, P. K., Rhodes, E. G. H., and Duguid, J. K. M. (1991) Continuous flow cell separator use for bone marrow processing. *Transfus. Sci.* **12**, 183–187.
15. Baker, P. K., Rhodes, E. G. H., and Duguid, J. K. M. (1991) Effective concentration of bone marrow mononuclear cells using density gradient 'separation within an automated cell separator. *Transfus. Sci.* **12**, 353–356.
16. Rhodes, E. G. H., Baker, P. K., Rhodes, J. M., Davies, J. M., and Duguid, J. K. M. (1994) Long-term follow up in patients with myeloma treated with PNA lectin/CD19 MAb purged autograft. *Bone Marrow Transplantation* **13**, 795–799.
17. Hagenbeek, A. and Martens, A. C. (1989) Cryopreservation of autologous marrow grafts in acute leukemia: survival of in vivo clonogenic leukemic cells and normal hemopoietic stem cells. *Leukemia* **3**, 535–537.
18. Fauser, A. A. and Messner, H. A. (1978) Granuloerythropoietic colonies in human bone marrow, peripheral blood and cord blood. *Blood* **52**, 1243–1248.
19. Ching, C. K. and Rhodes, J. M. (1988) Identification and partial characterization of a new pancreatic cancer related glycoprotein by SDS-PAGE and lectin blotting. *Gastroenterology* **95**, 137–142.

VIII

EFFECTS OF LECTINS ON MAMMALIAN CELLS

30

Mechanisms and Assessment of Mitogenesis

An Overview

David C. Kilpatrick

1. Historical Introduction

When Nowell *(1)* used a preparation of phytohemagglutinin (PHA) to separate red and white blood cells (a procedure described a decade earlier *[2])*, the consequences were far-reaching for studies of both lectins and lymphocytes. The key observation was that, in addition to the expected agglutination of erythrocytes, lymphocytes were induced to grow and divide.

Younger readers may need help in appreciating the state of immunological knowledge at that time. Lymphocytes were thought of as fully differentiated cells, the function of which was uncertain. T- and B-cells, lymphokines and T-cell receptors had still to be discovered. The role and significance of the major histocompatibility complex was obscure. "The exact functions of lymphocytes are not clearly understood. Some investigators believe that they are responsible for the formation of antibodies ... They appear to have a protective function but the mechanism has yet to be determined" *(3)*.

Nowell's discovery soon led to the discoveries of several other mitogenic lectins, most notably concanavalin A, for the latter was the first mitogen to be specifically and reversibly inhibited by simple sugars. This indicated, of course, that both hemagglutinating and mitogenic activities of plant lectins were consequences of their carbohydrate binding ability and, incidentally, the first evidence that cell surface glycoconjugates were involved in the mitogenic process. Unlike typical antigens, which perhaps stimulate 0.01–0.1% of the lymphocyte population, mitogenic lectins can stimulate 20% or more. Consequently, such lectins were invaluable as tools to study the biochemical changes associated with lymphocyte activation and proliferation. Immunological knowledge from these in vitro studies expanded enormously, and interest in lectins, too, further excited by the differential reactivity of wheatgerm agglutinin toward normal and tumor cells *(4,5),* increased in tandem.

Shortly after the discovery of the mitogenic activity of PHA, another fortuitous observation led to some highly significant investigations in this field. A little girl died after eating pokeberries, the fruit of the American pokeweed, *Phytolacca americana.* Postmortem examination revealed lymphoblasts similar in appearance to those observed after culture with PHA. This led to the discovery of pokeweed mitogen, shown to be mitogenic both in vitro and in vivo *(6).* These were the first indications that mitogenic lectins could be active both within the gut and also systemically after entry by either alimentary or parenteral routes. It is now clear that one of the many remarkable properties of lectins is their resistance to degradation within the alimentary canal; there is clear evidence for this in humans *(7–9),* as well as in rodents. In rats, various dietary lectins act as powerful growth stimulators for the small intestine, and after absorption can even affect remote organs like the pancreas *(10).*

The early work on lectins as mitogens has been summarized in reviews by Lis and Sharon *(11,12).* More recent work has been included in brief *(13)* and comprehensive *(14)* reviews of human lymphocyte interactions with lectins by the present author. What follows is an overview of the cellular and molecular interactions occurring during lectin-induced mitogenesis and a brief discussion of the measurements that may be used to quantify a mitogenic response.

2. Mitogenic and Antimitogenic Lectins

At present, an appreciable number of purified lectins have been found to have mitogenic activity that can be inhibited by low concentrations of sugars with a specificity identical to that exhibited by their hemagglutinating activity. Nevertheless, most purified or crude lectin preparations tested are found not to be mitogenic. Twenty years after Nowell's original discovery, several lectins were surprisingly found to be antimitogenic; that is to say, they act to antagonize the stimulative activity of mitogens with which they are cocultured. The first examples of this were wheatgerm agglutinin *(15)* and the tomato lectin, which inhibited chicken, but not mouse, lymphocytes *(16).* Before long, the lectin from the mushroom *Agaricus bisporus* was reported as antimitogenic for human lymphocytes *(17),* and later the tomato lectin was shown to have the same property toward human cells *(18).* There was some evidence that wheatgerm agglutinin, *A. bisporus* lectin and tomato lectin all bound to the same receptor on the T-cell surface *(18,19).*

Mitogenic lectins have little in common regarding saccharide specificity, but that would not exclude the possibility that different mitogenic lectins could bind to different saccharide structures on the same glycoprotein. Unfortunately, most lectins can bind to several different surface glycoproteins, so it is not obvious which interaction is functionally important.

Wheatgerm agglutinin binds to many receptors, but tomato lectin binds principally to only three *(18)*, and the *Helix pomatia* lectin to only one *(20)*. It is likely that the last two lectins bind to the leukocyte common antigen, CD45. If so, the nonmitogenic nature of those lectins could be explained by their blocking the natural function of CD45, a tyrosine phosphatase with an essential role in T-cell activation *(21)*. One of the other glycopeptides that the tomato lectin binds is the major T-lymphocyte sialoglycoprotein, CD43, another molecule implicated in T-cell activation *(22)*. It is entirely plausible therefore that nonmitogenic lectins are not lectins that simply do not bind to the mitogen receptor(s), but rather may act positively by binding to one or more subsidiary molecules that regulate activation signals.

It seems likely that the "mitogen receptor" is the T-cell receptor and associated molecules (hereafter referred to as the T-cell receptor complex), and that lectin activation is similar to antigen-induced activation *(23)*. The major difference is that, whereas antigens stimulate a few clones of lymphocytes corresponding to the precise recognition of the epitopes on the antigen molecule, lectins bring about a polyclonal activation via common saccharide structures. It would not be necessary for the lectin to bind the T-cell receptor directly, for the latter is associated on the lymphocyte surface with other glycosylated molecules (especially CD3, but also to some extent with CD2, CD4, CD5, and CD8 *[24–26]*).

The simplest model arising from these considerations is that mitogenic lectins/antigens bind to the T-cell receptor complex, whereas nonmitogenic lectins either do not bind to the complex or bind to other, accessory, molecules essential to the transmission of activation signals. Those binding to accessory molecules might be expected to demonstrate antimitogenic activity. Mitogenic lectins cause signals leading to the synthesis of interleukin-2 (IL-2) and interleukin 2 receptors (IL-2R); non (anti)-mitogenic lectins do not.

Lectins binding to both T-cell receptor complex and accessory molecules would be less predictable. They might enhance or inhibit the mitogenic process and might be particularly susceptible to additional influences. Perhaps that is why wheatgerm agglutinin *(27)* and *Datura* lectin *(28)* may be mitogenic or antimitogenic, depending on the lectin concentration used and other experimental conditions. A key step appears to be the activation of protein kinase C, an event induced physiologically by the hydrolysis of membrane phospholipid to yield diacylglycerol *(29),* and experimentally by the use of phorbol esters like the tumor promoter, TPA (12–0–tetradecanoyl-phorbol-13-acetate). TPA usually enhances lectin-mediated lymphocyte activation and, with wheatgerm agglutinin and *Datura* lectin, a very significant synergistic response is evident *(30)*. Membrane phospholipid hydrolysis induced by mitogenic lectin receptor occupancy concomitantly leads to inositol triphosphate production, which induces an increase in intracellular calcium concentration. This mediates several cellular events and also helps stimulate protein kinase C. The latter catalyzes a series of phosphorylation reactions, which stimulate metabolism and mitotic activity.

The mitogenic process is mediated in part by the synthesis and action of cytokines. It is now accepted that both human and rodent helper T-cells can be divided into two major subsets based on their pattern of lymphokine secretion. One subset (Th-1) secretes IL-2, interferon-γ and tumor necrosis factor, whereas the other (Th-2) secretes interleukins 4, 5, 6, and 10 *(31)*. As a crude simplification, the former might be thought of as helper cells for T-lymphocytes and the latter as B-cell helpers. In reality, both T- and B-lymphocytes (and other cells) respond to subtle changes in complex mixtures of cytokines. Nevertheless, it remains a helpful simplification that the growth and cell division of T-cell lymphocytes depends principally on the synthesis of IL-2 and IL-2R. Mitogenic lectins promote IL-2/IL-2R synthesis; nonmitogenic lectins do not. The pivotal events in the mitogenic process discussed so far are illustrated in Fig. 1.

3. Accessory Cells and Molecules

The model outlined in Fig. 1 does not specify the accessory molecules involved and is concerned only with the responding lymphocytes. In fact, as first shown convincingly by Habu and Raff *(32)*, lectin-induced mitogenesis has a clear requirement for accessory cells, i.e., cells other than T-lymphocytes. This is analogous to antigen-induced mitogenesis, in which presentation of an antigen on macrophages (and/or B-cells, dendritic cells, and so on), in association with MHC class-2 molecules is required. The accessory cell dependency of mitogenic lectins is very variable, ranging from the great sensitivity of wheatgerm agglutinin to the minimal requirement of PHA *(33)*.

To some extent, accessory cells provide a purely physical advantage, presumably by increasing the effective local concentration of the lectin and facilitating effective crosslinking of surface receptors *(34)*. To some extent also, monokines including interleukin-1 enhance the mitogenic process. The strong accessory cell dependence of WGA, for example, depends entirely on soluble factors which can be recovered from the conditioned medium *(33)*. IL-1 does not substitute for conditioned medium in this system. Purified T-cells did respond, however, to wheatgerm agglutinin in the presence of IL-2 (which, of course, is not produced by monocytes or other non-T-cells). Therefore, it would seem that the binding of WGA to the lymphocyte surface may be enough to induce

Overview of Mitogenesis

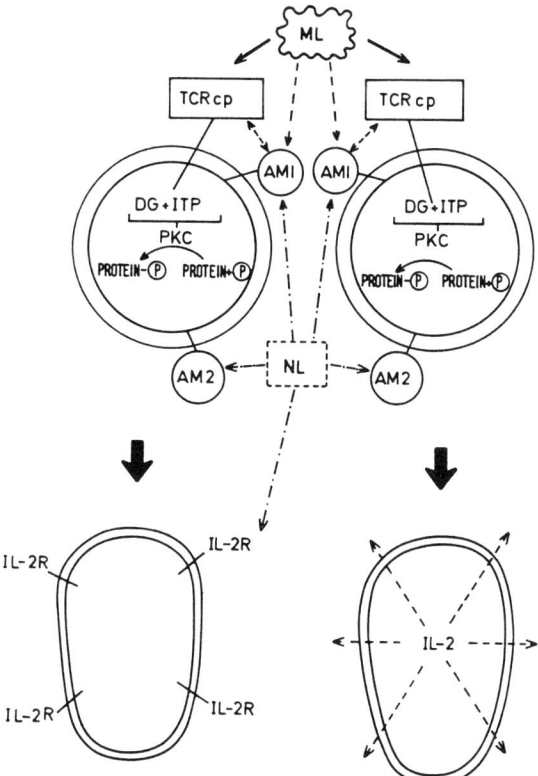

Fig. 1. Interactions of mitogenic and nonmitogenic lectins with the lymphocyte surface. Mitogenic (ML) and nonmitogenic (NL) lectins differ in their ability to bind to one or more molecules of the T-cell receptor complex (TCRcp) and relevant accessory molecules (AM).

the synthesis of IL-2R, but the synthesis if IL-2 itself requires soluble mediators provided by monocytes.

Soluble suppressor factors are undoubtedly sometimes produced in stimulated mononuclear cell cultures *(19,35)* and contribute to antimitogenic activity. However, the study of the regulation of *Datura* lectin-mediated mitogenesis (mitogenic when few monocytes are present, but readily inhibited by more monocytes) clearly indicated that a cell-mediated mechanism could inhibit mitogenesis *(33)*. Direct accessory cell-lymphocyte interactions as well as soluble mediators therefore control the mitogenic response. Indeed, accessory molecules on the cell surface appear to be the key players in the regulation of lymphocyte activation. Investigations with MAbs of precise specificity have identified many cell surface glycoproteins and implicated a surprisingly large number of them in mitogenesis, many of which are thought of as adhesion molecules *(13)*.

T-cell triggering via the sheep red blood cell receptor (E-R or CD2) was first described as an "alternative pathway" by analogy with complement activation *(36)*. In addition, a second separate pathway of T-cell activation has been proposed by Yachie and coworkers *(37)* to account for their results with wheatgerm agglutinin. The sheep red blood cell receptor may be regarded as an adhesion molecule or an endogenous lectin *(38)*. Engagement of its ligands, CD58 *(39)*, CD59 *(40)*, and CD48 *(41)* enhances cell adhesion. Random cell adhesion between lymphocytes and antigen presenting cells (APC) is probably a necessary first step in the induction of the immune response, but is a transient event unless recognition occurs (through the T-cell receptor) when adhesion is further enhanced. CD2 engagement, with its ligands, differs from the effects of other accessory molecules in vitro in being equivalent to stimulation by simultaneous TCR crosslinking to CD4 or CD8 on antigen presenting cells *(42)*. The in vivo significance of this phenomenon is unclear, particularly given the wide distribution of CD58 (LFA-3), and the partial membrane association of CD2 with the T-cell receptor complex *(24)*, but a regulatory role seems plausible. The relevance of CD2 in vivo is indicated by a single case report of immunodeficiency associated with absence of CD2 expression on T-cells *(43)*.

Various pieces of evidence support in outline a two-signal model of immune induction. The first signal consists of a specific recognition of antigen/MHC by the T-cell receptor. This leads to the synthesis of IL-2R, but not IL-2, and if nothing more happens, a state of anergy (tolerance) results. However, a second signal may be produced via accessory molecules, leading to cytokine secretion and ultimately cell proliferation and development of effector function.

These costimulatory molecules (which do not initiate, but rather amplify T-cell responses), are therefore critical in determining the outcome of immune cellular interaction. Guinan and coworkers *(44)* have argued that the involvement of CD28 is especially important. If adhesion or specific recognition events are blocked, a reversible suppression of the immune response occurs; if the CD28/B7 signal is blocked, an irreversible long-term state of unresponsiveness (anergy, tolerance) results *(43)*. In fact, CD28 reacts with at least two ligands, CD80 (B7.1) and CD86 (B7.2), both of which are found on activated B-cells and monocytes, although the latter appears first on the cell surface after activation. Both B7 molecules can also bind CTLA-4, a homologous molecule to CD28 that might have arisen via gene duplication. CTLA-4 is not able to substitute for CD28 and may function to enhance the signal delivered via CD28.

The function of CD28 and its ligands may be pivotal, but numerous other surface molecules have been implicated in mitogenesis. These include CD26 *(45)*, CD27 *(46)*, CD44 *(47)*, and CD49/29 *(48)* as well as those previously men-

Overview of Mitogenesis

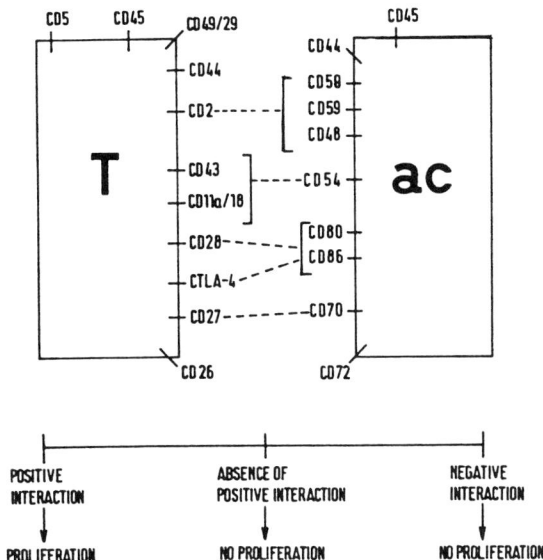

Fig. 2. Postrecognition events following lectin interaction with human T-lymphocytes. The listed structures on T-cells or accessory cells have all been implicated as accessory molecules in the regulation of human T-lymphocyte mitogenesis.

tioned. This apparent complexity implies that control of lymphocyte activation and proliferation may be exerted at different points and that a varying spectrum of outcomes may be possible when T-cells and accessory cells interact.

The mechanisms by which lectins activate lymphocytes have not been fully elucidated, but much can be understood within the general framework of this two-signal model (Fig. 2). Potentially mitogenic lectins are those which can bind to the T-cell receptor complex, and might be particularly effective if they can also bind MHC molecules on accessory cells since mitogenicity correlates well with MHC binding ability *(49)*. This would be enough to activate the T-cells and induce the synthesis of IL-2R. What happens next depends on the interactions of accessory molecules with each other and with the lectin. Lectin binding to CD2 might enhance the recognition response and, in some circumstances, obviate the need for the T-cell receptor altogether. Overall, one set of interactions leads to a positive response and the production of IL-2. A simple failure of the accessory molecules to function in this way results in no response, but that can be overcome by exogenous application of IL-2. An example might be provided by wheatgerm agglutinin: An absence of monocytes deprives the T-cells of complementary accessory molecules, but monocytes can be replaced by conditioned medium, TPA, or exogenous IL-2 *(27,33)*. A separate negative signal may result from the positive inhibitory action of complementary accessory molecules perhaps via lectin bridging. This kind of interaction is required to explain the strong inhibitory effect of monocytes when *Datura* lectin is used *(33)*; this effect can nonetheless be readily reversed by the addition of submitogenic amounts of TPA, indicating that a signal preceding the activation of protein kinase C is blocked by *Datura* lectin-influenced monocytes. Tomato and potato lectins are not comitogenic with TPA and their antimitogenic activity cannot be overcome with exogenous IL-2. The *Datura* lectin may differ from the tomato lectin in the recognition step, if the former can bind to the T-cell receptor complex. The tomato lectin is not simply nonmitogenic, but has an antimitogenic property, which could be related to its ability to bind CD43 and/or CD45.

4. In Vivo Relevance of Mitogenic Lectins

Mitogenic lectins are useful as reagents, and the study of their interactions with mononuclear cells in vitro has been invaluable in understanding lymphocyte activation and its control. Lectins are undoubtedly active toward lymphocytes in vivo also, as was first suggested by the cases of pokeweed poisoning mentioned earlier, and convincingly demonstrated by the induction of transplantation tolerance by iv injection of purified lectins *(50)*. Yet these effects are surely accidental. It is difficult to understand any advantage to a plant to have a protein that can stimulate the lymphocytes, or agglutinate the erythrocytes, of higher animals.

The same may not be true of bacterial or viral lectins. Bacteria and viruses colonize animals and have much to gain by doing so. The surface lectins on some micro-organisms, which can bind to animal membrane receptors and thereby mediate infection, have an obvious role.

Bacteria and viruses also have immunostimulatory molecules that act as polyclonal activators for animal lymphocytes, some of which have been implicated in several human diseases *(51)*. These "superantigens" may stimulate 5–30% of the T-cell population and do so by binding simultaneously to part of the variable region of the T-cell receptor β-chain (or γ-chain) and MHC class-2 molecules on accessory cells. The specificity is not for the antigen combining site, but for the product of a particular Vβ gene family. Like lectins, superantigens do not require processing and are not MHC-restricted; despite the MHC-class-2 dependence, superantigens do not lie within peptide binding grooves.

Licastro et al. *(52)*, comparing lectins and superantigens conclude that, despite their similarities, the two belong to separate families of molecules with important differences between them. Yet these crucial differences are becoming blurred. The mitogenic lectin from the nettle *(Urtica dioica)* satisfies the criteria for both lectins and superantigens *(53)*. Wheatgerm agglutinin has an homologous primary structure and a similar carbohydrate binding specificity, so might be similar. Various observations are inconsistent with the accepted model of superantigenic mechanism of action, leading Woodland and Blackman *(54)* to propose that there is direct contact between the T-cell receptor and MHC on accessory cells during superantigen engagement. Although superantigens may normally utilize the CD28 pathway *(55)*, superantigen responses are not necessarily dependent on B7 costimulation *(56)*, a circumstance more lectin-like than antigen-like. So, too, is the fact that superantigens can stimulate the small subset of T-cells that are CD4 and CD8 negative *(57)*. Staphylococcal enterototoxins can even stimulate T-cells in MHC class-2 negative mice *(58)*. Finally, superantigens may differ from true antigens in the mechanism through which T-cell deletion takes place *(59)*.

I suggest the only fundamental difference is that lectins bind to the saccharide moieties of glycoproteins, whereas superantigens bind to peptide structures. Superantigens, which may well function to help their sources overcome host defense mechanisms, may be the natural factors that plant lectins inadvertently mimic when used in the unnatural settling of in vitro culture.

5. Assessment of Mitogenesis

The original and most direct means of establishing whether lymphocytes have been stimulated is by microscopic examination. Resting lymphocytes (which are small and round with a very thin rim of cytoplasm) enlarge, become more oval-shaped, and acquire an increased area of cytoplasm on mitogenic stimulation. Simple examination of cells, after harvesting with a cytospin centrifuge and staining, is probably the best way to estimate the number (proportion) of cells that have been stimulated. However, it is only semiquantitative and, above all, is a laborious means of assessing mitogenic activity.

Other methods of assessment give an average measure for the whole cell preparation. To achieve this, various steps in the activation process can be exploited. Table 1 outlines the sequence of major events following the interaction of a mitogen with lymphocytes; a similar pattern occurs with other cell types. Various stages in the sequence can be targeted to measure stimulation. Examples include changes in amino acid transport *(60)*, and the production of γ-interferon *(61)*. Sometimes such methods can be particularly helpful in discriminating between activation and proliferation.

Commonly measured correlates of cellular proliferation are increases in DNA, RNA, and protein synthesis. Assaying simple increases in total DNA, RNA, or protein is a satisfactory means of quantifying growth of whole organs in vivo, e.g., for estimating the growth factor activity of dietary lectins on the small intestine *(10)*. Usually, more sensitive techniques are required for cell

**Table 1
Major Events During Mitogenic Stimulation of Lymphocytes**

1. Binding of mitogen to lymphocyte cell surface
2. Altered cell permeability and transport systems
3. Protein phosphorylation
4. Synthesis of IL-2 receptors
5. Synthesis of lymphokines
6. DNA synthesis
7. Mitosis

culture involving the use of radiolabelled precursors viz. ^{14}C-leucine incorporation into protein, ^{3}H-uridine into RNA or ^{3}H-thymidine into DNA.

In practice, the rate of change of DNA synthesis, which correlates well with the number of cells stimulated, is the most commonly used measure of lymphocyte activation and proliferation. Measurement of ^{3}H-thymidine incorporation into DNA is a highly sensitive method, and has become the standard technique. The major disadvantage is the necessity to adhere to the regulations and safety procedures required of radioisotope usage. To provide a nonradioactive substitute, 5-bromo-2'-deoxyuridine (BrdU) has been used in conjunction with immunoassay *(62,63)*. A commercial kit is now available based on an enzyme-linked immunosorbant assay in which BrdU-incorporated DNA forms the solid phase and the key reagent is a monoclonal anti-BrdU antibody.

An alternative nonradioactive method exploits the color change in a tetrazolium salt (MTT) when enzymatically reduced by metabolically active cells *(64)*. This colorimetric assay measures the number and metabolic activity of viable cells in the culture and is suitable for processing large numbers of samples. Various improvements to the original procedure have since been published *(65–67)*.

It seems likely that one or more of the nonradioactive methods will ultimately replace the standard ^{3}H-thymidine incorporation assay described in Chapter 32.

References

1. Nowell, P. (1960) Phytohemagglutinin: an initiator of mitosis in cultures of normal human leucocytes. *Cancer Res.* **20,** 462–464.
2. Li, J. G. and Osgood, E. E. (1949) A method for the rapid separation of leukocytes and nucleated erythrocytes from blood or marrow with a phytohemagglutinin from red beans *(Phaseolus vulgaris). Blood* **4,** 670–675.
3. Leeson, C. R. and Leeson, T. S. (1966) Histology, Saunders, Philadelphia, PA.
4. Aub, J. C., Tieslau, C., and Lankester, A. (1963) Reaction of normal and tumor cell surfaces to enzymes I. Wheatgerm lipase and associated mucopolysaccharides. *Proc. Natl. Acad. Sci. USA* **50,** 613–619.

5. Burger, M. M. and Goldberg, A. R. (1967) Identification of a tumor-specific determinant on neoplastic cell surfaces. *Proc. Natl. Acad. Sci. USA* **57,** 359–366.
6. Barker, B. E. (1969) Phytomitogens and lymphocyte blastogenesis. *In Vitro* **4,** 64–79.
7. Kilpatrick, D. C., Pusztai, A., Grant, G., Graham, C., and Ewen, S. W. B. (1985) Tomato lectin resists digestion in the mammalian alimentary canal and binds to intestinal villi without deleterious effects. *FEBS Lett.* **185,** 299–305.
8. Brady, P. G., Vannier, A. M., and Banwell, J. G. (1978) Identification of the dietary lectin, wheatgerm agglutinin, in human intestinal contents. *Gastroenterology* **75,** 236–239.
9. Freed, D. L. J. and Buckley, C. H. (1978) Mucotractive effect of lectin. *Lancet* **1,** 585,586.
10. Pusztai, A. (1991) *Plant Lectins,* Cambridge University Press, Cambridge, UK.
11. Lis, H. and Sharon, N. (1977) Lectins: their chemistry and application to immunology, in *The Antigens,* vol. 4 (Sela, M., ed.), Academic, New York, pp. 429–529.
12. Lis, H. and Sharon, N. (1986) Biological properties of lectins, in *The Lectins: Properties, Functions, and Applications in Biology and Medicine* (Liener, I. E., Sharon, N., and Goldstein, I. J., eds.), Academic, London, pp. 265–291.
13. Kilpatrick, D. C. (1991) Lectin interactions with human leukocytes: mitogenicity, cell separation, clinical applications, in *Lectin Reviews,* vol. 1 (Kilpatrick, D. C., Van Driessche, E., and Bøg-Hansen, T.-C., eds.), Sigma, St. Louis, MO, pp. 69–80.
14. Kilpatrick, D. C. (1995) Lectins in immunology, in *Lectins-Biomedical Perspectives* (Pusztai, A. and Bardocz, S. eds.), Taylor and Francis, Bazingstoke, pp. 155–182.
15. Greene, W. C. and Waldmann, T. A. (1980) Inhibition of human lymphocyte proliferation by the nonmitogenic lectin wheatgerm agglutinin. *J. Immunol.* **124,** 2979–2987.
16. Nachbar, M. S., Oppenheim, J. D., and Thomas, J. O. (1980) Lectins in the US diet. Isolation and characterization of a lectin from the tomato (Lycopersicon escalentum). *J. Biol. Chem.* **255,** 2056–2063.
17. Greene, W. C., Fleisher, T. A., and Waldmann, T. A. (1981) Suppression of human T- and B-lymphocyte activation by *Agaricus bisporus* lectin. *J. Immunol.* **126,** 580–586.
18. Kilpatrick, D. C., Graham, C., and Urbaniak, S. J. (1986) Inhibition of human lymphocyte transformation by tomato lectin. *Scand. J. Immunol.* **24,** 11–19.
19. Greene, W. C., Fleisher, T. A., and Waldmann, T. A. (1981) Soluble suppressor supernatants elaborated by concanavalin A-activated human mononuclear cells. *J. Immunol.* **126,** 1185–1197.
20. Axelsson, B., Kimura, A., Hammarström, S., Wigzell, H., Nilsson, K., and Mellstedt, H. (1978) *Helix pomatia* A hemagglutinin: selectivity of binding to lymphocyte surface glycopeptides on T-cells and certain B-cells. *Eur. J. Immunol.* **8,** 757–764.
21. Trowbridge, I. S. (1994) CD45: an emerging role as a protein tyrosine phosphatase required for lymphocyte activation and development. *Annu. Rev. Immunol.* **12,** 85–116.

22. Park, J. K., Rosenstein, Y. J., Remold-O'Donnell, E., Bierer, B. E., Rosen, F. S., and Burakoff, S. J. (1991) Enhancement of T-cell activation by the CD43 molecule whose expression is defective in Wiskott-Aldrich syndrome. *Nature* **350**, 706–709.
23. Chilson, O. P. and Kelly-Chilson, A. E. (1989) Mitogenic lectins bind to the antigen receptor on human lymphocytes. *Eur. J. Immunol.* **19**, 289–296.
24. Brown, M. H., Cantrell, D. A., Brattsand, G., Crumpton, M. J., and Gullberg, M. (1989) The CD2 antigen associates with the T-cell antigen receptor CD3 antigen complex on the surface of human T-lymphocytes. *Nature* **339**, 551–553.
25. Suzuki, S., Kupsch, J., Eichmann, K., and Saizawa, M. K. (1992) Biochemical evidence of the physical association of the majority of CD3 chains with the accessory/coreceptor molecules CD4 and CD8 on nonactivated T-lymphocytes. *Eur. J. Immunol.* **22**, 2475–2479.
26. Osman, N., Ley, S. C., and Crumpton, M. J. (1992) Evidence for an association between the T-cell receptor/CD3 antigen complex and the CD5 antigen in human lymphocytes. *Eur. J. Immunol.* **22**, 2995–3000.
27. Kilpatrick, D. C. and McCurrach, P. M. (1987) The wheatgerm agglutinin is mitogenic, nonmitogenic and antimitogenic for human lymphocytes. *Scand. J. Immunol.* **25**, 343–348.
28. McCurrach, P. M. and Kilpatrick, D. C. (1988) *Datura* lectin is both an antimitogen and a comitogen acting synergistically with phorbol ester. *Scand. J. Immunol.* **27**, 31–34.
29. Altman, A., Coggeshall, K. M., and Mustelin, T. (1990) Molecular events mediating T-cell activation. *Adv. Immunol.* **48**, 227–360.
30. Kilpatrick, D. C., Peumans, W. J., and Van Damme, E. J. M. (1990) Mitogenic activity of monocot lectins, in *Lectins—Biology, Biochemistry, Clinical Biochemistry*, vol. 7 (Kocourek, J., ed.), Sigma Chemical Co., St. Louis, MO, pp. 259–263.
31. Paul, W. E. and Seder, R.-A. (1994) Lymphocyte responses and cytokines. *Cell* **76**, 241–251.
32. Habu, S. and Raff, M. C. (1977) Accessory cell dependence of lectin-induced proliferation of mouse T-lymphocytes. *Eur. J. Immunol.* **7**, 451–457.
33. Kilpatrick, D. C. (1988) Accessory cell paradox: monocytes enhance or inhibit lectin mediated human T-lymphocyte proliferation depending on the choice of mitogen. *Scand. J. Immunol.* **28**, 247–249.
34. Gallagher, R. B., Whelan, A., and Feighery, C. (1986) Studies on the accessory requirement of T-lymphocyte activation by concanavalin A. *Clin. Exp. Immunol.* **66**, 118–125.
35. Grillon, C., Monsigny, M., and Kieda, C. (1991) Soluble human lymphocyte sugar binding proteins with immunosuppressive activity. *Immunol. Lett.* **28**, 47–56.
36. Meurer, S. C., Hussey, R. E., Fabbi, M., Fax, D., Acuto, O., Fitzgerald, K. A., Hodgdon, J. C., Protentis, J. P., Schlossmann, S. F., and Reinherg, E. L. (1984) An alternative pathway of T-cell activation: a functional role for the 50 kd T11 sheep erythrocyte receptor protein. *Cell* **36**, 897–906.

Overview of Mitogenesis

37. Yachie, A., Hernandez, D., and Blaese, R. M. (1987) T3–T-cell receptor (Ti) complex-independent activation of T-cells by wheatgerm agglutinin. *J. Immunol.* **138,** 2843–2847.
38. Boldt, D. H. and Armstrong, J. P. (1976) Rosette formation between human lymphocytes and sheep erythrocytes. Inhibition of rosette formation by specific glycopeptides. *J. Clin. Invest.* **57,** 1068–1078.
39. Selvaraj, P., Plunkett, M. L., Dustin, M., Sanders, M. E., Shaw, S., and Springer, T. A. (1987) The T-lymphocyte glycoprotein CD2 binds the cell surface ligand LFA-3. *Nature* **326,** 400–402.
40. Deckert, M., Kubar, J., Zoccola, D., Bernard-Pomier, G., Angelisova, P., Horejsi, V., and Bernard, A. (1992) CD59 molecule: a second ligand for CD2 in T-cell adhesion. *Eur. J. Immunol.* **22,** 2943–2947.
41. Sandrin, M. S., Mouhtouris, E., Vaughan, H. A., Warren, H. S., and Parish, C. R. (1993) CD48 is a low affinity ligand for human CD2. *J. Immunol.* **451,** 4606–4613.
42. Meuer, S. C., Schraven, B., and Sanstag, Y. (1994) An "alternative" pathway of T-cell activation. *Int. Arch. Allergy Immunol.* **104,** 216–221.
43. Sneller, M. C., Eizenstein, E. M., Baseler, M., Lane, H. C., Donoghue, E. T., and Falloon, J. (1994) A unique syndrome of immunodeficiency and autoimmunity associated with absent T-cell CD2 expression. *J. Clin. Immunol.* **14,** 359–367.
44. Guinan, E. C., Gribben, J. G., Boussiotis, V. A., Freeman, G. J., and Nadler, L. M. (1994) Pivotal role of the B7:CD28 pathway in transplantation tolerance and tumor immunity. *Blood* **84,** 3261–3282.
45. Torimoto, Y., Dang, N. H., Vivier, E., Tanaka, T., Schlossman, S. F., and Morimoto, C. (1991) Coassociation of CD26 (dipeptidyl peptidase N) with CD45 on the surface of human T-lymphocytes. *J. Immunol.* **147,** 2514–2517.
46. Kobata, T., Agernatsu, K., Kameoka, J., Schlossman, S. F., and Marimoto, C. (1994) CD27 is a signal-transducing molecule involved in CD45RA+ naive cell costimulation. *J. Immunol.* **153,** 5422–5432.
47. Pierres, A., Lipcey, C., Mawas, C., and Olive, D. (1992) A unique CD44 MAb identifies a new T-cell activation pathway. *Eur. J. Immunol.* **22,** 413–417.
48. Yamada, A., Nojima, Y., Sugita, K., Dang, N. H., Schlossman, S. F., and Morimoto, C. (1991) Crosslinking of VLA/CD29 molecule has a comitogenic effect with antiCD3 on CD4 cell activation in serum-free culture system. *Eur. J. Immunol.* **21,** 319–325.
49. Kimura, A. and Ersson, B. (1981) Activation of T-lymphocytes by lectins and carbohydrate-oxidizing agents viewed as an immunological recognition of cell surface modifications seen in the context of self major histocompatibility complex antigens. *Eur. J. Immunol.* **11,** 475–783.
50. Hilgert, I., Horejsi, V., Angelisova, P., and Kristofova, H. (1980) Lentil lectin effectively induces allotransplantation tolerance in mice. *Nature* **284,** 273–275.
51. Drake, C. G. and Kotzin, B. L. (1992) Superantigens: biology, immunology and potential role in disease. *J. Clin. Immunol.* **12,** 149–162.
52. Licastro, F., Davis, L. J., and Morini, M. (1993) Lectins and superantigens: membrane interactions of these compounds with T-lymphocytes affect immune responses. *Int. J. Biochem.* **25,** 845–852.

53. Galelli, A. and Truffa-Bachi, P. (1993) *Urtica dioica* agglutinin. A superantigenic lectin from stinging nettle rhizome. *J. Immunol.* **151,** 1821–1831.
54. Woodland, D. L. and Blackman, M.-A. (1993) How do T-cell receptors, MHC molecules and superantigens get together? *Immunol Today* **14,** 208–212.
55. Ohnishi, H., Tanaka, T., Takahara, J., and Kotb, M. (1993) CD28 delivers costimulating signals for superantigen-induced activation of antigen-presenting cell-depleted human T-lymphocytes. *J. Immunol.* **150,** 3207–3214.
56. Damle, N. K., Klussman, K., Leytye, G., and Linsley, P. S. (1993) Proliferation of human T-lymphocytes induced with superantigens is not dependent on costimulation by the CD28 counter-receptor B7. *J. Immunol.* **150,** 726–735.
57. Quarantino, S., Murison, G., Knyba, R. E., Verhoef, A., and Londei, M. (1991) Human CD4⁻CD8⁻αβ⁺ T-cells express a functional T-cell receptor and can be activated by superantigens. *J. Immunol.* **147,** 3319–3323.
58. Avery, A. C., Markowitz, J. S., Grusby, M. J., Glimcher, L. H., and Cantor, H. (1994) Activation of T-cells by superantigen in class II-negative mice. *J. Immunol.* **153,** 4853–4861.
59. Gonzalo, J. A., Baixeras, E., González-Garcia, A., George-Chandy, A., Rooijen, N. V., Martinez-A, C., and Kroemer, G. (1994) Differential in vivo effects of a superantigen and an antibody targeted to the some T-cell receptor. *J. Immunol.* **152,** 1597–1608.
60. Udey, M. C., Chaplin, D. D., Wedner, H. J., and Parker, C. (1980) Early activation events in lectin-stimulated human lymphocytes, evidence that wheatgerm agglutinin and mitogenic lectins cause similar early changes in lymphocyte metabolism. *J. Immunol.* **125,** 1544–1550.
61. Peumans, W. J., De Ley, M., and Broekaert, W. F. (1984) An unusual lectin from stinging nettle *(Urtica dioica)* rhizomes. *FEBS Lett.* **177,** 99–103.
62. Porstmann, T., Ternynck, T., and Avrameas, S. (1985) Quantitation of 5-bromo-2-deoxyuridine incorporation into DNA: an enzyme immunoassay for the assessment of the lymphoid cell proliferation response. *J. Immunol. Methods* **82,** 169–179.
63. Huong, P. L. T., Kolk, A. H. J., Eggelte, T. A., Verstijnen, C. P. H. J., Gilis, H., and Hendriks, J. T. (1991) Measurement of antigen specific lymphocyte proliferation uzing 5-bromo-deoxyuridine incorporation. *J. Immunol. Methods* **140,** 243–248.
64. Mosmann, T. (1983) Rapid colorimetric assay for cellular growth and survival: application to proliferation and cytotoxicity assays. *J. Immunol. Methods* **65,** 55–63.
65. Denizot, F. and Lang, R. (1986) Rapid colorimetric assay for cell growth and survival. Modification to the tetrazolium dye procedure giving improved sensitivity and reliability. *J. Immunol. Methods* **89,** 271–277.
66. Tada, H., Shiho, O., Kuroshima, K., Koyama, M., and Tsukamato, K. (1986) An improved colorimetric assay for interleukin 2. *J. Immunol. Methods* **93,** 157–165.
67. Hansen, M. B., Neilsen, S. E., and Berg, K. (1989) Re-examination and further development of a precise and rapid dye method for measuring cell growth/cell kill. *J. Immunol. Methods* **119,** 203–210.

31

Mitogenic Effects of Lectins on Epithelial Cells

Lu-Gang Yu and Jonathan M. Rhodes

1. Introduction

The specific binding of lectins to the carbohydrate moieties of glycoproteins and glycolipids on the cell surface has been shown to result in a variety of signal transduction processes *(1,2)*. One of the most dramatic effects of the interaction of lectins with cells is mitogenic stimulation, e.g., the triggering of nondividing lymphocytes into a state of growth and proliferation *(3)*, which has led to the extensive and successful use of lectins in the analysis of events occurring during lymphocyte activation and proliferation *(4,5)*.

Epithelial cells are amongst the principal cell types in which cancers commonly arise *(6)*. Changes in surface carbohydrate expression are common in epithelial neoplasia and often represent neoexpressions of oncofetal antigens *(7)*. There is increasing evidence that such changes in carbohydrate expression may play a key role in determining the metastatic behavior of tumors *(8,9)*. Among the most commonly demonstrated abnormalities in malignant and hyperplastic epithelia is the increased expression of the TF-antigen (Thomsen Friedenreich antigen), which has the structure galactosyl β1-3 NAcetyl-galactosamine. This structure, which is the type I core structure in *O*-linked oligosaccharides, is recognized by several lectins including the peanut and mushroom lectins. Recently it has been shown that peanut lectin (which does not bind sialyl-TF) stimulates the proliferation of both colon cancer cell line HT29 *(10)* and normal human colonic explants *(11)*, whereas mushroom lectin (which binds both the sialylated and unsialylated TF antigen) markedly inhibits the proliferation of a range of epithelial cells without cytotoxcity *(12)*. The interaction of lectins with epithelial cells could therefore offer a very useful tool in the analysis of events occurring during tumor cell proliferation.

From: *Methods in Molecular Medicine: Vol. 9: Lectin Methods and Protocols*
Edited by: J. M. Rhodes and J. D. Milton Humana Press Inc., Totowa, NJ

Cell proliferation is generally studied by measuring the rate of DNA synthesis. Measurement of [^{3}H]thymidine incorporation into DNA synthesis still remains the most popular method because of the rapidity and convenience of the procedure, but care should be taken when interpreting the results. There are a number of potential artifacts in correlating thymidine uptake to DNA synthesis, in particular the competition between the labeled thymidine taken up by the salvage pathway from the medium and the endogenous thymidine synthesized by the normal cellular pathway *(13)*.

2. Materials

1. Sterile, 24-multiwell tissue culture plates, 9-cm tissue culture dishes (Nunc, Kamstrup, Denmark), 10-mL plastic pipets, Pasteur glass pipets, and 30-mL plastic universals.
2. Eppendorf 4780 multipipeter and 12.5-mL combitips.
3. Hemocytometer or Coulter counter.
4. Sterile 10 mM phosphate-buffered saline (PBS) pH 7.2 (Gibco-BRL Life Technologies, Paisley, UK).
5. Dulbecco's modified Eagle's medium (DMEM), heat-inactivated fetal calf serum (FCS), L-glutamine, penicillin, streptomycin, 1:5000 Versene, and 2.5% trypsin solution (Gibco-BRL).
6. Standard culture medium: DMEM supplemented with 5% FCS (v/v), 100 U/mL penicillin, 100 µg/mL streptomycin, 4 mM glutamine and 0.375% sodium bicarbonate (Gibco-BRL). Store at 4°C.
7. Trypsin solution for cell release: 0.5 mL of 2.5% trypsin solution in 25 mL of 1:5000 Versene. Store at 4°C.
8. Bovine serum albumin (BSA) solution for cell culture (10 mg/mL): dissolve 200 mg BSA (Sigma, Poole, UK) in 20 mL distilled water and sterilize by passing through a 0.2-µm filter. The solution can be stored at 4°C for up to 2 mo.
9. Serum-free medium: DMEM supplemented with 100 U/mL penicillin, 100 µg/mL streptomycin, 4 mM glutamine, 0.375% sodium bicarbonate, and 0.25 mg/mL BSA. To make 100 mL serum-free medium, add 2.5 mL of 10 mg/mL BSA solution to 100 mL standard culture medium without the presence of fetal calf serum. Store at 4°C.
10. 40 µCi/mL [methyl-^{3}H]thymidine in 0.8 µM thymidine (Amersham, Little Chalfont, UK) *(see* Note 1).
11. 95% Ethanol (v/v).
12. 5% Trichloroacetic acid (TCA) (w/v).
13. 0.2M NaOH.
14. Scintillation cocktail, measuring tubes, and Packard scintillation counter (Packard, Pangbourne, UK).
15. HT29 colon cancer cells (European Cell Culture Collection at the Public Health Laboratory Service, Porton Down, UK).
16. Mushroom lectin (ABL) from common edible mushroom *Agaricus bisporus* and peanut lectin (PNA) from *Arachis hypogaea* (Sigma).

3. Methods

1. Culture HT29 cells in 9-cm tissue culture dishes in a 5% CO_2/95% air incubator at 37°C (*see* Note 2).
2. When the cells become 70–80% confluent, wash twice with 8 mL PBS (*see* Note 3).
3. Add 1 mL trypsin solution to the dish, swirl over the surface of the cells, and leave the dish at 37°C for 2–5 min to release the cells. Check under the microscope after 2 min (*see* Note 4).
4. When the cells are detached, add 6 mL standard culture medium to the dish and aspirate up and down gently using a narrow-aperture pipet to disperse the cells (*see* Note 5).
5. Take 0.5 mL of the cell suspension into 10 mL PBS and measure the cell number by Coulter counter (*see* Note 6).
6. Make a 2–5 × 10^4 cells/mL suspension by dilution of the trypsin-released cell suspension with standard culture medium. Seed 0.5 mL/well into 24-multiwell plates with a multipipeter, and incubate the plates at 37°C.
7. After 2 d culture when the cells have reached the log phase growth (*see* Note 7), wash twice with 0.5 mL/well PBS (*see* Note 8), add 0.5 mL serum-free medium to each well, and culture the cells for a further 24 h in this medium at 37°C (*see* Note 9).
8. Add the lectin under test to the relevant wells, and culture for a further 24 h at 37°C (*see* Note 10).
9. Add 20 µL of 40 µCi/mL [methyl-^{3}H]thymidine to each well, and incubate for 1 h at 37°C (*see* Note 11).
10. Wash the cells twice with PBS, precipitate the proteins and DNA with 5% ice-cold TCA, and leave the plate at 4°C for at least 30 min.
11. Wash once with 5% ice-cold TCA to remove the attached and unincorporated [^{3}H]thymidine and twice with ice-cold 95% ethanol to remove the remaining TCA.
12. Dry at room temperature for 2 h.
13. Solubilize the precipitate with 0.5 mL/well 0.2M NaOH, and leave the plate at room temperature for at least 2 h (*see* Note 12).
14. Transfer the dissolved precipitate to scintillation vials, add 1 mL Optima Gold MV scintillation cocktail, and count in a Packard scintillation counter (*see* Fig. 1).

4. Notes

1. A mixture of a small quantity (40 µCi/mL) of [^{3}H]thymidine in a larger amount (0.8 µM) of cold thymidine is used to label the cells. This has proved to be an effective way of regulating thymidine cellular pool size and so allowing reliable measurement of DNA syntheis *(14)*.
2. All cells are cultured in a 5% CO_2/95% air humidified incubator at 37°C to avoid rapid medium loss. Cells are routinely passaged when they become 80–90% confluent at a 1:6–1:8 subculture ratio and narrow passage cells are used for all studies. All cell culture medium and solutions are warmed to 37°C before use to avoid shock to the cells.

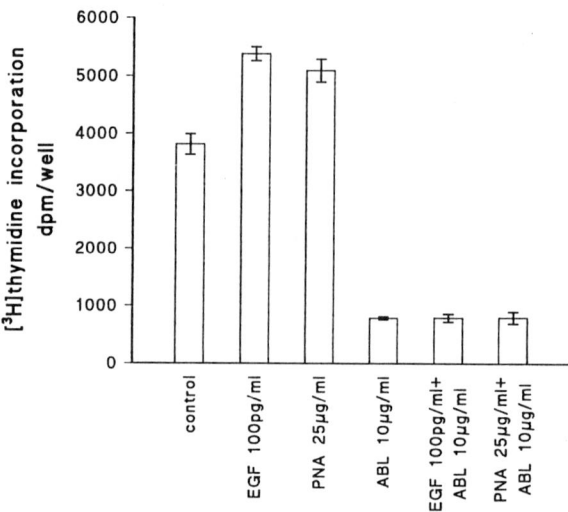

Fig. 1. Stimulation of proliferation of HT29 cells by peanut lectin (PNA) and epidermal growth factor (EGF) with inhibition by mushroom lectin (ABL). 100 pg/mL EGF, 25 µg/mL PNA, and 10 µg/mL ABL. Reprinted with the permission from ref. *12*.

3. 70–80% Confluent cells are used for all the studies. More confluent cells produce less reproducible results.
4. Gently shaking the plates after trypsin treatment helps to release the cells from the plates.
5. FCS is included in the medium to inhibit further activity of trypsin.
6. Cell number is measured either with a Coulter counter or a hemocytometer. If a Coulter counter is employed, it is essential that there is a single cell suspension. Poorly separated cells will result in a lower count and may block the aperture. If the cell number needs to be measured after interaction with lectins, the resulting agglutination means that it is usually necessary to add some hapten sugars into the medium to help to separate the cells. If the hapten addition is not successful at separating the cells, then using a hemocytometer instead of a Coulter counter to measure the cell number is suggested.
7. Cells are grown to log-phase prior to synchronization in serum free medium as in this stage cell growth is at its maximum. The time of log-phase growth varies between cell lines; for the HT29 cells this is between d 2 and 5 after plating.
8. Special attention needs to be paid to avoid washing the cells off the plate during the washing procedures, especially when using poorly adherent cell lines. Large aperture plastic pipets are used in all washing procedures. After each wash, check the plates under the microscope to make sure that the cells have not been washed off. Low radioactive counts at the end of experiments are most likely to be caused by loss of cells during washing.

9. The main advantage of using serum-free medium is that the system is simple and "clean" without the potential interference of the components of fetal calf serum (e.g., fetuin in fetal calf serum is a strong inhibitor of mushroom lectin).
10. To avoid dilutional effects, different doses of the lectins should preferably be added in the same volume, but usually the effect of tiny volume differences can be neglected if the volume of lectin solution added to the culture is <2–3% of the volume of the culture medium.
11. We use 1 h flash labeling with [^{3}H]thymidine to label the cells instantaneously in the synthetic phase. This is the most common method to label the cells in this system. Continuous labeling/longer labeling time, in which the thymidine is available for an interval exceeding cell cycle time is also used by some investigators.
12. The solubilization time can be shortened to 15–30 min by incubating the plates at 37°C. Leaving plates at room temperature overnight is convenient for solubilization.

References

1. Liener, I. E., Sharon, N., and Goldstein, I. J. (1986) *The Lectins; Functions, and Applications in Biology and Medicine.* Academic, New York, pp. 270–277.
2. Carraway, K. L. and Carraway, C. A. C. (1989) Membrane-cytoskeleton interactions in animal cells. *Biochim. Biophys. Acta* **988,** 147–171.
3. Sharon, N. (1976) *Mitogens in Immunobiology.* Academic, New York, pp. 31–34.
4. Peacock, J. S., Colsky, A. S., and Pinto, V. B. (1990) Lectins and antibodies as tools for studying cellular interactions. *J. Immunol. Methods* **126,** 147–157.
5. Bates, L. G., Grove, D. S., and Mastro, A. M. (1995) Mechanisms of activation and suppression in rat Nb 2 lymphoma cells: a model for interactions between prolactin and the immune system. *Exp. Cell Res.* **218,** 567–572.
6. Wright, N. and Alison, M. (1984) *The Biology of Epithelial Cell Populations.* Clarendon Press, Oxford, UK, pp. 1–20.
7. Feizi, T. (1985) Demonstration by monoclonal antibodies that carbohydrate structures of glycoproteins and glycolipids are oncodevelopmental antigens. *Nature* **314,** 53–57.
8. Raz, A. and Lotan, R. (1987) Endogenous galactoside-binding lectins: a new class of functional tumor cell surface molecules related to metastasis. *Cancer Metastasis Rev.* **6,** 433–452.
9. Nicolson, G. L. (1982) Cancer metastasis, organ colonization and cell-surface properties of malignant cells. *Biochim. Biophys. Acta* **95,** 113–176.
10. Ryder, S. D., Smith, J. A., and Rhodes, J. M. (1992) Peanut lectin: a mitogen for normal human epithelium and human HT29 colorectal cancer cells. *J. Natl. Cancer Inst.* **84,** 1410–1416.
11. Ryder, S. D., Parker, N., Ecclestone, D., Haqqani, M. T., and Rhodes, J. M. (1994) Peanut lectin stimulates proliferation in colonic explants from patients with inflammatory bowel dizease and colon polyps. *Gastroenterology* **106,** 117–124.

12. Yu, L.-G., Fernig, D. G., Smith, J. A., Milton, J. D., and Rhodes, J. M. (1993) Reversible inhibition of proliferation of epithelial cell lines by *Agaricus bisporus* (edible mushroom) lectin. *Cancer Res.* **53,** 4627–4632.
13. Boulton, R. A. and Hodgson, H. J. F. (1995) Assessing cell proliferation: a methodological review. *Clin. Sci.* **88,** 119–130.
14. Smith, J. A., Winslow, D. P., and Rudland, P. S. (1984) Appearance of basic growth factor stimulate cell division of rat mammary epithelial, myoepithelial and stromal cell lines in culture. *J. Cell. Physiol.* **119,** 320–326.

32

Use of Lectins as Mitogens for Lymphocytes

David C. Kilpatrick

1. Introduction

Resting lymphocytes can be induced to undergo DNA synthesis and subsequently cell division and proliferation by a wide variety of agents, but undoubtedly lectins constitute the most convenient generic group of mitogens to pick off the shelf as reagents. Under the influence of strongly mitogenic lectins, a high proportion of T-lymphocytes (irrespective of T-cell receptor specificity) will differentiate into active cytokine-synthesizing cells (helper cells, predominantly with a CD4-positive phenotype) and/or into cells with a functional suppressor or cytotoxic (both predominantly CD8-positive) capacity. Some resting B-lymphocytes will also become activated and subsequently proliferate and differentiate as fully as plasma cells. Although most work has been done in vitro, there is good evidence that some lectins are also mitogenic in vivo *(1)*.

To describe a lectin as mitogenic is not to indicate an absolute quality, as a particular lectin may be mitogenic with certain cells under certain conditions, but inactive toward a different source of lymphocytes or toward the same lymphocyte source under inappropriate conditions. Usually, lectins are evaluated for activity with human peripheral blood mononuclear cells, or with peripheral blood, spleen, or thymus cell preparations from rodents. A mitogenic lectin is one which displays mitogenic activity toward a specified cell preparation under optimal experimental conditions. Often a particular lectin is mitogenic (or nonmitogenic) for both murine and human cells. Some of the lectins that display strong (and largely species nonspecific) mitogenic activity are listed in Table 1.

The lectins most commonly used as mitogens are *Phaseolus vulgaris* lectin (PHA), concanavalin A (con A), and pokeweed mitogen (PWM). Con A stimu-

Table 1
Some Mitogenic Lectins

Name	Source	Ref.
Black locust lectin	*Robinia pseudoacacia*	6
Common vetch lectin	*Vicia sativa*	7
Con A	*Canavalia ensiformis* (jack bean)	8
Crotalaria lectin	*Crotalaria juncea*	9
H. crepitans lectin	*Hura crepitans*	10
Lentil lectin	*Lens culinaris*	8
Limulin-III	*Limulus polyphemus* (Horseshoe crab)	8
Meadow saffron lectin	*Colchicum autumnale*	11
P. coccineus lectin	*Phaseolus coccineus*	9
PHA	*Phaseolus vulgaris* (kidney bean)	8
PA-I & PA-II	*Pseudomonas aeruginosa*	12
PWM	*Phytolacca americana* (pokeweed)	8
Rice lectin	*Oryza sativa*	13
Shoriku mitogen	*Phytolacca esculenta*	14
Sweet pea lectins	*Lathyrus* spp	15
Tulip lectin	*Tulipa* spp	11

lates mainly T-lymphocytes, whereas PWM stimulates B-cells but in a T-cell dependent manner. I am not aware of any lectin which exclusively stimulates human B-cells.

Lectins are used as mitogenic agents for two routine purposes in pathology, karyotyping and assessment of immune competence. Within a day of culturing with PHA, there are sufficient lymphocytes in metaphase at any moment to permit ready assessment of chromosomal number and gross structure. The use of PHA revolutionized karyotyping, and there is no reason why it should be superceded in the foreseeable future. The assessment of immune function may be performed without lectins, and the impairment of the lymphocyte proliferation response to many stimuli is secondary to many diseases, rendering such assays nondisease specific and difficult to interpret. Nevertheless, the response to mitogenic lectins is still considered to be an essential investigation into suspected primary immune deficiency *(2)* (Table 2). In this context, the ability to mount a delayed-type hypersensitivity response is measured, and the lymphocyte proliferation assay is the in vitro equivalent of the intradermal skin test. It should be noted that, whereas the use of recall antigens (whether in vitro or in skin tests) requires previous sensitization, the PHA test is independent of antigenic recognition, and probably correlates simply with the ability to respond with an appropriate secretion of cytokines.

Table 2
Primary Immune Deficiency Diseases Associated with Low or Absent Proliferative Response to Mitogenic Lectins

Severe combined immune deficiency (all forms)
Di George syndrome (congenital thymic aplasia)
X-linked lymphoproliferative (Duncan's) syndrome
Nezelof's syndrome
LFA-1/Mac-1 glycoprotein deficiency
Immunodeficiency with ataxia-telangiectasia
Immunodeficiency with short-limbed dwarfism
Immunodeficiency with purine nucleoside phosphorylase deficiency
Immunodeficiency with adenosine deaminase deficiency

The research uses of lymphocyte proliferation assays are many and various. It is advisable when characterizing a new lectin to determine whether it is mitogenic or nonmitogenic. Conversely, newly discovered lectins and other materials can be tested for antimitogenic or comitogenic activity by coculture with a known mitogenic lectin. Lectins are useful tools to investigate aspects of the immune response (e.g., accessory cell dependence, synthesis of a particular metabolite or cytokine, and so on) and this may involve similar methodology. Lymphocyte proliferation assays can also be used to monitor response to immunosuppressive or immunostimulatory treatments, although particular care has to be taken regarding the inevitable day-to-day variation in activity.

Lymphocyte blastogenesis was originally assessed by microscopic examination of stained blood smears, looking for the presence of lymphoblasts. This laborious procedure was soon replaced by measurement of DNA synthesis using radioactive (^{3}H)thymidine. The main drawback of this method is the inconvenience of radioisotopes, with the attendant codes of practice and legal requirements. Despite the development of some nonradioactive proliferation assays, the detection of radioactive thymidine incorporation remains the standard and most widely used assay. What follows is a detailed description of the basic method, which can be modified for specific purposes as indicated in Section 4. The methodology is the same regardless of the species or the tissue source of the lymphocytes, although I chose to describe, as an example, the use of human peripheral blood mononuclear cells. It is essential that all procedures are performed under strictly aseptic conditions (*see* Note 1). The variables chosen are broadly based on the procedure of Penhale and coworkers *(3)*, but it should be noted that miniatured assays have been described that use approx one-tenth the number of cells *(4)*.

2. Materials

1. McCartney bottles containing around 25 glass balls (2-mm diameter [Mackay & Lynn Ltd, Edinburgh, Scotland]); or preservative-free lithium heparin blood collection tubes.
2. Dextran solution: 10% (w/v) dextran (mol wt 150,000–200,000) in 0.9% NaCl.
3. Serum (*see* Note 2).
4. RPMI-1640 culture medium. This should be supplemented with penicillin (final conc. 100×10^3 U/L) and streptomycin (100 mg/L) immediately before use.
5. Antibiotics: A mixture of penicillin and streptomycin should be purchased in lyophilized form and reconstituted immediately before use.
6. Ficoll solution, S.G. 1.077 g/L (*see* Note 3).
7. Trypan blue: 1% (w/v) in phosphate-buffered saline.
8. 6-^{3}H-thymidine, specific activity 307 MBq/mg (8.3 mCi/mg) diluted 1 in 4 vol of RPMI culture medium.
9. Glass fiber filter paper for the cell harvester.
10. Scintillation cocktail (*see* Note 4).

3. Methods

3.1. Preparation of Cells

1. Pour a fresh venous blood sample (20 mL) into a McCartney bottle with glass balls and defibrinate by shaking gently by hand for 10 min. Alternatively, anticoagulated blood may be used as starting material (*see* Note 5).
2. Divide the blood sample equally into two, sterile 30-mL plastic containers. Add dextran solution (2 mL) and make up to a final volume of 20 mL with RPMI medium. Mix well by inversion, and leave in a water bath at 37°C for 15–20 min, by which time the red blood cells will have settled to the bottom leaving a leukocyte-rich upper phase.
3. Place ficoll solution (7 mL/tube) into 15-mL sterile clear plastic centrifuge tubes. Layer an equal volume (per tube) of the leukocyte-rich supernatant from the dextran sedimentation step on top. (This is best done by tilting the ficoll-containing tube at a 45° angle and gently allowing the leukocyte suspension to run down the side).
4. Centrifuge at 700g for 20–25 min, at room temperature. Remove the mononuclear cell layer formed at the interface of the solution by aspiration into a separate centrifuge tube.
5. Pellet cells by centrifugation at 500g for 10 min, then wash the pellet twice by resuspension in RPMI and harvesting by centrifugation at 400g for 8 min. Finally, resuspend the cells in 5 mL of RPMI containing 15% serum.
6. Count the cells in a hemocytometer chamber, after first removing a small aliquot and adding trypan blue solution to a final concentration of 0.1% (*see* Note 6).
7. Adjust the cell density to 1.5×10^6 viable mononuclear cells/mL by addition of an appropriate extra volume of 15% serum in RPMI.

3.2. Culture of Cells with Mitogen

1. Place mitogen solution (25 µL) in triplicate into round-bottomed tissue culture trays. Usually a range of concentrations of mitogen in medium will be used, and

certainly a negative control (culture medium only) must always be included (*see* Note 7).
2. Make the volume/culture well up to 50 µL with RPMI, or an additional factor if, for example, an unknown is to be tested for antimitogenic activity.
3. Add 100 µL of the mononuclear cell suspension to each culture well.
4. Place the culture plate in an incubator set at 37°C and 5% CO_2, and leave for 2 d (40–48 h) (*see* Note 8).
5. Add 4 µL (1 µCi) of ^{3}H-thymidine solution to each culture well, and return to the CO_2 incubator for a further 4 h (*see* Note 9).

3.3. Measurement of DNA Synthesis

1. Use a cell harvester (Skatron, UK; Newmarket, UK) to transfer the cultures onto a glass fiber filter mat, and wash thoroughly with distilled water (*see* Note 10).
2. Allow the glass fiber filters to dry completely, e.g., at 60°C for 2 h.
3. Place each filter disk in a scintillation vial and add an appropriate scintillation cocktail (*see* Note 11).
4. Count vials for radioactivity using a scintillation counter set for appropriate low energy β-emissions (*see* Note 11).

3.4. Analysis of Results

1. The means and standard deviations of the triplicates should be calculated. This alone might be a satisfactory way to present the data and undertake any statistical significance testing required.
2. Additionally, and depending on the appropriateness of the circumstances, express the data as stimulation indices, SI: (experimental cpm ÷ control cpm), or as relative proliferation indices, RPI ($\triangle$ cpm of test individual ÷ $\triangle$ cpm of normal individual) (Table 3) (*see* Note 12).

4. Notes

1. All operations should take place within a microbiological safety cabinet. All solutions should be autoclaved or filter sterilized. All plastics and other materials that cannot be autoclaved should be purchased in a sterile condition.
2. Autologous serum is often most suitable. Alternatively, a pool of AB human sera or fetal calf serum may be used. Nonautologous sera should always be checked in advance to ensure it is neither excessively inhibitory or stimulatory.
3. Various commercial preparations are available for example, "Lymphocyte Separation Medium" (Nycomed), consisting of ficoll and sodium metrizoate or "Histopaque 1077" (Sigma) containing sodium diatrizoate and ficoll. Alternatively, Percoll (Pharmacia) can be used.
4. The choice of scintillation fliud depends not only on the isotope but on the scintillation counter available to detect it. The cocktail recommended by the counter's manufactures (usually available ready for use) should be most suitable and certainly a high counting efficiency is to be expected and can be calculated by spiking with a known amount of radioisotope.

Table 3
Typical Results of Lymphocyte Proliferation Assays and Their Presentation[a]

	PHA, μg/mL	cpm, mean ± 1SD	SI	RPI
A. Patient (pretransplant)	0	105 ± 20		
	0.5	130 ± 15	1.2	0.002
	1.0	176 ± 27	1.7	0.002
	2.0	180 ± 45	1.7	0.001
B. Patient (posttransplant)	0	123 ± 18		
	0.5	2328 ± 210	18.9	0.14
	1.0	9573 ± 450	77.8	0.28
	2.0	27,388 ± 3362	223	0.56
C. Control 1	0	153 ± 21		
	0.5	8529 ± 995	55	
	1.0	28,166 ± 3410	184	
	2.0	68,019 ± 1020	445	
Control 2	0	295 ± 141		
	0.5	15,481 ± 5210	52	
	1.0	46,263 ± 8675	157	
	2.0	61,730 ± 5920	209	
Control 3	0	190 ± 25		
	0.5	16,223 ± 3548	85	
	1.0	33,640 ± 1376	177	
	2.0	48,840 ± 2426	256	

[a]The example is given of a child with congenital immune deficiency whose response to PHA was simultaneously compared to that of two normal controls (1 and 2). The mean values from these controls were used to calculate the relative proliferation index (RPI) for each concentration of PHA. The patient was treated by bone marrow transplantation with her HLA-identical brother as a donor. About a month later, the patient's lymphocyte proliferation response was again assessed and simultaneously compared to a third normal individual (control 3). Ideally, the RPI should be calculated from mean values derived from at least three individuals previously shown to have proliferative responses typical of the healthy population.

5. A mononuclear cell suspension prepared as described is the simplest and most commonly used target for mitogens. However, some purposes require the isolation of particular cellular constituents. A detailed practical description of how to separate the major cellular components of blood is given elsewhere (5).
6. In the presence of trypan blue, viable mononuclear cells appear clear, bright, and refractile under phase contrast microscopy.
7. This negative (no added stimulus) control, reflecting spontaneous DNA synthesis, is also a valuable quality assurance indicator. Typically, one would expect a few hundred counts per min. Less than 100 cpm should raise suspicion about the postculture viability of the cells, whereas greater than 1000 cpm may be an indication of microbial contamination.

8. A 2-d culture period is suitable for commonly used mitogenic lectins like PHA, but, if testing an unknown for mitogenic activity varying periods, up to 6–8 d should be tried.
9. A 4-h pulse is optimal for strong mitogens, but for weak mitogens a longer pulse may be required and up to 18 h is permissible.
10. Cell transfer is unacceptably tedious without some kind of automatic or semiautomatic cell harvester. The glass fiber filter mats used in conjunction with the harvester would normally be those recommended by the manufacturer of the harvester.
11. The settings of the scintillation counter, like the choice of scintillation cocktail, depend on the nature of the counter available. Known amounts of ^{3}H-thymidine can be spotted on to the glass fiber filters to check the counting efficiency.
12. Standardization of lymphocyte proliferation assays is a major problem and results can sometimes be difficult to interpret, especially if the responses of different individuals to the same mitogenic lectin are being compared. For that purpose, the SI is unsuitable as a minor change in the negative control value can give a misleadingly high alteration to the SI. The RPI is generally more appropriate for comparing individuals, especially if the control lymphocyte donor(s) is chosen because his response is known to be typical of normal individuals. For a more complete discussion, see ref. 4.

References

1. Kilpatrick, D. C. (1995) Lectins in immunology, in *Lectins-Biomedical Perspectives* (Pusztai, A. and Bardocz, S., eds.), Taylor & Francis, Basingstoke, UK, pp. 155–182.
2. IUIS/WHO Report (1988) Laboratory investigations in clinical immunology: methods, pitfalls, and clinical indications. *Clin. Exp. Immunol.* **74,** 494–503.
3. Penhale, W. T., Farmer, A., Maccuish, A. C., and Irvine, W. J. (1974) A rapid micro-method for the phytohemagglutinin-induced human lymphocyte transformation test. *Clin. Exp. Immunol.* **18,** 155–167.
4. Maluish, A. E. and Strong, D. M. (1986) Lymphocyte proliferation, in *Manual of Clinical Laboratory Immunology,* 3rd ed. (Rose, N. R., Friedman, H., and Fahey, J. L., eds.), American Society for Microbiology, Washington, DC, pp. 274–281.
5. Kilpatrick, D. C. (1993) Isolation of lymphocytes and accessory cells for assays of mitogenicity, in *Lectins and Glycobiology* (Gabius, H.-J. and Gabius, S., eds.), Springer-Verlag, Berlin, pp. 369–375.
6. Wantyghem, J., Goulut, C., Frénoy, J.-P., Turpin, E., and Goussault, Y. (1986) Purification and characterization of *Robinia pseudoacacia* seed lectins. *Biochem. J.* **237,** 483–489.
7. Falasca, A., Franceschi, C., Rossi, C. A., and Stirpe, F. (1979) Purification and partial characterization of a mitogenic lectin from *Vicia sativa. Biochim. Biophys. Acta* **577,** 71–81.
8. Lis, H. and Sharon, N. (1981) Lectins in higher plants, in *The Biochemistry of Plants,* vol. 6 (Marcus, A., ed.), Academic, New York, pp. 371–447.

9. Licastro, F., Chiricolo, M., Barbieri, L., and Stirpe, F. (1983) Effect of 32 purified animal and plant lectins on human T-lymphocytes, in *Lectins—Biology, Biochemistry, Clinical Biochemistry,* vol. 3 (Bøg-Hansen, T.-C. and Spengler, G. A., eds.), Walter deGruyter, Berlin, pp. 293–302.
10. Falasca, A., Franceschi, C., Rossi, C. A., and Stirpe, F. (1980) Mitogenic and hemagglutinating properties of a lectin purified from *Hura crepitans* seeds. *Biochim. Biophys. Acta* **632,** 95–105.
11. Kilpatrick, D. C., Peumans, W. J., and Van Damme, E. J. M. (1990) Mitogenic activity of monocot lectins, in *Lectins—Biology, Biochemistry, Clinical Biochemistry,* vol. 7 (Koucourek, J., ed.), Sigma, St. Louis, MO, pp. 259–263.
12. Avichezer, D. and Gilboa-Garber, N. (1987) PA-II, the L-fucose and D-mannose binding lectin of *Pseudomonas aeruginosa* stimulates human peripheral lymphocytes and murine splenocytes. *FEBS Lett.* **216,** 62–66.
13. Shen, Z.-W., Sun, C., Zhu, Z., Tang, X.-H., and Shen, R.-J. (1984) Purification and properties of rice germ lectin. *Can. J. Biochem. Cell Biol.* **62,** 1027–1032.
14. Tokuyama, H. (1973) Isolation and characterization of pokeweed mitogen-like phytomitogens from shoriku, *Phytolacca esculenta. Biochim. Biophys. Acta* **317,** 338–350.
15. Borrebaeck, C. A. K. and Rougé, P. (1986) Mitogenic properties of structurally related *Lathyrus* lectins. *Arch. Biochem. Biophys* **248,** 30–34.

33

Effect of Lectins on Uptake of Polyamines

Susan Bardocz and Ann White

1. Introduction

The aliphatic polyamines (putrescine, cadaverine, spermidine, and spermine) are flexible polycations, since under physiological conditions they exhibit 2, 3, or 4 positive charges, respectively, and are able to rotate around the carbon-carbon or carbon-nitrogen bonds, imparting conformational flexibility (see Fig. 1).

Polyamines fulfil an array of roles in cellular metabolism (1,2) and are involved in many steps of protein, RNA, and DNA synthesis, from the control and initiation of translation (3); regulation of its fidelity (4); stimulation of ribosome subunit association (5), through enhancement of RNA (6) and DNA syntheses (7); stabilization of the structure of tRNA (8) and reduction of the rate of RNA degradation (9); and help to condense DNA (10) to the covalent modification of proteins (11). Although some of the biological effects of polyamines are specific, they interact nonspecifically with negatively charged structures of cells (12). In these nonspecific roles, polyamines can be replaced by metal ions, most often by Mg^{2+} or Ca^{2+}.

Putrescine, spermidine, and spermine are found in different quantities in all mammalian cells: putrescine is present at usually <0.1 µmol/g fresh weight in resting tissue and hardly ever exceeds 0.3–0.4 µmol/g; spermidine and spermine concentrations in most tissues lie within the range of 0.1–2 µmol/g fresh weight. Exceptions are the pancreas and the prostate gland, in which spermidine concentrations are higher (11). Polyamine content tends to decrease with age in many tissues and after hormone deprivation in hormone-dependent organs. It is interesting that the only types of mammalian cells that contain little or no polyamines are those with no (functional) nuclei, such as mature erythrocytes or platelets, or those in which nuclear DNA replication, RNA transcription,

From: *Methods in Molecular Medicine: Vol. 9: Lectin Methods and Protocols*
Edited by: J. M. Rhodes and J. D. Milton Humana Press Inc., Totowa, NJ

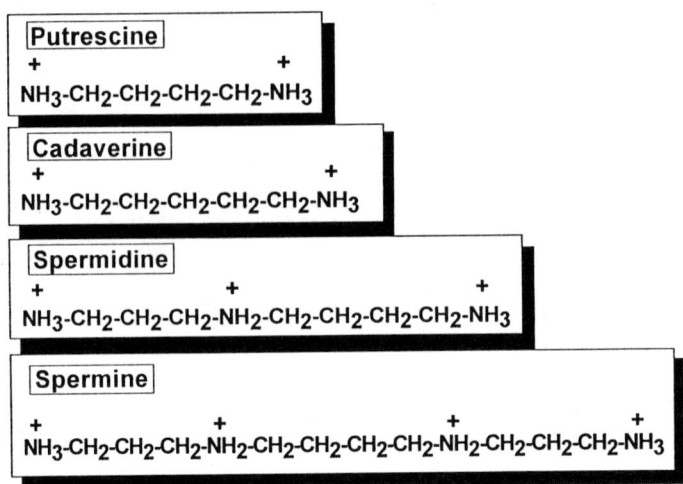

Fig. 1. The structure of the most common polyamines.

and extramitochondrial protein synthesis are irreversibly switched off, such as spermatozoa *(11)*. Although polyamines are also found in blood plasma and many other body fluids, their concentrations are extremely low, in the pmol/mL range.

Polyamines are essential for growth and cell proliferation. However, their most important function, in which they cannot be replaced by any of the other positively charged molecules, is as mediators of the action of all known hormones and growth factors. Similarly, because by binding to and interacting with cell surface receptors of the brush border membrane, plant lectins such as PHA (phytohaemagglutinin, *Phaseolus vulgaris*), SBA (soya bean, *Glycine max*), WGA (wheatgerm, *Triticum aestivum*), DSA (thorn apple, *Datura stramonium*), and UDA (nettle, *Urtica dioica*), can induce extensive proliferation and changes in the metabolism of epithelial cells via activation of second messenger pathways, this hyperplastic growth of the gut also requires large amounts of polyamines, particularly spermidine (Fig. 2).

In model experiments with PHA (the lectin from the kidney bean, *Phaseolus valgaris*), it was shown that lectins binding avidly to the mucosal surface induce dose- and time-dependent and fully reversible hyperplastic and hypertrophic growth of the small bowel, which is accompanied by the accumulation of polyamine in the tissue *(13)*.

Since all cells are able to synthesize some polyamines, it was believed that polyamines are synthesized *in situ* according to the cells' need. In the case of the gut, its metabolic activity is greatly influenced by the diet and particularly by antinutritional factors, including some lectins, tannins, and saponins, which are present in everyday foodstuffs of plant origin. When the metabolic activity

Lectin Effects on Polyamine Uptake

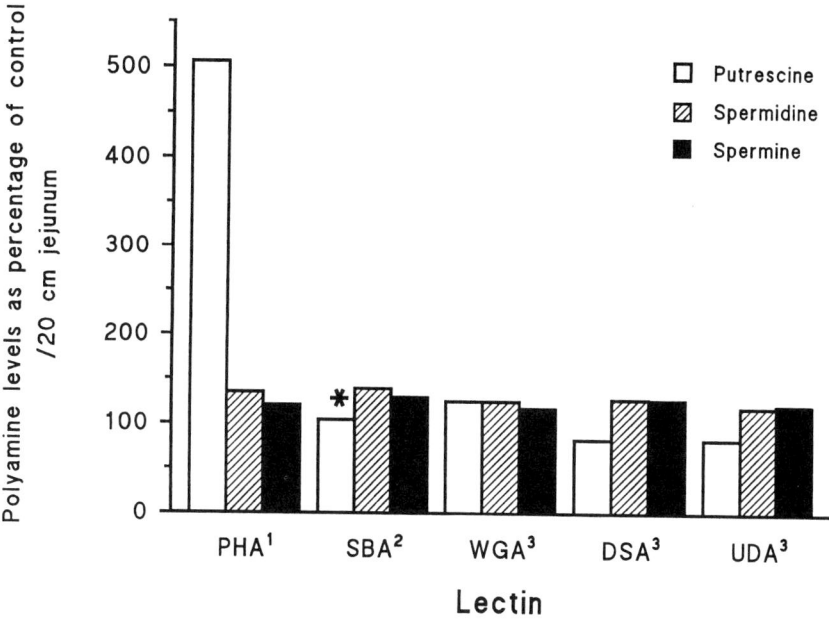

Fig. 2. Polyamine content of the rat jejunum fed either lactalbumin (control, semisynthetic diet) or lectin containing diets (42 mg lectin/rat/d for 10 d) expressed as a percentage of the control (pair-fed on semisynthetic [lactalbumin] diet) values. All polyamine levels of lectin-fed rats were significantly different than than the corresponding control (except for *) to at least $p < 0.05$. PHA, phytohemagglutinin (lectin from kidney beans, *Phaseolus vulgaris)*; SBA, soya bean *(Glycine max)*, agglutinin; WGA, wheatgerm *(Triticum aestivum)*, agglutinin; DSA, thorn apple *(Datura stramonium)*, agglutinin; UDA, nettle *(Urtica dioica)* rhizome agglutinin. [1]ref. *13*; [2]ref. *14*; [3]ref. *15*.

of the gut is increased, its polyamine requirement is also enhanced and must be met if the structural integrity and the proper functioning of the gut are to be maintained. However, in many models of induced, adaptive gut growth, such as after fasting-refeeding *(16)*, or consumption of PHA *(17)*, the tissues' capacity for biosynthesis of putrescine is not sufficiently elevated to account for all polyamines that accumulate in the tissue, despite a significant increase in ornithine decarboxylase (ODC; EC 4.1.1.17) activity. Therefore, as more and more information has gathered, the importance of polyamines from extracellular source has become obvious. Thus, one of the first effects of the PHA signal is to stimulate the basolateral polyamine uptake system for the sequestration of polyamines from the circulation in sufficient amounts to sustain the growth of the tissue *(18)*. The data indicate that the main source of polyamines for the gut is the circulating blood and that growth of the gut is preceded by the accumulation of large amounts of polyamines, mostly spermidine, in the tissue, without a major increase in its ODC activity *(17–20)*. This was confirmed in studies using the substrate analog, α-difluoromethylornithine (DFMO; *21*), an enzyme-activated, irreversible inhibitor of ODC *(22)*. As this drug did not block PHA-induced gut growth, it appears that ODC induction is not a prerequisite for polyamine accumulation *(21)*. The large increase in tissue polyamine pool size, despite only a slight elevation in intestinal ODC activity and the lack of effect of DFMO on gut growth, further confirmed that there must be an extracelluar source of polyamines and that the small bowel may also acquire polyamines independent of *de novo* synthesis.

Finally, as it has been suggested that the polyamine content of an organ correlates with its metabolic activity *(23)*, measurement of the content and uptake of polyamines by a tissue can be used as an indicator of the effectiveness of the lectins as growth factors *(24)*.

1.1. Quantification of Polyamines by HPLC Analysis

In spite of the undoubted importance of polyamines in growth, little information is available on the polyamine content and composition of the gut tissue under different metabolic conditions. In order to determine the polyamine content and composition of a particular tissue and correlate these with its hyperplastic or hypertrophic growth, reliable sample preparation methods, suitable for application to HPLC, have been devised for the identification of individual polyamines and standards. The HPLC method is based on that of Seiler and Knödgen *(25)*. The routine estimation of protein *(26)*, RNA *(27)*, and DNA *(28)* concentrations on these samples is also carried out.

1.2. Luminal and Basolateral Uptake of Labeled Polyamines

The origin of polyamines is of importance. To establish the extent of luminal polyamine uptake, rats are fed either a lactalbumin diet (control) or lactalbumin diet supplemented with PHA and after 3 d they are given a mixture of labeled and unlabeled polyamines by gastric intubation. In a separate experiment, rats are fed the same two diets for 3 or 7 d, after which they are injected intraperitoneally with labeled polyamine to determine the basolateral polyamine uptake and compare this with the luminal uptake *(18,23)*.

1.3. Variation in Ornithine Decarboxylase Activity in Relation to Polyamine Levels

Cellular response to stress is characterized by a marked induction of ornithine decarboxylase, the key regulatory enzyme in the polyamine biosynthetic pathway. It's activity can be determined by measuring the rate of formation of $^{14}CO_2$ from labeled ornithine hydrochloride *(29)*.

2. Materials
2.1. Quantification of Polyamines by HPLC Analysis, and Protein, RNA, and DNA
2.1.1. Tissue Preparation

1. Ultra-Turrax (Janke & Kinkel GmbH & Co., IKA Labortechnik, D-79219 Staufen, Germany) variable speed homogenizer with an 8-mm shaft was used throughout.
2. 10% Perchloric acid (PCA).
3. Stock internal standard (1,7-diaminoheptane): dissolve 13 mg in 100 mL distilled water. This will keep at 4°C for several weeks.
3. 500 mL 10% PCA containing 10 mL stock internal standard.
4. 0.3M NaOH.
5. 0.05M NaOH.

2.1.2. Polyamine Analysis

1. In this section, all reagents, including water, should be of HPLC grade.
2. 1% PCA.
3. Standard solution containing: putrescine, spermidine, spermine, acetyl-spermine, cadaverine, histamine, N^1-acetylspermidine, N^8-acetylspermidine and 1,7-diaminoheptane. Individual polyamines made up to a concentration of 10 mM in 1% PCA. Add 100 μL of each to a 100 mL volumetric flask and make up to the mark with 1% PCA (*see* Note 1 for storage details).
4. Column: Dynamax Microsorb C18 reverse-phase column (25 cm × 4.6 mm) packed with 5 μm octadecasilane (*see* Note 2 for storage details).
5. Guard column: 1 cm × 2 mm column packed with 30–40 μm octadecasilane (*see* Note 3).
6. Waters 600E pump and system controller for gradient.
7. Gilson model 231 automated sample injector fitted with 200 μL loop.
8. Pye Unicam PU4010 pump for postcolumn derivatization.
9. Waters 470 scanning fluorescence detector.
10. Phillips PU 6000 chromatography data package for integration of chromatograms.
11. *o*-Phthaldehyde reagent: stock reagent: dissolve 50 g boric acid, 44 g potassium hydroxide and 3 mL 30% Brij solution in 1 L water (this can be made up in 5-L batches and stored at room temperature for about 1 wk). Just before use, add 400 mg *o*-phthaldehyde dissolved in 5 mL methanol to 1 L of stock reagent. Immediately before use, add 2 mL 2-mercaptoethanol. **Note:** As soon as the mercaptoethanol is added, the solution begins to deteriorate and must be discarded at the end of the day.
12. Elution solution A: 0.1M sodium acetate (adjusted to pH 4.5 with acetic acid) containing 10 mM octane sulfonic acid.
13. Elution solution B: 769 mL 0.2M sodium acetate (adjusted to pH 4.5 with acetic acid) and 231 mL acetonitrile containing 10 mM octane sulfonic acid.
14. For gradient details, *see* Note 4.

2.1.3. Protein

1. 0.3M NaOH.
2. 0.05M NaOH.
3. Bovine serum albumin (BSA) standard: dissolve 100 mg in distilled water (*see* Note 5). Make up to 100 mL with distilled water and store at +1°C for a maximum of 1 mo.
4. Copper alkali reagent *(11)*.
    a. 20 g NaOH;
    b. 100 g Sodium carbonate;
    c. 1.0 g Potassium sodium (+)-tartrate; and
    d. 0.5 g Copper (II) sulfate (0.5 H_2O).

    Dissolve a–c in sequence in distilled water. Dissolve (d) separately in distilled water then add to the a–c solution. The volume is made up to 1 L, and stored at room temperature.
5. Folin Ciocalteau reagent (BDH) is diluted 1:17 with distilled water.

2.1.4. RNA

1. 5% PCA.
2. Dissolve 50 mg RNA (from yeast; Koch-Light) in 0.05M NaOH and make up to 100 mL. Store at 1°C for a maximum of 1 mo.
3. 0.5 g Ferric chloride ·6H_2O is dissolved in <100 mL concentrated HCl. Add 1.0 g orcinol and when dissolved, make up to 100 mL with concentrated HCl. This is made up immediately before use.

2.1.5. DNA

1. 0.05M NaOH.
2. 5% PCA.
3. Dissolve 50 mg DNA (Na-salt, Type III from salmon testes: Sigma, St. Louis, MO) in 0.05M NaOH and make up to 100 mL. Store a 1°C for a maximum of 1 mo.
4. Diphenylamine reagent: 1.5 mg diphenylamine is dissolved in 100 mL glacial acetic acid. To this 100 mL add 1.5 mL concentrated sulfuric acid and 0.5 mL 1.6% acetaldehyde. This is made up fresh and the acetaldehyde added immediately before use (*see* Note 6).

2.2. Uptake of Labeled Polyamines

1. Phosphate-buffered saline (PBS).
2. Ice-cold 0.15M NaCl.
3. 2% PCA.
4. [1,4-^{14}C]putrescine; [^{14}C]spermidine·3HCl; [^{14}C]spermine·4HCl; NCS tissue solubilizer fluid (Amersham International, Amersham, UK).
5. Unlabeled putrescine, spermidine, and spermine (Sigma).
6. Intragastric intubation (final volume 0.5 mL): Intubate a mixture of a standard amount of the appropriate ^{14}C-labeled polyamine (8.5 nmol, 2.4×10^6 dpm/rat,

^{14}C-labeled putrescine; 9.9 nmol, 3.3 × 10^6 dpm/rat, ^{14}C-labeled spermidine; 9.9 nmol, 3.0 × 10^6 dpm/rat, ^{14}C-labeled spermine) and either 0.1, 1.0, or 5.0 mg of the individual unlabeled polyamine. These quantities (0.1, 1.0, or 5.0 mg) correspond to 630, 6200, or 31,000 nmol putrescine; 400, 3900, or 19,600 nmol spermidine; 300, 2900, or 14,400 nmol spermine (see Note 7).
7. Intraperitoneal injection (per rat): dilute stock solutions with PBS to 0.2 mL so that there is 8.5 nmol (2.4 × 10^6 dpm) ^{14}C-labeled putrescine; 9.9 nmol (3.3 × 10^6 dpm) labeled spermidine; 9.9 nmol (3.0 × 10^6 dpm) labeled spermine.
8. Diethyl ether as anesthetic.
9. NE 265 scintillation fluid (NE Technology, Edinburgh, UK).

2.3. Ornithine Decarboxylase Activity

1. ODC buffer (make up fresh): 0.05M phosphate buffer (adjust to pH 7.2 with NaOH: approx 34 mL 1M NaOH/L) containing 5 mM dithiothreitol and 1 mM pyridoxal-5-phosphate.
2. ODC substrate (make up fresh): 50 μL (2.5 μCi) L-[1-^{14}C]ornithine hydrochloride made up to 2.5 mL with 2 mM D,L-ornithine (see Note 8).
3. 50% Trichloroacetic acid (TCA).
4. 0.5M NaOH.
5. 1 g Hyamine 10-X (methylbenzethonium chloride) in 10 mL 0.5M NaOH.
6. NE 265 scintillation fluid (NE Technology).

3. Methods

3.1. Quantification of Polyamines by HPLC Analysis, and Protein, RNA, and DNA

3.1.1. Tissue Preparation

1. Everything is kept on ice, including all reagents.
2. Homogenize 300–500 mg (wet weight) or 40–50 mg (dry weight) tissue with 2.5 mL cold 10% PCA at 20,500 rpm for 60 s.
3. Precool the centrifuge to 2°C.
4. Centrifuge for 25 min at 10,000g.
5. Remove supernatant and keep in the refrigerator (4°C).
6. Rehomogenize the pellet with another 2.5 mL of cold 10% PCA for 60 s at 20,500 rpm.
7. Centrifuge for 25 min at 10,000g.
8. Remove supernatant and combine with the first supernatant (stock polyamine).
9. The combined supernatant is stored at –20°C until required for polyamine analysis.
10. The insoluble pellet is then used for the remaining analysis (12).
11. For analysis, dilute 1-mL aliquot of the stock polyamine with 4 mL of distilled water.
12. The PCA-insoluble pellet from above is homogenized with 5 mL 0.3M NaOH for 60 s at 20,500 rpm.
13. Incubate sample(s) in a water bath at 37°C for 1 h.

14. Take a 1-mL aliquot and add 4 mL of 0.05M NaOH. This is used for protein estimation and stored at –20°C.
15. The remaining 4 mL is used for the RNA analysis.
16. To the remaining 4 mL, add 4 mL of 10% PCA.
17. Vortex the sample and leave sitting on ice for 1 h.
18. Centrifuge for 10 min at 1000g, the centrifuge does not need to be precooled.
19. Remove supernatants and store at –20°C for RNA estimation.
20. Homogenize the insoluble pellet for 60 s at 9500 rpm with 5 mL 5% PCA.
21. Heat samples in a water bath at 80°C for 1 h.
22. Leave the samples to cool on the bench before placing them into the refrigerator (4°C) overnight.
23. Next day: Centrifuge samples for 10 min at 1000g.
24. Remove supernatants and store at –20°C for DNA estimation.
25. Discard the insoluble pellet.

3.1.2. Polyamine Analysis

1. Equilibrate column with Eluent A (*see* Note 9).
2. Inject 200 µL of sample or standard.
3. The eluent is mixed with *o*-phthaldehyde for postcolumn derivatization.
4. Peaks are detected by fluorescence (340-nm excitation; 455-nm emission).
5. Calculate the polyamine concentration from peak areas and correct for the internal standard.
6. Values are expressed as nmol/g or /mL.

3.1.3. Protein

1. Standards: Take 0, 10, 25, 50, 75, 100, 125, 150, 200, and 300 µL from stock standard solution and make up to 1 mL with distilled water.
2. Samples: Take 50 µL and make up to 1 mL with distilled water.
3. Add 1 mL of copper alkali reagent.
4. Add 4 mL of Folin reagent.
5. Vortex.
6. Place marbles or stoppers onto top of tubes.
7. Incubate for 15 min in water bath at 55°C.
8. Leave tubes to cool for approx 10 min.
9. Read on spectrophotometer, wavelength 740 nm.
10. The color change ranges from clear to blue.
11. Use standard absorbance readings to give calibration curve in order to calculate protein concentration of unknowns.

3.1.4. RNA

1. Standards: Take 0, 10, 25, 50, 75, 100, 125, 150, 200, and 300 µL from stock standard solution and make up to 1.5 mL with 5% PCA.
2. Sample: 100 or 200 µL made up to 1.5 mL with 5% PCA.
3. Add 1.5 mL Orcinol reagent.

Lectin Effects on Polyamine Uptake

4. Vortex.
5. Place marbles or stoppers onto top of tubes.
6. Heat in boiling water for 30 min.
7. Leave tubes to cool down for approx 15 min.
8. Read on a spectrophotometer, wavelength 660 nm.
9. Use standard absorbance readings to give calibration curve in order to calculate RNA concentration of unknowns.
10. The color change ranges from yellow to a bottle green.

3.1.5. DNA

1. Standards: Take 0, 10, 25, 50, 75, 100, 125, 150, 200, and 300 µL from stock standard solution and make up to 1 mL with 5% PCA.
2. Samples: 200 µL made up to 1 mL with 5% PCA.
3. Add 2 mL of diphenylamine reagent.
4. Vortex.
5. Place marbles or stoppers onto top of tubes.
6. Incubate in water bath at 37°C for 16 h.
7. Leave tubes to cool for approx 10 min.
8. Read on a spectrophotometer, wavelength 600 nm.
9. Use standard absorbance readings to give calibration curve in order to calculate DNA concentration of unknowns.
10. Samples will develop a light blue color within approx 30 min, whereas with the standards this coloration is not apparent for several hours.

3.2. Uptake of Labeled Polyamines

1. Kill rats exactly 1 h (or at different times) after gastric intubation or intraperitoneal injection.
2. Remove small intestine, wash with ice-cold saline, and measure the length.
3. Collect the luminal contents of the stomach and the small intestine by washing it with 5–10 mL of cold saline. Measure volume.
4. All other body organs, blood, carcass, and fur can be collected if necessary.
5. Cut each small intestine into sections of known length (*see* Note 10).
6. Weigh each 2-cm section and place into separate scintillation vials containing 4 mL scintillation fluid.
7. Leave extracting for 2 d at room temperature in a fume cupboard in the dark (extraction for this time gives the maximum counts).
8. Count for 10 min.
9. Weigh all other sections of the small bowel. Calculate intestinal polyamine uptake from the amount of labeled polyamines incorporated into the six 2-cm long sections in relation to other sections of small intestine.
10. Homogenize the 18-cm sections of small bowel in 5 mL of 2% PCA at 20,500 rpm for 60 s.
11. Precool the centrifuge to 2°C.
12. Centrifuge for 25 min at 10,000g.

13. Remove supernatant and store at –20°C until required for polyamine analysis.
14. Tissues such as the gastrocnemius muscle, stomach, liver, kidney, or dried and ground carcass require treatment with tissue solubilizer fluid (*see* Section 2.2., item 4).
15. Add 1 mL tissue solubilizer to not more than 100 mg dry wt/200 mg wet wt of weighed material (up to 2 mL may be used).
16. Heat to 45–60°C and leave overnight with constant shaking.
17. Add 30 µL glacial acetic acid to neutralize the solubilized tissue.
18. Add 1 mL of this mixture to 10 mL scintillation fluid.
19. Count for 10 min.
20. Calculate the absorption of labeled polyamine per organ.

3.3. Ornithine Decarboxylase Activity

1. Sample preparation: Homogenize on ice 100–150 mg (wet weight) of tissue section or about 10^6 cells at 20,500g for 45 s in 2 mL ice-cold ODC buffer.
2. Precool centrifuge to 4°C.
3. Centrifuge the homogenate at 50,000g for 60 min at 4°C, and remove the supernatant for the ODC estimation (keep samples on ice, 1 h at the most at 1°C). This supernatant can also be used to determine the protein concentration if required.
4. Pipet 0.4 mL of ice-cold sample and 0.1 mL of ice-cold substrate into a 50 × 12-mm reaction tube.
5. Place a Durham tube (Fischer, Loughborough, Leicestershire, UK) containing 0.2 mL hyamine hydroxide (for collection of CO_2) inside the reaction tube.
6. Seal the reaction tube with a Suba seal rubber stopper (Gallenkamp, Loughborough, Leicestershire, UK).
7. Place the reaction mixture in a shaking water bath for 30 min at 37°C.
8. Inject 0.2 mL TCA through the rubber stopper to stop the enzyme reaction mixture. (Take care that no acid gets into the Durham tubes!)
9. Leave tubes in the shaking water bath for a further 45 min at 37°C to allow complete absorption of the liberated CO_2 by the hyamine hydroxide.
10. Remove the Durham tube and place 0.1 mL of its contents in a scintillation vial containing 5 mL scintillation fluid.
11. For the blank add 100 µL of ODC buffer instead of sample (same volume); for the standard add 100 µL substrate instead of sample.
12. Count radioactivity for 10 min.
13. ODC activity is expressed as nmol putrescine/h/organ (or /mg protein) (*see* Note 11 for reaction equation).

4. Notes

1. Individual standard solutions (10 mM) are stored at –20°C. The mixture of polyamine standards (100 mL) is aliquoted to 5 × 20 mL. Working standard solution is stored at 4°C and the remaining 4 vials are stored at –20°C. Once thawed, these must not be refrozen.
2. If column is not being used, it cannot be left in buffer. It can be stored in acetonitrile (or methanol): it takes about 1 h for all the liquid in the column to be dis-

Table 1
Gradient for Polyamine Analysis

Time	Eluent A	Eluent B
0	100	0
13	100	0
17	60	40
41	40	60
61	0	100
84	0	100
89	100	0

Load next sample at 89 min.

placed. All the salt solution must be removed. If you use methanol, you must use a gradient of 100% distilled water to 100% methanol and pump until pure methanol comes through the column.

3. Samples are best filtered through a 0.2-μm spin filter. These, however, are expensive and it is cheaper to run the guard column until the pressure starts to increase and then replace.
4. Gradient for polyamine analysis (*see* Table 1).
5. Add approx 50 mL distilled water to a 100-mL beaker. Scatter weighed BSA on top of the water and leave to dissolve. **Note:** Do not try to speed things up by stirring, as you will end up with a frothy mixture, which is impossible to make up accurately to 100 mL.
6. Diluted (1.6%) acetaldehyde is stored in a dark bottle at 1°C indefinitely.
7. Intragastric intubation is carried out by passing a narrow plastic tube down into the stomach of the rat. As the rat cannot vomit, there is no risk of the polyamine mixture being regurgitated.
8. To get rid of all ^{14}C-CO and/or -CO_2 absorbed during the isotope production process, keep ^{14}C-ornithine in an open vial for about 30 min during constant shaking at 37°C.
9. Before using a column which has been stored, the acetonitrile or methanol must be displaced before samples are applied. Flush column with Eluent A for about 5 min if column is stored in acetonitrile. If column is stored in methanol, flush with gradient of 100% methanol to 100% water and then Eluent A for about 60 min.
10. From the duodenum cut lengths of 3, 2, 2, 18, 2, 18, 2, and 18 cm. The remainder is cut into 2, 2, and 3 cm, but this will vary depending on what is left.
11. Ornithine $\xrightarrow{ODC}$ Putrescine + CO_2.

References

1. Tabor, C. W. and Tabor, H. (1984) Polyamines. *Annu. Rev. Biochem.* **53,** 747–790.
2. Pegg, A. E. (1986) Recent advances in the biochemistry of polyamines in eukaryotes. *Biochem. J.* **234,** 249–262.

3. Konecki, D., Kramer, G., Pinphanichakarn, P., and Hardesty, B. (1975) Polyamines are necessary for maximum in vitro synthesis of globin peptides and play a role in chain initiation. *Arch. Biochem. Biophys.* **169,** 192–198.
4. Abraham, A. K., Olsner, S., and Phil, A. (1979) Fidelity of protein synthesis in vitro is increased in the pancreas in the presence of spermidine. *FEBS Lett.* **101,** 93–96.
5. Kyner, D., Zabros, P., and Levin, D. H. (1973) Inhibition of protein chain initiation in eukaryotes by deacetylated transfer RNA and its reversibility by spermine. *Biochim. Biophys. Acta* **324,** 386–396.
6. Barbiroli, B., Corti, A., and Caldarera, C. M. (1971) The pattern of synthesis of ribonucleic acid species under the action of spermine in the chick embryo. *Biochem. J.* **123,** 123,124.
7. Fillingame, R. H., Jorstad, C. M., and Morris, D. R. (1975) Increased cellular levels of spermidine or spermine are required for optimal DNA synthesis in lymphocytes activated by concanavalin A. *Proc. Natl. Acad. Sci. USA* **72,** 4042–4046.
8. Cohen, S. S. (1978) What do the polyamines do? *Nature* **274,** 209,210.
9. Fausto, N. (1972) RNA metabolism in isolated perfused normal and regeneration livers: polyamine effects. *Biochim. Biophys. Acta* **281,** 543–553.
10. Anderson, N. G. and Norris, C. B. (1960) The effects of amines on the structure of isolated nuclei. *Exp. Cell Res.* **19,** 605–618.
11. Williams-Ashman, H. G. and Canellakis, Z. N. (1979) Polyamines in mammalian biology and medicine. *Perspect. Biol. Med.* **22,** 421–453.
12. Schuber, F. (1989) Influence of polyamines on membrane function. *Biochem. J.* **260,** 1–10.
13. Bardocz, S., Grant, G., Ewen, S. W. B., Duguid, T. J, Brown, D. S., Englyst, K., and Pusztai, A. (1995) Reversible effect of phytohaemagglutinin on the growth and metabolism of rat gastrointestinal tract. *Gut* **37,** 353–360.
14. Gelencser, E. Hajos, G., Ewen, S. W. B., Pusztai, A., Grant, G., Brown, D. S., and Bardocz, S. (1994) Biological effects and survival of soya bean agglutinin (SBA) in the gut of the rat, in *Lectins: Biology, Biochemistry, Clinical Biochemistry,* vol. 10 (Van Driessche, E., Fischer, J., Beekmans, S., and Bøg-Hansen, T. C., eds.), Textop, Hellerup, Denmark, pp. 305–311.
15. Pusztai, A. Ewen, S. W. B., Grant, G., Brown, D. S., Stewart, J. C., Peumans, W. J., Van Damme, E. J. M., and Bardocz, S. (1993) Antinutritional effects of wheat germ agglutinin and other N-acetylglucosamine specific lectins. *Br. J. Nutr.* **70,** 313–321.
16. Tabata, K. and Johnson, L. R. (1986) Ornithine decarboxylase and mucosal growth in response to feeding. *Am. J. Physiol.* **251,** G270–G274.
17. Pusztai, A., Grant, G., Brown, D. S., Ewen, S. W. B., and Bardocz, S. (1988) *Phaseolus vulgaris* lectin induces the growth and increases the polyamine content of rat small intestine *in vivo. Med. Sci. Res.* **16,** 1283,1284.
18. Bardocz, S., Brown, D. S., Grant, G., and Pusztai, A. (1990) Luminal and basolateral polyamine uptake by rat small intestine stimulated to grow by *Phaseolus vulgaris* lectin phytohaemagglutinin *in vivo. Biochim. Biophys. Acta* **1034,** 46–52.

19. Bardocz, S., Grant, G., Brown, D. S., Ewen, S. W. B., and Pusztai, A. (1989) Involvement of polyamines in *Phaseolus vulgaris* lectin induced growth of rat pancreas *in vivo. Med. Sci. Res.* **17,** 309–311.
20. Pusztai, A. (1991) *Plant Lectins,* Cambridge University Press, Cambridge, UK.
21. Bardocz, S., Grant, G., Brown, D. S., Wallace, H. M., Ewen, S. W. B., and Pusztai, A. (1989) Effect of α-difluoromethylornithine on *Phaseolus vulgaris* lectin-induced growth of the rat small intestine. *Med. Sci. Res.* **17,** 143–145.
22. Metcalf, B. W., Bey, P., Danzin, C., Jung, M. J., Casara, P., and Vevert, J. P. (1978) Catalytic irreversible inhibition of mammalian ornithine decarboxylase (E.C.4.1.1.17) by substrate and product analogs. *J. Am. Chem. Soc.* **100,** 2551–2553.
23. Bardocz, S., Brown, D. S., Grant, G., Pusztai, A., Stewart, J. C., and Palmer, R. M. (1992) Effect of the β-adrenoceptor agonist clenbuterol and phytohaemagglutinin on growth, protein synthesis and polyamine metabolism of tissues of the rat. *Br. J. Pharmacol.* **106,** 476–482.
24. Bardocz, S., Grant, G., Brown, D. S., Duguid, T. J., Ewen, S. W. B., Peumans, W. J., Van Damme, E. J. M., and Pusztai, A. (1994) Lectin-induced polyamine-dependent growth of the small intestine, in *Lectins: Biology, Biochemistry, Clinical Biochemistry,* vol. 10 (Van Driessche, E., Fischer, J., Beekmans, S., and Bøg-Hansen, T. C., eds.), Textop, Hell rup, Denmark, pp. 289–294.
25. Seiler, N. and Knödgen, B. (1980) High-performance liquid chromatography procedure for the simultaneous determination of natural polyamines and their monoacetyl derivatives. *J. Chromatog.* **221,** 227–235.
26. Schachterle, G. R. and Pollack, R. L. (1973) A simplified method for the quantitative assay of small amounts of protein in biological material. *Anal. Biochem.* **51,** 654,655.
27. Sneider, W. C. (1957) Determination of nucleic acids in tissue by pentose analysis. *Methods Enzymol.* **3,** 680–684.
28. Løvtrup, S. and Roos, K. (1961) Observation on the chemical determination of deoxyribonucleic acid in animal tissues. *Biochim. Biophys. Acta* **53,** 1–10.
29. Russell, D. H. and Snyder, S. H. (1968) Amine synthesis in rapidly growing tissues: ornithine decarboxylase activity in regenerating rat liver, chick embryo, and various tumors. *Proc. Natl. Acad. Sci. USA* **60,** 1420–1427.

34

Effects of Lectins on Cytoskeletal Organization in Mammalian Cells

Paolo Carinci, Ennio Becchetti, and Maria Bodo

1. Introduction
1.1. Cytoskeleton Organization

The cytoskeleton of mammalian cells is a complex assembly of microfilaments, microtubules, and intermediate filaments. The thicker fibers are the microtubules which have a diameter of approx 25 nM and an apparently hollow core; microfilaments have a diameter of 6 nM; but the third type of filaments have a diameter of approx 10 nM and are known as intermediate filaments. These filamentous structures are commonly referred to as a cytoskeletal framework; however, this description is misleading, since their functions certainly include not only "skeletal properties" but also force generation and movement, cell division, uptake of materials, and transport of exocytotic and endocytotic vesicles.

Remarkable advances in the study of cytoskeletal structures were made with the introduction of immunofluorescence microscopy. The use of anti-actin antibodies revealed an elaborate network of microfilament bundles along the cytoplasm, in the leading lamella actively involved in mobility and in perinuclear regions (1,2). In a variety of different cell types, actin has been localized in stress fibers, ruffles, attachment plaques, and cell junctions (3). In stress fibers, actin is generally distributed continuously along the length of the fibers (4) although, in a few cases, its distribution is discontinuous (5). A great number of actin-associated proteins have been identified, among them myosin, tropomyosin, filamin, alpha-actinin, vinculin, and many others. Together they form the molecular architecture of cellular mechanochemical machinery (6). Tropomyosin is distributed along stress fibers and is absent in areas where

From: *Methods in Molecular Medicine: Vol. 9: Lectin Methods and Protocols*
Edited by: J. M. Rhodes and J. D. Milton Humana Press Inc., Totowa, NJ

alpha-actinin is present—for example, stress fiber densities, ruffles, cell junctions, and attachment plaques *(7)*. Alpha-actinin and tropomyosin are localized in adjacent bands. There are differences in the stress fibers between fibroblasts and epithelial cells. In epithelial cells (Ptk2, bovine lens) and in fibroblasts (Gerbil fibroma, WI-38, primary human), the spacing between sites of alpha-actinin localization differs by a factor of approx 1.6 as determined by indirect immunofluorescence. It has been suggested that actin filaments and perhaps myosin filaments could be longer in fibroblast cells than in epithelial cells, and that the different alpha-actinin periodicities could be owing to variation in filament lengths *(8)*.

The actin network is involved in a force generating mechanism, with locomotory and adhesive functions. For example, chemotactic activation of neutrophils is accompanied by a dramatic reorganization of actin filaments with pseudopodia enriched in F-actin, stabilized by alpha-actinin *(9)*. During the process of neutrophil adhesion, in the area of strong adhesion, pseudopodia are particularly rich in F-actin with a peripheral distribution of vinculin *(10)*. Several models for the attachment of actin filaments to membrane have been proposed. For example, vinculin may interact directly with actin in focal contacts *(11)*, and alpha-actinin colocalizes with actin and vinculin in type II plaques *(12)*.

Vinculin and alpha-actinin, bound together, have been implicated in the linkage of actin to the cytoplasmic domain of integrin in focal contacts of stationary cells *(13)*, and it has been shown that talin binds vinculin and integrin in vitro *(14,15)*. Thus, the evidence suggests that integrins are transmembrane proteins that interact with the actin cytoskeleton via links mediated by proteins, such as talin, vinculin, alpha-actinin, tensin, and paxillin *(16–19)*.

Myosins are components of many actin structures and are essential for their motility. They represent a diverse group of actin-binding proteins with ATPase activity. The list of new myosins is constantly lengthening, and most of these proteins have not been functionally characterized. In nonmuscle cells, myosin is found only on linear actin filament bundles that are aligned with the cell's long axis in cortical and subcortical areas, while it is absent from actin filaments perpendicular to these bundles and from actin filaments having a complex geometric configuration *(20)*. The myosins may be found up to the very edge of the cortex both at the sides and at the ends of cells.

The other main type of cytoskeletal fibril is the microtubule. Several microtubule-associated proteins have been described; they appear to promote the assembly of microtubules and their interaction with other cellular organelles *(21)*. The microtubule network extends to the periphery of the cell from a perinuclear microtubule organizing center (MOC), has a morphogenetic role, controls intracellular traffic of organelles, is the prime cytoskeletal element determining cell polarity *(22)*, reacts to the substratum topography, and is sensitive to mechanical stress and tensile forces *(23)*. The microtubule network may affect the alignment of other elements, thereby influencing both the distribution of intermediate filaments and actin cortex *(24)* and stretching of the microfilament network *(25)*.

The intermediate filaments (IF) are the most insoluble filaments within cells. A characteristic of these filaments is their resistance to detergent and to buffers of low and high ionic strength *(26)*. Intermediate filaments, whose protein subunit composition and biochemical properties differ in a cell type-specific manner, are a remarkable part of the cytoskeleton. About 40 IF proteins are known and have been grouped into six different types on the basis of sequence homology: type I (acidic keratin), type II (neutral-basic keratin), type III (vimentin, desmin, glial fibrillary acid protein, peripherin), type IV (neurofilament, alpha-internexin), type V (nuclear laminin), type VI (nestin) *(27)*.

Each species of IF protein is encoded by a unique gene but, since the positions of intron sequences segmenting the coding portions are highly similar, all types of IF protein genes could have a common ancestor *(28)*. A structural support role has been attributed to these cytoskeletal elements because of their apparent involvement in nuclear anchorage in cells *(29)* and in disposition of organelles, but they may also play a role in signalling function *(30)*.

IF interact with the nuclear surface and, from this region, they radiate toward the cell-surface, where they are closely associated with the plasma membrane. Thus, IF represents a cytoskeletal system interconnecting the cell-surface with the nucleus. IFs also appear to be crosslinked to each other and may be physically connected to the microtubule network *(31)* and to the microfilament system *(31)*. Staining with antibodies against tubulin and vimentin has shown that most intermediate filaments are coaligned with the radiating microtubules. The effects of depolymerizing microtubule drugs, which induce the collapse of IF, are fully reversible after removal of the drug *(33)*.

The crosslinking of different cytoskeletal systems is mediated by cytoskeletal-associated proteins. For instance, microtubule-associated protein (MAPs) bind IFs and actin filaments. Interrelationships between microtubules and actin structures are complex. Actin bundles are not coaligned with microtubules, but drug-induced depolymerization of microtubules causes contraction of the actin cortex and loss of parallel orientation of actin bundles. Spectrin is an actin crosslinking protein, binding microtubules and microfilaments. Recently, it was observed that filamin serves to interconnect the microfilaments with IF *(34)*.

Thus, the three systems of cytoskeleton network interact with one another and an alteration regarding one system leads to structural reorganization of other systems.

1.2. Cytoskeleton and Cellular Organelles

Cellular organelles have a regular anisotropic distribution within cells; their position and orientation are correlated to the organization of cytoskeletal structures. For example, both lysosomes and mitochondria are aligned along the microtubules *(35)*. The Golgi apparatus is normally formed by numerous stacks of membrane-bound cisternae, grouped near the centrosome. Depolymerization of microtubules leads to dispersal of these stacks over the entire cytoplasm. The integrity of the radiating system of microtubules is essential for the exact organization and localization of various organelles.

Unconventional myosins are thought to function as organelle motors *(36)*. Axoplasmic organelles have a myosin-like motor on their surface and actin and microtubules are implicated in its function. The role and the exact nature of links between the various elements of cytoskeleton and the organelles is still not clear. In all probability, control of the distribution of cellular organelles by the cytoskeleton is dynamic and may be considered the result of the intracellular movement of organelles by cytoskeleton. It has been suggested that distinct "translocators" are necessary for organelle movement. Kinesin is one of the microtubule-associated ATPases *(37)*; gelsolin and brevin are proteins associated with actin filaments that are responsible for axon transport *(38)*.

The cytoskeleton is also important in the maintenance of the differentiated state, since it may induce posttranscriptional regulation of mRNA by modulating the interaction of the polyribosomes with various elements of the cytoskeleton *(39)*. The association of both hnRNA and mRNA with the structural networks of the cell may be essential features of gene expression.

1.3. Lectin Effects on Cytoskeleton

Lectins are specific carbohydrate binding proteins. Local crosslinkage of cell-surface glycoproteins by lectins induces a propagated change in the lateral mobility of a variety of cell-surface receptors. These receptors, originally dispersed uniformly over the surface, are redistributed to a limited extent, in small patches, or collected in a few large patches or in a single "cap" on the cell-surface *(40,41)*. The capped regions are then internalized by endocytosis *(42)*. The capping process induced by several different lectins resulted in changes in cytoskeletal organization, since submembranous cytoskeleton is actively involved in anchoring specific proteins to particular membrane domains *(43)*.

The actin network appears to be modified from its original uniform distribution in the cellular cytoplasm and organized into patches. Membrane-associated myosin shows the same distribution. Actin and myosin are so closely associated in restricted region of the cell-surface following lectin-receptor binding as to suggest a strong correlation between cytoskeleton and cell-surface receptors *(44)*.

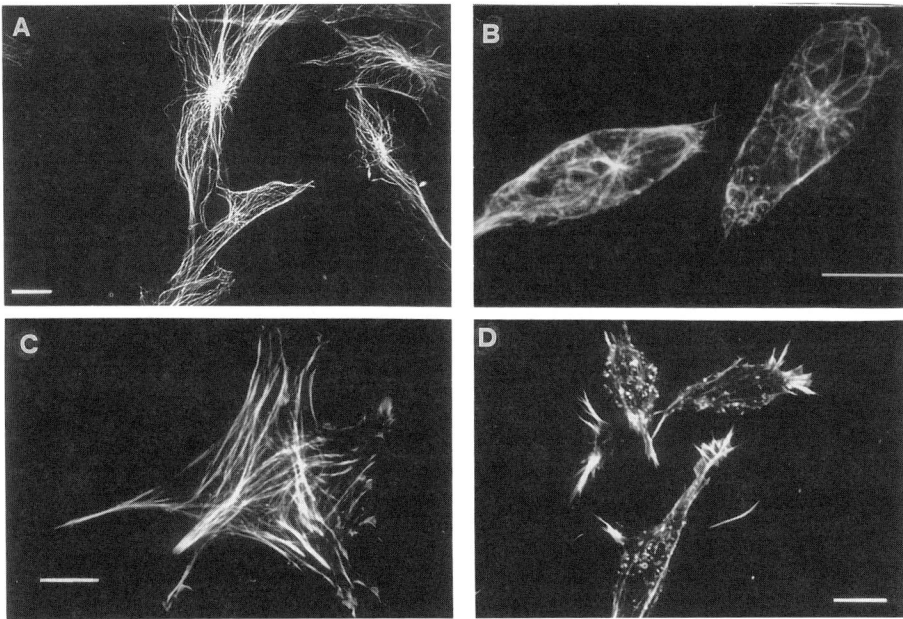

Fig. 1. (A,B) Immunofluorescence staining of microtubule pattern in mouse embryo fibroblasts pretreated or not with Con A for 24 h in 199 serum-free medium. Cells grown on glass cover slips were fixed in 1.5% glutaraldehyde + 0.75% paraformaldehyde in PBS, permeabilized with 0.2% Triton X-100 in PBS, and labeled with antibodies to tubulin in indirect immunofluorescence assay. Micrographs were taken using a Bio-Rad MRC-500 confocal microscope. (A) Microtubule distribution was normal in control cells. Microtubules rayed out from a perinuclear area to reach the cell membrane and then run parallelly beneath it. (B) 20 µg/mL Con A provoked an evident modification in cell shape that became rounded and contained a thickened tubulin network. (C,D) Actin pattern in mouse embryo fibroblasts. After fixation and permeabilization, cells were labeled with phalloidin-TRITC that binds to F-actin. (C) Numerous stress fibers are arranged in a regular diffuse pattern throughout the cell body. (D) 20 µg/mL Con A caused the filamentous actin to convert to aggregates or a diffusely stained form. Bars = 10 µm.

In concanavalin A (Con A)-capped ovarian granulosa cells, a redistribution of microtubules in cytoplasmic regions underlying the cap has also been shown *(45,46)*. Interaction between cell-surface components and the underlying microtubule cytoskeleton was also evident when the binding between wheatgerm agglutinin (WGA) lectin and the neonatal rat photoreceptor connecting cilium was studied *(47)*. Personal observations confirm that modulation of membrane receptors by lectin alters the cytoskeleton organization (*see* Figs. 1 and 2). Addition of exogenous peanut agglutinin (PNA) and soybean

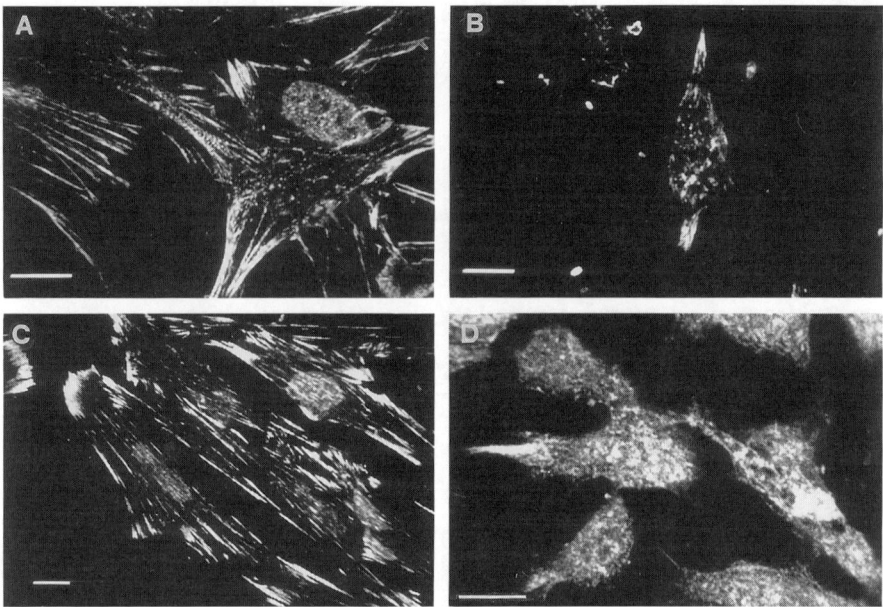

Fig. 2. **(A,B)** Immunofluorescence staining of alpha-actinin pattern in mouse embryo fibroblasts pretreated or not with Con A for 24 h in 199 serum-free medium. Cells grown on glass cover slips were fixed, permeabilized, and then labeled with antibodies to alpha-actinin. (A) Control cells showed alpha-actinin localized in the cytoplasm, extended into the fine prolongations and periodically punctated by bundles of microfilaments. Instead, in Con A-treated cells (B), the alpha-actinin appeared as spots scattered throughout the cell body. **(C,D)** Immunofluorescence staining of vinculin pattern in mouse embryo fibroblasts pretreated or not with Con A for 24 h in 199 serum-free medium. (C) In controls, vinculin is found closely associated with stress fibers endings and the protein forms a possible linking between the end of the bundles of actin filaments and the plasma membrane. (D) In Con A-treated cells, was evident a disorganization of normal vinculin pattern. Cells appeared diffusely stained over the entire cytoplasm. Bars = 10 µm.

agglutinin (SBA) to chick embryo fibroblasts in vitro affects myosin, actin, and microtubule network and cellular morphology in a dose-dependent manner *(48)*. PNA and SBA, at higher doses, cause disorganization of the myosin filaments and depolymerization of actin. Microtubules appear thickened, highly condensed, and interrupted in PNA-treated fibroblasts. Since lectin treatments also cause rounding of the cells, we suggested that the lectin-membrane interaction induced spatial modification of cytokinetic structure responsible for altered adhesion to substrate.

Reversible alterations in cell phenotype and microtubule architecture are induced by Con A treatment in the same cell system *(49–53)*. These data support the hypothesis that microtubules are indirectly responsible for the modulation of the mobility of the receptors for lectins by a cooperative mechanism involving interactions with microfilaments.

Conversely, Con A treatment does not alter the phenotype or cytoskeletal organization of mouse macrophages, which show a rather uniform distribution of tubulin, actin, and alpha-actinin *(54)*.

Lectin-induced modifications in cytoskeletal pattern and cell morphology were also found by others. For example, the binding of Con A to amebae of *Dictyostelium discoideum* caused a dramatic change in cell shape with cellular rounding, a 20-fold increase in the concentration of myosin, and a threefold increase in actin associated with Triton X-100 insoluble cytoskeleton *(55)*.

A short time exposure (60 min) of rat intestinal epithelium to Con A or WGA appears to be sufficient to achieve rearrangement of the cytoskeleton *(56–59)*. In differentiated human colon carcinoma cells, actin cytoskeleton lesions were evident after exposure to SBA lectin *(60)*. In particular, SBA treatment caused an increase in the amount of G-actin and a decrease in F-actin, which correlated with a shortening of the microvilli of enterocytes. In fact, SBA contains multiple binding sites for specific carbohydrate residues such that a crosslinking of the membrane glycoconjugates of neighboring microvilli causes their aggregation. It has been suggested that actin cytoskeleton changes caused by the interaction between SBA and its specific sugar on the cell-surface membrane may play a role in the pathogenesis of microvillous abnormalities.

Studies of lectin effects in blood cells show that the redistribution of crosslinked surface immunoglobulin (sIg) on B-lymphocytes is inhibited by Con A treatment. The inhibition of sIg mobility is exerted indirectly through the cytoskeleton, since the effect is diminished by colchicine and cytochalasin treatment. Crosslinking of Con A receptors on lymphocyte surfaces also causes a large increase in cellular resistance to compression. Elastic and viscous resistance to deformation and mechanical properties are determined by the cytoskeletal cortex, which lies immediately beneath the plasma membrane *(61)*.

Treatment of human lymphocytes with other lectins, such as phytohemagglutinin (PHA), WGA, and Con A, caused a dose-dependent decrease in the amount of G-actin and conversion of G-actin to F-actin *(62)*. According to the authors of this study, actin polymerization following WGA stimulation is a signal required for initiation of DNA synthesis.

Con A treatment of platelets induces a specific transmembranous interaction between the glycoprotein IIb-IIIa complex, which forms the fibrinogen-binding site, and the cytoskeleton, in particular the polymerized actin. Thus, Con A triggers a cytoskeletal assembly consisting of polymerized actin, which interacts with myosin and alpha-actinin both of which have key roles in platelet functions. Another effect of Con A on platelets is a slight reduction in cytoskeletal core actin content with a significant decrease of myosin. In particular, the platelet pseudopodal cytoskeletal elements (actin crosslinked via actin-binding protein and alpha-actinin) form a precipitable matrix in the presence of Con A *(63)*.

Binding of WGA to erythrocyte membranes causes membrane rigidity owing to intermolecular linkage of the lectin to glycophorin on the outside of the membrane and to the cytoskeleton network that lies on the inner face of the membrane *(64,65)*.

All of the data cited above demonstrate lectin effects on the cytoskeleton components. Conformational changes in the cytoskeletal network are involved not only in cell shape but also in transmitting and distributing information among the major cellular domains: extracellular matrix-cell-surface, cytoplasm-nuclear surface, and nuclear surface-nuclear matrix. In fact, reorganization of the cytoskeleton after lectin binding and uptake on the surface induces a reprogramming of many metabolic reactions. This correlation suggests that lectins, by altering the distribution of membrane molecules, generate signals responsible for a reorganization of the cytoskeletal network, which act as the transducer of these signals. Further work in this field may better analyze this intriguing hypothesis.

2. Materials

1. Cells cultured on glass cover slips.
2. 1X PBS (phosphate-buffered saline): $0.14M$ NaCl, 2.7 mM KCl, 1.5 mM KH_2PO_4, 8.1 mM Na_2HPO_4, pH 7.4. Dissolve chemicals in bidistilled water and sterilize by autoclaving.
3. PBS with Ca^{2+} and Mg^{2+}: PBS containing 1 mM $CaCl_2$ and 1 mM $MgCl_2$.
4. PBS with Mg^{2+} and EGTA: PBS containing 2 mM $MgCl_2$ and 2 mM EGTA ([ethylene-dioxy] diethylene dinitrilo-tetraacetic acid).
5. Fixation buffer: 0.5% (v/v) glutaraldehyde, 0.25% (w/v) paraformaldehyde in PBS.
6. Permeabilization buffer: 0.1% (v/v) Triton X-100 in PBS.
7. Sodium borohydride ($NaBH_4$): 1 mg/mL in PBS.
8. Dilution buffer for antibodies (BSA/PBS): 0.05% bovine serum albumine (BSA) in PBS.
9. Blocking solution: 1% BSA in PBS. This solution acts to block nonspecific binding sites.
10. Primary antibody raised against the antigen of interest. Polyclonal antibodies are purified from serum, MAbs are purified from ascites fluid or supernatant from cultured hybridomas.
11. Suitable control antibodies (preimmune sera, ascites fluid from nonimmune mice, Ig fractions from nonimmune animals).

Lectins and Cytoskeletal Organization 415

12. Secondary antibody (preferably affinity purified and fluorescently conjugated) raised against immunoglobulins of the same species as the primary antibody.
13. Mounting medium: 90% glycerol in PBS supplemented with an antifading agent (for example, *p*-phenylenediamine).

3. Methods
3.1. Indirect Immunofluorescence Labeling of Cultured Cells

1. Rinse cells cultured on cover slips twice with PBS warmed at 37°C (*see* Note 1).
2. Fix cultures in fixation buffer for 5 min at 37°C (*see* Notes 2–4).
3. Wash three times with PBS for 10 min each.
4. Permeabilize cells with permeabilization buffer for 2 min (*see* Note 5).
5. Rinse as in step 3.
6. Reduce aldehyde groups caused by fixatives (paraformaldehyde and glutaraldehyde) incubating cover slips with freshly prepared sodium borohydride solution. Shake on ice for 5 min.
7. Wash three times with PBS for 5 min each, monitoring for the complete disappearance of autofluorescence due to residual aldehyde via fluorescence microscopy.
8. Incubate in 1% BSA in PBS for 30 min. Alternatively, use 1:10 or 1:30 dilutions of preimmune serum in PBS. Serum should be from the same species as the primary antibody.
9. Incubate with primary antibody (*see* Note 6). Add a drop of diluted antibody over a glass slide wrapped with Parafilm. Each antibody works at a well defined dilution depending on various factors. Thus, the ideal working diluition must be empirically determined by titration for each antibody. In our system, for example, anti-tubulin is diluted 1:400. Invert cover slip onto the drop and incubate for 60 min at room temperature or for 30 min at 37°C in a humidified chamber. Alternatively, incubate overnight at 4°C.
10. Wash thoroughly in BSA/PBS (three times, for 10 min each); if primary antibody is incubated overnight, increase the wash time up to 1 h.
11. Incubate each cover slip with a drop of secondary antibody (FITC or TRITC conjugated) diluted in 0.05% BSA in PBS for 60 min at room temperature or for 30 min at 37°C in a humidified chamber. Cover slips must be kept in the dark, because fluorochromes fade when exposed to light (*see* Note 8).
12. Rinse as in step 8.
13. Mount cover slip on a glass slide using a drop of 90% glycerol in PBS.
14. Observe the samples under a microscope equipped with epifluorescence.

3.1.1. Indirect Immunofluorescence Protocol for Paraffin-Embedded Tissues

A specific example of a lectin experiment in organ cultures is reported.

1. Remove embryo explants under sterile conditions.
2. Wash three times in PBS.
3. Plate on isopore membranes (0.4–μm pore size, Millipore, Bedford, MA) floating in a chemically-defined medium at 37°C in a humidified 5% CO_2 incubator

(composition of a standard medium: 1:1 mixture of Ham's F12 and DMEM with 1 μg BAS/mL, 10 μg transferrin/mL, and 50 μg gentamicin/mL).
4. Add the lectin (Con A, SBA, PNA) at doses ranging from 5–20 μg/mL for at least 24 h.
5. Wash in PBS for 5 min.
6. Fix 2–16 h in 4% paraformaldehyde in PBS.
7. Dehydrate in increasing grades of ethanol (50, 70, 90%, 3 min each).
8. Embed in parafin wax.
9. Dewax sections (1–5 μm) in xylene bath for 15 min (2 changes). Rehydrate tissues in decreasing grades of ethanol: absolute, 96, 70, 50%, 3 min each.
10. Wash in PBS for 5 min.
11. Reduce aldehyde groups caused by fixatives (formaldehyde, paraformaldehyde, glutaraldehyde): cover the sections with 1 mg/mL sodium brorohydride ($NaBH_4$) in PBS for 5 min; rinse in three changes of PBS (3 min each).
12. Block nonspecific binding sites: incubate with 1% BSA in PBS for 30 min at room temperature (alternatively, preimmune serum diluted 1:10, 1:30 in PBS may be used).
13. React with primary antibody (diluted in PBS or 0.05% BSA in PBS): cover the section with primary antibody (about 200 mL) at working dilution and incubate for 30–40 min at room temperature in a humidified chamber.
14. Wash with 0.05% BSA in PBS three times for 5 min each.
15. React with secondary antibody (FITC or TRITC conjugated): cover the section with secondary antibody (about 200 mL) and incubate for 35–45 min at room temperature in the dark in a humidified chamber.
16. Wash with 0.05% BSA in PBS (or PBS alone) three times for 5 min each.
17. Mount the section with a drop of 90% glycerol in PBS. Keep in the dark in a humidified chamber at 4°C until observed. View under a fluorescence and then observe at fluorescent microscope. Specimens can be stored at –20°C.

4. Notes

1. To preserve cell junctions, it is useful to employ PBS with Ca^{2+} and Mg^{2+}. Washes and fixation may be performed at 37°C.
2. Fixation protocols must prevent antigen leakage and preserve the antigen in a form recognizable by antibody. In general, fixative must preserve the cytoarchitecture. The choice of method depends on the antigen type and antibody characteristics. Two classes of fixatives are useful: either organic solvents (alcohol or acetone), or chemical compounds (formaldehyde or glutaraldehyde), which are able to make intermolecular links (mostly between free amino-groups) forming a space-lattice of associated antigens.
3. Fixing solution containing paraformaldehyde and glutaraldehyde in PBS may be substituted by methanol cooled to –20°C. In such a case, the permeabilization step should be omitted.
4. Formaldehyde vapors are toxic and solutions containing formaldehyde should be prepared in a chemical hood. Fixatives that crosslink proteins (paraformalde-

hyde, glutaraldehyde) preserve cell ultrastructure better than organic solvents but can reduce the antigenicity of some cell components. Glutaraldehyde causes a considerable autofluorescence owing to residual free aldehyde. Autofluorescence may be completely quenched by treating the samples with 1 mg/mL $NaBH_4$ in PBS for 5 min. Sometimes, fixation modifies antigen so that it is no larger recognized by the antibody. If this is the case, fixation is performed after primary antibody incubation. Cells are pre-equilibrated with 2% saccharose to reduce possible osmotic shock damage and to better preserve cytoskeleton structures and adhesion plaques.
5. To detect a membrane antigen, permeabilization (Section 3.1., step 4) is obviously omitted.
6. Before incubating cells with antibodies, is useful to drain excess PBS from cover slips by touching the edge to filter paper.
7. It is important not to allow the cover slip to dry out at any stage.
8. FITC- or TRITC-labeled cover slips must be kept at −20°C protected from light. Under such conditions, fluorescence is preserved for approx 1 mo.
9. To maintain microtubule structures and their antigenicity, it is necessary to use PBS containing 2 mM EGTA and 2 mM $MgCl_2$ and without Ca^{2+}.
10. Keep diluted antibodies on ice or at 4°C.
11. It is useful to centrifuge antibody solutions in order to pellet precipitates that may alter results.
12. Adjust the pH of PBS to 7.4 with HCl and/or NaOH.
13. Method requires 3–4 h.

Acknowledgments

This work was supported by M. U. R. S. T. 60%. We would like to thank Dr. Tiziano Baroni for helpful discussions and for assistance in the preparation of the manuscript and the figures.

References

1. Theriot, J. A. and Mitchison, T. J. (1991) Actin microfilament dynamics in locomoting cells. *Nature* **352,** 126–131.
2. Theriot, J. A. and Mitchison, T. J. (1992) Comparison of actin and cell-surface dynamics in motile fibroblasts. *J. Cell Biol.* **118,** 367–377.
3. Amato, P. A. and Taylor, D. L. (1986) Probing the mechanism of incorporation of fluorescently labeled actin into stress fibers. *J. Cell Biol.* **102,** 1074–1084.
4. Sanger, J. W. (1975) Intracellular localization of actin with fluorescently labeled heavy meromyosin. *Cell Tissue Res.* **161,** 432–444.
5. Gordon, W. E. (1978) Immunofluorescent and ultrastructural studies of "sarcomeric" units in stress fibers of cultured nonmuscle cells. *Exp. Cell Res.* **117,** 253–260.
6. Heath, J. P. and Holifield, B. F. (1991) Cell locomotion: new research tests old ideas on membrane and cytoskeletal flow. *Cell Motil. Cytoskel.* **18,** 245–257.
7. Sanger, J. W., Mittal, B., and Sanger, J. M. (1984) Interaction of fluorescently-labeled contractile proteins with the cytoskeleton in cell models. *J. Cell Biol.* **99,** 918–928.

8. Sanger, J. W., Sanger, J. M., and Jockusch, B. M. (1983) Differences in the stress fibers between fibroblasts and epithelial cells. *J. Cell Biol.* **96,** 961–969.
9. Zigmond, S. H. (1989) Chemotactic response of neutrophils. *Respir. Cell Mol. Biol.* **1,** 451–453.
10. Yürüker, B. and Niggli, V. (1992) alpha-actinin and vinculin in human neutrophils: reorganization during adhesion and relation to the actin network. *J. Cell Sci.* **101,** 403–414.
11. Burridge, K., Nuckolls, G., Otey, C., Pavalko, G., Simon, K., and Turner, C. (1990) Actin-membrane interaction in focal adhesions. *Cell Differ. Dev.* **32,** 337–342.
12. Drenckhahn, D. and Franz, H. (1986) Identification of actin-, alpha-actinin-, and vinculin-containing plaques at the lateral membrane of epithelial cells. *J. Cell Biol.* **102,** 1843–1852.
13. Geiger, B. (1989) Cytoskeleton-associated cell contacts. *Curr. Opinion Cell Biol.* **1,** 103–109.
14. Pavalko, F. M., Othey, C. A., Simon, K. O., and Burridge, K. (1991) Alpha-actinin: a direct link between actin and integrins. *Biochem. Soc. Trans.* **19,** 1065–1068.
15. Khoory, E., Wu, E., and Hartford Svoboda, K. K. (1993) Intracellular relationship between actin and alpha-actinin in a whole corneal epithelial tissue. *J. Cell Sci.* **106,** 703–717.
16. Turner, C. E., Glenney, J. R. Jr., and Burridge, K. (1990) Paxillin: a new vinculin-binding protein present in focal adhesions. *J. Cell Biol.* **111,** 1059–1068.
17. Luna, E. J. and Hitt, A. L. (1992) Cytoskeleton-plasma membrane interactions. *Science* **258,** 955–964.
18. Burridge, K., Turner, C. E., and Romer, L. H. (1992) Tyrosine phosphorilation of paxillin and pp125FAK accompanies cell adhesion to extracellular matrix: A role in cytoskeletal assembly. *J. Cell Biol.* **119,** 893–903.
19. Wood, C. K., Turner, C. E., Jackson, P., and Critchley, D. R. (1994) Characterization of the paxillin-binding site and the C-terminal focal adhesion targeting sequence in vinculin. *J. Cell Sci.* **107,** 277–290.
20. Lawson, D. (1986) Myosin distribution and actin organization in different areas of antibody-labeled quick-frozen fibroblasts. *J. Cell Sci.* **5,** 45–54.
21. Collot, M., Louvard, D., and Singer, S. J. (1984) Association between lysosomes and microtubules in cultured fibroblasts, as studied by double immunofluorescence labeling. *J. Submicrosc. Cytol.* **16,** 65–67.
22. Oakley, C. and Brunette, D. M. (1993) The sequence of alignment of microtubules, focal contacts and actin filaments in fibroblasts spreading on smooth and grooved titanium substrata. *J. Cell Sci.* **106,** 343–354.
23. Hill, T. L. and Kirschner, M. W. (1982) Bioenergetic and kinetics of microtubule and actin filament assembly-disassembly. *Int. Rev. Cytol.* **78,** 1–12.
24. Bershadsky, A. D., Ivanova, O. Y., Lyass, L. A., Pletyushkina, O. Y., Vasiliev, J. M., and Gelfand, I. M. (1990) Cytoskeletal reorganizations responsible for the phorbol ester-induced formation of cytoplasmic processes: possible involvement of intermediate filaments. *Proc. Natl. Acad. Sci. USA* **87,** 1884–1888.
25. Vasiliev, J. M. (1991) Polarization of pseudopodial activities: cytoskeletal mechanisms. *J. Cell Sci.* **98,** 1–4.

26. Virtanen, I., Lehto, V. P., Lehtonen, E., Vartio, T., and Stenman, S. (1981) Expression of intermediate filaments in cultured cells. *J. Cell Sci.* **50,** 45–63.
27. Skalli, O. and Goldman, R. D. (1991) Recent insights into the assembly, dynamics, and function of intermediate filament networks. *Cell Motil. Cytoskel.* **19,** 67–79.
28. Capetanaki, Y. C., Ngai, J., and Lazarides, E. (1984) Regulation of the expression of genes coding for the intermediate filament subunits vimentin, desmin, and glial fibrillary acidic protein, in *Molecular Biology of the Cytoskeleton* (Barisy, G. G., Cleveland, D. W., and Murphy, D. B., eds.), Cold Spring Harbor Laboratory, Cold Spring Harbor, New York, pp. 415–434.
29. Sangiorgi, F., Woods, C. M., and Lazarides, E. (1990) Vimentin downregulation is an inherent feature of murine erythropoiesis and occurs independently of lineage. *Development* **110,** 85–96.
30. Goldman, R. D., Goldman, A. E., Green, K. J., Jones, J. C. R., Jones, S. M., and Yang, H.-Y. (1986) Intermediate filament networks: organization and possible functions of a diverse group of cytoskeletal elements. *J. Cell Sci.* **(Suppl. 5),** 69–67.
31. Dráberová, E. and Dráber, P. (1993) A microtubule-interacting protein involved in coalignment of vimentin intermediate filaments with microtubules. *J. Cell Sci.* **106,** 1263–1273.
32. Green, K. J., Geiger, B., Jones, J. C. R., Talian, J. C., and Goldman, R. D. (1987) The relationship between intermediate filaments and microfilaments before and during the formation of desmosomes and adherens-type junction in mouse epidermal keratinocytes. *J. Cell Biol.* **104,** 1389–1402.
33. Geuens, G., De Bradander, M., Nuydens, R., and De Mey, J. (1983) The interaction between microtubules and intermediate filaments in cultured cells treated with taxol and nocodazole. *Cell Biol. Int. Rep.* **7,** 35–47.
34. Brown, K. D. and Binder, L. I. (1992) Identification of the intermediate filament-associated protein gyronemin as filamin. Implications for a novel mechanism of cytoskeletal interaction. *J. Cell Sci.* **102,** 19–30.
35. Mithieux, G. and Rousset, B. (1989) Identification of a lysosome membrane protein which could mediate ATP-dependent stable association of lysosomes to microtubules. *J. Biol. Chem.* **264,** 4664–4668.
36. Langford, G. M., Kuznetsov, S. A., Johnson, D., Cohen, D. L., and Weiss, D. G. (1994) Movement of axoplasmic organelles on actin filaments assembled on acrosomal processes: evidence for a barbed-end-directed organelle motor. *J. Cell Sci.* **107,** 2291–2298.
37. Gyoeva, F. and Gelfand, V. I. (1991) Coalignment of vimentin intermediate filaments with microtubules depends on kinesin. *Nature* **353,** 445–448.
38. Brandy, S. T., Lasek, R. J., and Allen, R. D. (1984) Gelsolin inhibition of fast axonal transport indicates a requirement for actin microfilaments. *Nature* **310,** 56–58.
39. Fey, E. G., Ornelles, D. A., and Penman, S. (1986) Association of RNA with the cytoskeleton and the nuclear matrix. *J. Cell Sci.* **5,** 99–119.
40. Inbar, M. and Sachs, L. (1973) Mobility of carbohydrate containing sites on the surface membrane in relation to the control of cell growth. *FEBS Lett.* **32,** 124–128.

41. Asch, J. F. and Singer, S. J. (1976) Concanavalin-A induces transmembrane linkage of concanavalin A surface receptors to intracellular myosin-containing filaments. *Proc. Natl. Acad. Sci. USA* **73,** 4575–4579.
42. Bourguignon, L. Y. W. and Bourguignon, G. J. (1984) Capping and the cytoskeleton. *Int. Rev. Cytol.* **87,** 195–224.
43. André, P., Gabert, J., Benoliel, A. M., Capo, C., Boyer, C., Schmitt-Verhulst, A. M., Malissen, B., and Bongrand, P. (1991) Wild type and tailless CD8 display similar interaction with microfilaments during capping. *J. Cell Sci.* **100,** 329–337.
44. Katsumoto, T. and Kurimura, T. (1988) Ultrastructural localization of concanavalin A receptors in the plasma membrane: association with underlying actin filaments. *Biol. Cell* **62,** 1–10.
45. Albertini, D. F. and Clark, J. I. (1975) Membrane-microtubule interactions: Concanavalin A capping induced redistribution of cytoplasmic microtubules and colchicine binding proteins. *Proc. Natl. Acad. Sci. USA* **72,** 4976–4980.
46. Albertini, D. F. and Anderson, E. (1977) Microtubule and microfilament rearrangements during capping of concanavalin A receptors on cultured ovarian granulosa cells. *J. Cell Biol.* **73,** 111–127.
47. Horst, C. J., Forestner, D. M., and Besharse, J. C. (1987) Cytoskeletal-membrane interactions: a stable interaction between cell-surface glycoconjugates and doublet microtubules of the photoreceptor connecting cilium. *J. Cell Biol.,* **105,** 2973–2987.
48. Arena, N., Bodo, M., Baroni, T., Alia, F. A., Gaspa, L., and Becchetti, E. (1990) Effects of lectins on cytoskeleton and morphology of cultured chick embryo fibroblasts. *Cell. Mol. Biol.* **36,** 317–328.
49. Bodo, M., Becchetti, E., Pezzetti, F., Baroni, T., Calvitti, M., Alia, F. A., and Arena, N. (1990) Cytoskeletal and DNA synthesis modification by concanavalin A in embryonic fibroblasts mantained in serum-added medium. *Cell. Mol. Biol.* **36,** 673–687.
50. Bodo, M., Arena, N., Pezzetti, F., Baroni, T., Calvitti, M., Alia, F. A., and Becchetti, E. (1990) Restoration of a normal phenotype, microtubular pattern and DNA synthesis in embryonic fibroblasts concanavalin A pretreated. *Cell. Mol. Biol.* **36,** 689–703.
51. Carinci, P., Becchetti, E., Locci, P., Bodo, M., and Evangelisti, R. (1991) Lectins: cell morphology, cytoskeleton and GAG metabolism. *Acta Embriol. Morphol. Exp.* **12,** 93.
52. Evangelisti, R., Becchetti, E., Locci, P., Bodo, M., Arena, N., Lilli, C., De Matteo, M., and Carinci, P. (1993) Coordinate effects of concanavalin A on cytoskeletal organization, cell shape, glycosaminoglycans accumulation and exoglycosidase activity in chick embryonic cultured fibroblasts. *Eur. J. Histochem.* **37,** 161–172.
53. Evangelisti, R., Becchetti, E., Baroni, T., Rossi, L., Arena, N., Valeno, V., Carinci, P., and Locci, P. (1995) Modulation of phenotypic expression of fibroblasts by alteration of the cytoskeleton. *Cell Biochem. Funct.* **13,** 41–52.
54. Bodo, M., Becchetti, E., Baroni, T., Mocci, S., Merletti, L., Giammarioli, M., Calvitti, M., and Sbaraglia, G. (1994) Internalization of Candida albicans and cytoskeletal organization in macrophages and fibroblasts treated with concanavalin A. *Cell. Mol. Biol.* **41,** 247–305.

55. Fukui, Y., De Lozanne, A., and Spudich, J. A. (1990) Structure and function of the cytoskeleton of a dictyostelium myosin-defective mutant. *J. Cell Biol.* **110,** 367–378.
56. King, T. P., Pusztai A., and Clarke, E. M. W. (1980) Immunocytochemical localization of ingested kidney bean *(Phaseolus vulgaris)* lectins in rat gut. *Histochem. J.* **12,** 201–208.
57. King, T. P., Pusztai, A. and Clarke, E. M. W. (1982) Kidney bean (*Phaseolus vulgaris*) lectin-induced lesions in rat small intestine. *J. Comp. Pathol.* **92,** 357–373.
58. King, T. P., Begbie, R., and Cadenhead, A. (1983) Nutritional toxicity of raw beans in pigs. Immunocytochemical and cytopathological studies on the gut and the pancreas. *J. Sci. Food Agric.* **34,** 1004–1412.
59. Sjölander, A., Magnusson, K. E., and Latkovic, S. (1986) Morphological changes of rat small intestine after short-time exposure to concanavalin A or wheatgerm agglutinin. *Cell Struct. Funct.* **11,** 285–293.
60. Draaijer, M., Koninkx, J., Hendriks, H., Kik, M., Van Dijk, J., and Mouwen J. (1989) Actin cytoskeletal lesions in differentiated human colon carcinoma Caco-2 cells after exposure to soybean agglutinin. *Biol. Cell.* **65,** 29–35.
61. Pasternak, C. and Elson, E. L. (1985) Lymphocyte mechanical response triggered by crosslinking surface receptors. *J. Cell Biol.* **100,** 860–872.
62. Murali, K. R. (1984) Lectin-induced actin polymerization in human lymphocytes: a possible signal for mitogenesis. *Cell. Immunol.* **83,** 181–188.
63. Wheeler, M., E., Gerrard, J. M., and Carroll, R. C. (1985) Reciprocal transmembranous receptor-cytoskeleton interactions in concanavalin A-activated platelets. *J. Cell Biol.* **101,** 993–1000.
64. Smith, L. and Hochmuth, R. M. (1982) Effect of wheatgerm agglutinin on the viscoelastic properties of erythrocyte membrane. *J. Cell Biol.* **94,** 7–11.
65. Evans, E. and Leung, A. (1984) Adhesivity and rigidity of erythrocyte membrane in relation to wheatgerm agglutinin binding. *J. Cell Biol.* **98,** 1201–1208.

35

Effect of Lectins on Protein Kinase Activity

Kiyonao Sada and Hirohei Yamamura

1. Introduction

Protein kinases have critical roles for the amplification and distribution of the signals that are sent from the cell surface to the cytoplasm after ligand binding. They are generally classified into two groups: protein-serine/threonine kinases, such as cAMP-dependent protein kinase or casein kinases, and protein-tyrosine kinases (PTKs), such as EGF receptor kinase or p60^{src}. Although the amount of protein-tyrosine phosphorylation is much less than that of protein-serine/threonine phosphorylation, PTKs are particularly important in the control of cell proliferation and differentiation. PTKs have themselves been divided into two groups: the transmembrane receptor group and nonreceptor group.

Lectins can induce a dramatic increase of protein-tyrosine phosphorylation in T-cells, neutrophils, mast cells, splenocytes, and platelets. The increase in protein-tyrosine phosphorylation by lectins usually involves crosslinking of cell-surface glycoproteins. Since these receptors often have no intrinsic protein-kinase activity, nonreceptor PTKs, such as p56^{lck}, p60^{fyn}, and p72^{syk} are implicated in participating in the lectin-induced signaling pathway. Therefore, lectins are useful agents for investigating intracellular signal transductions during cellular activation involving PTKs.

1.1. Immunoblotting of Phosphotyrosine

Mitogenic lectins induce cytoplasmic protein phosphorylation, especially on tyrosine residues in cells of hematopoietic origin. In lectin stimulated human lymphocytes, a rapid increase in tyrosine phosphorylation of several proteins, such as T-cell receptor zeta chain is observed *(1–4)*. These phenomena are widely observed in hematopoietic cells, e.g., splenocytes, human neutrophil, and platelets *(4–7*, Fig. 1). Protein serine phosphorylation is also induced by lectin stimulation. Lectins lead to activation of several protein-serine/threo-

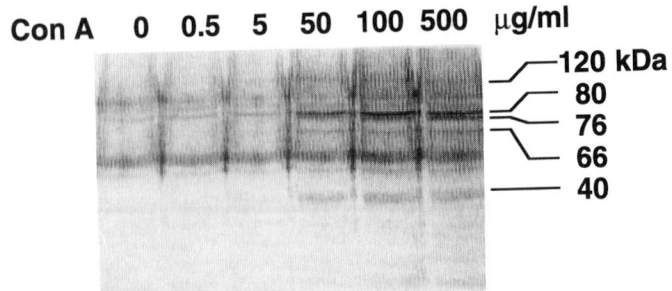

Fig. 1. Immunoblotting of protein-tyrosine phosphorylation induced by Con A in human neutrophils. Neutrophils are incubated with various amount of con A for 4 min and tyrosine phosphorylation of 120, 80, 76, 66, and 40 kDa proteins are observed after stimulation. In this figure, rabbit polyclonal antiphosphotyrosine antibodies are employed.

nine kinases *(8,9)*, phosphorylation of CD3 gamma chain *(10,11*, or retinoblastoma (RB) protein *(12)*.

Alterations in the response of tyrosine phosphorylation to lectins have been found in acute myeloblastic leukemic cells *(13)*. The component of the T-cell receptor that is phosphorylated after lectin stimulation is constitutively phosphorylated in lymphoproliferative mutant mice T-cells *(10)*. A similar phenomenon is observed in T-lymphoma cells *(4)*.

For detection of the increase of protein-tyrosine phosphorylation, immunoblotting is most useful procedure. After addition of lectin to cells, the stimulated cell lysate is separated by SDS-polyacrylamide gel electrophoresis (SDS-PAGE) followed by transfer to polyvinylidene difluoride membrane and probing with antiphosphotyrosine antibody. Monoclonal antiphosphotyrosine antibodies are available commercially. Enhanced chemiluminescence (ECL) can be used to increase the sensitivity of immunoblotting.

1.2. Immunoprecipitation Kinase Assay

Lectins induce protein phosphorylation through the activation of cytoplasmic protein kinases. The candidate protein kinases responsible for protein-tyrosine phosphorylation are nonreceptor-type PTKs. In hematopoietic cells that react to lectin, expression of several nonreceptor-type PTKs has been reported, and increase of PTK activity has been observed *(1,13)*. Pretreatment of cells with genistein and tyrphostin, PTK inhibitors, reduces protein-tyrosine phosphorylation and suppresses the specific cellular functions induced by lectins *(4,10)*.

An in vitro kinase assay using specific antibody is required to determine the increase of specific activities of PTKs. $p72^{syk}$, a nonreceptor type PTK which

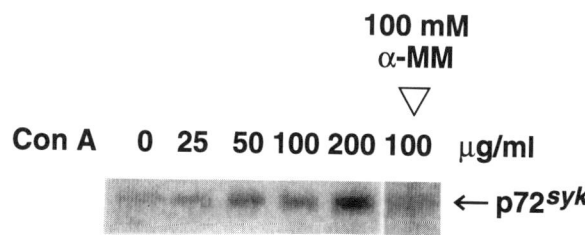

Fig. 2. Immunoprecipitation kinase assay reveals dose-dependent activation of PTK by con A measured by autophosphorylation activity. The arrow indicates the position of p72^{syk}. Coincubation with methyl α-mannopyranoside (α-MM) prevented this activation of p72^{syk}.

possesses two Src homology 2 (SH2) domains instead of SH3 and shows limited expression in hematopoietic cells, increases its kinase activity both for exogenous substrate and for autophosphorylation in lectin stimulated cells (5,7,14–16, Fig. 2). In thymocytes, the association of some membrane molecules with nonreceptor type PTK, p56^{lck} and p60^{fyn}, and the increase of kinase activities associated with CD 48 have been observed after concanavalin A (Con A) stimulation (17).

Stimulated cell lysate is mixed with specific antibody, then the immunoprecipitate is incubated with bivalent cation and [γ-^{32}P]ATP with or without exogenous substrates. H2B histone and enolase are frequently used as exogenous substrates. The immunoprecipitate is separated by SDS-PAGE gel and followed by autoradiography.

To determine simply the increase of tyrosine phosphorylation of all substrates, alkaline treatment of the gel is sometimes performed. Since this procedure is not specific, the result of autoradiography includes the increase of protein-threonine and -tyrosine phosphorylations.

To detect the increase of phosphotyrosine, an immunoprecipitation kinase assay using "cold" ATP can be performed, but the increase of phosphoserine and phosphothreonine is not determined by this procedure. The amount of phosphorylated proteins is confirmed by immunoblotting.

1.3. Phosphoamino Acid Analysis

Phosphoamino acid analysis reveals the state of phosphorylation on serine, threonine and tyrosine residues. To determine the activation of PTK, the phosphoamino acid analysis of exogenous substrate or autophosphorylated kinase should be performed (Fig. 3). The proteins that are phosphorylated with [γ-^{32}P]ATP in the immunoprecipitation kinase assay are eluted from the gel and precipitated with trichloroacetic acid. Determination of phosphoamino acid is performed using high voltage electrophoresis.

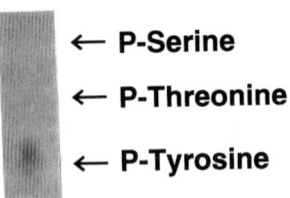

P: Phospho

Fig. 3. Phosphorylated 72-kDa band is excised from the gel and analyzed by phosphoamino acid analysis. This figure shows only phosphotyrosine is detected.

2. Materials
2.1. Immunoblotting of Phosphotyrosine After Lectin Stimulation

1. Hanks balanced salt solution (HBSS): 136.7 mM NaCl, 5.4 mM KCl, 0.81 mM MgSO$_4$, 1.3 mM CaCl$_2$, 0.33 mM Na$_2$HPO$_4$, 0.44 mM KH$_2$PO$_4$, 5.6 mM dextrose, and 4.2 mM NaHCO$_3$, pH 7.4, at 4°C.
2. Con A is purchased from Sigma (St. Louis, MO), dissolved in HBSS, and stored at 4°C for up to 1 mo.
3. 3X Sample buffer: 6% SDS, 187.5 mM Tris–HCl, pH 6.8, 30% glycerol, 15% 2-mercaptoethanol, and 0.003% bromophenol blue.
4. Polyvinylidene difluoride membrane: Immobilon P is obtained from Millipore (Bedford, MA).
5. Methanol.
6. Whatman 3MM paper.
7. Anode 1 solution: 300 mM Tris, 20% methanol.
8. Anode 2 solution: 25 mM Tris, 20% methanol.
9. Cathode solution: 25 mM Tris, 20% methanol, 40 mM ε-amino-n-caproic acid.
10. BSA: Albumin fraction V (Bovine) is purchased from Seikagaku (Tokyo, Japan).
11. Skim milk.
12. TPBS: PBS (pH 7.4) containing 0.05% Tween-20.
13. Antiphosphotyrosine antibody (anti-PY Ab): Monoclonal anti-PY Ab 4G10 is purchased from UBI (Lake Placid, NY).
14. Horseradish peroxidase-conjugated goat antimouse IgG Ab is purchased from Bio-Rad (Hercules, CA).
15. PBS.
16. POD staining kit is obtained from Wako Pure Chemicals (Osaka, Japan).

2.2. Immunoprecipitation Kinase Assay
2.2.1. Immunoprecipitation Kinase Assay with [γ-^{32}P]ATP

1. Lysis buffer: 2% Triton X-100, 50 mM Tris–HCl, pH 7.5, 150 mM NaCl, 5 mM EDTA, 100 μM Na$_3$VO$_4$, 1 mM phenylmethylsulfonyl fluoride and 1 μg/mL of leupeptin.

Lectin Effects on Protein Kinase Activity

2. Specific antibody for the target PTK, for example, rabbit polyclonal anti-Syk antibodies are raised against the synthtic partial polypeptide of p72^{syk}.
3. Protein A sepharose 4FF is obtained from Pharmacia (Uppsala, Sweden).
4. Washing buffer: 0.5M NaCl, 10 mM HEPES/NaOH, pH 8.0.
5. 10 mM HEPES/NaOH, pH 8.0.
6. 3X Reaction mixture: 300 mM HEPES/NaOH, pH 8.0, 30 μM Na$_3$VO$_4$, 150 mM magnesium acetate, 15 mM MnCl$_2$, and 3 μM [γ-^{32}P] ATP (200 cpm/fmol).
7. 3X Sample buffer (*see* Section 2.1., step 3).

2.2.2. Alkaline Treatment of SDS-PAGE Gel

1. 1M KOH.
2. 40% Methanol and 10% acetic acid.

2.2.3. Immunoprecipitation Kinase Assay and Immunoblotting of Protein-Tyrosine Phosphorylation

1. 3X modified reaction mixture: 300 mM HEPES/NaOH, pH 8.0, 30 μM Na$_3$VO$_4$, 150 mM magnesium acetate, 15 mM MnCl$_2$, and 150 μM ATP.
2. Reprobe solution: 62.5 mM Tris-HCl, pH 6.8, 2% SDS, 100 mM 2-mercaptoethanol.
3. ECL staining kit (Amersham, Amersham, UK).

2.3. Phosphoamino Acid Analysis

1. Homogenizing buffer: 50 mM NH$_4$HCO$_3$ and 0.1% SDS.
2. 2-Mercaptoethanol.
3. 1% Human γ-globulin.
4. 100% Trichloroacetic acid (TCA).
5. Ethanol, at –20°C.
6. Ethanol-ether (1:1) mixture, at –20°C.
7. Standard phosphoamino acid solution: each contains 1 mg/mL of phosphoserine, phosphothreonine, and phosphotyrosine, pH 3.5.
8. Whatman 3MM paper.
9. Electrophoresis buffer, pH 3.5: 0.5% pyridine and 5% glacial acetic acid.
10. 0.2% Ninhydrin in acetone.

3. Methods

3.1. Immunoblotting of Phosphotyrosine

1. Incubate the neutrophils (1 × 10^7 cells/mL), which are prepared from freshly obtained blood for various times in the presence of 50 μg/mL of con A at 37°C. The neutrophils are isolated by Ficoll-Conray density centrifugation *(16)* and suspended in HBSS at 4°C.
2. Terminate the reaction by boiling with 1/2 vol of 3X sample buffer for 3 min.
3. Separate the cell lysates by SDS-PAGE gel using 8% gel at 30 mA constant current.
4. Soak the polyvinylidene difluoride membrane in methanol for 20 s (wetting).

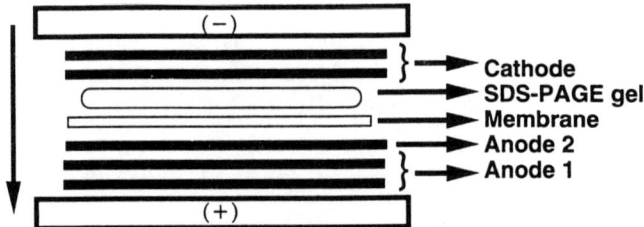

Fig. 4. Separated proteins on SDS-PAGE gel are electrophoretically transferred to polyvinylidene difluoride membrane with Whatman 3MM papers, which are soaked in several electrode solutions. This figure shows the scheme.

5. Gently soak this membrane in water for 1 h.
6. Soak two Whatman 3M*M* papers in anode 1 solution, another in anode 2 solution and two in cathode solution.
7. Place two papers soaked in anode 1 solution on the anodic electrode plate.
8. Next, place one paper soaked in anode 2 solution on the papers of anode 1, then polyvinylidene difluoride membrane on anode 2 papers.
9. Place the SDS-PAGE gel on this membrane.
10. Place the two papers soaked in cathode 2 solution on the gel, and cathodic electrode plate on top of the papers (*see* Note 1, Fig. 4).
11. Electrophoretically transfer the separated proteins on SDA-PAGE gel to polyvinylidene difluoride membrane with current up to 4 mA/cm^2 of gel for 30 min.
12. Block the membrane with TPBS containing BSA (1%) or skim milk (5%) for 1 h at room temperature.
13. Wash the membrane with 10 mL of TPBS, 3 × 5 min (*see* Note 2).
14. Incubate the membrane with 10 mL of TPBS containing 1 µL of monoclonal anti-PY Ab 4G10 (10,000-fold dilution) for 1 h at room temperature.
15. Wash the membrane with 10 mL of TPBS, 3 × 5 min.
16. Incubate the membrane with 5 mL of TPBS containing 1 µL of horseradish peroxidase-conjugated goat antimouse IgG antibody (5000-fold dilution).
17. Wash the membrane with 10 mL of TPBS, 3 × 5 min.
18. Rinse the membrane with 10 mL of PBS for 5 min.
19. Visualize by POD staining kit (*see* Section 2.1., item 16, and Note 3).

3.2. Immunoprecipitation Kinase Assay

3.2.1. Immunoprecipitation Kinase Assay with [γ-^{32}P]ATP

1. Incubate 1 mL of isolated neutrophils (2 × 10^7 cells/mL) with 100 µg/mL con A for various times (e.g., 0–2 min) at 37°C.
2. Collect the cells by a rapid centrifugation and aspirate the supernatants immediately (*see* Note 4).
3. Dissolve the resulting pellets with 500 µL of lysis buffer and keep on ice for 30 min, and if necessary disrupt the cells by sonication.

Lectin Effects on Protein Kinase Activity

4. Centrifuge the lysates at 100,000g for 10 min to remove the detergent insoluble fraction.
5. Transfer the supernatants to Eppendorf tubes and then add 0.3 µg of polyclonal anti-Syk antibodies to each tube (*see* Note 5).
6. Add 40 µL of protein A sepharose 4FF as an antibody–anchor to each tube.
7. Mix the supernatant with antibodies and Protein A for 1 h at 4°C.
8. Centrifuge at 10,000g for 2 min, and aspirate the supernatant (*see* Note 6).
9. Add 1 mL of lysis buffer, disperse the pellets, centrifuge, and aspirate the supernatants.
10. Wash the pellets with lysis buffer.
11. Add the 1 mL of washing buffer containing 0.5M NaCl and disperse the pellet.
12. Centrifuge again, and aspirate the supernatant.
13. Add 1 mL of 10 mM HEPES/NaOH and wash the pellets once more.
14. Finally, suspend the pellets in 40 µL of 10 mM HEPES/NaOH, pH 8.0.
15. Incubate with 20 µL of 3X reaction mixture for 10 min at 30°C with or without exogenous substrate, such as 10 µg of H2B histone (Sigma) (*see* Note 7).
16. Terminate the reaction by boiling with ½ vol of 3X sample buffer for 3 min, then centrifuge the tubes at 10,000g for 2 min to remove the anchor.
17. Separate the immunoprecipitates by SDS-PAGE.
18. Autoradiography (overnight using medical X-ray film) (*see* Note 8).

3.2.2. Alkaline Treatment of SDS-PAGE Gel

Phosphotyrosine and phosphothreonine are stable to heat treatment, whereas phosphoserine is not.

1. To determine the increase of protein phosphorylation by alkaline treatment gently shake the gel prepared from 3.2.17 in 500 mL of 1M KOH for 1 h at 56°C.
2. Gently shake the gel in 500 mL of 40% methanol and 10% acetic acid solution for 2 h at room temperature, until the size of the gel becomes normal again (*see* Note 9).
3. Autoradiography (*see* Note 8).

3.2.3. Immunoprecipitation Kinase Assay and Immunoblotting of Protein-Tyrosine Phosphorylation

1. To determine the lectin–induced increase of protein-tyrosine phosphorylation using anti-PY Ab, incubate the washed immunoprecipitate prepared from Section 3.2.1., step 14 with 20 µL of 3X modified reaction mixture for 10 min at 30°C with or without an exogenous substrate, such as 10 µg H2B histone.
2. Terminate the reaction, and separate the immunoprecipitate by SDS-PAGE as in steps 16 and 17 of Section 3.2.1.
3. Electrophoretically transfer the separated immunoprecipitates on SDS-PAGE gel to polyvinylidene difluoride membrane, and probe with MAb anti-PY as in steps 4–18 of Section 3.1.

4. Soak the membrane with the mixture of solution 1 and 2 in ECL staining kit and exposure to the X-ray films (medical X-ray film from Fuji, Japan) to visualize the results (*see* Note 10).
5. Soak the membrane with reprobe solution at 50°C for 30 min.
6. Block the membrane with skim milk solution as step 12 of Section 3.1.
7. Wash the membrane as in step 13 of Section 3.1.
8. Incubate the membrane with 3 mL of TPBS containing the same antibody as is used for immunoprecipitation, such as 1000-fold diluted polyclonal anti-Syk Ab for 1 h at room temperature.
9. Wash the membrane with 10 mL of TPBS, three times for 5 min.
10. Incubate the membrane with 5 mL of TPBS containing second antibody, such as 1 µL of horseradish peroxidase-conjugated goat antirabbit IgG antibody (5000-fold dilution).
11. Wash the membrane with 10 mL of TPBS, three times for 5 min.
12. Rinse the membrane with 10 mL of PBS for 5 min.
13. Visualize the results using ECL staining kit and exposure to the X-ray films.

3.3. Phosphoamino Acid Analysis

1. Excise the phosphorylated protein band on gel from step 18 of Section 3.2.1.
2. Homogenize the gel in 1 mL of homogenizing buffer using a Potter-Elvehjem homogenizer.
3. Add 50 µL of 2-mercaptoethanol and boil the samples for 5 min.
4. Shake and incubate the sample overnight at 37°C.
5. Centrifuge the sample at 1600g for 10 min.
6. Add 7.5 µL of 1% human γ-globulin and mix with the supernatant.
7. Add 100% of TCA solution to a final concentration of 20%, and place on ice for 4 h.
8. Centrifuge the samples at 10,000g for 15 min, and aspirate the supernatants.
9. Mix the resulting pellets with 1 mL of ethanol (−20°C), then centrifuge the samples again.
10. Mix the pellets with 100 mL of ethanol-ether (1:1) mixture (−20°C), then centrifuge again.
11. Evaporate ethanol-ether from eluted proteins using N_2 gas.
12. Add 100 µL of 6N HCl and dissolve the dried protein, then transfer the sample to the small heat stable glass vial (*see* Note 11).
13. Perform acid hydrolysis of sample at 120°C for 2 h.
14. Add 500 µL water and lyophilize the sample to remove the HCl and water.
15. Dissolve the dried proteins in 10 µL of standard phosphoamino acid solution.
16. Apply the sample to Whatman 3MM paper (*see* Note 12), then dry the paper.
17. Wet Whatman 3MM paper with electrophoresis buffer, pH 3.5.
18. Electrophorese at 1500 V for 2.5 h in the electrophoresis buffer, pH 3.5.
19. Dry the Whatman 3MM paper.
20. Spray 0.2% ninhydrin on the paper to visualize the position of each phosphoamino acid, then dry the paper at 60°C for 15 min.
21. Autoradiography (*see* Note 8).

4. Notes

1. Squeeze out the bubbles between the layers by hand with a plastic glove.
2. For blocking, soaking, and incubation of membrane, a gel destain rotator (Marisol) is useful tool.
3. The results of POD staining will deteriorate in light, so we keep the membranes in the dark and record the results by digital sampling using a computer scanner. Alternatively, ECL staining is more useful because the result is more sensitive, stable, and immunoblotting is probed by at least two different antibodies using the same membrane by reprobing method.
4. We use high speed refrigerated microfuge model TOMY MRX 151 (Tokyo, Japan).
5. The amount of the antibodies should be optimized for each antibody, cell, and set of experimental conditions.
6. Since protein A does not form a firm sediment, aspiration of the supernatant should be done carefully. We use a pipetman and do not aspirate the supernatant completely.
7. Instead of [γ-^{32}P]ATP we sometimes use 50 μM ATP, and perform an immunoprecipitation kinase assay to determine the increase of protein-tyrosine phosphorylation by immunoblotting. In this case, the immunoprecipitate is purified, incubated with 3X modified reaction mixture and separated by SDS-PAGE. Then proteins are electrophoretically transferred to polyvinylidene difluoride membrane following immunoblotting using anti-PY MAb. To confirm the amount of precipitate in each lane, ECL and reprobe procedure are performed as in Section 3.2.3.
8. The results of autoradiography are analyzed by a digital optical scanning of gels on a BAS-2000.
9. The size of the gel is increased after soaking in KOH solution, so handle the gel carefully because of its weakness.
10. To reduce the background and pseudopositive band on X-ray film, the residual ECL solution should be removed by sandwiching the membrane between two Whatman 3MM papers.
11. Close the cap of this vial tightly.
12. We gradually apply the sample to the paper with drying to minimize the diameter of the sample spot.

References

1. Fischer, S., Fagard, R., Gacon, G., Genetet, N., Piau, J. P., and Blaineau, C. (1984) Stimulation of tyrosine phosphorylation in lectin treated human lymphocytes. *Biochem. Biophys. Res. Commun.* **124,** 682–689.
2. Wedner, H. J. and Bass, G. (1986) Tyrosine phosphorylation of a 66,000 Mr soluble protein in lectin-activated human peripheral blood T lymphocytes. *J. Immunol.* **136,** 4226–4231.
3. Trevillyan, J. M., Lu, Y. L., Atluru, D., Phillips, C. A., and Bjorndahl, J. M. (1990) Differential inhibition of T-cell receptor signal transduction and early activation events by a selective inhibitor of protein-tyrosine kinase. *J. Immunol.* **145,** 3223–3230.

4. Stanley, J. B., Gorczynski, R., Huang, C. K., Love, J., and Mills, G. B. (1990) Tyrosine phosphorylation is an obligatory event in IL-2 secretion. *J. Immunol.* **145,** 2189–2198.
5. Yamada, T., Taniguchi, T., Nagai, K., Saitoh, H., and Yamamura, H. (1991) The lectin wheatgerm agglutinin stimulates a protein-tyrosine kinase activity of p72^{syk} in porcine splenocytes. *Biochem. Biophys. Res. Commun.* **180,** 1325–1329.
6. Inazu, T., Taniguchi, T., Ohta, S., Miyabo, S., and Yamamura, H. (1991) The lectin wheatgerm agglutinin induces rapid protein-tyrosine phosphorylation in human platelets. *Biochem. Biophys. Res. Commun.* **174,** 1154–1158.
7. Ohta, S., Inazu, T., Taniguchi, T., Nakagawara, G., and Yamamura, H. (1992) Protein-tyrosine phosphorylations induced by concanavalin A and N-formyl-methionyl-leucyl-phenylalanine in human neutrophils. *Eur. J. Biochem.* **206,** 895–900.
8. Iyer, A. P., Pishak, S. A., Sniezek, M. J., and Mastro, A. M. (1984) Visualization of protein kinases in lymphocytes stimulated to proliferate with concanavalin A or inhibited with a phorbol ester. *Biochem. Biophys. Res. Commun.* **121,** 392–399.
9. Evans, G. A., Linnekin, D., Grove, S., and Farrar, W. L. (1992) Specific protein kinases modulated during T-cell mitogenesis. Activity of a 55-kDa serine kinase is associated with growth arrest in human T-cells. *J. Biol. Chem.* **267,** 10,313–10,317.
10. Samelson, L. E., Davidson, W. F., Morse, H. C., III, and Klausner, R. D. (1986) Abnormal tyrosine phosphorylation on T-cell receptor in lymphoproliferative disorders. *Nature* **324,** 674–676.
11. Davies, A. A., Canrell, D. A., Hexham, J. M., Parker, P. J., Rothbard, J., and Crumpton, M. J. (1987) The human T3 γ chain is phosphorylated at serine 126 in response to T lymphocyte activation. *J. Biol. Chem.* **262,** 10,918–10,921.
12. Evans, G. A., Wahl, L. M., and Farrar, W. L. (1992) Interleukin-2-dependent phosphorylation of the retinoblastoma-susceptibility-gene product p110-115RB in human T-cells. *Biochem. J.* **282,** 759–764.
13. Fischer, S., Fagard, R., Piau, J. P., Genetet, N., Blaineau, C., Reibel, L., Le, Prise, Y., and Gacon, G. (1985) Acute myeloblastic leukemia with active tyrosine protein kinase. *Leuk. Res.* **9,** 1345–1351.
14. Ohta, S., Taniguchi, T., Asahi, M., Kato, Y., Nakagawara, G., and Yamamura, H. (1992) Protein-tyrosine kinase p72^{syk} is activated by wheatgerm agglutinin in platelets. *Biochem. Biophys. Res. Commun.* **185,** 1128–1132.
15. Takeuchi, F., Taniguchi, T., Maeda, H., Fujii, C., Takeuchi, N., and Yamamura, H. (1993) The lectin concanavalin A stimulates a protein-tyrosine kinase p72^{syk} in peripheral blood lymphocytes. *Biochem. Biophys. Res. Commun.* **194,** 91–96.
16. Asahi, M., Taniguchi, T., Hashimoto, E., Inazu, T., Maeda, H., and Yamamura, H. (1993) Activation of protein-tyrosine kinase p72^{syk} with concanavalin A in polymorphonuclear neutrophils. *J. Biol. Chem.* **268,** 23,334–23,338.
17. Garnett, D., Barclay, A. N., Carmo, A. M., and Beyers, A. D. (1993) The association of the protein tyrosine kinases p56^{lck} and p60^{fyn} with the glycosyl phosphatidylinositol-anchored proteins Thy-1 and CD48 in rat thymocytes is dependent on the state of cellular activation. *Eur. J. Immunol.* **23,** 2540–2544.

36

Lectin-Induced Calcium Mobilization in Human Platelets

Use of Fluorescent Probes

Giuseppe Ramaschi and Mauro Torti

1. Introduction

The cytosolic calcium concentration, $(Ca^{2+})_c$, represents in many types of cells a versatile regulatory system involved in several signal transduction pathways of cell activation (1), and is tightly regulated in human platelets. Under basal conditions, platelets maintain the $(Ca^{2+})_c$ at about 80 nM by actively sequestering calcium into the intracellular stores or by extruding it into the extracellular medium. The binding of specific agonists to their receptors induces a rapid increase of the $(Ca^{2+})_c$, which can reach the micromolar range, depending on the type and the dose of agonist used. This effect is owing to Ca^{2+} influx from the extracellular medium to the cytosol and to the release of Ca^{2+} from internal stores (1,2). This latter effect is mediated by a phospholipase C-dependent mechanism; phospholipase C hydrolyzes plasma membrane phosphatidylinositol 4,5-bisphosphate and produces diacylglycerol and inositol 1,4,5-trisphosphate (3). Inositol 1,4,5–trisphosphate causes the release of Ca^{2+} from the dense tubular system by interacting with a specific receptor that behaves as a Ca^{2+} channel by itself. Much less is known about the Ca^{2+} influx across the plasma membrane. Several studies suggested a role for membrane glycoproteins, such as glycoprotein IIb-IIIa complex (GPIIb-IIIa) in controlling calcium homeostasis in platelets. GPIIb-IIIa is the platelet receptor for fibrinogen and other adhesive proteins and belongs to the integrin superfamily (4). It represents a high affinity binding site for Ca^{2+} in the plasma membrane (5) and may regulate the Ca^{2+} homeostasis in human platelets (6), by acting as a channel by itself (7) or by interacting with a closely adjacent channel (8).

From: *Methods in Molecular Medicine: Vol. 9: Lectin Methods and Protocols*
Edited by: J. M. Rhodes and J. D. Milton Humana Press Inc., Totowa, NJ

Moreover, some authors reported a role for GPIIb-IIIa occupation in the regulation of phospholipase C activation *(9)*.

1.1. Platelet Activation and Lectins

Lectins are valuable tools for studying the role of membrane glycoprotein occupancy and/or clustering in the signal transduction mechanisms leading to cell activation. Two lectins are well known as platelet agonists: concanavalin A (Con A) *(10)* and wheatgerm agglutinin *(11)*. The interaction of Con A with the platelet surface promotes patching/capping of the glycoprotein GPIIb-IIIa complex *(12)*, that is the main Con A-receptor on the platelet surface *(13)*. Moreover, Con A induces full platelet aggregation and secretion *(10)*, activation of phospholipase C *(14)*, and Ca^{2+} ions movement as a consequence of receptor clustering *(15)*. Recently, it has been shown that both Con A and wheat germ agglutinin are able to induce strong tyrosine protein phosphorylation in human platelets *(16,17)*.

1.2. Fluorescent Indicators of Calcium: Fura-2

Indicators of Ca^{2+} like fura-2 are tetracarboxilic acid calcium chelators. Chromophore moieties are incorporated into the structures which give fluorescent signals varying with $(Ca^{2+})_c$ *(18)*. Cells are loaded with the tetraacetoxymethyl esters of the Ca^{2+} indicators that permeate the plasma membrane and are hydrolyzed by aspecific esterases in the cytosol. The entrapped molecules do not influence functional responses of platelets. Fura-2 binds Ca^{2+} with a K_D of 224 nM *(19)* and undergoes a marked wavelength shift of its fluorescence spectrum after the binding of Ca^{2+}. Fura-2 binds Mg^{2+} only weakly, but its fluorescence is strongly quenched by Mn^{2+} ions; this property is useful to follow the influx of cations through the membrane using Mn^{2+} as a calcium surrogate *(20)*.

2. Materials

2.1. Platelet Preparation

1. Siliconized or plastic labware (*see* Note 1).
2. Gel filtration column and sepharose CL-2B resin (*see* Note 2).
3. ACD anticoagulant: 152 mM sodium citrate, 130 mM citric acid, 112 mM glucose. Use it at $^1/_{10}$ dilution in whole blood (*see* Note 3).
4. HEPES-Tyrode buffer: 10 mM HEPES, 137 mM NaCl, 2.9 mM KCl, 12 mM NaHCO$_3$, 0.5% (w:v) bovine serum albumin, pH 7.4 (*see* Note 4).

2.2. Fluorimetric Determination of Free Calcium Concentration

1. Fura-2-AM (the tetraacetoxymethyl ester of fura-2) (Calbiochem, San Diego, CA), dissolved in dimethylsulphoxide (Sigma, St. Louis, MO) (Fura-2-AM stock solution: 1 mM in DMSO; *see* Note 5).

2. Con A, wheat germ agglutinin, or other lectins to be tested (Sigma) (1–10 mg/mL stock solutions) dissolved in buffer or in distilled water.
3. 10% (v:v) Triton X-100.
4. 100 mM $CaCl_2$.
5. 400 mM Tris base.
6. 200 mM Ethyleneglycolbis-(aminoethylether)tetra-acetic acid (EGTA) (*see* Note 6).
7. A fluorescence spectrometer to monitor fluorescence continuously by using settings of 340 nm (excitation) and 500 nm (emission). Quartz cuvets have to be mounted in a thermostatically controlled holder at 37°C and with a magnetic stirring device (*see* Note 7).

3. Methods
3.1. Platelet Preparation

1. Centrifuge about 20 mL of fresh whole blood (*see* Note 3) anticoagulated with 2% of ACD (152 mM sodium citrate, 130 mM citric acid, 112 mM glucose) at 120g for 15 min at room temperature.
2. Collect the platelet-rich plasma in a 10-mL plastic tube and incubate it with 3 µM Fura-2-AM for 40 min at 37 °C. The tube has to be well mixed every 10 min.
3. Centrifuge the platelet-rich plasma at 340g in order to obtain a platelet pellet.
4. Resuspend cell pellet by gently shaking with a drop of autologous plasma.
5. Load resuspended platelets (<1 mL) on the top of the HEPES-Tyrode-equilibrated column and let the sample enter the separation gel. When the gel is almost dry, add the buffer to elute cells.
6. Follow the turbidity of the eluate, and collect cells when the eluate is cloudy. The platelet preparation will be almost free of contaminating plasma proteins.
7. Adjust platelet count to 2×10^8/mL with HEPES-Tyrode buffer.

3.2. Cytosolic Calcium Movements

1. Prepare samples of 400 µL (8×10^7 cells) in the quartz cuvets and place in the fluorescence spectrometer at 37°C. Set 340 nm excitation and 500 nm emission.
2. Equilibrate cells for 3 min at 37°C with 1 mM $CaCl_2$. To exclude Ca^{2+} influx add 1 to 5 mM EGTA to the medium instead of $CaCl_2$ (*see* Note 8). Begin recording the fluorescence traces with a pen recorder. The fluorescence value at this point is the basal level (Fb).
3. Add the desired amount of lectin (final concentration usually ranges from 10–200 µg/mL). If the lectin works as an agonist there will be a fluorescence peak. The fluorescence at the top of the peak is called Fs (fluorescence after stimulus).
4. Lyse the cells by adding 0.1% Triton X-100 (4 µL of the 10% stock solution) to obtain the maximal fluorescence of the system (F_{max}).
5. Basify the solution with Tris base (20 µL of the 400 mM stock solution-final concentration 20 mM). This will improve the action of EGTA.
6. Add EGTA (20 µL of the 200 mM stock solution-final concentration 10 mM) to the suspension in order to chelate all the free Ca^{2+} and to obtain F_{min}, the minimal fluorescence of the system (*see* Fig. 1).

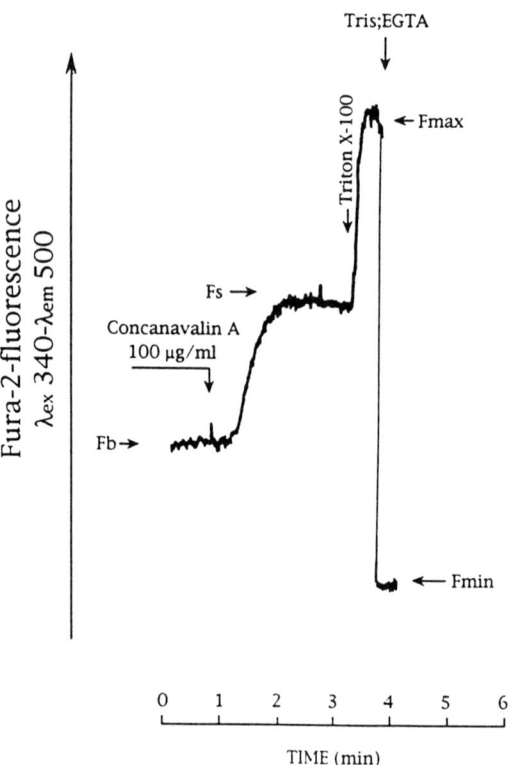

Fig. 1. Intracellular Ca^{2+} increase induced by 100 μg/mL concanavalin A in human platelets in the presence of 1 mM $CaCl_2$. The addition of the reagents is reported (for details, *see* Chapter 16). In this example, Fb = 363, Fmax = 712, F_{ConA} = 530, and Fmin = 222. The calculated Ca^{2+} values are (Ca^{2+})basal = 90 nM and (Ca^{2+})Con A = 380 nM.

7. Calculate the calcium concentration from fluorescence values by using the general formula *(21)*: $(Ca^{2+}) = K_d$ (F-Fmin)/(Fmax-F), where K_d = 224 nM (*see* Note 9).

3.3. Mn^{2+} as Bivalent Cation Influx Indicator

1. Prepare the platelet suspension as reported in Section 3.2. and use fluorescence settings of 360 nm excitation and 500 nm emission (*see* Note 10).
2. Do not add $CaCl_2$ or EGTA. The system is nominally Ca^{2+}-free.
3. Add 200 μM $MnCl_2$ (4 μL of a 20 mM $MnCl_2$ stock solution) to the cell suspension. Follow the entry of Mn^{2+} recording the fall of Fura-2 fluorescence for 1 min.
4. Add the agonist and record the change of the slope of the basal influx of Mn^{2+} into the cells. The change in slope indicates the opening of membrane cation channels due to the lectin action (*see* Fig. 2).

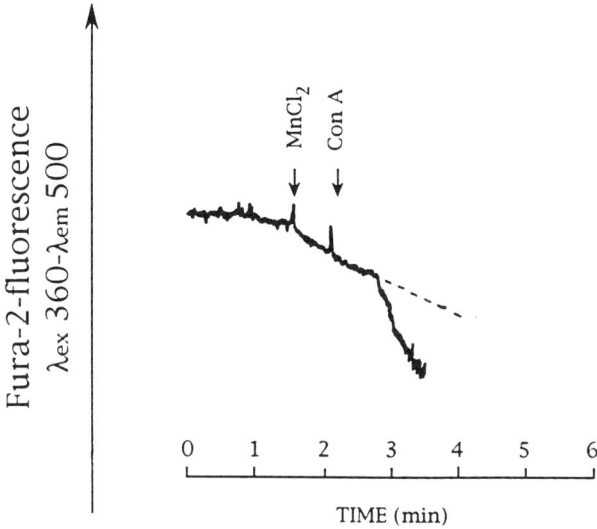

Fig. 2. Effect of concanavalin A on membrane cation channels: Fura-2-loaded platelets are incubated with 200 µM $MnCl_2$ and then stimulated with 100 µg/mL concanavalin A. The broken line shows the decline of fluorescence in control samples, without addition of agonist.

4. Notes

1. It is important to avoid the contact of platelets with glass surfaces because the cells will be activated by glass or other anionic surfaces.
2. Prepare the resin in water and degas it for 1 h under vacuum; pack a column of about 15 cm with a diameter of 1–2 cm. Wash the column with several beds of deionized water, and then equilibrate it with buffer.
3. Take fresh blood from healthy volunteers, who have not taken drugs for 2 wk before venipuncture. Mix the blood well with ACD directly in the 20-mL syringe. Use gloves when manipulating blood! Remember to dispose of blood-contaminated waste properly. Centrifugation of whole blood should be done using a swinging bucket rotor and not a fixed angle rotor.
4. HEPES-Tyrode buffer can be stored for 1 wk at 4 °C; it is important to check and adjust pH to 7.40 every time: platelets are inhibited when pH is lower than 7.2 and can easily aggregate if pH is higher than 7.5.
5. Carefully dissolve Fura-2-AM in dimethylsulfoxide in the dark and make 20-µL aliquots to be stored in the dark at –20°C. The stock solution is stable for about 5 mo. Dimethylsulphoxide is a harmful compound: avoid contact with skin.
6. EGTA is almost insoluble in water at pH <7.0. To prepare EGTA 200 mM, dissolve the powder in ¾ of the final volume, and add 5M NaOH drop-to-drop mixing with a stirring bar. When the pH is about 7.0, the solution will be clear.

7. Teflon-coated stirring bars for aggregometer cuvets should fit well into the spectrofluorimeter cuvets.
8. Remember that most lectins depend on cations for their binding properties. In the presence of EGTA there could be a underestimation of the potency of the lectin as agonist.
9. Data obtained using Fura-2 on platelets suspensions after lectin- or, in general, agonist-induced activation come from the summed fluorescence of all the cells and this reports some average value for calcium concentration. However, it is possible to challenge with lectins the single fura-2-loaded platelet, as done with rabbit cells *(22)*. These data confirm and complete our results *(14,15)*.
10. The use of excitation at 360 nm allows the selective study of Mn^{2+} influx without the interference caused by changes in the concentration of calcium *(20)*.

References

1. Siess, W. (1989) Molecular mechanism of platelet activation. *Physiol. Rev.* **69**, 58–178.
2. Rink, T. J. and Sage, S. O. (1990) Calcium signalling in human platelets. *Annu. Rev. Physiol.* **52**, 431–449.
3. Berridge, M., J. and Irvine, R. F. (1984) Inositol triphosphate, a novel second messenger in cellular signal transduction. *Nature* **312**, 315–321.
4. Phillips, D. R., Charo, I. F., Parise, L. V., and Fitzgerald, L. (1988) The platelet membrane glycoprotein IIb-IIIa complex. *Blood* **71**, 831–843.
5. Brass, L. F. and Shattil, S. J. (1984) Identification and function of the high affinity binding sites for Ca^{2+} on the surface of platelets. *J. Clin. Invest.* **73**, 626–632.
6. Brass, L. F. (1985) Ca^{2+} transport across the platelet plasma membrane. A role for membrane glycoproteins IIb and IIIa. *J. Biol. Chem.* **260**, 2231–2236.
7. Rybak, M., E., Renzulli, L. A., Bruns, M., J., and Cahaly, D. P. (1988) Platelet glycoproteins IIb and IIIa as a calcium channnel in liposomes. *Blood* **72,** 714–720.
8. Powling, M., J., and Hardisty, R. M. (1985) Glycoprotein IIb-IIIa complex and Ca^{2+} influx into stimulated platelets. *Blood* **66**, 731–734.
9. Sinigaglia, F., Torti, M., Ramaschi, G., and Balduini, C. (1989) The occupancy of glycoprotein IIb-IIIa complex modulates thrombin activation of human platelets. *Biochim. Biophys. Acta* **984**, 225–230.
10. Wheeler, M. E., Gerrard, J. M., and Carroll, R. C. (1985) Reciprocal transmembraneous receptor-cytoskeleton interactions in Concanavalin A-activated platelets. *J. Cell. Biol.* **101,** 993–1000.
11. Higashihara, M., Takahata, K., Ohashi, T., Kariya, T., Kume, S., and Oka, H. (1985) The platelet activation induced by wheatgerm agglutinin. *Febs Lett.* **183**, 433–438.
12. Kakaiya, R. M., Kiraly, T. L., and Cable, R. G. (1988) Concanavalin A induces patching/capping of the platelet membrane glycoprotein IIb/IIIa. *Thromb. Haemost.* **59**, 281–283.
13. McGregor, J. L., Clemetson, K. J., James, E., Greencond, T., and Dechavanne, M. (1979) Identification of human platelet glycoproteins in SDS-polyacrylamide gels using ^{125}I-labeled lectins. *Thromb. Res.* **16**, 825–831.

14. Torti, M., Balduini, C., Ramaschi, G., and Sinigaglia, F. (1992) Stimulation of human platelets with Concanavalin A involves phospholipase C activation. *Cell Biochem. Funct.* **10**, 53–59.
15. Ramaschi, G., Torti, M., Sinigaglia, F., and Balduini, C. (1993) Intracellular calcium mobilization is triggered by clustering of membrane glycoproteins in Concanavalin A-stimulated platelets. *Cell Biochem. Funct.* **11**, 241–249.
16. Torti, M., Ramaschi, G., Sinigaglia, F., and Balduini, C. (1995) Dual mechanism of protein-tyrosine phosphorylation in Concanavalin A-stimulated platelets. *J. Cell. Biochem.* **57**, 30–38.
17. Inazu, T., Taniguchi, T., Ohta, S., Miyabo, S., and Yamamura, H. (1991) The lectin wheat germ agglutinin induces rapid protein-tyrosine phosphorylation in human platelets. *Biochem. Biophys. Res. Comm.* **174**, 1154–1158.
18. Hallam, T. J. and Rink, T. J. (1987) Insights into platelet function gained with fluorescent Ca^{2+} indicators, in *Platelets in Biology and Pathology III* (Macintyre, D. E. and Gordon, J. L., eds.), Elsevier, Amsterdam, 353–372.
19. Grynkiewicz, G., Poenie, M., and Tsien, R. Y. (1985) A new generation of Ca^{2+} indicators with greatly improved fluorescence properties. *J. Biol. Chem.* **260**, 3440–3450.
20. Sage, S. O., Merritt, J. E., Hallam, T. J., and Rink, T. J. (1989) Receptor-mediated calcium entry in fura-2-loaded human platelets stimulated with ADP and thrombin. *Biochem. J.* **258**, 923–926.
21. Pollock, W. K., Rink, T. J., and Irvine, R. F. (1986) Liberation of [^{3}H]arachidonic acid and changes in cytosolic free calcium in fura-2-loaded human platelets stimulated by ionomycin and collagen. *Biochem. J.* **235**, 869–877.
22. Ikegami, Y., Nishio, H., Fukuda, T., Nakata, Y., and Segawa, T. (1991) Effect of concanavalin A on intracellular calcium concentration in single blood platelets. *Japan. J. Pharmacol.* **57**, 233–241.

37

Lectin-Triggered Superoxide/H_2O_2 and Granule Enzyme Release from Cells

Alexander V. Timoshenko, Klaus Kayser, and Hans-Joachim Gabius

1. Introduction

The modulatory potency of lectins on cellular activities deserves attention from a cell biological and a clinical point-of-view. In addition to serving as tools to delineate signaling processes that follow carbohydrate-dependent cell binding, plant lectins can apparently affect certain characteristics of the host immune system in vitro and in vivo with potential clinical benefit *(1,2)*. With respect to defense mechanisms the generation of reactive oxygen compounds such as the superoxide anion radical (O_2^-) and H_2O_2 or the release of granule enzymes are supposed to play a notable role *(3–6)*. Deliberate enhancement of these activities by lectins may increase the host's capacity to control growth of infectious organisms or malignant cells. Therefore, rigorous testing of plant agglutinins and endogenous lectins that have been isolated from human tissues may enable one to devise a rational lectin-mediated treatment modality. This chapter describes the protocols for the respective assay procedures used to determine the effects of lectins on production of reactive oxygen compounds and release of granule enzymes.

2. Materials

2.1. Isolation of Human Neutrophils and Rat Thymocytes

1. Monovette tubes with Na-citrate (Sarstedt, Nuembrecht, Germany).
2. Gauze bandage.
3. Phosphate-buffered saline (PBS): 10 mM Na_2HPO_4/KH_2PO_4, 137 mM NaCl, 2.7 mM KCl, pH 7.3.
4. PBS, pH 7.3, containing 5.55 mM D-glucose, 0.9 mM $CaCl_2$, and 0.5 mM $MgCl_2$ (PBSG; *see* Note 1).

From: *Methods in Molecular Medicine: Vol. 9: Lectin Methods and Protocols*
Edited by: J. M. Rhodes and J. D. Milton Humana Press Inc., Totowa, NJ

5. Lymphoprep from Nycomed (Oslo, Norway; *see* Note 2).
6. 6% (w/v) solution of dextran (70 kDa) in 0.9% NaCl (*see* Note 3).
7. 0.6M solution of NaCl.

2.2. Release of $O_2^{\cdot -}/H_2O_2$ from Cells

1. PBS and PBSG, pH 7.3.
2. Cytochrome-c (Sigma, Munich, Germany): 16 mg/mL in PBS (freshly prepared).
3. Superoxide dismutase (SOD) (Sigma): 3 mg/mL in PBS, can be stored frozen in aliquots at –20°C for at least 2 mo without notable loss of activity.
4. Scopoletin solution: 0.2 mM in PBS. This solution is prepared freshly from a 2 mM stock solution of scopoletin (*see* Note 4).
5. Horseradish peroxidase: 1 mg/mL in PBS (freshly prepared).
6. Menadione: 0.02M in dimethyl sulfoxide (DMSO). The solution should be protected from light and can be stored in aliquots at –20°C for at least 1 mo.
7. 0.1M NaN$_3$ in water.

2.3. Release of Granule Enzymes from Neutrophils

1. PBS and PBSG, pH 7.3 (*see* Section 2.1.); 0.1M PBS, pH 6.2.
2. HEPES/Brij 35 solution: 0.1M HEPES, pH 7.5, 0.5M NaCl, 0.1% Brij 35.
3. Suspension of lyophilized cells of *Micrococcus lysodeikticus* (Sigma): 0.2 mg/mL in 0.1M PBS, pH 6.2.
4. *O*-dianisidine solution: 5 mg/mL in water; *o*-dianisidine is carcinogenic and should be handled with care.
5. 0.05% (v/v) H$_2$O$_2$ (freshly prepared).
6. 10% (v/v) Triton X-100.
7. 20 mM Methoxysuccinyl-ala-ala-pro-val-*p*-nitroanilide in DMSO.
8. 1% (w/v) NaN$_3$ in water.
9. 2% (v/v) Acetic acid.

3. Methods
3.1. Isolation of Cells
3.1.1. Isolation of Human Neutrophils

1. Mix 20 mL of blood from Na-citrate-containing Monovette tubes with 10 mL of 6% dextran ($M_r \sim$ 70 kDa) solution in a 30-mL plastic syringe and keep the syringe vertical for 1 h at room temperature (*see* Note 3).
2. Collect the upper neutrophil-rich layer in two 10-mL plastic tubes and centrifugate for 10 min at 300g (e.g., in Hettich Universal 2S).
3. Both pellets are carefully suspended in a total volume of 10 mL PBS and centrifugation is repeated (step 2).
4. Add 3 mL ice-cold water to the cell pellet, cautiously suspend the cells by using a 1-mL Gilson-type pipet to produce a solvent flow, and finally add 1 mL ice-cold 0.6M NaCl to the suspension to restore the isotonicity (*see* Note 5).
5. Centrifugate the suspension for 4 min at 200g.

6. Repeat steps 4 and 5.
7. Suspend the pellet in 6 mL of PBS and place the solution on 4 mL of Lymphoprep in a 10-mL plastic tube. Do not mix the Lymphoprep and the neutrophil-rich suspension.
8. Centrifuge the sample for 20 min at 600g.
9. Wash the pellet twice in 8 mL of PBSG for 5 min at 300g.
10. Resuspend the pellet in 5 mL ice-cold PBSG and keep the suspension of human neutrophils on ice. Usually the suspension contains more than 98% granulocytes (staining with hematoxylin); the cell viability is usually more than 95% in routine assays with trypan blue.

3.1.2. Isolation of Rat Thymocytes

1. Remove the thymus gland from a decapitated rat, 6–8 wk of age, and transfer the organ into a 10-mL plastic tube (*see* Note 6).
2. Cut the tissue into small pieces for 15–20 s using a small pair of scissors, mix the resulting homogenate with 10 mL ice-cold PBSG, and filter the suspension through a piece of gauze bandage.
3. Centrifuge the suspension for 5 min at 300g.
4. Suspend the pellet in 8 mL ice-cold PBSG and repeat the centrifugation (step 3).
5. Resuspend the pellet in 5 mL ice-cold PBSG and keep the suspension of rat thymocytes on ice. Usually the suspension contains more than 95% viable cells that are not stained with trypan blue.

3.2. Assay of Superoxide/H_2O_2 Release from Cells

3.2.1. Lectin-Induced Release of Superoxide from Human Neutrophils

1. Prepare in Eppendorf tubes 800-µL aliquots of the reaction mixture containing 2×10^6 cells/mL, 80 µM (1 mg/mL) cytochrome-c, and 10–50 µg lectin/mL in PBSG at 4°C (*see* Note 7). Do not forget to include a control sample that additionally contains 200 U SOD/mL to eliminate superoxide anions in the reaction mixture.
2. Place the tubes in a waterbath at 37°C for 15 min.
3. Stop the reaction by placing the tubes on ice.
4. Centrifuge the cooled samples for 1 min at 2000g (e.g., in Eppendorf centrifuge 5415C).
5. Measure the absorbance of supernatants at 550 nm relative to control specimens in the presence of SOD or to the standard solution (1 mg cytochrome-c/mL) that lacks any lectin.
6. Calculate the lectin-induced release of superoxide in nmol using the formula: O_2^- (nmol) = 47.7 × ΔA, where ΔA is the difference in absorbance of samples with and without lectin *(7)*.

3.2.2. Lectin-Induced Release of H_2O_2 from Human Neutrophils

1. 1.9 mL of reaction mixture containing (2–4) × 10^6 cells, 2 nmol scopoletin, 2 µmol NaN_3, and 30 µg horseradish peroxidase in PBSG are prepared at room tempera-

ture directly in a fluorimeter cuvet, which is then placed into the holder of a fluorimeter (e.g., SLM 4800) that is maintained at 37°C for 3 min with constant stirring (*see* Note 8).
2. Start the reaction by adding 100 µL of lectin containing solution (final concentration of lectin: 5 µg/mL or any suitable value) and record the kinetics of scopoletin oxidation (decrease in fluorescence).
3. Calculate the rate of H_2O_2-mediated scopoletin oxidation as the maximal slope of the measured kinetic curve (*see* Note 9).

3.2.3. Menadione-Dependent Release of H_2O_2 from Rat Thymocytes in the Presence of Lectins

1. 2 mL of the reaction mixture containing 2×10^7 cells/mL, 10–15 µg horseradish peroxidase/mL, 1 µM scopoletin, 1 mM NaN_3, and 1–15 µg lectin/mL in PBSG are prepared at room temperature directly in a fluorimeter cuvet, which is then placed into the holder of fluorimeter that is maintained at 37°C for 10 min with constant stirring (*see* Note 10).
2. Start the reaction by adding 4 µL of a 0.02M menadione solution in DMSO and record the kinetics of scopoletin oxidation (decrease in fluorescence) (*see* Note 11).
3. Calculate the rate of H_2O_2-mediated scopoletin oxidation as the maximal slope of the measured kinetic curve (*see* Note 9).

3.3. Assays of Lectin-Induced Granule Enzyme Release from Human Neutrophils (see Note 12)

3.3.1. Preparation of the Enzyme-Containing Supernatant

1. Incubate the suspension of neutrophils (6.25×10^6 cells/mL in PBSG) for 5 min at 37°C in the presence of 5 µg cytochalasin B/mL (Sigma, Munich, Germany) (*see* Note 13).
2. Place aliquots of cell containing solution (400 µL) in Eppendorf tubes on ice and add 100 µL of lectin containing solution, e.g., in PBS, to reach the desired concentration of lectins (10–50 µg/mL). Add 100 µL of buffer instead of the lectin containing solution to a blank specimen. Concomitant addition of lectin and haptenic sugar to ascertain the assumed carbohydrate specificity as well as addition of the sugar to exclude any metabolic or osmolarity effects are also required as controls.
3. Place the tubes in a waterbath at 37°C for 20 min.
4. Stop the reaction by placing the tubes on ice.
5. Centrifugate the cooled samples for 5 min at 2000g (e.g., in Eppendorf centrifuge 5415C).
6. Place supernatants on ice or store frozen at –20°C.
7. Process a control sample with cells by adding 5 µL of a 10% Triton X-100 solution to 500 µL of the suspension of neutrophils (5×10^6 cells/mL) and thorough mixing for cell lysis to determine the total amount of enzymatic activity.

3.3.2. Assay of Lysozyme Activity

1. Add 50 µL of the supernatant to 1 mL of a suspension of *Micrococcus lysodeikticus* cells (0.2 mg lyophilized cells/mL 0.1M PBS, pH 6.2) directly in a spectrophotometer cuvet that is kept at 25°C.
2. Measure the decrease in the absorbance at 450 nm at 30 s intervals for 4 min (*see* Note 14).
3. Calculate the velocity of the reaction and express results as the percentage of total enzyme activity, determined in the control sample containing 0.1% Triton X-100 for cell lysis.

3.3.3. Assay of Elastase Activity

1. Add 50 µL of the supernatant to 200 µL of ice-cold substrate solution containing 1 mM methoxysuccinyl-alanyl-alanyl-prolyl-valyl-*p*-nitroanilide in 0.1M HEPES, pH 7.5/0.1% Brij 35 solution (*see* Section 2.3., item 2; Note 15).
2. Place the aliquots in a waterbath, and incubate for 15 min at 40°C.
3. Stop the reaction by adding 200 µL of 2% acetic acid.
4. Transfer 100–200 µL of the mixture into individual wells of 96-well plates with F-shaped bottom and determine the absorbance of each sample at 405 nm in any suitable type of plate reader (*see* Note 16).
5. Calculate the release of enzyme as the percentage of total enzyme activity, determined in the control sample containing 0.1% Triton X-100 for cell lysis.

3.3.4. Assay of Myeloperoxidase Activity

1. Add 50 µL of the supernatant to 600 µL of ice-cold substrate solution (150 µg *o*-dianisidine/mL in a solution containing 6 parts 10 mM PBSG, pH 7.3, 5 parts 0.1M PBS, pH 6.2, and 1 part 0.05% H_2O_2).
2. Place the samples into a waterbath, and incubate for 15 min at 26°C.
3. Stop the reaction by adding 50 µL of 1% NaN_3.
4. Transfer 200-µL aliquots from each assay mixture into individual wells of 96-well plates with F-shaped bottom and determine the absorbance of each sample at 450 nm in any suitable type of plate reader (*see* Note 17).
5. Calculate the release of enzyme as the percentage of total enzyme activity, determined in the control sample containing 0.1% Triton X-100 for cell lysis.

4. Notes

1. The final concentrations of calcium and magnesium salts are reached by additions of 1 mL from respective stock solutions (66.2 mg $CaCl_2 \cdot 2H_2O$/mL and 50 mg $MgCl_2 \cdot 6H_2O$/mL) to 500 mL of PBS with vigorous stirring.
2. Instead of Lymphoprep from Nycomed (Oslo, Norway), other media with a density of 1.077 g/mL, for instance Ficoll-Paque from Pharmacia (Freiburg, Germany), can also be successfully used, unless they unfavorably affect cell viability.
3. 6% (w/v) Solution of dextran is required for sedimentation of erythrocytes. Our experience shows that dextran from *Leuconostoc* species with a M_r ~70 kDa from

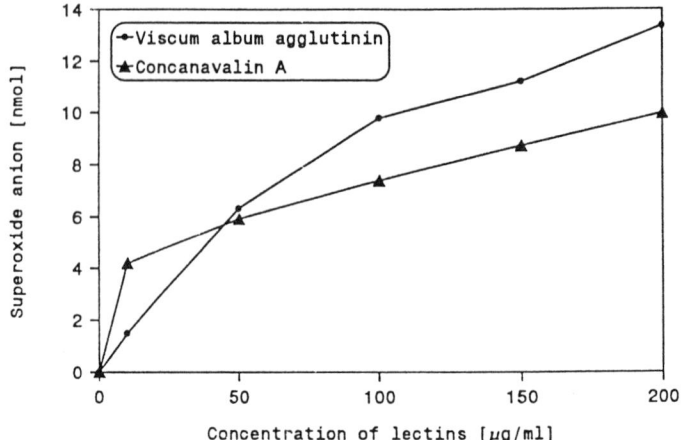

Fig. 1. Effect of lectin concentration on superoxide release from human neutrophils. The cells (2×10^6/mL) were incubated at 37°C in Eppendorf tubes containing 80 μM cytochrome-c with various concentrations of *Viscum album* agglutinin and concanavalin A. The values of superoxide release from untreated control cells were subtracted from the results, obtained in the presence of lectin. The illustrated data are a representative of 4–6 experiments on different cell batches, each series was done at least in duplicate.

Fluka (Neu-Ulm, Germany) and dextran T70 from Pharmacia (Freiburg, Germany) yield comparable results. Other types of dextran with an increased mol-wt of up to 500 kDa can also be used for this purpose.
4. Scopoletin has a limited solubility in aqueous solutions *(8)*. To prepare the stock solution of scopoletin (5–10 mL, 2 mM) the mixture is placed into a waterbath, set to 37–40°C, for at least 1 h under light protection with constant stirring. The solution is then kept at 4°C for up to 1 mo and should always be also warmed up to 37–40°C prior to preparation of the assay mixture of scopoletin (0.2 mM).
5. An increase of incubation time of cells in water may reduce the viability of neutrophils.
6. When removing the thymus, this step should be performed very carefully to avoid any damage of blood vessels. If contamination with blood has occurred, the thymus gland should be rinsed thoroughly with PBS and the processing continued, as described.
7. The rate of reduction of cytochrome-c, indicative for the generation of superoxide by neutrophils, depends on the lectin concentration *(9)*. As shown in Fig. 1, a lectin concentration of 50 μg/mL appears to be sufficient for preliminary assays to test this activity of lectins. Several lectins like *Viscum album* agglutinin (VAA), concanavalin A (Con A), or *Ricinus communis* agglutinin can obviously activate the plasma membrane NADPH-oxidase at lectin concentrations below this level

Superoxide and Granule Enzyme Release from Cells 447

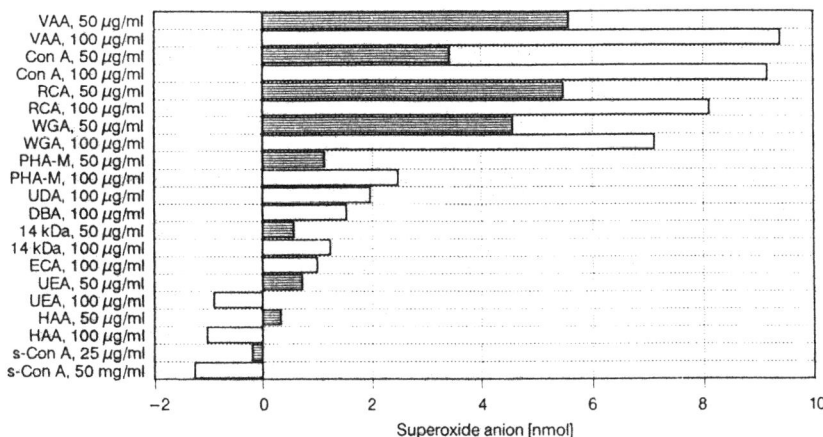

Fig. 2. Effect of various lectins on superoxide release from human neutrophils. VAA, *Viscum album* agglutinin; Con A, concanavalin A; RCA, *Ricinus communis* agglutinin; WGA, wheat germ agglutinin; PHA-M, phytohemagglutinin M; UDA, Urtica dioica agglutinin; DBA, *Dolichos biflorus* agglutinin; 14 kDa, galectin-1, β-galactoside-specific lectin from human placenta; ECA, *Erythrina cristagalli* agglutinin; UEA, *Ulex europaeus* agglutinin; HAA, *Helix aspersa* agglutinin; s-Con A, succinyl-concanavalin A.

(9). Presence of synergistic agents such as phorbol-myristate-acetate or chemotactic peptides, too, reduce the required amount of lectin. However, lectin binding does not always trigger a greater response at higher concentration (Fig. 2). Following preliminary assays, it will thus be indispensable to determine the optimal lectin concentration for each tested substance. The specificity of lectin-dependent generation of superoxide can be checked by addition of haptenic sugar, e.g., lactose for VAA and α-methyl-D-mannoside for Con A (Fig. 3). It should also be noted for clinical considerations that the level of lectin-dependent production of superoxide can be altered in cancer patients *(10)*. The interindividual heterogeneity precludes a clear-cut interpretation (Fig. 4).

8. In this step the recorder of the fluorimeter is set to the maximal value of fluorescence, because the lectin-dependent oxidation of scopoletin leads to a decrease in fluorescence intensity. Among the individual components, peroxidase should be added in the last step to control the purity of specimens with respect to presence of peroxides as well as to adjust the basal level of the H_2O_2 generating system by recording the fluorescence signal during the period of temperature equilibration. Control samples show a nonspecific decrease in the intensity of scopoletin fluorescence that does not exceed 4–6% of the initial level. The presence of NaN_3 in the reaction mixture is necessary to block the cellular catalase activity. In the absence of NaN_3 the lectin-dependent reaction can still be monitored with decreased rate.

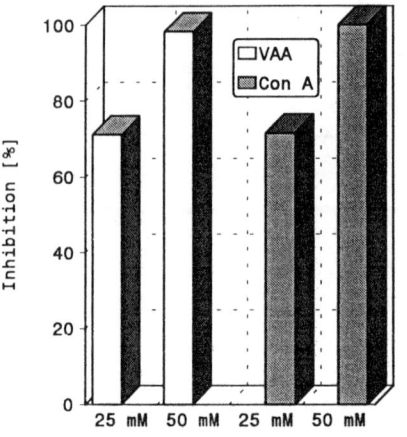

Fig. 3. Inhibitory efficiency of sugars on superoxide release from human neutrophils that was dependent on the presence of *Viscum album* agglutinin (VAA) or concanavalin A (Con A) at a concentration of 50 µg/mL.

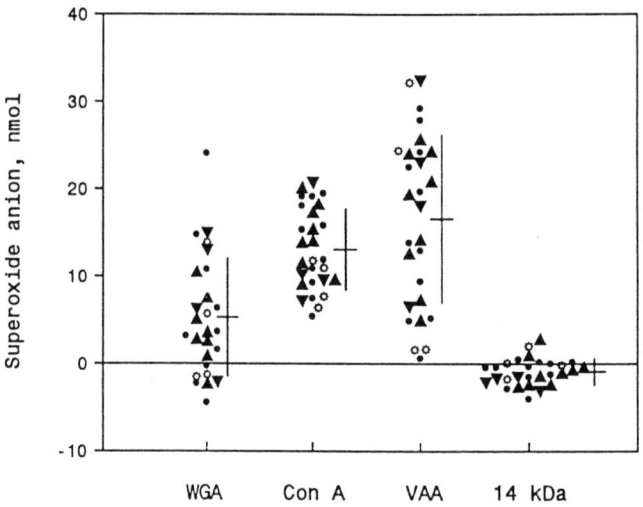

Fig. 4. Effect of wheat germ agglutinin (WGA), concanavalin A (Con A), *Viscum album* agglutinin (VAA), and galectin-1 from human placenta (14 kDa) on superoxide production by human neutrophils in patients with bronchial carcinoma, who received no chemotherapy. The lectin concentration was 50 µg/mL. The production of superoxide in the absence of lectins is subtracted from each value. Bars: mean ± 1 SD; patients with various types of nonsmall cell carcinoma: squamous cell carcinoma (●), large cell anaplastic carcinoma (▼), and adenocarcinoma (▲); patients with small cell carcinoma (○).

Superoxide and Granule Enzyme Release from Cells

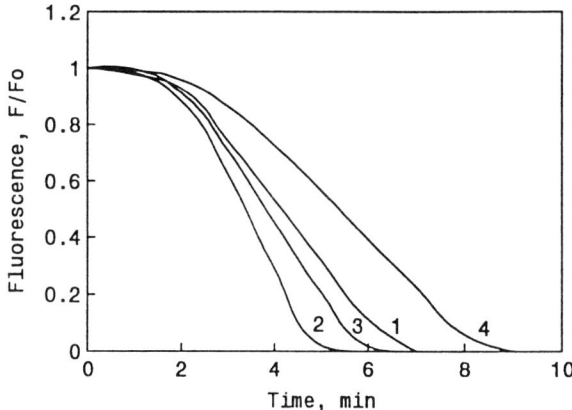

Fig. 5. Continuous assessment of menadione-dependent oxidation of scopoletin by rat thymocytes at 37°C in the absence or in the presence of *Viscum album* agglutinin (VAA), its carbohydrate-binding B-chain or its toxic A-subunit: 1–control or 1.25 μg VAA/mL + 25 mM lactose, 2–1.25 μg VAA/mL, 3–1.25 μg VAA-B/mL, 4–12.5 μg VAA-A/mL. The cells were preincubated with the lectin or its subunits for 10 min prior to addition of menadione. The sample contains 2×10^7 cells/mL, 1 μM scopoletin, 1 mM NaN$_3$, 10 μg peroxidase/mL, 40 μM menadione in PBSG. A representative of five experiments with cells from different animals is shown.

9. An example for the typical kinetics of scopoletin oxidation is presented in Fig. 5. To calculate precisely the lectin-dependent production of H_2O_2 in nmol, it is necessary to calibrate the reaction in control experiments without cells using a series of solutions with known concentrations of H_2O_2. Such solutions can freshly be prepared from a 30% H_2O_2 stock solution. The exact concentration is determined spectrophotometrically with the molar extinction coefficient of H_2O_2 in the UV field *(8)*: $\varepsilon^{240-260} = 26.41 \times mol^{-1} \times cm^{-1}$.
10. The lectin concentration should be chosen in a range that is not sufficient to induce aggregation of cells.
11. This method allows one to reveal drastic differences in the activity of the toxic A-subunit and the carbohydrate-binding B-chain of VAA *(11)*. Whereas the B-chain induces a lactose-inhibitable acceleration in the menadione-dependent oxidation of scopoletin like the hololectin, the A-subunits suppress this activity, probably due to their inherent toxicity (Fig. 5).
12. For unequivocal interpretation of the results of lectin-dependent enzyme release, it is essential to exclude damage of the plasma membrane within processing by measuring the activity of a strictly plasma-localized enzyme like lactate dehydrogenase (LDH) in cell supernatants. Two kinds of measurement of LDH-activity may be used: oxidation of NADH in the presence of pyruvate and reduction of NAD$^+$ in the presence of lactate *(12)*. Because the sensitivity of these methods for application on cell suspensions may not always be sufficient, the application of proce-

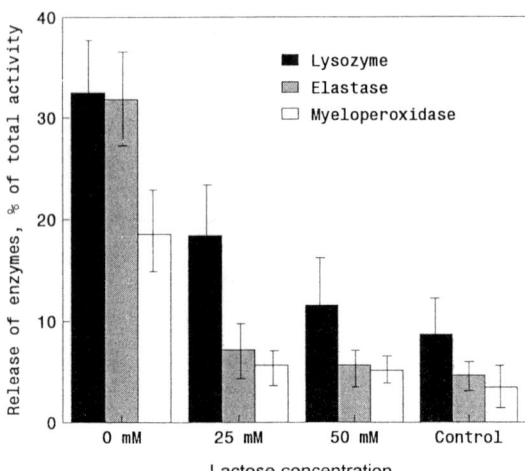

Fig. 6. Effect of increasing concentrations of lactose (25 mM, 50 mM) on VAA-induced release of enzymes from human neutrophils in comparison with the control value in the absence of the lectin. The cells (5 × 10^6 cells/mL) were incubated for 20 min in the presence of lectin (25 µg/mL) at 37°C and the supernatants after centrifugation were analyzed. Enzyme release is expressed as percentage of total activity of neutrophils that were treated with 0.1% Triton X-100. Results are presented as mean ± SD (bars), $n = 3$.

dures with increased level of detection like the cytotoxicity analysis kit from Boehringer (Mannheim, Germany) may be considered an alternative.

13. In contrast to particulate stimulators, which induce enzyme release from specific as well as from azurophil granules of neutrophils, soluble activators apparently affect only specific granules *(3)*. Preincubation with cytochalasin B is required to reach a pronounced level of release of enzymes from azurophil granules in response to soluble stimulators *(13)*. Similar to the lectin-triggered oxidative burst, the release of granule enzymes from cytochalasin B-treated human neutrophils is inhibitable by the specific sugar *(14)*. An illustrative example with the mistletoe lectin is shown in Fig. 6, underscoring the specificity of the protein-carbohydrate interaction.
14. The rate of decrease in absorbance of the bacterial suspension should be in the range of 0.015–0.050/min. If the rate of reaction exceeds a reduction of 0.050 U of absorbance/min, the assay solution should be diluted for accurate assessment.
15. The presence of the nonionic detergent Brij 35 in this solution is needed to prevent the nonspecific adsorption of elastase to the tube wall *(15)*.
16. Elastase cleaves the nitroanilide substrate with high specificity, yielding free 4-nitroanilide *(16)*. The product of this reaction is very stable and the plate with the samples can be stored at 4°C for at least 1 d without changes in absorbance at 405 nm.
17. The plate should be read within 1 h after completing the experiment and can not be stored for a long time.

References

1. Gabius, H.-J., Gabius, S., Joshi, S. S., Koch, B., Schroeder, M., Manzke, W. M., and Westerhausen, M. (1994) From ill-defined extracts to the immunomodulatory lectin: will there be a reason for oncological application of mistletoe? *Planta Med.* **60,** 2–7.
2. Gabius, H.-J. (1994) Lectinology meets mythology: oncological future for the mistletoe lectin? *Trends Glycosci. Glycotechnol.* **6,** 229–238.
3. Boxer, L. A. and Smolen, J. E. Neutrophil granule constituents and their release in health and disease. *Hematol. Oncol. Clin. North Am.* **2,** 101–134.
4. Weiss, S. J. (1989) Tissue destruction by neutrophils. *N. Engl. J. Med.* **320,** 365–380.
5. Segal, A. W. and Abo, A. (1993) The biochemical basis of the NADPH oxidase of phagocytes. *Trends Biochem. Sci.* **18,** 43–47.
6. Timoshenko, A. V. and Cherenkevich, S. N. (1994) Glycobiological aspects of the activation of phagocytes' respiratory chain. *Biopolimery i Kletka* **10,** 58–66 (in Russian).
7. Markert, M., Andrews, P. C., and Babior, B. M. (1984) Measurements of $O_2^{·-}$ production by human neutrophils. The preparation and assay of NADPH oxidase-containing particles from human neutrophils. *Methods Enzymol.* **105,** 358–365.
8. Corbett, J. T. (1989) The scopoletin assay for hydrogen peroxide. A review and a better method. *J. Biochem. Biophys. Methods* **18,** 297–308.
9. Timoshenko, A. V. and Gabius, H.-J. (1993) Efficient induction of superoxide release from human neutrophils by the galactoside-specific lectin from *Viscum album*. *Biol. Chem. Hoppe Seyler* **374,** 237–243.
10. Timoshenko, A. V., Kayser, K., Drings, P., Kolb, G., Havemann, K., and Gabius, H.-J. (1993) Modulation of lectin-triggered superoxide release from neutrophils of tumor patients with and without chemotherapy. *Anticancer Res.* **13,** 1789–1792.
11. Timoshenko, A. V. and Gabius, H.-J. (1995) Influence of the galactose-specific lectin from *Viscum album* and its subunits on cell aggregation and selected intracellular parameters of rat thymocytes. *Planta Med.* **61,** 130–133.
12. Bergmeyer, H. U. (1974) *Methoden der enzymatischen Analyse,* vol. 1, Verlag Chemie, Weinheim, Germany, pp. 607–623.
13. Fittschen, C. and Henson, P. M. (1994) Linkage of azurophil granule secretion in neutrophils to chloride ion transport and endosomal transcytosis. *J. Clin. Invest.* **93,** 247–255.
14. Timoshenko, A. V., Kayser, K., Drings, P., André, S., Dong, X., Kaltner, H., Schneller, M., and Gabius, H.-J. (1995) Carbohydrate-binding proteins (plant/human lectins and autoantibodies from human serum) as mediators of release of lysozyme, elastase, and myeloperoxidase from human neutrophils. *Res. Exp. Med.* **195,** 153–162.
15. Barrett, A. J. (1981) Leukocyte elastase. *Methods Enzymol.* **80,** 581–588.
16. Nakajima, K., Powers, J. C., Ashe, B. M., and Zimmerman, M. (1979) Mapping the extended substrate binding site of cathepsin G and human leukocyte elastase. Studies with peptide substrates related to the α1-protease inhibitor reactive site. *J. Biol. Chem.* **254,** 4027–4032.

38

Cytotoxic Effects of Lectins

Elieser Gorelik

1. Introduction

The interaction of lectins with various normal or malignant cells not infrequently results in their proliferation or death *(1–3)*. Incubation of normal T-lymphocytes with some lectins, such as Con A and PHA stimulates their proliferation and renders them highly cytotoxic and capable of lysing various cell targets *(4)*. This phenomenon is known as lectin-dependent cell cytotoxicity (LDCC). However, several lectins have been found to be able to kill various normal and malignant cells in the absence of lymphocytes. This type of lectin-mediated cell lysis is obviously distinguishable from LDCC and can be termed direct lectin cytotoxicity. The cytotoxic properties have been found with some lectins such as Con A, PHA, WGA, $GS1A_4$, $GS1B_4$, *lens culinaris* (LCA), ricin (RIC), and abrin lectins *(2,3,5)*. Cells that survived after exposure to the cytotoxic lectins manifested resistance to cytotoxic activity of the same lectin. Numerous lectin-resistant sublines were isolated from the original lectin-sensitive cell lines. These lectin-resistant variants were widely used for the analysis of mechanisms of lectin resistance or sensitivity, and of glycoprotein and glycolipid biosynthesis in mammalian cells, the investigation of the biological role of cell surface carbohydrates in cell-to-cell interactions, and in regulation of metastatic behavior of malignant cells *(2,6)*. For many cultured cell lines, the frequency of lectin-resistant variants was found to be about 10^{-5}–10^{-6}. Some cell lines require mutagenization in order to obtain lectin-resistant cell variants *(2)*. Analysis of mechanisms responsible for resistance to cytotoxic action of lectins revealed that, in some cases, lectin-resistant variants appeared as a result of loss of a specific glycosyltransferase activity with a consequent loss of the particular cell surface carbohydrate necessary for lectin binding *(2)*. However, some lectin-resistant variants showed no loss of cell-surface carbohydrates, lectin binding, or cell agglutination *(2)*. This might indicate that resistance of tumor cells to lectin cytotoxicity is not solely based on the loss of cell surface carbohydrates reacting with this lectin and that some postbinding events are probably involved in determining cell sensitivity to lectin cytotoxicity.

Several methods can be utilized for evaluation of the cytotoxic activity of lectins or cell sensitivity to lectin-mediated cytotoxicity. Basically, all tests used for evaluation of cytotoxic activity of chemicals, biological products, or cytotoxic cells can be helpful for assessment of cytotoxic activity of lectins. These assays are based on direct count of viable and dead cells using trypan blue exclusion test, ^{51}Cr release from the damaged cells, measurement of ^{3}H-thymidine incorporation into DNA by viable cells, total protein determination or evaluation of activity of intracellular enzymes capable of converting some fluorochromic substrates with changes of their color. The colorimetric MTT assay has become very popular as a simple and inexpensive method able to compare the total numbers of cultured cells in control and treated cultures.

1.1. MTT Assay for Evaluation of Cytotoxic Properties of Lectins

A quantitative colorimetric assay was initially developed by Mosmann et al. *(7)*, and numerous modifications have been subsequently introduced. The assay is based on the reduction of the tetrazolium salt MTT (3-[4,5-dimethylthiazol-2-yl]-2,5-diphenyl tetrazolium bromide) by the mitochondrial dehydrogenase of viable cells to form blue formazan crystals. The crystals are solubilized by the addition of an acidic alcohol. The intensity of staining is measured by determining the optical density, which is usually proportional to the number of living cells in culture and provides a possibility of evaluating the cytotoxic and cytostatic effects of the tested agents. Solubilization of the formazan crystals with alcohol might result in precipitation of proteins in culture media, therefore, several other solubilizing agents were used. We found that utilization of DMSO for solubilization of the formazan crystals is most suitable and does not require removal of media. Furthermore, MTT assay in the 24-well, in comparison to 96-well plate gives more reproducible data.

1.2. Lectin-Induced Programed Cell Death (Apoptosis)

The precise mechanisms of cytotoxic action of lectins mostly remains obscure *(8)*. It is not known how binding of some lectins to cell surface carbohydrates results in cell lysis, whereas binding of other lectins does not lead to cell death. It is believed that cells may be killed by various modalities via induction of necrosis or programmed cell death (apoptosis) *(9)*. Necrotic cell death is a result of pore formation in cell membrane, with a balloon-like increase in cell volume and cell disruption. Cell lysis by antibodies and complement is an example of necrotic cell death. Apoptosis is a common mechanism of cell elimination in the multicellular organism and occurs during embryonic development, as well as in a variety of adult normal and malignant tissues *(9,10)*. Apoptosis can be induced by various signals, such as exposure to corticosteroids, cytotoxic T-lymphocytes, tumor necrosis factor (TNF), and some cytotoxic drugs. In addition, apoptosis can be triggered in some cells by hormone or lymphokine deprivation *(9,10)*. It was recently demonstrated that some cytotoxic lectins such as WGA and $GS1B_4$ kill tumor cells by triggering apoptosis *(5)*.

Apoptosis is characterized by induction of plasma and nuclear membrane blebbing, chromatin condensation, and DNA fragmentation that is a result of activation of an endogenous endonuclease that cleaves cell DNA at the internucleosomal linker regions and produces oligonucleosome-length fragments (multiple of 180 bp) *(9,10)*. Various techniques have been developed to identify apoptotic type of cell death. We will detail only some of them.

1.2.1. Quantitation of Apoptotic Cells Using Fluorescent Dyes

To assess the ability of lectins to induce apoptosis, tumor cells, after incubation with lectins, can be analyzed for aberration in chromatin organization using acridine orange staining and fluorescent microscopy *(11)*. Acridine orange intercalates into DNA and makes nuclei appear green, whereas the cytoplasm due to RNA staining is red–orange. Apoptotic nuclei in contrast to normal have highly condensed chromatin that is uniformly and strongly stained by acridine orange. This can take the form of crescents around the periphery of the nucleus, or the nucleus contains one or a group of featureless, bright spherical beads. In advanced apoptosis, the cells will have lost DNA or became fragmented into "apoptotic bodies" with associated reduction in the brightness of nucleus staining *(11)*. Analysis of 200 total cells allows determination of the percentage of apoptotic cells.

1.2.2. Flow Cytometric Analysis of Apoptotic Cells

This assay is rapid, simple, and reproducible and provides a possibility to determine the percentage of apoptotic cells after any treatment. This method has mostly been applied to the evaluation of apoptotic death of normal diploid cells *(12)*. Apoptosis is associated with DNA fragmentation and DNA loss, resulting in a reduction of total level of DNA with increased number of hypodiploid cells that can be detected by flow cytometric analysis *(12)*.

1.2.3. Analysis of DNA Fragmentation

The presence of a "ladder" of oligonucleosome-length degraded DNA in agarose gel electrophoresis is indicative for apoptotic type of death. Agarose gel electrophoresis helps to identify the nonfragmented high-mol-wt DNA on the top of the gel, whereas fast migrated fragmented DNA forms an easily identified apoptotic DNA ladder consisting of multiples of 180 bp *(9,10)*.

2. Materials
2.1. MTT Assay For Evaluation of Cytotoxic Properties of Lectins
1. 24-Well plates.
2. MTT (Sigma, St. Louis, MO).
3. DMSO (Sigma).
4. Lectins (WGA, GS1B$_4$) (Vector Laboratories, Burlingame, CA).
5. Cell culture media: RPMI-1640 media supplemented with 10% fetal bovine serum, L-glutamine, antibiotics (Gibco-BRL, Gaithersburg, MD).
6. Tumor cells (B16 melanoma, MCA105 fibrosarcoma, 3LL lung carcinoma, and so on).

2.2. Quantitation of Apoptotic Cells Using Fluorescent Dyes
1. Acridine orange (Sigma). Stock solution in PBS at concentration 100 µg/mL.
2. Glass tubes (12 × 75-mm) (Fisher Scientific, Pittsburgh, PA).
3. Microscopic slides and 22-mm^2 cover slips.
4. Fluorescence microscope equipped with fluorescein filter set.

2.3. Flow Cytometric Analysis of Apoptotic Cells
1. Polypropylene tubes (12 × 75 mm).
2. Hypotonic fluorochrome solution (propidium iodide (50 µg/mL in 0.1% sodium citrate plus Triton X-100, all from Sigma).

2.4. Analysis of DNA Fragmentation
1. Tumor cells in T-75 culture flask (about 10 × 10^6 cells).
2. Lectins (WGA, GS1B$_4$ or others).
3. Lysis buffer: 100 mM EDTA, 0.5% SDS, 10 mM Tris-HCl, pH 7.5 (Sigma).
4. Proteinase K stock solution: 20 mg/mL.
5. RNase stock solution (10 mg/mL), electrophoresis-grade agarose.
6. Phenol, chloroform, isoamyl alcohol, sodium acetate (3M), absolute ethanol.
7. TE: 10 mM Tris-HCl, pH 7.6, 1 mM EDTA.
8. TPE: 90 mM Tris-phosphate, 2 mM EDTA *(13)*.
9. Gel loading buffer (15% Ficoll 400 containing bromophenol blue and bromocresol green).
10. DNA mol-wt markers (*Hae*II-digested phageX 174 DNA) (Boehringer-Mannheim, Indianapolis, IN).
11. Horizontal gel electrophoresis apparatus and power supply.

3. Methods
3.1. MTT Assay for Evaluation of Cytotoxic Properties of Lectins
1. Harvest plastic adherent tumor cells after short (1 min) incubation with trypsin (0.025%) and EDTA (0.53 mM) solution (Gibco-BRL, Gaithersburg, MD). Prepare tumor cell suspension in complete RPMI-1640 media (1 × 10^6 cells/mL).

2. Distribute 1 mL of cell suspension into each well of 24-well plate.
3. After overnight incubation, add tested lectins into wells in triplicates for each concentration (*see* Note 1).
4. After 1–2 d of incubation, prepare MTT solution in complete culture media at a final concentration of 1 mg/mL (*see* Note 2).
5. Add 0.66 mL of MTT solution into each well without removing the culture media in the well.
6. Incubate at 37°C for 1.5–4 h depending on cell line ability to form the formazan crystals.
7. Carefully aspirate the medium from the wells.
8. Add 0.66 mL of DMSO to each well to dissolve crystals.
9. Remove 0.2 mL from each well, and place it into 96-well flat-bottomed plate suitable for reading.
10. Read optic density at 540 nm wavelength using Microplate Reader.

% cytotoxicity can be determine as
(OD control–OD experimental group) × 100/OD control

3.2. Quantitation of Apoptotic Cells Using Fluorescent Dyes

1. Culture tumor cells with the tested lectins for 24 h. Prepare cell suspension at final concentration 5×10^5–5×10^6 cell/mL.
2. Place 1 µL of acridine orange in bottom of glas tube, add 25 µL of cell suspension, and mix gently by hand.
3. Place 10 µL of this mixture on a microscope slide and cover with a 22-mm^2 cover slip. Use of a microscope slide with cover slip is preferable to a hemocytometer since it gives a greater flattening of the cells and better visualization of the possible apoptotic changes.
4. Under a fluorescence microscope with an appropriate filter, count 200 cells and record the number of cells with apoptotic nuclei and calculate the percentage of apoptotic cells.

3.3. Flow Cytometric Analysis of Apoptotic Cells

1. Incubate cells with the test lectins for 24 h. Prepare cell suspension of treated and nontreated with lectins (*see* Note 4).
2. Distribute 0.5×10^6 cells/tube. Pellet cells by centrifugation (200*g*, 10 min).
3. Resuspend pellet in 1.5 mL of hypotonic fluorochrome solution.
4. Incubate tubes at 4°C in the dark overnight.
5. Analyze the red fluorescence of individual nuclei using a FACScan flow cytometer. Determine the percent of diploid and hypodiploid cells in the control and lectin-treated cultures.

3.4. Analysis of DNA Fragmentation

1. Pellet 5–10 × 10^6 tumor cells after 24 h exposure to lectin in an Eppendorf tube using low speed (200–300*g*, 5 min). Discard the supernatant.

2. Add 292.5 µL of lysis buffer and 7.5 µL of proteinase K stock solution to the pelleted cells. Vortex the tube, and incubate for 1 h at 50°C.
3. Spin down condensate (in an Eppendorf centifuge at maximum speed for 2 min), add 10 µL of RNase stock solution, and incubate for an additional 1 h at 50°C.
4. Purify the DNA by sequential extractions with phenol, phenol:chloroform:isoamyl alcohol (25:24:1), and chloroform:isoamyl alcohol (24:1).
5. Precipitate the DNA with the addition of 1/10 volume of $3M$ sodium acetate and 2 vol of absolute ethanol. Place the samples at −70°C for 15 min and then spin at 14,000g for 15 min to pellet DNA. Remove all traces of ethanol by aspiration and allow samples to stand for several minutes open.
6. Resuspend the DNA in 20 µL of TE buffer and load the samples into the wells of a 2% agarose gel with gel loading buffer.
7. Run gel for 2 h at 40 V in TPE buffer.
8. Stain with ethidium bromide by incubation of the gel in TPE buffer containing 0.5 µg/mL of ethidium bromide for 30–45 min and photograph using Polaroid 667 film.

4. Notes

1. In all experiments the specificity of lectin-mediated cytotoxicity has to be evaluated. For this purpose, cultured cells should be incubated with test lectin and inhibitory sugar at concentrations that abrogate cytotoxic effect of lectin.
2. Lectin-induced cytotoxicity is usually observed after 18–24 h incubation. It should be noted that during this period, highly proliferative malignant cells could double their number in the control group, whereas their proliferation in the presence of lectin might be inhibited. Thus, differences between lectin-treated and nontreated control groups might be a result of cytolytic and/or cytostatic effects of lectins. MTT assay actually evaluates both the cytolytic and cytostatic effects of lectins. Cytolytic effect of lectins are manifested by an increased number of dead cells. To determine the cytolytic effects of lectins, slow growing cells should be used or other tests such as ^{51}Cr release assay should be employed.
3. Acridine orange is highly mutagenic and should be handled with care.
4. Flow cytometric analysis of apoptotic cells was mostly applied for testing normal diploid cells such as lymphocytes or fibroblasts by defining the appearance of hypodiploid cells as a result of DNA fragmentation. Malignant cells usually show a wide range of heteroploidy from hypodiploid to tetraploid cells and this variation may complicate flow cytometric identification of apoptotic lectin-treated tumor cells.

References

1. Sharon, N. and Lis, H. (1989) Lectins as cell recognition molecules. *Science* **246**, 227–234.
2. Stanley, P. (1981) Surface carbohydrate alterations of mutant mammalian cells selected for resistance to plant lectins, in *The Biochemistry of Glycoproteins and Proteoglycans* (Lennarz, W. J., ed.), Plenum Press, NY., pp. 161–190.

3. Stanley, W., Peters, B., Blake, D., Yep, D., Chu, E., and Goldstein, I. (1979) Interaction of wild-type and variant mouse 3T3 cells with lectins from *Bandeiraea simplicifolia* seeds. *Proc. Natl. Acad. Sci. USA* **76**, 303–307.
4. Berke, G. (1989) Functions and mechanisms of lysis induced by cytotoxic T-lymphocytes and natural killer cells, in *Fundamental Immunology* (Paul, W., ed.), Raven, NY, pp. 735–763.
5. Kim, M., Rao, V., Tweardy, D., Prakash, M., Galili, U., and Gorelik, E. (1993) Lectin-induced apoptosis of tumor cells. *Glycobiology* **3**, 447–453.
6. Nicolson, G. (1982) Cancer metastasis. Organ colonization and the cell-surface properties of malignant cells. *Biochem. Biophys. Acta* **695**, 113–176.
7. Mosmann, T. (1983) Rapid colorimetric assay for cellular growth and survival, application to proliferation and cytotoxicity assays. *J. Immunol. Methods* **65**, 55–63.
8. Gorelik, E. (1994) Mechanisms of cytotoxic activity of lectins. *Trends Glycosci. Glycotech.* **6**, 435–445.
9. Arends, M., Morris, R., and Wyllie, A. (1990) Apoptosis. The role of the endonuclease. *Am. J. Pathol.* **136**, 593–607.
10. Gerschenson, L. and Rotello, R. (1992) Apoptosis: a different type of cell death. *FASEB J.* **6**, 2450–2455.
11. Kerr, J., Wyllie, A., and Currie, A. (1972) Apoptosis: a basic biological phenomenon with wide-ranging implications in tissue kinetics. *Br. J. Cancer* **26**, 239–257.
12. Nicoletti, I., Migliorati, G., Pagliacci, M., Grignani, F., and Riccardi, C. (1991) A rapid and simple method for measuring thymocyte apoptosis by propidium iodide staining and flow cytometry. *J. Immunol. Methods* **139**, 271–279.
13. Sambrook, J., Fritsch, E., and Maniatis, T. (1989) *Molecular Cloning. A Laboratory Manual*, vol. 1, Cold Spring Harbor Laboratory Press, Cold Spring Harbor, NY.

39

Application of the Lectin-Dependent Cell-Mediated Cytotoxicity Assay to Bronchoalveolar Lavage Fluid and Venous Blood Samples Collected from Canine Lung Allografts

Allan G. L. Lee and Hani A. Shennib

1. Introduction

One of the major obstacles in postoperative management of lung transplant recipients is differentiating between rejection and infection episodes. In addition, there are no reliable methods routinely to monitor lung allografts to ascertain that they are well tolerated by the host. Conventional noninvasive methods such as chest roentgenographs, radio nuclide perfusion scans, and pulmonary function tests used in conjunction with clinical assessment have been shown to be nonspecific *(1,2)*. Other invasive methods used to facilitate the differentiation of rejection from infection are transbronchial biopsy (TBB) and bronchoalveolar lavage (BAL) *(3–6)*. The technique of BAL offers a unique opportunity for the safe and repetitive harvesting of large quantities of graft infiltrating immunocompetent cells from the transplanted lung. The supernatant fluid collected may also contain microorganisms, soluble cytokines, and other mediators that may reflect the changes occurring in the allograft due to infection or rejection.

Several researchers have applied the lectin-dependent cell-mediated cytotoxicity assay (LDCMC) to BAL fluid and venous blood samples obtained from canine lung allografts, in order to determine if grafts are well tolerated or if rejection or infection exists *(7–11)*. The LDCMC assay is believed to detect the total T-cell-mediated cytotoxicity of a lymphocyte population since lectin substitutes for alloantigens in the binding phase of the lytic reaction *(11–13)*.

The population of cytotoxic T-cells mediating LDCMC and specific cell-mediated cytotoxicity have been shown to be the same *(14–16)*. The LDCMC assay may be useful for assessing rejection in outbred species since numerous antigenic differences may evoke a wide variety of specific cytolytic cell responses. As well, LDCMC assays utilize standard cell lines as targets, which may eliminate the need for extensive storage of frozen donor cells. Although, the LDCMC assay is capable of detecting in vitro peripheral blood lymphocytes (PBL) isolated from humans *(17)*, it has not yet been accepted as a clinical test to differentiate episodes of rejection from infections in transplanted patients.

Nguyen et al. *(9)* reported that rejection of the single lung allograft results in alterations in BAL lymphocyte cytotoxicity that differ from those associated with bacterial infection. They show that BAL lymphocytes from infected allografts showed an elevation of both natural killer (NK) and LDCMC activity, whereas those from rejecting allografts only expressed a signifigant increase in LDCMC. Normal allografts on the other hand, possessed extremely low but detectable NK and LDCMC levels *(9)*. Cellular events following warm ischemia (1 h) of lung tissue were also assessed using the LDCMC assay because the effects of warm ischemia are an important factor in early allograft dysfunction *(7)*. LDCMC was noted to be increased in the BAL from the ischemic lung, but decreased in BAL and PBL at 72 h after injury. Norin et al. used the LDCMC assay to examine the development of intragraft and peripheral blood cytolytic T-lymphocyte activity during cyclosporine A (CsA) dose tapering in canine single lung transplant recipients *(10)*. Results demonstrated a correlation between increased intragraft LDCMC and clinical evidence of lung allograft rejection following CsA dose tapering. Peripheral blood LDCMC did not correlate well with rejection episodes. These results are supported by Emeson et al. *(11)* but are in contrast to the findings of Nguyen et al. *(9)*.

1.1. Principles of Lectin Dependent Cell-Mediated Cytotoxicity

The principle behind the LDCMC assay is to measure nonspecific Tc activity against a population of target cells when induced by lectin addition in vitro. The target cell line for the LDCMC assay in canine and human studies are Raji-cells, derived from human Burkitt's lymphoma. Nonspecific Tc activity can be induced in vitro with lectins phytohemagglutinin (PHA) or concanavalin A (Con A) *(14,18)*. The LDCMC identifies functionally mature cytotoxic T-lymphocytes. These include $CD8^+$ and $CD4^+$ cells as described by Meuer et al. in studies involving cloned $OKT4^+$ and $OKT8^+$ CTL lines *(19)*. Perl et al. noticed higher responsiveness in the $OKT8^+$ T-cell subsets in LDCMC against HEp-2 targets *(20)*. Further studies on murine models by Phillips et al. demonstrate that LDCMC is mediated preferentially by low density $CD3^+$ T-lymphocytes coexpressing the Leu-7 antigen when Con A is added to the medium. In contrast, high-density $CD3^+$ T-lymphocytes and $CD16^+$ NK cells did not mediate LDCMC *(21)*. Furthermore the depletion of NK cells or complement does not affect LDCMC *(10)*. Other cytotoxic cells such as lymphokine-activated killer cells (LAK) and more specifically the $Th1^+CD4^+$ T-cells are also suggested to be detected by the LDCMC assay *(22)*.

Mizushima et al. investigated the role of cytokines interleukin-2 (IL-2) OK-432 and interferon-γ (IFN-γ) on induction of LDCMC in a murine model *(23)*. Con A cultured effector spleen cells isolated from C57BL/6 mice displayed nonspecific cytotoxicity against NK sensitive YAC-1 and NK-resistant EL-4 target cells. Effector cells cultured with Con A and IL-2 displayed increased cytotoxicity against YAC-1 EL-4 target cells. Con A and IFN-γ cultured effector cells had no effect on LDCMC. Combinations of IL-2 with OK-432 or IL-2 with IFN-γ had no synergistic effects on nonspecific lectin-dependent cytotoxicity.

1.1.1. The Role of Lectin Con A in LDCMC

The majority of studies describing the mechanisms of LDCMC have been performed in murine models. In order to determine the role of Con A and other lectins, Green et al. used effector spleen cells isolated from alloimmune C57BL/6 mice and EL4 lymphoma cells syngeneic to the effector as targets *(24)*. Results demonstrate that agglutination of target cells to T-effector cells is necessary for cytotoxicity, but on its own is insufficient. This was demonstrated by the fact that agglutinating agents that are nonmitogenic, e.g., wheatgerm agglutinin or antisera to interacting cells did not elicit cytotoxicity. Only lectins Con A and phytohemagglutinin (PHA) were able to promote cytolysis. Second, mitogen activated T-cells under conditions that precluded cell–cell bridging did not lead to lysis. Results from Parker et al. also demonstrate that nonlethal adhesion occurs between cytotoxic T-lymphocytes and its target cell line when adhesive contact occurs in the absence of lectin *(25)*. These observations lead Green et al. to conclude a dual role for lectin in mediating cytotoxicity: activation of the effector cell to express its cytolytic potential, and concomitant bridging of effector cells to susceptible target cells. Similar conclusions were drawn by Bonavida et al. *(14)*.

The effects of adding B-cell mitogens lipopolysaccharide (LPS) and purified protein derivative (PPD) was shown to be ineffective in mediating LDCMC *(14)*. This supports the hypothesis that only lectins that are T-cell mitogens and agglutinating will stimulate LDCMC. Addition of Con A inhibitor α-methyl-D-mannoside (α-MDM), at a optimal concentration of $10^{-1}M$, inhibited LDCMC activity against target cells. This suggests the presence of specific Con A and PHA receptors on the surface of the cells and supports the "binding" mechanism as described by Green et al. *(24).*

Further studies by Ballas et al. set out to determine the site of lectin action on the effector cell *(26).* Debate centered around whether lectin acted solely on the surface membrane of the effector cell or needed be localized intracellularly (in a similar fashion to steroid hormones) to activate the effector cells. To answer this question, Ballas et al. insolubilized Con A by coupling it to sepharose beads and then tested for its ability to support LDCMC. Results demonstrated that Con A-sepharose was able to support LDCMC in a fashion identical with soluble Con A, although a higher concentration was needed as compared to soluble Con A. sepharose alone did not elicit a cytotoxic response. Addition of α-MDM to either soluble Con A assays or Con A-sepharose assay had the same effect of inhibiting LDCMC activity. Therefore, evidence suggests that the activation signal leading to target cell destruction in the presence of effector cells and mitogen can occur exclusively at the effector cell membrane and does not require intracellular presence of lectin.

1.1.2. The Role of Target Cell Structures in Mediating LDCMC

Bradley and Bonavida, using both the ^{51}Cr LDCMC assay and the single cell assay, demonstrate that lectins act on specific sites of target cells when revealed by neuraminidase (N'ase) treatment and mediate nonspecific LDCMC activity *(27).* In this study, it was shown that the lectins soybean agglutinin (SBA) and peanut agglutinin (PNA) were able to bind effector cells (isolated from C57BL/6 mice) to EL-4 or P815 target cells but were unable to stimulate cytotoxicity. Target cells pretreated with N'ase followed with SBA and effector cell addition displayed LDCMC activity. No cytotoxicity was observed when effector cells were pretreated with N'ase. The N'ase treatment revealed the penultimate galactosyl residues on the target cells for interaction with lectin and effector cells.

2. Materials

2.1. Preparation and Passing of Target Cells for LDCMC Assay

1. Raji target cells (ATCC, Rockville, Maryland). Raji-cells are effective targets for both human and canine effector lymphocytes.
2. 37°C water bath.
3. Rosewell Park Memorial Institute (RPMI)-complete: RPMI-1640 medium (Gibco Laboratories, Gaithersburg, MD; Life Technology, Grand Island, NY) enriched with 10% fetal calf serum, penicillin (100 I.U./mL), streptomycin (100 µg/mL), and glutamine (2 mM). Store at 4°C for up to 1 mo.
4. Sterile 5% CO_2 incubator.
5. Improved Neubauer hemacytometer.
6. Tissue culture flasks with breathable caps.

2.2. Isolation and Differentiation of Peripheral Blood Lymphocytes From Venous Blood Samples

1. Heparinized 20-mL syringes.
2. Ficoll-Paque cushion (specific gravity 1.077) (Pharmacia Biotech, Montreal, Quebec, Canada) cushion in a 50-mL conical tube.
3. Turks solution: 0.01% gentian violet in 3% acetic acid in water. This is the counting medium.
4. RPMI-complete.
5. Leukostat Stain Kit (Fisher Scientific, Orangeburg, NY).
6. Improved Neubauer hemacytometer.
7. Sterile incubator and CO_2 tank.
8. Standard microscope slides.
9. Cytospin (Shandon Southern Instruments, Camberley, Surrey, UK).

2.3. Isolation and Differentiation of Bronchoalveolar Lavage Lymphocytes

1. Bronchoscopy unit: We use the Schott Fiber Optics Scope (Model VFS-2A) and Illuminator (Model 11858A) (Southbridge, MA) for small animals.
2. 8×50-mL aliquots of sterile, phosphate-buffered saline.
3. RPMI-complete.
4. Turks solution (*see* Section 2.2.3.).
5. Leukostat stain kit.
6. Nylon wool column: 1 g of nylon wool (Fenwall Laboratories, Morton Grove, IL), $0.2N$ HCl, 10-mL autoclavable syringe.
7. Neubauer hemocytometer.
8. Standard microscope slides.
9. $0.16M$ Tris-NH_4Cl solution to lyse red blood cells. Better results can be obtained with fresh solution. Solution can be kept up to 2 wk at room temperature. $0.16M$ Tris-NH_4Cl is prepared by mixing 900 mL of $0.16M$ NH_4Cl (a) with 100 mL of $0.17M$ Tris (b). Adjust to 7.2.
    a. $0.16M$ NH_4Cl (ammonium chloride, Fisher) is prepared by dissolving 8.56 g of ammonium chloride in 100 mL of distilled water. Once the NH_4Cl is thoroughly dissolved, add distilled water up to 1000 mL.
    b. $0.17M$ Tris ($C_4H_{11}NO_3$) is prepared by dissolving 2.06 g of Tris-base (Sigma, Oakville, Ontario, Canada) with 10 mL of distilled water. Adjust pH to 7.65 with $1N$ HCl. Add distilled water up to 100 mL.
10. Distilled water.
11. Hanks' balanced salt solution (BSS).
12. Cytospin.

2.4. Plating of Effector Cells and Target Cells

1. 96-Well, flat-bottomed tissue culture plate (Fisher).
2. 0.5% Triton X-100 detergent.
3. RPMI-complete.

2.5. Lectin-Dependent Cell-Mediated Cytotoxicity

1. Turks solution.
2. RPMI-complete.
3. Lectin concanavalin A (20 µg/mL) (Sigma). Store at –20°C.
4. Tris-phosphate buffer solution, 10 mM, pH 7.4.
5. Five µCi of sodium chromate ^{51}Cr. Specific activity 150 Ci/mg Cr. Store at 4°C (*see* Note 1) (I.C.N. Pharmaceuticals, Montreal, Quebec, Canada).
6. Cobra auto gamma counter (Packard Instrument, Meriden, CT).

3. Methods (*see* Note 2)

3.1. Preparation and Passing of Target Cells for Assay

1. Remove 1-mL aliquot of Raji cells from the –80°C freezer and thaw for approx 5 min in a 37° C water bath.
2. Place Raji cell suspension solution in a sterile flask, and add 10 mL of RPMI-complete drop-by-drop slowly. Mix solution by pipeting (*see* Note 3).
3. Incubate in tissue culture flask for 24 h at 37°C in 5% CO_2-humidified air (*see* Note 4).
4. Place culture in a conical tube and centrifuge at 435g for 10 min at room temperature.
5. Discard supernatant. Add 10 mL of RPMI-complete to the pellet. Mix by pipeting.
6. Prepare three Raji-cell dilutions of 2:10, 3:10, and 5:10 concentration. Store in sterile plastic culture flasks. For the 2:10 dilution, mix 2 mL of Raji cell solution from step 5 with 8 mL of RPMI-complete. For the 3:10 dilution, mix 3 mL of Raji cell solution from step 5 with 7 mL of RPMI-complete. For the 5:10 dilution, mix 5 mL of Raji cell solution from step 5 with 5 mL of RPMI-complete.
7. Incubate for 3 d at 37°C in 5% CO_2 humidified air.

3.1.1. Passing of Target Cells (see Note 5)

1. After 3 d of incubation, take the 2:10 dilution and mix well with a pipet.
2. Label and date two new sterile culture flasks 3:10 and 5:10.
3. For the 3:10 dilution, mix 3 mL of 2:10 Raji-cell medium with 7 mL of RPMI-complete. For the 5:10 dilution, mix 5 mL of 2:10 Raji-cell medium with 5 mL of RPMI-complete. Add 8 mL of RPMI-complete to the remaining 2 mL of 2:10 Raji cell culture and mix with a pipet.
4. Check that target cells in the 2:10 Raji cell culture are still viable with the Turks exclusion dye test. Target cells do not have to be adjusted to 2×10^6 cells/mL yet (*see* Note 6).
5. Continue passing cells twice a week.

3.2. Procedure for Isolating and Differentiating Peripheral Blood Lymphocytes (PBL) (see Note 7)

1. Add 5 mL of Ficoll-Paque solution to a 50-mL conical tube.
2. Draw 10 mL of venous blood into a heparinized 20-mL syringe.

3. Add blood to the tube containing the Ficoll-Paque, but avoid mixing the two. This can be done by touching the needle of the syringe to the wall of the tube.
4. Centrifuge immediately for 30 min at 435g and 19°C. The braking mechanism of the centrifuge must be turned off in order to prevent mixing the blood with the Ficoll-Paque.
5. Four layers will be present after centrifugation. Use a sterile Pasteur pipet to draw out the second layer from the top and place in a sterile 50-mL conical tube. This cloudy white layer contains the mononuclear cells (*see* Note 8).
6. Wash mononuclear cell layer with 40 mL of RPMI-complete. Centrifuge for 10 min at 193g, 19°C and brakes engaged.
7. Resuspend pellet in 1–4 mL of RPMI-complete. Begin by adding 1 mL of RPMI-complete. Label as solution A (*see* Note 9).
8. To count cells in solution A, mix 50 µL of the solution A with 250 µL of RPMI-complete. Label as solution B. Mix by pipeting, 50 µL of solution B with 50 µL of Turks solution. Count all nonstained cells in the four corner chambers of the hemocytometer (*see* Notes 10 and 11).
9. To prepare microscopic slides for differential PBL cell counts, cytospin remaining 200 µL of solution B onto a standard slide at 34g for 5 min. Stain slides with Leukostat stain kit. Wash with tap water. Air dry and differentiate under light microscopy using oil immersion lens. Lymphocyte count should be greater than 90% for the LDCMC assay to be valid.
10. Plate effector cells only when they have been adjusted to 5×10^6 cells/mL and are greater than 90% of all cytotoxic cells. *See* Section 3.4. Incubate at 37°C and in 5% CO_2 humidified air until BAL lymphocytes and Raji-cells are ready.

3.3. Procedure for Isolating and Differentiating Lymphocytes from Bronchoalveolar Lavage Fluid (see Note 12)

1. Prepare the nylon wool column the day before the assay. Wash 1 g of nylon wool with 0.2N HCl and multiple large volumes of distilled water. Dry at 60°C and pack 0.3 g into a 10-mL plastic syringe. Seal the plunger free syringe with a stopcock and autoclave. When cooled, saturate the column with 50 mL of balanced salt solution (BSS) and discard solution. Then rinse with 10 mL of RPMI-complete.
2. Filter BAL lavage fluid through two layers of gauze to remove mucus debris.
3. Centrifuge for 10 min at 302g at room temperature with the braking mechanism engaged.
4. Remove BAL supernatant and store in separate micro tubes –80°C (*see* Note 13).
5. If red blood cells are present in the pellet, add 10–20 mL of Tris-NH_4Cl to the pellet and vortex. Mix by pipeting and centrifuge for 10 min at 302g and at room temperature. Repeat step 5 if necessary.
6. Resuspend RBC free pellet with 1 mL of RPMI-complete (solution C) in a pipet and count cells (*see* step 8 of Section 3.2.).
7. Load solution C into the column, seal with aluminum foil, and incubate for 45 min at 37°C and 5% CO_2.

8. Elute nonadherent cells by adding 50 mL of RPMI-complete over the column and adjust the outflow with a three-way stopcock at a rate of 1 drop/s until 50 mL are recovered (*see* Note 14).
9. Centrifuge for 10 min at 435*g*. Discard supernatant. Resuspend pellet with 0.5 mL of RPMI-complete (solution D). Mix 50 µL of solution D with 250 µL of RPMI-complete (solution E).
10. To count BAL cells, mix 50 µL of solution E with 50 µL of Turks solution. Count cells in the four corner chambers of the hemacytometer. Adjust cell count to 5×10^6 effector cells/mL (*see* Note 11).
11. To prepare slides for differential BAL cell counts, cytospin remaining 200 µL of solution E onto a standard microscopic slide at 34*g* for 5 min. Stain slides with Leukostat stain kit. Wash with tap water. Air-dry and count under light microscopy using oil immersion lens. The BAL effector population should be >90% for the LDCMC assay to be valid.
12. Plate BAL effector T-lymphocytes only when it has been adjusted to 5×10^6 effector cells/mL and is confirmed to be greater than 90% of cell population. *See* Section 3.4.

3.4. Plating of PBL and BAL Effector Cells

1. Label 96-well, tissue culture plate as described:
   a. Label the first six columns as 1–6. Columns 1–3 will be reserved for plating PBL effector cells in triplicate. The remaining columns (4–6) will be reserved for plating BAL effector cells in triplicate.
   b. Label the first eight rows as A–H. Rows E and G will not be used.
2. Add 100 µL of RPMI-complete to rows B, C, D, and F.
3. Add 100 µL of Triton X-100 (0.5%) to row H.
4. Plate in triplicate 100 µL of 5×10^6 PBL/mL in rows A and B. (A1–A3 and B1–B3). Mix by pipeting. Transfer 100 µL of B1 to C1. Mix and transfer 100 µL of C1 to D1. Mix and discard 100 µL from D1. *See* Note 15. Repeat procedures for wells B2 and B3.
5. Incubate tissue culture plate at 37°C in 5% CO_2 humidified air until BAL lymphocytes are ready.
6. Plate in triplicate 100 µL of 5×10^6 BAL lymphocytes/mL in rows A and B. (A4–A6 and B4–B6). Mix by pipeting. Transfer 100 µL of B4 to C4. Mix and transfer 100 µL of C4 to D4. Mix and discard 100 µL from D4. Repeat procedures for wells B5 and B6.
7. Incubate tissue culture plate at 37°C in 5% CO_2 humidified air until Raji-target cells are ready.

3.5. Lectin-Dependent Cell-Mediated Cytotoxicity Assay

1. Mix 5:10 Raji-cell medium by pipeting.
2. Mix 50 µL of 5:10 Raji cell medium with 50 µL of Turks Solution. Count all nonstained cells in the two corner chambers of the hemacytometer (*see* Note 16).
3. When the 5:10 Raji-cell medium has been adjusted to 2×10^6 cells/mL, centrifuge for 10 min at 193*g*.

Cell-Mediated Cytotoxicity Assay

4. Discard supernatant. Add 250 µL of RPMI-complete and 250 µL of phosphate buffer (pH 7.4) to the pellet (solution F). Mix by pipeting.
5. Add 5 µCi of sodium chromate ^{51}Cr (specific activity 150 Ci/mg Cr) to Raji-target cells in solution F.
6. Incubate with lid off for 1 h at 37°C and in 5% CO_2 humidified air. Shake medium every 15 min.
7. Add 5 mL of RPMI-complete. Centrifuge for 5 min at 109g. Discard supernatant appropriately. Repeat washing process three times.
8. Add 10 mL of RPMI-complete and 0.8 mL of Concanavalin A (20 µg/mL) to the pellet. Resuspend.
9. Plate Raji-target cells with PBL and BAL effector cells. Add 100 µL of Raji-cell medium to PBL and BAL wells in rows A–D, F, and H.
10. Incubate plate for 6 h at 37°C and in 5% CO_2 humidified air.
11. Centrifuge plate for 5 min at 537g.
12. Extract 100 µL of supernatant from the 36-wells and place into separately labeled tubes and read ^{51}Cr emittence in a γ-counter.
13. Calculate LDCMC (*see* Note 17).

4. Notes

1. Radioactive Na_2CrO_4 or $CrCl_3$ has a half-life of 27.8 d and should be stored at 4°C. Electron capture is 100% with VX-rays of 0.320 MeV(9%).
2. In order to prevent contamination of the assay, **all procedures** must be carried out with **sterile techniques**. When target cell lines are established, the LDCMC assay must be done in 1 d.
3. Do not mix Raji cells by vortexing. Cells are sensitive to harsh environments.
4. If breathable caps are not available, normal flask caps can be used. Caps are placed on flasks loosely.
5. The process of passaging cells must be done twice a week.
6. Turks solution only enters cells that are nonviable and stains them blue in color.
7. Do not carry out lymphocyte isolation from venous blood and BAL fluid samples until target cells are prepared and adjusted to the appropriate concentration.
8. The procedure of extracting the mononuclear layer should be done immediately after centrifugation in order to prevent sedimentation.
9. Solution A will be used for the LDCMC assay when effector cells have been adjusted to 5×10^6 effector cells/mL.
10. Solution B will be used for counting effector cells and preparation of microscopic slides to differentiate the population of lymphocytes.
11. Calculation for number of effector cells/mL in PBL or BAL is as follows:

    no. cells/mL = total number of cells in the four corner chambers $\times 2 \times 6 \times 10^4/4$

    Where 2 is the dilution factor when 50 µL of suspension was added to 50 µL of Turks solution; 4 is the number of corner chambers; and 6 is the dilution factor when 50 µL of suspension was added to 250 µL of RPMI-complete.

Adjust PBL and BAL effector cells to 5×10^6 cells/mL. If cell count is significantly greater than expected, dilute Solution A with appropriate amount of RPMI-complete (e.g., 12×10^6 cells were counted when 1 mL of RPMI-complete was added to the lymphocyte pellet). We need the count to be approx 5×10^6 effector cells/mL.

$$C1 \times V1 = C2 \times VTotal$$

$$12 \times 10^6 \text{ cells} \times 1.0 \text{ mL of RPMI} = 5 \times 10^6 \text{ cells} \times VTotal$$

$$VTotal = C1 \times V1/C2$$
$$VTotal = 2.4$$
$$V2 = VTotal - V1$$
$$= 2.4 - 1$$
$$= 1.4$$

Therefore, add an additional 1.4 mL of RPMI-complete to solution A.

12. Animals should be gently anesthetized with iv pentobarbital 10 mg/kg body weight in order maintain spontaneous breathing and weak cough reflex. This aids in return of high-quality BAL fluid. Lavages and handling of BAL cells must be performed with strict aseptic techniques. Under bronchoscopic guidance, inject 8×50-mL aliquots of sterile PBS into selected lung segments. Wait several minutes to allow lavage fluid to reach distal alveoli. Gently suction the BAL fluid. When performing serial lavages, choose different pulmonary segments to avoid BAL cellular changes secondary to repeated lavage that have been previously described in dogs. BAL fluid should be processed immediately for best results. BAL fluid can be kept at 4°C for a maximum of 2 h.
13. The BAL supernatant can be stored at –80°C freezer for future studies on cytokine expression and concentration.
14. When BAL cells are passed through the nylon wool column, alveolar macrophages and polymorphonuclear leukocytes (PML) are removed, thus enriching the lymphocyte population. Natural killer cells are not removed with this purification step.
15. Row B contains the 50:1 effector:target (ET) ratio, row C the 25:1 ET ratio, and row D the 12.5:1 ET ratio. Row F will be used to calculate spontaneous release of ^{51}Cr and row H will be used to calculate total release of ^{51}Cr in the assay.
16. The calculation for number of Raji target cells in 5:10 dilution/mL is as follows:

    no. cells/mL = total number of cells in the four corner chambers $\times 2 \times 10^4/4$

    Where 2 is the dilution factor when 50 µL of suspension was added to 50 µL of Turks solution and 4 is the number of corner chambers. Adjust Raji-target cells to 2×10^6 cells/mL.
17. Calculation of LDCMC.
    % cytotoxicity = counts/min experimental–counts/min spontaneous release/
    counts/min total release–counts/min spontaneous release
    Spontaneous release is determined in wells containing target cells and medium alone. Total release is determined by incubating target cells with 0.1 mL of 0.5%

Triton X-100 detergent. Spontaneous release should be >5% of total release. Plot effector-to-target cell ratio (*x*-axis) vs % cytotoxicity (*y*-axis). The slope of the graph should be –1.

Acknowledgments

The authors acknowledge the expert technical support in preparation of the manuscript from Nada Chikhani.

References

1. Kamholtz, S. L. (1988) Current perspectives on clinical and experiment single lung transplantation. *Chest* **94**, 390–396.
2. Millet, B., Higenbottam, T. W., and Flower, C. D. R. (1989) Radiographic appearances of infection and acute rejection on the lung after heart-lung transplantation. *Am. Rev. Respir Dis.* **140**, 62–67.
3. Higenbottam, T., Stewart, S., Penketh, A., and Wallwork, J. (1988) Transbronchial lung biopsy for the diagnosis of rejection in heart-lung transplant patients. *Transplantation* **46**, 532–539.
4. Higenbottam, T., Hutter, J. A., Stewart, S., and et al. (1988) Transbronchial biopsy has eliminated the need for endocardial biopsy in heart-lung recipients. *J. Heart Trans.* **7**, 435–439.
5. Hutter, J. A., Stewart, S., Higenbottam, T., and et al. (1988) Histologic changes in heart lung transplant recipients during rejection episodes and at routine biopsy. *J. Heart Trans.* **7**, 440–444.
6. Shennib, H. and Nguyen, D. (1991) Bronchoalveolar lavage in lung transplantation. *Ann. Thorac. Surg.* **51**, 335–340.
7. Nguyen, D., Mulder, D. S., and Shennib, H. (1991) Warm ischemia induces alterations in lung immune cell functions. *J. Thorac. Cardiovasc. Surg.* **101**, 1030–1036.
8. Nguyen, D., Mulder, D. S., and Shennib, H. (1991) Altered cellular immune function in the atelectatic lung. *Ann. Thorac. Surg.* **51**, 76–80.
9. Nguyen, D., Mulder, D. S., and Shennib, H. (1993) Lectin dependent cell-mediated cytotoxicity and natural killer function in rejecting and infected lung allografts. *Transplantation* **55**, 1250–1256.
10. Norin, A., Kamholtz, S., Pinsker, K., Emeson, E. E., and Veith, F. J. (1986) Concanavalin A dependant cell-mediated cytotoxicity in bronchoalveolar lavage fluid. *Transplantation* **42(5)**, 466–472.
11. Emeson, E., Norin, A., and Veith, F. (1982) Lectin dependant cell-mediated cytoxicity. *Transplantation* **33(4)**, 365–369.
12. Bonavida, B. Bradley, T., Fan, J., Hiserodt, J., Effros, R., and Wexler, H. (1983) Molecular interactions in T-cell-mediated cytotoxicity. *Imm. Rev.* **72**, 119–141.
13. Henkart, P. A. (1985) Mechanism of lymphocyte mediated cytotoxicity. *Ann. Rev. Immunol.* **3**, 31–58.
14. Bonavida, B. and Bradley, T. P. (1976) Studies on the induction and expression of T-cell-mediated immunity. V. Lectin-induced nonspecific cell-mediated cytoxicity by alloimmune lymphocytes. *Transplantation* **21**, 94–102.

15. Gately, M. K. and Martz, E. (1977) Comparative studies on the mechanisms of nonspecific, Con A-dependent cytolysis and specific T-cell-mediated cytolysis. *J. Immunol.* **119,** 1711–1722.
16. Bradley, T. and Bonavida, B. (1981) Mechanism of cell-mediated cytotoxicity at the single cell level. III. Evidence that cytotoxic T-lymphocytes lyse both antigen-specific and nonspecific targets pretreated with lectins or periodate. *J. Immunol.* **126,** 208–213.
17. Persson, U. and Johansson, G. (1984) The appearance of cytotoxic lymphocytes in unstimulated and concanavalin A-activated human peripheral mononuclear cells. *Scand. J. Immunol.* **20,** 209–217.
18. Forman, J. and Moller, G. (1973) Generation of cytotoxic lymphocytes in mixed lymphocyte reactions. I. Specificity of the effector cells. *J. Exp. Med.* **138,** 672–685.
19. Meuer, S. C., Hussey, R. E., Hodgdon, J., et al. (1982) Surface structures involved in target recognition by human cytotoxic T-lymphocytes. *Science* **218,** 471–473.
20. Perl, A., Gonzalez-Cabello, R., Lang, I., and Gergely, P. (1984) Effector activity of okt4+ and okt8+ T-cell subsets in lectin-dependent cell-mediated cytotoxicity against adherent hep-2 cells. *Cell. Immunol.* **84(1),** 185–193.
21. Phillips, J. H. and Lanier, L. L. (1986) Lectin dependent and anti-cd3 induced cytotoxicity are preferentially mediated by peripheral blood cytotoxic T-lymphocytes expressing leu-7 antigen. *J. Immunol.* **136(5),** 1579–1585.
22. Grimm, E., Mazumder, A., Zhang, H., and Rosenberg, S. (1982) Lymphokine-activated killer cell phenomena. Lysis of natural killer-resistant fresh solid tumor cells by interleukin-2 activated autologous human peripheral blood lymphocytes. *J. Exp. Med.* **155,** 1823–1841.
23. Mizushima, Y., Iwata, M., Sato, M., and Yano, S. (1986) Effects of interleukin-2, ok-432 and interferon-gamma on in vitro induction of non-specific killer cells by concanavalin A in mice. *Tohoku J. Exp. Med.* **148(1),** 79–85.
24. Green, W., Ballas, Z., and Henney, C. (1978) Studies on the mechanism of lymphocyte-mediated cytolysis. XI. The role of lectin in lectin-dependent cell-mediated cytotoxicity. *J. Immunol.* **121(4),** 1566–1572.
25. Parker, B. M. (1982) Lectin-induced nonlethal adhesions between cytolytic T-lymphocytes and antigenically unrecognizable tumor cells and nonspecific "triggering" of cytolysis. *J. Immunol.* **11(5),** 387–400.
26. Ballas, Z., Green, W., and Henney, C. (1981) Studies on the mechanism of T-cell-mediated lysis. XIII. Lectin-dependent T-cell-mediated cytotoxicity is supported by con a-coupled sepharose beads. *Cell. Immunol.* **59,** 411–418.
27. Bradley, T. P. and Bonavida, B. (1982) Mechanisms of cell-mediated cytotoxicity at the single cell level. V. The importance of target cell structures in cytotoxic T-lymphocyte-mediated antigen nonspecific lectin dependent cellular cytotoxicity. *J. Immunol.* **129(6),** 2352–2356.

IX

EFFECTS OF LECTINS IN ORGAN CULTURE

40

The Effect of Lectins on Crypt Cell Proliferation in Organ Culture

Stephen D. Ryder

1. Introduction
1.1. Dietary Lectins and Intestinal Proliferation

Lectins are highly active biological molecules that are present in large quantities in the human diet. Changes in colonic epithelial glycoconjugates, such as the enhanced expression of the Thomsen-Friedenreich (TF) antigen, are commonly seen in hyperplastic, premalignant, and malignant colorectal epithelium *(1,2)*. As lectins, such as peanut agglutinin would be expected to bind to TF expressed on the epithelial cell *(3)*, it is possible to hypothesize that this interaction could have profound biological effects. This hypothesis has been confirmed by the finding that PNA stimulates proliferation in colonic cancer cell lines *(4)* and in colonic explants from both normal *(4)* and diseased colonic epithelium *(5)*. These findings suggest that many lectins may have effects on growth and oncogenesis in the colon; there is therefore a need for laboratory techniques to quantitate changes in proliferation in the human colonic epithelium.

1.2. Assessment of Intestinal Proliferation

There are a large number of techniques that have been utilized, but in vitro organ culture techniques with dynamic measurement of epithelial proliferative rates is the closest possible approximation to the situation in vivo.

The relatively high proliferation rates and the restriction of proliferation to anatomically discrete zones in the colon facilitates accurate cell population kinetic measurements. A large number of techniques have been used for the asessment of epithelial cell proliferation *(6);* the most frequently used methods either use radioisotopes to label dividing cells or use direct counts of mitotic cells. The use of radioisotopes, such as tritiated thymidine, is clearly difficult in human subjects, in whom ethical considerations limit its use and has technical drawbacks concerning the source of thymidine used for DNA synthesis *(7)*. Techniques based on direct microscopy and counting of mitoses are therefore preferable. These techniques rely on microdissection of individual colonic crypts and counting highlighted mitotic figures with the unit of proliferation being the entire crypt.

There are a number of techniques available to provide means of identifying cells in division in colonic epithelial preparations, such as immunostaining for nuclear proliferation markers, such as Ki67 *(8)*, but the simplest and most robust method available is histochemical staining using the Feulgen reaction to highlight cells in an active phase of the cell cycle *(9)*.

Another major factor to be taken into account when attempting to quantitate proliferation is that cell proliferation is a dynamic process that procedes at a variable rate, but all labeling methods require fixation of cells and hence calculation of the proportion of cells in cycle is possible only at a single, fixed, timepoint. No technique will currently allow the calculation of a dynamic measure of cell turnover in vivo, as these would require the use of DNA-toxic agents to allow calculation of the rates of accumulation of metaphases. For a dynamic measure to be used, an in vitro technique has been developed, based on the use of an organ culture system and metaphase arrest agents with calculation of a dynamic measure of proliferation, the crypt cell proliferation rate (CCPR).

1.3. Crypt Cell Proliferation Rate

Crypt cell proliferation rate (CCPR) provides the best in vitro method to study the kinetics of colonic epithelial cell proliferation. The technique relies on the accumulation of metaphase arrests with time in culture and is a useful system to estimate the effects of any mitogen under study in human tissue. It is relatively simple to perform and provides a robust measurement that has been shown to closely approximate to that seen in vivo *(10)*. There are a number of elements involved in the calculation of crypt cell proliferation rate: the basic organ culture of colonic tissue, considerations of the lectin and metaphase-arrest agent concentrations used in the experiments, and the mechanics of CCPR calculation itself.

The calculation of CCPR is based on the ability of the metaphase arrest agent vincristine sulfate to stop cells progressing through the cell cycle. Vincristine is a mitotic spindle poison which arrests cells in metaphase and will not allow further progress through the cycle *(11)*. These arrested metaphases are stained and counted in individual crypts in biopsies harvested over a 3-h time period after addition of vincristine. Detection is via nonspecific DNA staining with the Feulgen reaction followed by microdissection of individual colonic crypts and crypt cell counting *(12)*.

Crypt Cell Proliferation

Fig. 1. Culture plate ready for incubation. The central well is filled with 1 mL of culture medium plus the mitogen of interest. The colonoscopic biopsies are placed on an alloy mesh that is suspended on the top of the medium so that the cut biopsy surface is in full contact with the medium. The biopsies must not be submerged. The outer well contains distilled water.

2. Materials

1. There are a wide variety of different unconjugated lectins available from many suppliers, such as Sigma, Dorset, UK. When assessing the proliferative effects of lectins in organ culture, it is important to be aware that both standard medium and fetal calf serum will contain carbohydrates.
2. Alloy wire mesh (Plastic Padding, UK).
3. Dual well culture plates (Falcon, UK) (Fig. 1).
4. Any general purpose cell culture medium, such as Roswell Park Memorial Institute 1640 (Sigma) will provide an adequate substrate for culturing colonic biopsies.
5. Fetal calf serum provides essential growth factors and is used at concentrations of 5–10% (Gibco, Berks, UK).
6. Antibiotics and antifungals are added as required. It is not essential for the media to be supplemented with L-glutamine for short-term (24-h cultures).
7. Carnoys solution: Six parts ethanol, 1 part acetic acid, 3 parts chloroform.
8. Feulgen reagent!
9. The Feulgen reagent can be bought commercially (Sigma or BDH [Poole, UK]), but there is some variation in staining from batch to batch and you can make your own. Boil 1 L distilled water. Cool for 30 s. Add 5 g basic Fuchsin (Hopkins and Williams, Essex, UK). Cool to 50°C. Add 15 g sodium metabisulfite.
10. Vincristine sulfate (*see* Section 3.2., step 1).

3. Methods
3.1. Organ Culture (13)

1. Obtain colonic biopsies (after obtaining ethical committee consent for the study and signed informed consent from the patient); *see* Note 1).
2. Transfer biopsies from the patient to incubator (*see* Note 2).
3. Align biopsies on alloy wire mesh on culture plate with the mucosal surface uppermost (*see* Note 3).
4. Fill inner well with 1 mL culture medium (Fig. 1).
5. Fill the outer well of the culture plate with sterile distilled water to prevent the biopsies from drying during culture (Fig. 1).
6. Add the lectin under study (e.g., peanut lectin 25 µg/mL) to the culture medium of one plate (*see* Note 4).
7. Prepare an identical parallel plate as control without lectin.
8. Incubate the culture plates in 95% oxygen, 5% carbon dioxide for 24 h (*see* Notes 5 and 6).
9. Ensure viability of culture (*see* Note 7).

3.2. Crypt Cell Production Rate Calculation (14)

1. Add vincristine sulfate to culture medium after 21 h of culture (*see* Note 8).
2. Remove biopsies from culture plates at 1, 2, and 3 h after addition of vincristine sulfate.
3. Fix in Carnoy's for 2–6 h.
4. Store in 70% ethanol if required (not recommended to store for more than 2–3 d).
5. If stored, rehydrate in 50% ethanol 10 min, 25% ethanol 10 min.
6. Hydrolyze for 1–2 min in $1N$ HCl at 60°C (*see* Note 9).
7. Stain in Feulgen reagent for 1 h.
8. Microdissect whole colonic crypts under binocular microscope (*see* Note 10).
9. Count the number of metaphase arrests per crypt (*see* Note 11).
10. Plot metaphase number per crypt against time of culture.
11. Regression line gives CCPR in cells per crypt per hour (*see* Note 11; Fig. 2).

4. Notes

1. Obtaining biopsy samples: Standard rectal biopsies obtained at rigid sigmoidoscopy or at flexible colonoscopy are used. Biopsy size is important since, if too small a biopsy is used, the number of crypts present will be inadequate; if too large, there will be a large variation in relative oxygenation and hence proliferative rate of the biopsy *(10)*. If standard rectal biopsies are used, it is possible to cut these into appropriate-sized explants prior to culture. A single rectal biopsy will usually provide three culturable explants. We have also observed that the viability in organ culture is also markedly altered by biopsy size, small biopsies showing much lower viability *(11)*. The probable explanation for this is that small biopsies contain a higher proportion of traumatized tissue and that there is a greater tendency for the mucosal surface to become submerged in the culture medium with resulting hypo-

Crypt Cell Proliferation

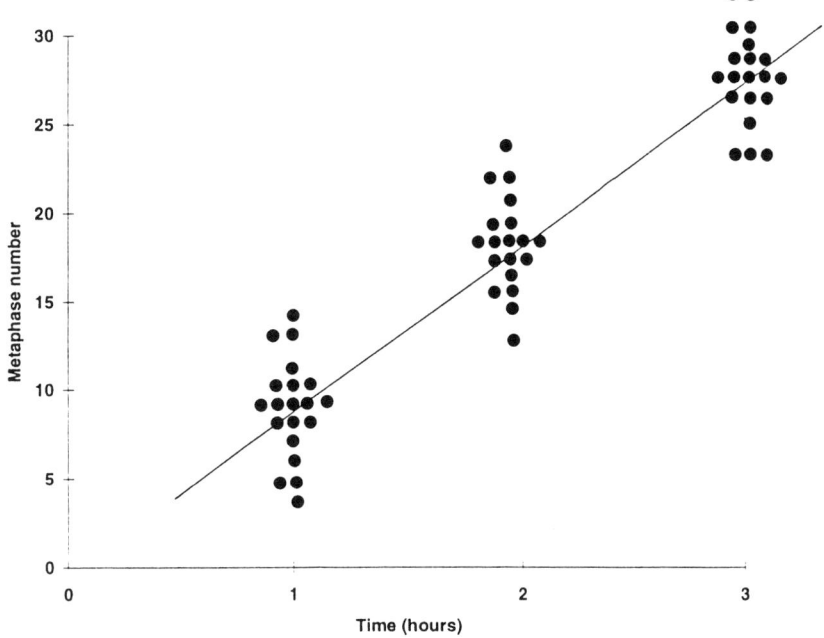

Fig. 2. Calculation of crypt cell production rate. After microdissection, the number of metaphase figures was counted in 20 crypts at 1, 2, and 3 h after the addition of vincristine sulfate to the culture medium. Crypt cell production rate is the slope of the regression line drawn through the points thus obtained. In this case the CCPR is 9.3 cells per crypt per hour.

oxygenation. The ideal-sized biopsy for organ culture is probably 8–12 µg in wet wt. At this weight proliferation indices and oxygenation give reliable estimates of proliferation when repeated measures are made *(11)*.

2. Transfer of biopsies: It is important to transfer biopses from the patient to the culture plates as rapidly as possible. If transport is required, the biopsies should be gassed with 95% oxygen and kept on ice. If this is undertaken, viability and proliferation indices are not affected compared to samples transferred immediately from patient to incubator.

3. Placing biopsies on culture plates: The biopsy must be aligned correctly with the mucosal surface uppermost and the cut surface resting in medium. This takes more practice than one would expect and is one of the major reasons for culture failure.

4. Addition of lectin to the culture system: The major potential problem with measures of CCRP in response to potential luminal mitogens is that the lectin is present in the medium exposed to the submucosal cut surface rather than the luminal (oxygen-exposed) surface. In practice, this is not a major problem since,

if biopsies are cultured with peroxidase-tagged lectins, they can be shown to reach all epithelial cells after 4 h in culture. The concentration of lectin used in the system will depend on two factors; the likely concentration found in colonic contents and the presence of inhibitory carbohydrates in the culture medium and fetal calf serum. Using peanut lectin, there is little inhibitory carbohydrate present, but this may not be the case with lectins with other carbohydrate specificities. It is possible to obtain medium with a specified carbohydrate content from the major manufacturers (Gibco).

5. Incubator design: Specially designed incubators (e.g., Jencons, London, UK) are required to allow the use of such high oxygen concentrations.
6. Culture additives: In our experience, viability is very well maintained over this time and it is possible to prolong culture for another few hours if required. The limiting factor appears to be infection of the medium by fecal organisms transferred to the plates with the tissue samples. This can be reduced by the addition of antibiotics to the culture medium, such as gentamicin. With a 24-h culture period, this is not essential as infection is rarely a major problem with this short-term culture.
7. Ensuring viability: The ideal method is to calculate CCPR in a series of biopsies at a variety of times in culture, a linear growth rate is seen after an initial slow phase relating to transfer of biopsies from patient to culture system. Many lectins have marked toxic effects on epithelial cells, and it is important to establish viability of the colonic explants over the culture period by calculation of CCRP or mitotic index (the static, single time-point equivalent) at a variety of time-points during the proposed culture period.
8. Addition of vincristine to medium: There is an important relationship between vincristine dose and the detection of metaphase arrests. If too little vincristine is used not all cells entering metaphase will be arrested; if too much is used vincristine toxicity may cause degradation of metaphases and an apparent fall in proliferation causing an underestimate of CCPR. A dose–response curve should be calculated for each culture system and for each lectin used (Fig. 3).
9. Hydrolysis time: Hydrolysis time is important; if too long, poor staining will result, and so prior study is essential.
10. Microdissection and crypt cell counting: Microdissection requires the use of a low-power binocular microscope and two dental probes that have been blunted. The technique requires practice, but is essentially simple. The biopsy is gently teased apart so that individual crypts are separated. When this has been achieved, the crypts are gently sqashed by placing them on a standard microscope slide and squashing them gently with a cover slide. The aim is to produce a preparation in which individual cells can be readily identified under a low-power microscope.
11. Crypt cell counting: The technique requires an accurate count of metaphase arrests per crypt. It is easy to detect darkly stained metaphase figures in the colonic crypt, the whole crypt provides a useful unit and results are conventionally expressed as metaphase arrests per crypt. Hyperproliferation in colonic epithelium results in a change in the proliferative compartment, normally cycling cells are restricted to the lower one-third of the crypt, but one of the earliest changes is

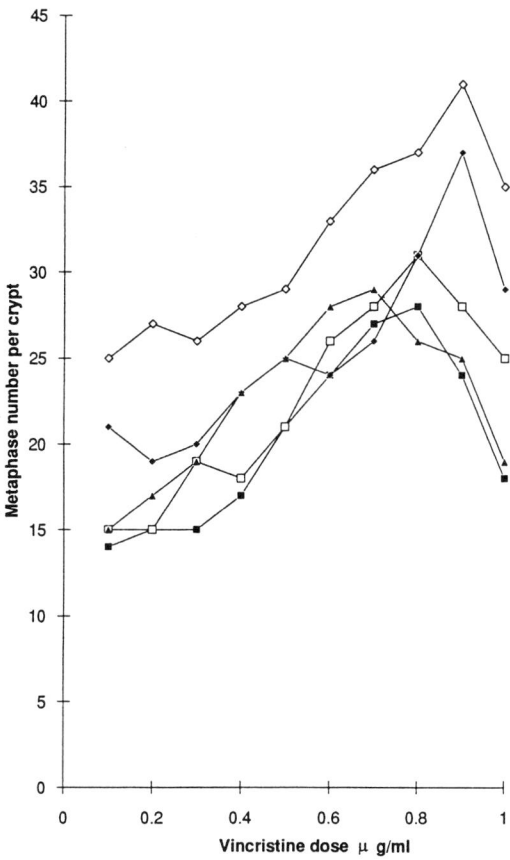

Fig. 3. Dose–response curve for vincristine sulfate in organ culture of colonic epithelium in normal and diseased epithelia. This shows that the optimal dose may be different for the same tissue in disease states and it emphasizes the need for dose–response experiments to be performed in any new tissue under study. ■, normal; □, prior polyps; ●, ulcerative colitis (inactive); ◇, ulcerative colitis (active); ▲, Crohn's disease.

a shift of cycling cells up the crypt. This phenomenon can be quantified but is not essential for assessing response to a mitogen, a count of cycling cells per crypt provides an adequate measure. The more crypts counted, the better. Practicality determines a limit. The process of crypt counting is laborious and time consuming, but a substantial number is required as there is considerable variation from crypt to crypt. It is also important to Note that there is a difference in CCRP in crypts at the edge of biopsies compared to the center *(10)*. Using standard colonoscopic biopsies, at least 50 viable crypts are present; I would recommend counting 40 as a minimum. Crypts at the outer edge of the biopsy should be excluded.

12. Calculation of CCPR: A graph of number of metaphases against time in culture after addition of vincristine is produced. CCPR is the slope of a regression line drawn through the data points and approximates to the cell birth rate in cells per crypt per hour (Fig. 2). Comparisons are then possible between explants cultured with and without a mitogen present.

References

1. Rhodes, J. M., Black, R. R., and Savage, A. (1986) Glycoprotein abnormalities in colonic carcinomata, adenomata and hyperplastic polyps shown by lectin peroxidase histochemistry. *J. Clin. Pathol.* **39**, 1331–1334.
2. Rhodes, J. M., Black, R. R., and Savage, A. (1988) Altered lectin binding by colonic epithelial glycoconjugates in ulcerative colitis and Crohn's disease. *Dig. Dis. Sci.* 1359–1363.
3. Bird, G. W. G. (1964) Anti-T in peanuts. *Vox Sang.* **9**, 748–749.
4. Ryder, S. D., Smith, J. A., and Rhodes, J. M. (1992) Peanut lectin is a mitogen for normal human epithelium and HT29 colon cancer cells. *J. Natl. Cancer Inst.* **84**, 1410–1416.
5. Ryder, S. D., Parker, N., and Rhodes, J. M. (1994) Peanut lectin stimulates proliferation in colonic explants from patients with ulcerative colitis, Crohn's disease and colonic polyps. *Gastroenterology* **106**, 117–124.
6. Aherne, W. A., Camplejohn, R. S., and Wright, N. A. (1977) An introduction to cell population kinetics. Edward Arnold, London.
7. Al Mukhtar, M. Y. T., Polak, J. M., Bloom, S.R., and Wright, N. A. (1982) The search for appropriate measurements of proliferative and morphologic status in intestinal adaptation. MTP, Lancaster.
8. Gerdes, J., Schwab, U., and Stein H. (1983) Production of a mouse monoclonal antibody reactive with a human nuclear antigen associated with cell proliferation. *Int. J. Cancer* **31**, 13–20.
9. Wright, N. A. and Appleton, D. R. (1980) The metaphase arrest technique. A critical review. *Cell Tissue Kinet.* **13**, 643–663.
10. Finney, K. J., Ince, P., Appleton, D. R., Sunter, J. P., and Watson, A. J. (1986) A comparison of crypt cell proliferation in rat colonic mucosa in vivo and in vitro. *J. Anat.* **149**, 177–188.
11. Tannock, I. F. (1967) A comparison of the relative effectiveness of various metaphase arrest agents. *Exp. Cell Res.* **47**, 344–356.
12. Goodlad, R. A. and Wright, N. A. (1982) *Techniques in the Life Sciences. Physiology,* vol. 2. Quantitative studies on epithelial replacement in the gut (Titchen, D. A., ed.), Elsevier Biomedical, Ireland, pp. 212–223.
13. Trier, J. S. (1976) Organ-culture methods in the study of gastrointestinal-mucosal function and development. *N. Engl. J. Med.* **295**, 150–155.
14. Ryder, S. D. (1994) The effect of dietary lectins on colonic epithelial proliferation. DM Thesis, University of Nottingham, UK, pp. 89–93.

X

EFFECTS OF LECTIN INGESTION

41

Effects of Lectin Ingestion on Animal Growth and Internal Organs

Arpad Pusztai

1. Introduction

Lectins are essential and omnipresent plant constituents. As many foods are of plant origin, the daily ingestion of lectins by both humans and animals is appreciable. For example, in an *ad hoc* survey, 53 edible plants were shown to contain lectins and approx 30% of fresh and processed food regularly consumed by humans had significant hemagglutinating activity *(1)*. The situation is potentially even more acute in animal nutrition because animal diet is less diverse than that of humans, and in most instances foodstuffs are not thoroughly heat-treated. This is particularly significant in the light of our finding a correlation between lectin activity and antinutritional effects *(2)*. As in evolution, the mammalian gut has been regularly exposed to lectins, they must have played an important part in the development of the digestive system. Although based on experience, most overtly toxic plants have been eliminated from the diet, many plants with appreciable lectin content are still consumed because it has not been easy to relate growth retardation and antinutritional, mild allergic or other subclinical symptoms to the food consumed or a particular component of it. As some lectins are at least partially heat stable and most survive the passage through the gut in functionally and immunologically intact form, their interaction with the gut surface epithelium *(3)* can damage the gut at high dietary intakes and this may lead to digestive disorders/diseases in some instances. However, it is not generally appreciated that not all lectins are antinutrients and indeed some may have beneficial effects and be of potential value in nutritional practice. Accordingly, it is of considerable importance to establish whether a lectin has deleterious or potentially beneficial effects for mammals. Unfortunately at present there are no adequate in vitro methods to do this reliably and it is usually necessary to carry out in vivo animal feeding studies, despite their relatively cumbersome nature, particularly for large-scale screening.

Because of the lack of suitable biomarkers of lectin toxicity, the best macroindicators of antinutritional effects are obtained by measuring the changes in body weight and the absolute and relative weights of internal organs of young, actively growing animals (usually rats) fed test diets containing the appropriate lectin for variable lengths of time, usually 10 d. As all conditions are rigorously standardized, the results with the test diets are directly comparable with those obtained with animals fed control diets of identical composition, except for the absence of the lectin. The value of the biomarkers can be further refined by carrying out compositional and (optionally) histological analyses on affected tissues. To make further comparisons and establish antinutritional ranking, the lectins should be included in the diet at identical concentrations. We normally use a diet of 7 g lectin/kg. In order to maintain the total protein content of the diet at a constant level, usually 100 g/kg, the lectin is included at the expense of the basal protein, lactalbumin. Feeding studies are carried out by strict pair feeding. It is well-established that rats survive without fatalities when fed for 10 d on diets containing 7 g/kg of one of the most antinutritional lectins, the kidney bean lectin, phytohemagglutinin (PHA). Because, at best, a rat will eat about 6 g/d under these conditions, the intake of all rats is restricted to this amount in all comparative studies. This satisfies their requirements for both protein and energy although it is probably only about half of the amount of the control diet that they would eat under *ad lib* feeding conditions.

2. Materials

2.1. Diets for Rat Experiments

1. The composition of the semisynthetic control diet is given in Table 1:
   a. Lactalbumin is purchased from commercial sources (e.g., Sigma, Dorset, UK). The nitrogen content of these preparations needs to be checked by Kjeldahl N measurements (or other suitable method), because some preparations may be contaminated by nonprotein matter and the concentration of lactalbumin has to be exactly 100 g protein/kg (calculated on the basis of 6.25 × nitrogen content).
   b. Starch: One of the main ready sources of energy of the diet is uncooked maize starch (obtained from commercial suppliers) and is usually included in the control diet at a level of 400 g/kg. However, if adjustments are needed, such as when additional lactalbumin preparation has to be used to reach a true 100 g protein concentration of 100 g/kg, this is done at the expense of maize starch (e.g., Table 1). All diets also contain uncooked potato starch (from commercial suppliers) at 100 g/kg as a source of undigestible fiber. This starch is seldom used for dietary adjustments.

Table 1
Composition of Diets

	Control (lactalbumin)	Lectin diet
Maize starch[a]	373	374
Potato starch	100	100
Glucose	150	150
Corn oil	150	150
Vitamin mix[b]	50	50
Mineral mix[c]	50	50
Lactalbumin[d]	127	118
Lectins[e]	0	7
L-methionine	0	1
L-tryptophan	0	0
Silicic acid	0.4	0.4

[a]Dietary concentrations are given as g ingredient per kg.
[b]Composition of vitamin mix is given in Table 2.
[c]Composition of mineral mix is given in Table 3.
[d]Based on N × 6.25 value of a particular commercial sample.
[e]As an example, this is given for a sample of kidney bean lectin.

c. Glucose (from commercial sources) is used as the most instantly available energy source in the diet and is included at 150 g/kg concentration.
d. Corn oil (Mazzola or other equivalents) is the long acting energy source in the diet; it is incorporated at a concentration of 150 g/kg.
e. Vitamin mix is made up in advance in convenient quantities, and stored at 0–4°C. All essential vitamins, as given in Table 2, are added gradually to maize starch and dispersed in it by thorough mixing.
f. Mineral mix is made up in advance, and stored until needed. It contains all essential minerals and its composition is given in Table 3.
g. Silicic acid is also incorporated in the diet in small amounts to mimic normal plant-based diets.

Basal control diet is usually made up in 2-kg lots. Appropriate amounts of the ingredients are thoroughly mixed in a suitable kitchen food mixer (e.g., Kenwood KM250 "Major Chef" or other equivalents) in the following order of addition: first lactalbumin, starches, glucose, vitamin and mineral mixes, silicic acid, and, finally, the corn oil are blended into a homogeneous mix, making sure that the oil is fully dispersed. Diets not used immediately are best kept in plastic bags with a minimum of residual air at 0–4°C. When lectins are used, it is best to make up only a few hundred grams of the diet (or the precise amount needed for a 10-d feeding experiment) by hand mixing the ingredients including the lectins to avoid losses; alternatively, the lectin-addition is carried out daily by blending in appropriate quantities of lectin with a control diet whose lactalbumin content has been reduced to 93 g/kg diet to allow for the addition of the lectin protein. Although

Table 2
Composition of Vitamin Mix[a]

Ingredient	g
Thiamine	1.00
Pyridoxine (B_6)	1.00
Riboflavin	1.00
p-Amino benzoic acid	1.00
Nicotinic acid	3.00
Calcium pantothenate	2.00
Folic acid	0.50
Biotin	0.50
Inositol	40.00
Vitamin A	1.20
Vitamin D	0.25
Vitamin E	6.00
Vitamin K	0.01
Vitamin B_{12}	2.50
Choline chloride	80.00
Maize starch	4,870.00

[a]Made up in 5-kg batches.

Table 3
Composition of Mineral Mix[a]

Minerals	g
$CuSO_4 \cdot 5H_2O$	2.00
$FeSO_4 \cdot 7H_2O$	25.00
$MnSO_4 \cdot 4H_2O$	20.00
$ZnSO_4 \cdot 7H_2O$	18.00
KIO_3	0.20
KI	0.20
NaF	0.60
NH_4VO_3	0.05
$NiCl_2 \cdot 6H_2O$	0.40
$SnCl_4 \cdot 5H_2O$	0.60
$NaSeO_3$	0.03
$CrK(SO_4)2.1_2H_2O$	4.80
$CaCO_3$	2,100.00
KH_2PO_4	1,570.00
KCl	110.00
$MgSO_4 \cdot 7H_2O$	510.00
Na_2HPO_4	710.00

[a]Made up in 5-kg batches.

Table 4
Target Patterns of Essential Amino Acids for Rat Diets

Amino acid	Target[a]
Threonine[b]	4.0
Valine	5.5
Isoleucine	5.0
Leucine	8.0
Tyrosine	4.0
Phenylalanine	5.0
Lysine	6.0
Histidine	2.5
Arginine	5.0
Methionine + Cystine	4.5
Tryptophan	1.5

[a]Taken from ref. 4.
[b]Amino acid values are expressed as g amino acid/16 g nitrogen.

the amino acid composition of the lactalbumin used in the control diet is ideal and needs no supplementation with essential amino acids, the same may not apply to the lectins. Their amino acid composition must therefore be determined after hydrolysis with $6N$ HCl. To make up for deficiencies in essential amino acid content, diets containing lectins may have to be supplemented with appropriate amino acids to reach optimum target levels. For example, target essential amino acid requirements of the rat diet (4) are given in Table 4. The small amounts of supplementary amino acids are best dispersed in maize starch at the beginning of the diet mixing process.

2. Lectins obtained from commercial sources should be checked for purity according to supplier's instructions before use. This should include an SDS-polyacrylamide gel electrophoresis run, particularly when the subunit patterns of the lectin are known.
3. Individual rats are housed in "Techniplast" metabolism cages whose minimum floor space for different sizes of rats are specified by Home Office regulations. These can be obtained from Stephen Clark Fabrications, Alva, Clackmannanshire, UK. They are equipped with plastic separators that enable feces and urine to be collected separately and contain feeding tunnels and pots of appropriate dimension for different size of rats to minimize feed spillage.
4. Halothane-M&B volatile anesthetic is obtained from Rhone Mérieux, Essex, UK. For terminal anesthesia, the concentration of halothane is 4% (v/v) in pure oxygen and this gas mixture is administered using anesthetic equipment purchased from International Market Supply, Cheshire, UK.
5. Phosphate-buffered saline solution (PBS), pH 7.6: 8.0 g/L NaCl, 0.2 g/L KCl, 1.15 g/L Na_2HPO_4, 0.2 g/L KH_2PO_4.

2.2. Chemical Analysis

1. Kjeldahl N digester mix; 1.5 mL concentrated sulfuric acid, 1.0 mL HgO catalyst (5 g HgO dissolved in 100 mL 10% w/v H_2SO_4) and 1.5 g K_2SO_4 per sample.
2. Perchloric acid: 5 or 10% (w/v).
3. $0.3M$ NaOH.
4. Composition of Copper-alkali reagent (Folin): $0.5M$ NaOH (20 g/L); 10% Na_2CO_3 (100 g/L); 1 g/L potassium sodium (+)–tartrate, 0.5 g/L $CuSO_4 \cdot 5H_2O$. The first three components are dissolved in water in sequence. Copper sulfate is dissolved separately, and then added to the first mixture and made up to 1 L with water.
5. Folin-Ciocolteau reagent (BDH) is kept at 0–4°C and diluted 1 part to 17 parts of distilled water fresh before use.
6. Bovine serum albumin (BDH) standard solution: 0.1% (w/v), kept at 0–4°C for a maximum of 1 mo.
7. Orcinol reagent (RNA determination): 0.5 g $FeCl_3 \cdot 6H_2O$ is dissolved in 100 mL concentrated HCl followed by the addition of 1.0 g of Orcinol immediately before use.
8. RNA (from yeast; Koch-Light) standard: 0.05 g is dissolved in 100 mL $0.05M$ NaOH.
9. Diphenylamine reagent (DNA determination): 1.5 mg diphenylamine is dissolved in 100 mL glacial acetic acid containing 1.5 mL concentrated sulfuric acid. To this solution 0.5 mL 1.6% (v/v) acetaldehyde is added immediately before use (the 1.6% acetaldehyde solution, if stored in a dark bottle at 0–4°C, can be kept indefinitely).
10. DNA (Na-salt, Type III from salmon testes; Sigma) standard: 0.05 g is dissolved in 100 mL $0.05M$ NaOH. This is stored under refrigeration for a maximum of 1 mo.

3. Methods

3.1. Rat Feeding Experiments

1. Male Hooded Lister (Rowett) rats, which are usually weaned at 19 d of age.
2. Feed a stock diet or control (lactalbumin) diet for 10 d, then select into groups of at least four rats of average weight 80 ± 1 g for the feeding studies, which should preferably be done in full nitrogen balance experiments.
3. House the rats are individually in metabolism cages (*see* Note 1) containing plastic separators to obtain feces and urine samples separately.
4. Feed for 10 d (6 g/rat/d) on fully balanced semisynthetic diets, containing 93 g/kg lactalbumin protein and 7 g/kg individual pure lectin samples. Control rats are fed the same amount of diet containing 100 g/kg lactalbumin protein. The composition of the diets is given in Table 1. Water should be made available *ad lib*.
5. Weigh each animal daily and collect urine and feces daily, and store at –20°C until required.
6. Freeze-dry fecal samples and grind in a mortar before analysis.
7. On the morning of the 10th d, rats are given only 2 g of the respective diet (14 mg of lectin) and kill by halothane anesthesia 2 h later (*see* Note 2).

Animal Growth and Internal Organs

8. Cut open the abdomen and remove the entire gastrointestinal tract, together with the pancreas and spleen.
9. Separate the stomach and small intestine from the other organs and rinse with saline. If needed, the remaining immunochemically intact lectin can be determined in the washings by rocket immunoelectrophoresis or some other suitable method (e.g., ELISA).
10. Wash the stomach contents with approx 10 mL PBS to stop futher proteolysis. The buffer and the small intestine with PBS contains 0.1 mg/mL aprotinin (Sigma).
11. Take a weighed section of 2 cm from the small intestine (5 cm from the pylorus) for histology (optional) and the following 18 cm for chemical analyses (*see* Note 3).
12. Excise, rinse, blot dry, and weigh tissues, including pancreas, spleen, cecum, colon, liver, kidneys, thymus, heart, lungs, adrenals, testes, prostate, and hind leg muscles of soleus, plantaris, and gastrocnemius.
13. Freeze-dry the tissues and the remainder of the carcasses to constant weight and weigh (*see* Notes 4 and 5).

3.2. Chemical Analysis

1. Protein content: Analyze diets, feces, urine, carcasses, and required tissues for total Kjeldahl N after digestion with concentrated sulfuric acid in the presence of an HgO catalyst mixture *(5)* (*see* Note 6).
2. Polyamine determination (*see* Note 7): Extract samples of pancreas and small intestine (and other tissues of interest) in the presence of an internal standard of 1,7-diamino heptane with 10% (w/v) perchloric acid (15 mg tissue/mL) for 30 min at 0°C, centrifuge, and determine individual polyamines in the supernatant by HPLC *(6)* (*see* Chapter 33, Notes 7 and 8).
3. The protein determination is performed on the residue insoluble in perchloric acid by a modified Lowry method *(7)*. Homogenize the pellets with 5 mL $0.3M$ NaOH for 30 s at a speed of 20,500 rpm in a Janke-Kunkel homogenizer and then incubate in a waterbath at 37°C for 1 h. To each 1-mL aliquot, add 4 mL $0.05M$ NaOH and use these solutions for the protein estimations.
4. Dilute these samples, usually 50-µL aliquots of the solutions obtained after solubilization in $0.3M$ NaOH, and the bovine serum albumin standards to 1 mL with distilled water.
5. Add 1 mL copper-alkali reagent and 4 mL Folin-Ciocolteau reagent.
6. After vortexing, incubate in a water bath at 55°C for 15 min, leave to cool for approx 10 min, and read on a spectrophotometer at 740 nm.
7. RNA estimation *(8)*: Neutralize the remaining 4 mL of the $0.3M$ NaOH solution from the above protein assays are neutralized by vortexing with 4 mL 10% (w/v) perchloric acid, left to stand on ice for 1 h, and centrifuged on a bench-top centrifuge for 10 min at 3500 rpm. Dilute supernatants and RNA standards to a final volume of 1.5 mL with 5% (w/v) perchloric acid. Add 1.5 mL orcinol reagent, vortex the tubes, stopper with marbles, and heat in a boiling water bath for 30 min. Cool to room temperature and read OD on a spectrophotometer at 660 nm.

8. DNA estimation *(9)* (*see* Note 8): Homogenize the pellets obtained after precipitation of the 0.3*M* NaOH solutions with 10% (w/v) perchloric acid in the previous (RNA) step for 60 s at 9500 rpm (Janke-Kunkel) with 5 mL 5% (w/v) perchloric acid and the mixtures are heated in a water bath at 80°C for 1 h. Cool at room temperature, and stored in a refrigerator overnight after centrifuging at 3500 rpm for 10 min. Dilute samples (usually 200 µL each) and standards to 1 mL final volume with 5% (w/v) perchloric acid and vortex after the addition of 2 mL diphenylamine reagent. Stopper the tubes with marbles and incubate in a water bath at 37°C for 16 h. Cool at room temperature and read on a spectrophotometer at 600 nm.
9. Lipid estimation: Extract the dried carcass or tissue samples in chloroform-methanol (2:1, v/v) for 24 h. Remove the solvent by filtration followed by drying under reduced pressure in a desiccator. Calculate the lipid content from the difference in weight before and after extraction.
10. Statistical analysis: The results are best subjected to one-way analysis of variance (ANOVA). Significant differences between means can then be determined by using Student's *t*-test.

3.3. Morphology

Fix sections of the small intestine immediately with 4% buffered, pH 7, paraformaldehyde, embed in paraffin wax, section at 3 µm, and, for histological measurements, stain with H&E. Select 10 properly oriented villi and crypts at random from each animal and measure their length and count the number of cells (*see* Note 9).

4. Notes

1. The use of high-quality cages is necessary to meet Home Office regulations and also to ensure that full nitrogen balance experiments are carried out accurately. These cages must be designed to minimize diet spillage by the rats and to separate urine clearly from feces.
2. For terminal anesthesia, halothane is the method of choice. This method is relatively humane and induces the least changes in body composition. Previously used methods, such as ether anesthesia are no longer advised by the Home Office.
3. All antinutrient lectins cause growth retardation in comparison with controls, and the extent of this may be used as an index of their antinutritional activity. In contrast, the same lectins are potent growth factors for the rat digestive tract, inducing mainly hyperplastic growth of the intestines. The growth is particularly dramatic in the proximal part of the small intestine. Most of these lectins also induce hypertrophic growth of the pancreas, involution of the thymus, and partial atrophy of skeletal muscle.
4. A valid comparison of the antinutritional potency of lectins can only be obtained from the measurement of the growth of animals and their tissues if the diets and their composition and all experimental conditions are rigorously standardized.

Further safeguards to the validity of comparisons can be obtained if, in addition to the usual lectin-free control group, a standard lectin group is included. The rats in this group should be given a diet containing a highly antinutritive lectin, such as kidney bean lectin. Under these conditions, the reproducibility of changes in body and organ weights is of the order of ±10% or better.

5. Standardization of rats, particularly their starting weight, and a rigorous adherence to the animal management protocols are of crucial importance for obtaining valid comparisons.
6. As antinutritional activity is strictly dependent on the concentration of lectins in the diet, the lectin-protein concentration of samples to be tested must be precisely determined. This may be best done by measuring the optical density at 280 nm of a 1 mg/mL solution of the lectin in an appropriate buffer and comparing this with a reference value in the literature if known. If this is not available, the lectin sample may have to be subjected to micro-Kjeldahl. Other protein determinations, such as color reactions, appropriate immunochemical procedures (if antibodies are available), or other suitable techniques may also be used. However, to obtain absolute protein-lectin values, these need to be compared to a sample of standard high-purity lectin, which may not be readily available.
7. The sequence of chemical steps used for the analysis (polyamines, proteins, RNA, and DNA) needs to be adhered to. This protocol is based on long standing experience with tissue analysis. The values are not only useful for comparing the difference in the effects of the different lectins on rat tissues but also for establishing whether the compositional changes in protein, RNA, and DNA contents are parallel or not. For example, as increases in DNA content signify cellular proliferation, higher DNA concentrations in the tissue usually indicate hyperplastic growth. Increased protein concentration without higher cellular DNA levels usually suggest hypertrophic growth of the tissue.
8. Increased cellular polyamine, particularly spermidine, levels are usually indicative of increased metabolic activity in the tissue. As a large proportion of the spermine content of cells is to be found in the nucleus, increases in spermine levels are well correlated with increased DNA synthesis in cells and are therefore useful indicators of hyperplasia.
9. Sections may also be examined for the presence of lectins bound to the brush border by antibody-peroxidase-antiperoxidase (PAP) staining *(3)* After inhibition of endogenous peroxidase, antigenic sites are unmasked with trypsinization and the sections are reacted with appropriate antibody solutions, followed successfully by the link antiserum and PAP serum. The label is visualized with 3,3'-diaminobenzidine and the sections are counterstained with hematoxylin.

References

1. Nachbar, M. S. and Oppenheim, D. J. (1980) lectins in the United States diet: a survey of lectins in commonly consumed foods and a review of the literature. *Am. J. Clin. Nutr.* **33,** 2338–2345.

2. Grant, G., More, L. J., McKenzie, N. H., and Pusztai, A. (1983) A survey of the nutritional and hemagglutination properties of legume seeds generally available in the UK. *Br. J. Nutr.* **50,** 207–214.
3. Pusztai, A., Ewen, S. W. B., Grant, G., Peumans, W. J., Van Damme, E. J. M., Rubio, L., and Bardocz, S. (1990) The relationship between survival and binding of plant lectins during small intestinal passage and their effectiveness as growth factors. *Digestion* **46(Suppl. 2),** 308–316.
4. Coates, M. E., O'Donoghue, P. N., Payne, P. R., and Ward, R. J. (eds.) (1969) *Laboratory Animal Handbooks. 2. Dietary Standards for Laboratory Rats and Mice,* London Laboratory Animals, London, p. 15.
5. Davidson, J., Mathieson, J., and Boyne, A.W. (1970) The use of automation in determining nitrogen by the Kjeldahl method with final calculations by computer. *Analyst* **95,** 181–193.
6. Seiler, N. and Knödgen, B. (1981) HPLC procedure for the simultaneous determination of the natural polyamines and their monoacetyl derivatives. *J. Chrom.* **221,** 227–235.
7. Schachterle, G. R. and Pollack, R. L. (1973) A simplified method for the quantitative assay of small amounts of protein in biological material. *Anal. Biochem.* **51,** 654,655.
8. Sneider, W. C. (1957) Determination of nucleic acids in tissue by pentose analysis. *Methods Enzymol.* **3,** 680–684.
9. Lovtrup, S. (1962) Chemical determination of DNA in animal tissues. *Acta Biochim. Polon.* **9,** 411–424.

42

Lectin Ingestion

*Changes in Mucin Secretion
and Bacterial Adhesion to Intestinal Tissue*

Howard Ceri, John G. Banwell, and Rixun Fang

1. Introduction

Phytohemagglutinin (PHA), the lectin derived from red kidney beans *(Phaseolus vulgarus)* causes reduced growth rates in several animal species, when incorporated at 0.5–5% of dietary protein. Lectin feeding results in diarrhea, impaired nutrient absorption, growth rate inhibition, and can even lead to the eventual death of PHA-fed animals. These effects are believed to result from changes in the autochthonous microflora induced by the presence of PHA in the diet, as germ-free animals do not display the same changes seen in conventional animals *(1–11)*. Jayne-Williams' *(1,2)* now classical studies in the quail clearly demonstrate this point. It is established that these adverse effects are the result of PHA-induced changes in the normal endogenous flora and are not due to lectin selection of specific pathogenic bacteria *(9,10)*. The major change appears in the levels of facultative aerobes, which increase in PHA-fed animals without an increase in obligate anaerobes *(2,6)*. These observations separate the effects of PHA feeding from blind loop syndrome or other stress-related changes in bacterial flora, in which anaerobic bacterial overgrowth is observed *(12,13)*. Phytohemagglutinin feeding has also been observed to affect small intestinal growth *(14)*.

1.1. The Mechanism of Lectin-Induced Changes

The mechanism by which PHA induces changes in the facultative aerobic population of the intestine remains a mystery. It is known that the changes, over the time-course of these observations, are independent of dietary protein or energy content *(10)*. It is also known that PHA is able to bind directly to the mucosal surface of the intestine and brush border membranes *(8,15,16)*, but did not to bind directly to the LPS of *E. coli (17)*. We have demonstrated that PHA also does not bind to or stabilize the exopolysaccharides of *E. coli* or of the majority of Gram-positive organisms isolated from PHA-induced bacterial overgrowths of the rat intestine *(9)*. It was further demonstrated that both the Gram-positive and Gram-negative isolates, from PHA-induced bacteria overgrowths, bound both intestinal tissue or brush border cells that were washed free of mucin *(9)*. This binding was not enhanced in the presence of PHA *(9)*. The overgrowth of the intestine induced by PHA could be inhibited by the addition of a second lectin activity specific for mannose *(18)*. These findings would suggest that PHA does not function as a direct ligand for bacterial adhesion to the mucosal surface, but may induce changes to the intestine that facilitate bacterial colonization. Scanning electron microscopy of the mucosal surfaces of PHA-fed rat intestines, suggested that bacteria found on the mucosal surface may gain access through windows in the mucous blanket *(10)*. Methods for assessing PHA effects on mucin secretion and bacterial binding to mucosal surfaces will be presented.

2. Materials
2.1. Mucin Purification and Antibody Production

1. Mucin scrapings are collected in 10 vol (w/v) of ice-cold $0.2M$ NaCl containing 0.05% (w/v) NaN_3, 1 mM PMSF, 5 mM EDTA, and 1 mM benzamidine to prevent degradation of the mucin.
2. Chromatography is carried out in $0.2M$ NaCl containing 0.02% (w/v) NaN_3.
3. Antigen for immunization should contain 100 µL of purified intestinal mucin in water (approx 400 µg), 100 µL of Pertussus vaccine (Sigma, St. Louis, MO), and 400 µL of complete Freund's adjuvant. Boosters of antigen were given as above, except that incomplete Freund's adjuvant was substituted (*see* Note 1).

2.2. Mucin Secretion Assays

1. Krebs Ringer Buffer: 118 mM NaCl, 5 mM KCL, 3 mM $CaCl_2$, 1.4 mM KH_2PO_4, 1.4 mM $MgSO_4$, 29 mM $NaHCO_3$, 8 mM L-glutamine, pH 7.4.
2. Carbogen (95% O_2, 5% CO_2).
3. ELISA plates coated with a total of 500 ng of mucin (in 50 µL), and stored at 4°C overnight.
4. A stock of quail mucin at 1 mg/mL in PBS (75 mM NaH_2PO_4/K_2HPO_4, pH 7.2, 75 mM NaCl) is prepared, and 100 µL is used to prepare 16 serial dilutions to which are added an equal volume of antibody to mucin (1:1000 dilution of Ab in PBS, as determined by ELISA). The mucin-antibody preparations were left overnight at 37°C.
5. PBS-0.1% Tween-80.
6. 10% Skim milk powder in PBS.

7. Horseradish peroxidase-conjugated protein A (1/1000 dilution of Sigma stock in PBS).
8. Chromagen: 0.6 mg o-phenylenediamine and 1 μL H_2O_2/mL of substrate buffer (25 mM citric acid, 50 mM NaH_2PO_4, pH 5.2).
9. 2.5M H_2SO_4.
10. Carnoy's fixative: (30% [v/v] chloroform, 10% [v/v] glacial acetic acid, and 60% ethanol).
11. 80, 90, and 100% Ethanol.
12. 1% Periodic acid in water.
13. Schiff reagent and Gill hematoxylin (Fisher, Edmonton, Alberta, Canada).

2.3. Bacterial Adhesion

1. Defined minimal medium M63 for incorporation of ^{35}S-methionine *(9,19,20)*.
2. 100 μCi L-[^{35}S]methionine (1000 Ci/mM).
3. 10 mM Tris-HCl, pH 7.6.
4. PBS or PBS +0.3M lactose.

3. Methods

3.1. Mucin Purification

1. Fast animals, in our case quails *(Coturnix coturnix japonica),* overnight, decapitate, and remove intestine.
2. Cut the intestine longitudinally and flush with ice-cold PBS containing protease inhibitors.
3. Scrape the mucosa with a glass slide and collect the scrapings into approx 10 vol of 0.2M NaCl containing the protease inhibitors.
4. Homogenize the scrapings for 30 s at high speed in a Warring Blender at 4°C, and then centrifuge for 30 min at 30,000g at 4°C.
5. Subject the supernatant to two sequential density-gradient centrifugations in CsCl as described previously *(21,22)*. Add solid CsCl to the supernatant to a final concentration of 60% (w/v) to reach a starting density of 1.43 g/mL.
6. Place the solution in a quick seal centrifuge tube, seal, and centrifuge at 4°C for 48 h at $1.5 \times 10^5 g$ in a Beckman Ti60 rotor.
7. Collect eight equal fractions with a Pasteur pipet. Determine the density of each and assay for glycoprotein, protein, and nucleic acid after dialysis against distilled water (*see* Note 2).
8. Pool the major glycoprotein peak and recentrifuge (Fig. 1).
9. Dialyze the resulting carbohydrate-rich peak (density 1.38–1.50 g/mL) against 0.2M NaCl and 0.02% (w/v) NaN_3 to remove CsCl and concentrate by Amicon (Danvers, MA) ultra filtration prior to chromatography on a Sepharose CL-2B (Pharmacia, Toronto, Ontario) column equilibrated in the same buffer.
10. The void volume fractions, following elution in the same buffer, contain the purified mucin (*see* Note 3).

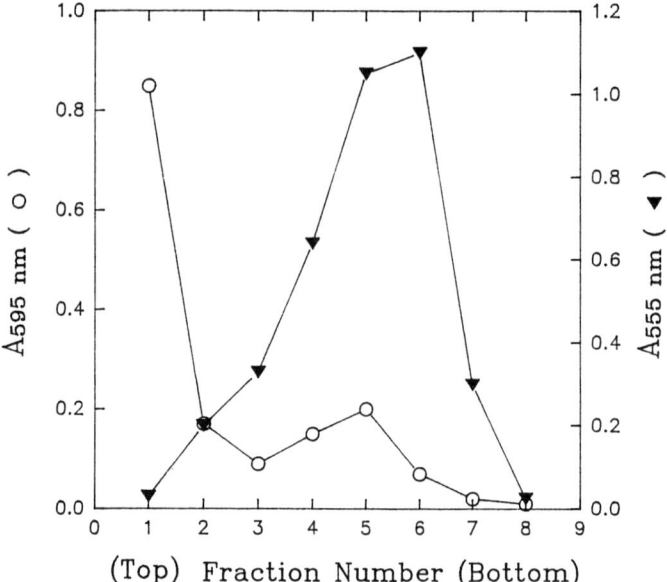

Fig. 1. Glycoprotein and protein analysis for different fractions from cesium chloride equilibrium density gradient centrifugation. Eight equal fractions were collected and assayed for glycoprotein (▼) and protein (○). Protein was measured using the Bradford reagent (Bio-Rad) and glycoprotein by the periodic acid/Schiff assay.

3.2. Mucin Secretion

1. Fast animals overnight and sacrifice by cervical dislocation.
2. Remove the jejunum and ileum segments and place in warm oxygenated Krebs Ringer buffer. Open the intestine, rinse clean with the same buffer, and cut into 2-cm length slices, that are randomly allocated to control or test groups *(22)*.
3. Mount sections on plastic screens in scintillation vials and keep in KRB or KRB containing lectin (stock of 1 mg/mL PHA; *see* Note 4).
4. Shake sections gently, for a prescribed period of time, in a water bath at 37°C, under an atmosphere of carbogen.
5. Following incubation, collect the culture media by aspiration and wash the tissue three times in a total vol of 1 mL of buffer (*see* Note 5). Pool these fractions as the secreted fraction. The tissue is divided into two sections, fix one in Carnoy's fixative for light microscopy (*see* Note 6) and homogenize the other, using a Polytron homogenizer for 1 min in 2 mL PBS containing protease inhibitors.
6. Determine mucin concentration in the secreted fraction and the tissue homogenate by quantitative ELISA *(23)*.

3.3. Mucin Assay by ELISA

1. Prepare serial dilutions of purified mucin stock (1 mg/mL in distilled water) in a volume of 100 µL (*see* Note 7).

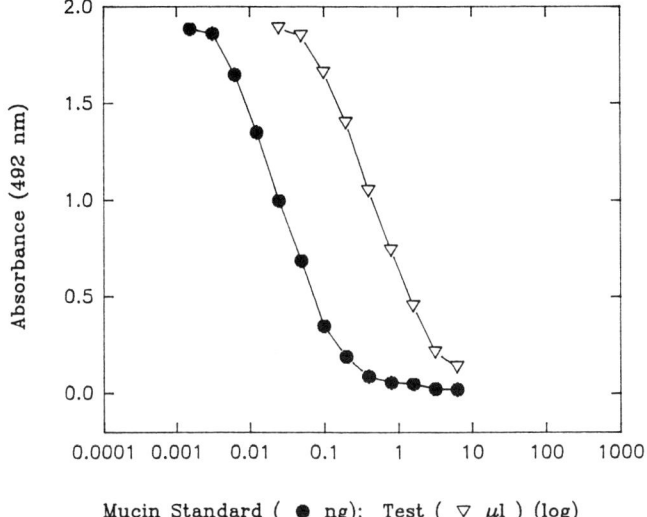

Fig. 2. Standard curve for mucin secretion derived by incubation of serial dilutions of a stock (1 mg/mL) of purified quail mucin (100 µL) with an equal volume of quail mucin specific antibody (1/1000 dilution) overnight at 4°C and conducting an ELISA assay of free antibody as compared to readings derived from dilutions of secreted mucin from control and test secretion assays.

2. Add an equal volume of quail mucin specific antibody (*see* Note 6) (1/1000 dilution) to each sample and incubate overnight at 37°C.
3. Add 50 µL of the dilution series, following the incubation, to ELISA wells coated with mucin (500 ng/well) and block with skim milk protein in PBS.
4. After washing three times with PBS-Tween-80, add 50 µL of a 1/1000 dilution of HRP-conjugated Protein A (Sigma) in PBS and incubate the plate for 1 h at 37°C.
5. Wash the wells in PBS-Tween-80 four times, add 150 µL of substrate per well, and incubate for 30 min at 37°C.
6. Read plates at 492 nm on an ELISA plate reader.
7. Plot results to form a standard curve by plotting optical density against the log of mucin concentration in the preincubation mixture (Fig. 2).
8. Determine protein concentration using the Bradford reagent (Bio-Rad, Hercules, CA), and report mucin secretion/mg of protein (Fig. 3).

3.4. Mucin Histochemistry

1. Cut tissue fixed in Carnoy's fixative to 1-mm thick sections and dehydrate.
2. Embed tissue in plastic using an immunobed kit as per manufacturers instructions.
3. Cut sections of 2–3 µm on a microtome, pick the sections up on a clean glass slide, and heat dry on a hot plate (40°C) for 1 h.

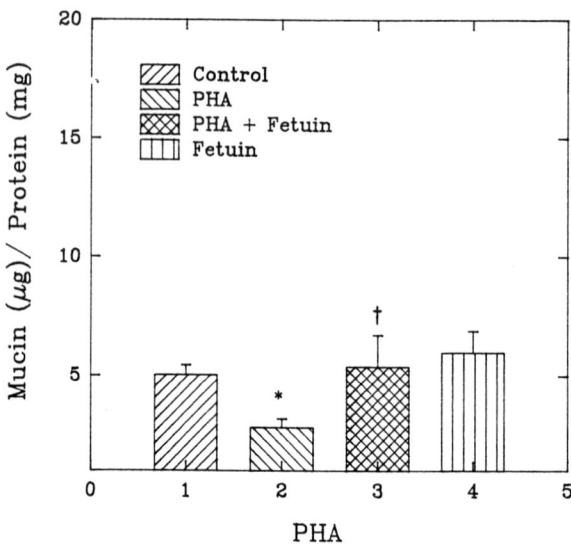

Fig. 3. The effect of phytohemagglutinin on baseline mucin secretion in quail jejunal tissue. Jejunal tissue was incubated with medium containing no additives, PHA (1 mg/mL), PHA plus fetuin (0.3*M*), or fetuin alone at the same concentration. *$p < 0.05$ compared with control, †$p < 0.05$ when compared to PHA-treated tissue (Student's *t*-test for unpaired data).

4. Incubate tissue in 1% periodic acid for 10 min and then rinse with water.
5. Immerse tissue in Schiff Reagent (Fisher) for 8 min and dry in a microwave oven at high (600 W output) for 1 min.
6. Counter stain sections in Gill Hematoxylin for 8 min and wash prior to soaking in ammonia hydroxide for 20 s, rinse in deionized water, dry, and view.

3.5. Bacterial Adhesion

1. Label midexponential phase bacteria by adding 100 mCi of L-[^{35}S] methionine to 10 mL of cells and incubating for 5 min.
2. Harvest bacteria by centrifugation at 3000g for 5 min and wash in Tris-HCl.
3. Resuspend bacteria at approx $5 \times 10^8/100$ μL and determine the specific activity.
4. Prepare intestinal sections as described above and wash gently in either PBS or in PBS + 0.3*M* lactose to remove the mucus layer (*see* Note 8).
5. Distribute sections of PBS washed or PBS-lactose washed tissue randomly in wells of a 96-well tissue culture dish and cover with 100 μL of PBS.
6. Add 1 mL of the labeled bacterial suspension to the tissue and incubate for 15 min at 37°C and shake gently.
7. Wash sections five times in PBS, dry, and count by liquid scintillation.

4. Notes

1. Antibody to mucin was raised by sc injection of rabbits with purified quail intestinal mucin in complete Freund's adjuvant. Boosters in incomplete Freund's adjuvant were given three times at 2-wk intervals and every 6 wk thereafter. Antibody titer was determined by ELISA and the specificity measured by Western Blot as previously described *(21)*.
2. The CsCl gradient is fractionated, and each fraction is read at both 595 nm and 555 nm to determine the crude distribution of the mucin peak in the gradient. Assays for DNA, protein, and carbohydrate can be carried out to verify these peaks, but these readings give an excellent first evaluation of separation in the gradient.
3. Characterization of the purified mucin should be carried out to ensure the purity of the product. Gel electrophoresis under reduced and nonreduced conditions, amino acid composition, and sugar analysis should be performed as described *(21)*; however, the exact nature of these assays is beyond the scope of this review of methods.
4. The secretion vessel was designed to optimize the collection of the secreted mucin and to maintain the organ culture in as healthy a state as possible.
5. The volume into which the mucosal scrapings are resuspended becomes a critical factor. A balance between mucin solubility, the ease of homogenization, and the need to keep the volume small for CsCl gradient centrifugation must be achieved.
6. Tissue sections from each assay are collected for microscopy to verify the integrity of the tissue sample, and this must be done for each sample to ensure validity of the assay.
7. To quantitate mucin by ELISA, all assays are carried out in duplicate and read against preincubation reagent blanks to control for nonspecific Protein A conjugate binding to substrate, and against antibody control wells containing no antigen for antibody binding. The appropriate dilution of antibody was determined from a titer curve, and assay conditions were optimized for time, temperature, and relative concentration of antigen, antibody, and protein A conjugate.
8. Bacterial adhesion to the mucosal surface differs from the association of bacteria with the mucus blanket that covers the intestine. A gentle wash of the mucosal surface with PBS alone will leave the mucin intact, however, adding lactose to the buffer results in the dissociation of the mucin leaving the naked mucosal surface.

References

1. Jayne-Williams, D. T. and Hewitt, D. (1972) The relationship between the intestinal microflora and the effects of diets containing raw navy beans *(Phaseolus vulgaris)* on the growth of Japanese quails *(Coturnix coturnix japonica). J. Appl. Bacteriol.* **35,** 331–344.
2. Jayne-Williams, D. T., and Burgess, C. D. (1974) Further observations on the toxicity of navy beans *(Phaseolus vulgaris)* for Japanese quails *(Coturnix coturnix japonica) J. Appl. Bacteriol.* **37,** 149–169.

3. Hewitt, D., Coates, M. E., Kakade, M. L. and Liener, l. E. (1973) A comparison of fractions prepared from navy (Haricot) beans *(Phaseolus vulgaris)* for germ free and conventional chicks. *Br. J. Nutr.* **29**, 423–435.
4. Evans, R. J., Pusztai, A., Watt, W. B. and Bauer, D. H. (1973) Isolation and properties of protein fractions from navy beans *(Phaseolus vulgarus)* which inhibit growth in rats. *Biochim Biophy. Acta* **303**, 175–184.
5. Wilson, A. B., King, T. P., Clarke, E. M. W., and Pusztai, A. (1980) Kidney bean *(Phaseolus vulgarus)* lectin-induced lesions in rat small intestine. 2. Microbial studies. *J. Comp. Pathol.* **90**, 597–602.
6. Banwell, J. G., Boldt, D. H., Meyers, J., and Weber, F. L., Jr. (1983) Phytohemagglutinin derived from red kidney beans *(Phaseolus vulgaris)*: a cause for intestinal malabsorption with bacterial overgrowth in the rat. *Gastroenterology* **84**, 506–515.
7. Greer, F., Brewer, A. C., and Pusztai, A. (1985) Effect of kidney bean *(Phaseolus vulgarus)* toxin on tissue weight and composition and some metabolic functions of rats. *Br. J. Nutr.* **54**, 95–103.
8. Banwell, J. G., Howard, R., Cooper, D., and Costerton, J. W. (1985) Intestinal microbial flora after feeding phytohemagglutinin lectins *(Phaseolus vulgarus)* to rats. *Appl. Environ. Microbiol.* **50**, 68–80.
9. Ceri, H., Falkenberg-Anderson, K., Fang, R., Costerton, J. W., Howard, R., and Banwell, J. G. (1988) Bacterial-lectin interactions in Phytohemagglutinin (PHA)-induced bacterial overgrowth of the small intestine. *Can. J. Microbiol.* **34**, 1003–1008.
10. Banwell, J. G., Howard, R., Kabir, I., and Costerton, J. W. (1988) Bacterial overgrowth by indigenous microflora in the phytohemagglutinin-fed rat. *Can. J. Microbiol.* **34**, 1009–1013.
11. Banwell, J. G., Howard, R., Costerton, W., Ceri, H., Falk, P., and Larson, G. (1988) Chronic intestinal microbial infection in the rat induced by dietary intake of phytohemagglutinin lectin, in *Inflammatory Bowel Disease, Current Status and Future Approach* (MacDermott, R. P. ed.), Elsevier, pp. 609–614.
12. Tannock, G. W. and Savage, D. C. (1974) Influences of dietary and environmental stress on microbial populations in the murine intestinal tract. *Infect. Immunol.* **9**, 591–598.
13. Savage, D. C. (1977) Microbial ecology of the digestive tract. *Ann. Rev. Microbiol.* **31**, 107–133.
14. Banwell, J. G., Howard, R., Kabir, I., Adrian, T. E., Diamond, R. H., and Abromowsky, C. (1993) Small intestinal growth caused by feeding red kidney bean phytohemagglutinin lectin to rats. *Gastroenterology* **104**, 1669–1677.
15. Boldt, D. H. and Banwell, J. G. (1886) Binding of isolectins from red kidney beans *(Phaseolus vulgaris)* to purified rat brush boarder membranes. *Biochim. Biophys. Acta* **843**, 230–237.
16. Pusztai, A., Clarke, E. M. W., and King, T. P. (1979) The nutritional toxicity of *Phaseolus vulgaris* lectins. *Proc. Nutr. Sci.* **38**, 115–120.
17. Pistole, T. G. (1981) Interaction of bacteria and fungi with lectins and lectinlike substances. *Annu. Rev. Microbiol.* **35**, 85–112.

18. Pusztai, A., Grant, G., Spencer, R. J., Duguid, T. J., Brown, D. S., Eiven, S. W., Peumans, W. J., Van Damme, E. J., and Bardocz, S. (1993) Kidney bean lectin-induced *Escherichia cold* overgrowth in the small intestine is blocked by GNA, a mannose-specific lectin. *J. Appl. Bacteriol.* **75,** 360–368.
19. Herzenberg, L. A. (1959) studies on the induction of β-galactosidase in a cryptic strain of *Esherichia cold Biochim. Biophys. Acta* **31,** 525–539.
20. Whitfield, C., Vimr, E. R., Costerton, J. W., and Troy, F. A. (1985) Membrane proteins correlated with expression of the polysialic acid capsule in *Escherichia cold* K1. *J. Bacteriol.* **161,** 743–749.
21. Fang, R., Mantle, M., and Ceri, H. (1993) Characterization of quail intestinal mucin as a ligand for endogenous quail lectin. *Biochem. J.* **293,** 867–872.
22. Mantle, M. and Thakore, E. (1988) Rabbit intestinal mucins, isolation, partial characterization and measurements using enzyme-linked immunoassay. *Biochem. Cell Biol.* **66,** 1045–1054.
23. Forstner, J. F., Roomi, N. W., Fahim, R. E. F., and Forstner, G. G. (1981) Cholera toxin stimulates secretion of immunoreactive intestinal mucin. *Am. J. Physiol.* **240,** G10–G16.

43

Assessment of Lectin Inactivation by Heat and Digestion

Arpad Pusztai and George Grant

1. Introduction

Proteins/glycoproteins from plants, particularly lectins, are more resistant to heat denaturation than animal proteins *(1,2)*. With legume seeds, whose lectin content is appreciable, this presents potentially serious problems in nutritional practice. Therefore, before they can be used safely, legume-based food/feeds usually require thorough and expensive heat processing to inactivate antinutritive components. Indeed, dry or moist heating of seeds at 70°C for several h has little or no effect on their lectin activity (Fig. 1) and treatment at much higher temperatures is needed to inactivate the biological and antinutritional effects of legume lectins *(1,2)*. The safety aspect is even more serious with some monocot lectins, such as wheatgerm agglutinin or a number of oilseed lectins, such as peanut agglutinin and many others because they are extremely heat stable and normal cooking or other conventional heat treatments may fail to inactivate them *(3)*. Thus, the best way to avoid potential harmful effects of these heat-resistant lectins is to limit their dietary intake to a minimum.

A wide range of procedures have been used to eliminate lectin activity in legume-based or other plant products including dry roasting or toasting, autoclaving, microwaving, and infrared heating treatments *(4,5)*. However, these processes generally require expensive equipment and are more suited to large-scale processing units. Furthermore, to be effective, a number of variables including temperature, the duration of the heating, particle size, and the moisture content of the meal have to be precisely controlled. This can be difficult when large quantities or numerous batches of material have to be processed

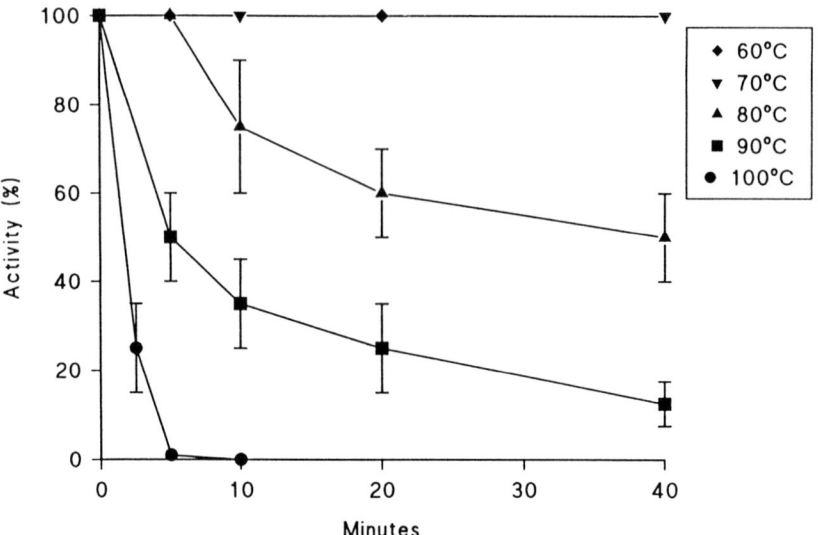

Fig. 1. Loss of lectin activity during aqueous heat treatment of soybean at various temperatures.

and may explain why small but significant amounts of active lectins can frequently be detected in some processed seed products *(3,6)*.

The most practical, effective, and commonly used method to abolish lectin activity is aqueous heat treatment. Seeds are first fully soaked in water, and then heated in water at or close to 100°C. Under these conditions, the lectin activity in fully hydrated soya beans *(Glycine max)* (Fig. 1), kidney beans *(Phaseolus vulgaris)*, faba beans *(Vicia faba)*, and lupinseeds *(Lupinus angustifolius)* could be eliminated by heating at 95°C for 1 h or at 100°C for 10 min *(2,7)*.

Orally administered native, undenatured lectins are extremely resistant to proteolytic breakdown by pancreatic and small intestinal proteases during passage through the mammalian digestive tract *(8)*. As most of these lectins can be recovered from the feces, they must also be resistant to bacterial proteases in the large intestine. It has been suggested that this resistance is mainly the result of stabilization of the conformation of the lectin molecule by its avid binding to carbohydrate moieties of gut epithelial membrane glycoconjugates and this may also shield peptide bonds, which in the absence of the saccharides would be open to protease attack *(9)*. However, as some lectins, like the mannose-specific snowdrop *(Galanthus nivalis)* bulb agglutinin (GNA), which do not bind to the gut wall can also be recovered in high amounts, binding induced stabilization is not always essential for stability against proteolytic breakdown *(8)*.

The high stability of lectins to proteolysis can be fully abolished by denaturation with appropriate heat treatments. Thus, the substantial improvement in the nutritional quality of legume meals after processing is most likely owing not only to the inactivation of lectin activity by denaturation, but also to the increased digestibility of the lectin protein whose component amino acids can then be absorbed in the gut and fully utilized by the body.

1.1. Indirect Noncompetitive ELISA

ELISA assay methods *(10)* are illustrated with the description of the protocols used for the measurement of the lectin content in kidney bean lectin, PHA.

In this assay, the antigen, PHA, is first immobilized on the solid-phase, ELISA microtiter plate, and then reacted with a suitably diluted specific rabbit antilectin antibody solution. The antigen–antibody complex formed is detected by a goat immunoglobulin preparation specific for the rabbit antibody, and labeled with ExtrAvidin peroxidase, whose quantity is measured by a colorimetric reaction. The reaction rate, i.e., the increase in absorbance per unit time, A/t, is directly related to the quantity of antigen present in the sample. The first step in this method is to establish the optimum dilution of the rabbit antilectin antibody. For this, the wells in each horizontal row are coated with a serially diluted PHA solution followed by the addition of different dilutions of the rabbit antibody so that the antibody concentration is halved in the wells of each successive horizontal row. From the results of the color reaction obtained at different dilutions of the antibody, calibration curves are drawn on a semilog graph paper. The curve showing the best sigmoidal shape (high sample dilution = low value of A/t, and low sample dilution = high value of A/t, S-shaped) indicates optimal parameters.

1.2. Indirect Competitive ELISA

This assay is based on the competitive inhibition of the reaction between a known amount of antigen coated to the plate and its specific antibody by free antigen present in test or calibration samples.

2. Materials
2.1. Inactivation by Heat

1. Seed samples should be purchased from a reliable source to ensure that the individual seeds are from the same cultivar. This is particularly important with kidney beans, in which the lectin content of different cultivars may differ considerably. To obtain seed meals containing uniform size particles and to minimize exposure to heat, seeds are ground in a high precision hammermill, such as a Glen Creston Hammermill (Glen Creston, Stanmore, Herts, UK) fitted with a 1-mm mesh, taking care that the mill is not overheated but remains at room temperature.

2. Phosphate-buffered saline solution, pH 7.6 (PBS): 8.0 g/L NaCl, 0.2 g/L KCl, 1.15 g/L Na_2HPO_4, 0.2 g/L KH_2PO_4.
3. Saline: 9 g/L NaCl in water.
4. Collect blood into a heparinized tube (30 U/mL blood), quickly dilute 20-fold with saline, and store at 1°C. The cells are stable for up to 1 wk if kept at this temperature. This diluted and untreated blood is used directly in the assays.
5. Trypsin-treatment of blood cells increases the sensitivity of the hemagglutination assay. Centrifuge diluted rat or cattle blood at $500g_{max}$ for 5 min and wash the cells twice with saline. Resuspend the cells in saline to their original diluted volume, add trypsin (0.1 mg/10 mL of diluted blood cells) and incubate for 45 min at 25°C. Centrifuge the trypsin-treated cells, wash four times with saline, and finally resuspend in saline to their original diluted volume. Store cells at 1°C, but do not keep for more than 1 d.

2.2. Inactivation by Proteolysis In Vivo

1. ELISA 96-well microtiter plates (TITERTEK, Flat Bottom, Labsystems, Basingstoke, UK): Incubator set at 37°C; pipet tips (Labsystems Finntips, 0.5–300 µL; Labsystems); microdensitometer (ELISA reader; Uniskan II, Labsystems).
2. Coating buffer (stored at 4°C in the refridgerator): $0.05M$ carbonate-bicarbonate buffer, pH 9.8, 2.69 g $NaHCO_3$, and 1.91 g Na_2CO_3 dissolved in 1 L distilled water.
3. Blocking solution (prepared fresh): 0.5% (w/v) gelatin in PBS diluted 10X with distilled water. Dissolve 0.15g gelatin in 30 mL of PBS with heating to 45°C and stirring, then dilute to 300 mL with distilled water.
4. Washing solution (prepared fresh): 0.1% (v/v) Tween-20 in PBS. To 1000 mL PBS add 1 mL Tween-20.
5. Dilution buffer: T-G-PBS (prepared fresh). To 1000 mL PBS add 1 mL Tween-20 and 0.5 g gelatin.
6. Substrate working buffer solution: $0.05M$ citric acid-Na_2HPO_4 buffer, pH 5.0.
    a. Solution A: $0.1M$ citric acid; 2.10 g citric acid monohydrate dissolved in 100 mL water.
    b. Solution B: $0.2M$ Na_2HPO_4; 2.84 g Na_2HPO_4 dissolved in 100 mL water. Stock substrate buffer: $0.1M$ citric acid-Na_2HPO_4, pH 5.0, is made up by mixing 48.5 mL of solution A with 51.5 mL of solution B. For the substrate working buffer, the stock substrate buffer is diluted 1:1 with distilled water.
7. OPD-H_2O_2 reagent, 20 mg o-phenylenediamine free base (1 tablet, Sigma) is dissolved in 50 mL substrate working buffer and 20 µL 30% (v/v) H_2O_2 is added 10 min before use.
8. Stopping solution: $3M$ sulfuric acid.
9. Antilectin (rabbit) antibodies are obtained from commercial sources (e.g., Sigma) or raised in rabbits locally.
10. Kit reagents for ELISA: rabbit extravidin peroxidase staining kit Extra-3 (Sigma) containing 1 vial biotinylated goat antirabbit immunoglobulin, affinity-purified in buffer containing preservative (diluted 1:1000), and 1 vial extravidin peroxidase in buffer containing preservative (diluted 1:500).

3. Methods
3.1. Inactivation by Heat

For whole seeds, carry out steps 1–8. For seed meals, use steps 9–10.

1. Sort seed samples visually to ensure purity.
2. Soak seeds in water (1:4 w/v) at 20°C for up to 16 h (a minimum of 8 h).
3. Remove any seeds floating on the surface of the water and discard.
4. Pour off the soaking water and discard.
5. Sort the seeds visually and discard any that are not fully hydrated (*see* Note 1).
6. Pour the seeds into a container (cooking pot or waterbath) of preheated water (300 g wet wt/L) and boil for 10 min. In localities in which the boiling temperature of water is <100°C, the cooking time needs to be extended accordingly (*see* Note 2).
7. Pour off the cooking water and wash the seeds with fresh water, drain, and check for uniform softness. Discard abnormal seeds (*see* Note 3).
8. The seeds may then be safely consumed without further treatment or be added to other dishes. Alternatively, they can be dried, ground, and tested by hemagglutination or by rat feeding experiments to establish the degree of inactivation of the seed lectin.
9. The recommended method for the heat treatment of seed meals is simpler. Weigh suitable amounts of ground seed meal into a stoppered flask, followed by the addition of 3–4 vol of water. There is no need for presoaking, place the stoppered flasks in a preheated waterbath and keep at the appropriate temperature for the required time. After the heating, recover the contents of the flasks by freeze-drying and test by hemagglutination and/or rat feeding experiments.
10. For heat treatment of purified lectins, the procedure is similar to that recommended for seed meals. Dissolve the lectins in PBS and heat in stoppered flasks at appropriate temperatures for the required time and test the degree of inactivation by hemagglutination (*see* Note 4).

3.2. Inactivation by Proteolysis in the Gut

1. Feed rats with 1 mL of lectin solution (10 mg/mL) by intragastric intubation.
2. Kill by halothane overdose precisely 1 h later.
3. Ligate stomach and small intestine, excise, and rinse with PBS containing aprotinin (1000 kIU/mL) to recover free, unbound lectins from the lumen of these tissues.
4. Homogenize the washed stomach and small intestinal tissues with PBS (10 mL), also containing aprotinin and the monosaccharide (1 g/L) appropriate for the specificity of the lectin (Table 1), in a Janke-Kunkel homogenizer (20,000 rpm, 30 s).
5. Estimate the amounts of immunoreactive lectins in both the luminal washings and tissue homogenates by a suitable technique, such as ELISA.
6. Calculate the degree of survival as percentage of the lectin originally administered (*see* Note 5).

Table 1
Survival and Binding of Pure Lectins to the Small Intestinal Mucosa

Lectins	Specificity	Binding	Recovery, %
PHA *(Phaseolus vulgaris)*	Complex	+++	>90
Con A *(Canavalia ensiformis)*	Man/Glc	+	>90
GNA *(Galanthus nivalis)*	Man	–	>90
SNA–I *(Sambucus nigra)*	α-2,6-NeuAc–Gal	+	50–60
SNA–II	GalNAc	+++	>60
SBA *(Glycine max)*	GalNAc/Gal	++	40–50
LEL *(Lycopersicon esculentum)*	GlcNAc	+	40–50
WGA *(Triticum vulgare)*	GlcNAc	++	50–60
PSL *(Pisum sativam)*	Man/Glc	±	30–40
VFL *(Vicia faba)*	Man/Glc	±	20–30
DGL *(Dioclea grandiflora)*	Man/Glc	±	18–20

The results are taken from ref. 8. Rats were intragastrically intubated with 10 mg of individual lectins. The amounts of lectin surviving in the stomach and small intestine were estimated from luminal washings and supernatants of the tissues homogenized with $0.1M$ solution of the appropriate specific carbohydrate in phosphate-buffered saline, pH 7.6. The strength of binding is marked on an arbitrary scale: +++, strong binding; –, represents no binding at all.

7. Fecal samples are freeze-dried, ground to a fine powder, extracted with PBS containing 0.02% (w/v) NaN_3 (feces:PBS ratio is 1:10; w/v), and centrifuged (50,000g max for 30 min); the clear supernatants are used for measurement of lectin concentration by ELISA.

3.3. Hemagglutination Assay

1. Twelve small tubes (possibly 24 or 36 if treated cells are used) each containing 150 µL of saline are set up for each sample.
2. Add 150 µL of sample to the first tube, and mix the contents well.
3. Remove 150 µL from this first tube and transfer to tube 2. Continue this serial dilution to the last tube, mix the diluted samples with 150-µL aliquots of blood cells, and leave for 2–3 h at room temperature.
4. Resuspend the cells by agitation, and assess the degree of agglutination by eye or, preferably, by a microscope.
5. The degree of clumping is expressed as follows: $3+^v$, large clumps visible by eye; 3+, 80–100% of cells are clumped (by microscope); 2+, 40–60% of cells are clumped (by microscope); 1+, 10–20% of cells are aggregated (by microscope) and tr, < 10% of cells are clumped (by microscope) (*see* Note 6).
6. One unit of hemagglutinating activity (HU) is defined as the amount of material (µg/mL) in the last dilution in which 50% of the cells are agglutinated (*see* Notes 7 and 8).

3.4. ELISA Method

3.4.1. Indirect Noncompetitive ELISA

1. To the first well in each horizontal row of a 96-well ELISA plate add 200-µL aliquots of a solution of PHA (500 µg/mL) (in coating buffer), and into all others add 100 µL coating buffer.
2. Serially dilute the PHA solution in a horizontal direction so that the amount of PHA-antigen decreases from 50 µg in the first well to 0.025 µg in the last.
3. Incubate the plate for 60 min at 37°C.
4. Remove the unbound PHA by draining and washing the wells three times with washing solution and twice with distilled water (250 µL each time).
5. Add 200 µL of blocking solution to each well and incubate the plate for a minimum of 30 min at 37°C.
6. After the removal of the blocking solution, wash the plate three times with washing solution and twice with distilled water (250 µL each time).
7. Since the conditions of the immune reaction are best optimized in a checkerboard design, add 100 µL of anti-PHA antibody in twofold dilutions in dilution buffer in each successive row from top to bottom of the plate. Thus, although the dilution of the antibody is constant for the 12 wells in each horizontal row, overall it changes from 1:250 in the first to 1:32,000 in the eighth row.
8. Cover the plate with parafilm and aluminium foil and incubate for 60 min at 37°C in a humid atmosphere.
9. Wash the plate three times with washing solution and twice with distilled water (250 µL each time) and dry by turning upside down and banging it against filter paper.
10. Pipet 100 µL of prediluted (1:1000) conjugate of biotinylated goat antirabbit immunoglobulin into all wells and incubate the plate for 60 min at 37°C in a humid atmosphere.
11. Drain the wells, wash three times with washing solution and twice with distilled water (250 µL each time) and dry.
12. Pipet 100 µL of 1:500 diluted extravadin peroxidase into each well and incubate for 60 min at 37°C in a humid atmosphere.
13. Wash the plate five times with washing solution (250 µL each time) and dry.
14. Develop the color reaction by adding 100 µL OPD reagent to each well, leaving the plate to stand for 16 min and stopping the reaction by adding 50 µL stopping reagent.
15. Read the color with an ELISA reader at 492 nm.

3.4.2. Indirect Competive ELISA

1. Two plates are used. Coat the first plate with a known optimal concentration of purified PHA, 100 µL, dissolved in the coating buffer; incubate at 37°C for 60 min.
2. Remove the unbound PHA, wash three times with washing solution, and twice with distilled water (250 µL each time) and dry.
3. Incubate the second plate with 200 µL blocking solution overnight at 37°C.

4. Wash the plate three times with washing solution and twice with distilled water (250 µL each time) and dry.
5. Into all wells, except those in the first vertical column which contain 200 µL of a PHA standard solution, add 100 µL of dilution buffer in which the PHA solution is serially diluted in a horizontal direction.
6. Transfer aliquots of the serially diluted PHA solutions (50 µL each) from wells on plate 2 to equivalent wells on the first plate, then add 50 µL of suitably diluted anti-PHA antibody (previously determined to be optimal by indirect noncompetitive ELISA).
7. Cover the plate with parafilm and aluminium foil, incubate at 37°C for 60 min in a humid atmosphere.
8. Drain the plate and wash three times with washing solution and twice with distilled water (250 µL each time) and dry.
9. Immunoreaction, color development and reading at 492 nm are done as before.

4. Notes

1. It is important to ensure that seeds are fully hydrated. Thus, at 100°C, the cooking time required to eliminate lectin activity in partially hydrated seeds can be more than sixfold greater than that necessary for fully hydrated seeds.
2. The aqueous heating procedure described can effectively eliminate the lectin activity from most plant materials consumed in human or animals diets. However, wheatgerm agglutinin or gluten-associated lectins are not inactivated under these conditions *(11)*. Taro tuber lectin is also reported to require prolonged heating at high temperature *(12)*.
3. Many beans develop hard-shell or hard-to-cook characteristics during long-term storage under conditions of high humidity and temperature *(13)* and cooking times necessary to eliminate antinutritional factors, such as lectins become very extended regardless of whether dry or moist heating is used. If at stages 2, 5, or 7 of the aqueous heat treatment procedure a high proportion of the seeds have to be thrown away, this is a clear indication that there are problems with that batch of seeds and it would be wise to reject all seeds in the batch.
4. Purified lectins in aqueous solution, in plant extracts or seed meal suspensions appear to be inactivated under the same temperature conditions as those required for intact seeds *(2)*.
5. The degree of total lectin survival is estimated by measuring the amounts of immunoreactive lectins in the feces of rats and comparing them to the total dietary input over the entire period of feeding. To allow for slow stomach emptying and increased intestinal passage time owing to the presence of lectins in the diet, collection of fecal samples is continued for at least 48 h after the last meal containing lectins. The extent of the survival of lectins during small intestinal passage is always significant but quite variable. An example is given in Table 1.
6. Occasionally in hemagglutination assays with enzyme-treated cells, the resuspended cells exhibit an unacceptably variable degree of background clumping. If this is the case, the stock suspension of the cells should be left to stand for 20–30

min at room temperature to allow the majority of the clumps to settle out. The upper layer is then decanted and used in the assay.
7. A lectin standard is included in each assay. For comparison purposes, the results are best expressed as HU/100 g bean meal or as lectin equivalents/100 g bean meal *(1)*.
8. Some examples: A kidney bean cultivar of high lectin content, e.g., "Processor" using native blood cells gave the following values: 1 HU = 49 µg or 2040×10^6 HU/100 g meal. A low lectin kidney bean cultivar, Pinto III, using native blood cells gave: 1 HU = 12,500 µg or 8×10^3 HU/100 g meal. The titers estimated with trypsin-treated cells are far higher.

References

1. Pusztai, A. (1991) *Plant lectins.* Cambridge University Press, Cambridge, UK.
2. Grant, G. and van Driessche, E. (1993) Legume Lectins: physicochemical and nutritional properties, in *Recent Advances of Research in Antinutritional Factors in Legume Seeds* (van der Poel, A. F. B., Huisman, J., and Saini, H. S., eds.), Wageningen Pers, Wageningen, The Netherlands, pp. 219–234.
3. Nachbar, M. S. and Oppenheim, D. J. (1980) Lectins in the United States diet: a survey of lectins in commonly consumed foods and a review of the literature. *Am. J. Clin.Nutr.* **33,** 2338–2345.
4. Melcion, J.-P. and van der Poel, T. F. B. (1993) Process technology and antinutritional factors: principles, adequacy and process optimization, in *Recent Advances of Research in Antinutritional Factors in Legume Seeds* (van der Poel, A. F. B., Huisman, J., and Saini, H. S., eds.), Wageningen Pers, Wageningen, The Netherlands, pp. 419–434.
5. Rackis, J. J., Wolf, W. J., and Baker, E. C. (1986) Protease inhibitors in plant foods: content and inactivation, in *Nutritional and Toxicological Significance of Enzyme Inhibitors in Foods* (Friedmann, M., ed.), Plenum, NY, pp. 299–347.
6. Ryder, S. D., Smith, J. A., and Rhodes, J. M. (1992) Peanut lectin: a mitogen for normal human colonic epithelium and human HT29 colorectal cancer cells. *J. Natl. Cancer Inst.* **84,** 1410–1416.
7. Grant, G., More, L. J., McKenzie, N. H., and Pusztai, A. (1982) The effect of heating on the hemagglutinating activity and nutritional properties of bean *Phaseolus vulgaris* seeds. *J. Sci. Food Agric.* **33,** 1324–1326.
8. Pusztai, A., Ewen, S. W. B., Grant, G., Peumans, W. J., Van Damme, E. J. M., Rubio, L., and Bardocz, S. (1990) The relationship between survival and binding of plant lectins during small intestinal passage and their effectiveness as growth factors. *Digestion* **46(Suppl. 2),** 308–316.
9. Pusztai, A., Begbie, R., Grant, G., Ewen, S. W. B., and Bardocz, S. (1991) Indirect effects of food antinutrients on protein digestibility and nutritional value of diets, in *In Vitro Digestion for Pig and Poultry* (Fuller, M., ed.), CAB International, Wallingford, Oxon, UK, pp. 45–61.
10. Tijssen, P. (1985) Practice and theory of enzyme immunoassays, in *Laboratory Techniques in Biochemistry and Molecular Biology,* 5th ed. (Burdon, R. H. and Van Knippenberg, P. H., eds.), Elsevier, Amsterdam, The Netherlands, pp. 329–355.

11. Liener, I. E. (1986) Nutritional significance of lectins in the diet, in *The Lectins* (Liener, I. E., Sharon, N., and Goldstein, I. J. eds.), Academic, NY, pp. 527–552.
12. Seo, Y.-J., Satsuki, U., Tsukamoto, I., and Miyoshi, M. (1990) The effect of lectin from taro tuber *(Colocasia antiquorum)* given by force-feeding on the growth of mice. *J. Nutr. Sci. Vitaminol.* **36,** 277–285.
13. Aguilera, J. M. and Stanley, D. W. (1985) A review of textural defects in cooked reconstituted legumes–the influence of storage and processing. *J. Food Proc. Preserv.* **9,** 145–169.

XI

LECTINS IN THE INVESTIGATION OF NEURONAL TRAFFICKING

44

Use of Lectins as Transganglionic Neuronal Tracers in the Study of Unmyelinated Primary Sensory Neurons

Mark B. Plenderleith and Peter J. Snow

1. Introduction

One of the major advances in neurobiology in the last two decades has been the development of neural tracing techniques that allow the investigator to unequivocally establish the precise pattern of connections between different populations of neurons. Early neuronal tracers were relatively nonspecific agents transported by all neurons at the injection site. More recently however, the affinity of plant lectins for the membrane-associated glycoconjugates expressed by neurons has been utilized to develop a class of neuronal tracers that are taken up and transported by specific populations of neurons. The effectiveness of plant lectins as neuronal tracers is dependent on the distribution of the membrane-associated glycoconjugates to which they bind. If the glycoconjugate to which the lectin binds is present on the majority of neurons, then this lectin will be suitable for use throughout the nervous system. However if the lectin binds to a relatively discrete group of neurons, it may be used to examine the connections of this specific population of neurons.

In this chapter, we describe the use of a plant lectin that is taken up and transported specifically by unmyelinated primary sensory neurons. This tracer enables us to distinguish the unmyelinated fiber input to the dorsal horn of the spinal cord from other inputs and provides a very concise picture of the pattern of synaptic connections that these neurons make in the superficial dorsal horn. This technique has proved to be a very powerful tool in the analysis of synaptic transmission at the first synapse in the pain pathway.

1.1. Lectin Binding Sites on Unmyelinated Primary Sensory Neurons

The presence of a unique membrane-associated carbohydrate expressed by unmyelinated sensory neurons was suspected when it was shown that the plant lectins soybean agglutinin (SBA) and *Bandeiraea simplicifolia* I-isolectin B_4 (BSI-B_4), both of which exhibit an affinity for terminal galactose residues, bind to the plasma membrane and Golgi apparatus of a subpopulation of small diameter sensory neurons in the rat and cat *(1–10)*. Ultrastructural analysis of these binding sites revealed that they were expressed by the axolemma of some (but not all) of the unmyelinated axons in the dorsal roots of the rat and cat, whereas myelinated axons exhibit no binding *(6,7)*. Examination of lectin binding in the spinal cord revealed that both SBA and BSI-B_4 bind to the superficial dorsal horn (laminae I and II) in both species *(6,7)*. This binding was found to be associated with axons in Lissauer's tract and with terminal-like varicosities in the superficial dorsal horn. Binding of these lectins in the superficial dorsal horn was abolished by dorsal rhizotomy in the rat and cat *(7)* and by neonatal treatment of rats with the unmyelinated axon neurotoxin capsaicin (Plenderleith, unpublished observations). The presence of SBA and BSI-B_4 binding sites in the Golgi apparatus and the plasma membrane as well as the uptake of tritiated galactose by small diameter sensory neurons *(11)* suggests that the lectin binding site may be a galactose containing glycoconjugate that undergoes glycosylation in the Golgi cisternae before inserting into the plasma membrane. The binding of both SBA and BSI-B_4 to sensory neurons (and their central processes) survives wax embedding *(6,7)* and lipid extraction (Plenderleith, unpublished observations) suggesting that the binding site is a glycoprotein rather than glycolipid. As this galactose containing glycoconjugate is expressed by a subpopulation of unmyelinated axons we recently attempted to establish whether this subpopulation had a discrete function. One possibility we considered was that this glycoconjugate is expressed by unmyelinated axons which innervate a particular target tissue (e.g., skin, muscle, or viscera). To examine this, we retrogradely labeled sensory neurons from the saphenous nerve (innervating the skin), the lateral gastrocnemius nerve (innervating muscle) or the greater splanchnic nerve (which innervates abdominal viscera) with the fluorescent tracer Diamidino yellow. We then screened these neurons for BSI-B_4 binding. The results revealed that BSI-B_4 binding was associated with 39% of cutaneous sensory neurons but only 0.7–3.4% of either muscle or visceral neurons *(12)*. From these results, we concluded that BSI-B_4 binding sites are located almost exclusively on a subpopulation of sensory neurons that innervate the skin. The small proportion of muscle and visceral neurons labeled with BSI-B_4 may be associated with that small number of primary sensory neurons having an axon that branches to innervate the skin and either muscle or viscera. Our findings indicated that a membrane-associated, galactose-containing glycoconjugate is expressed by a subpopulation of unmyelinated fibers that innervate the skin and raised the possibility that, by virtue of their affinity for this binding site, these lectins may be very effective anterograde tracers for unmyelinated sensory neurons. This hypothesis was supported by the recent observation that injection of BSI-B$_4$ into the saphenous nerve of the rat produces Golgi-like labeling of the central terminals of unmyelinated fibers in the spinal cord 72 h later *(13)*.

1.2. Summary of Technique

The technique for visualizing the central terminals of unmyelinated primary sensory neurons using transganglionic tracing with the plant lectin BSI-B$_4$ at the light microscopic level essentially involves three steps. The first stage involves microinjection of the lectin conjugated to the reporter enzyme horseradish peroxidase (HRP) and subsequent transport of the tracer. The second stage involves immobilization of the tracer by aldehyde fixation. Finally the tracer is visualized by a very sensitive modification of the horseradish peroxidase histochemistry procedure. If the resolution of light microscope is inadequate (as it is in our studies) then following visualization, the reaction product will have to be stabilized and processed for electron microscopy. In the following section, each of these stages will be described in detail. The procedure described is for the injection of a single peripheral nerve of a rat but should be applicable for other species with appropriate adjustments for transport times and perfusion volumes.

2. Materials
2.1. Tracer Injection

1. Peroxidase-labeled *Bandeiraea simplicifolia* I-isolectin B$_4$ (BSI-B$_4$–HRP) is dissolved in 0.1*M* PBS, pH 7.4., to a final concentration of 1%. This is made up in a small, glass conical vial, and mixed on a vortex mixer. There are various commercial suppliers of lectin conjugates, we routinely use Sigma (St. Louis, MO). This solution is kept at 4°C and we have noticed little attenuation of transport efficacy after 2–3 mo of storage (*see* Note 1).
2. Micropipets used to inject the tracer are prepared from 1.5-mm OD filamented glass capillary tubing (Clark Electromedical, Reading, UK) and pulled on an electrode puller (Campden, Loughborough, UK). The puller is setup to manufacture electrodes with a long shank to permit insertion of the electrode several millimetres into the nerve with minimal damage. Prior to filling, the electrodes are visualized under a dissecting microscope and broken back to a tip diameter of 50–250 µm.

3. Because the lectin is injected by pressure, some form of easily controllable pressure source must be attached to the shaft of the pipet. The easiest technique is to glue (using cyanoacrylate adhesive) the pipet to the needle of a 1-µL Hamilton syringe (*see* Note 2) and mount the syringe/pipet assembly onto a micromanipulator. This allows the pipet to be easily maneuvered into the nerve under microscopic control.

2.2. Fixation

1. Heparin sodium (5000 I.U./mL) 0.25 mL/rat.
2. Phosphate-buffered saline (PBS; $0.1M$, pH 7.4), 250 mL/rat at room temperature.
3. Glutaraldehyde (2.5% in $0.1M$ phosphate-buffer [PB], pH 7.4), 500 mL/rat at 4°C.

2.3. Horseradish Peroxidase Histochemistry

1. Ammonium heptamolybdate 0.25% in $0.1M$ PB at pH 6.0.
2. Tetramethylbenzidine (TMB): 0.2% in absolute ethanol (*see* Note 3).
3. Hydrogen peroxide: 0.3% in distilled water.
4. Solution containing 0.05% 3,3' diaminobenzidine (DAB), 0.02% cobalt acetate, and 0.01% hydrogen peroxidase (*see* Note 4).

2.4. Electron Microscopy

1. Osmium tetroxide (2%) in $0.1M$ PB, pH 7.4.
2. Dilution series of (AR grade) ethanol: 50, 70, 90, 95 (in distilled water), and 100%.
3. Acetone (AR grade).
4. Resin embedding kit for preferred resin. We routinely use Epon Araldite (Probing and Structure, Thuringowa, Australia).
5. Copper grids (200 mesh).
6. Uranyl acetate (1% in distilled water) and Reynolds lead citrate *(14)*.

3. Methods

3.1. Tracer Injection

1. The rat is deeply anesthetized (Nembutal 60 mg/Kg ip) and placed on a heated blanket the temperature of which is regulated by feedback control from a thermistor monitoring rectal temperature. This allows us to clamp the body temperature of the rat at 37.5°C throughout the procedure.
2. Make a small incision in the skin and expose the nerve of choice using aseptic procedures.
3. Place a loose silk ligature around the distal portion of the nerve and apply a small amount of tension to gently stretch the nerve. Then swab the area surrounding the nerve to remove all fluids (*see* Note 5).
4. Fill the micropipet with 1 µL of $BSI\text{-}B_4$–HRP and position it parallel to, but above the nerve, with the help of a micromanipulator. The pipet is then pushed through

the epineurium (just proximal to the ligature) and into the nerve for a reasonable distance (3–6 mm). Take care to keep the pipet parallel to the axons (*see* Note 6).
5. Expel approx one-fifth of the total volume of the tracer from the pipet by the Hamilton syringe or pressure injection system (*see* Note 7). Following injection, leave the pipet in place for 2–3 min to prevent leakage of the tracer as the electrode is removed. At this time, withdraw the electrode move it across the width of the nerve slightly, reinsert it make another injection. Repeat this process until the total 1 µL of tracer has been injected.
6. Close the wound in layers, apply topical antibiotic powder, and monitor the animal during recovery from the anesthetic.
7. The survival time is dependent on transport distance for the nerve and will have to be determined empirically. As a guide, we routinely use a survival time of 72 h for transport into the lumbar spinal cord after injection into the sciatic nerve at the level of the popliteal fossa of a rat. This is an approximate transport distance of 80 mm and correlates with an effective transport rate of 1.1 mm/h.

3.2. Fixation

1. Following the appropriate survival time, reanesthetize the animal (Nembutal 60 mg/kg ip) and open the thoracic cavity to expose the heart.
2. Inject the Heparin into the left ventricle, and (after a few heart beats to permit distribution of the solution) make a nick in the wall of the left ventricle just lateral to the apex of the heart.
3. Insert a small steel cannula into the left ventricle and push it up into the aorta. Then clamp the cannula in place with a pair of artery forceps and cut the right atrium to allow fluids to leave the systemic circulation.
4. Use the cannula to perfuse the circulatory system (at a pressure of 100 mmHg) with the PBS followed immediately by the fixative. Perfuse the PBS and first half of the fixative at the maximum flow rate permitted by the circulatory system, then reduce the flow rate such that the remaining volume of fixative takes 10–15 min to flow through the circulatory system.
5. Following perfusion, remove the appropriate segments of spinal cord and place them in 2.5% glutaraldehyde in PB, pH 7.4 for 4 h at 4°C.
6. Then transfer the tissue to PBS and store at 4°C. Although we have not evaluated the stability of the tracer systematically, we have successfully visualized the tracer after 5 wk storage with no noticeable effect on the intensity of the reaction product.

3.3. Sectioning

In order to visualize the tracer, cut serial 80-µm transverse sections of the appropriate segments of spinal cord. For light microscopy we use a freezing microtome and for electron microscopy an Oxford vibratome. In either case, collect the sections into 10-mL wells containing PB, pH 7.4.

3.4. Horseradish Peroxidase Histochemistry

All histochemistry is carried out in plastic multiwell trays. Each well holds 10 mL of solution and one section. Transfer of sections between wells is carried out with the aid of a small paint brush. Note that both DAB and TMB are suspected carcinogens and should only be handled in a fume cupboard, wearing appropriate protective clothing.

1. Wash sections in PB, pH 7.4 (3 changes of 5 min each), in order to remove any fixative remaining in the sections.
2. Incubate the sections in a 97.5:2.5 ratio mixture of 0.25% ammonium heptamolybdate (in $0.1M$ PB at pH 6.0) and 0.2% TMB (in absolute ethanol).
3. Add 0.1 mL of 0.3% hydrogen peroxide (in distilled water)/10 mL of presoak, every 5 min for 1 h or until the blue reaction product is observed (under a dissecting microscope; Fig. 1A; see Note 8).
4. Wash section for 60 s in PB, pH 7.4 (see Note 9).
5. Stabilization of the reaction product is achieved by incubating sections in PB containing 0.05% 3,3' diaminobenzidine (DAB), 0.02% cobalt acetate, and 0.01% hydrogen peroxide for 10–30 min (see Note 10; Fig. 1B).
6. Wash sections in distilled water (3 changes of 5 min each).

3.5. Light Microscopy

1. For light microscopic analysis, mount frozen sections on subbed slides, allow them to dry overnight, and then dehydrate by placing the slides for 5 min in each of the following ethanol solutions: 50, 70, 90, 95% (in distilled water), followed by two, 10-min washes in 100% ethanol.
2. Then clear the sections by two, 5-min washes in xylene before "coverslipping" with DePeX. The labeled terminals appear as dark brown punctate granules in the superficial dorsal horn of the spinal cord.

3.6. Electron Microscopy

For electron microscopic analysis, osmicate, dehydrate, and then embed the sections in resin. In our laboratory, sections are flat embedded in Epon-Araldite. Throughout the dehydration and embedding process, the contents of vials containing the sections are thoroughly mixed by placing them on a rotary mixer.

1. Incubate sections in 2% osmium tetroxide in PB, pH 7.4, for 60 min at room temperature.
2. Wash the sections in distilled water (5 changes of 5 min each).
3. Dehydrate sections by washing for 10 min in each of the following solutions of ethanol: 50, 70, 90, 95% (in distilled water) and 2 × 100%. This is followed by 2, 10-min washes in 100% acetone.
4. Infiltrate sections in 1:1 mixture of Epon-Araldite and acetone for 8 h.
5. Infiltrate sections in three changes (4 h each) of 100% Epon-Araldite.

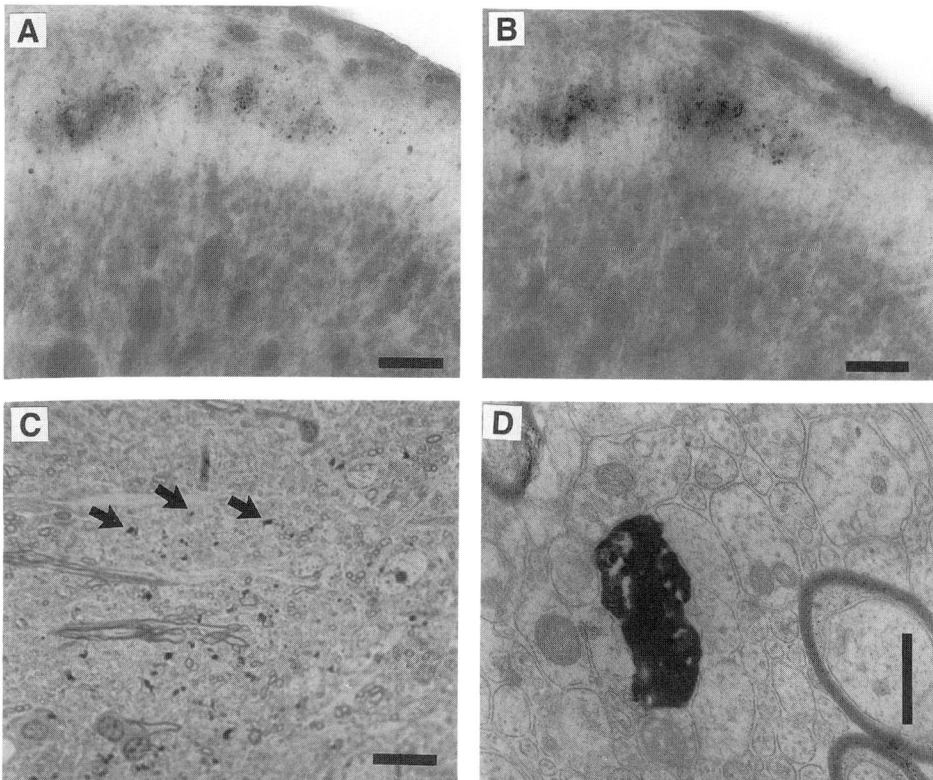

Fig. 1. (A) Photomicrogaph of a transverse section through the dorsal horn of the lumbar spinal cord of a rat following transganglionic transport of BSI-B_4. The dark punctate granules are HRP reaction product in the superficial dorsal horn as they appear following the initial stage of the histochemistry (after step 3 of Section 3.4.). The reaction product is blue in color. Bar = 50 μm. (B) Photomicrograph of HRP reaction product in the superficial dorsal horn of the same section as shown in A following stabilization (after step 5 of Section 3.4.). The reaction product is generally more intense and dark brown in color. Bar = 50 μm. (C) Photomicrograph of semithin section (counterstained with toluidine blue) through the superficial dorsal horn showing a large number of punctate granules of HRP reaction product, three of that are indicated by arrows. Bar = 10 μm. (D) Electron micrograph of the superficial laminae of the dorsa horn showing a single axon terminal containing HRP reaction product. Bar = 0.5 μm.

6. Mount sections between two thermanox cover slips with 100% Epon-Araldite, and then polymerize resin at 60°C for 72 h. In order to keep the sections flat in the oven, the thermanox cover slips are placed between two plates of glass. Each sheet of glass is separated from the thermanox cover slips by an acetate sheet (overhead projector transparency) to prevent the resin gluing the plates of glass together.

7. Following polymerization, examine the thermanox cover slips containing the sections under a light microscope and cut out the region containing labeled terminals and mount on a blank Epon-Araldite block (prepared by polymerization of excess resin in a BEEM capsule).
8. From this block, use glass knives to cut semithin sections (Fig. 1C) for orientation prior to ultrathin sectioning with a diamond knife. Mount ultrathin sections on 300 mesh copper grids.
9. Counterstain the ultrathin sections with uranyl acetate and lead citrate before viewing on a transmission electron microscope.
10. Using this procedure, labeled axon and axon terminals are clearly visible by the electron-dense reaction product associated with such profiles in the superficial dorsal horn (Fig. 1D).

4. Notes

1. Prior to injection the conjugate is removed from the refridgerator, allowed to come to room temperature, and then thoroughly mixed on a vortex mixer.
2. The use of a Hamilton syringe has the disadvantage that if the pipet is damaged, while maneuvering near the animal then a new syringe/pipet assembly has to be prepared, which can be time consuming. An alternative solution is to glue a standard 25-gage hypodermic needle inside the shaft of the pipet (again with cyanoacrylate adhesive), and then connect the pipet/needle assembly to a 5-mL syringe via a 20–30-cm length of polyethylene tubing with a luer connector at each end. Providing the luer connector can be firmly attached to the micromanipulator, then the pipet/needle assembly can be easily replaced if damaged. A three-way tap near the syringe makes it easy to release the pressure in the system at any time. This system is certainly more convenient (and cheaper) than the Hamilton syringe assembly but does require a lot more manual dexterity to control the pressure in the pipet.
3. In order to dissolve the TMB, it is usually necessary to either sonicate it or gently warm it (<60°C), and then shake it vigorously.
4. In order to dissolve the DAB and prevent precipitation of the cobalt acetate, we dissolve both of these in half the final volume of distilled water, and then make up to the final volume with $0.2M$ PB. This solution must be freshly made up shortly before required, and the hydrogen peroxidase added to the final concentration immediately before use.
5. It is very important that the nerve and surrounding tissue are dry. If significant amounts of fluid are present near the nerve, then these will be drawn up into the pipet by capillary action and replace the tracer in the tip.
6. Depending on the anatomical position of the nerve to be injected (or the surgical approach adopted) it may be necessary to raise the nerve above the surrounding tissue. This is easily achieved by placing a small cotton wool swab under the nerve. This has the added advantage of keeping the area surrounding the injection site dry. In some cases, it may be necesarry to make a small cut in the epineurium to permit entry of the pipet. The easiest way to do this is with a glass pipet

(similar to those used for the injection), which has been broken back to give a small cutting edge.
7. During the injection process, it is relatively easy to observe the flow of the tracer into the nerve by monitoring the meniscus formed by the solution in the shank of the pipet with the aid of a dissecting microscope. After application of pressure to the injection system, the flow of tracer usually only begins after the pipet has been withdrawn very slightly.
8. Although the reaction product is usually apparent under the dissecting microscope after 60 min, we have on occasions failed to observe anything at this time. In some instances, we have left the incubation running overnight (without further addition of hydrogen peroxide) and have successfully located the reaction product in the morning.
9. The TMB reaction product appears to be particularly unstable at this point, so the PB wash is kept as short as possible.
10. In order to optimize the time of the stabilization step, it is best to watch the reaction of a few sections under a dissecting microscope. The blue TMB reaction product will change to dark brown and the reaction should be stopped once the intensity of this stabilized reaction product plateaus.

References

1. Streit, W. J., Schulte, B. A., Balentine, J. D., and Spicer, S. S. (1985) Histochemical localization of galactose containing glycoconjugates in sensory neurons and their processes in the central and peripheral nervous system of the rat. *J. Histochem. Cytochem.* **33**, 1042–1052.
2. Streit, W. J., Schulte, B. A., Balentine, J. D., and Spicer, S. S. (1986) Evidence for glycoconjugate in nociceptive primary sensory neurons and its origin from the Golgi complex. *Brain Res.* **377**, 1–17.
3. Streit, W. J. and Kreutzberg, G. W. (1987) Lectin binding by resting and reactive microglia. J. *Neurocytology* **16**, 249–260.
4. Silverman, J. D. and Kruger, L. (1988) Lectin and neuropeptide labeling of separate populations of dorsal root ganglion neurons and associated "nociceptor" thin axons in rat testis and cornea whole-mount preparations. *Somatosensory Res.* **5**, 259–267.
5. Tajti, J. Fischer, J. Knyihar-Csillik, E., and Csillik, B. (1988) Transganglionic regulation and fine structural localization of lectin-reactive carbohydrate epitopes in primary sensory neurons of the rat. *Histochemistry* **88**, 213–218.
6. Plenderleith, M. B., Cameron, A. A., Key, B., and Snow, P. J. (1988) Soybean agglutinin binds to a subpopulation of primary sensory neurons in the cat. *Neuroscience Letts.* **86**, 257–262.
7. Plenderleith, M. B., Cameron, A. A., Key, B., and Snow, P. J. (1989) The plant lectin soybean agglutinin binds to the soma, axon and central terminals of a subpopulation of small diameter primary sensory neurons in the rat and cat. *Neuroscience* **3**, 683–695.
8. Plenderleith, M. B. and Snow, P. J. (1990) The effect of peripheral nerve section on lectin binding in the superficial dorsal horn of the rat spinal cord. *Brain Res.* **507**, 146–150.

9. Plenderleith, M. B. and Snow, P. J. (1991) Plant lectins: new tools in neurobiology. *Today's Life Science* **3,** 34–40.
10. Plenderleith, M. B., Wright, L., and Snow, P. J. (1992) Expression of lectin binding in the superficial dorsal horn of the rat spinal cord during pre and postnatal development. *Dev. Brain Res.* **68,** 103–109.
11. Droz, B. L. (1967) L'appareil de Golgi comme site d'incorporation du galactose-H^3 dans les neurons ganglionaires spinaux chez le rat, *J. Microsc.* **6,** 419–424.
12. Plenderleith, M. B. and Snow, P. J. (1993) The plant lectin *Bandeiraea simplicifolia* I-B_4 identifies a subpopulation of small diameter primary sensory neurons which innervate the skin in the rat. *Neurosci. Lett.* **159,** 17–20.
13. Kitchener, P. D., Wilson, P., and Snow, P. J. (1993) Selective labeling of primary sensory afferent terminals in lamina II of the dorsal horn by injection of *Bandeiraea simplicifolia* isolectin B_4 into peripheral nerves. *Neuroscience* **54,** 545–551.
14. Reynolds, E. S. (1963) The use of lead citrate at high pH as an electron opaque stain in electron microscopy. *J. Cell Biol.* **17,** 208–215.

XII

USE OF LECTINS IN THE INVESTIGATION OF PATHOGEN–HOST INTERACTIONS

45

Lectin Inhibition of Bacterial Adhesion to Animal Cells

Murray W. Stinson and Jen Ren Wang

1. Introduction

Adhesion of bacteria to epithelial cells of the respiratory, gastric, and genitourinary mucosa is generally considered to be the initial step in the pathogenesis of many bacterial infections (1). Adhesion enables the bacteria to localize near a food source and to resist being washed away by the fluids that constantly bathe mucosal surfaces. Bacteria that persist at the site of attachment can proliferate and thus establish a stable colonization. If the bacteria produce the necessary exoenzymes and/or exotoxins to overcome other host defenses, they may invade deeper into the tissues and cause clinical symptoms.

The specificity of bacterial adhesion usually resides in proteins on the surface of the bacteria that recognize particular glycoproteins and glycolipids on the host cell membrane. In some cases, adhesion is inhibited by extraneous monosaccharides or oligosaccharides indicating that the adhesins are lectins (1–4). These lectin-like adhesins exhibit high-affinity binding and can be found at the tips of long filamentous appendages known as pili or fimbriae; e.g., the mannose-specific type I fimbriae, the Galα1,4Gal-specific type P fimbriae (1,3) and the NeuAcα2,3Gal-specific type S fimbriae of *Escherichia coli* (5). In other bacterial genera, the adhesins may reside in shorter, densely arrayed, protein fibrils called fibrillae. These include the fucose-specific M6 protein of *Streptococcus pyogenes* (6,7) and the sialic acid adhesin of *Streptococcus sanguis* (8). The filamentous configuration of bacterial adhesins may enable the bacterium to make physical contact with host cells by circumventing the problem of anionic repulsion by opposing cell surfaces and steric interference by bacterial capsular polysaccharides or other cell wall components.

In addition to inhibition by sugars, the binding specificity of bacterial adhesins may be revealed by a reduction in the amount of adhesion following treatment of host cell surfaces with selected glycosidases or lectins *(7–11)*. The use of lectins to block carbohydrate groups, which may serve as receptors for bacteria, is sometimes hindered by the presence of complex carbohydrates (polysaccharides, lipopolysaccharides, teichoic acids, or teichuronic acids) on the bacterial cell wall that can bind the same lectins *(4,12,13)*. If this occurs, the multivalent lectins may crosslink the two cell surfaces and obscure the blocking effects of the lectin on the receptor. These complications may be minimized by studying bacterial adhesion to artificial surfaces coated with a purified host receptor or with a structural analog, e.g., fetuin-coated latex beads for adhesion of *Helicobacter pylori (14)*, or by studying the binding of a purified bacterial adhesin to a native animal cell, e.g., M6 protein of *Streptococcus pyogenes (7)*.

1.1. Adhesion of Streptococci to Human Epithelial Cells

The adhesion of *Streptococcus pyogenes* to human epithelial cells (HEp-2) is believed to involve several adhesins: lipoteichoic acid, M protein *(6,15,16)*, and fibronectin binding protein *(17,18)*. Of these, only the M protein has been shown to have carbohydrate binding properties *(7)*. As might be expected, the relative contribution of each type of adhesin to the attachment of streptococci to a particular surface is difficult to measure experimentally. Selective inhibition of one adhesin in such a redundant system may not result in a proportional loss in the amount of cell adhesion because the alternative mechanisms may remain fully functional. The problem of discriminating between multiple adhesions has been addressed genetically in streptococci by inactivating the structural gene for M6 protein and testing the mutant for adhesion activity *(7,18)*. When compared with the isogenic parental strain, the M6-deficient mutant lost 78% of its adhesion activity for HEp-2 cells (Fig. 1). The residual activity was a result of the alternative adhesins on the bacterial cell walls. Although the genetic evidence indicates that the presence of M6 protein is necessary for adhesion, it does not prove that it directly mediates attachment. It can be argued that the loss of M6 protein from the cell surface perturbs the function of another protein that is the actual adhesin. Therefore, characterization of the binding mechanisms of streptococcal adhesins also requires the use of purified adhesin instead of intact bacteria. The binding of streptococcal M6 protein to HEp-2 cell membranes has been determined and shown to be inhibited by pretreatment of the HEp-2 components with α-L-fucosidase or with *Ulex europaeus* (UEA-1) lectin (Fig. 2) *(7)*.

Studies of bacterial adhesion can be further complicated by artifacts induced by the constituents of the nutrient medium used to grow the bacteria. Strepto-

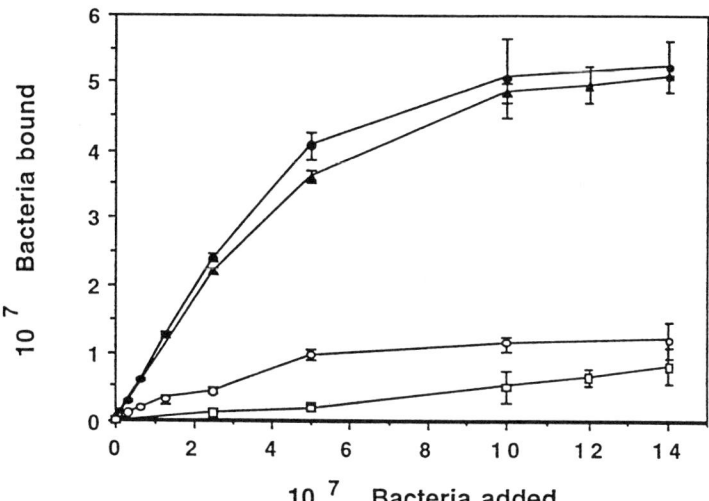

Fig. 1. Adhesion of radiolabeled *S. pyogenes* S43 (M$^+$) (▲), JRS4 (M$^+$) (●), and JRS75 (M$^-$) (○) to HEp-2 monolayers and adhesion of radiolabeled S43 cells to tissue culture medium treated control wells (□). Strains JRS4 andJRS75 showed the same amounts of adhesion to the control wells and thus results for them are not presented. Results are expressed as mean ± SD of triplicate data from a single representative experiment. Reprinted with permission from ref. 6.

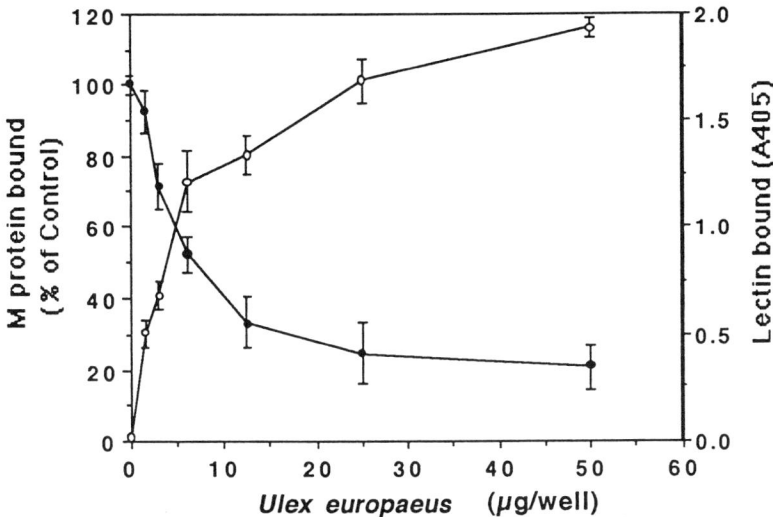

Fig. 2. Enzyme immunoassay showing UEA-1 binding to HEp-2 cell membranes and the resulting blockage of receptors for streptococcal M6 protein. Symbols: (○), binding of UEA-1 as detected by rabbit antibodies to UEA-1; (●), dose-dependent inhibition of purified M6 protein binding as detected by rabbit antiscrum to *S. pyogenes*. Reprinted with permission from ref. 7.

cocci and many other pathogenic bacteria grow best in complex media comprized of meat infusions, acid-hydrolyzed casein, enzyme digests of soybean meal, and blood components, which may contain a variety of glycoconjugates with sugar sequences similar to the oligosaccharide receptors being studied on animal cell surfaces. Binding of culture medium components to bacterial adhesins can be expected to modulate bacterial adhesion in subsequent assays since bound medium components may persist on the bacteria through numerous washing steps (20). It is suspected that tissue components in Todd-Hewitt Broth (a bovine heart infusion) causes spontaneous clumping of S. pyogenes JRS4. These bacteria do not clump when grown in a semisynthetic medium; but clump quickly when resuspended in THB. The isogenic mutant strain JRS75 that does not produce the M6 protein does not clump when grown in THB (6). Consequently, it is desirable to grow bacteria on a chemically defined medium, whenever possible, prior to their use in adhesion assays. It is also recommended that freshly grown bacteria be used for each adhesion experiment. Suspensions of bacteria that are stored at 4°C often give inconsistent results because they can lose viability and shed surface proteins. Because the expression of surface proteins may vary during growth, bacteria from both logarithmic and early stationary (within 2 h of maximum growth) phases should be evaluated in adhesion assays to determine the time of maximum adhesion activity.

Three methods have been widely used for enumerating bacteria that have adsorbed to animal cell surfaces:

1. Direct microscopic counting of bacteria on individual host cells.
2. Colony count after dissociation of the bacteria from host cells and growth on agar medium.
3. Quantitation of radiolabeled bacteria on host cells.

The method of choice is influenced by the number of individual assays being conducted and on the nature of the bacterium and animal cells. Generally, the microscopic counting is technically tedious and is hindered by spontaneous clumping of some bacteria and by the inability to view all host cell surfaces. Although colony counting methods are rapid and sensitive, they are subject to complications owing to the inherent clumping activities of some bacterial genera (e.g., *Staphylococcus* and *Streptococcus*) and to loss of cell viability. Radioactivity assays detect all adherent bacteria regardless of clumping, viability, and position on the host cell surface.

2. Materials
2.1. Adhesion of Streptococci to Epithelial Cell Monolayers

1. *Streptococcus pyogenes* strain S43 (serotype M6); ATCC 12348, American Type culture Collection, Rockville, MD (*see* Note 1).

Lectin Inhibition of Bacterial Adhesion

2. Streptococcal growth medium *(21)* supplemented with 0.2% ultrafiltered yeast extract (SMY). To prepare SMY, a stock solution of 20% yeast extract (Difco, Detroit, MI) is ultrafiltered through a Centriprep 3000 (Amicon, Beverly, MA) and added to the medium (1–100 mL). Sterilize the medium by filtration (*see* Note 2).
3. ^{3}H-adenine; ICN, Irvine, CA (*see* Note 3).
4. Sterile phosphate-buffered saline: 150 mM sodium chloride, 10 mM sodium phosphate, pH 7.2.
5. Human epithelial cells (HEp-2 cells): ATCC–CCL 23, American Type Culture Collection, Rockville, MD.
6. Dulbecco's modified Eagle's medium (DMEM) containing 5% fetal bovine serum, penicillin (100 U/mL), streptomycin (100 µg/mL), and amphotericin B (0.25 µg/mL) (Gibco, Grand Island, NY).
7. Sterile 0.05% Trypsin-0.53 mM EDTA.
8. Sterile 24-well plastic culture plate; local vendor.
9. 1% Triton X-100.

2.2. Inhibition of Streptococcal Adhesion Using UEA-1 Lectin

1. All materials listed in Section 2.1.
2. Eagle's minimal essential medium (EMEM); Gibco.
3. *Ulex europeus* lectin UEA-1; local vendor. Prepare a stock solution (1 mg/mL) in EMEM.

3. Methods
3.1. Adhesion of Streptococci to Epithelial Cell Monolayers

1. Harvest a monolayer (75-cm^2 flask) of HEp-2 cells 48 h prior to the adhesion assay. Detach the HEp-2 cell monolayer from the flask using 2 mL of trypsin-EDTA for 5 min at room temperature.
2. Transfer the cell suspension promptly to a sterile test tube containing 5 mL of DMEM-5% fetal calf serum (FCS). Sediment the cells at 200g for 5 min and wash them with a fresh vol of DMEM-5% FCS.
3. Resuspend the final cell pellet to 1.6×10^5 cells/mL DMEM-5% FCS. Transfer 1 mL of this suspension to each well of a sterile 24-well plate and incubate statically at 37°C in air with 5% CO_2. After 48 h, examine the monolayer in each well to confirm cell confluency.
4. Grow *S. pyogenes* statically in 10 mL of SMY containing 2 µCi/mL of ^{3}H adenine at 37°C for 12–16 h (*see* Note 4).
5. Harvest the bacteria by centrifugation (10,000g for 15 min) and wash the cells three times with 0.5 mL of PBS to remove soluble radiolabel.
6. Resuspend the final pellet in DMEM to a density of 4×10^8 bacteria/mL; this corresponds to an optical density 1.0 at 600 nm (*see* Note 5).
7. Count the radioactivity in several volumes of this suspension by scintillation spectrometry. The specific radioactivity of the bacteria grown under these conditions should be approx 2×10^5 cpm/10^8 bacteria.

8. Dilute the radiolabeled bacteria in DMEM in separate test tubes to give a range of cell concentrations: e.g., 2, 4, 8, 12, 16, 20, 30, and 40×10^7/mL.
9. Remove the culture medium from the 48 h confluent monolayers by inverting the 24-well culture dish and replace it with 0.5 mL of the selected bacterial suspension. Test each dilution of bacteria in triplicate.
10. To monitor possible adhesion to exposed plastic surfaces, an identical series of bacterial suspensions should be added to plastic wells devoid of HEp-2 cells but pretreated for 60 min with DMEM-5% FCS (see Note 6).
11. Centrifuge the culture plate for 10 min at 200g at room temperature using a swinging rotor and microtiter plate buckets (see Note 7).
12. Remove the culture medium containing free bacteria using a vacuum line connected to a radiation-waste trap and wash each monolayer three times with 2 mL of PBS.
13. Dissociate the HEp-2 monolayer and adherent bacteria from the plastic wells by extraction with 400 µL of 1% Triton X-100 for 30 min at room temperature.
14. Transfer the contents of each well to a scintillation vial and count the radioactivity. The number of bacteria that adhered to the HEp-2 cell monolayer can be calculated using the specific radioactivity of the original bacterial suspension (see Note 8). The number of bacteria required to saturate the surface receptors on the HEp-2 monolayers is often indicated by a plateau in the amount of adhesion when the number of bacteria bound to HEp-2 cells is plotted against the total number of bacteria in each culture well (see Note 9). The bacteria that adsorbed to the control wells may be plotted separately or subtracted from the numbers obtained in corresponding wells containing HEp-2 cell monolayers.

3.2. Inhibition of Streptococcal Adhesion by UEA-1 Lectin

1. Block terminal α L-fucose receptors on the surface of HEp-2 cells by treating the monolayers with *Ulex europaeus* lectin (UEA-1) dissolved in EMEM. Several concentrations of UEA-1 should be tested in order to determine the dose response, e.g., 10, 25, 50, 100, 150, 200, 250, and 300 µg/monolayer. After 1 h at 37°C, wash the monolayers three times with EMEM to remove unbound lectin. For positive adhesion controls, leave 3–6 monolayers untreated with lectin.
2. Add 5×10^7 radiolabeled bacteria to each well. This number of bacteria should be below the saturating dose for a monolayer of this size; verify using the data in Section 3.1. (see Note 10).
3. Follow assay steps 11–14 in Section 3.1. to complete the adhesion assay.

4. Notes

1. Because *S. pyogenes* is a human pathogen, aseptic procedures must be followed when handling this bacterium.
2. Although *Streptococcus pyogenes* has been reported to grow in the chemically defined medium *(21)* without nutrient supplements, we find that bacterial growth ceases after two passages. The unidentified nutrient deficiency is satisfied by the low molecular fraction of yeast extract.

3. It is necessary to select a radiolabeled compound that will be taken up and assimilated into macromolecular constituents and remain in the cytoplasm. Although much higher specific radioactivity can be achieved in bacteria with radiolabeled amino acids or sugars, the labeled products are often released into the cell environment during growth and during the adhesion assay. Less than 1% of the adenine radioactivity is released by viable streptococci during this assay period.
4. For maximum uniformity, the culture should be harvested during logarithmic phase of growth. A small inoculum of 5–10 µL of a fresh starter culture is recommended.
5. Because *S. pyogenes* grows in chains, the most accurate enumeration is achieved microscopically with a Petroff-Hausser counting chamber. For safety considerations, streptococci grown in nonradioactive medium should be used to determine the number of cells/mL at OD 1 (600 nm). Vortex the suspension vigorously for 1 min to break up cell chains before counting. Low-intensity ultrasonification can also be used to disperse Gram-positive bacteria; however, it is not suitable for dispersing the more fragile Gram-negative bacteria. Vortexed streptococcal suspensions will contain single cells, pairs, and short chains; count all cells in the preparation. The resulting number of bacteria/mL should be used for all subsequent calculations pertaining to the radiolabeled bacteria (OD 1) in the adhesion assay. Colony count assays may be an alternative procedure when using bacteria that do not form autologous clumps.
6. Some bacteria exhibit a large amount of nonspecific adhesion to exposed plastic surfaces (walls) that can obscure their adhesion to animal cells in radioactivity assays. In some cases, this adsorption to plastic can be depressed to an insignificant level by pretreating the empty control wells with growth medium containing 5% FCS. If this does not prevent nonspecific adhesion to plastic, the bacteria can be suspended in 1% bovine serum albumin.
7. The adhesion assay may be conducted without centrifugation but the incubation time with the bacteria must be extended considerably to obtain similar numbers of adherent bacteria, e.g., 90 min at 37°C for *S. pyogenes*. Centrifugation shortens the duration of the experiment and thus minimizes complications resulting from bacterial growth and damage to the monolayer by bacterial toxins (e.g., hemolysin) and exoenzymes (e.g., protease) that may be produced during the incubation period.
8. The average number of HEp-2 cells in a confluent monolayer can be determined microscopically. Detach HEp-2 cells from 10 wells using trypsin-EDTA and suspend the cells in 10 mL of DMEM. Transfer the suspension to a hemocytometer for enumeration.
9. Other streptococci (e.g., *Streptococcus* mutans) do not appear to saturate host cell surfaces in radioactivity assays because of their ability to form homologous clumps. To minimize bacterial cohesion, this assay should be conducted with the lowest detectable numbers of bacteria.
10. UEA-1 does not bind to *S. pyogenes* cell surfaces; however, other lectins bind to a variety of bacterial genera and species and cause them to agglutinate. The latter lectins are unsuitable as potential blocking agents for bacterial adhesion since

they can crosslink bacteria and animal cell surfaces. Relevant lectins should be pretested for their abilities to agglutinate the bacteria of interest on glass microscope slides.

References

1. Beachey, E. H. (1981) Bacterial adherence: adhesin-receptor interactions mediating the attachment of bacteria to mucosal surfaces. *J. Infect. Dis.* **143,** 325–345.
2. Wadstrom, T. and Tylewska, S. K. (1982) Glycoconjugates as possible receptors for *Streptococcus pyogenes*. *Curr. Microbiol.* **7,** 343–346.
3. Leffler, H. and Svanborg-Eden, C. (1986) Glycolipids as receptors for *Escherichia coli* lectins, in *Microbial Lectins and Agglutinins, Properties and Biological Activity* (Mirelman, D., ed.), Wiley, NY, pp. 83–112.
4. Lis., H. and Sharon, N. (1986) Lectins as molecules and tools. *Annu. Rev. Biochem.* **55,** 35–67.
5. Parkkinen, J., Rogers, G. N., Korhonen, T. K., Dahr, W., and Finne, J. (1986) Identification of the o-linked sialyloligosaccharides of glycophorin A as the erythrocyte receptors of S-fimbriated *Escherichia coli*. *Infect. Immunol.* **54,** 37–42.
6. Wang, J. R. and Stinson, M. W. (1994) M protein mediates streptococcal adhesion to HEp-2 cells. *Infect. Immunol.* **62,** 442–448.
7. Wang, J. R. and Stinson, M. W. (1994) Streptococcal M6 protein binds to fucose-containing glycoproteins on cultured human epithelial cells. *Infect. Immunol.* **62,** 1268–1274.
8. Murray, P. A., Levine, M. J., Tabak, L. A., and Reddy, M. S. (1982) Specificity of salivary-bacterial interactions. II. Evidence of a lectin on *Streptococcus sanguis* with specificity for a NeuAcα2,3Galβ1,3GalNAc sequence. *Biochem. Biophys. Res. Commun.* **106,** 390–396.
9. Falk, P., Roth, K. A., Boren, T., Westblom, T. U., Gordon, J. I., and Normark, S. 1993. An in vitro adherence assay reveals that *Helicobacter pylori* exhibits cell lineage-specific tropism in the human gastric epithelium. *Proc. Natl. Acad. Sci. USA* **90,** 2035–2044.
10. Ensgraber, M., Genitsariotis, R., Storkel, S. and Loos, M. (1992) Purification and characterization of a Salmonella typhimurium agglutinin from gut mucus secretions. *Microbiol. Pathogen.* **12,** 255–266.
11. Sauter, S. L., Rutherfurd, S. M., Wagener, C., Shively, J. E. and Hefta, S. A. (1993) Identification of specific oligosaccharide sites recognized by type I fimbriae from *Escherichia coli* on nonspecific cross-reacting antigen: a CD66 cluster granulocyte glycoprotein. *J. Biol. Chem.* **268,** 15,510–15,516.
12. Pistole, T. G. (1981) Interaction of bacteria and fungi with lectins and lectin-like substances. *Annu. Rev. Microbiol.* **35,** 85–112.
13. Doyle, R. and Keller, K. (1984) Lectins in diagnostic microbiology. *Eur. J. Clin. Microbiol.* **3,** 4–9.
14. Lelwala-Guruge, J., Ascencio, F., Ljungh, A. and Wadstrom, T. (1993) Rapid detection and characterization of sialic acid-specific lectins of *Helicobacter pylori*. *APMIS* **101,** 695–702.

15. Courtney, H. S., von Hunolstein, C., Dale, J. B., Bronze, M. W., Beachey, E. H., and Hasty, D. L. (1992) Lipoteichoic acid and M protein: dual adhesins of group A streptococci. *Microbial Pathogen.* **12,** 199–208.
16. Hasty, D. L., Ofek, I., Courtney, H. S., and Doyle, R. J. (1992) Multiple adhesins of streptococci. *Infect. Immunol.* **60,** 2147–2152.
17. Hanski, E., Horwitz, P. A., and Caparon, M. (1992) Expression of Protein F, the fibronectin-binding protein of *S. pyogenes* JRS4, in heterologous streptococcal and enterococcal strains promotes their adherence to respiratory epithelial cells. *Infect. Immunol.* **60,** 5119–5125.
18. Hanski, E. and Caparon, M. (1992) Protein F, a fibronectin binding protein, is an adhesin of group A streptococci. *Proc. Natl. Acad. Sci. USA* **89,** 6172–6176.
19. Caparon, M. G., Stephens, D. S., Olsen, A., and Scott, J. R. (1991) Role of M protein in adherence of group A streptococci. *Infect. Immunol.* **59,** 1811–1817.
20. Stinson, M. W. and Jones, C. A. (1983) Binding of Todd-Hewitt broth antigens by *Streptococcus* mutans. *Infect. Immunol.* **40,** 1140–1145.
21. van de Rijn, I. and Kessler, R. E. (1980) Growth characteristics of group A streptococci in a new chemically defined medium. *Infect. Immunol.* **27,** 444–448.

46

Inhibition of HIV Infection by Lectin Binding to CD4

Jean Favero and Virginie Lafont

1. Introduction

The progression of AIDS all over the world has led research laboratories to focus their energy on the identification of molecules that can impair the infection process. Every step of the viral replicative cycle can be considered as a potential target to block infection. One of the first events in the infection of T-cells by the human immunodeficiency virus (HIV) is the binding of the viral envelope glycoprotein gp120 to the differentiation antigen CD4 expressed on the host cell *(1–3)*. Among the therapeutic strategies aimed at controlling HIV disease progression, one was to inhibit or perturb this interaction.

The high glycosylated character of gp120 *(4,5)* has led to the hypothesis that the oligosaccharides present on the virus envelope could play a fundamental role in the infection mechanisms (*see* ref. *6* for a review). This hypothesis is still a matter of controversy *(7–9);* however, lectins with binding specificities for gp120 saccharides have been studied and tested as potential inhibitors for HIV infection (*see* ref. *10* for a review). Thus, a considerable number of lectins (mostly mannose-specific lectins) have been described that display an inhibitory effect on HIV infection. Although the mechanism by which some lectins inhibit infection is not yet defined (blocking of virus binding to the host cell by steric hindrance or alteration of the conformation of gp120 causing destabilization of the complex formed by gp120 and the transmembrane glycoprotein gp41), it is clear that a perturbation of oligosaccharide residues of gp120 by lectin binding leads to impairment of in vitro HIV infection.

As an alternative to considering gp120 as a target to block essential events in the infection process, one can consider the viral receptors on the host cell. In lymphocytes and probably in cells of the monocyte/macrophage lineage, the CD4 molecule is one of these receptors for HIV. The human CD4 antigen is a

From: *Methods in Molecular Medicine: Vol. 9: Lectin Methods and Protocols*
Edited by: J. M. Rhodes and J. D. Milton Humana Press Inc., Totowa, NJ

55-kDa glycoprotein containing two asparagine-linked glycosylation sites *(11)*. Despite the presence of oligosaccharide chains on CD4, lectin binding to cell viral receptor has not been considered a valuable therapeutic strategy for AIDS. To begin with, the HIV envelope glycoprotein interacts with the V1 region of the CD4 molecule (particularly involving the Phe-43 amino acid in the immunoglobulin-CDR2-like domain) (*see* ref. *12* for a review), far removed from the two glycosylation sites (Asn-271 and Asn-300) *(11)*. Moreover, those lectins that appeared to block or retard syncytia formation (a cell fusion phenomenon involving the interaction between gp120 expressed on the surface of infected cells and CD4 present on uninfected cells) are ineffective when preincubated with CD4 bearing uninfected cells *(13,14)*. Furthermore, it has been shown that carbohydrates of recombinant soluble CD4 (sCD4) are not necessary for virus binding *(15)*. Even though these results strongly suggested that the carbohydrate moieties of CD4 are not involved in the infection process, and that lectins susceptible to recognize these oligosaccharides do not inhibit infection, we studied the effect of jacalin, a lectin purified from jackfruit seeds, on in vitro HIV infection. Indeed this lectin was shown specifically to stimulate T-lymphocytes bearing the CD4 antigen *(16)* i.e., the T-lymphocyte population susceptible to be infected by the virus. This result led us to hypothesize that jacalin could interact with CD4 *(17)*, act through CD4 *(18,19)* and, as a consequence, have an effect on virus infection *(17)*. However, the high specificity of jacalin for the β-D-Gal(1→3)D-GalNAc motif *(20)* in *O*-linked glycans, i.e., a structure not found in CD4, led us to suggest that a possibility existed that jacalin-CD4 binding could be the result of a sugar-independent protein–protein interaction.

1.1. Analysis of Jacalin-CD4 Interaction

Lectins are sugar binding molecules and, in contrast to antibodies that generally recognize one well-defined molecule on the cell surface, they can interact with several membrane antigens. This has been a real problem, for example, in the determination of the actual molecules involved in lectin-induced mitogenic stimulation of T-lymphocytes. In the case we are dealing with, the problem is different since the goal is to demonstrate that jacalin interacts with CD4, a well-characterized antigen. To resolve this problem, the cell surface molecules bound by the lectin were separated from CEM cell lysates (a CD4 positive T-cell line) by adsorption on jacalin-coated agarose beads. Of course, this technique would not discriminate CD4 from the other putative glycoproteins recognized by the lectin. After SDS-PAGE of the jacalin-bound molecules and electroblotting on nitrocellulose sheet or nylon sheet, CD4 has been detected using an anti-CD4 MAb labeled with ^{125}I followed by autoradiography (Fig. 1) or using a biotin-conjugated anti-CD4 antibody followed by

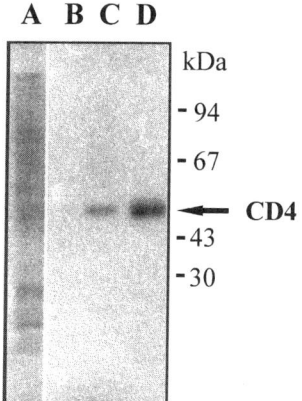

Fig. 1. Precipitation of CEM cell antigens by anti-CD4 MAb and jacalin. Lane A shows cell lysate adsorbed to jacalin agarose and revealed by ^{125}I-labeled-jacalin. Lane B shows the precipitation of CEM cell lysate with protein A-sepharose alone (non specific binding), lane C, precipitation with jacalin immobilized on agarose beads and lane D, immunoprecipitation with anti-CD4 MAb and protein A-sepharose. The blot (lanes B,C and D) was revealed by incubation with ^{125}I-labeled-anti-CD4 MAb. (Reprinted from ref. *17* with permission.)

peroxidase-conjugated streptavidin and the chemiluminescence technique. As a control, CD4 was immunoprecipitated from CEM cell lysates using anti-CD4 MAb and protein A sepharose. The profile of the total molecules actually bound by the lectin was visualized by incubating the blot in a solution containing ^{125}I-labeled jacalin. Before incubating the blot with either anti-CD4 MAb or jacalin, it had to be saturated to avoid nonspecific binding. This technique allowed us to show that among the molecules adsorbed on jacalin-coated agarose beads, one was recognized by an anti-CD4 MAb suggesting that jacalin does interact with CD4.

1.2. Effect of Jacalin on In Vitro HIV Infection of Human Lymphoid Cells

CEM cells, a CD4$^+$ T-cell line highly susceptible to in vitro HIV infection was used in this study. The effect of jacalin was tested on the infection of these cells by HIV-1$_{GER}$, a virus isolated from an AIDS patient. The effect of several concentrations of the lectin was compared to that of OKT4A, an anti-CD4 MAb known to block infection *(21)*. The HIV infection related to viral production was monitored twice a week by measuring reverse transcriptase (RT) activity. OKT4A has been described to bind a CD4 epitope located in the CDR2 domain of the molecule, i.e., close to the virus binding site. This antibody blocks infec-

Table 1
Inhibition of HIV-1 Infection by Jacalin or OKT4A

	Days after infection			
	3	6	10	13
Additives				
None	0.5	20	250	150
Jacalin, 0.1 µg/mL	0.8	40	360	280
Jacalin, 1 µg/mL	0.5	0.8	250	440
Jacalin, 10 µg/mL	0.7	0.7	0.5	0.6
OKT4A, 2 µg/mL	0.8	0.6	0.6	0.9

Viral production was evaluated by measuring reverse transcriptase activity.
RT activity; cpm × 10^{-3}/mL.

tion by sterically preventing virus binding. On the contrary, OKT4, another anti-CD4 MAb that recognizes an epitope far from the virus binding site, has no effect on infection. The results presented (Table 1) show that jacalin, like OKT4A, inhibits in vitro HIV infection of CEM cells. We, however, brought evidence that jacalin does not prevent virus binding suggesting, as already described for certain anti-CD4 MAb *(22)*, that the inhibition of infection observed in the presence of jacalin is not owing to an inhibition of HIV binding.

1.3. Evaluation of CD4 as a Cell Signaling Receptor in Jacalin Stimulation

Because jacalin blocks in vitro HIV-1 infection of lymphoid cells without preventing virus binding, we hypothesized that jacalin could deliver signals inside the cell, which would impair the infection process. It is indeed well documented that cellular signaling is associated with gp120-CD4 interaction *(23–27)*. Several experiments have shown that cellular signals triggered through CD4 play a critical role in T-cells within the context of the CD3-T-cell receptor (TCR) activation pathway; however, it is conceivable that the CD4-binding lectin jacalin may induce or affect cellular signaling directly through this antigen. The experiments described here led us to demonstrate that jacalin can trigger a signal (increase of free intracellular calcium concentration) directly via CD4 in lymphoid cells independently of the CD3-TCR complex. The only example of such a stimulation via CD4 in the absence of coexpression of CD3-TcR was that described by Carrel et al. using a special antihuman CD4 antibody, B66 *(28)*.

The experiments reported here have been performed using Jurkat cell variants (T-cell hybridoma line), which have been immunoselected from several

HIV Inhibition by Binding to CD4 543

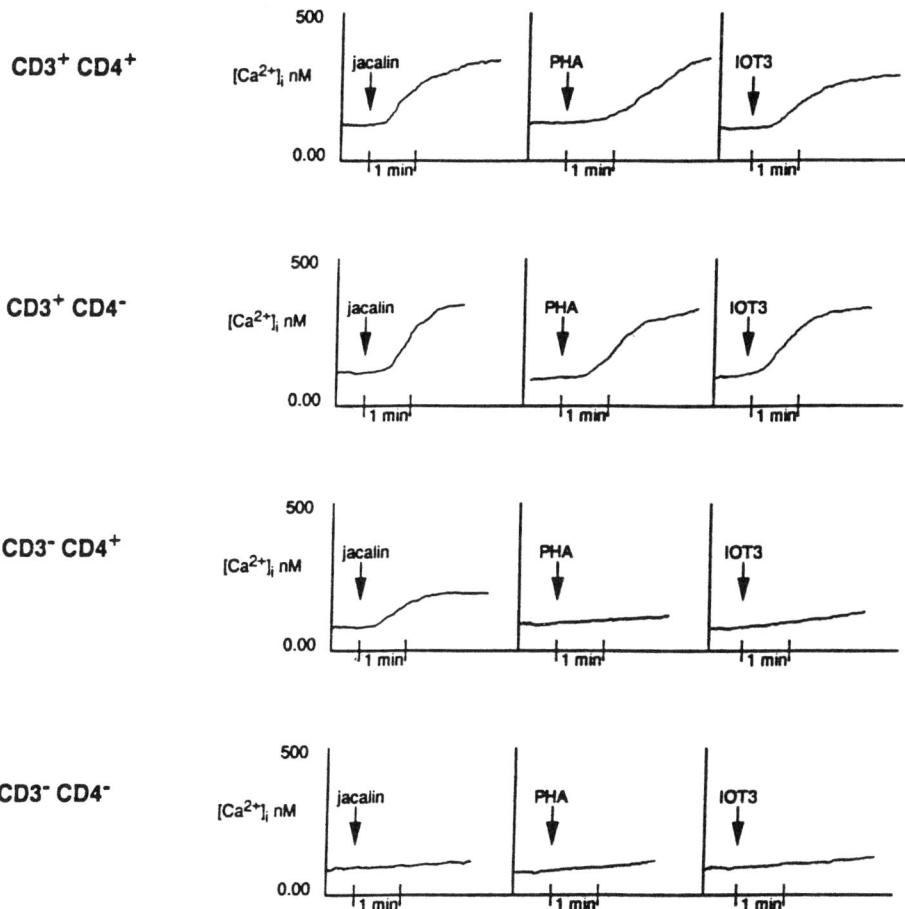

Fig. 2. Intracellular calcium concentration was measured in the four different jurkat variants on stimulation with jacalin, PHA or anti-CD3 MAb (IOT3). (Reprinted from ref. *19* with permission.)

laboratory strains in terms of their expression of CD3 and/or CD4. The immunoselection was achieved using magnetic beads coated with the corresponding monoclonal antibodies. Phenotype analysis of the resulting selected variants was checked by flow cytometry. The Jurkat variants were loaded with Fura-2AM (a dye commonly used for intracellular calcium measurement), and stimulated with jacalin, anti-CD3 MAb, or phytohemagglutinin (PHA); the respective induced calcium increase was measured in a spectrofluorometer. As shown in Fig. 2, a calcium increase was induced by the three mitogens in $CD3^+$ cells, whether CD4 was expressed or not. However, when CD3 was absent, a

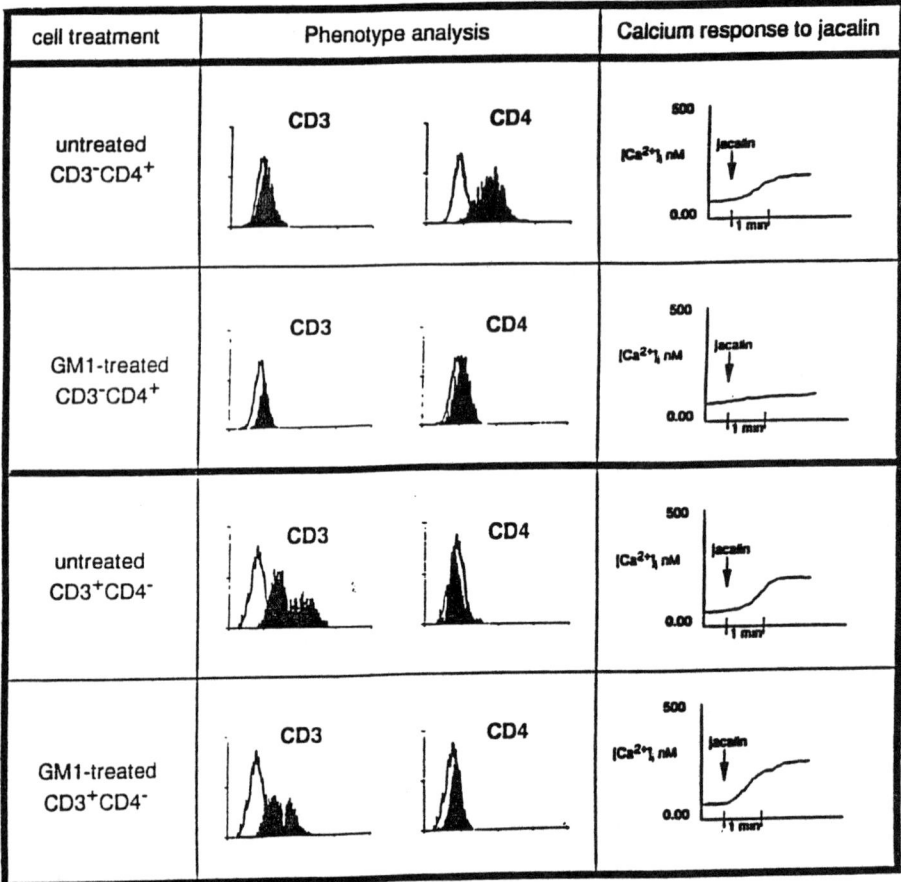

Fig. 3. Phenotype analysis and measurement of calcium response in CD3⁻CD4⁺ and CD3⁺CD4⁻ cells before and after GM1 treatment. (Reprinted from ref. *19* with permission.)

calcium rise was observed only on jacalin stimulation provided that CD4 was expressed. To determine whether jacalin is able to induce intracellular calcium increase via CD4 in cells lacking the CD3-TCR complex, CD4 was modulated from the surface of these cells using the monosialoganglioside GM1. Indeed, GM1 induces a specific internalization and degradation of CD4 *(29)*. Figure 3 shows that modulation of CD4 from CD3⁻CD4⁺ Jurkat cells abolishes the calcium response normally induced by jacalin. A control experiment shows that, GM1 does not modify the calcium response to jacalin in CD3⁺CD4⁻ cells ruling out the possibility of a direct GM1-induced inhibiting effect.

2. Materials
2.1. Analysis of Jacalin-CD4 Interaction

1. CEM cells (American Type Culture Collection, Rockville, MD).
2. Culture medium: 500 mL RPMI containing 10% heat inactivated (56°C, 45 min) fetal calf serum (FCS), 500 µL gentamicin (50 mg/mL), 5.5 mL glutamine (200 mM).
3. Lysis buffer (LB): 10 mM Tris-HCl, 140 mM NaCl, 0.025% NaN$_3$, pH 8.0, 1% Nonidet P40 (NP40), and 1 mM phenyl methyl sulfonyl fluoride (PMSF; *see* Note 1).
4. 0.01M Phosphate-buffered saline (PBS), pH 7.2.
5. Sample buffer (2X): 62 mM Tris-HCl pH 6.8 containing 2% sodium dodecyl sulfate (SDS), 10% glycerol, 5% 2-mercaptoethanol and bromophenol blue.
6. Nonspecific site saturation solutions: 5% dry milk in PBS or PBS-0.05% Tween-20 (*see* Note 2).
7. 10% Polyacrylamide gel.
8. 50 mM Tris-HCl, pH 6.8.
9. Nitrocellulose membrane or polyvinyl difluoride (PVDF) transfer membrane (slightly larger than the gels).
10. Electroblotting buffers: anode buffer 1: 0.3M Tris-HCl, 20% methanol, pH 10.4; anode buffer 2: 0.025M Tris-HCl, 20% methanol, pH 10.4; cathode buffer: 0.025M Tris-HCl, 0.04M glycine, 20% methanol, pH 9.4.
11. Ponceau solution (0.2%) in distilled water for nitrocellulose membrane staining (destained in distilled water) or Coomassie brilliant blue solution (227 mL methanol, 227 mL H$_2$O, 45 mL acetic acid, and 0.5 g Coomassie blue) for PVDF membrane staining and 45% MeOH/7% AcOH solution for destaining.
12. Protein A sepharose (50 mg/mL).
13. OKT4A (IgG2a), OKT4 (IgG2b) anti-CD4 MAb (Ortho Diagnostic, Raritan, NJ); biotinylated anti-CD4 MAb, horseradish peroxidase conjugated streptavidin (Immunotech, Marseilles, France).
14. Jacalin from Lectinola (Prague, Czech Republic) or Pierce (Rockville, MD).
15. Jacalin-coated agarose beads (Pierce).
16. ^{125}INa (100 mCi/mL); G25 sephadex columns (PD10).
17. Western blot chemiluminescence reagent (ECL).
18. Iodogen (1,3,4,6-tetrachloro-3α,6α diphenyl glycouril).
19. Dichloro-methane.
20. Kodak X-Omat AR films.

2.2. Effect of Jacalin on In Vitro HIV Infection of Human Lymphoid Cells

1. CEM cells, RPMI, containing fetal calf serum, glutamine, and gentamicin.
2. Jacalin, OKT4A.
3. HIV-1GER a virus isolated from an AIDS patient (supplied by Dr. Devaux, Montpelier, France) (solution from a chronically infected CEM cell supernatant).

4. Methyl-[^{3}H] deoxythymidine triphosphate (^{3}H-TTP), specific activity: 10–25 Ci/mmol (1 mCi/mL).
5. Poly(rA)·p(dT)$_{12-18}$ template primer (Pharmacia, Piscataway, NJ) (25 U/vial; make a 10 U/mL solution in universal buffer and store as 0.5-mL aliquots at –20°C. Aliquots can be frozen and thawed several times).
6. Poly(dA)·p(dT)$_{12-18}$ template primer (Pharmacia) (same solution as for Poly[rA]·p[dT]$_{12-18}$).
7. Yeast tRNA (Sigma, St. Louis, MO) (make a 5 mg/mL solution in universal buffer and store at –20°C).
8. Glass fiber filters (2.4 cm) (Millipore, Bedford, MA).
9. Buffer A: $1M$ Tris-HCl, pH 7.8 (12.5 mL), $0.1M$ EDTA (1.25 mL), 10% Triton X-100 (1.25 mL), glycerol (250 mL), dithiothreitol (DTT) (0.77 g), KCl (3.72 g), double-distilled water (ddH$_2$O) to 500 mL.
10. Solution 1: 10% Triton X-100 (45 mL), KCl (1.63 g), ddH$_2$O to 500 mL.
11. 10% Trichloroacetic acid (TCA) solution: TCA (50 g), sodium pyrophosphate (4.46 g), ddH$_2$O to 500 mL.
12. 5% TCA solution: TCA (100 g), sodium pyrophosphate (17.8 g), ddH$_2$O to 2 L.
13. Universal buffer: $0.01M$ Tris-HCl, pH 8, $0.015M$ NaCl.
14. Reaction mixture (rA cocktail per sample, 90 µL): $1.0M$ Tris-HCl, pH 7.8 (4 µL), $0.2M$ DDT in universal buffer (4 µL), $0.2M$ MgCl$_2$, ddH$_2$O (47 µL), universal buffer (22.5 µL), ^{3}H-TTP (2.5 µL), Poly(rA)·p(dT) (5 µL).
15. Reaction mixture (dA cocktail per sample, 90 µL): identical as rA cocktail with 5 µL of Poly(dA)·p(dT) instead of Poly(rA)·p(dT).

2.3. Evaluation of CD4 as a Cell Signaling Receptor in Jacalin Stimulation

1. Jurkat cells (Dr. Alcover, Institut Pasteur, Paris, France).
2. Mouse antihuman CD4 MAb-coated magnetic beads (*see* Note 3); goat antimouse IgG Ab-coated magnetic beads, A magnet, special reagent to detach cells from beads (Detachabead) (Dynal).
3. 2% Bovine serum albumin (BSA).
4. Culture medium: RPMI containing FCS, gentamicin, glutamine.
5. Phytohemagglutinin (PHA), anti-CD3 MAb, jacalin.
6. Monosialoganglioside (GM1).
7. Fura 2-AM (1 mg/mL in DMSO) (store protected from light), RPMI without phenol red.
8. FITC-conjugated anti-CD3 and anti-CD4 MAb.
9. 1% Paraformaldehyde solution in PBS (PFA).
10. Spectrofluorometer.

3. Methods

3.1. Analysis of Jacalin-CD4 Interaction

1. Culture CEM cells in medium, in 250-mL culture flasks, at 37°C in 5% CO$_2$ humidified atmosphere until a concentration of 10^6 cells/mL is reached. Centri-

fuge the CEM cells at 400g for 5 min, remove the supernatant, and resuspend the pellet (10^8 cells) in 1 mL prechilled (4°C) lysis buffer (LB).
2. Leave at 4°C for 1 h with occasional mixing.
3. Centrifuge for 10 min at 13,000g at 4°C and collect the supernatant.
4. Preclear the supernatant by adding swollen protein A sepharose CL4B (50 μL/mL) for 2 h at 4°C, then centrifuge at 400g for 5 min and remove the supernatant to a fresh tube. Treat the pellet as described in steps 7 and 8.
5. Add 100 μL of immobilized jacalin on beaded agarose to 1 mL supernatant and leave for 12 h at 4°C under gentle rocking agitation.
6. In parallel, add 10 μg of a mixture of IgG2a and IgG2b anti-CD4 MAb (OKT4A + OKT4) to 1 mL supernatant. After 2 h incubation at 4°C under gentle mixing, wash the cells and add 100 μL swollen protein A sepharose CL4B and leave for 12 h at 4°C under gentle rocking agitation.
7. Centrifuge both tubes and wash the pellets successively with LB, PBS pH 7.2, and 50 mM Tris-HCl buffer pH 6.8.
8. After centrifugation (400 g for 5 min) resuspend the pellet in 50 μL of sample buffer and heat at 100°C for 5 min. Centrifuge at 13,000g for 1 min and remove the supernatant to a fresh tube.
9. Prepare a 10% SDS-PAGE gel with enough lanes (20 lanes) to allow duplicates. The molecular weight standards will be run in the first lane (*see* Note 4); in the other lanes will be the lysate adsorbed onto jacalin-immobilized agarose (two lanes), the lysate incubated with protein A sepharose alone (nonspecific binding), and the immunoprecipitate with anti-CD4 antibody and protein A sepharose.
10. Electroblotting in Semidry Graphite Electrobloter system.
    a. Place the gel in cathode buffer for 5 min at room temperature.
    b. Wet PVDF membrane for 15 s in methanol then 15 s in deionized (milli-Q) water; then equilibrate the membrane in anode buffer no. 2 for 5 min.
    c. Soak two sheets of filter paper (Whatmann 3MM) in anode buffer no. 1. Place these two sheets on the graphite anode electrode plate.
    d. Soak one sheet of filter paper with anode buffer no. 2. Place this sheet on top of the two sheets already placed on the electrode plate.
    e. Place the transfer membrane equilibrated in anode buffer no. 2 on the three filter papers (manipulate the membrane with gloves).
    f. Place the equilibrated gel without allowing any air bubbles between the gel and the membrane (*see* Notes 5 and 6).
    g. Place three filter papers previously soaked in the cathode buffer on above the gel and recover with the cathode electrode plate.
    h. Electroblot for 1 h at 1.2 mA/cm^2.
    i. Peal off the transfer membrane with forceps then, to reveal the position of the bands, stain membrane with staining solution (Coomassie blue) for 5 min, and destain with the destaining solution.
11. Completely destain the membrane in 100% methanol. Nonspecific site saturation is then done in PBS containing 0.05% Tween-20 or milk (*see* Note 2), overnight at room temperature under gentle shaking agitation.

12. Incubate the membrane for 1 h in 10 mL PBS containing 150 μL of either ^{125}I-jacalin or ^{125}I-anti-CD4 MAb. Jacalin and anti-CD4 MAb were iodinated using the iodogen method (*see* step 13).
13. ^{125}I-labeling of proteins using the iodogen method (in restricted and protected area).
    a. Prepare 1 mL iodogen solution in CH_2Cl_2 (100 μg/mL).
    b. Pour 100 μL of the iodogen solution in glass hemolysis tube and evaporate solvent, allowing iodogen to coat the bottom inner side of the tube.
    c. In an iodogen-coated tube, add 25 μL of jacalin or antibody solution, 121 μL of PBS, and 4 μL of ^{125}INa (100 mCi/mL), leave for 30 min at room temperature with occasional mixing.
    d. During the reaction time, equilibrate a PD10 column with 25 mL PBS pH 7.2.
    e. After 30 min, put the reaction mixture on the column and eluate with 500 μL fractions of PBS. Each fraction is gathered in an hemolysis tube. The iodinated protein is generally eluted between tube 6 and tube 9 (check 2 μL in a γ-counter). The following radioactive elution peak represents residual ^{125}INa.
14. Wash the membrane liberally with PBS, then wash twice (5 min each wash) with PBS/0.1% Tween-20 and three times (5 min each) with PBS.
15. Dry the membrane and put with a Kodak X-OMAT-AR film in a cassette for at least 24 h exposure.
16. Develop the film and put a new one for a longer exposure if necessary.
17. The membrane can be revealed with biotinylated-anti-CD4 MAb instead of ^{125}I-anti-CD4 MAb. In that case:
    a. Incubate the membrane in PBS containing 1 μg/mL biotinylated anti-CD4 for 1 h at room temperature.
    b. After extensive washing with PBS (five times in 5 min incubation in PBS under gentle shaking agitation) incubate the membrane in 1/2000 dilution of horseradish peroxidase-conjugated streptavidin for 30 min at room temperature.
    c. After extensive washing (three times in PBS, one in PBS/0.1% Tween-20 and one time in PBS) incubate the membrane in western blot chemiluminescence reagent for 1 min and put the membrane between sealed transparent nylon foil.
    d. Put the membrane on an X-OMAT-AR film in a cassette and develop the film after a few minutes exposure (1–10 min).

3.2. Effect of Jacalin on In Vitro HIV Infection of Human Lymphoid Cells

These experiments must be conducted in a P3 facility using P3 practices and equipment.

3.2.1. Cell Infection and Culture

1. Prepare a 5×10^6 CEM cell suspension/mL in culture medium.
2. Put 100 μL of this cell suspension in wells of a flat-bottomed 96-microwell plate and add various concentrations of jacalin (0.1–10 μg/mL) or a saturating amount

HIV Inhibition by Binding to CD4

of anti-CD4 MAb OKT4A (2 µg/mL). Cells with no additives will be the control. Incubate for 1 h at 4°C. Each point is done in duplicate.
3. Expose the cell suspension for 1 h at 4°C to 100 µL of a diluted stock of HIV-1 from a chronically infected CEM cell supernatant corresponding to 100 $TCID_{50}$ (*see* Note 7). In cell culture, the TCID50 (50% tissue culture infective dose) is the dilution of virus suspension that infects 50% of the cell cultures as measured by visible virus-induced cytopathic effect (multinucleated giant cell formation-syncytium formation with ballooning cytoplasm).
4. Wash the cells five times. The cells are then cultured in 24-well culture plate (5×10^5 cells/2 mL culture medium) in the presence (or the absence) of the various abovementioned concentrations of jacalin or OKT4A.
5. Twice a week, 1 mL cell culture is removed from the culture wells for reverse transcriptase assay; the remaining cell suspension is washed (centrifugation at 400g for 5 min), counted, and resuspended (5×10^5 cells/2 mL) in culture medium containing the original additive concentrations.

3.2.2. Reverse Transcriptase Assay

1. Centrifuge the supernatants at 400g for 5 min.
2. Pellet virions in the clarified supernatants by ultracentrifugation at 100,000g for 2 h.
3. Resuspend pellet from 1 mL supernatant in 50 µL of Tris buffer, pH 7.8 (buffer A) and 25 µL of 10% Triton X-100 (solution 1) (*see* Note 8). The samples can be kept frozen at –20°C until assayed. From this step the virus is inactivated, and the following steps do not need to be strictly carried out in a biohazard hood.
4. Prepare sufficient volumes of rA and dA cocktails and distribute 90 µL of each cocktail in several Eppendorf tubes according to the number of samples to be assayed (1 rA and 1 dA for each sample), and place tubes in ice bath.
5. Add 10 µL of sample to rA and dA tubes. Mix and incubate for 1 h in a 37°C water bath.
6. Stop the reaction by adding 10 µL of yeast RNA followed by 1 mL cold 10% TCA. Vortex well and leave in ice bath for 20–30 min.
7. Soak glass fiber filters in 5% TCA before placing in a Millipore sampling manifold attached to a vacuum source. Put the content of each Eppendorf tube on a separate filter. Rinse each tube four times with 5% TCA and wash filters twice with 5% TCA. Dry filters by turning off vacuum and filling each well with 70% ethanol. After 15 s reapply vacuum.
8. Dry filters at 80–100°C for 15 min in an oven.
9. Count each filter in a liquid scintillation counter.
10. The results are calculated by subtracting dA counts per min (cpm) from rA (*see* Note 9) and multiplying the result by a coefficient to convert to net cpm incorporated per mL of original culture supernatant fluid (here, 10 µL [from the 75 µL prepared from 1 mL supernatant] were essayed, thus the coefficient will be 7.5).

3.3. Evaluation of CD4 as a Cell Signaling Receptor in Jacalin Stimulation

3.3.1. Immunoselection of Jurkat Variants

1. In a 15-mL tube resuspend the Jurkat cells to be immunoselected in term of CD4 expression at 10^7 cells/mL in PBS containing 2% BSA. Add 100 µL of anti-CD4-coated magnetic beads/mL and leave the cells for 1 h at 4°C under gentle shaking agitation.
2. Put the tube in a magnet and leave for 5 min to allow beads bound to CD4$^+$ cells to gather on the inner side part of the tube against the magnet.
3. Remove medium containing free cells and add new medium (PBS/BSA). Resuspend the beads in the medium and act as in step 2. This operation is performed five times.
4. After the last wash, resuspend the beads in PBS/BSA and add 50 µL of "detachabead" solution/mL. Leave the cells at room temperature for 1 h under gentle shaking agitation.
5. Remove the free beads with the magnet and remove cell suspension to a new tube.
6. Wash the cells three times with RPMI-1640 and incubate the cells in culture medium (*see* Note 10).
7. To immunoselect the cells in term of CD3 expression:
   a. Centrifuge the cells and resuspend at 10^6 cells/100 µL RPMI.
   b. Add 0.5 µg anti-CD3 MAb and leave 30 min at 4°C with occasional mixing.
   c. Wash the cells with RPMI and treat the cells as in steps 1–6 using antimouse IgG-coated magnetic beads instead of CD4-coated beads.
8. Label the immunoselected cells by adding 0.5 µg/mL anti-CD3 or anti-CD4 MAb and leave 30 min at 4°C with occasional mixing.
9. After washing, resuspend the pellet and fix the cells in 1% PFA. Check the presence of the corresponding antigen (CD3 or CD4) by flow cytometry.

3.3.2. Measurement of Intracellular Calcium Increase in Stimulated Cells

1. Centrifuge the immunoselected purified Jurkat cells and resuspend the cell pellet in a 15-mL tube at 5×10^6 cells/mL in RPMI-1640 not containing phenol red. Treat a maximum of 15×10^6 cells.
2. Add 5 µL Fura-2AM/mL and incubate the cells at 37°C for 30 min with occasional mixing.
3. Wash the cells with RPMI-1640 without phenol red and resuspend the cells at 10^6 cells/mL.
4. Use a spectrofluorimeter set at 340 and 380 nm excitation wavelength and 560 nm emission wavelength.
5. Evaluate Ca^{2+} increase by adding in the cuvet containing 2 mL cell suspension, 10 µL PHA, or 20 µL Jacalin (1 mg/mL) or 20 µL anti-CD3 MAb (20 µg/mL).

3.3.3. GM1 Treatment of the Cells

1. Incubate the cells (10^6 cells/mL) at 37°C for 4 h with 100 μg/mL GM1 in RPMI-1640 without FCS (*see* Note 11).
2. Wash the cells three times with RPMI-1640.
3. Check the resulting phenotype by flow cytometry.
4. Measure Ca^{2+} increase in GM1 treated cells on stimulation with jacalin or anti-CD3 MAb as described in Section 3.3.2.

4. Notes

1. Prepare PMSF as a 100 mM stock solution in ethanol. Add PMSF from the stock solution and NP-40 extemporaneously in the TBS buffer.
2. Do not saturate the blotting membrane with milk when it has to be revealed with jacalin (it contains glycoconjugates which are be bound by this lectin); use preferentially PBS/Tween-20. In contrast, when it has to be revealed with anti-CD4 MAb, it is better to use milk. Tween-20 may impair CD4/anti-CD4 interaction.
3. Use these beads for positive or negative cell selection.
4. It is better to leave one or two empty wells on each side of the gel to avoid a "smiling" migration.
5. Remove air bubbles every time you put a sheet (paper filter or membrane or gel) with a roll.
6. Cut a small piece of gel and membrane at the same corner to mark the orientation of the membrane with respect to the gel.
7. In cell culture, the TCID50 (50% tissue culture infective dose) is the dilution of virus suspension that infects 50% of the cell cultures as measured by visible virus-induced cytopathic effect (multinucleated giant cell formation-syncytium formation with ballooning cytoplasm).
8. Resuspend pellet by placing a Pasteur pipet in the tube and vortexing. The Pasteur pipet is necessary to break up the hard pellet.
9. The samples can be kept frozen at –20°C until assayed. From this step the virus is inactivated, and the following steps do not need to be strictly carried out in a biohazard hood.
10. Each sample yields a cpm value for both the poly rA oligo $(dT)_{12-18}$ template primer and the poly dA oligo$(dT)1_{2-18}$ template primer. The dA value (which reflects nonspecific cellular polymerase activity) is then subtracted from the rA value (which reflects HIV-specific RT polymerase activity).
11. After immunoselection, leave the cell in culture medium for one night before calcium assay.
12. It is important to remove the FCS from culture medium before GM1 treatment.

References

1. Dalgleish, A. G., Beverly, P. C. L., Clapham, P. R., Crawford, D. H., Greeves, M. F., and Weiss, R. A. (1984) The CD4 (T4) antigen is an essential component of the receptor for the AIDS retrovirus. *Nature* **312,** 763–766.

2. Sattentau, Q. J. and Weiss, R. A. (1988) The CD4 antigen: physiological ligand and HIV receptor. *Cell* **52**, 631–633.
3. Klatzmann, D., McDougal, J. S., and Maddon, P. J. (1990) The CD4 molecule and HIV-1 infection. *Immunodeficiency Rev.* **2**, 43–66.
4. Lasky, L. A., Groopman, J. E., Fennie C. W., Benz, P. M., Capon, D. J., Dowbenko, D. J., Nakamura, G. R., Nunes, W. M., Renz, M. E., and Berman, P. W. (1986) Neutralization of the AIDS retrovirus by antibodies to a recombinant envelope glycoprotein. *Science* **233**, 209–212.
5. Leonard, C. K., Spellman, M. W., Riddle, L., Harris, R. J., Thomas, J. N., and Gregory, T. J. (1990) Assignment of intrachain disulfide bonds and characterization of potential glycosylation sites of type 1 recombinant human immunodeficiency virus envelope glycoprotein (gp120) expressed in Chinese hamster ovary cells. *J. Biol. Chem.* **265**, 10,373–10,383.
6. Feizi, T. and Larkin, M. (1990) AIDS and glycosylation. *Glycobiology* **1**, 17–23.
7. Robinson, W. E., Jr., Montefiori, D. C., and Mitchell, W. M. (1987) Evidence that mannosyl residues are involved in human immunodeficiency virus type 1 (HIV-1) pathogenesis. *Aids Res. Hum. Retrovir.* **3**, 265–282.
8. Fenouillet, E., Gluckman, J. C., and Barhaoui, E. (1990) Role of N-linked glycans of envelope glycoproteins in infectivity of human immunodeficiency virus type 1. *J. Virol.* **64**, 2841–2848.
9. Dirck, L., Lindemann, D., Ette, R. Manzoni, C., Moritz, D., and Mous, J. (1990) Mutation of conserved N-glycosylation sites around the CD4-binding site of human immunodeficiency virus type 1 gp120 affects viral infectivity. *Virus Res.* **18**, 9–20.
10. Favero, J. (1994) Lectins in *AIDS* research. *Glycobiology* **4**, 387–396.
11. Carr, S. A., Hemmling, M. E., Folena-Wasserman, G., Sweet, R. W., Anumula, K., Barr, J. R., Huddleston, M. J., and Taylor, P. (1989) Protein and carbohydrate structural analysis of a recombinant soluble CD4 receptor by mass spectrometry. *J. Biol. Chem.* **264**, 21,286–21,295.
12. Levy, J. A. (1993) Pathogenesis of human immunodeficiency virus infection. *Microbiol. Rev.* **57**, 183–289.
13. Lifson, J., Coutré, S., Huang, E., and Engleman, E. (1986) Role of envelope glycoprotein carbohydrate in human immunodeficiency virus (HIV) infectivity and virus-induced cell fusion. *J. Exp. Med.* **164**, 2101–2106.
14. Hansen, J. E. S., Nielsen, C. M., Nielsen, C., Heegaard, P., Tathieson, L. R., and Nielsen, J. O. (1989) Correlation between carbohydrate structures on the envelope glycoprotein gp120 of HIV-1 and HIV-2 and syncytium inhibition with lectins. *AIDS* **3**, 635–641.
15. Fenouillet, E., Clerget-Raslain, B., Gluckman, J. C., Guetard, D., Montagnier, L., and Bahraoui, E. (1989) Role of N-linked glycans in the interaction between the envelope glycoprotein of human immunodeficiency virus and its CD4 cellular receptor. *J. Exp. Med.* **169**, 807–822.
16. Pineau, N., Aucouturier, P., Brugier, J. C., and Preud'homme, J. L. (1990) Jacalin: a lectin mitogenic for human CD4 T-lymphocytes. *Clin. Exp. Immunol.* **80**, 420–425.
17. Favero, J., Corbeau, P., Nicolas, M., Benkirane, M., Travé, G., Dixon, J. F. P., Aucouturier, P., Rasheed, S., Parker, J. W., Liautard, J. P., Devaux, C., and

Dornand, J. (1993) Inhibition of human immunodeficiency virus infection by the lectin jacalin and by a derived peptide showing a sequence similarity with gp120. *Eur. J. Immunol.* **23,** 179–185.
18. Taimi, M., Dornand, J., Nicolas, M., Marti, J., and Favero, J. (1994) Involvement of CD4 in interleukin-6 secretion by U937 monocytic cells stimulated with the lectin jacalin. *J. Leuk. Biol.* **5,** 214–220.
19. Lafont, V., Dornand, J., Dupuy d'Angeac A., Monier, S., Alcover, A., and Favero, J. (1994) Jacalin, a lectin that inhibits in vitro HIV-1 infection, induces intracellular calcium increase via CD4 in cells lacking the CD3/TcR complex. *J. Leuk. Biol.* **56,** 521–524.
20. Sastry, M. V. K., Banarjee, P., Patanjali, S. R., Swamy, M. J., Swarnalatha, G. V., and Surolia, A. (1986) Analysis of saccharide binding to Artocarpus integrifolia lectin reveals specific recognition of T-antigen (β-D-Gal($1\rightarrow 3$)D-GalNac). *J. Biol. Chem.* **261,** 11,726–11,733.
21. Corbeau, P., Devaux, C., Kourilsky, F., and Chermann, J. C. (1990) An early postinfection signal mediated by monoclonal anti-$\beta 2$ microglobulin antibody is responsible for delayed production of human immunodeficiency virus type 1 in peripheral blood mononuclear cells. *J. Virol.* **64,** 1459–1464.
22. Hasunuma, T., Tsubota, H., Watanabe, M., Chen, Z. W., Lord, C. I., Burkly, L. C. Daley, J. F., and Letvin, N. L. (1992) Regions of the CD4 molecule not involved in virus binding or syncytia formation are required for HIV-1 infection of lymphocytes. *J. Immunol.* **148,** 1841–1846.
23. Juszczak, R. J., Turchin, H., Truneh, A., Culp, J., and Kassis, S. (1991) Effect of human immunodeficiency virus gp120 glycoprotein on the association of the protein tyrosine kinase p56lck with CD4 in human T-lymphocytes. *J. Biol. Chem.* **266,** 11,176–11,183.
24. Cohen, D. I., Tani, Y., Tian, H., Boone, E., Samelson, L. E., and Lane, H. C. (1992) participation of tyrosine phosphorylation in the cytopathic effect of human immunodeficiency virus-1. *Science* **256,** 542–545.
25. Yoshida, H., Koga, Y., Moroi, Y., Kimura, G., and Nomoto, K. (1992) The effect of P56lck, a lymphocyte specific protein kinase, on syncytium formation induced by human immunodeficiency virus envelope glycoprotein. *Internal. Immunol.* **4,** 233–242.
26. Soula, M., Fagard, R., and Fisher, S. (1992) Interaction of human immunodeficiency virus glycoprotein gp160 with CD4 in jurkat cells increases p56lck autophosphorylation and kinase activity. *Int. Immunol.* **4,** 295–299.
27. Tremblay, M., Meloche, S., Graton, S., Wainberg, M. A., and Sekaly, R. P. (1994) Association of p56lck with the cytoplasmic domain of CD4 modulates HIV-1 expression. *EMBO J.* **13,** 774–783.
28. Carrel, S., Salvi, S., Gallay, P., Rapin, C., and Sekaly, R. P. (1991) Positive signal transduction via surface CD4 molecules does not need coexpression of the CD3/TcR complex. *Res. Immunol.* **142,** 97–108.
29. Saggioro, D., Sorio C., Calderazzo, F., Callegaro, L., Panozzo, M., Berton, G., and Chieco-Bianchi, L. (1993) Mechanism of action of the monosialoganglioside GM1 as a modulator of CD4 expression. *J. Biol. Chem.* **268,** 1368–1375.

47

Inhibition of HIV Infection by Lectin Binding to gp120

Theresa Animashaun and Naheed Mahmood

1. Introduction

Human immunodeficiency virus (HIV) is the causative agent of AIDS (acquired immunodeficiency syndrome). The polypeptide precursor gp160 of HIV-1 forms the external glycoprotein, gp120 and the transmembrane glycoprotein, gp41 *(1)*. Sequence variability is a feature of HIV viruses that have been classified into several subtypes *(2)*. There are 22–31 potential N-linked glycosylation sites on gp120 depending on the HIV-1 isolate and thus, approximately half of its molecular weight is composed of carbohydrate. Gp120 oligosaccharides are a mixture of high mannose-, hybrid-, and complex-type N-glycans *(3–5)*. The proportion of these N-glycan substituents on the envelope glycoprotein varies on different HIV-2 isolates propagated in different cell lines *(6)*. The less-processed oligosaccharides are primarily located on conserved N-linked glycosylation sites on the recombinant gp120s produced in CHO cells and by a baculovirus expression system *(4,7)*. This points to the high mannose and hybrid-type oligosaccharides as being important in the structure and/or function of the envelope glycoprotein.

1.1. Screening of Lectins as Antiviral Agents and Examination of Their Mode of Action

Lectins have been examined for their potential as antiviral agents and to shed light on the mechanism of HIV pathogenesis *(8–10)* (Table 1). Early experiments by Lifson et al. *(8)* established that the mannose binding lectin, Con A, inhibited HIV-1 induced cell fusion but at relatively high concentrations. Some mannose binding lectins with different oligosaccharide specifici-

Table 1
Effect of Lectins at 10 μg/mL on Cell Viability and Syncytium Formation of C8166 Cells Infected with HIV-1

Plant	Lectin[a]	Syncytia	Cell viability % control	
			Infected	Uninfected
Mannose-specific lectins				
Machaerium biovulatum	MBA	–[b]	79	75
Machaerium lunatus	MLA	–[b]	88	88
Galanthus nivalis	GNA	–	100	100
Bowringia mildbraedii	BMA	–	100	100
Canavalia ensiformis	ConA	–	100	100
Lablab niger	LNA	–	100	100
Narcissis lobularis	NLA	±	58	100
Vigna racemosa	VRA	±[b]	48	100
Pterocarpus rhorii	PRA	±	55	100
Galactose-specific lectins				
Tetracarpidium conophorum	TCA	toxic	25	30
Sphenostylis stenocarpa	SSA	+	24	100
Erythrina senegalensis	ESA	+	33	100
Artocarpus altilis	AAA	+	22	100
AZT control	–	–	100	100
Control	–	+	21	100

[a]Each lectin is called after the plant from which it is purified, e.g., MBA (*Machaerium biovulatum* agglutinin) was isolated from *M. biovulatum* seeds *(9)*. The other lectins MLA, PRA, SSA, ESA *(9)*, BMA *(15)*, GNA *(13)*, LNA, VRA *(16)*, TCA *(17)*, AAA *(18)*, and NLA *(19)* were similarly named.

[b]Control cells were fused.

ties have been shown to be effective at much lower concentrations *(9,11)*. However, lectins binding to galactose residues on *N*-glycans are ineffective at halting the progress of HIV-1 infection of cells *(9)*. The mannose-specific lectins MBA and MLA are the most potent reported so far. Maximum protection against HIV-1 infection is obtained at low concentrations of MBA when the virus is preincubated with lectin before addition to the cells suggesting that MBA acts by binding to envelope glycoprotein *(9)* (Fig. 1). The following protocols explain how the potential of lectins as inhibitors of HIV infectivity can be examined. The antiviral activity and cytotoxicity of the lectins is assessed in the XTT-Formazan assay and the cells are also examined microscopically. Further confirmatory experiments can be conducted, e.g., measurement of progeny virus production and p24 antigen levels *(12)*.

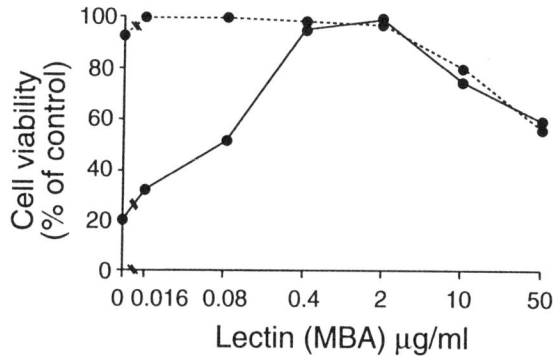

Fig. 1. Anti-HIV-1 activity of MBA: HIV was preincubated with a range of concentrations of MBA. After 30 min, the mixture was added to C8166 cells and the cell viability in infected (—) and uninfected cells (·····) was determined as described in Section 3.1.

1.2. Determination of gp120 Captured by GNA on ELISA Plates

Envelope glycoprotein, gp120, can be determined in the cell culture fluids of infected cells by an ELISA method in which the plates are coated with GNA to capture gp120. Most mannose binding lectins recognize numerous serum and cellular glycoproteins. However, GNA binds to a restricted number of glycoproteins bearing terminal mannose (α 1→3) mannose units *(13)*. Thus, when GNA is coated onto ELISA plates, it efficiently captures gp120 *(14)*, which is not diluted by the binding of other glycoproteins.

1.3. Binding of gp120 to Lectin–Agarose Matrices

In order to ascertain whether gp120 binds directly to a lectin, the viral lysate is mixed with lectin coupled to a matrix. The gp120 bound to lectin-matrix is detected by Western blotting. This protocol can also be performed to decide on the suitability of a lectin-affinity matrix for glycoprotein purification.

2. Materials

2.1. Screening of Lectins as Antiviral Agents and Examination of Their Mode of Action

1. Containment level 3 laboratory fully equipped with Class I cabinets.
2. HIV viral stock (MRC ADP Reagent Repository, Herts, UK), e.g., HIV-1$_{IIIB}$ or HIV-1$_{MN}$ supernatant from clinically infected H9 cells (*see* Note 1).
3. C8166 cells (*see* Note 2).
4. RPMI 10%: RPMI-1640 (ICN Flow, Thane, UK) containing 10% fetal calf serum, 2m*M* glutamine, 100 U/mL penicillin, 100 µg/mL streptomycin and kanamycin.

5. 96-well microtiter tissue culture plates.
6. Lectin stocks at 1 mg/mL stored at –20°C (*see* Note 3).
7. Azido thymidine (AZT).
8. 1 mg/mL XTT (2,3-bis[2-methoxy-4-nitro5-sulfophenyl]-5-[phenylamino carbonyl]-2H-tetrazolium hydroxide) and 0.02 mM N-methylphenazonium methosulfate in water (*see* Note 4).

2.2. Determination of gp120 Captured by GNA on ELISA Plates

1. Supernatant from d 5 cultures (*see* Section 3.1., step 8).
2. GNA at 10 μg/mL in PBS, freshly diluted.
3. RPMI 10%.
4. RPMI 10% Empigen: RPMI 10% containing 0.5% Empigen BB, Alyl dimethyl betaine (Calbiochem, Beeston, UK).
5. RPMI 10% Tween: RPMI 10% containing 0.1% Tween-20 (polyoxyethylene-sorbitan monolaurate).
6. PBST: PBS containing 0.1% Tween-20.
7. Antihuman Ig-horseradish peroxidase-conjugated sera (Amersham, Amersham, UK).
8. Human anti-HIV sera, pooled from three patients (a gift from Professor Karpas, Cambridge).
9. OPD, O-phenylenediamine, 1 mg/mL in 0.1M sodium citrate buffer pH 5.0, containing 0.03% hydrogen peroxide.
10. 1M sulfuric acid.

2.3. Binding of gp120 to Lectin–Agarose Matrices

1. H9 cells (1×10^7) uninfected and chronically infected with HIV-1.
2. Laemmli buffer: 1 mL 0.5M Tris-HCl, pH 6.8, 0.8 mL glycerol, 1.6 mL 10% SDS, 0.4 mL β-mercaptoethanol, 0.2 mL 0.05% (w/v) bromophenol blue.
3. Triton-X100.
4. Buffer A: 10 mM Tris-HCl, pH 7.5, 0.1M NaCl, 1 mM MgCl$_2$, 1 mM CaCl$_2$, 0.02% sodium axide.
5. Lectin–agarose beads (e.g., Vector Labs, Burlingame, CA; or Sigma, St. Louis, MO).
6. 7.5% SDS-PAGE gels prestained protein molecular weight markers.
7. Nitrocellulose filters of 0.45-μm porosity; Whatman 3MM paper.
8. Transfer buffer: 25 mM Tris-HCl, 192 mM glycine, 0.1% SDS, 20% methanol at pH 8.3.
9. PBST: PBS containing 0.05% Tween-20.
10. Marvel-PBST: 5% Marvel (nonfat dried milk) from the local supermarket in PBST.
11. Sheep anti-gp120: ADP 401 from Cell-tech donated by the MRC AIDS Reagent Repository.
12. Antisheep IgG alkaline phosphatase conjugate (Sigma).
13. Stock-solutions for alkaline phosphatase detection:
    a. NBT: Dissolve 0.5 g of NBT in 10 mL of 70% dimethylformamide (nitroblue tetrazolium).

b. BCIP: Dissolve 0.5 g of BCIP (bromochloroindolyl phosphate disodium salt) in 10 mL of 100% dimethylformamide.
c. Alkaline phosphatase buffer: 100 mM NaCl, 5 mM MgCl$_2$, 100 mM Tris-HCl, pH 9.5. All stocks are stable at 4°C for at least 1 yr.

3. Methods

3.1. Screening of Lectins as Antiviral Agents and Examination of Their Mode of Action

1. Make six serial fivefold dilutions of each lectin in 50 μL of RPMI–10% in microtiter plate wells using a multichannel pipet. The initial concentration of each lectin is 200 μg/mL to give a final concentration of 100 μg/mL. Also include a positive control of AZT. Use only the central 60-wells and fill the outer row wells with 50 μL of medium only.
2. Add 25 μL of a dilution of virus HIV-1$_{111B}$ or HIV-1$_{MN}$, supernatant from chronically infected H9s (clarified by centrifugation at 170g for 5 min) to a triplicate set of dilutions. The dilution of virus which kills 70–80% infected cells in 5 d is predetermined.
3. Add 25 μL of medium to a triplicate set of lectin dilutions to measure cytotoxicity. Include untreated wells as controls for both infected and uninfected cells. Leave for 30 min.
4. Centrifuge C8166 cells at 170g for 5 min and resuspend in fresh medium at a concentration of 1.6×10^6/mL.
5. Add 25 μL of the resuspended cells to each well.
6. Incubate the plates at 37°C in humidified boxes with 5% CO$_2$.
7. Examine the cell cultures for evidence of virus-induced cytopathic effects 2–3 d after infection (*see* Table 1).
8. Add 25 μL of a mixture of XTT and N-methylphenazonium methosulfate to infected and uninfected cultures 5 d postinfection.
9. Incubate plates for 3–4 h and replace plastic covers with adhesive plate sealers.
10. Agitate the mixture using a plate shaker for 1 h.
11. Read the optical densities at test wavelength of 450 nm and a reference wavelength of 650 nm.
12. The data from the plate reader is collected via a remote link on a computer located outside the containment laboratory. The proportions of viable cells, both infected and uninfected are readily calculated using a spread sheet (e.g., Lotus 123 or Excel), in order to assess antiviral activity and cytotoxicity. The data are expressed as the percentage of the untreated uninfected control. The results of cell viability and syncytia formation for several lectins tested at 10 μg/mL are presented in Table 1 (*see* Notes 6 and 7). The protective effect of the mannose binding lectin, MBA, at low concentrations against HIV-1 infection is illustrated in Fig. 1.

3.2. Determination of gp120 Captured by GNA on ELISA Plates

1. Coat microtiter plate (Falcon, Los Angeles, CA) wells with 50 μL GNA at room temperature overnight.

2. Wash three times with PBS.
3. Add 100 µL RPMI 10% to block at room temperature for 1–6 h until used.
4. Repeat step 2.
5. Add 25 µL of RPMI 10% Empigen to all wells.
6. Add 25 µL of supernatant from lectin-treated wells and 25, 12.5, and 6.25 µL from untreated wells for standard curve (1:2, 1:4, and 1:8 dilutions are made by adding 12.5 and 18.75 µL of RPMI 10% to second and third wells).
7. Incubate at 37°C for 3 h or at 4°C overnight.
8. Remove unbound antigen by washing three times with PBST.
9. Add 50 µL of human anti-HIV sera (1:200) in diluted RPMI 10%-Tween, incubate at 37°C for 4 h.
10. Wash three times with PBST.
11. Add 50 µL of a 10^3 dilution of antihuman Ig peroxidase conjugate in RPMI 10%-Tween.
12. Incubate at 37°C for 90 min.
13. Repeat step 10.
14. Add 50 µL of OPD solution.
15. Stop the reaction as soon as sufficient color is produced by the addition of 50 µL of $1M$ sulfuric acid.
16. Measure the absorbance at 492 nm in an ELISA plate reader.
17. The amount of gp120 is calculated from linear logarithmic plots, e.g., using three dilutions of untreated infected cultures as standards (using WIACALC, Pharmacia LKB, Sweden).

3.3. Binding of gp120 to Lectin–Agarose Matrices

1. Spin down uninfected and chronically infected H9 cells (1×10^7). Resuspend the cell pellet in 150 µL Laemmli buffer containing 0.1% Triton-X100 and boil for 5 min. The treated lysate may now be removed from the Containment level 3 laboratory.
2. Make the lysate to 1 mL with buffer A.
3. Prepare one tube of lectin-sepharose beads for each sample. Place a suspension of lectin-sepharose beads (80 µL) in an eppendorf tube and add buffer A to 1.5 mL. Vortex the tube, spin for 30 s, and aspirate the buffer. Wash the beads twice more with buffer A.
4. Add 100 µL diluted cell lysate to the washed lectin beads and mix on a rotator at 4°C overnight.
5. Spin the beads 1 min at 4°C. Remove approx 50 µL of the supernatant (nonbound material) and save (*see* Note 8).
6. Add 1.4 mL buffer A to the beads which are mixed on the vortex, then spin down and aspirate the buffer. Repeat this washing process four more times.
7. Suspend the washed beads in 100 µL Laemmli buffer and boil for 5 min. Spin down the beads and load 30 µL supernatant onto a track of a 7.5% SDS-PAGE gel. Also boil a 10-µL sample of infected and uninfected cell lysate (stage 2) with Laemmli buffer (20 µL). Treat 10 µL of prestained protein molecular markers similarly.

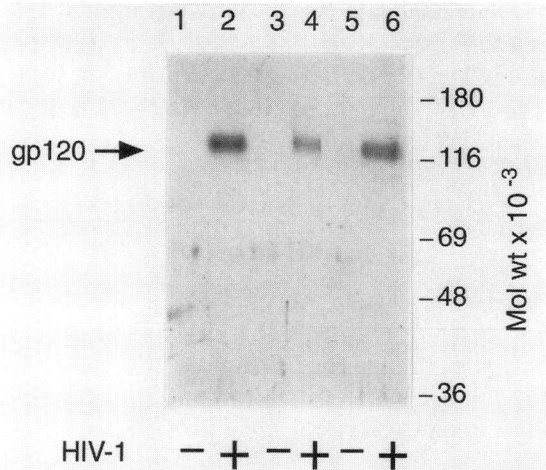

Fig. 2. Binding of mannose-specific lectins to gp120. Lysates were prepared from infected (lanes 2, 4, 6) and uninfected (lanes 1, 3, 5) H9 cells. The glycoproteins bound to GNA-agarose (lanes 1, 2) or MBA-sepharose (lanes 3, 4) were run on SDS-PAGE, blotted, and detected with anti-gp120. *See* Section 3.3.

8. When the electrophoresis is completed, transfer the gel to nitrocellulose membrane using a semidry blotter. Snip off the top left-hand corner of the membrane.
9. Block protein binding sites on the nitrocellulose sheet with Marvel PBST at 4°C overnight (*see* Note 9).
10. Wash twice with PBST at room temperature.
11. Incubate the membrane for 2 h at room temperature with sheep anti-gp120 (500-fold dilution in Marvel-PBST; *see* Note 10).
12. Incubate the nitrocellulose sheet for 1 h with antisheep immunoglobulin alkaline phosphatase conjugate (500-fold dilution in Marvel-PBST; *see* Note 11).
13. Wash the blot for 4 × 5 min with PBST.
14. Prepare fresh substrate solution just prior to developing the blot. Add 66 µL of NBT stock to 10 mL of alkaline phosphatase buffer. Mix well, and add 33 µL of BCIP stock.
15. Add fresh substrate solution to the washed blot. Develop the blot at room temperature with agitation until the bands are suitably dark (*see* Note 12).
16. Stop the reaction by rinsing with PBS. The binding of gp120 to GNA agarose and MBA-Sepharose beads is shown in Fig. 2.

4. Notes

1. Although C8166 cells are sensitive to infection by most HIV-1, HIV-2, and SIV strains, HIV-1$_{111B}$ or HIV-1$_{MN}$ are preferentially used for primary screening. The virus stocks are maintained in the culture supernatant of the chronically infected H9 cell line. The titer is determined in C8166 cells.

2. It is extremely important to use cells that are sensitive to HIV infection and its lytic effect. C8166 cells are human T-lymphoblastoid cells which are ideal for these experiments. Infected C8166 cells produce large, visible syncytia with a ballooning effect resulting in cell death within 4–6 d of infection. All cultures should be passaged twice a week in RPMI 10%.
3. Lectins are commercially available (e.g., from Vector, Peterborough, UK) but they can also be prepared from seeds and other plant materials (*see* Table 1) *(9)*.
4. The soluble-Formazan assay is based on the metabolic reduction of XTT (2,3-bis[2-methoxy-4-nitro-5-sulfophenyl]-5-[{phenylamino}carbonyl]-2H-tetrazolium hydroxide) to a soluble brown product (XTT-Formazan) in surviving cells *(20)*; it replaces 3-(4,5-dimethylthiazol-2-yl)-2,5-di-Phenyl-tetrazolium bromide (MTT), which has been used routinely in our laboratory. The MTT-Formazan product is insoluble and requires vigorous and time-consuming mixing with isopropanol and Triton X-100 prior to colorimetric determination. Particularly in the case of adherent cell lines, the XTT-Formazan assay is easier to perform.
5. The importance of lectin-only controls is shown here. TCA is toxic to cells at 10 µg/mL (Table 1), so the decrease in virus at higher concentrations is due to the toxicity of the lectin. Occasionally the cells develop an abnormal appearance in the presence of lectin, e.g., the lectin VRA fused the cells (Table 1).
6. Selected active lectins can also be tested against SIV in C8166 cells and HIV-2 in MT2 cells (HIV-2 produces a better cytopathic effect in MT2 cells). The isolation and maintenance of either peripheral blood lymphocytes or peripheral blood monocytes/macrophages is costly and labor intensive so will only be considered for selected lectins *(12)*.
7. To examine the mode of action of the inhibitory lectins, the order of addition of the original components is altered. In one alternative protocol, the cells are added to the lectin (i.e., reverse steps 4 and 5 with step 2) and left for 30 min. Virus is then added and the procedure continued. We observed that the inhibitory effect of MBA was reduced using this protocol, suggesting the lectin inhibits by binding to virus rather than cells *(9)*. The mechanism of action of GNA may differ from that of MBA *(21)*.
8. Care should be taken not to remove any of the beads. Subsequent extensive washing can compensate for inefficient removing of the supernatant at the first stage.
9. The blocking time can be reduced to 1 h at room temperature.
10. Several different detecting antibodies are available, e.g., MAb 221 (MRC AIDS Reagent Repository) diluted 1:1500 also gives clear bands with gp120. The dilution of antibody may be increased if there is evidence of nonspecific results and decreased if the signal is too weak.
11. Anti-immunoglobulin-hydrogen peroxidase can be used instead of the alkaline-phosphatase-conjugate. The ECL™ kit supplied by Amersham is based on a luminescent technique, and provides sensitive results with hydrogen-peroxidase. After stage 13 the ECL™ reagents are mixed and incubated with the blot for 1 min. The blot is wrapped in Saran Wrap and placed in a X-ray film cassette. A 1-cm strip of Tracker Tape™ is placed alongside the blot to line up the blot on

the film. In a dark room, a sheet of Hyperfilm™-ECL is placed on the blot for 1 min. The film is developed and meanwhile another sheet of film is placed on the blot for between 1 min and 1 h according to the amount of signal seen on the first film. This method provides a hard copy of the result.
12. Gp120 is cleaved in the V3 loop to a 70K (amino-terminus) and to a 50K fragment, the carboxy-terminus. Polyclonal antibodies will bind to both these fragments, whereas monoclonals (unless they recognize the cleavage region in the V3 loop) will only bind to the whole molecule and a fragment of gp120.

References

1. Earl, P. K., Dos, R. W., and Moss, B. (1990) Oligomeric structure of the human immunodeficiency virus type I envelope glycoprotein. *Proc. Natl. Acad. Sci USA* **87**, 648–652.
2. Sharp, P. M., Robertson, D. L., Gao, F., and Hahn, B. H. (1994) Origins and diversity of human immunodeficiency viruses. *AIDS* **8(suppl. 1)**, S27–S42.
3. Mizuochi, T., Spellman, M. W., Larkin, M., Solomon, J., Basa, L. J., and Feizi,T. (1988) Carbohydrate structures of the human-immunodeficiency-virus (HIV) recombinant envelope glycoprotein gp120 produced in Chinese hamster ovary cells. *Biochem J.* **254**, 599–603.
4. Leonard, C. K., Spellman, M. W., Riddle, L., Harris, R. J., Thomas, J. N., and Gregory, T. J. (1990) Assignment of intrachain disulphide bonds and characterization of potential glycosylation sites of the Type 1 recombinant human immunodeficiency virus envelope glycoprotein (gp120) expressed in Chinese hamster ovary cells. *J. Biol. Chem.* **265**, 10,373–10,382.
5. Feizi, T. and Larkin, M. (1990) AIDS and glycosylation. *Glycobiology* **1**, 17–23.
6. Liedtke, S., Adamski, M., Geyer, R., Pfutzner, A., Rubsamen-Waigmann, H., and Geyer, H. (1994) Oligosaccharide profiles of HIV-2 external envelope glycoprotein: dependence on host cells and virus isolates. *Glycobiology* **4**, 477–484.
7. Yeh, H.-C., Seals, J. R., Murphy, C. I., van Halbeek, H., and Cummings, R. D. (1993) Site-specific *N*-glycosylation and oligosaccharide structures of recombinant HIV-1 gp120 derived from a baculovirus expression system. *Biochemistry* **32**, 11,087–11,099.
8. Lifson, J., Coutre, S., Huang, E., and Engelman, E. (1986) Role of envelope glycoprotein carbohydrate in human immunodeficiency virus (HIV) infectivity and virus-induced cell fusion. *J. Exp. Med.* **164**, 2101–2106.
9. Animashaun, T., Mahmood, N., Hay, A. J., and Hughes, R. C. (1993) Inhibitory effects of novel mannose binding lectins on HIV-infectivity and syncytium formation. *Antiviral Chem. Chemother.* **4**, 145–153.
10. Muller, W. E. G., Renneisen, K., Kreuter, M. H., Schroder, H. C., and Winker, I. (1988) The D-mannose-specific lectin from *Gerardia savaglia* blocks binding of human immunodeficiency virus type 1 to H9 cells and human lymphocytes in vitro. *J. AIDS* **1**, 453–458.
11. Favero, J. (1994) Lectins in AIDS research. *Glycobiology* **4**, 387–396.

12. Mahmood, N. (1995) Cellular assays for antiviral drugs, in *HIV. A Practical Approach*, vol. 2 (Karn, J., ed.), IRL, Oxford University Press, pp. 271–287.
13. Shibuya, N., Berry, J. E., and Goldstein, I. J. (1988) One-step purification of murine IgM and human alpha-2-macroglobulin on immobilized snowdrop bulb lectin. *Arch. Biochem. Biophys.* **267,** 676–680.
14. Mahmood, N. and Hay, A. J. (1992) An ELISA utilising immobilized snowdrop lectin GNA for the detection of envelope glycoproteins of HIV and SIV. *J. Immunol. Methods* **151,** 9–13.
15. Animashaun, T. and Hughes, R. C. (1989) *Bowringia milbraedii* agglutinin, specificity of binding to early processing intermediates of asparagine-linked oligosaccharide and use as a marker of endoplasmic reticulum glycoproteins. *J. Biol. Chem.* **264,** 4657–4663.
16. Animashaun, T. (1981) Survey of haemagglutinating activity in extracts of Nigerian seeds. *Nigerian J. Natl. Sci.* **3,** 75–83.
17. Sato, S., Animashaun, T., and Hughes, R. C. (1991) Carbohydrate binding specificity of *Tetracarpidium conophorum* lectin. *J. Biol. Chem.* **266,** 11,485–11,494.
18. Pekelharing, J. M. and Animashaun, T. (1989) Binding of *Artocarpus* lectins to plasma glycoproteins. *Biochem. Soc. Trans.* **17,** 131,132.
19. Kaku, H., Van Damme, E. J. M., Peumans, W. J., and Goldstein, I. J. (1990) Carbohydrate-binding specificity of the Daffodil *(Narcissus pseudonaricissus)* and Amaryllis *(Hippeastrum hybr)* bulb lectins. *Arch Biochem Biophys.* **279,** 298–304.
20. Weislow, O. S., Kiser, R., Fine, D. L., Bader, J., Shoemaker, R. H., and Boyd, M. R. (1989) New soluble-formazan assay for HIV-1 cytopathic effects: application to high-flux screening of synthetic and natural products for AIDS. *J. Natl. Cancer Inst.* **81,** 577–586.
21. Hammar, L., Hirsch, I., Machado, A., DeMareid, J., Baillon, J., and Chermann, J. C. (1994) Lectin effects on HIV-1 infectivity. *Ann. NY Acad. Sci.* **724,** 166–169.

XIII

USE OF LECTINS FOR DRUG DELIVERY

48

Absorption Enhancement by Lectin-Mediated Endo- and Transcytosis

Ellen Haltner, Gerrit Borchard, and Claus-Michael Lehr

1. Introduction

The rapid scientific progress in the areas of gene technology and biotechnology has provided an increasing number of biologically active compounds, raising the hope that better cures will be available in the near future for a number of severe diseases. However, before such novel biopharmaceuticals (e.g., peptides, proteins, oligonucleotides, gene vectors) can be used for drug therapy in humans, novel ways of drug delivery have to be developed in order to ensure the safe and efficient administration of these compounds to the patient (For general references, see refs. *1–4*). Certainly, convenient and easy routes of drug administration—ideally by simply swallowing a tablet or capsule—would be preferable to parenteral injections, which are painful and require trained healthcare personnel. However, because of their relatively large size, poor lipid solubility, and delicate structure—in contrast to many conventional small drug molecules—the transport of modern biopharmaceuticals across biological barriers, such as epithelial tissues or the membranes of individual cells themselves, is usually very slow. Moreover, the residence time of drugs or drug delivery systems at a given mucosal absorption barrier may be relatively short. This limits the time available for drug absorption before the compound is either metabolically destroyed or swept away by mechanical clearance processes (e.g., mucus secretion and transport, bowel movements, coughing, sneezing, and so on). Both problems, namely to control and prolong the residence time of the compound to be delivered at a given biological barrier, and to increase the rate at which the compound is transported across this barrier, might be solved by employing bioadhesive drug delivery systems *(6–8)*. The first such systems were based on mucoadhesive polymers, which "stick" to wet mucous surfaces

From: *Methods in Molecular Medicine: Vol. 9: Lectin Methods and Protocols*
Edited by: J. M. Rhodes and J. D. Milton Humana Press Inc., Totowa, NJ

by nonspecific, physicochemical mechanisms (e.g., hydrogen bonding, swelling, polymer chain interpenetration, and surface energy effects). More recently, however, lectins have attracted the interest of pharmaceutical scientists, because of their ability to accomplish specific binding to membrane-bound receptors located at the luminal surface of epithelial cells *(9)*.

1.1. Lectins as Bioadhesives

Though most known lectins are of plant origin, lectins are also produced by bacteria and animals *(10)*. Lectins produced by higher animals, e.g., mammals, are sometimes referred to as endogenous or reverse lectins. They are located, for example, on the surface of epithelial cell membranes, and are also responsible for the binding of bacteria to the epithelial cell surface, in some cases thereby initiating adsorptive endocytosis *(11–13)*. Isolated and fluorescently tagged lectins can be bound and transcytosed by epithelial cell monolayers and intestinal sacs in vitro *(14–17)*. By binding lectins to drug molecules or carriers, this effect might be exploited for the intestinal absorption of protein drugs through the transcytotic pathway. The coupling of drugs to lectin molecules might yield a higher metabolic stability of the drug during absorption and transcytosis, e.g., regarding pH changes and enzymatic activity in endo-/lysosomes. This protective effect may be enhanced when the drug is incorporated in colloidal drug carriers, as nanoparticles. One big advantage of these systems is their versatility: They can be designed to include a range of drugs and their surfaces can be functionalized with binding moieties, such as lectins through covalent coupling. In addition, the release of the drug from these carrier systems can occur in a time-controlled fashion.

1.2. In Vitro Studies of Lectin-Induced Drug Transport

To study the binding and endocytosis of lectins as well as lectin-labeled nanoparticles to intestinal epithelial cells under controlled conditions in vitro, we are using the Caco-2 cell model, which is derived from a human tumor cell line *(18)*. Although originating from a colonic tumor, these cells grow to monolayers with typical enterocyte-like differentiation, expressing microvilli and tight junctions. The epithelial properties of these cells have been characterized by various authors *(19–21)*, and they are presently used in many laboratories to study intestinal drug transport. The interaction of lectins and nanoparticles with these cells can conveniently be studied by fluorescence-based methods. Incubation at reduced temperature gives an indication whether or not membrane binding is followed by endocytosis, which as an active, energy-dependent process will not take place at 5°C. As an indicator for possible cytotoxic effects of lectins, we routinely measure the release of intracellular lactate dehydrogenase

(LDH-assay) *(22)* or the activity of mitochondrial dehydrogenases (MTT-assay *(23)*). The transepithelial electrical resistance (TEER) of the cell monolayer and the transport of a marker substance impermeable to cell membranes (fluorescein-Na) serve as indicators for epithelial permeability and possible lectin-induced changes. Finally, the noninvasive optical method of confocal laser scanning microscopy (CLSM) *(24)* can be employed for the three-dimensional scanning of the cell monolayer at high resolution. This allows visualization of the extent of binding, transport pathways, and the intracellular localization of the particulate carriers.

2. Materials
2.1. Cell Culture of Caco-2 Cells

1. "Falcon" (Becton Dickinson, Rutherford, NJ) or "Nunc" (Gibco-BRL, Gaithersburg, MD) plastic ware (96-well plates, cell culture flasks 75-cm^2), "Costar" (Costar, Cambridge, MA) Transwells with a pore size of 0.4 µm.
2. Dulbecco's modified Eagle's medium (DMEM) high glucose supplemented with 10% fetal calf serum (FCS), 1% nonessential amino acids, 100 U/mL penicillin, and 100 µg/mL streptomycin. Medium supplemented as described above will be called culture medium in this chapter.
3. Trypsin/EDTA solution: 0.25% trypsin, 0.02% EDTA (*see* Note 1).
4. Phosphate-buffered saline (PBS) pH 7.4 without calcium and magnesium (Note 2).
5. MTT (Sigma, St. Louis, MO); (3-[4,5-Dimethylthiazol-2-yl]-2,5-diphenyltetrazolium bromide).
6. LDH-Testkit No. 500 (Sigma).
7. Triton X-100
8. Equipment: tissue culture incubator, set at 37°C, 5% carbon dioxide, humidified atmosphere; inverted microscope; low-speed benchtop centrifuge, microplate reader.

2.2. Assessment of Caco-2 Cell Monolayers by Measuring Transepithelial Electrical Resistance or Paracellular Transport of Fluorescein-Na

2.2.1. Measurement of Transepithelial Electrical Resistance

1. Caco-2 cells grown on polycarbonate filters.
2. EVOM epithelial voltohmmeter (Millipore, Bedford, MA).

2.2.2. Transport of a Water-Soluble Paracellular Transport Marker Impermeable to the Membrane

1. Krebs-Ringer-Buffer (KRB) (freshly prepared) (*see* Note 3; Table 1).
2. Fluorescein-Na 1 mg/mL in KRB.
3. 96-Well plate fluorescence reader (e.g., Millipore Cytofluor II).

Table 1
Composition of Krebs-Ringer Buffer (KRB)

Stock solution no.	Substance	Amount needed to prepare 200 mL stock solution, g	Concentration after final dilution dilution buffer, mM
1	MgCl$_2$ · 6 H$_2$O	0.94	1.10
	CaCl$_2$ · 2 H$_2$O	0.734	1.25
2	NaCl	26.64	114
	KCl	1.5	5
3	NaHCO$_3$	6.72	20
4	Na$_2$HPO$_4$ · 7 H$_2$O	1.768	1.65
	NaH$_2$PO$_4$ · H$_2$O	0.166	0.30
5	HEPES	9.532	10
6	Glucose	18.016	25

2.3. Binding Studies with Fluorescently Labeled Lectins

1. Fluorescently (usually FITC-) labeled lectins (various commercial suppliers, e.g., Sigma, Vector, EY Labs, San Mateo, CA) dissolved in KRB. Some lectins are mitogenic and/or toxic, so they should be handled with care. Store the lectin solutions at high concentrations (1–10 mg/mL) in aliquots at –20°C. Thaw and dilute to the appropriate concentration immediately prior to use. Suggested highest lectin concentration is 50 µg/mL.
2. Nonlabeled lectins dissolved in KRB as described for fluorescently labeled lectins.
3. Krebs-Ringer-Buffer (described in Section 2.2.2.).
4. Bovine serum albumin (BSA): 10 mg/mL in KRB (*see* Note 4).
5. Sugars that are specific for the different lectins.
6. Confluent Caco-2 cell monolayers grown on 96-well plates.

2.4. Measurement of Changes in the Permeability of Caco-2 Cell Monolayers During Incubation with Different Lectins

1. Lectins dissolved in KRB (*see* Section 2.3.).
2. Caco-2 cell monolayers grown on Transwell cell culture inserts.
3. Fluorescein-Na 1 mg/mL in KRB.

2.5. Confocal Imaging of Lectin-Mediated Endocytosis

1. Caco-2 cells grown on polycarbonate filter inserts in 24-well plates.
2. Fluorescently labeled control and lectin particles, 0.3% (w/v) in KRB. (Fluorescently labeled polystyrene particles and methods to couple proteins,

e.g., lectins, to their surface are available from Polysciences, St. Goar, Germany.
3. Mannose solution in KRB (3 μmolar).
4. Confocal microscope: MRC-600 Lasersharp System (Bio-Rad), Zeiss IM 35 inverted microscope equipped with a Zeiss Neofluar 63, NA 1.25 oil-immersion lens (both Carl Zeiss, Oberkochen, Germany), Video Graph Printer UP-850 (Sony, Tokyo, Japan).

3. Methods
3.1. Cell Culture of Caco-2 Cells
3.1.1. Cell Culture of Caco-2 Cells in Flasks

1. Plate out cells at a concentration of 60,000 cells per cm^2 in 15 mL culture medium in a 75-cm^2 flask. Cell cultures grow at 37°C, 5% carbon dioxide in a humidified atmosphere.
2. Change the culture medium every second day (*see* Note 5).
3. Subculture the cells when they reach confluence. Monolayer formation occurs in approx 7 d. Wash three times with phosphate-buffered saline (PBS) and detach cells with 1 mL trypsin/EDTA at 37°C (*see* Note 6). Add 10 mL culture medium and resuspend the cell suspension five times using a pipet. Centrifuge cell suspension (5 min), decant supernatant, and resuspend pellets in 10 mL culture medium. One 75-cm^2 flask is split into three.

3.1.2. Culture of Caco-2 Cells on 96-Well Plates

1. Seed cells at a density of 17,000 cells in 0.2 mL (60,000 cells/cm^2) culture medium.
2. Subsequently change the medium every second day. Cells become differentiated 5–7 d after reaching confluence.

3.1.3. Culture of Polarized Caco-2 Cell Monolayers on Membrane Filters

Caco-2 cells may be grown on polycarbonate filters inserts (12-mm diameter), which fit into a 12-well plate. Once confluence has been reached and tight junctions are formed, the cell monolayer represents an artificial epithelium and can be used to study transport processes from the apical to the basolateral compartment (Fig. 1).

1. Seed Caco-2 cells at a density of 67,800 cells per 0.5 mL culture medium in the upper chamber of each filter insert. Add 1.5 mL culture medium to the lower chamber.
2. Subsequently, change the medium every second day (*see* Note 7) After 10–12 d the cells form a monolayer. For transport experiments they are usually used at d 21–25 after seeding.

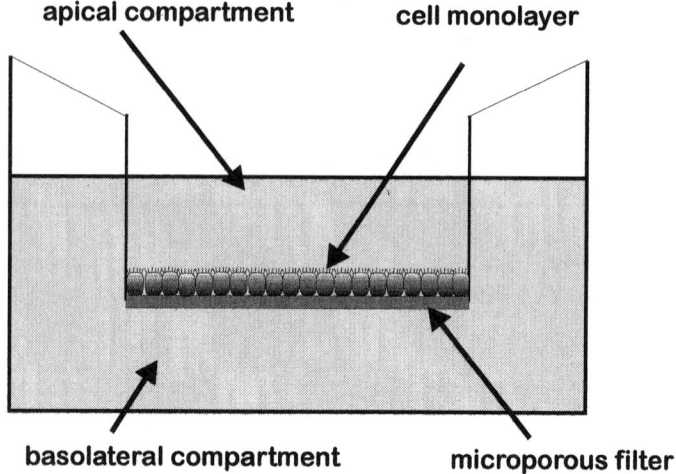

Fig. 1. Caco-2 cells growing on a polycarbonate filter membrane in a COSTAR Transwell.

3.1.4. Colorimetric Assays for Chemosensitivity Testing
3.1.4.1. MTT-Assay

Intracellular esterase activity is a parameter of cell viability and cell proliferation. One colorimetric method measures the metabolization of MTT (3-[4,5-dimethyldiazol-2-yl]-2,5-diphenyl tetrazolium bromide) through mitochondrial LDH in living cells to a water-insoluble formazan product. After dilution in isopropanol, the color changes to violet and the amount of formazan can be read with a scanning multiwell spectrophotometer at 570 nm.

1. Incubate 50,000 cells per well in flat-bottomed, 96-well plates in 0.2 mL culture medium for 48 h.
2. Replace culture medium with 0.1-mL lectin solutions of different concentrations in DMEM suppplemented with 0.5% BSA. Every experiment should be done in quadruplicate.
3. After 24 h incubation at 37°C, add 25 µL MTT (5 mg/mL) in PBS and allow the cells to incubate for another 2 h.
4. Then aspirate 0.1 mL solution and add 0.15 mL isopropanol to each well, to dissolve the formazan salt (*see* Note 8).
5. Read absorbance at 570 nm with a multiwell spectrophotometer.
6. Use Triton X-100 in a serial dilution from 1% to 3.9×10^{-2}% in DMEM to damage the cells and to determine 0% and 100% damage (*see* Note 9).

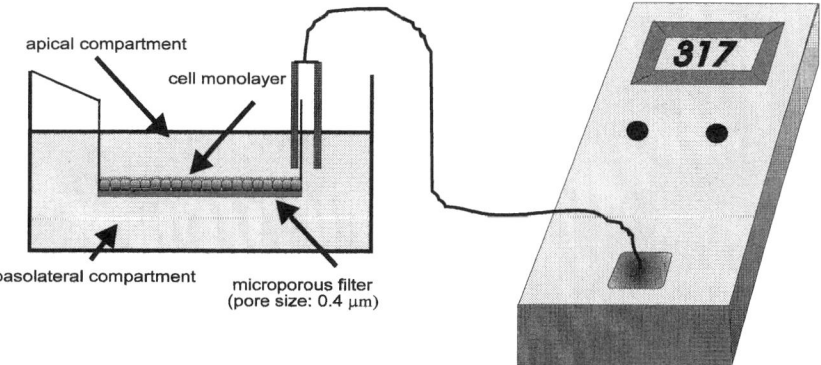

Fig. 2. Schematic view of measurement of transepithelial electrical resistance (TEER) across Caco-2 cell monolayers grown on a polycarbonate filter.

3.1.4.2. LDH-Assay

To determine possible damage of Caco-2 cells after incubation with lectin solutions or other chemicals it is useful to measure LDH release at the end of each experiment.

1. Take 50 µL solution out of each well and dilute to an appropriate concentration.
2. Pipet these solutions into a 96-well plate and add 50 µL LDH reagent.
3. After incubation for 30 min at 37°C, add 50 µL color reagent and incubate at 37°C for another 20 min.
4. After adding of 100 µL $1N$ NaOH, measure the absorbance at 492 nm in a multiwell spectrophotometer.
5. Controls: fresh KRB; medium from Caco-2 cells, which were incubated with 0.1% Triton X-100 (*see* Note 10).

3.2. Assessment of a Caco-2 Cell Monolayer Permeability by Measuring Transepithelial Electrical Resistance (TEER) and Marker Transport

3.2.1. Measurement of Transepithelial Electrical Resistance

Caco-2 cells grown on filters form a monolayer that separates the apical from the basolateral compartment. The measurement of electrical resistance across the monolayer is thus, a measure of passive ion transport. The transepithelial electrical resistance (TEER) is a criterion for the tightness of the cell monolayer. In practice, the resistance is measured by an epithelial voltohmmeter (*see* Fig. 2; Note 11).

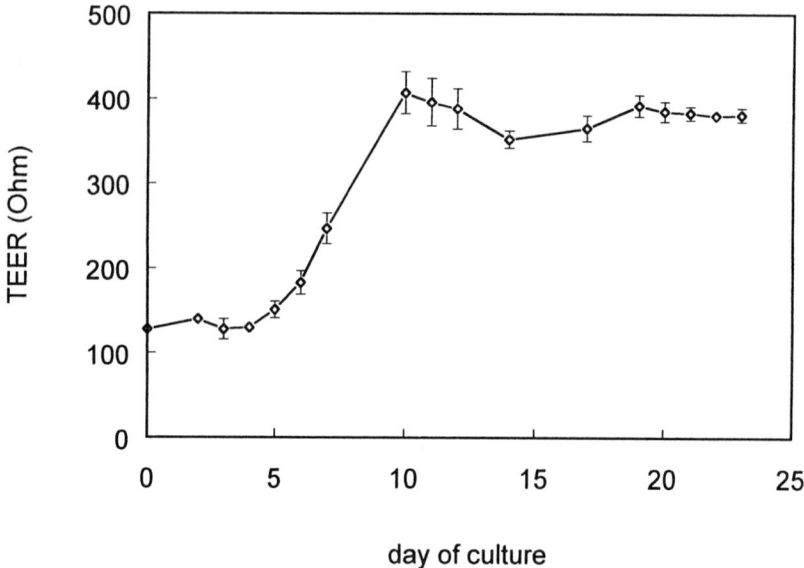

Fig. 3. Typical run of transepithelial electrical resistance (TEER) from seeding (d 0) to d 24 (bar means SD, $n = 12$).

The cellular resistance across the monolayer steadily increases as the cells differentiate, reaches a maximum, and decreases slightly to a plateau when the cells are fully differentiated. 21–25 d after seeding, the cells form a tight monolayer and can be used for experiments. A decrease of TEER is indicative for the disruption of monolayers or membrane damage. Figure 3 shows the typical time-course of TEER during culture of Caco-2 cells.

3.2.2. Transport of a Water-Soluble Transport Marker Impermeable to the Cell Membrane

1. Wash Caco-2 cell monolayers grown on Transwell cell culture systems three times with freshly prepared KRB.
2. Add 0.5 mL KRB containing 1 mg/mL fluorescein-Na to the upper chamber and 1.5 mL KRB to the lower chamber of the Transwell system. Start the experiment when TEER remains stable (*see* Note 12).
3. Incubate Transwells at 37°C in the carbon dioxide incubator. Measure TEER before taking out 50 µL solution from the lower chamber to determine the concentration of transported fluorescein-Na. Add 50 µL KRB to the lower chamber to compensate for the solution taken out.
4. Determine the amount of fluorescein-Na at each time-point with a 96-well plate fluorescence reader.

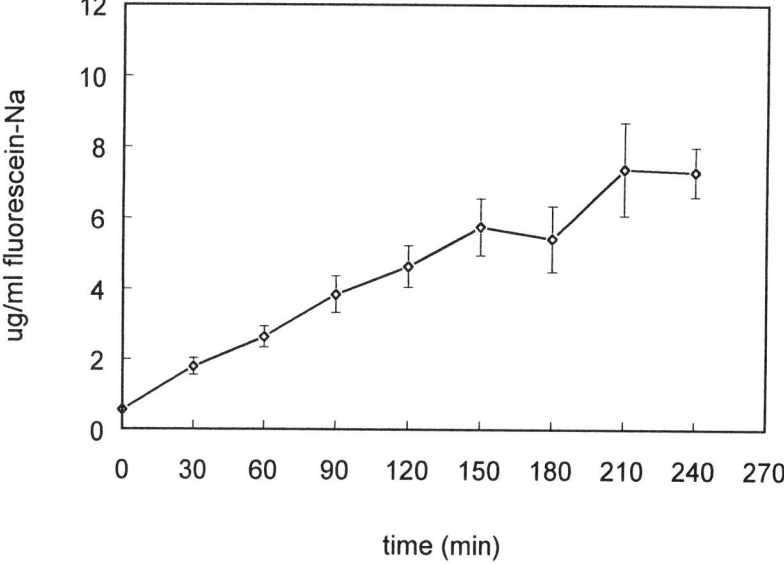

Fig. 4. Typical transport experiment with fluorescein-Na, resulting in linearly increased concentrations in the basolateral compartment.

5. Concentration of the fluorescein-Na in the lower chamber should steadily increase during the experiment. The rate of fluorescein increase can be used to describe epithelial permeability (Fig. 4).

3.3. Binding Studies with Fluorescently Labeled Lectins

1. Wash the confluent cells three times with 0.2 mL KRB. Then add to each well 0.15 mL solution of fluorescently labeled lectin in varying concentrations to the cell monolayers and determine the fluorescence with the fluorescence reader.
2. Each experiment should be done in triplicate. Use cell monolayers incubated with KRB as the control.
3. After incubation for 1 h at 37°C, then aspirate the lectin solution, wash the monolayers three times with 0.2 mL KRB, and measure fluorescence again (see Note 13).
4. To exclude binding by unspecific protein–protein interactions, repeat incubation with lectin solution of a constant concentration in the presence of different concentrations of BSA. Start at a concentration of 10 mg/mL BSA.
5. Specific binding of lectin to cell surface structures should be saturable and specifically inhibited by the unlabeled lectin or the appropriate inhibitory sugar. To determine specific binding, incubate the cell monolayers with a solution of constant concentration of fluorescently labeled protein in the presence of free lectin in varying concentrations. The highest concentration should be the 100-fold of

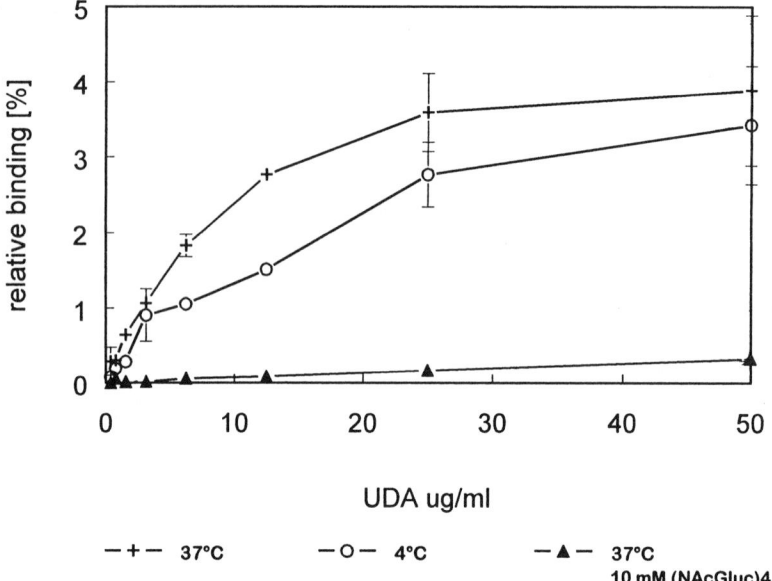

Fig. 5. Relative binding of FITC labeled *Urtica dioica* agglutinin to Caco-2 cells grown on a 96-well plate under different conditions: +, Binding after 1 h incubation at 37°C; ○, binding after 1 h incubation at 4°C; ▲, binding after 1 h incubation at 37°C in presence of 10 mM (NAcGluc)$_4$, the specific inhibitory sugar for UDA.

the fluorescently labeled lectin. Perform respective experiments with the specific sugar instead of free lectin.
6. In contrast to the uptake by fluid-phase pinocytosis, binding of a substrate to cell surface structures does not require metabolic energy. Uptake by fluid-phase pinocytosis as a process which needs energy for the movement and fusion of membrane vesicles should be inhibited at low temperatures. An energy-dependent uptake mechanism is indicated by decreased fluorescence bound after incubation at 4°C (Fig. 5).

3.4. Measurement of Changes in the Permeability of Caco-2 Cell Monolayers During Incubation with Different Lectins

1. Wash Caco-2 cell monolayers grown on Transwell cell culture systems three times with freshly prepared KRB.
2. Add 0.5 mL KRB containing 1 mg/mL fluorescein-Na and lectin solution at different concentrations to the upper chamber and 1.5 mL KRB to the lower chamber of the Transwell system. Start the experiment when TEER is stable at 37°C (*see* Note 12).
3. Incubate Transwells at 37°C in a carbon dioxide incubator. Take out 50 μL solution from the lower chamber to measure the concentration of paracellularly trans-

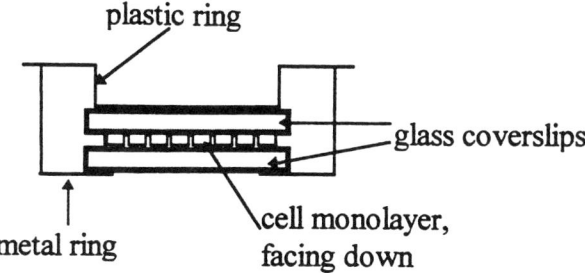

Fig. 6. Holder for confocal microscopy. Cells on polycarbonate filter membrane are "sandwiched" between two glass cover slides.

ported fluorescein-Na. Add 50 µL KRB to the lower chamber to compensate for the solution taken out.
4. Measure TEER at each timepoint you take out an aliquot.
5. Determine the amount of fluorescein-Na at each time-point with a 96-well plate fluorescence reader.

3.5. Confocal Imaging of Lectin-Mediated Endocytosis

3.5.1. Confocal Imaging of Lectin-Mediated Endocytosis

1. From filter-grown confluent monolayers, remove culture medium and wash twice with KRB.
2. Sonicate particle suspension for 5 min prior to incubation and add 200 µL to the apical side of the monolayer.
3. Incubate for 1 h at 37°C, 5% CO_2.
   a. For mannose inhibition studies, preincubate the particle suspension with the mannose solution (3 µmolar).
   b. For suppression of energy consuming uptake mechanisms, incubate 1 h at 4°C.
4. Remove particle suspension from the apical side, wash with KRB once, and remove insert from the culture plate, using a pair of tweezers.
5. Remove filter following a procedure as described by Hurni et al. *(25):*
   a. Position the insert on a glass cover slip lying (0.15-mm-thick, ø 12 mm) on an inverted film-plastic pack (30-mm diameter; *see* Note 14).
   b. Put a 17-mm diameter tube (e.g., from a 10-mL plastic syringe) on the apical side of the filter membrane.
   c. While pressing the tube downwards, the insert is lifted upward, removing the filter from the supporting insert.
   d. Lay another glass cover slide on top of the filter.
6. Adjust the "sandwich" in a specially designed holder (Fig. 6). This holder is made up of a metal ring, on which the sandwich is centered with the apical surface of the monolayer facing down (for inverted microscopes). A second plastic ring is fitted into the metal ring, holding the sandwich in a proper position for microscopy, preventing the cells from touching the bottom cover slip.

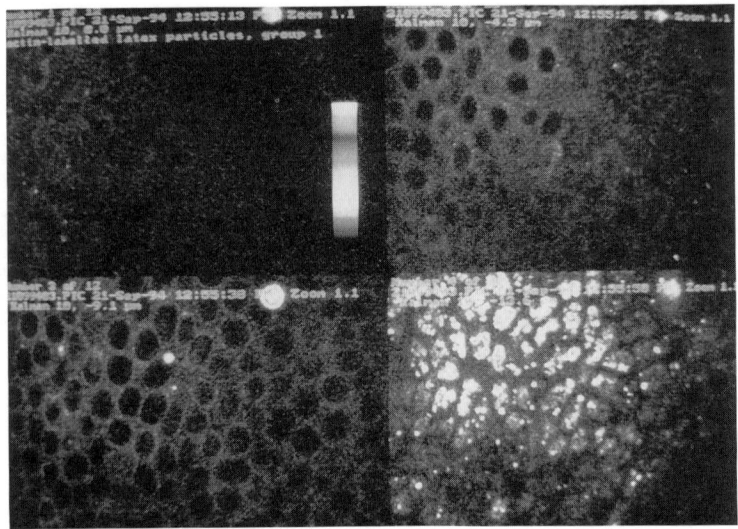

Fig. 7. Series of confocal Sections through Caco-2 cell monolayer incubated for 1 h with fluorescently labeled nanoparticles, which were coated with Con A. The labeled material has obviously been endocytosed.

7. Place holder on the stage of the confocal microscope.
8. Try to record images within 10–15 min after removal of the filter from the culture plate (Figs. 7 and 8; Note 15).

4. Notes

1. Trypsin/EDTA should be stored frozen in 1-mL aliquots in Eppendorf cups at –40°C.
2. The absence of Mg^{2+} and Ca^{2+} facilitates the loosening of tight intercellular junctions.
3. KRB should be freshly prepared. To avoid precipitation of unsoluble salts, add 50 mL of each stock solution to 200 mL water into a 1000-mL volumetric flask in the following order: 1, 2, 4, 3, 5, 6. Finally, fill up with water to 1000 mL.
4. Use freshly prepared BSA solution only, because it is easily susceptible to microbiological overgrowth.
5. An appropriate schedule is changing medium on Monday, Wednesday, and Friday. It is possible to leave the cells during the weekends.
6. Detachment of cells will be achieved within 5–15 min.
7. To avoid damage of cell monolayers by buoyancy, first aspirate the medium from the basolateral compartment, then from the apical compartment. Fill up the apical compartment with fresh medium first, then the basolateral compartment.
8. The formazan salt formed has a very low solubility in water. We, therefore, found it useful to replace water by other solvents like isopropanol or DMSO.

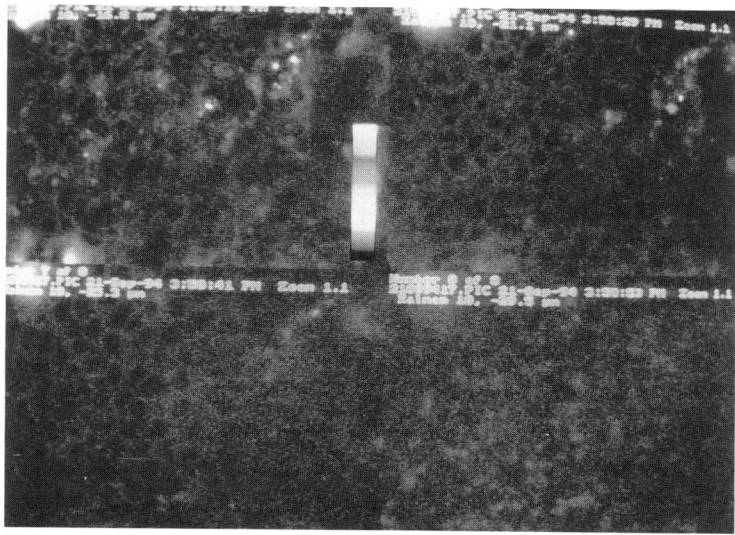

Fig. 8. Same experiment as in Fig. 7, but using fluoresecently labeled nanoparticles without lectin coating. No binding or endocytosis can be observed in the absence of lectins, illustrating their role as mediators of endocytosis.

9. Triton X-100 leads to lysis of cells and inactivation of intracellular dehydrogenases which form the formazan. Minimal absorbance read at 570 nm after such treatment is considered as maximal (100%) cytotoxic damage.
10. Triton X-100 leads to lysis of cells and release of cytosolic LDH. The LDH release is calculated relatively to maximal LDH-release after incubation with 0.1% Triton X-100.
11. Handle the electrodes with care, and never touch the cell monolayers! The TEER is dependent on temperature and pH, so keep the time the cells are outside the incubator as short as possible.
12. Usually, the TEER will have stabilized stable after 1 h.
13. For most lectins we have studied, binding is a fast process and complete after 1 h incubation time. However, this may be different for other lectins. Therefore it is advisable to determine the influence of incubation time for each new lectin as well.
14. Put the filter on the glass slide using a pair of tweezers. Be sure you know which side of the filter membrane the cells are on. Try not to move the filter already adhering to the glass surface, as this will certainly damage your cells and lead to artifacts.
15. Both decreased temperature and altered CO_2 content will shorten the viability of the cells considerably. First the cells will swell, forming so-called domes, which can easily be detected when running an xz-scan. After about 15 min, they will begin to die, indicated by the massive uptake of fluorescent markers. So, if you do not have the luxury of a heatable microscopic stage complete with incubation chamber, you had better be fast!

References

1. Lee, V. H. L. (ed.) (1991) *Peptide and Protein Drug Delivery.* Marcel Dekker, New York.
2. Audus, K. L. and Raub, T. J. (eds.) (1993) *Biological Barriers to Protein Delivery.* Plenum, New York.
3. Boer, A. G. de (ed.) (1993) *Drug Absorption Enhancement.* Harwood, Chur.
4. Zhou, X. H. (1994) Overcoming enzymatic and absorption barriers to non-parenterally administered protein and peptide drugs. *J. Control. Rel.* **29,** 239–252.
5. Lee, V. H. L., Hashida, M., and Mizushuma, Y. (eds.) (1995) *Trends and Future Perspectives in Peptide and Protein Drug Delivery.* Harwood, Chur.
6. Junginger, H. E. (1990) Bioadhesive polymer systems for peptide delivery. *Acta Pharm. Technol.* **36,** 110–126.
7. Harris, D. and Robinson, J. R. (1990) Bioadhesive polymers in peptide drug delivery. *Biomaterials* **11,** 652–658.
8. Lehr, C.-M. (1994) Bioadhesion Technologies for the delivery of peptide and protein drugs to the GI tract. *CRC Crit. Rev. Therap. Drug Carrier Syst.* **11(2 and 3),** 119–160.
9. Lehr, C.-M. and Pusztai, A. (1995) The potential of bioadhesive lectins for the delivery of peptide and protein drugs to the gastrointestinal tract, in *Lectins, Biomedical Perspectives* (Pusztai, A. and Bardocz, S., eds.), Taylor and Francis, London, pp. 117–140.
10. Goldstein, I. J., Hyghes, R. C., Monsigny, M., Osawa, T., and Sharon, N. (1980) What should be called a lectin? *Nature* **285,** 66.
11. Izhar, M, Nuchamowitz, Y., and Merelman, D. (1982) Adherence of Shigella Flexneri to guinea pig intestinal cells is mediated by mucosal adhesin. *Infect. Immunol.* **35,** 1110–1118.
12. Ashkenazi, S. (1986) Adherence of nonfimbriated entero invasive *Escherichia coli* O 124-guinea pig intestinal tract in vitro and in vivo. *J. Med. Microbiol.* **21,** 117–123.
13. Hoepelman, A. I. M. and Tuomanen, E. I. (1992) Consequences of microbial attachment: directing host cell functions with adhesins. *Infect. Immunol.* **60,** 1729–1733.
14. Lehr, C.-M., Bouwstra, J. A., Kok, W., Noach, A. B. J., de Boer, A. G., and Junginger, H. E. (1992,) Bioadhesion by means of specifically binding tomato lectin. *Pharm. Res.* **4,** 547–553.
15. Lehr, C.-M. and Lee, V. H. L. (1993) Binding and transport of some bioadhesive plant lectins across Caco-2 cell monolayers. *Pharm. Res.* **12,** 1796–1799.
16. Naisbett, B. and Woodley, J. (1994) The potential use of tomato lectin for oral drug delivery. 1. Lectin binding to rat small intestine in vitro. *Int. J. Pharm.* **107,** 223–230.
17. Naisbett, B. and Woodley, J. (1994) The potential use of tomato lectin for oral drug delivery. 2. Mechanism of uptake in vitro. *Int. J. Pharm.* **110,** 127–136.
18. Fogh,. J., Fogh, J. M., and Orfeo, T. (1977) One hundred and twenty seven cultured human tumor cell lines producing tumors in nude mice. *J. Natl. Cancer Inst.* **59,** 221–226.

19. Pinto, M., Robine-Leon, S., Appay, M. D., Kedinger, M., Triadou, N., Dussaulx, E., Lacroix, B., Simon-Assmann, P., Haffen, K., Fogh, J., and Zweibaum, A., (1983) Enterocyte-like differentiation and polarization of the human colon carcinoma cell line Caco-2 in culture. *Biol. Cell.* **47**, 323–330.
20. Hilgers, A. R., Conradi, R. A., and Burton, P. B. (1990) Caco-2 cell monolayers as a model for drug transport across intestinal mucosa. *Pharm. Res.* **7**, 902–910.
21. Artursson, P. (1990) Epithelial transport of drugs in cell culture. I: a model for studying the passive diffusion of drugs over intestinal absorptive (Caco-2) cells. *J. Pharm. Sci.* **79**, 476–482.
22. Schasteen, C. S., Donovan, M. G., and Cogburn, J. N. (1992) A novel in vitro screen to discover agents which increase the absorption of molecules across intestinal epithelium. *J. Contr. Rel.* **21**, 49–62.
23. Mosman, T. (1983) Rapid colorimetric assay for cellular growth and survival: application to proliferation and cytotoxic assays. *J. Immunol. Res.* **65**, 55–63.
24. Matsumoto, B., ed. (1993) Cell Biological Applications of confocal microscopy (1993) *Methods in Cell Biology,* vol. 38, Academic, San Diego, CA.
25. Hurni, M. A., Noach, A. B. J., Blom-Roosemalen, M. C. M., de Boer, A. G., Nagelkerke, J. F., and Breimer, D. D. (1993) Permeability enhancement in Caco-2 cell monolayers by sodium salicylate and sodium taurohydrofusidate: Assessment of effect-reversibility and imaging of transepithelial transport routes by confocal laser scanning microscopy. *J. Pharmacol. Exp. Ther.* **267**, 942–950.

49

The Use of Lectins for Liposome Targeting in Drug Delivery

Michael Kaszuba and Malcolm N. Jones

1. Introduction

Most mammalian cell surfaces are covered with a layer of oligosaccharides covalently linked to glycoproteins and glycolipids that form part of the surface structure, or "glycocalyx," of the plasma membrane. In bacteria, the cells' outer walls contain sugar-amino heteropolymers (peptidoglycans) in the case of Gram-positive strains or lipopolysaccharides in the case of Gram-negative strains. These cell surface sugar (monosaccharide) units are potential binding sites or receptors for many lectins and offer a means of targeting materials, such as drugs to the cell.

Liposomes or vesicles generally consist of a single bilayer (unilamellar liposomes) encapsulating an aqueous space or several concentric bilayers (multilamellar liposomes) and can be used for the delivery of both hydrophilic (water-soluble) and hydrophobic (bilayer-soluble) drugs *(1,2)*. There are numerous ways of exploiting carbohydrate interactions in the targeting of liposomes to cells for drug delivery *(3)*, but one of the most direct methods is to conjugate a lectin to the liposome surface to produce what have been described as proteoliposomes. Lectins have been used on the surface of liposomes to target liposomes to chicken erythrocytes *(4)*, mouse spleen cells *(4)*, Hela cells *(5)*, mouse fibroblasts *(6)*, and various oral and skin-associated bacteria *(7)*. The procedures required to produce and characterize liposomes with covalently attached lectins are described in detail below. They are readily adaptable to liposomes prepared from a wide range of bilayer forming lipids and, in principle, any lectin, but will be described specifically for phospholipid liposomes with the surface-conjugated lectins, wheatgerm agglutinin and concanavalin A.

1.1. Principles of Proteoliposome Preparation

The essential steps in the production of lectin-bearing liposomes are as follows.

1.1.1. Lipid Activation

A reactive lipid must be prepared to be incorporated into the liposome formulation for subsequent coupling to the lectin. The most useful phospholipids that can be activated for use as lectin anchors are the phosphatidylethanolamines (PE). They can be activated by reaction with a "double agent," having a succinimidyl group that reacts readily with $-NH_3^+$ in the PE head group and a second group which will react with either a disulfide or sulfhydryl group in the lectin. The method described in Section 3.1. uses m-maleimidobenzoyl-N-hydroxy-succinimide (MBS) (reactive to free -SH groups). An alternative reagent is N-succinimidylpyridyl-thiopropionate (SPDP) (see Note 1).

1.1.2. Liposome Preparation

The preparation of liposomes can be done using a variety of methods. The method used determines the size and size distribution of the liposomes. Small unilamellar liposomes (SUV) can be prepared by sonication *(8)*, larger liposomes can be prepared by extrusion (VETs or Vesicles by Extrusion Technique *[9]*), or by reverse-phase evaporation (REV *[10]*). The method described in Section 3.2. uses vesicles produced by extrusion (VETs) having a diameter of approx 100 nm. These are easily produced, stable, and convenient to use.

1.1.3. Lectin Activation

Derivatization of the lectin to introduce free sulfhydryl groups that will react with the m-maleimide$N$-hydroxysuccinimide derivative of PE (PEMBS) can be brought about using N-succinimidyl-S-acetylthioacetate (SATA). This reagent introduces thioacetate groups at lysine residues that can be deacetylated by reaction with hydroxylamine to yield free –SH groups.

1.1.4. Conjugation

The final step in the production of lectin-bearing liposomes is the coupling of the free –SH groups introduced in the lectin to the reactive lipid in the liposomes. This reaction goes easily over 2 h at room temperature or approx 12 h at 4°C. The overall reaction scheme is shown in Fig. 1.

2. Materials

2.1. Synthesis of the MBS Derivative of DPPE (11)

1. L-α-dipalmitoylphosphatidylethanolamine (DPPE) (e.g., Sygena, Liestal, Switzerland; Sigma, Poole, Dorset, UK).

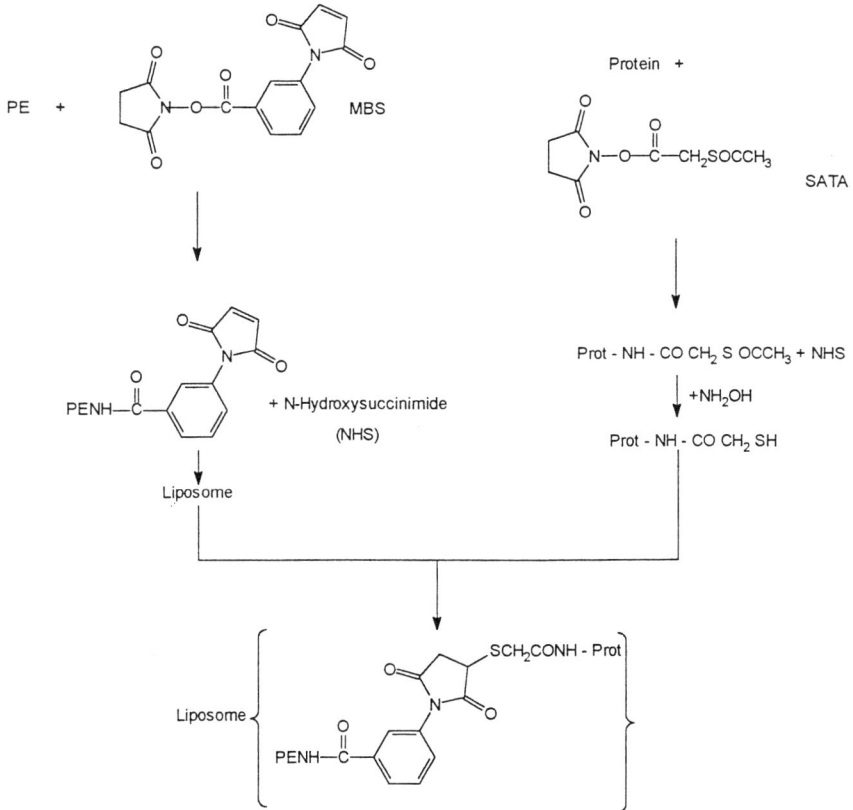

Fig. 1. Flowchart showing the derivatization of phosphatidylethanolamine and protein (lectin) and the conjugation of lectin to PEMBS containing liposomes.

2. *m*-Maleimidobenzoyl-*N*-hydroxysuccinimide ester (MBS) (e.g., Pierce Warriner, Chester, UK).
3. Dry chloroform (16 mL), dry methanol (2 mL), and dry triethylamine (20 mg) (dried over molecular sieves type 4A, BDH, Poole, Dorset, UK).
4. Phosphate-buffered saline (PBS) buffer, pH 7.3.
5. Silica gel TLC plates.
6. TLC solvent: chloroform, methanol, glacial acetic acid (65:25:13 v/v).
7. Zinzade reagent: 1.3% molybdenum oxide in $4.2M$ sulfuric acid.
8. Nitrogen gas supply.

2.2. Liposome Preparation

1. L-α-dipalmitoylphosphatidylcholine (DPPC), phosphatidylinositol (PI) from wheatgerm (e.g., Sygena; Lipid Products, South Nutfield, UK), [^{3}H] dipalmitoyl phosphatidylcholine (e.g., Amersham, Amersham, UK; *see* Note 2).

2. Distilled chlorofom/methanol (4:1 v/v).
3. PBS, pH 7.3.
4. 0.1-μm pore diameter polycarbonate membranes (e.g., Poretics, Livermore, CA; Nucleopore, Pleasanton, CA).

2.3. Derivatization of Lectins with SATA (12)

1. Succinyl concanavalin A (sConA), wheatgerm agglutinin (WGA) (e.g., Sigma).
2. N-succinimidyl-S-acetylthioacetate (SATA) (e.g., Pierce Warriner), dimethylformamide.
3. 40 mM phosphate, 1 mM EDTA, pH 7.5.
4. Sephadex G-50 column (15 × 2 cm) set up in phosphate EDTA buffer.
5. Lowry reagents: Reagent A: 2% w/v Na_2CO_3 in 0.1M NaOH, Reagent B: 1% w/v $CuSO_4 \cdot 5H_2O$, Reagent C: 2% w/v NaK tartrate.

2.4. Conjugation of Lectins to Liposomes

1. Hydroxylamine solution: 40 mM NH_2OH plus 2 mM EDTA plus solid Na_2HPO_4, pH 7.5.
2. Sepharose 4B column (30 × 2 cm) made up in PBS buffer.
3. Ecoscint A scintillation fluid (e.g., National Diagnostics, Atlanta, GA).
4. Wang and Smith modified Lowry reagents [13]: Reagent A: Folin-Ciocalteau's reagent diluted 1:1 with water (prepared freshly on day of assay). Reagent B: 250 mg CuEDTA, 100 mL of 20% Na_2CO_3 (w/v), 10 mL of 10M NaOH made up to 1 L. Reagent C: 10% SDS (w/v).

2.5. Equipment

1. Rotary evaporator for removal of organic solvents.
2. Liposome extruder for liposome preparation (e.g., Lipex Biomembranes, Vancouver, British Columbia, Canada).
3. Autosizer (e.g., Malvern Instruments, Malvern, UK) for measuring the size of the liposomes at various stages of the preparation.
4. Scintillation counter for measurement of radiolabeled lipids.
5. Fraction collector for gel chromatography with Sepharose 4B and Sephadex G50 columns.
6. UV/VI spectrophotometer for assay of protein derivatization with SATA and Lowry/Wang and Smith modified Lowry protein assays.

3. Methods

3.1. Synthesis of the MBS Derivative of DPPE

1. Dissolve 40 mg of DPPE in a mixture of dry chloroform (16 mL), dry methanol (2 mL), and dry triethylamine (20 mg).
2. Add 20 mg of MBS to the mixture and stir under nitrogen at room temperature for 24 h.

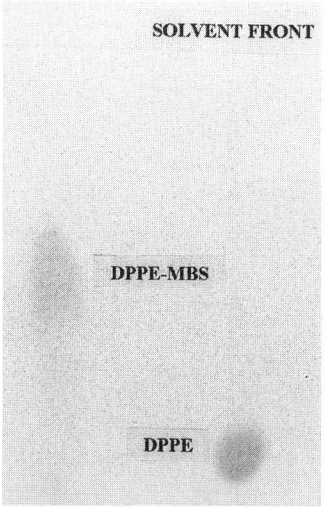

Fig. 2. Typical TLC plate showing Zinzade-stained phosphatidylethanolamine and its m-maleimidobenzoyl-N-hydroxysuccinimide derivative.

3. Wash the organic phase three times with PBS buffer (pH 7.3, 4°C) to remove unreacted MBS (*see* Note 3).
4. Rotary evaporate the organic phase to recover the DPPE-MBS derivative.
5. Analyze the DPPE-MBS derivative by thin layer chromatography (TLC) using a silica plate and a solvent mixture consisting of chloroform, methanol, and glacial acetic acid (65:25:13 v/v).
6. After running, spray the air-dried TLC plate with Zinzade reagent to develop the lipid spots (Fig. 2).

3.2. Liposome Preparation

1. Dissolve DPPC (27 mg), PI (3 mg), DPPE-MBS (3 mg) and 1 µCi [^{3}H] DPPC in chloroform/methanol (4:1 v/v) in a 50 mL round bottom flask.
2. Remove the organic solvent on a rotary evaporator (60°C) to obtain a thin, uniform lipid film (*see* Note 4).
3. Add 3 mL of PBS buffer (pH 7.3, 60°C) to the lipid film and vigorously mix to produce a milky suspension of multilamellar vesicles (MLVs).
4. Take 3 × 10-µL aliquots of the MLVs for scintillation counting (*see* Note 5).
5. Extrude the MLV suspension through two stacked polycarbonate membranes (pore size of 0.1 µm) 10 times at 60°C to produce unilamellar vesicles (VETs).
6. Take 3 × 10-µL aliquots of the VET dispersion for scintillation counting and size the VETs using photon correlation spectroscopy or electron microscopy (*see* Note 6; Fig. 3).

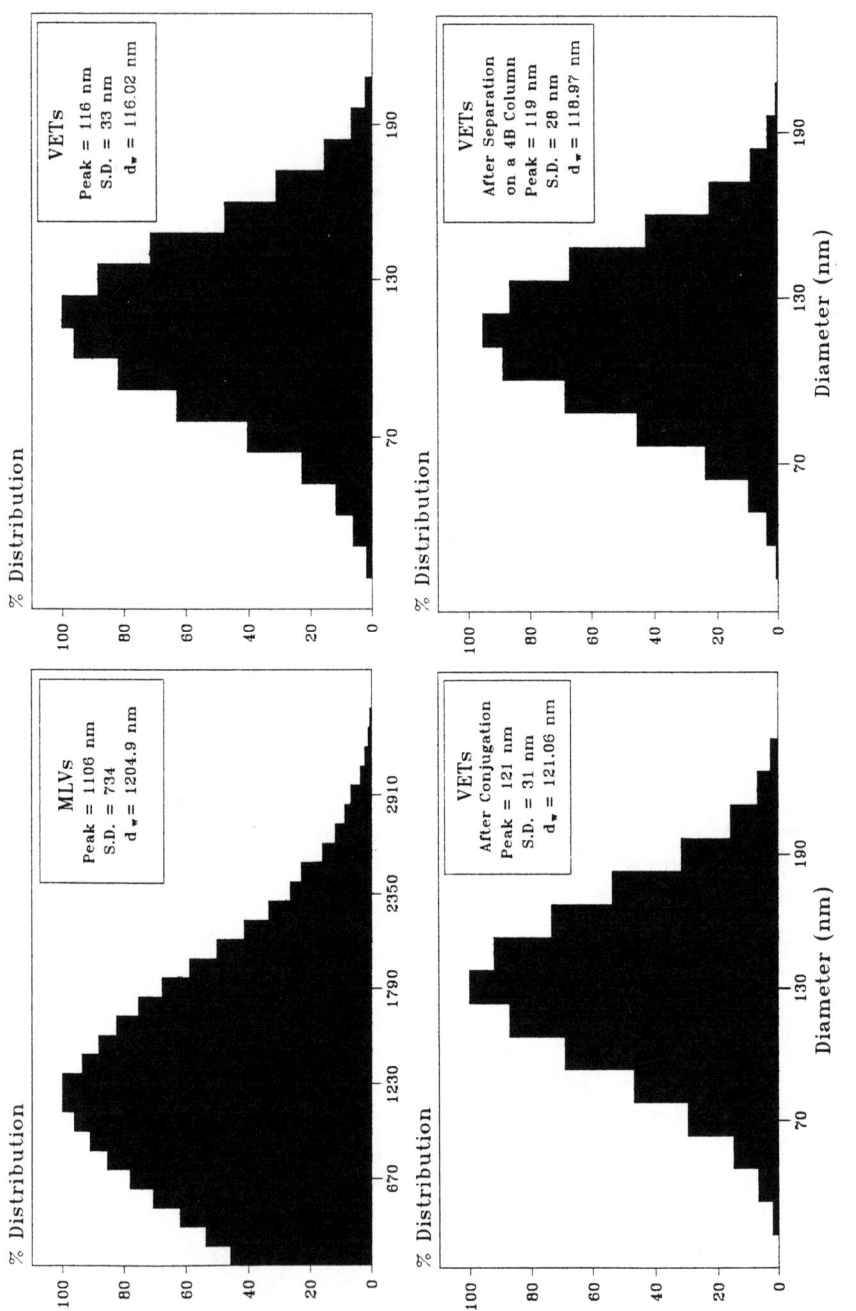

Fig. 3.

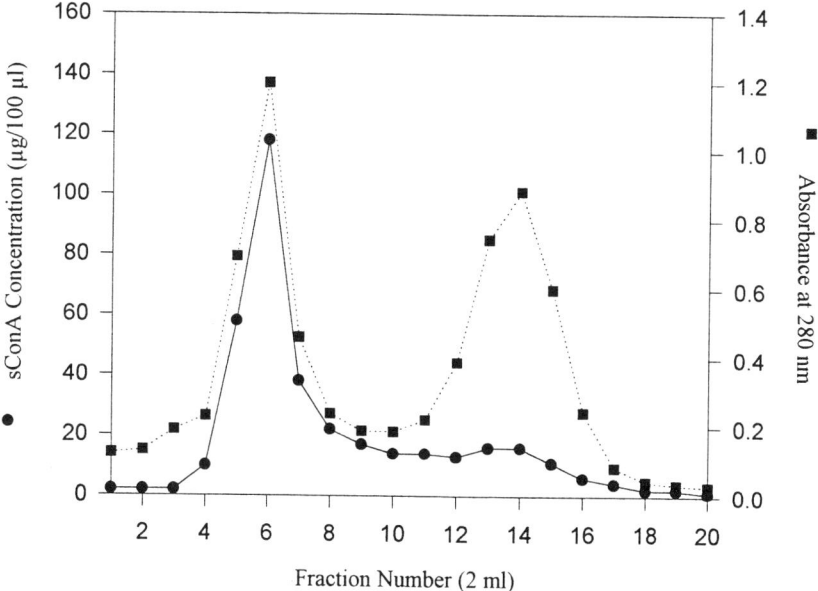

Fig. 4. Separation of the SATA derivative of sConA from the unreacted SATA on a Sephadex G50 gel filtration column. Both free and reacted SATA (derivatized sConA) absorb at 280 nm. sConA is assayed using the Wang and Smith modified Lowry assay. The peak at fraction 14 corresponds to unreacted SATA.

3.3. Derivatization of Lectins with SATA

1. Prepare a stock solution of SATA (9.08 mg) in dimethyl formamide (50 µL).
2. Add a required volume of the stock SATA solution to the lectin (0.2 µM in 2.5 mL of phosphate (40 mM)-EDTA (1 mM) buffer (pH 7.5) at room temperature (*see* Note 7).
3. After the reaction time (15 min), apply the mixture to a Sephadex G-50 column (15 × 2 cm) to separate the unreacted SATA from the derivatized lectin (Fig. 4). Collect 2-mL fractions.
4. Measure the absorbance at 280 nm to detect both derivatized lectin and unreacted SATA.
5. Assay the protein using a Lowry microassay with the appropriate lectin as the standard.

Fig. 3. *(previous page)* Size distributions from photon correlation spectroscopy (PCS) of multilamellar liposomes (MLV) (composition DPPC/PI/PEMBS), extruded liposomes (VETs), liposomes after conjugation with the SATA derivative of sConA and after gel filtration to separate the unreacted sConA. The size of the VETs is not changed by conjugation. For the VETs shown the number of lectin molecules per liposome ($\overline{P}_W$) was 14.3.

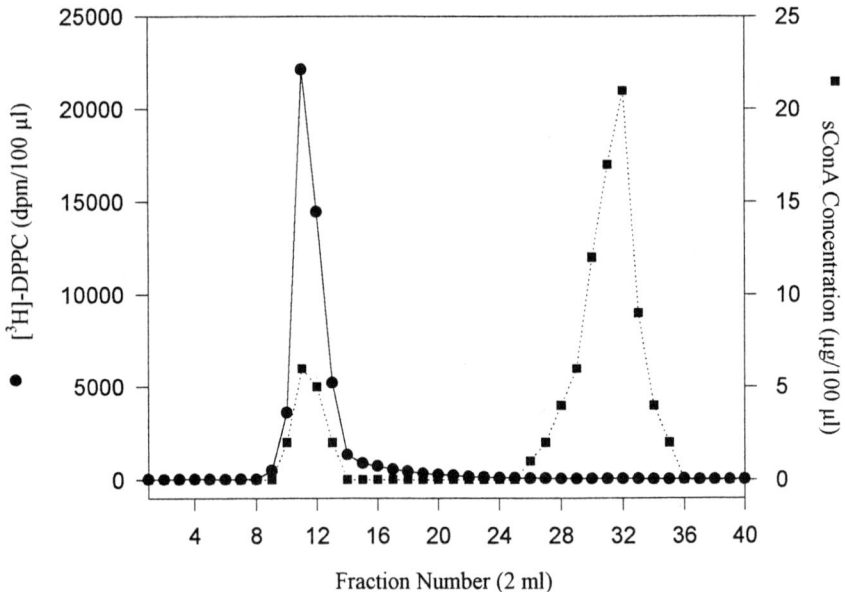

Fig. 5. Separation of unconjugated sConA from liposome conjugated sConA on a Sepharose 4B column. The coelution of radiolabeled liposomes (●) and sConA (■) confirms conjugation of the lectin to the liposome surface. In this experiment, $\overline{P}_W = 14.3$.

3.4. Lowry for Protein Determination

1. Freshly prepare the Lowry reagent by mixing 100 mL reagent A (2% w/v Na_2CO_3 in $0.1M$ NaOH), 1 mL reagent B (1% w/v $CuSO_4 \cdot 5H_2O$), and 1 mL reagent C (2% w/v NaK tartrate).
2. Add 1 mL of the Lowry reagent to 100 µL of standard or sample in glass tubes. Mix well, and leave at room temperature for 15 min (standards are usually 0.2–2 mg lectin/100 µL buffer).
3. Add 100 µL Folin-Ciocalteau's reagent to all the tubes, and mix well.
4. Leave for at least 1.5 h at room temperature before reading absorbance at 700 nm.

3.5. Conjugation of Lectins to Liposomes

1. Add 100 µL of hydroxylamine hydrochloride (40 mM hydroxylamine in 2.5 mM EDTA with sufficient solid Na_2HPO_4 added to bring the pH to 7.5) for each milliliter of the derivatized lectin solution and leave for 1 h to deacetylate the bound SATA from the lectin. The sulfhydryl content of the derivatized lectin can be determined by the method of Ellman *(16)* (*see* Note 8).
2. After deacetylation, mix and incubate the lectin solution with the liposome suspension at room temperature for 2 h or at 4°C overnight.
3. Apply the mixture to a Sepharose 4B column (30 × 2 cm made up in PBS buffer) to separate the proteoliposomes from the unconjugated lectin and collect 2-mL fractions (Fig. 5).

4. Take 100-µL aliquots from each fraction for scintillation counting.
5. Assay the lectin content of the proteoliposome fractions using a Wang and Smith modified Lowry assay *(13)* with the appropriate lectin as the standard (*see* Section 3.6.).
6. Measure the size of the peak proteoliposome fractions using photon correlation spectroscopy (Fig. 3).
7. From the measurements of the size of the proteoliposomes and their protein content, the number of lectin molecules per liposome may be calculated (*see* Note 9).

3.6. Wang and Smith Modified Lowry Assay for Protein Determination

1. Add 500 µL of reagent B to 100 µL of the appropriate solution (proteoliposome fraction, lectin standard or blank) in glass tubes, mix thoroughly and leave for 15 min (standards are usually 4–20 µg lectin/100 µL buffer).
2. Add 500 µL of reagent C to the tubes, and mix thoroughly.
3. Add 50 µL of reagent A to each tube, mix thoroughly, and leave for at least 30 min.
4. Measure the absorbance of each sample at 700 nm.

4. Notes

1. In the SPDP method, both lipid (PE) and lectin are activated by reaction of $-NH_3^+$ groups with the succinimidyl ester with *N*-hydroxysuccinimide (NHS) as by product according to the reactions:

    $PE(NH_2) + SPDP \longrightarrow PE\text{-}(NHCO(CH_2)_2\text{-S-S-Pyr}) + NHS$

    $Prot(NH_2) + SPDP \longrightarrow Prot\text{-}(NHCO(CH_2)_2\text{-S-S-Pyr}) + NHS$

    The protein is then reduced with dithiothreitol to give a sulfhydryl group, which undergoes an exchange reaction with the PE derivative. Because, at this stage, the PE and protein derivatives are in different oxidation states, there is little chance for the formation of crosslinking between protein–protein and liposome–liposome. It is thus essential that the dithiothreitol is rigorously removed form the reduced protein to avoid unwanted crosslinking.
2. The choice of radiolabel depends on what other materials are to be carried by the liposomes. If a drug is to be carried that is only available in a tritiated form then [^{14}C]-lipid would have to be used instead of [^{3}H]-lipid. Where both radiolabeled lipids are available, the [^{3}H]-lipids are much cheaper.

    The incorporation of phosphatidylinositol (PI) gives the liposomes a negative charge at neutral pH, which stabilizes them, inhibiting fusion and aggregation. Any negatively charged phospholipid may be used, such as phosphatidic acid or phosphatidylserine.
3. Separation of the organic and aqueous phases can be achieved more rapidly by centrifugation at 4°C. This normally results in the formation of a white, frothy layer between the two phases which can be removed, with the aqueous (upper) phase, carefully using a glass Pasteur pipet attached to a vacuum line.

4. The temperature of the buffer used in the preparation of the liposomes is governed by the phase transition temperature (T_c) of the lipid mixture. In the case of the lipid mixture used in Section 3.2., the T_c of the main lipid constituent, DPPC, is 41°C. Therefore a temperature normally 20°C in excess of T_c is required (i.e., so that the lipid is in the liquid crystalline state) to ensure that the liposomes form properly. If an unsaturated lipid is used (e.g., eggPC or soyaPC) then a much lower temperature is required (approx 20°C). The temperature of the lipid suspension should be maintained above T_c during the whole of the liposome preparation; once the liposomes are formed, the temperature can be reduced below T_c.
5. Aliquots of MLVs and VETs are taken for scintillation counting to ensure that no lipid loss occurs on the polycarbonate membranes during extrusion. This can be checked by comparison of the scintillation counts before (MLVs) and after (VETs) extrusion.
6. Photon correlation spectroscopy (PCS) measures the diffusion coefficient of the liposomes from the scattering of light (normally a He/Ne [red] laser beam is used). The scattered light is analyzed and fitted to an equivalent normal weight distribution of particle sizes assuming that the liposomes are spherical. From the size distribution, the peak particle size (diameter) may be taken as a measure of the liposome size (diameter) or a weight average diameter ($\bar{d}_w$) may be calculated from the distribution *(14)*. If PCS equipment is not available, electron microscopy of negatively stained (phosphotungstic acid) liposomes gives a measure of size although it should be remembered that the liposomes flatten on EM grids.
7. Changing the molar ratio of SATA:lectin results in a change in the number of sulfhydryl groups introduced per lectin molecule. For example, a molar ratio of 10:1 SATA:sConA results in 1.21 ± 0.05 –SH groups per sConA *(15)*. The lectin/SATA solution should always contain <10 µL of DMF/mL. In general, the number of sulfhydryl groups introduced is not large *(12)*. Increasing the SATA: sConA molar ratio to 25:1 only raises the number of sulfhydryl groups to 1.76 ± 0.1 *(15)*.
8. The sulfhydryl content of derivatized lectins can be determined by the method of Ellman *(16)*. Ellman buffers: Buffer I: 77.9 mg NaH_2PO_4 and 1.1342 g Na_2HPO_4 plus 90 mL water, pH adjusted to 8.0 and made up to 100 mL. Buffer II: 0.5377 g NaH_2PO_4 and 0.782 g Na_2HPO_4 plus 90 mL water, pH adjusted to 7.0 and made up to 100 mL. Ellman's reagent: 39.6 mg 5,5'-dithiobis(2-nitrobenzoic acid) (DTNB) in 10 mL Ellman's buffer II.
   a. Make up the following mixture: Lectin fraction (1.5 mL), Ellman's buffer I (pH 8.0, 1 mL) and water (2.5 mL).
   b. Add 20 µL of Ellman's reagent (39.6 mg DTNB in 10 mL Ellman's buffer II) to 3 mL of the prepared mixture in a cuvet and measure the absorbance at 412 nm until it becomes stable (approx 30 min).

Here is a sample calculation for the determination of sulfhydryl content for sConA (molecular mass 52,000).
A sConA fraction contains 380 µg protein/mL (determined from the Lowry assay). The Ellman's assay gave an absorbance of 0.018 at 412 nm.

$$\text{If } C_o = (A/E) \times D$$

where A is the absorbance at 412 nm, E is the extinction coefficient of the thiolate anion of 2-nitro 5-mercaptobenzoic acid (= $13,600 M^{-1} \text{ cm}^{-1}$), and D is the dilution factor.

$$\text{Therefore } C_o = (0.018/13,600) \times (5/1.5) \times (3.02/3)$$
$$= 4.44 \times 10^{-6} \text{ moles sulfhydryl/L}$$

$$\text{Moles SH groups} = (4.44 \times 10^{-6}) \times (1.5/1000)$$
$$= 6.66 \times 10^{-9}$$

$$\text{Moles of sConA} = (380 \times 10^{-6}/52,000) \times (1.5/2.1)$$
$$= 5.22 \times 10^{-9}$$

Therefore (moles SH/moles protein) = $(6.66 \times 10^{-9}/5.22 \times 10^{-9})$ = 1.28

9. It has been found that the surface density of lectin molecules on the liposome surface is determined only by the level of PEMBS initially incorporated into the bilayers during the liposome preparation, i.e., the surface density of lectin is not a function of liposome size at least in the diameter range 65–250 nm *(14,15)*. This being so the distribution of the number of lectin molecules per liposome (P_i) can be calculated from the equivalent normal weight distribution and a weight-average number of lectin molecules per liposome ($\bar{P}_w$) can be calculated from the relationship:

$$\bar{P}_W = (\Sigma_i\ P_i W_i / \Sigma_i W_i) = (\Sigma_i\ P_i W[d_i] / \Sigma_i W[d_i])$$

where W_i is the weight fraction of liposomes having P_i lectin molecules per liposome and $W(d_i)$ is the normal weight distribution of liposome diameters (d_i). A computer program is available on request to carry out the computation of $\bar{P}_w$ from the particle diameter distribution curves (Fig. 3) and the molar ratio of lectin to lipid.

References

1. Ostro, M. J., ed. (1987) *Liposomes: From Biophysics to Therapeutics.* Marcel Dekker, New York.
2. Jones, M. N. and Chapman, D. (1995) *Micelles, Monolayers and Biomembranes,* vol. 5, Wiley, New York, pp. 117.
3. Jones, M. N. (1994) Carbohydrate-mediated liposomal targeting and drug delivery. *Adv. Drug Delivery Rev.* **13,** 215–250.
4. Carpenter-Green, S. and Huang, L. (1983) Incorporation of acetylated wheatgerm agglutinin into liposomes. *Anal. Biochem.* **135,** 151–155.
5. Liautard, J. P., Vidal, M., and Philipott, J. R. (1985) Controlled binding of liposomes to cultured cells by means of lectins. *Biol. Int. Rep.* **2,** 1123–1137.
6. Bogdanov, A. A, Gordeeva, L. V., Torchilin, V. P., and Margolis, L. B. (1989) Lectin-bearing liposomes: differential binding to normal and to transformed mouse fibroblasts. *Exp. Cell Res.* **181,** 362–374.

7. Jones, M. N., Kaszuba, M., Hill, K. J., Song, Y.-H., and Creeth, J. E. (1994) The use of phospholipid liposomes for targeting to oral and skin-associated bacteria. *J. Drug Targeting* **2,** 381–389.
8. New, R. R. C., ed. (1990) *Liposomes: A Practical Approach,* Oxford University Press, Oxford, UK, pp. 44–48.
9. Mayer, L. D., Hope, M. J., and Cullis, P. R. (1986) Vesicles of various sizes produced by a rapid extrusion procedure. *Biochim. Biophys. Acta* **858,** 161–168.
10. Szoka, F. and Papahadjopoulos, D. (1978) Procedure for preparation of liposomes with large internal aqueous space and high capture by reverse-phase evaporation. *Proc. Natl. Acad Sci. USA* **75,** 4194–4198.
11. Hutchinson, F. J. and Jones, M. N. (1988) Lectin-mediated targeting of liposomes to a model surface. *FEBS Lett.* **234,** 493–496.
12. Duncan, R. J. S., Weston, P. D., and Wrigglesworth, R. (1983) A new reagent which may be used to introduce sulfhydryl groups into proteins and its use in the preparation of conjugates for immunoassay. *Anal. Biochem.* **132,** 68–73.
13. Wang, C. and Smith, R. L. (1974) Lowry determination of protein in the presence of Triton X-100. *Anal. Biochem.* **63,** 414–417.
14. Hutchinson, F. J., Francis, S. E., Lyle, I. G., and Jones, M. N. (1989) The characterization of liposomes with covalently attached proteins. *Biochim. Biophys. Acta* **978,** 17–24.
15. Francis, S. E., Hutchinson, F. J., Lyle, I. G., and Jones, M. N. (1992) The control of protein surface concentration on proteoliposomes. *Colloids and Surfaces* **62,** 177–184.
16. Ellman, G. L. (1959) Tissue sulfhydryl groups. *Arch. Biochem. Biophys.* **82,** 70–77.

Appendix: Lectin-Binding Specifications

Proper name	Common name		Specificity
Gal-GalNAc group			
Agaricus bisporus	Common edible mushroom	ABL	Galβ1–3GalNAc or Sial2–3Gal-GalNAc not Gal-(Sial2–6)GalNAc
Amaranthus caudatus		AML	Galβ1–3GalNAc or Sial2–3Gal-GalNAc
Arachis hypogaea	Peanut	PNA	Galβ1–3GalNAc or Gal-(Sial2–6)GalNAc not Sial2–3Gal-GalNAc
Artocarpus integrifolia	Jackfruit	JAC	Galβ1–3-GalNAc or Galβ or Sial2–3Gal-GalNAc
Bauhinia purpurea		BPA	Gal-GalNAc (not Sial)
Maclura pomifera	Osage orange	MPA	Galβ1–3GalNAc or GalNAcα (2 sites)
Gal or GalNAc group			
Artocarpus altilis			Gal
Abrus precatorius	Jequirity bean	Abrin	Galα or Galα-Galβ
Colchicum autumnale	Meadow saffron	CA	Gal or GalNAc
Crotalaria junceae	Sun hemp		Galβ
Dolichos bifloris	Horse gram	DBA	GalNAc-α1 ±-3GalNAcβ
Erythrina cristagelli	Coral tree	ECA	Galβ1–4GlcNAc
Erythrina senegalensis		ESA	Gal
Griffonia simplifolia		GS1-A4	GalNAc
(*Bandeiraea simplifolia*)		GS1-B4	Galα1–3Gal or Galα
Glycine max	Soya bean	SBA	GalNAcα
Hura crepitans	Sand box tree		GalNAc or Gal
Helix aspersi	Garden snail	HAA	GalNAcα
Helix pomatia	Roman snail	HPA	GalNAcα (A substance)
Lablab niger		LNA	Gal
Phaseolus lunatus		LBA	GalNAc (A substance)
Pseudomonas aeruginosa		PA-I	Galα (Blood gp B)
Ricinus communis	Castor bean	RCA	Galα or Gal-Gal-GlcNAc

(continued)

Proper name	Common name		Specificity
Sophora japonica	Pagoda tree	SJA	Galβ, GalNAcβ
Sphenostylis stenocarpa		SSA	Gal
Tetracarpidium conophorum	Nigerian walnut	TCA	Gal
Vicia villosa	Hairy vetch	VVA-A4	Gal-Gal-GalNAc
		VVA-B4	GalNAcα Tn antigen
Viscum album	Mistletoe	ML1	Gal-Gal-GalNAc
			Sial-Gal-Gal?
Wisteria floribunda		WFA	GalNAc(α- or β-, or 6Gal)
Fucose group			
Aleuria aurantia			Fucα-±GlcNAc
Anguilla anguilla	Eel	AAA	Fucα
Euonymus europa	Spindle tree	EEA	Fucα (A type 1 H chain)
Lotus tetragonolobus		LTA	Fucα
Pseudomonas aeruginosa		PA-II	Fucα Blood gp H, A, or B
Ulex europa	Gorse	UEA1	Fuc
Sialic acid group			
Allomyrina dichotoma	Japanese beetle		Sial2-6Galβ or Galβ
Limax flavus	Slug	LFA	Sial
Limulus polyphemus	Horse shoe crab	LPA	Sial
Maackia amurensis		MAA	Sial2–3Galβ1–4Glc/GlcNAc
Sambucus nigra	Elderberry	SNA	Sial-2–6-Gal
			Sial-2–6-GalNAc?
Sambucus sieboldiana	Japanese elderberry	SSA	As SNA
Mannose/Glucose/complex group			
Bowringia mildbraedii		BMA	Man
Canavalia ensiformis	Jack bean	ConA	Manα or Glcα or better oligo-Man
Galanthus nigalis	Snowdrop	GNA	Manα or better (Man)$_3$
Lathrus odoratus	Sweet pea		Man, Glc
Lens culinaris	Lentil	LHA	Manα (not terminal), Glc
Machaerium biovulatum		MBA	Man
Machaerium lunatus		MLA	Man
Narcissus lobularis		NLA	Man
Phaseolus coccineus			As PHA
Phaseolus vulgaris	Kidney bean	l-PHA	Man complex
		e-PHA	Man complex

Appendix

Proper name	Common name		Specificity
Pisum Sativa	Pea	PSA	Manα, Glc
Pterocarpus rhorii		PRA	Man
Vicia faba	Broad bean	VFA	Man, Glc
Vicia sativa	Common vetch	VSA	Man, Glc
Vigna racemosa		VRA	Man

GLcNAc group

Datura stramonium	Thorn apple	DSA	$(GlcNAc1–4)_n$ or Gal-β1–4GlcNAc or Galβ
Griffonia simplifolia		GSF-2	GlcNAcα or β
Lycopersicon esculentum	Tomato		$(GlcNAc1–4)_n$
Oryza sativa	Rice	OSA	$(GlcNAc1–4)_3$
Phytolacca americana	Pokeweed	PWM	GlcNAc
Psathyrella velestina		PVL	GlcNAc1–2Man Better than GlcNAc1–4 or 6
Solanum tuberosum	Potato	STA	GlcNAc
Triticum vulgaris	Wheat germ	WGA	GlcNAc also reacts with sialic
Ulex europa		UEA2	GlcNAc, Fuc-Gal-Glc
Urtica dioica	Stinging nettle	UDA	$(GlcNAc)_n$

Complex–uncertain

Robinia pseudoacacia	Black locust		Complex
Tulipa spp	Tulip		Man or GalNAc (2 sites)

Index

A

ABO(H) blood group antigens,
  expression on plasma glycoproteins,
    enzyme-linked immunosorbent assay,
      immunoprecipitation of proteins, 238
      materials, 237, 242
      plasma preparation, 237, 238
      reaction conditions, 241
      sensitivity, 243
    lectin blotting,
      electrophoresis and blotting, 238
      immunoprecipitation of proteins, 238
      materials, 237, 242
      plasma preparation, 237, 238
      reactivity of lectins on blots, 236, 243
    types of proteins, 235
  lectin specificity, 235
Abrin, toxicity, 9, 30
α-1 Acid glycoprotein (AGP), lectin-affinity electrophoresis of glycoforms,
  controls, 231
  dissociation constant determination for lectin–glycoprotein complex, 232
  materials, 228, 229
  one-dimensional electrophoresis, 227–229
  two-dimensional electrophoresis, 227, 228, 231
Acridine orange, apoptosis assay, 455–457
Adrenal, lectin histochemistry and cancer, 84, 85
Affinity chromatography, *see* Immunoaffinity chromatography; Lectin affinity chromatography
Agalactosyl immunoglobulin G,
  biosynthesis, 196, 197
  enzyme-linked immunosorbent assay, 197
  Psathyrella velutina lectin assay,
    biotinylation of lectin, 200, 201
    controls, 203
    enzyme-linked lectin binding assay, 200, 202
    immunoglobulin purification, 200
    lectin blotting, 200, 201
    reduction and alkylation of oligosaccharide chains, 202
  receptor affinities, 196, 197
  rheumatoid arthritis association, 195–197
  structure of sugar chain, 195
AGP, *see* α-1 Acid glycoprotein
Alkaline phosphatase, lectin blot detection, 160, 162
α-1-Antitrypsin,
  carbohydrate structure assessment by lectin reactivity, 210, 211
  crossed immunoaffinoelectrophoresis of variants with *Lens culinaris* agglutinin,
    electrophoresis conditions, 208, 211, 212
    lectin-reactive species in hepatocellular carcinoma, 208, 209
    staining of gels, 212
  mutation and disease, 207
  structure, 207
Apoptosis,
  assays of lectin effects,
    acridine orange staining, 455–457
    DNA fragmentation, 455, 457, 458
    flow cytometry, 455–458
    materials, 456

comparison to necrosis, 454, 455
triggers, 455
Avidin, detection of biotinylated lectins, 7, 15, 27, 28, 30, 31, 34

B

Bandeiraea simplicifolia I-isolectin B_4 (BSI-B_4),
  binding sites on unmyelinated primary sensory neurons, 518, 519
  neuronal tracing,
    electron microscopy, 520, 522–524
    fixation, 520, 521
    histochemistry, 520–522, 525
    injection of lectin–horseradish peroxidase conjugate, 519–521, 524, 525
    materials, 519, 520
    sectioning, 521
Bandeirea simplicifolia isolectin I (BS-I),
  endothelial cell staining, 9, 10, 76
B-cell, *see* Mitogenesis
Binding assay,
  epithelial cell lectin binding, 570, 575, 576
  gp120 and lectins, 557–562
  histochemistry applications, 19, 28, 29, 44, 45, 102, 108
  jacalin binding to CD4, 540, 541, 545–548, 551
  lectin monomer binding assay, 304, 305
Binding site, lectins,
  multivalency, 4, 22
  size, 22, 23
  subsites, 29
  unmyelinated primary sensory neurons, 518, 519
Biotinylation, lectins,
  enzyme-linked lectin binding assay, 200, 201
  histochemistry applications, 7, 11, 15, 17, 18, 30, 31, 34, 43, 46, 47, 49
Bladder, lectin histochemistry and cancer, 62
Blood group antigens, *see* ABO(H) blood group antigens
Bone, lectin histochemistry and cancer, 85, 86

Bone marrow transplantation,
  graft-vs-host disease, 329, 338, 339
  hematopoietic progenitor cell assay, 355, 357, 360
  hematopoietic progenitor cell enrichment,
    bone marrow, 339, 340
    human umbilical cord blood, 340–342
  purging for autologous transplantation,
    breast cancer, 333–336
    Burkitt's lymphoma, 337
    magnetic affinity cell sorting, 342–344
    neuroblastoma stage IV, 331–333
    principle, 330, 331
  T-cell depletion for allogeneic transplantation,
    haploidentical transplant, 338, 339
    siblings, 337, 338
Breast cancer,
  lectin histochemistry,
    electron microscopy,
      cell maintenance, 137
      fixation, 135, 136, 140
      postembedding technique, 138, 139
      pre-embedding technique, 137, 138, 140
    light microscopy, 10, 13, 16, 55, 56, 84
  purging for autologous bone marrow transplantation, 333–336, 342–344
BS-I, *see Bandeirea simplicifolia* isolectin 1
BSI-B_4, *see Bandeiraea simplicifolia* I-isolectin B_4
Burkitt's lymphoma, purging for autologous bone marrow transplantation, 337, 342–344

C

Cadaverine, *see* Polyamine uptake assay
Calcium flux,
  assay of jacalin binding to CD4, 543–546, 550, 551
  lectin effects on platelets, fura-2 assay,
    fluorescence measurement, 434–436, 438
    lectin selection, 434

Index

manganese quenching, 436
materials, 434, 435, 437
platelet preparation, 434, 435, 437
principle, 434
regulation, 433
Cartilage,
components, 65
lectin histochemistry,
degradative enzymes, 66–70
detection systems, 66
disease states, 86
fluorescein isothiocyanate conjugates and detection, 68
lectin selection, 66, 68
sample preparation, 69, 70
CD4,
gp120 binding, 167, 279, 539, 540
jacalin binding,
analysis of interaction, 540, 541, 545–548, 551
calcium flux assay, 543–546, 550, 551
effect on human immunodeficiency virus infection in vitro,
cell infection, 548, 549, 551
materials, 545, 546
mechanism, 541, 542
reverse transcriptase assay, 549, 551
structure, 540
Cell sorting, see Fluorescence-activated cell sorting; Magnetic affinity cell sorting
Central nervous system, lectin histochemistry and cancer, 86, 87
Cervix, lectin histochemistry and cancer, 83
Chondroitinase ABC, cartilage degradation and lectin histochemistry, 66–70
CIAE, see Crossed immunoaffinoelectrophoresis
Colloidal gold,
history of development, 121
lectin histochemistry,
electron microscopy,
lectin–gold conjugates after embedding, 114–119
particle diameter selection, 122–124
silver intensification, 121–129

light microscopy,
digoxigenin complex visualization, 42, 43
lectin staining of ultrathin sections, 34–36, 41
silver intensification, 41, 42, 44, 46, 50, 121, 122
Colon,
crypt cell proliferation assay,
biopsy, 478, 479
calculation of rate, 476, 478, 482
lectin incubations, 478–480
materials, 477
microdissection and crypt cell counting, 480, 481
organ culture, 478, 480
overview of assays, 475, 476
dietary lectins and epithelial proliferation, 475
lectin histochemistry,
cancer, 79, 80
inflammatory bowel disease, 79
normal tissue, 79
Con A, see Concanavalin A
Concanavalin A (Con A),
binding specificity, 3, 4, 159, 232
cytoskeleton effects, 411, 413, 414
histochemistry applications, 8, 9
lectin-affinity electrophoresis of α-1 acid glycoprotein glycoforms, 228–232
mitogenesis, 365, 385, 386
Confocal microscopy, imaging of lectin-mediated endocytosis, 570, 571, 577–579
Crossed immunoaffinoelectrophoresis (CIAE),
α-1-antitrypsin variants with *Lens culinaris* agglutinin,
electrophoresis conditions, 208, 211, 212
lectin-reactive species in hepatocellular carcinoma, 208, 209
staining of gels, 212
α-fetoprotein variants, 215
Crypt cell, see Colon
Cytoskeleton,

lectin effects,
   concanavalin A, 411, 413, 414
   immunofluorescence microscopy,
      cultured cell labeling, 415, 416
      materials, 414, 415
      paraffin-embedded tissue labeling, 415–417
   peanut agglutinin, 411, 412
   phytohemagglutinin, 413
   receptor mediation, 410–414
   soybean agglutinin, 411–413
   wheatgerm agglutinin, 411, 413, 414
organelle maintenance, 410
organization in mammalian cells,
   actin, 408
   crosslinking, 409
   intermediate filaments, 407, 409
   microtubules, 407–409
   myosin, 408, 410
Cytotoxicity, *see* Lectin-dependent cell cytotoxicity

D

DAB, *see* 3,3'-Diaminobenzidine tetrahydrochloride
Degranulation,
   assay of lectin effects,
      elastase assay, 445, 450
      enzyme supernatant preparation, 444, 450
      lactate dehydrogenase control, 449
      lysozyme assay, 445
      materials, 441, 442
      myeloperoxidase assay, 445
      neutrophil isolation, 441–443, 445, 446
   host defense, 441
3,3'-Diaminobenzidine tetrahydrochloride (DAB), *see* Peroxidase
Diet, *see also* Heat processing; Proteolytic digestion,
   antinutritional effects of lectins, 485, 486, 505
   lectin effects in rats,
      chemical analysis of tissues, 490–493
      composition of diet, 486, 487, 489
      controls, 493
      feeding protocol, 490, 491
      lectin quantification, 493
      small intestine morphology, 492, 493
DIG, *see* Digoxigenin
Digestion, *see* Proteolytic digestion
Digoxigenin (DIG),
   advantages over biotin systems, 42
   lectin conjugates in histochemistry, 42, 46, 47
Dolichos biflorus agglutinin, blood group antigen recognition, 235
Drug delivery,
   lectin-mediated transcytosis,
      Caco-2 cell culture, 569, 571, 578
      confocal imaging of endocytosis, 570, 571, 577–579
      cytotoxicity assays,
         lactate dehydrogenase, 568, 569, 573, 579
         MTT, 569, 572
      fluorescein-sodium assay, 569, 574–577
      lectin binding analysis, 570, 575, 576
      principle, 568
      transepithelial electrical resistance assay, 569, 573, 574, 576, 577, 579
   liposome–lectin conjugates in drug targeting,
      applications, 583
      proteoliposome preparation,
         conjugation, 584, 586, 590, 591
         lectin activation, 584, 586, 589, 592, 593
         lectin density determination, 593
         lipid activation, 584–587, 591
         liposome preparation, 584–587, 592
         materials, 584–586
         protein determination, 590, 591
   residence time of drugs, 567

E

EEA, *see Evonymus europaeus* agglutinin
Elastase, degranulation assay, 445, 450
Electron microscopy, *see* Histochemistry, electron microscopy
β-Elimination,
   glycan structure analysis, 180, 184, 187

lectin blots, 163, 164
ELISA, *see* Enzyme-linked immunosorbent assay
Ellman's reagent, quantification of lectin sulfhydryl groups, 592, 593
Endocytosis,
   analysis with fluorescein isothiocyanate conjugates, 58, 59
   lectin-mediated endocytosis in drug delivery,
      Caco-2 cell culture, 569, 571, 578
      confocal imaging of endocytosis, 570, 571, 577–579
      cytotoxicity assays,
         lactate dehydrogenase, 568, 569, 573, 579
         MTT, 569, 572
      fluorescein-sodium assay, 569, 574–577
      lectin binding analysis, 570, 575, 576
      principle, 568
      transepithelial electrical resistance assay, 569, 573, 574, 576, 577, 579
Endometrium, lectin histochemistry and cancer, 84
Endothelial cell,
   fluorescence-activated cell sorting, 319
   lectin staining,
      neoplastic vessels, 77
      normal vessels, 9, 10, 12, 13, 76, 77
   magnetic affinity cell sorting,
      cell isolation from liver, 320, 322–324
      immunofluorescent staining, 323
      lectin selection, 319, 320
      magnetic bead,
         preparation, 322–324, 326
         removal, 322, 324, 326
         separation, 324
      materials, 321, 322
      principle, 320
      purity assay, 322–324
Enzyme-linked immunosorbent assay (ELISA),
   ABO(H) blood group antigens expressed on plasma glycoproteins,
      immunoprecipitation of proteins, 238

   materials, 237, 242
   plasma preparation, 237, 238
   reaction conditions, 241
   sensitivity, 243
agalactosyl immunoglobulin G, 197
gp120, 157–160, 162, 281, 283, 284, 286, 557–560, 562
lectin monomer binding assay, 304, 305
mucins,
   intestinal mucins, 256–258, 260, 498, 499, 501
   lectin/antibody sandwich assay,
      calibration, 252, 253
      materials, 250, 251
      monoclonal antibody, 249, 250
      pancreatic cancer diagnosis, 249, 250
      protocols, 251, 252
      wheatgerm agglutinin capture, 249, 250
phytohemagglutinin,
   indirect competitive assay, 507, 511, 512
   indirect noncompetitive assay, 507, 511
   monitoring of proteolytic digestion in gut, 507–510
Epithelial cell,
   lectin-mediated transcytosis for drug delivery,
      Caco-2 cell culture, 569, 571, 578
      confocal imaging of endocytosis, 570, 571, 577–579
      cytotoxicity assays,
         lactate dehydrogenase, 568, 569, 573, 579
         MTT, 569, 572
      fluorescein-sodium assay, 569, 574–577
      lectin binding analysis, 570, 575, 576
      principle, 568
      transepithelial electrical resistance assay, 569, 573, 574, 576, 577, 579
mitogenesis,
   lectin stimulation,
      cell culture and incubation conditions, 381–383

materials, 380
tritiated thymidine assay, 380, 381, 383
tumors, 379
polyamine uptake assay in small intestine,
DNA analysis, 398, 401
high-performance liquid chromatography analysis of polyamines, 396, 397, 399, 400, 402, 403
luminal versus basolateral uptake, 396
materials, 397–399
ornithine decarboxylase activity during polyamine accumulation, 395, 396, 399, 402
polyamine synthesis, 394, 395
protein analysis, 398, 400
RNA analysis, 398, 400, 401
uptake conditions, 398, 399, 401–403
Streptococcus pyogenes adhesion,
adhesins, 529, 530
assays,
artifacts, 530, 532, 535
cell counting, 532, 535
monolayer adhesion, 532–534
lectin blocking, 530, 533, 534–536
M6, 530
Escherichia coli, adhesins, 529
Evonymus europaeus agglutinin (EEA), endothelial cell marker, 320

F

FACS, *see* Fluorescence-activated cell sorting
α-Fetoprotein,
expression in development and cancer, 215
lectin-affinity electrophoresis,
antibody–affinity blotting, 218–220
immunoenzymatic amplification and visualization, 218, 220–223
lectin selection, 218, 223
materials, 218–220
one-dimensional electrophoresis, 216, 221, 222

sensitivity, 216
two-dimensional electrophoresis, 216, 218, 222
FITC, *see* Fluorescein isothiocyanate
Fixation, *see* Histochemistry
Flow cytometry, *see also* Fluorescence-activated cell sorting,
apoptosis assay, 455–458
Golgi elements,
flow cytometry parameters, 309, 310, 313, 314
functional Golgi regions, 308
isolation of Golgi subfractions, 309, 310, 312, 313
lectin selection and preparation, 308, 309, 312
software for analysis, 314
titration of lectin-specific binding, 310, 312
intracellular glycoconjugates, 307, 308
monomers of lectins,
agglutination prevention, 301
cell culture, 303, 304
flow cytometry, 303
lectin,
enzyme-linked immunosorbent assay, 304, 305
labeling with fluorescein isothiocyanate, 303, 305
monomer preparation, 301, 303, 304
purification, 302
materials, 302, 303
overview, 307
Fluorescein isothiocyanate (FITC), lectin conjugates,
advantages in detection, 56
cartilage analysis, 68
cytophotometric analysis, 58, 59, 63
fluorescence-activated cell sorting analysis, 55, 59, 60, 62, 63
histochemistry of sectioned material, 57, 63
Fluorescence-activated cell sorting (FACS), lectin conjugates, 55, 59, 60, 62, 63
Fura-2,
assay of jacalin binding to CD4, 543–546, 550, 551

Index

assay of lectin effects on platelet calcium flux,
  fluorescence measurement, 434–436, 438
  lectin selection, 434
  manganese quenching, 436
  materials, 434, 435, 437
  platelet preparation, 434, 435, 437
  principle, 434

G

Galanthus nivalis agglutinin (GNA),
  binding specificity, 280
  gp120,
    capture enzyme-linked immunosorbent assay, 557–560, 562
    lectin affinity chromatography,
      column preparation, 283, 284, 287
      materials, 283, 284
      running conditions, 283–287
      sample preparation, 283–285
gC-1,
  lectin specificity, 177
  molecular organization, 176
  O-glycosylation sites, 176
  purification,
    immunoaffinity chromatography, 185, 186, 190
    lectin affinity chromatography, 178, 179, 183–185
  radiolabeling, 183–185, 189
  structure analysis,
    acid hydrolysis and thin-layer chromatography, 181, 184, 188–190
    β-elimination, 180, 184, 187
    gel filtration,
      peptide fragments, 180
      O-linked glycans, 181, 187, 188, 190
    hydrazinolysis, 180, 187
    lectin affinity chromatography of fragments, 180, 184, 186
    pronase treatment, 179, 180, 184, 186, 190
Gel filtration,
  glycopeptide fragments, 180
  O-linked glycans, 181, 187, 188, 190
Glycan,
  detection, *see* ABO(H) blood group antigens; Agalactosyl immunoglobulin G; α-1-Antitrypsin; α-Fetoprotein; Histochemistry
  disease roles, overview, 73, 74, 147
  functions, overview, 21, 147
  linkage discrimination by alkali hydrolysis and lectin histochemistry, 29, 35, 36
  structure analysis, *see* gC-1; gp120; Lectin blot; Nuclear magnetic resonance
Glycosidase,
  lectin blot treatment, 163, 164
  substrate specificity, 29
  treatment of sections, 35–37
GNA, *see Galanthus nivalis* agglutinin
Gold, *see* Colloidal gold
Golgi apparatus,
  flow cytometry with fluorescent lectins,
    flow cytometry parameters, 309, 310, 313, 314
    isolation of Golgi subfractions, 309, 310, 312, 313
    lectin selection and preparation, 308, 309, 312
    software for analysis, 314
    titration of lectin-specific binding, 310, 312
  functional Golgi regions, 308
gp120,
  affinity chromatography with *Galanthus nivalis* agglutinin,
    column preparation, 283, 284, 287
    materials, 283, 284
    running conditions, 283–287
    sample preparation, 283–285
  detection and quantification with lectin enzyme-linked immunosorbent assay, 281, 283, 284, 286, 557–560, 562
  glycosylation,
    CD4 binding role, 167, 279, 539
    sites, 167, 279, 555
  lectin binding,

effect on virus infectivity, 167, 279, 280, 539, 555, 556
screening for antiviral agents,
  binding assay, 557–562
  controls, 562
  gp120 quantification, 557–560, 562
  materials, 557–559
specificity of lectins, 279, 280
lectin blotting,
  glycosidase degradation, 169–171
  lectin selection, 172
  materials, 169, 170
  peroxidase detection, 168–170
preparation from cell lines, 168, 169
processing, 279, 555, 562
Western blotting, 169, 170, 557, 558, 560–562
Griffonia simplicifolia lectin (GSL),
  endothelial cell marker, 320
GSL, *see Griffonia simplicifolia* lectin

H

Heat processing,
  aqueous heat inactivation of lectins,
    efficiency, 506
    hemagglutination assay, 510, 512, 513
    materials, 507, 508
    protocol, 509, 512
  resistance of lectins, 505
  techniques, 505, 506
Helix pomatia lectin (HPA),
  binding specificity, 3, 10
  blood group antigen recognition, 235
  cancer staining, 10, 13, 55, 56
Hematopoietic progenitor cell, *see* Bone marrow transplantation
Hepatocellular carcinoma,
  α-1-antitrypsin variants, 208, 209
  α-fetoprotein expression, 215
Herpes simplex virus type 1 (HSV-1), *see also* gC-1,
  glycoprotein types, 175, 176
High-performance liquid chromatography (HPLC), polyamine uptake assay in small intestine, 396, 397, 399, 400, 402, 403

Histiocyte, lectin histochemistry and disease, 86
Histochemistry, electron microscopy,
  advantages over light microscopy, 133
  colloidal gold detection,
    advantages over peroxidase detection, 134, 135
    breast cancer cell lines,
      cell maintenance, 137
      fixation, 135, 136, 140
      postembedding technique, 138, 139
      pre-embedding technique, 137, 138, 140
    lectin–gold conjugates after embedding, 114–119, 134
    particle diameter selection, 122–124
    silver intensification,
      buffer, 126
      fixer, 126–128
      glutaraldehyde postfixation, 127, 128
      incubation conditions, 126, 128, 129
      lugol iodine pretreatment, 124
      osmium effects, 127, 129
      overview, 121–124
      protective colloid, 125, 126
      reducing agent, 125
      silver ions, 124, 125
      water purity, 126
  control incubations, 102, 108, 118
  embedding, 102, 103, 107, 111–113, 115, 116, 118, 134, 135
  fixation,
    postembedded gold staining, 111, 114, 115, 135, 136
    pre-embedded peroxidase staining, 98, 99, 101–103, 105
  lectin incubations, 102, 108, 116, 117, 119
  lectin selection, 103, 113, 114
  materials, 100, 101
  neuronal tracing with lectins, 520, 522–524
  peroxidase detection,
    diffusion and resolution, 99
    osmium black conversion of 3,3'-diaminobenzidine

Index

tetrahydrochloride product, 99, 106, 107
principle, 97, 98
reaction conditions, 102, 105
substrate preparation, 105, 106
sectioning, 107, 113, 114, 116, 118
Histochemistry, light microscopy,
adrenal, 84, 85
availability of lectins, 4
avidin–biotin methods, 7, 27, 28, 30, 31, 34
binding conditions, 15, 19, 32–35, 46, 70
bladder, 62
bone, 85
bone, 86
breast, 10, 13, 16, 55, 56, 84
cancers, 10, 13, 16, 76, 87
cartilage, 66–70, 86
central nervous system, 86, 87
cervix, 83
detection systems, see Colloidal gold; Fluorescein isothiocyanate; Peroxidase
direct methods, 5, 12, 15, 16
endometrium, 84
endothelial cells, 9, 10, 16, 76, 77
experimental design, 8, 9
fibrous tissue, 85
glycan linkage discrimination, 29, 35, 36
glycoprotein storage disorder analysis, 76
immune cells, 86
indirect antibody method, 5, 12, 13, 16, 18
indirect sandwich method, 5, 6
kidney, 81, 82
kits, 7, 8
large intestine, 79, 80
lectin buffer, 10, 17, 36
lectin conjugates, see Biotinylation; Colloidal gold; Digoxigenin; Fluorescein isothiocyanate; Peroxidase
lectin selection, 8, 9, 23, 24, 48, 66, 73–75
liver, 80
microbiological applications, 87
mouth, 78
neuronal tracing with lectins, 520–522, 525
ovary, 84
pancreas, 80
placenta, 83
prostate, 83
resin embedding,
coating of slides, 32
controls, 28, 29
dewaxing of paraffin samples, 27, 31, 32, 47
embedding reaction, 30, 45
fixation, 24
fixation, 30, 31
glycosidase treatment of sections, 35–37
lectin staining, 32–35
resins, 24–26
staining intensity, 26, 27
respiratory tract, 80, 81
salivary glands, 78
sample preparation,
cell smears, 10, 13, 14
cryostat sections, 11, 14, 55, 75
fixation, 11, 14, 18, 42, 45, 50, 55, 56, 75
overview, 4
paraffin-embedded sections, 11, 14, 18, 45, 75
tissue imprints, 10, 13
trypsinization, 11, 14, 17
skeletal muscle, 85
skin, 77
small intestine, 79
stomach, 78
testis, 84
thyroid, 85
tissue selection, 9
toxic lectin handling, 9, 30
HIV, see Human immunodeficiency virus
Horseradish peroxidase, see Peroxidase
HPA, see Helix pomatia lectin
HPLC, see High-performance liquid chromatography
HSV-1, see Herpes simplex virus type 1
Human immunodeficiency virus (HIV), see CD4; gp120

Hydrazinolysis, glycan structure analysis, 180, 187
Hydrogen peroxide, *see* Oxidative burst

I

IgA, *see* Immunoglobulin A
IgG, *see* Immunoglobulin G
Immunoaffinity chromatography, gC-1, 185, 186, 190
Immunofluorescence microscopy, lectin effects on cytoskeleton,
    cultured cell labeling, 415, 416
    materials, 414, 415
    paraffin-embedded tissue labeling, 415–417
Immunoglobulin A (IgA), secretory proteins,
    anion-exchange chromatography, 267–269
    cation exchange chromatography of IgA2, 268, 270, 273
    characterization of isotypes, 268, 271
    colostrum protein purification, 268, 269, 271, 272, 275
    gel filtration, 269, 272, 274
    isoforms, 265
    jacalin lectin affinity chromatography of IgA1
        binding of IgA1, 265–267
        history of application in purification, 266
        materials, 267–269
        running conditions, 269, 272, 276
        serum preparation, 267, 269
        structure, 265
Immunoglobulin G (IgG), *see* Agalactosyl immunoglobulin G
Immunoprecipitation, protein-tyrosine kinase assay, 424–430
Intestine, *see* Colon; Small intestine

J

Jacalin,
    binding specificity, 265–267
    CD4 binding,
        analysis of interaction, 540, 541, 545–548, 551
        calcium flux assay, 543–546, 550, 551
        effect on human immunodeficiency virus infection in vitro,
            cell infection, 548, 549, 551
            materials, 545, 546
            mechanism, 541, 542
            reverse transcriptase assay, 549, 551
    lectin affinity chromatography of IgA1
        binding of IgA1, 265–267
        history of application in purification, 266
        materials, 267–269
        running conditions, 269, 272, 276
        serum preparation, 267, 269
        structure, 266

K

Keratinase, cartilage degradation and lectin histochemistry, 66–70
Kidney, lectin histochemistry,
    glomeruli, 82
    tubules, 81, 82

L

Lactate dehydrogenase (LDH), cytotoxicity assay for lectin-mediated transcytosis, 568, 569, 573, 579
Large intestine, *see* Colon
LCA, *see* Lens culinaris agglutinin
LDCC, *see* Lectin-dependent cell cytotoxicity
LDH, *see* Lactate dehydrogenase
Lectin affinity chromatography,
    gC-1, 178, 179, 183–185
    glycopeptide fragments, 180, 184, 186
    gp120, lectin affinity chromatography with *Galanthus nivalis* agglutinin,
        column preparation, 283, 284, 287
        materials, 283, 284
        running conditions, 283–287
        sample preparation, 283–285
    IgA1 purification with jacalin,
        binding of IgA1, 265–267
        history of application in purification, 266
        materials, 267–269
        running conditions, 269, 272, 276
        serum preparation, 267, 269
    T-cell receptor, 292–294, 297

Lectin-affinity electrophoresis,
  α-1 acid glycoprotein glycoforms,
    controls, 231
    dissociation constant determination for lectin–glycoprotein complex, 232
    materials, 228, 229
    one-dimensional electrophoresis, 227–229
    two-dimensional electrophoresis, 227, 228, 231
  α-fetoprotein,
    antibody–affinity blotting, 218–220
    immunoenzymatic amplification and visualization, 218, 220–223
    lectin selection, 218, 223
    materials, 218–220
    one-dimensional electrophoresis, 216, 221, 222
    sensitivity, 216
    two-dimensional electrophoresis, 216, 218, 222
  information obtainable for glycoproteins, 227
Lectin blot,
  ABO(H) blood group antigens expressed on plasma glycoproteins,
    electrophoresis and blotting, 238
    immunoprecipitation of proteins, 238
    materials, 237, 242
    plasma preparation, 237, 238
    reactivity of lectins on blots, 236, 243
  acid hydrolysis, 154, 155, 163
  agalactosyl immunoglobulin G, 200, 201
  alkaline phosphatase detection, 160, 162
  detection system overview, 160
  electrophoresis and blotting of glycoproteins, 154–156, 160, 162, 164
  β-elimination, 163, 164
  glycosidase treatment, 163, 164
  gp120,
    glycosidase degradation, 169–171
    lectin selection, 172
    materials, 169, 170
    peroxidase detection, 168–170
  incubation conditions for lectin blot, 155, 162
  materials, 148–153, 161, 162
  mucins from intestine, 256–260
  peroxidase detection, 155, 156
  sample preparation, 163, 164
  Smith degradation, 148, 155
  specificity controls, 160–163
  storage of blots, 156
  structure analysis of glycans, principle, 148
Lectin-dependent cell cytotoxicity (LDCC),
  assays,
    lung transplantation evaluation,
      applications, 462
      bronchoalveolar lavage lymphocyte isolation, 461, 465, 467–469
      calculations, 469, 470
      cell plating, 465, 468
      concanavalin A, 463, 464
      incubation conditions, 468, 469
      materials, 464–466
      peripheral blood lymphocyte isolation, 465–467
      target cell preparation, 464, 466
    MTT assay, 454, 456, 457
    overview, 454, 458
  cytokine role, 463
  lectin types, 453
  mechanism, 462–464
  neuramidase treatment effects, 464
  resistant cell variants, 453, 454
Lectin-mediated transcytosis, *see* Drug delivery
Lens culinaris agglutinin (LCA),
  crossed immunoaffinoelectrophoresis of α-1-antitrypsin variants,
    electrophoresis conditions, 208, 211, 212
    lectin-reactive species in hepatocellular carcinoma, 208, 209
    staining of gels, 212
  lectin-affinity electrophoresis of α-fetoproteins, 218, 223
Liposome, *see* Drug delivery
Liver, lectin histochemistry and cancer, 80
Lung transplantation,
  lectin-dependent cell cytotoxicity assay,

applications, 462
bronchoalveolar lavage lymphocyte isolation, 461, 465, 467–469
calculations, 469, 470
cell plating, 465, 468
concanavalin A, 463, 464
incubation conditions, 468, 469
materials, 464–466
peripheral blood lymphocyte isolation, 465–467
target cell preparation, 464, 466
rejection differentiation from infection, 461, 462
Lung, lectin histochemistry and cancer, 80, 81
Lycopersicon esculentum lectin, endothelial cell marker, 320
Lysozyme, degranulation assay, 445

M

α_2M, *see* α_2-Macroglobulin
Maackia arnurensis leukoagglutinin (MAL),
flow cytometry application,
cell culture, 303, 304
flow cytometry parameters, 303
lectin,
enzyme-linked immunosorbent assay, 304, 305
labeling with fluorescein isothiocyanate, 303, 305
monomer preparation, 301, 303, 304
purification, 302
materials, 302, 303
specificity, 301
Machaerium biovulatum agglutinin (MBA), inhibition of human immunodeficiency virus infection, 556
Maclura pomifera lectin, binding specificity, 29
α_2-Macroglobulin (α_2M), blood group antigen expression,
enzyme-linked immunosorbent assay,
immunoprecipitation of proteins, 238
materials, 237, 242

plasma preparation, 237, 238
reaction conditions, 241
sensitivity, 243
lectin blotting,
electrophoresis and blotting, 238
immunoprecipitation of proteins, 238
materials, 237, 242
plasma preparation, 237, 238
reactivity of lectins on blots, 236, 243
Macrophage, lectin histochemistry and disease, 86
Magnetic affinity cell sorting,
bone marrow purging and hematopoietic progenitor cell enrichment in transplantation,
hematopoietic progenitor cell enrichment,
bone marrow, 339, 340
human umbilical cord blood, 340–342
overview, 329, 330
purging for autologous transplantation,
breast cancer, 333–336
Burkitt's lymphoma, 337
magnetic affinity cell sorting, 342–344
myeloma, 351–360, 354–356
neuroblastoma stage IV, 331–333
principle, 330, 331
T-cell depletion for allogeneic transplantation,
haploidentical transplant, 338, 339
siblings, 337, 338
endothelial cells,
cell isolation from liver, 320, 322–324
immunofluorescent staining, 323
lectin selection, 319, 320
magnetic bead,
preparation, 322–324, 326
removal, 322, 324, 326
separation, 324
materials, 321, 322

Index

purity assay, 322–324
principle, 320
MAL, *see Maackia arnurensis* leukoagglutinin
Mast cell, lectin histochemistry and disease, 86
MBA, *see Machaerium biovulatum* agglutinin
Mitogenesis,
   antimitogenic lectins, 366, 372
   epithelial cells,
      lectin stimulation,
         cell culture and incubation conditions, 381–383
         materials, 380
         polyamine uptake, *see* Polyamine uptake assay
         tritiated thymidine assay, 380, 381, 383
      tumors, 379
   lymphocytes,
      accessory molecules, 368–372
      assays of lectin response,
         cell preparation and incubation, 388, 389
         macromolecule synthesis and tritiated thymidine assay, 373, 374, 387, 389–391
         materials, 388
         microscopy, 373, 387
         MTT assay, 374
         standardization, 389, 391
      cytokine role, 367–371
      enhancement by TPA, 367, 368
      history of lectin studies, 365, 366
      lectin response in immune deficiency, 386, 387
      major events, 373, 374
      mitogenic lectins, 365, 385, 386
      pathological applications of lectins, 386
      receptor binding by lectins, 367, 370
      signal transduction, 367, 368, 372
      superantigens, lectins comparison, 372, 373
Mouth, lectin histochemistry and cancer, 78
M protein, *see Streptococcus pyogenes*

MTT,
   cytotoxicity assay for lectin-mediated transcytosis, 569, 572
   lectin cytotoxicity assay, 454, 456, 457
   mitogenesis assay, 374
Mucin,
   intestinal mucins,
      antibody production, 496, 501
      histochemistry, 499, 500
      purification, 496, 497, 501
      quantification,
         enzyme-linked immunosorbent assay, 256–258, 260, 498, 499, 501
         lectin slot blotting, 256–260
         lectin versus antibody methods, 255
      secretion assays, 496–499
      structure, 255
   lectin/antibody sandwich enzyme-linked immunosorbent assay,
      pancreatic cancer diagnosis, 249, 250
      calibration, 252, 253
      materials, 250, 251
      monoclonal antibody, 249, 250
      protocols, 251, 252
      wheatgerm agglutinin capture, 249, 250
Muscular dystrophy, lectin histochemistry, 85
Mushroom, lectin extraction, 11, 14, 15, 18
Myeloma, purging for autologous bone marrow transplantation,
   melphalan therapy, 351
   monoclonal antibodies, 352, 353, 355–357
   peanut agglutinin purging,
      bone marrow harvest, 354, 355
      cell recovery and efficiency, 358–360
      cryopreservation, 355, 357
      hematopoietic progenitor cell assay, 355, 357, 360
      large-scale development, 353, 354, 358
      magnetic affinity cell sorting,
         bead preparation, 354, 356
         purging, 355–357
      materials, 354, 355

overview, 352, 353
Myeloperoxidase, degranulation assay, 445

N

Neuramidase, cartilage degradation and lectin histochemistry, 66–70
Neuroblastoma, purging for autologous bone marrow transplantation, 331–333, 342–344
Neuronal tracing, lectins,
 binding sites on unmyelinated primary sensory neurons, 518, 519
 electron microscopy, 520, 522–524
 fixation, 520, 521
 histochemistry, 520–522, 525
 injection of lectin–horseradish peroxidase conjugate, 519–521, 524, 525
 materials, 519, 520
 overview, 517
 sectioning, 521
Neutrophil,
 degranulation, assay of lectin effects,
  elastase assay, 445, 450
  enzyme supernatant preparation, 444, 450
  lactate dehydrogenase control, 449
  lysozyme assay, 445
  materials, 441, 442
  myeloperoxidase assay, 445
  neutrophil isolation, 441–443, 445, 446
 oxidative burst, assay of lectin effects,
  hydrogen peroxide assay, 443, 444, 447, 449
  materials, 441, 442
  neutrophil isolation, 441–443, 445, 446
  superoxide assay, 443, 446, 447
NMR, see Nuclear magnetic resonance
Nuclear magnetic resonance (NMR), glycan structure analysis, 148

O

ODC, see Ornithine decarboxylase
Organ culture, see Colon
Ornithine decarboxylase (ODC), activity during polyamine accumulation in small intestine, 395, 396, 399, 402
Orosomucoid, see α-1 Acid glycoprotein
Osmium black,
 avoidance before silver enhancement, 127
 peroxidase product enhancement for electron microscopy, 99, 106, 107
Ovary, lectin histochemistry and cancer, 84
Oxidative burst,
 assay of lectin effects,
  hydrogen peroxide assay, 443, 444, 447, 449
  materials, 441, 442
  neutrophil isolation, 441–443, 445, 446
  superoxide assay, 443, 446, 447
  thymocyte isolation, 441–443
 host defense, 441

P

Pancreatic cancer,
 diagnosis with lectin/antibody sandwich assay of mucin,
  calibration, 252, 253
  materials, 250, 251
  monoclonal antibody, 249, 250
  protocols, 251, 252
  sensitivity and specificity, 249, 250
  wheatgerm agglutinin capture, 249, 250
 lectin histochemistry, 80
Peanut agglutinin (PNA),
 cytoskeleton effects, 411, 412
 endothelial cell binding in histochemistry, 10, 12
 purging for autologous bone marrow transplantation,
  bone marrow harvest, 354, 355
  cell recovery and efficiency, 358–360
  cryopreservation, 355, 357
  hematopoietic progenitor cell assay, 355, 357, 360
  large-scale development, 353, 354, 358
  magnetic affinity cell sorting,
   bead preparation, 354, 356
   purging, 355–357
  materials, 354, 355

overview, 352, 353
Peroxidase,
  lectin blot detection, 155, 156, 168–170
  lectin histochemical staining detection,
    3,3'-diaminobenzidine
      tetrahydrochloride substrate
      preparation, 105, 106
    electron microscopy,
      diffusion and resolution, 99
      osmium black conversion of 3,3'-
        diaminobenzidine
        tetrahydrochloride product,
        99, 106, 107
      principle of detection, 97, 98
      reaction conditions, 102, 105
    light microscopy, 5, 11, 14, 15, 17,
      27, 28, 30, 31, 34, 43
    neuronal tracing with lectin conjugates,
      520–522, 525
PHA, *see* Phytohemagglutinin
Phosphotyrosine, *see* Protein-tyrosine
  kinase
Phytohemagglutinin (PHA),
  cytoskeleton effects, 413
  effects on intestinal microbes, 495, 496
  enzyme-linked immunosorbent assay,
    indirect competitive assay, 507, 511,
      512
    indirect noncompetitive assay, 507,
      511
    monitoring of proteolytic digestion,
      507–510
  growth retardation effects, 495
  karyotyping, 386
  mitogenic activity, 365, 366
PKC, *see* Protein kinase C
Placenta, lectin histochemistry, 83
Platelet, lectin effects on calcium flux, fura-
  2 assay,
  fluorescence measurement, 434–436,
    438
  lectin selection, 434
  manganese quenching, 436
  materials, 434, 435, 437
  platelet preparation, 434, 435, 437
  principle, 434
PNA, *see* Peanut agglutinin
Pokeweed mitogen, discovery, 366

Polyamine uptake assay,
  physiological roles of polyamines, 393,
    394
  small intestine,
    DNA analysis, 398, 401
    high-performance liquid
      chromatography analysis of
      polyamines, 396, 397, 399,
      400, 402, 403
    luminal versus basolateral uptake,
      396
    materials, 397–399
    ornithine decarboxylase activity
      during polyamine
      accumulation, 395, 396, 399,
      402
    polyamine synthesis, 394, 395
    protein analysis, 398, 400
    RNA analysis, 398, 400, 401
    uptake conditions, 398, 399, 401–
      403
  structures of polyamines, 393, 394
  tissue concentrations of polyamines,
    393, 394
Programmed cell death, *see* Apoptosis
Pronase, hydrolysis and structural analysis
  of gC-1, 179, 180, 184, 186, 190
Prostate, lectin histochemistry and cancer,
  83
Protein kinase C (PKC), lymphocyte
  mitogenesis role, 367, 368, 372
Protein-tyrosine kinase (PTK),
  immunoprecipitation kinase assay, 424–
    430
  lectin-induced signaling, 423
  phosphoamino acid analysis, 425, 427,
    430
  phosphotyrosine immunoblotting, 423,
    424, 426–431
Proteolytic digestion,
  monitoring in gut by enzyme-linked
    immunosorbent assay,
    indirect competitive assay, 507, 511,
      512
    indirect noncompetitive assay, 507,
      511
    tissue preparation, 507–510
  resistance of lectins, 506, 507

Psathyrella velutina lectin (PVL),
  agalactosyl immunoglobulin G assay,
    biotinylation of lectin, 200, 201
    controls, 203
    enzyme-linked lectin binding assay, 200, 202
    immunoglobulin purification, 200
    lectin blotting, 200, 201
    reduction and alkylation of oligosaccharide chains, 202
PTK, *see* Protein-tyrosine kinase
Putrescine, *see* Polyamine uptake assay
PVL, *see* Psathyrella velutina lectin

R

Reed-Sternberg cell, lectin histochemistry and disease, 86
Resin embedding, *see* Histochemistry
Ricin,
  affinity chromatography of T-cell receptor, 292–294, 296, 297
  toxicity, 9, 30, 296, 297

S

Salivary gland, lectin histochemistry and cancer, 78
Sambucus sieboldiana agglutinin (SSA),
  flow cytometry application,
    cell culture, 303, 304
    flow cytometry, 303
  lectin,
    enzyme-linked immunosorbent assay, 304, 305
    labeling with fluorescein isothiocyanate, 303, 305
    monomer preparation, 301, 303, 304
    purification, 302
    materials, 302, 303
    specificity, 301
SBA, *see* Soybean agglutinin
Selectivity, definition, 23
Sialic acid, lectin binding specificity, 159
Sialidase, lectin specificity assays, 49, 50
Silver acetate, intensification of colloidal gold signal,
  electron microscopy, 121–129
  light microscopy, 41, 42, 44, 46, 50
Simian immunodeficiency virus (SIV), *see* gp120

SIV, *see* Simian immunodeficiency virus
Skeletal muscle, lectin histochemistry and disease, 85
Skin, lectin histochemistry and cancer, 77
Small intestine,
  bacterial adhesion assay, 497, 500, 501
  dietary lectin effects,
    bacteria, 495, 496
    rat morphology, 492, 493
  lectin histochemistry and disease, 79
  mucins,
    antibody production, 496, 501
    histochemistry, 499, 500
    purification, 496, 497, 501
    quantification,
      enzyme-linked immunosorbent assay, 256–258, 260, 498, 499, 501
      lectin slot blotting, 256–260
      lectin versus antibody methods, 255
    secretion assays, 496–499
    structure, 255
  polyamine uptake assay,
    DNA analysis, 398, 401
    high-performance liquid chromatography analysis of polyamines, 396, 397, 399, 400, 402, 403
    luminal versus basolateral uptake, 396
    materials, 397–399
    ornithine decarboxylase activity during polyamine accumulation, 395, 396, 399, 402
    polyamine synthesis, 394, 395
    protein analysis, 398, 400
    RNA analysis, 398, 400, 401
    uptake conditions, 398, 399, 401–403
Smith degradation, lectin blotting in structure determinations, 148, 155, 161
Solid-phase sugar, lectin binding affinity, 4
Soybean agglutinin (SBA),
  binding sites on unmyelinated primary sensory neurons, 518, 519

Index

bone marrow purging and hematopoietic
progenitor cell enrichment in
transplantation,
hematopoietic progenitor cell
enrichment,
bone marrow, 339, 340
human umbilical cord blood, 340–342
overview, 329, 330
purging for autologous
transplantation,
breast cancer, 333–336
Burkitt's lymphoma, 337
magnetic affinity cell sorting,
342–344
neuroblastoma stage IV, 331–333
principle, 330, 331
T-cell depletion for allogeneic
transplantation,
haploidentical transplant, 338,
339
siblings, 337, 338
cell binding specificity, 329
cytoskeleton effects, 411–413
structure, 330
Specificity,
assay, 28, 29, 44, 45, 47, 49, 50, 102,
108
definition, 23
lectin comparison to monoclonal
antibodies, 76
Spermidine, *see* Polyamine uptake assay
Spermine, *see* Polyamine uptake assay
SSA, *see Sambucus sieboldiana* agglutinin
Stomach, lectin histochemistry and cancer,
78
Storage disorder, lectin histochemistry
analysis, 76, 133
Streptococcus pyogenes, epithelial cell
adhesion,
adhesins, 529, 530
assays,
artifacts, 530, 532, 535
cell counting, 532, 535
monolayer adhesion, 532–534
lectin blocking, 530, 533, 534–536
M6, 530
Streptococcus sanguis, adhesins, 529

Superantigen, comparison to mitogenic
lectins, 372, 373
Superoxide, *see* Oxidative burst

T

T-cell, *see also* Mitogenesis,
depletion for allogeneic bone marrow
transplantation,
haploidentical transplant, 338, 339
siblings, 337, 338
T-cell receptor (TCR),
assembly, 291
glycosylation, 292
immunoprecipitation, glycosidase
digestion, and gel electrophoresis,
294–296
intracellular transport, 291, 292
lectin affinity chromatography,
materials, 292–294
running conditions, 294, 297
mitogenesis role, 367
subunits, 291
TCR, *see* T-cell receptor
TEER, *see* Transepithelial electrical
resistance
Testis, lectin histochemistry and cancer,
84
Thin-layer chromatography (TLC), glycan
structure analysis following acid
hydrolysis, 181, 184, 188–190
Thyroid, lectin histochemistry and cancer,
85
TLC, *see* Thin-layer chromatography
Transepithelial electrical resistance
(TEER), evaluation of lectin-
mediated transcytosis, 569, 573, 574,
576, 577, 579
Trypsinization, samples for histochemical
analysis, 11, 14, 17

U

UEA-I, *see Ulex europaeus* I lectin
UEA-II, *see Ulex europaeus* II lectin
Ulex europaeus I lectin (UEA-I),
binding specificity, 9
blocking of bacterial adhesion, 530, 533,
534
blood group antigen recognition, 235
cartilage staining, 66

endothelial cell staining, 9, 10, 12, 13, 16, 76, 77. 320
Ulex europaeus II lectin (UEA-II), binding specificity, 9

V

von Willebrand factor (vWF), blood group antigen expression,
  enzyme-linked immunosorbent assay,
    immunoprecipitation of proteins, 238
    materials, 237, 242
    plasma preparation, 237, 238
    reaction conditions, 241
    sensitivity, 243
  lectin blotting,
    electrophoresis and blotting, 238
    immunoprecipitation of proteins, 238
    materials, 237, 242
    plasma preparation, 237, 238
    reactivity of lectins on blots, 236, 243
vWF, *see* von Willebrand factor

W

Western blotting,
  glycoproteins, 154–156, 160, 164
  gp120, 169, 170, 557, 558, 560–562
  phosphotyrosine proteins, 423, 424, 426–431
WGA, *see* Wheatgerm agglutinin
Wheatgerm agglutinin (WGA),
  cytoskeleton effects, 411, 413, 414
  endothelial cell binding in histochemistry, 10
  uptake assay in activated lymphocytes, 58, 59